Handbook of Research on Bioeconomy and Economic Ecosystems

Ana Pego
Universidade Nova de Lisboa, Portugal

IGI Global
PUBLISHER of TIMELY KNOWLEDGE

A volume in the Practice, Progress, and Proficiency in Sustainability (PPPS) Book Series

Published in the United States of America by
IGI Global
Business Science Reference (an imprint of IGI Global)
701 E. Chocolate Avenue
Hershey PA, USA 17033
Tel: 717-533-8845
Fax: 717-533-8661
E-mail: cust@igi-global.com
Web site: http://www.igi-global.com

Library of Congress Cataloging-in-Publication Data

Names: Pego, Ana, 1969- editor.
Title: Handbook of research on bioeconomy and economic ecosystems / edited by Ana Pego.
Description: Hershey, PA : Engineering Science Reference, [2023] | Includes bibliographical references and index. | Summary: "Handbook of Research on Bioeconomy and Economic Ecosystems is a critical resource that explores the conservation of ecosystems and their biodiversity and discusses potential new challenges in terms of the economic, social, and environmental path for Europe and other regions of the world. Featuring research on topics such as bioeconomy, circular economy, and economic and social analysis, this book is ideally designed for city authorities, experts, officers, business representatives, economists, politicians, academicians, and researchers"-- Provided by publisher.
Identifiers: LCCN 2023014901 (print) | LCCN 2023014902 (ebook) | ISBN 9781668488799 (hardcover) | ISBN 9781668488805 (ebook)
Subjects: LCSH: Sustainable development--Europe. | Circular economy--Europe. | Environmental policy--Europe. | Biodiversity--Europe.
Classification: LCC HC240.9.E5 H35 2023 (print) | LCC HC240.9.E5 (ebook) | DDC 333.7/094--dc23/eng/20230420
LC record available at https://lccn.loc.gov/2023014901
LC ebook record available at https://lccn.loc.gov/2023014902

This book is published in the IGI Global book series Practice, Progress, and Proficiency in Sustainability (PPPS) (ISSN: 2330-3271; eISSN: 2330-328X)

British Cataloguing in Publication Data
A Cataloguing in Publication record for this book is available from the British Library.

For electronic access to this publication, please contact: eresources@igi-global.com.

Practice, Progress, and Proficiency in Sustainability (PPPS) Book Series

Ayman Batisha
International Sustainability Institute, Egypt

ISSN:2330-3271
EISSN:2330-328X

Mission

In a world where traditional business practices are reconsidered and economic activity is performed in a global context, new areas of economic developments are recognized as the key enablers of wealth and income production. This knowledge of information technologies provides infrastructures, systems, and services towards sustainable development.

The **Practices, Progress, and Proficiency in Sustainability (PPPS) Book Series** focuses on the local and global challenges, business opportunities, and societal needs surrounding international collaboration and sustainable development of technology. This series brings together academics, researchers, entrepreneurs, policy makers and government officers aiming to contribute to the progress and proficiency in sustainability.

Coverage

- Socio-Economic
- Outsourcing
- E-Development
- Knowledge clusters
- ICT and knowledge for development
- Global Business
- Green Technology
- Environmental informatics
- Innovation Networks
- Sustainable Development

IGI Global is currently accepting manuscripts for publication within this series. To submit a proposal for a volume in this series, please contact our Acquisition Editors at Acquisitions@igi-global.com or visit: http://www.igi-global.com/publish/.

The Practice, Progress, and Proficiency in Sustainability (PPPS) Book Series (ISSN 2330-3271) is published by IGI Global, 701 E. Chocolate Avenue, Hershey, PA 17033-1240, USA, www.igi-global.com. This series is composed of titles available for purchase individually; each title is edited to be contextually exclusive from any other title within the series. For pricing and ordering information please visit http://www.igi-global.com/book-series/practice-progress-proficiency-sustainability/73810. Postmaster: Send all address changes to above address.

Titles in this Series

For a list of additional titles in this series, please visit: www.igi-global.com/book-series

Hydrogen Fuel Cell Technology for Mobile Applications
Raluca Andreea Felseghi (Department of Mechanical Engineering, Faculty of Automotive, Mechatronics, and Mechanical Engineering, Technical University of Cluj-Napoca, Romania)
Engineering Science Reference • copyright 2023 • 313pp • H/C (ISBN: 9781668467213) • US $240.00 (our price)

Energy Transition in the African Economy Post 2050
Olayinka Ohunakin (Covenant University, Nigeria)
Engineering Science Reference • copyright 2023 • 300pp • H/C (ISBN: 9781799886389) • US $215.00 (our price)

Natural Resource Management Issues in Human-Influenced Landscapes
Dilini Hemachandra (Faculty of Agriculture, University of Peradeniya, Sri Lanka) and Jeevika Weerahewa (Faculty of Agriculture, University of Peradeniya, Sri Lanka)
Engineering Science Reference • copyright 2023 • 300pp • H/C (ISBN: 9781668470510) • US $225.00 (our price)

Multidisciplinary Approaches to Sustainable Human Development
Vishal Jain (Sharda University, India) and Vinay Kukreja (Chitkara University, India)
Information Science Reference • copyright 2023 • 390pp • H/C (ISBN: 9781668482230) • US $225.00 (our price)

5G, Artificial Intelligence, and Next Generation Internet of Things Digital Innovation For Green and Sustainable Economies
Patricia Ordóñez de Pablos (University of Oviedo, Spain) and Xi Zhang (Tianjin University, China)
Engineering Science Reference • copyright 2023 • 320pp • H/C (ISBN: 9781668486344) • US $280.00 (our price)

Operational Research for Renewable Energy and Sustainable Environments
Joshua Thomas (UOW Malaysia KDU, Malaysia) Gerhard Wilhelm Weber (Poznan University of Technology, Poland) Roman Rodríguez Aguilar (Universidad Panamericana, Mexico) Elias Munapo (North West University, South Africa) and Pandian Vasant (Modeling Evolutionary Algorithms Simulation and Artificial Intelligence (MERLIN), TDTU, Vietnam)
Engineering Science Reference • copyright 2023 • 300pp • H/C (ISBN: 9781668491300) • US $240.00 (our price)

Artificial Intelligence Tools and Technologies for Smart Farming and Agriculture Practices
Rajeev Kumar Gupta (Pandit Deendayal Energy University, India) Arti Jain (Jaypee Institute of Information Technology, India) John Wang (Montclair State University, USA) Santosh Kumar Bharti (Pandit Deendayal Energy University, India) and Samir Patel (Pandit Deendayal Energy University, India)
Engineering Science Reference • copyright 2023 • 303pp • H/C (ISBN: 9781668485163) • US $225.00 (our price)

IGI Global
PUBLISHER of TIMELY KNOWLEDGE

701 East Chocolate Avenue, Hershey, PA 17033, USA
Tel: 717-533-8845 x100 • Fax: 717-533-8661
E-Mail: cust@igi-global.com • www.igi-global.com

List of Contributors

Table of Contents

Detailed Table of Contents

The world is facing pressing environmental challenges, including climate change, biodiversity loss, and resource scarcity. These issues have led businesses and investors to concentrate on ESG factors, which are increasingly crucial for long-term sustainable growth. However, the existing ESG framework is not a panacea for urgent problems such as energy, climate, and biodiversity crises. This chapter explores the emerging role of the TNFD and the potential benefits of incorporating biodiversity credits in ESG reporting practices as a unified mechanism to combat biodiversity challenges and sustainability issues, highlighting the synergies among these concepts and the possible advantages of integrating them into business operations. By addressing nature-related risks and adopting market-based instruments like biodiversity credits, companies can better position themselves to achieve their ESG objectives and generate long-term value.

Chapter 4
Boosting Local Economies: The Impact of N-Contour Gesellian Money on Regional Growth 55
Rinat Galiautdinov, Independent Researcher, Italy

This article explores the practical use of Gesellian money in several local regions which allowed the authors to collect the data and research the influence of the Gesellian model on the local economies. The implementation of Gesellian money has resulted in a significant increase in the turnover of money within local economies, which has helped to save struggling businesses that were unable to pay their employees. By using Gesellian money, local businesses have been able to circulate more money within their communities, leading to a more sustainable and prosperous economy. The implementation of Gesellian money has proven to be an effective tool for stimulating local economies and supporting small businesses. The research also showed that introduction of Gesellian model allows to avoid crucial crises in economy. The article also provides the solution of introduction Gesellian model into big economies using n-contour which allows to consider different growth amplitude of local economies, and the instruments for interaction between local economies and local-to-international economies.

Chapter 5
Environmental Regulation and Incentives in Socio-Ecological and Green Technological Innovation 69
José G Vargas-Hernandez, Tecnológico Superior de Jalisco, Mexico
Omar C. Vargas-González, Tecnológico Nacional de México, Ciudad Guzmán, Mexico

This study intends to analyze the impact of environmental regulation and incentives in socio-ecological and green technological innovation. It departs from the assumption that environmental regulation affects the performance of socio-ecological and green technological innovation in organizations. The method employed is critical analysis and reflective based on the theoretical, conceptual, and methodological literature. It is concluded that the analysis confirms that there is a direct relationship between the incentives of environmental regulation and the socio-ecological and green technological innovation in organizations.

Chapter 6
Estimating the Impact of the Sustainable Development Goals on the Lebanese Society: Effects on Bio-Economic Indicators 88
Geoges Bellos, Lebanese International University, Lebanon
Mazen Muhieddine Kotob, Lebanese International University, Lebanon

The current project thesis aims to examine the influence of the sustainable development goals on Lebanese society and whether or not they enhance the Lebanese people's quality of life. For this purpose, via

The purpose of this chapter is to introduce research on the impact-driven productivity and innovation gains achieved through the adoption of unified communication and collaboration (UC&C) technologies as drivers of digital transformation across a global automotive organization. Against the background of research on the bioeconomy and economic ecosystems, indicators of cultural change in the workplace and other transformation factors shown to have an impact included the ease of use and adoption, legacy reduction, and elimination, increased organizational efficiency and generating savings.

The interrelationship between economic ecosystems and social impact refers to the way in which the economic environment of a community can impact its social well-being and vice versa. Economic ecosystems refer to the complex network of economic activities, resources, and institutions that exist within a community or region, including businesses, industries, financial institutions, government agencies, and other stakeholders. Social impact, on the other hand, refers to the effect that these economic activities have on the social and cultural fabric of a community, such as employment opportunities, income levels, access to resources and services, and quality of life. Examining the ripple effect on communities involves looking at the way in which economic activities can create both positive and negative effects on the social well-being of a community.

The aim of this chapter is to provide a systematic literature review, using bibliometric methods, on the topics of Maritime Cluster and Competitiveness, complemented with the combination of innovation, networks, and co-petition. Data were collected from the web of science and quantitative and qualitative analyses were applied to understand the relevance of the topics, the years and countries of production and the types of articles published. It was possible to identify three approaches: one that relates the theme from a strategic management perspective, another that associates the theme with regional development policies, and the third associated with the development of networks. It was also found that although co-petition does not appear explicitly, it appears as a fundamental element for the promotion of maritime clusters of the simultaneous combination of competition and collaboration of different actors within the cluster and as a mobilizing element of innovation networks.

The ocean economy is directly linked to the biodiversity of the ocean. Biodiversity is under great threat as a result of the effects of climate change and global warming. Therefore, the ocean economy and the marine economy are among the economies that will be most affected by the effects of climate change. Each change in ocean and sea levels causes changes in biodiversity. Considering its importance in terms of food health, economic and social development, the sustainability of the marine economy is an important issue. The aim of this study is to present indicators of ocean economy among OECD countries and show the importance of ocean and marine economy in the long term.

The global waste trade refers to the movement of waste materials across international borders for disposal or recycling. This trade has grown rapidly in recent years as countries seek to capitalize on the economic benefits of exporting waste while also addressing the challenges of managing increasing volumes of waste at home. This chapter will provide an overview of the global waste trade over the past decade, including key trends and developments, major players and destinations, and this rapidly growing industry's environmental and economic impacts. The chapter will also explore the challenges and opportunities presented by the global waste trade and offer insights into the future of this complex and dynamic field.

The financial stability of millions of older adults around the world rests on the shoulders of well-functioning pension systems. For this reason, the sustainability of such systems is one of the most important challenges facing states, especially in the face of low birth rates and increased life expectancy. Hence the growing importance of the debate over taxation and economic policy. This study presents the historical backgrounds of the pension system in Ecuador to give an all-encompassing view of its present configuration, and examines the relevant data of the factors that affect the pension system with the purpose of evaluating the sustainability of the Ecuadorian pension system in the long term. Both the historical and evaluative aspects of the study lead to an analysis of fiscal policies, the political relationships of the system and numerous exogenous processes that reveal aspects of the system that lack the expected solidarity.

The bioeconomy presents a promising opportunity for sustainable economic growth, but it also poses several challenges that require innovative solutions. This chapter provides an overview of the bioeconomy and its key components, including biotechnology, agriculture, and forestry. It then explores some of the challenges associated with the bio economy, such as the need to balance economic growth with

Preface

Bioeconomy is highly complex owing to the interaction between anthropogenic activities, economic vs. environmental goals, and the detrimental effects of climate change. To adequately act on and achieve effective and efficient bioeconomy management, a deep-rooted multi-actor innovation ecosystem must be in place. This requires harmonized governance and monitoring structure that coordinates and connects the activities of stakeholders (i.e., researchers, policy-makers and regulators, private sector, and consumers) and their initiatives. This will enable efficient and effective cooperation between the existing governments, projects, and actors in different regions, countries, and sectors. The Stakeholders' support is vital for various restoration and protection projects in many regional innovation systems. This implies innovators' access and use of knowledge, capacity building, technical, and funding assistance. Finally, societal awareness of these challenges needs to be raised. Bioeconomy will set the principles for the operation of a multi-actor and multi-level innovation for natural and sustainable management in regions with varying characteristics.

This book aims to address a range of approaches including, but limited to aspects of Bioeconomy. In the next paragraphs, the book structure is detailed, in respective chapters.

Chapter 1: Anthropocene, Urban and Antho-Socio-Ecology Planning Resilience

José Vargas, Selene Castañeda-Burciaga, Omar C. Vargas-González

This study aims to analyse some of the existing implications between urban anthropology and Anthropocene urban social-ecological resilience planning. Based on the assumption that urban anthropology provides support for the creation and development of urban planning based on Anthropocene social-ecological resilience. The staff's methods are analytical-descriptive and based on ethnographic interpretation and reflection of theoretical and empirical literature review. The analysis concludes that the foundations of urban anthropology help to strengthen the Anthropocene orientation of urban socio-ecological planning.

Chapter 2: Anthropological Ecology and Ecological Anthropology as a Framework for the Analysis of Socio-Ecological Sustainable Development

José G Vargas-Hernandez, María F. Higuera-Cota, Omar C. Vargas-González

This study aims to analyse anthropological ecology and ecological anthropology as a framework for the analysis of socio-ecological sustainable development. It starts from the assumption that sustainable development requires a socio-ecological approach underpinned by anthropology and ecology. The method used is analytical reflection based on a conceptual, theoretical and empirical literature review. The study concludes that anthropology and ecology combine elements to support the approaches of anthropological ecology and ecological anthropology and to develop the framework for the analysis of socio-ecological sustainable development.

Chapter 3: Biodiversity Initiatives: Theoretical and Practical Intersections Among TNFD, Biodiversity Credits, and ESG Reporting

Ruopiao Zhang, Carlos Noronha

The world faces pressing environmental challenges, including climate change, biodiversity loss and resource scarcity. These issues have led companies and investors to focus on ESG factors, which are increasingly important for long-term sustainable growth. However, the existing ESG framework is not a panacea for pressing issues such as energy, climate and biodiversity crises. This chapter explores the emerging role of TNFDs and the potential benefits of incorporating biodiversity credits into ESG reporting practises as a unifying mechanism to address biodiversity challenges and sustainability issues, highlighting the synergies between these approaches and the potential benefits of integrating them into business operations. By addressing nature-related risks and using market-based tools such as biodiversity credits, companies can better position themselves to achieve their ESG goals and create long-term value.

Chapter 4: Boosting Local Economies: The Impact of n-Contour Gesellian Money on Regional Growth

Rinat Galiautdinov

This article examines the practical application of Gesellian money in several local regions, which made it possible to collect data and study the impact of the Gesellian model on the local economy. The introduction of Gesellian money has led to a significant increase in money turnover in the local economy, which has helped to bail out struggling businesses that were unable to pay their employees. The use of gesellian money has enabled local businesses to circulate more money within their communities, leading to a more sustainable and prosperous economy. The introduction of Gesellian money has proven to be an effective tool in stimulating the local economy and supporting small businesses. Research has also shown that the introduction of the Gesellian model makes it possible to avoid crucial economic crises. The article also provides a solution for introducing the Gesellian model in large economies using the n-contour, which allows for different growth amplitudes of local economies, as well as the tools for interaction between local economies and local-international economies.

Chapter 5: Environmental Regulation and Incentives in Socio-Ecological and Green Technological Innovation

José G Vargas-Hernandez, Omar C. Vargas-González

This study aims to analyse the impact of environmental regulations and incentives on socio-environmental and green technological innovations. It starts from the assumption that environmental regulation influences the performance of socio-ecological and green technological innovations in organisations. The method used is the critical-analytical and reflective based on the theoretical, conceptual and methodological literature. The conclusion is that the analysis confirms that there is a direct relationship between the incentives of environmental regulation and social-ecological and green technological innovation in organisations.

Chapter 6: Estimating the impact of the Sustainable Development Goals on the Lebanese Society: Effects on Bio-Economic Indicators

Geoges Bellos, Mazen Muhieddine Kotob

This project work aims to investigate the impact of the Sustainable Development Goals on Lebanese society and to determine whether or not they improve the quality of life of the Lebanese population. To this end, the researcher intended to use sequential exploratory mixed research methods to interview an expert in the field of sustainability, followed by a survey sent to over 40 respondents. Our findings show that the shift from a service economy to a production economy is exactly what we need to build a strong economy, and that sustainability will indeed ultimately help us live healthier and better lives. Thus, a relationship was found with the above variables and indeed a positive relationship was found between the implementation of the SDGs and the improved quality of life of the Lebanese population. This shows the proportional relationship with both variables.

Chapter 7: Experiential Marketing: A Tool to Maximize the Contribution of Natural History Museums to the Bioeconomy

Hasan Hüseyin Erdoğan

Natural history museums are important actors in the bioeconomy. They contribute to the bioeconomy by increasing the bio ecological knowledge of the public. This chapter shows that their contribution can be maximised if they apply the right marketing practises. In this sense, experiential marketing, which aims to provide customers with personalised, entertaining and engaging educational experiences, seems to be the best approaChapter Accordingly, the three most visited natural history museums were studied to identify experiential marketing practises that can help maximise the contribution of natural history museums to the bioeconomy. The findings highlighted six marketing practises: holistic experiential design, storytelling boosters, sensory interactive elements, participatory interactive elements, co-creation and online interactive elements. The last section presents detailed practical suggestions on these practises for professionals in natural history museums.

Chapter 8: Green Knowledge in Green Roofs and Organizational Green Innovation

José G Vargas-Hernandez, Carlos Alberto A. Rodriguez Maillard, Omar C. Vargas-González

This study aims to analyse the impact of green knowledge and green technology on organisational green innovation, urban green innovation and green roofs. Green roofs can be an effective tool for cities to improve the thermal environment, save energy and combat climate change, and they are a suitable method to save energy. The analysis is based on the assumption that green technology is the foundation for green innovation practises, operations and activities in businesses and cities. Striking a balance between urban development's and environmental issues, taking into account the principle of sustainable development and its innovative green solutions. The methods used are based on analytical-reflective and descriptive methods supported by a review of theoretical and empirical literature. The analysis concludes that green knowledge sharing is important for the creation and development of green technologies, which has a positive impact on organisational green innovation, urban green innovation spaces and green roofs.

Chapter 9: Impact-Driven Productivity and Innovation with Unified Communication and Collaboration Technologies: Drivers to be Addressed Across Global Regions

Anthony D Bolton, Leila Goosen, Elmarie Kritzinger

The purpose of this chapter is to present research on the productivity and innovation gains achieved through the adoption of Unified Communication and Collaboration (UC &C) technologies as a driver of digital transformation in a global automotive company. Against the backdrop of research on the bioeconomy and economic ecosystems, indicators of cultural change in the workplace and other transformational factors were identified as ease of use and acceptance, reduction and elimination of legacy, increased organisational efficiency and achievement of savings.

Chapter 10: Interrelationship of Economic Ecosystems with Social Impact Examining the Ripple Effect on Communities

Nidhi Agarwal, Jocelyn B. Hipona, Ekansh Agarwal

The interrelationship between economic ecosystems and social impact refers to the ways in which a community's economic environment can affect its social well-being and vice versa. Economic ecosystems refer to the complex network of economic activities, resources and institutions that exist in a community or region, including businesses, industries, financial institutions, government agencies and other actors. Social impacts, on the other hand, refer to the effects that these economic activities have on the social and cultural fabric of a community, such as employment opportunities, income levels, access to resources and services, and quality of life. The study of community impact is about how economic activities can have both positive and negative impacts on the social well-being of a community.

Chapter 11: Systematic Literature Review of Maritime Clusters and Competitiveness

Maria Manuela Santos Natário, Elisabeth T. Pereira, João Pedro A. Couto

The aim of this chapter is to provide a systematic literature review using bibliometric methods on Maritime Clusters and Competitiveness complemented by the combination of Innovation, Networks and Co-Petition. Data were collected from the Web of Science and quantitative and qualitative analyses were conducted to understand the relevance of the topics, the years and countries of origin, and the types of articles published. Three approaches were identified: one that looks at the topic from a strategic management perspective, another which relates the topic to regional development policy, and the third that has to do with network development. It was also noted that although co-petition does not explicitly emerge, it appears as a fundamental element for the promotion of maritime clusters of the simultaneous combination of competition and cooperation between different actors within the cluster and as a mobilising element of innovation networks.

Chapter 12: The Climate Change and Ocean Economy: A Review on OECD's Ocean Economy

Seda Yildirim, Merve Kaplan, Durmus Cagri Yildirim

The marine economy is directly linked to the biodiversity of the ocean. Therefore, the marine and maritime economies are among the economies that will be most affected by the impacts of climate change. Biodiversity is highly threatened by the impacts of climate change and global warming. Any change in the oceans and sea levels will lead to a change in biodiversity. Given its importance for food health, economic and social development, the sustainability of the marine economy is an important issue. The aim of this study is to present indicators of the marine economy in OECD countries and to show the importance of the marine economy in the long term.

Chapter 13: The Global Waste Trade: A Decade in Review

Ahmed Yusuf Sarihan

The global waste trade refers to the movement of waste materials across international borders for disposal or recycling. This trade has increased rapidly in recent years as countries seek to capitalise on the economic benefits of exporting waste, while at the same time overcoming the challenges of managing growing waste volumes at home. This chapter provides an overview of global waste trade over the past decade, including key trends and developments, key players and destinations, and the environmental and economic impacts of this fast-growing industry. The chapter also explores the challenges and opportunities presented by global waste trade and offers insights into the future of this complex and dynamic field.

Chapter 14: The Pension System in the Ecuadorian Social Security Institute (IESS) Analysis and Perspectives in the Economic and Legal Field

Arturo Luque, Marcelo Centeno Ponce, Ana Corrales Mendoza

The financial stability of millions of older people around the world rests on the shoulders of well-functioning pension systems. For this reason, the sustainability of such systems is one of the most important challenges for states, especially in the face of low birth rates and higher life expectancy. Therefore, the debate on fiscal and economic policy is becoming increasingly important. This study presents the historical background of the pension system in Ecuador to provide a comprehensive overview of its current design, and examines the relevant data of the factors that influence the pension system with the aim of assessing the sustainability of the Ecuadorian pension system in the long run. Both the historical and evaluative aspects of the study lead to an analysis of fiscal policies, the political relations of the system and numerous exogenous processes that highlight aspects of the system that lack the expected solidarity.

Chapter 15: The Role of Innovation in Driving the Bioeconomy, the Challenges and Opportunities: Innovation in Driving the Bioeconomy

S M Nazmuz Sakib

The bioeconomy represents a promising opportunity for sustainable economic growth, but it also presents some challenges that require innovative solutions. This chapter provides an overview of the bioeconomy and its main components, including biotechnology, agriculture and forestry. It then discusses some of the challenges associated with the bioeconomy, such as the need to balance economic growth and environmental sustainability, the potential risks of genetically modified organisms and the need to ensure that benefits are shared equitably across society. Finally, the potential for new bio-based products and services, the creation of new jobs and economic opportunities, and the potential to contribute to global sustainability goals will be discussed.

Chapter 16: The Role of IoT an AI in Bioeconomy

Justino Lourenço

The new era brings with it a host of new challenges in terms of proactive practises to achieve a better ecosystem in Europe. A number of new trending technologies such as Artificial Intelligence (AI) and the Internet of Things (IoT) can and will bring a new approach to tackling our problems. AI can contribute to sustainability in a number of ways, such as optimising energy consumption, reducing waste, improving supply chain efficiency and enabling precision agriculture. AI can help drive scientific research and innovation for sustainable development, e.g. by developing new materials and technologies for clean energy generation and storage. On the other hand, the Internet of Things (IoT) can contribute to sustainability by enabling real-time monitoring and control of energy and resource consumption in buildings, transport systems and industrial processes. This chapter will start the discussion on the expected positive impact of the introduction of AI and IoT for a more sustainable planet.

Chapter 17: Towards Sustainable Development: A Multi-Criteria Assessment of the Circular Economy in the European Union

Milan Marković

In a sustainable society, which countries aspire to, the circular economy has a special place. The circular economy offers certain solutions that help to make development more sustainable by increasing resource productivity and reducing waste. The European Union is a world leader in implementing the principles and practises of the regenerative economy and the circular economy. The chapter aims to provide a multi-criteria analysis of the performance of the circular economy by ranking the countries of the European Union, taking into account the composite index created. The aggregation was carried out using the Grey Relational Analysis method, while the weighting coefficients were determined using the entropy method. The results of the study show that the Western European countries are the most successful in achieving the goals of the circular economy towards sustainable development and that the Netherlands ranks first among them.

Chapter 18: Transformational Innovation Technologies for Regenerative Bioeconomy: Case Study on Green Initiatives for Tourism Logistics Service Providers

Sejana Jose V, Bindi Varghese

The tourism industry has a negative impact on the environment and dwarfs the regenerative bioeconomy. Climate change, land degradation and resource depletion are major challenges. The overuse of non-biodegradable resources threatens the planet and requires bio-based resources. The shift to bio-based resources and their reuse is crucial. In this chapter, the regenerative bioeconomy has far-reaching implications for the achievement of SDGs 6, 8, 11 and 12, with a focus on the participation of the circular economy in tourism logistics. Investments in talent development, digital technologies and partnerships are needed to realise the potential of the bioeconomy. Engaging local communities and implementing sustainable business practises can reduce energy consumption and environmental impacts. Digital transformation requires technological advances, foreign investment and the active participation of all stakeholders. This chapter attempts to explain the complicated interplay between the regenerative bioeconomy, tourism logistics and sustainable practises.

Chapter 19: Manufacturing of Marketable Energy Sources from Agricultural Crops: A Spanish Sector-Based Study on Bioeconomy

Luis A. Millan-Tudela, Bartolomé Marco-Lajara, Eduardo Sánchez-García, Javier Martínez-Falcó

The development of societies has historically been linked to the production, consumption and ultimately availability of energy. The development of more efficient technologies has meant that various sources of energy have evolved and changed in recent centuries. However, these changes, along with a significant increase in energy consumption, have also brought with them a threat to ecosystems in the form of various pollutants. In this chapter, a study is conducted on the potential use of crops to produce marketable energy, using the Spanish agricultural sector as a study case, as it is a major contributor to crop production in the European Union.

Chapter 20: Understanding the Dynamics of Economic Development in Developing Countries

Puneet Kumar, Maricel C. San Diego, Maria Teresa R. Fajardo

For emerging economies to advance and prosper, their economies must develop in complex and diverse ways. The aim of this study is to provide readers with a thorough understanding of the forces driving economic growth in emerging economies. The chapter's introduction reviews the main ideas and theories of economic development, which include elements such as human capital, technological innovation, institutional quality and globalisation. It then analyses the unique problems that developing countries face, including poverty, income inequality and limited access to resources and infrastructure. In addition, this study examines how social, market and governmental elements interact to promote economic development. It examines how social capital, inclusive institutions and good governance help to promote long-term economic success. In addition to the potential benefits and difficulties of inclusion in the global economy, the impact of globalisation and trade on developing countries is also addressed.

Chapter 21: Bioeconomy and Ecosystem of China's Polar Silk Road

Maria João Tomás

The Polar Silk Road is one of the major projects of the People's Republic of China (PRC). As part of the grand plan of the "One Belt, One Road" initiative, it is the continuation and reinforcement of the country's Arctic strategy. In 2018, China officially reaffirmed its goals with the publication of the book China's Arctic Policy announcing the establishment of the Arctic Route, underscoring a more proactive policy towards the region. The exploration of the Arctic with the ongoing thaw raises big questions about the sustainability of a changing ecosystem. The concerns of the Arctic bioeconomy are growing, not only about navigability, but also about the exploitation of the riches beneath the ice. This article focuses on China's political and economic ambitions in the Arctic and how they are compatible with the bioeconomy and sustainability policies, as the North Polar Region is one of the most fragile ecosystems on Earth and one of the most dynamic resulting from the thawing of permafrost.

Ana Pego
Universidade Nova de Lisboa, Portugal

Chapter 1
Anthropocene, Urban, and Antho-Socio-Ecology Planning Resilience

José G. Vargas-Hernandez
Tecnológico Superior de Jalisco, Mexico

Selene Castañeda-Burciaga
Universidad Politécnica de Zacatecas, Mexico

Omar C. Vargas-González
Tecnológico Nacional de México, Ciudad Guzmán, Mexico

ABSTRACT

This study aims to analyze some of the existing implications between urban anthropology and Anthropocene urban socio-ecology planning resilience, beginning from the assumption that urban anthropology gives support to create and develop any urban planning based on the Anthropocene urban socio-ecology resilience. The methods employes are analytical-descriptive based on an ethnographic interpretation and reflection of the theoretical and empirical literature review. The analysis concludes that urban anthropology fundamentals give support to strengthen the Anthropocene orientation of the urban socioecological planning resilience.

INTRODUCTION

Urban anthropology is a scientific discipline epistemologically and intricately linked and interrelated with other scientific theories. Anthropology, like any social science, predicts various phenomena (Álvarez, 2016; Brady, 2019; Grimmer et al., 2021; Hindman, 2015; Hofman et al., 2017; Hofman et al., 2021; Taagepera, 2008). The scientific status of anthropology goes beyond the conception of a soft, interpretative, and humanistic phenomenon to interpret meanings, to use quantitative techniques, to apply formal methods in experimental designs, to postulate scientific theories predicting phenomena and formulating scientific laws linked and interrelated with other scientific theories.

DOI: 10.4018/978-1-6684-8879-9.ch001

On the other hand, the term resilience was introduced by Holling (1973) in the field of socio-ecological research to describe the ability of the socio-ecosystem to return to its original shape after having been deformed by the impact of external pressures from the external environment. Resilience influences urban planning and innovation, urban infrastructure construction and development leading to urban and regional economic resilience as one of the contents of research on resilience in urban planning and governance.

Likewise, the improvement of urban socio-ecological resilience contributes to the prevention, management, and reduction of catastrophes, which strengthens urban resilience processes (Yang, Lili & Hongchi, 2022). Some approaches to urban anthropology do not consider it a science and those who prefer theories or the particular over the universal, and those who choose to interpret rather than explain.

However, the anthropological approach to urban socio-ecological resilience has shorter interactions in time between social groups that aim at urban socio-ecological resilience planning and green innovations. The development of urban anthropological motivation on creative capacities generates the conditions of the local innovation economy based on urban socio-ecological planning and green innovation areas. Being that, currently, international institutions are dedicated to studying lateral work with groups of experts in urban socio-ecological planning and areas of green innovation with the support of urban anthropology (Hannerz, 2004; Holmes & Marcus, 2006, 2021; Ortner, 2010, 2013).

On the other hand, the changes from traditional urban anthropology to the urban sociology of the Anthropocene and resilience have been analyzed by Kuhn (1962/1996) who has focused on paradigm shifts, focusing on the epistemic cycles of the methods of research from its rise and decline. Thus, this study aims to analyze some of the relationships and implications between urban anthropology and urban socio-ecological planning for the resilience of the Anthropocene. To achieve this goal, this study first starts from the analysis through the study of organizational and urban anthropology cases, followed by urban socio-ecological resilience planning and urban green innovation areas. Finally, some conclusions are offered.

THEORETICAL BACKGROUND

The ancient Greeks developed critical thinking about urban socio-ecological resilience based on mutualistic relationships (Egerton, 2001). Currently it is recognized that the socio-ecological change of societies is caused by exogenous factors, such as globalization, colonialism, capitalism, etc. Some colonial societies are composed of heterogeneous elements with instability due to divergent interests (Radcliffe-Brown, 1940). Similarly, the phenomena of sociology, anthropology, and political science differ from individual behavior, according to Bunge (1999), who argues that predicting the behavior of an unknown individual is different from predicting the behavior of a system. However, the social sciences have translated small group prediction into complex social systems using sophisticated computational methods (Bunge, 1999; Kaplan, 1940).

Likewise, it is recognized that a scientific law is a logical-mathematical representation of the relationships between variables (Pfeifer, 2006). Scientific predictions arise from the verification and application of a theory capable of explaining specific phenomena. Research in urban anthropology and socio-ecology of resilience in the fields of urban planning and green innovation using interdisciplinary methodologies is changing people's societies, which is relevant for research on a global scale. However, anthropomorphic representations in research have not been convincing in favor of the functional hypothesis.

Science resolves testable hypotheses (Hirst, 2019). Research in the socioeconomic and organizational sciences focuses on sympathy for the population of subjects who move from the realm of an observer to become an active participant (Bilharz, 1998; Kolb, 1998). The interdisciplinarity and transdisciplinarity of urban anthropology applied to urban socio-ecological planning and green innovation areas are limited in space to conduct the development of disciplines in multidimensional crisis including ethical values and moral issues that drive to an ontological crisis of homo sapiens (Tarantino-Curseri, 2018).

It is recognized that urban anthropology leading to socio-ecological planning processes is evolving as a dialectical process from centralized, sequenced, and systematic rational integral planning to complete fragmentation and disjointed incrementalism (Lindblom, 1959). Strengthening the consolidation of national science, technology and innovation systems has effects on competitiveness derived from incremental socio-ecological and green innovations in urban planning. Therefore, research in urban anthropology and socio-ecology of resilience requires a comprehensive analysis, returning to a scientific perspective, without neglecting issues of human development in an environment that adapts to current demands, both of globalization and awareness of the environment.

METHOD

The analysis of the relations between the establishment of urban anthropology of action is framed in the reflexive dialectic actor and author with specific disciplinary orientation in urban socio-ecology resilience planning and areas of green innovation. The analysis of the relationships between the objectivity of the intersubjectivity of urban anthropology become methodological problems and dilemmas that need to be addressed specifically. The intersubjectivity of urban anthropology is an objective element of construction of reality, not with a relativistic ideology but with a methodological and practical one based on valid assumptions for urban socio-ecological resilience planning and green innovation.

In accordance with the above, a qualitative, analytical, and descriptive study is proposed on the state of urban anthropology and resilient urban socio-ecological planning of the Anthropocene; the foregoing, through the analysis of case studies related to organizational and urban anthropology, urban socio-ecological resilience planning and urban green innovation areas.

CASE STUDIES

The formal methods must meet the requirements of operationalization of terms, data analysis, logical consistency, justification of affirmations and replication of results, although neither the formal theories were dominant nor were the dominant theories ever (Reynoso, 1995b). The discipline of urban anthropology can formalize the knowledge produced. Formal computational methods used for ethnographic analysis in urban anthropology and archaeological data based on agent models, cellular automata, diffusion and percolation models, network analysis, etc., proposed a new form of research open to theoretical development and experimentation (Diaz et al. 2007). Urban anthropological experiments, predictions and theories can be compared with the natural and physical sciences to determine scientific status (Llanos, 2009), but they are not sufficient scientific criteria for good science if it is carried out without laws or predictions.

The most common experiments in urban anthropology and social sciences are natural experiments (Diamond & Robinson, 2010; Dunning, 2012), field experiments (Baldassarri & Abascal, 2017) and naturalistic experiments (Bernard, 2011). Accordingly, ethnographic writing, interpretivism, and perspectives in urban anthropology must be replaced by experimentation, theorizing, causation, prediction, mathematical modeling, and epistemology. Participant observation is an ethnographic methodology and one of the most popularized and important techniques developed by Malinowski (1961) in anthropological research that can be applied to urban anthropology and its impact on urban socio-ecological resilience planning and development (DeWalt, DeWalt, & Wayland, 1998; Kawulich, 2005; Jorgensen, 2015).

It should be noted that unstructured techniques without a specific direction led to collect deeper and broader answers and the original terms to know the native categories and perspectives of urban anthropology in urban socio-ecological planning (Pizarro, 2014). Interviews through defined questions are recommended as an ethnographic research instrument at the lowest level of structuring to extract a greater amount of information that links urban anthropology and urban socio-ecological resilience planning. The qualitative-quantitative division of methods and techniques in urban anthropology and urban socio-ecological resilience planning and innovation is objectionable (Aldrich, 2014; Reynoso, 1995b).

A recent investigation that addresses quantitative and qualitative approaches to reveal the consequences of the COVID-19 pandemic crisis in the theoretical framework of urban resilience, society and institutional governance, was carried out by Olimid, Georgescu & Gherghe, (2022). Quantitative methods and techniques in urban anthropology and urban socio-ecological resilience planning and innovation aim to collect and analyze numerical data (Roni et al., 2020).

ORGANIZATIONAL AND URBAN ANTHROPOLOGY

In this section, an analysis is carried out from a sociohistorical perspective of the interrelationships between urban anthropology and labor organizations, on anthropological practices in organizations and urban environments that have challenges that affect research and academic training and development. Whereas a multidisciplinary perspective on urban anthropology and green and socio-ecological innovation attempts to understand traditional technology in the production process. It should be noted that work on entrepreneurship and business anthropology dates to the study of ethnology in urban societies in the 1970s (Durrenberger, 2007). On the other hand, Warner's work ratified Durkheim's analysis from a sociohistorical perspective has an influence on anthropology and sociology related to the social structures of complex tribal and industrialized societies, advancing the analysis of the interrelationships between urban anthropology and organizations.

Industrial technology and social research acquire relevance and, consequently, can improve the understanding of technology in the anthropology of organizations and in urban socio-ecological planning and innovation (Roca, 2008). Organizational anthropology has close relationships with urban anthropology, and this includes demands for participatory actions in practical anthropology of urban imaginaries and organizational facilitation (Gravano, 1992). This situation integrates the organizational and urban anthropological perspectives, as well as the ethnological methodological tools to the analysis of organizational studies, transversal to the demands for research funding.

The social sciences can make possible scientific predictions about specific events and problems in urban socio-ecological planning and development, despite the arbitrariness of human behavior and the assumption that it is not useful to predict behavior (Nießen et al., 2020). In this same sense, the debates in the analysis of organizational and urban anthropology are building bridges between the studies of organizations and urban areas from an anthropological perspective, which seeks to analyze the interrelationships between urban anthropology and business organizations. and labor, from a socio-historical orientation to develop reflections that situate the analysis of urban areas and neighborhoods in post-industrial and global societies. Thus, urban anthropology can be analyzed from a sociohistorical orientation through its interrelationships with organizations and companies.

It should be noted that organizational and urban anthropology has conceptual, methodological and theoretical tools aimed at developing the interrelationships between organizations and urban settlements, which lead to creating the meanings that both organization workers and citizens have of their heritage. industrial. Regarding the above, the studies of Mayo (1945) and his collaborators Roethlisberger and Dickson (1939) in the Hawthorne Wester Electric Co. industrial plant stand out. They were pioneers in organizational anthropology research applied to work productivity. The design of jobs in settlements and urban labor organizations is characterized by the development of multiple skills based on the redundancy of a variety of functions rather than the redundancy of parts that make those of the job capable of performing.

The approach to the professional practice of organizational facilitation and participatory processes presents theoretical-methodological challenges in metropolitan urban environmental socio-ecological planning (Gravano, 2007b) supported by the work of the organizational culture of the organizational and economic agents and of the political actors. In this regard, the Manchester and Chicago schools have worked on urban anthropology (Cucó Giner, 2004; Gravano, 2005; Hannerz, 1986; Signorelli, 1999; Velho, 1978). Therefore, organizational performance reflects the problems and dilemmas of organizational practices linked to the state and urban anthropology in the socio-ecological planning of control and weighting of the anthropological ethnography method used in facilitating organization in professional practice.

Urban anthropology in socio-ecological planning is a process of negotiation and mobilization that leads to the application of the definition of incremental action (de Melo, 2002a, b). The urban neighborhood, as organizational anthropology in organizational culture, controls the symbolic culture that is manipulated as an ideology aimed at controlling urban anthropology or organizational life conditions and the interrelationships of social subjects.

In another order of ideas, the socio-technical approach has demonstrated its effectiveness in some studies carried out in coal mines and in a project in which operational units of Royal Dutch Shell in England participated, in a textile factory in India, with the participation of unions, employers and the government involved in improving the quality of working life in Norway and Sweden, and other socio-technical projects in Canada, the Netherlands and Australia. The execution of urban socio-technical projects is carried out in a continuous process of urban renewal to reach fullness because of revisions and modifications.

The urban planning of urban socio-technical and socio-ecological innovation in urban regions known as local industrial systems can be an example of an economy derived from the industrial location of self-reinforcing agglomerations of accumulated socio-technical skills, investments and venture capital, infrastructures, suppliers, spillovers of knowledge, services, etc. associated with informal information flows and proximity to research centers and universities (Saxenian, 1994). Likewise, urban techno-scientific models of relationality, such as radioecology, make it possible to visualize cross-border movements in urban areas. The socio-technical approach is based on semi-autonomous groups with roots in self-regulated projects and redundancy of functions (Trist, 1976; de Melo, 1997).

The socio-technical support or urban systems based on promotions and rewards, promotions, must be consistent with the activities and be aligned with the objectives of the institution. Communication across the urban metropolitan boundaries of the organization's departments links the levels of control and supervision. In accordance with the above, it is considered that radioecology offers something more than limited human relationships and delimits an eco-semiotics of ineffable interrelation of living, decomposing and dying things. The application of eco-semiotics considers that humans and other species co-constitute ecological phenomena (Maran & Kull 2014). Radioecology uses specific radioisotopes to investigate the tracking and mapping connections of radiological contaminants and their transboundary movements between the earth and the atmosphere (Odum, 1959).

The resulting production of cumulative nuclear waste that brings urban socio-ecological resilience to collective recovery. The socio-technical urban criterion fosters the autonomy and initiative of employees. However, the variation must be controlled in the sociotechnical urban criteria as the source for each person responsible for controlling job learning and failures. Urban movements reveal the state's limits to post-disaster urban socio-ecological resilience.

URBAN SOCIO-ECOLOGY RESILIENCE PLANNING AND URBAN GREEN INNOVATION AREAS

The theory of urbanism for green and socio-ecological innovation continued to be marginal during the 1970s until, recently, innovation acquired a multidisciplinary character, encompassing knowledge in fields such as sociology, economics, psychology, administration, geography, engineering etc. (Sundbo, 1998; Schlapfer & Marinova, 2001). Throughout the 1980s and 1990s, the economy experienced a socio-ecological restructuring of technological innovation and a process of economic internationalization and market globalization that led to the substitution of the scientific and technological system to focus on the competitiveness of the industrial sector. oriented to satisfy international markets (González Alvarez, 1997).

The urban planning of areas of socio-ecological and green innovation is the process that begins with the invention of a new element until its development for commercial use. The commercialization of green socio-ecological urban planning innovations generated in public universities and research institutions requires clear regulations to share and transfer intellectual property rights to private companies, new forms of partnership between public-private sectors for development technological. Green innovation must be practical, have commercial value and be socially acceptable (Sundbo, 1998).

The sources of knowledge for the planning of urban green areas are rooted in institutional and human forms subject to financial capital and less to geographic movement (Schlapfer & Marinova 2001). On the other hand, the conceptual framework of the institutional analysis model is proposed by René Lourau (1979) and Lapassade (1986) that can be applied to analyze urban anthropology in the Anthropocene of socio-ecological urban planning and green innovation areas.

The ethnomethodological urban anthropology perspective of urban socio-ecological resilience planning and green innovation areas points to the inherent transformation of the sociohistorical process that affects non-contemplative knowledge production that claims meaning and challenges deciphering without typologies. in an ahistorical way. Regarding the above, urban socio-ecological planning processes occur with institutional and community participation, addressing the circularity of management issues and exchanging institutional power relations based on the anthropological object of cultural otherness (Gravano, 2006).

However, anthropological contributions to urban socio-ecological and environmental planning are based on community participation in institutional and organizational processes. Individual and business benefits prevail over the vocations and values of the local socio-ecological communities, which leads to the sacrifice of the common good. The community participatory process imposed by law delimits the specific possibilities, difficulties and opportunities based on the asymmetric rationalities at stake and state action in the context or urban environmental socio-ecological planning. A specific situation of demand for anthropological contribution by the State facilitates the community participatory process. A well-structured State can develop a local system of green innovation and urban socio-ecological resilience to support the activities of companies, institutions, universities and research centers, the economic and private sector, etc.

Likewise, the process of urban planning and green innovation and green technology transfer begins with an idea that research turns into a technology that will be used and applied, followed by dissemination and evaluation of the potential economic value, as well as discussion and consultation with specialists in the urban areas concerned with the discovery, considering criteria based on scientific and technical relevance, existing patents, and market potential. From there, the registration of the patent and intellectual property follows. Therefore, the evolution of scientific knowledge and socio-ecological and technological green innovation in urban planning is driven by the challenges posed by the design of new goods, services, and processes by carrying out internal research and development activities in the companies and in cooperation with other agents, in efforts to improve the efficiency of regions with deficiencies in their scientific, technological and innovation systems.

However, companies that intend to innovate in socio-ecological urban planning face financial obstacles. Fiscal and market incentives leave the decision on which urban development projects to finance in the hands of companies (de Melo, 2001). Therefore, to stimulate investments in green science, green technology and socio-ecological innovation systems for urban planning, the necessary conditions are required to invest in research and development (de Melo, 2001). Unproven green technological innovation has high risks and difficulties in assessing the market for green high-tech socio-ecological goods and services, and green technology-based companies have difficulties in accessing sources of capital. Investment activities in innovation activities have a high level of risk and uncertainty in investments in credit financing and equity financing in research and development of urban green planning innovation in technologies.

Currently, universities offer technical training and development in urban planning and socio-ecological innovation. In addition, university-business interactions occur between socio-ecological and environmental urban planning research institutes and centers with companies interested in joint research programs sponsored by consortia or individual government agencies and corporations to exploit the results through patents and licenses. Thus, through business incubators, universities and research institutions, technology parks and agents of science, green technology and socio-ecological innovation in local university-based green innovation systems, early-stage start-ups may have the opportunity to access government venture capital.

In accordance with the above, government funding of research in green technological innovation for urban planning carried out by companies has a positive effect on business spending on research and development (Guellec & van Pottelsbergue, 2000, in Melo, 2001). Government authorities facilitate the formation of local green innovation and urban socio-ecological resilience systems from universities and research centers by increasing public investment in science foundations, green technology and socio-ecological innovation. Government intervention in university-oriented local urban green innovation can regulate incentives to promote green technology innovation training and strengthen green technology-based start-ups. The technological training of companies reinforces the convergence of scientific research on green innovation in universities.

Companies can share training in green technological innovation, fostered by public green investment. Technological and social capital determines the level of socio-ecological and urban anthropology in companies' green innovation. The accumulation of social, economic, and symbolic capital is relevant to undertake planning activities for urban socio-ecological resilience and green innovation (Reguera, 2009). On the other hand, local urban innovation systems for green socio-ecological areas integrate the definitions considered in Lundvall (1992) and the one used by MIT (Lester, 2005). Local innovation systems in urban planning have some elements that interact with the creation, diffusion, transfer and use of knowledge for economic value in specific regions (Lundvall, 1992). The context of the transfer of urban green technological innovation from the University to companies is a factor of competitiveness of the productive sectors of society.

Thus, government agencies and institutions of research centers and universities make massive investments in infrastructure and equipment for urban planning and green innovation (Porter, 2001). Creation of technical schools and incentive programs for the application of urban planning policies and innovative green practices in the industry. Therefore, urban socio-ecological resilience planning and green innovation networks constitute a related economy of connections, which can be effective in the future of work by fueling debates of ideas, thoughts, and feelings to discover the point of view. of the natives in the local perspective to obtain, analyze and interpret the data.

Some actors such as research laboratories, the network of students and researchers, incubator companies, angel investors and venture capital, etc., participate in urban environmental planning processes, although those who are not involved in the governance of local urban planning and socio-ecological innovation systems are not necessarily less influential. However, non-financial government institutions are hardly available to complement the green investment of private venture capitalists in green companies. Declining government involvement is a principle that applies to mature local urban green innovation systems (OECD, 1997b; de Melo, 2001).

Anchoring organizations and companies focus on urban planning of university-oriented local innovation systems to cooperate in joint efforts of R&D projects directing the focus of research cooperation towards socio-ecological and green areas with which they can gain a competitive advantage with a competent university. On the other hand, the group of green companies participates in an articulating task to support identification, selection and consolidation mechanisms of urban green innovation projects through articulating instruments (Pimienta-Bueno, 1999). The local urban socio-ecological concentration of green companies allows economies of scale, agglomeration and gains from externalities (Audretsch, 1998).

Green urban innovation technologies in technology-intensive emerging green companies can be compared to mature companies in terms of achievements and results. Green technology-based start-ups have higher costs related to their seeding and start-up phases, which are higher than those of small firms (OECD, 1997b).

DISCUSSION

The intersection between expanded urbanization and multilevel analysis in urban socio-ecology and resilience is a promising line of research (Connolly et al., 2021; Wolf, 2016). More recently, the theory and methods of multi-criteria decision analysis (MCDA) assess urban sustainability and socio-ecological resilience complemented with AHP and Electre III meta-analysis techniques forming a hierarchy of problems to make decisions that satisfy the criteria and indicators of urban communities in cities that suffer frequent natural disasters, for example (Kosova, Qendraj & Xhafaj, 2022).

The impact of specific risk disasters in urban areas, such as earthquakes, hurricanes, floods, public health emergencies, etc., leading to urban socio-ecological planning resilience to prevent the possible impact of these extreme events and provide a valuable diagnosis of the weak links in urban resilience and improve the resistance and resilience of the city of the future sustainable development of human society. Regarding the above, the case analyzed by Mu, Fang, Yang, & Guo, 2022 stands out) based on population mobility, centrality, and nodality between cities in China, which was significantly reduced, affecting the urban resilience within the network during the COVID-19 epidemic. Other natural phenomena are also causing urban ecological and resilience dysfunctions, such as strong storms that inundate urban communities, immobilize them, and disrupt urban transport, energy, water, and waste networks.

Cities and any urban areas are practically saturated with fossil fuels controlled to a large extent by corporations. The interactions of stakeholders and actors contribute in disarray to identify the complexity inherent in urban socio-ecological meta-problems and contribute to intensify the complexity that the intertwining process requires to address.

It should be noted that the objective of social ecosystems in urban areas is to maximize the benefits of human needs in the Anthropocene era. The concept of the Anthropocene is linked to the socio-ecological aspects of climate change and to the analysis of the causes of the collapse of civilization and of societies that can learn from archaeological research when dealing with crises (Tainter, 1988; Tainter, 2008; McIntosh, et al., 2000; Trawick & Alf, 2015). Western science as a mode of existence is a source of prescriptions for climate change complemented by interpretation and translation (Latour, 2013; Finucane, 2009). Communities merge global scientific explanations with regional and local knowledge that blame humans for global warming and climate change (Schnegg et al., 2021).

Likewise, the approach of the global framework for climate services has provoked critical reflections on initiatives related to ethical implications, epistemic politics, colonial legacies, etc. (Haines, 2019b; Krauß, 2020; Nost, 2019; Webber, 2019). International political frameworks underpinned discourses combined with the adverse effects of warming (Baer & Singer, 2014). Consistent with anthropogenic climate change perspectives, humans may be responsible for climate. People make sense of and respond to changing patterns of global warming and climate change (Barnes et al., 2013; Fiske et al., 2014; Peterson & Broad, 2009). The global transformation has at the center of the concerns that Anthropocene in climate change.

A sociological approach refers to unpredictability in Bauman and Beck's work research and social acceleration by Hartmut Rosa (Beck, 2009; Hartmut Resonanz, 2016; W. S. et al., 2007). The evolution of human activities has given rise to temporary urban landscapes perceived as natural but shaped by the evolution of generations (Meeus, 1995; Kuna and Dreslerová, 2007). The transformations of urban landscapes by human groups cause anthropogenic impacts on the local socio-ecological balance. The donut economy model proposes knowledge and practices for the resilience, sustainability and well-being of urban socio-ecology that should be applied everywhere (Strang, 2017).

Now, urban resilience is more tangible than invisible greenhouse gases, with serious implications. The isolation of homes is associated with investment and frugality, while organic food with risk and pastoral ideals. (Press and Arnould, 2014; Wilk and Wilhite, 1985). Furthermore, taking up the ability of human beings to make decisions to choose nutritious and healthy foods that increase the chances of survival (Birch et al., 1998; Cashdan, 1998; Wright, 1991).

At the same time, pleasure, holistic health, and purity are values of purchasing behavior in some European countries (Ditlevsen et al., 2019). Where, consumption is relational embedded in social relationships. Consumers of traditional marketing goods and services are alienated (Lambert, 2019). In accordance with the above, Marx's criticisms of consumer capitalism led to a naturalistic epistemology (Kramer, 2021; Stoner, 2021; Trujillo, 2021). In line with the above, social exchange theory is a dyadic approach to marketing that is extended to incorporate philanthropic actors and multiple stakeholders. Marketing theory and practices replace the consumer fetish with value creation by a collective of stakeholders (Arnould & Cayla, 2015; Tsing, 2015; Atanasova, 2021; Casey et al. 2020), Giesler, 2006, 2008; Figueiredo & Scaraboto, 2016).

Without replacing the concept of consumer, consumers can share, gift, barter, and other semiotic consumption processes (Belk, 2014; Wei et al., 2021). Consumption is socially learned and individually determined during primary socialization. The urban anthropology of consumption is the study, analysis and interpretation of consumption as a social phenomenon that incorporates the acquisition, consumption and above all the appropriation of products and brands that serve as a reinforcer of a person's identity (Páramo & Ramírez, 2017).

CONCLUSION

The analysis sought to find some implications and relationships between urban anthropology and the resilience of urban socio-ecological planning in the Anthropocene, based on the assumption that urban anthropology supports the creation and development of any urban planning based on socio-ecological resilience. urban Anthropocene. This analysis concludes that the foundations of urban anthropology serve as support to reinforce the anthropocenic orientation of the resilience of urban socio-ecological planning.

Urban anthropological research is positioned between the limits of global scope in connections and assemblies with urban anthropological socio-ecology and the areas of resilience and green innovation. Anthropoid dominance has resulted in the scarcity of the planet's resources that may be a more acute problem than climate change. The institutional support of urban anthropology to local urban planning and the socio-ecological innovation system is a relevant factor to achieve successful results. Urban planning and eco-innovation institutions are characterized by their ability to create and develop social capital to facilitate relationships of cooperation and conflict.

There are various levels of urban anthropology applied to urban green planning and socio-ecological innovation. The urban green innovation technology park provides green companies that seek the benefits of agglomeration. The Emergence of Innovative Green and Socio-Ecological Enterprises explores opportunities in urban innovation from the main sources and drivers that use the wealth of green knowledge. The generation of scientific green knowledge and technological socio-ecological innovation, extension and consulting activities should be structured through various mechanisms for planning and the formation of cooperative urban communities structured by interactions of deliberative processes to stimulate entrepreneurship and the innovation.

On the other hand, urban anthropology in the local dimension of the university-based socio-ecological urban planning and green innovation system must be supported by financial subsidies to guide the efforts of the governance structure. Another mechanism of urban socio-ecological planning and innovation is the indirect support of green infrastructures and green asset constitutions used by flow companies and a set of projects. Green infrastructure, socio-ecological physical facilities and green technologies in urban planning and innovation centers are green assets that can be financed. The foregoing is an effort that requires shared objectives and collaboration supported by a governance structure capable of synchronizing the mission and objectives of the agents and actors from the perspective of urban anthropology analysis to urban planning and urban centers and parks. socio-ecological innovation.

Deliberative actions on innovative urban green areas are triggered by normative and deliberative actions conducted by agents and actors. The stakeholders involved focused on the urban green innovation technology park which should have autonomy, flexibility and should avoid erosions caused by less naturally occurring regulatory issues. Similarly, research on the resilience mechanisms of socio-ecological urban planning based on cross-cutting linkages should be strengthened.

Finally, the reductionism of the works that address issues related to urban anthropology and urban socio-ecology of the Anthropocene are recognized as limitations of the present work, which present a diffuse causality about planning and resilience, therefore, For future research, an analysis of a greater number of case studies is established as recommendations, which allow a detailed understanding of urban anthropology as a support to create and develop urban planning based on the socio-ecological resilience of the Anthropocene. Regarding the above, the comprehensive analysis of these constructs is recognized as a contribution of this study.

REFERENCES

Aldrich, H. (2014). *Stand up and be counted: Why social science should stop using the qualitative/ quantitative dichotomy*. LSE. https://blogs.lse.ac.uk/impactofsocialsciences/2014/11/28/stand-up-and-becounted-social-science-qualitative-quantitative-dichotom

Alvarez, R. (2016). *Computational social science*. Cambridge University Press. doi:10.1017/ CBO9781316257340

Arnould, E. J., & Cayla, J. (2015). Consumer fetish: Commercial ethnography and the sovereign consumer. *Organization Studies*, *36*(October), 1361–1386. doi:10.1177/0170840615580012

Atanasova, A. (2021). Re-examining utopia in contemporary consumption: Conceptualization and implications for marketing. *AMS Review*, *11*(1-2), 23–29. doi:10.100713162-021-00193-0

Audretsch, D. B. (1998). *Agglomeration and the Location of Innovative Activity*. ULE.

Baer, H., & Singer, M. (2014). *The anthropology of climate change: An integrated critical perspective*. Routledge. doi:10.4324/9781315818702

Baldassarri, D., & Abascal, M. (2017). Field experiments across the social sciences. *Annual Review of Sociology*, *43*(1), 41–73. doi:10.1146/annurev-soc-073014-112445

Barnes, J., Dove, M., Lahsen, M., Mathews, A., McElwee, P., McIntosh, R., Moore, F., O'Reilly, J., Orlove, B., Puri, R., Weiss, H., & Yager, K. (2013). Contribution of anthropology to the study of climate change. *Nature Climate Change*, *3*(6), 541–544. doi:10.1038/nclimate1775

Barrett, S. (2009). *Anthropology*. University of Toronto Press. doi:10.3138/9781442697362

Beck, U. (2009). *World at Risk*. Polity.

Belk, R. W. (2014). You are what you can access: Sharing and collaborative consumption online. *Journal of Business Research*, *67*(August), 1595–1600. doi:10.1016/j.jbusres.2013.10.001

Bernard, H. (2011). *Research methods in anthropology: Qualitative and quantitative approaches*. AltaMira Press.

Bilharz, J. A. (1998). *The Allegany Senecas and Kinzua Dam: Forced Relocation through Two Generations*. University of Nebraska Press.

Birch, L. L., Gunder, L., Grimm-Thomas, K., & Laing, D. G. (1998). Infants' consumption of a new food enhances acceptance of similar foods. *Appetite*, *30*(3), 283–295. doi:10.1006/appe.1997.0146 PMID:9632459

Brady, H. (2019). The challenge of big data and data science. *Annual Review of Political Science*, *22*(1), 297–323. doi:10.1146/annurev-polisci-090216-023229

Bunge, M. (1999). *Buscar la filosofía en las ciencias sociales*. Siglo XXI.

Casey, K., Lichrou, M., & O'Malley, L. (2020). Prefiguring sustainable living: An ecovillage story. *Journal of Marketing Management*, *36*(17–18), 1658–1679. doi:10.1080/0267257X.2020.1801800

Cashdan, E. (1998). Adaptiveness of food learning and food aversions children. *Social Sciences Information. Information Sur les Sciences Sociales*, *37*(4), 613–632. doi:10.1177/053901898037004003

Connolly, C., Keil, R., & Ali, S. H. (2021). Extended urbanisation and the spatialities of infectious disease: Demographic change, infrastructure and governance. *Urban Studies (Edinburgh, Scotland)*, *58*(2), 245–263. doi:10.1177/0042098020910873

de Melo, A. (2001). *The Innovation Systems of Latin America and the Caribbean*. Inter-American Development Bank. Agosto.

de Melo, M. A. C. (1997). Processo de Planejamento e as Inovações Tecnológicas e Sociais: uma perspectiva sócio-ecológica. *Anais do 5o. Seminário de Modernização Tecnológica*. Maxwell.

de Melo, M. A. C. (2002a). *Enriquecendo a Atuação de Incubadora de Emrpesas. Tecnologia e Inovação: experiências de gestão nas micro e pequenas empresas*. PGT/USP.

de Melo, M.A.C. (2002b). Inovação e Modernização Tecnológica e Organizacional nas MPMEs: o domínio interorganizacional. *Seminário Internacional: Políticas*

DeWalt, K., DeWalt, B., & Wayland, C. (1998). Participant observation. In H. R. Bernard (Ed.), *Handbook of methods in cultural anthropology* (pp. 259–299). Altamira.

Diamond, J., & Robinson, J. (2010). *Natural experiments of history*. The Belknap Press of Harvard University Press.

Díaz, D., Kristoff, J., Castro, M., Miceli, J., Castro, D., Quinteros, R., & Guerrero, S. (2007). *Exploraciones en antropología y complejidad*. Editorial Sb.

Ditlevsen, K., Sandøe, P., & Lassen, J. (2019). Healthy food is nutritious, but organic food is healthy because it is pure: The negotiation of healthy food choices by Danish consumers of organic food. *Food Quality and Preference*, *71*(January), 46–53. doi:10.1016/j.foodqual.2018.06.001

Dunning, T. (2012). *Natural experiments in the social sciences*. Cambridge University Press. doi:10.1017/CBO9781139084444

Durrenberger, P. (2007) The Anthropology of Organized Labor in the United States. *Ann Review of Anthropology, 36*. Annual Reviews. https://www.annualreviews.org/doi/pdf/10.1146/annurev.anthro.36.081406.094433

Egerton, F. N. (2001). A history of the ecological sciences: Early Greek origins. *Bulletin of the Ecological Society of America*, *82*(1), 93–97.

Figueiredo, B., & Scaraboto, D. (2016). The systemic creation of value through circulation in collaborative consumer networks. *The Journal of Consumer Research*, *43*(4), 509–533. doi:10.1093/jcr/ucw038

Finucane, M. L. (2009). Why science alone won't solve the climate crisis: Managing climate risks in the Pacific. *Asia Pacific Issues*, *89*, 1–8.

Fiske, S. J., Crate, S. A., Crumley, C., Galvin, K., Lazrus, H., Lucero, L., Oliver-Smith, A., Orlove, B., Strauss, S., & Wilk, R. R. (2014). *Changing the atmosphere. Anthropology and climate change. Final report of the AAA global climate change task force*. American Anthropological Association.

Giesler, M. (2008). Conflict and compromise: Drama in marketplace evolution. *Journal of Consumer Research*, 34(April), Giesler, M. (2006). Consumer gift systems. *The Journal of Consumer Research*, *33*(September), 283–290. doi:10.1086/506309

Giner, C. (2004). *J. Antropología urbana*. Barcelona. *Ariel*.

González Álvarez, M. D. (1997). Processos de Planejamento nos Pólos Tecnológicos: um enfoque adaptativo. [Tese de Doutorado, Departamento de Engenharia Industrial, PUC-Rio, Rio de Janeiro].

Gravano, A. (1992). Antropología Práctica, muestra y posibilidades de la Antropología Organizacional. *Publicar en Antropología y Ciencias Sociales, Buenos Aires*, (1), 95–126.

Gravano, A. (2005). *El barrio en la teoría social*. Espacio Editorial.

Gravano, A. (2006). Imaginarios regionales y circularidad en la planificación: El caso del TOAR. *Intersecciones*, (7), 305–323.

Gravano, A. (2007b). Claves para la facilitación organizacional del proceso participativo en la planificación urbano-ambiental metropolitana. *Revista Regional de Trabajo Social*, *21*(40), 9-22.

Grimmer, J., Roberts, M., & Stewart, B. (2021). Machine learning for social science: An agnostic approach. *Annual Review of Political Science*, *24*(1), 395–419. doi:10.1146/annurev-polisci-053119-015921

Haines, S. (2019b). Managing expectations: Articulating expertise in climate services for agriculture in Belize. *Climatic Change*, *157*(1), 43–59. doi:10.100710584-018-2357-1

Hannerz, U. (1986). *Exploración de la ciudad*. Fondo de Cultura Económica.

Hannerz, U. (2004). *Foreign New: Exploring the World of Foreign Correspondents.* University of Chicago Press.

Hartmut Resonanz, R. (2016). *Eine Soziologie der Weltbeziehung*. Suhrkamp.

Hindman, M. (2015). Building better models: Prediction, replication, and machine learning in the social sciences. *The Annals of the American Academy of Political and Social Science*, *659*(1), 48–62. doi:10.1177/0002716215570279

Hirst, K. (2019). *Is anthropology a science?* ThoughtCo. https://www.thoughtco.com/isanthropology-a-science-3971060

Hofman, J., Sharma, A., & Watts, D. (2017). Prediction and explanation in social systems. *Science*, *355*(6324), 486–488. doi:10.1126cience.aal3856 PMID:28154051

Hofman, J., Watts, D., Athey, S., Garip, F., Griffiths, T., Kleinberg, J., Margetts, H., Mullainathan, S., Salganik, M., Vazire, S., Vespignani, A., & Yarkoni, T. (2021). Integrating explanation and prediction in computational social science. *Nature*, *595*(7866), 181–188. doi:10.103841586-021-03659-0 PMID:34194044

Holling, C. S. (1973). *Resilience and Stability of Ecological Systems. Annual Review of Ecology and Systematics* (Vol. 4). JSTOR. https://www.jstor.org/stable/2096802

Holmes, D. R., & Marcus, G. E. (2006). Fast capitalism: para-ethnography and the rise of the symbolic analyst. In M. Fisher & G. Downey (Eds.), *Frontiers of Capital: Ethnographic Reflections on the New Economy*. Duke University Press.

Holmes, D. R., & Marcus, G. E. (2021). How do we collaborate? An updated manifesto. In D. Boyer & G. Marcus (Eds.), *Collaborative Anthropology Today: A Collection of Exceptions*. Cornell University Press. doi:10.7591/cornell/9781501753343.003.0002

Jorgensen, D. (2015). Participant observation. In R. Scott & S. Kosslyn (Eds.), *Emerging trends in the social and behavioral sciences* (pp. 1–15). Wiley. doi:10.1002/9781118900772.etrds0247

Kaplan, O. (1940). Prediction in the social sciences. *Philosophy of Science*, *7*(4), 492–498. doi:10.1086/286658

Kawulich, B. (2005). La observación participante como método de recolección de datos. *Forum Qualitative Social Research*, *68*(2), 43.

Kolb, Ch. (1998). Review of Bilharz, Joy A., *The Allegany Senecas and Kinzua Dam: Forced Relocation through Two Generations*. H-AmIndian, H-Net Reviews. https://www.h-net.org/reviews/showrev.php?id=2366

Kosova, R., Qendraj, D.H., Xhafaj, E. (2022). Meta-Analysis ELECTRE III and AHP in Evaluating and Ranking the Urban Resilience. *Journal of Environmental Management* and Tourism, *3*(59): 756 - 768. doi:10.14505/jemt.v13.3(59).15

Kramer, J. M. (2021). From predator to parasite: On private property and our ecological disaster. *Journal of Economic Issues*, *55*(June), 416–422. doi:10.1080/00213624.2021.1908804

Krauß, W. (2020). Narratives of change and the co-development of climate services for action. *Climate Risk Management*, *28*, 100217. doi:10.1016/j.crm.2020.100217

Kuhn, T. S. (1962/1996). *The Structure of Scientific Revolutions* (3rd ed.). University of Chicago Press.

Kuna, M., & Dreslerová, D. (2007). Landscape archaeology and "community areas" in the archaeology of central Europe. D. Hicks, L. McAtackney, G.J. Fairclough (Eds.), Envisioning landscape: Situations and standpoints in archaeology and heritage. Left Coast Press.

Lambert, A. (2019). Psychotic, acritical and precarious? A Lacanian exploration of the neoliberal consumer subject. *Marketing Theory*, *19*(3), 329–346. doi:10.1177/1470593118796704

Lapassade, G. (1986). *Grupos, organizaciones e instituciones: la transformación de la burocracia*. Gedisa.

Latour, B. (2013). *An inquiry into modes of existence: An anthropology of the moderns* (C. Porter, Trans.). Harvard University Press.

Lester, R. K. (2005). *Universities, Innovation, and the Competitiveness of Local Economies A Summary Report from the Local Innovation Systems Project – Phase I.* Working Paper, Industrial Performance Center Massachusetts Institute of Technology.

Lindblom, C. E. (1959). The Science of Muddling through. *Public Administration Review*, *19*(2), 79–88. doi:10.2307/973677

Llanos, M. (2009). *Epistemología de las ciencias sociales*. UNMSM.

Lourau, R. (1979). *El análisis institucional*. Amorrortu.

Lundvall, B.-Å. (1992). (ed.). National Systems of Innovation: towards a theory of innovation and interactive learning. Printer Publisher. Londres.

Malinowski, B. (1961). *Argonauts of the Western Pacific*. Routledge.

Maran, T., & Kull, K. (2014). Ecosemiotics: Main principles and current developments. *Geografiska Annaler. Series B, Human Geography*, *96*(1), 41–50. doi:10.1111/geob.12035

Mayo, E. (1945). *The social problems of an Industrial Civilization*. Harvard University Press.

McIntosh, R. J., Tainter, J. A., & McIntosh, S. K. (Eds.). (2000). *The Way the Wind Blows: Climate Change, History, and Human Activity*. Columbia University Press.

Meeus, J. H. A. (1995). Pan-European landscapes. *Landscape and Urban Planning*, *31*(1-3), 57–79. doi:10.1016/0169-2046(94)01036-8

Mu, X., Fang, C., Yang, Z., & Guo, X. (2022). Impact of the COVID-19 Epidemic on Population Mobility Networks in the Beijing–Tianjin–Hebei Urban Agglomeration from a Resilience Perspective. *Land (Basel)*, *2022*(11), 675. doi:10.3390/land11050675

Nießen, D., Danner, D., Spengler, M., & Lechner, C. (2020). Big Five personality traits predict successful transitions from school to vocational education and training: A large-scale study. *Frontiers in Psychology*, *11*, 1827. doi:10.3389/fpsyg.2020.01827 PMID:32903700

Nost, E. (2019). Climate services for whom? The political economics of contextualizing climate data in Louisiana's coastal master plan. *Climatic Change*, *157*(1), 27–42. doi:10.100710584-019-02383-z

Odum, E. P. (1957). *Fundamentals of ecology*. W. B. Saunders Company.

OECD. (1997b). Government Venture Capital for Technology-Based Firms. Ducument OCDE/GD (97) 201. Paris: OECD, 1997b.

Olimid, A. P., Georgescu, C. M., & Gherghe, C. L. (2022). Influences of Covid-19 Crisis on Resilience Theories: An analysis of Community, Societal and Governance Resilience. *Revista de Ştiinţe Politice. Revue des Sciences Politiques*, *73*, 38–51.

Ortner, S. (2010). Access: Reflections on studying up in Hollywood. *Ethnography*, *11*(2), 211–233. doi:10.1177/1466138110362006

Ortner, S. (2013). *Not Hollywood: Independent Film at the Twilight of the American Dream*. Duke University Press.

Páramo, D., & Ramírez, E. (2017). *Etnomarketing, la dimensión cultural del marketing*. Klasse Editorial.

Peterson, N., & Broad, K. (2009). Climate and weather discourse in anthropology: From determinism to uncertain futures. In *Anthropology and climate change: From encounters to actions* (pp. 70–86). Left Coast Press.

Pfeifer, J. (2006). Laws of nature. In S. Sarkar & J. Pfeifer (Eds.), *The philosophy of science: An encyclopedia* (pp. 439–444).

Pimienta-Bueno, J.A. (1999). *Parque de Inovação Tecnológica e Cultural da Gávea: A Visão da PUC-Rio*. Documento Interno.

Pizarro, C. (2014). La entrevista etnográfica como práctica discursiva: Análisis de caso sobre las pistas meta-discursivas y la emergencia de categorías nativas. *Revista de Antropologia*, *57*(1), 461–496. doi:10.11606/2179-0892.ra.2014.87770

Porter, M. (2001). Clusters of Innovation: regional foundations of U.S. competitiveness. Council on Competitiveness. Washington, D.C: Monitor Group.

Press, M., & Arnould, E. J. (2014). Narrative transparency. *Journal of Marketing Management*, *30*(13–14), 1353–1376. doi:10.1080/0267257X.2014.925958

Reguera, A. (2009) Empresarios de ayer, de hoy y de siempre. Un recorrido latinoamericano por sus formas espacio-temporales *América Latina en la historia económica, 32*. https://www.scielo.org.mx/scielo.php?script=sci_arttext&pid=S1405- 22532009000200003&lng=es&nrm=iso

Reynoso, C. (1995b). Hacia la perfección del consenso: Los lugares comunes de la antropología. *Intersecciones*, *1*, 51–72.

Roca, A. (2008). *La comprensión de la tecnología en la antropología*. Actas VII ESOCITE UFRJ, Jornadas Latino-Americanas de Estudos Sociais das Ciências e das Tecnologías.

Roethlisberger, F., & Dickson, W. (1939). *Management and the Worker*. Harvard Univ. Press.

Roni, S., Merga, M., & Morris, J. (2020). *Conducting quantitative research in education.* Springer. doi:10.1007/978-981-13-9132-3

Saxenian, A. (1994). *Regional Advantage: culture and competition in Silicon Valley and Route 128.* Harvard Business Press.

Schlapfer, A., & Marinova, D. (2001). *Local Innovation Systems: nature, importance and role.* Conference Proceedings: International Summer Academy on Technological Studies: user involvement in technological innovation. Deutschlandsberg. Austria.

Schnegg, M., O'Brian, C. I., & Sievert, I. J. (2021). It's our fault: A global comparison of different ways of explaining climate change. *Human Ecology: an Interdisciplinary Journal*, *49*(3), 327–339. doi:10.100710745-021-00229-w

Signorelli, A. (1999). Antropología Urbana. Barcelona: Anthropos: UAM.

Stoner, A. M. (2021). Things are getting worse on our way to catastrophe: Neoliberal environmentalism, repressive desublimation, and the autonomous ecoconsumer. *Critical Sociology*, *47*(May), 491–506. doi:10.1177/0896920520958099

Strang, V. (2017). The Gaia complex: Ethical Challenges to an Anthropocentric 'Common Future. In B. Marc & L. Jerome (Eds.), *The Anthropology of Sustainability* (pp. 255–283). Palgrave. doi:10.1057/978-1-137-56636-2_12

Sundbo, J. (1998). *The Theory of Innovation: entrepreneurs, technology, and strategy*. Edward Elgar.

Taagepera, R. (2008). *Making social sciences more scientific: The need for predictive models.* Oxford University Press. doi:10.1093/acprof:oso/9780199534661.001.0001

Tainter, J. A. (1988). *The Collapse of Complex Societies*. Cambridge University Press.

Tainter, J. A. (2008). Collapse and Sustainability: Rome, the Maya, and the Modern World. *Archeological Papers of the American Anthropological Association*, *24*(1), 201–214. doi:10.1111/apaa.12038

Tarantino-Curseri, S. (2018). Antropología, etología, sociología, y psicología, 4 disciplinas útiles en el ámbito empresarial. *Management Review*, *3*(3). doi:10.18583/umr.v3i3.126

Trawick, P., & Alf, H. (2015). Revisiting the Image of Limited Good: On Sustainability, Thermodynamics, and the Illusion of Creating Wealth. *Current Anthropology*, *56*(1), 1–27. doi:10.1086/679593

Trist, E. L. (1976). *A Concept of Organizational Ecology: an invited address to the three Melbourne universities.*

Trujillo, M. L. N. (2021). Notes for a critical and ecological view of patriarchal capitalism in the web of life. *Capital and Class*, *45*(March), 21–32. doi:10.1177/0309816820929115

Tsing, A. L. (2015). *The mushroom at the end of the world*. Princeton University Press.

Velho, G. (1978). Observando o familiar. In: Nunez, E. de O. (Org.). An aventura sociológica: objetividade, paixao, improviso e método na pesquisa social. Rio de Janeiro: Zahar.

W. S. (2007). The Anthropocene: Are Human Beings Now Overwhelming the Forces of Nature? *Ambio, 36*, 614–621. doi:10.1579/0044-7447(2007)36[614:TAAHNO]2.0.CO;2 PMID:18240674

Webber, S. (2019). Putting climate services in contexts: Advancing multi-disciplinary understandings: Introduction to the special issue. *Climatic Change, 157*(1), 1–8. doi:10.100710584-019-02600-9

Wei, X., Lo, C. K. Y., Jung, S., & Choi, T.-M. (2021). From co-consumption to co-production: A systematic review and research synthesis of collaborative consumption practices. *Journal of Business Research, 129*(May), 282–294. doi:10.1016/j.jbusres.2021.02.027

Wilk, R. R., & Wilhite, H. L. (1985). Why don't people weatherize their homes? An ethnographic solution. *Energy, 10*(May), 621–629. doi:10.1016/0360-5442(85)90093-3

Wolf, M. (2016). Rethinking urban epidemiology: Natures, networks and materialities. *International Journal of Urban and Regional Research, 40*(5), 958–982. doi:10.1111/1468-2427.12381 PMID:32336869

Yang, Y., Lili, W., & Hongchi, Z. (2022). *Bibliometric research on the evolution of resilience theme from the perspective of Geographical Science*. 2022 29th International Conference on Geoinformatics, Beijing, China. 10.1109/Geoinformatics57846.2022.9963870

Chapter 2
Anthropological Ecology and Ecological Anthropology as a Framework for the Analysis of Socio-Ecological Sustainable Development

José G. Vargas-Hernandez
Tecnológico Superior de Jalisco, Mexico

María F. Higuera-Cota
Universidad Autónoma de Sinaloa, Mexico

Omar C. Vargas-González
Tecnológico Nacional de Mexico, Ciudad Guzmán, Mexico

ABSTRACT

This study is aimed at analyzing anthropological ecology and ecological anthropology as a framework approach for the analysis of socio-ecological sustainable development. It departs from the assumption that sustainable development requires a socio-ecological approach sustained by anthropology and ecology. The method used is analytical reflection based on the conceptual, theoretical, and empirical literature review. The study concludes that anthropology and ecology combine elements to sustain the anthropological ecology and ecological anthropology approaches to develop the framework for the analysis of a socio-ecological sustainable development.

INTRODUCTION

Ecology is adopted by various disciplines, geography, economics, psychology, behavioral ecology, sociology, anthropology, anthropological ecology, cultural ecology, urban ecology, political ecology, urban

DOI: 10.4018/978-1-6684-8879-9.ch002

political ecology, urban planning and design, ecological planning and design, landscape architecture (Lyle, 1985; McHarg, 1969; Ndubisi, 2002, Ndubisi, 2014; Palazzo, 1997; Steiner, 2001, Steiner, 1991, Steiner, 2016; Thompson & Steiner, 1997; Van der Ryn & Cowan 2007). The concept of human ecology evolved to embrace disciplines, such as economics, geography, land use, psychology, sociology, socioecology, socio-technology, sociobiology, anthropology, cultural ecology, anthropological ecology, etc. (Petterson, Kensek & Noble, 2017). This is due to the growing need to study the interaction between human beings and their environment with the aim of making this relationship sustainable.

In the same line of study of the relationships between the human being and its environment is anthropological ecology, which seeks to study the wide range of human responses to environmental problems. In turn, it has theoretical and practical implications on organizational sustainability through the integration of social and environmental dimensions. Sociology and anthropology are integrated into the organizational management aimed at co-evolution sustainability. The ecology of sustainability is a resource available to manage environmental changes in areas such as market forces, technology, geographical locations, natural resources, etc. Anthropology as well as psychology, medicine and social sciences have been interested in the study of diet in health and environmental effects of ecological adaptations (Montgomery, Bennett, & Scudder, 1973; Netting, 1974).

The engagements of urban political ecology with the public health crisis of Covid-19 led to important anthropological and biological insights. This is because through urban political ecology it is possible to explore the interactions between social, political, economic, and natural factors both in the production and reproduction of urban environments. As a result of the health crisis, it was possible to become aware not only of its effects on health, economy, and social development, but also the impact of human beings on earth, where industrialization, massive deforestation, extensive use of pesticides in the Agriculture, consumerism, the indiscriminate use of plastics, among other practices, are present and deepen the environmental disorders linked to climate change, which originate from human action.

Focus on climate change brings vulnerabilities with unsustainable development practices with impacts through material changes in socio-ecosystems driven by climate processes (Barnett & Adger, 2003). Anthropological and ethnographic ecology are used in policy making in climate change, considered a biophysical phenomenon. The ethnographic analysis of the human-centered efforts is directed towards a sustainable environment and ecological and relations. Comparative analysis of ecological anthropology uses cross-cultural datasets in ecological dynamics influencing human population size, land ownership, environmental impact, social behavior, social inequality, etc. (Bliege Bird & Codding, 2022).

Anthropological ecology framework of cultural ecology is used for analytical and theoretical approach of organizational phenomena (Weick, 1969). Biocultural perspectives in anthropology aim to analyze the deterministic bases of the economic, social, environmental, cultural, and ecological factors influencing human biology and wellbeing (McElroy, 1990; Elith, 2006; Leonard, 2020; Alland, 1966). Biocultural perspectives are necessary when analyzing organizational phenomena, since the organizational phenomenon is framed within a represented complexity, where the context is understood as a whole, in which inside, the individual, the Society, the State, shareholder, environment, are interrelated and interact in diverse ways. Therefore, conservation actions in a biocultural approach that sustain the biophysical and sociocultural components of dynamic and interdependent socio ecological systems are necessary.

Cultural studies refer to nature-culture more than ecocultures nurturing the linkages between the ecological and social environments in biocultural diversity, different concept of bio culture in biocultural studies (Daryl Slack, 2008; Giblett, 2012; Böhm, Bharucha & Pretty, (Eds.) 2015; Simberloff, 2018; Davis & Morris, 2007; Ehlers, & Krupar, (Eds.); Krupar & Ehlers, 2020). Biocultural diversity is all biological, environmental, cultural, and linguistic manifestations interrelated to life diversity adapted to a complex socio-ecological system. Biocultural perspectives in anthropology aim to analyze the deterministic bases of the economic, social, environmental, cultural, and ecological factors influencing human biology and wellbeing (McElroy, 1990; Elith, 2006; Leonard, 2020; Alland, 1966).

Biocultural studies in ecological anthropology are a term used in biocultural diversity to explain co-evolving socio-ecological systems and implications with cultural dimensions (Mercon, *et al.,* 2019). The term biocultural has overlapping descriptions used in biocultural diversity in the context of its usage being different from biocultural studies within anthropology. Biocultural studies in anthropology are a term used in biocultural diversity to explain co-evolving socio-ecological systems and implications with cultural dimensions.

Socio-ecological systems focus on international policy and planning of sustainable economic development to contribute to human wellbeing and health (European Commission, 2017; Food and Agriculture Organisation, 2017; United Nations Environment Program, 2012; United Nations General Assembly, 2017). The research design of biocultural studies includes human biology, medical ecology, and social, cultural, and behavioral variables (McElroy, 1990). Biocultural diversity is all biological, environmental, cultural, and linguistic manifestations interrelated to life diversity adapted to a complex socio-ecological system.

Biocultural studies within biological anthropology assumes humans as biological, social, and cultural beings (Elith, 2006) with connecting synergies to ecological and medical anthropology and political economy (Fletcher & Fortin, 2018; Fourcade, Engler, Rödder, & Secondi, 2014), evolutionary theory, political economic analysis and biocultural studies in anthropology (McElroy, 1990; Wiley & Cullin, 2016; Hoke & Schell, 2020; Hoke & Schell, 2020; Willermet, 2016; Farahmand, Nahrevanian, Shirazi, Naeimi, Farzanehnejad, 2011; Ghatee, Taylor & Karamian, 2020). Indigenous peoples have been unwilling subjects regarding science in anthropological, ecological and medical studies because they have been used in resource extraction (Schnarch, 2004).

Evolutionary anthropology is related to the abundance of resources, structure and distribution that have implications on the socio-ecological adaptive conditions leading to variability of resources on subsistence economies mediated by the interpersonal human violence (McCool, et al., 2022). Socio-ecological sources and methods of management policies to assess species recovery, restoration and conservation towards sustainable socio-ecological systems in operating social space (Raworth, 2012).

Development of new sustainable economic activities is an eco-restructuring process. These activities seek to increase social well-being by promoting responsible consumption through economic practices that are respectful of the environment and committed to society. It is in eco culturalism where the cultural and ecological dimensions are linked, such as biocultural diversity (Cajee, 2014). Issues such as economic development and social development are relevant for the exploitation of sustainable tourism and the regulation of territorial and urban development. Understanding sustainable tourism as the vehicle for regional and local socioeconomic development (Zan, Lusiani, 2011).

Previous research proceeds separately and must be integrated under sustainable development. Public welfare has been linked with private profits and was formulated on the assumptions of "trickle down" theory of the economics sustainability growth and development. Research and development of science and technology innovation is a fundamental factor in creating economic growth, social value, and environmentally sustainable development.

After this introduction to the study, the analysis continues delimiting the interrelationships between anthropological ecology and ecological anthropology as a conceptual, theoretical, and empirical framework leading to continuing with socio-ecological sustainable development. Finally, some conclusions are given.

ANTHROPOLOGICAL ECOLOGY AND ECOLOGICAL ANTHROPOLOGY

Anthropological wisdom considers humanity as an element of existence distinguishing from ecological wisdom that advocates an interdisciplinary approach to sustainable lifestyles development referring to the willingness of an embedded place in an ecosystem deciding the behavior in relation to the quality of life (Weston, 2009). Organizational sociology and ecological anthropology are at the forefront of ecology and natural resources management to integrate organizational sustainability development and conservation. The multidisciplinary approach combines the perspectives of organizational sociology and ecological anthropology in the integration of sustainability in organizational ecology.

Ecological anthropology is the foundation for sustainable development programs linking economic growth with sustainable community development. Ecological anthropology takes for granted these assumptions and categories to subject the analytical concept and current anthropological thinking and analyses of culture in organizations, providing a historical perspective of the development of organizational anthropology studies and the analytical methods used in research.

Organizations involved in local communities practice the anthropological perspective of sustainability and ownership, sponsoring public culture, and public recreational events, etc., always involved in local sustainability and social development programs. (Esty and Winston, 2006). The socio-ecological crisis of the communities supposes multiple challenges that go from the environment to the reproduction of patterns of social inequality, a situation that leads to biocultural crises. Ecocultural communities prioritize ethical principles to nurture ecological and sociological and economic wellbeing, considering humans in nature to sustain human and social capital (Böhm, Bharucha, Pretty (Eds.) 2015; Yongming, 2010).

A critique consistent with the current ecological crisis argues that the sustainable macromarketing entails the comprehensive consumption fostering corporate profit maximization in the market and a take-making waste ecology reliant un unsustainable demand of raw natural resources and energy (Borland & Lindgreen 2013; Kilbourne et al., 1997; Prothero et al., 2011; Slater & Tonkiss, 2001; O'Neil et al., 2018; Stiglitz, 2015; Temesgen et al., 2021; Wallace-Wells, 2019). In other words, the extractivist model is questioned, as well as its cultural patterns of consumption and entertainment. It is believed that both agents, articulations and States could be emissaries of change if their efforts are aimed at reducing poverty and guaranteeing quality development for present and future generations without compromising the planet's resources. Socio-ecological issues is the starting point of the sustainable marketing process and not as a set of constraints and externalities (Belz & Peattie, 2010)

The macro relational perspective treats socio ecological issues as the marketing mix process focusing on the transformational potential of institutional change for sustainable production, distribution, consumption and reproduction free of externalities (Belz & Peattie, 2010; Kemper & Ballantine, 2019).

Ecological anthropology assumes that sustainable accounting and reporting addresses natural resources management and conservation consistent with local cultures aimed to promote sustainable development. Resource management of communities are challenged to measure the socio-ecosystem changes with the acknowledgement of archaeological and anthropological data to identify changes of socio-ecological observations beyond the time scales (Pauly, 1995; Dayton and others 1998). Ecological anthropology and organizational disciplines benefit from sustainable accounting reporting (Kottack, 1999; Stone,

2003). Ecological and sustainable accounting reporting analysis incorporated into strategic planning and control aimed to address environmental factors and to manage and mitigate investments associated with financial risks.

Sustainability accounting focusing on the market includes process improvements, planning and control decisions, organizational growth and development, management and accounting costs in sustainable development, performance measures, etc. Ecological anthropology approach supports the economic and technological sustainability, the development dimensions of market and social components to protect natural resources and the environment. Long term sustainability interests require calculations for investments, production, and preservation of ecological and natural resources (Hopwood, 2009).

Ecocultures in cultural, environmental communication and psychology link ecology and culture and advocating sociocultural identity inseparable of ecology and recognizing local knowledge and memories in a dynamic lynk between the landscapes and ecocultural identities that promote resilience (Pretty, 2011; Berry, 2004; Nsamenang, 2015; *Chen, et al.* 2012; Daryl Slack, 2008; Milstein & Castro-Sotomayor, 2020; Hoffmann, 2020; Dahake, 2020; Zedler, & Stevens, 2018).

The eco-cultural perspective advocates sociocultural identity inseparable from ecology based on local memories and knowledge linked to landscapes and human culture shaping ecocultural identities and promoting resilience (Daryl Slack, 2008). Younger generations from villages and indigenous communities go away for opportunities of study and employment losing their best people. Migration can be explained by multiple issues, but the most recurrent point to issues of security, poverty, human rights violations, and climate change. Among the main consequences of migration is the loss of population, a situation that can reduce the productive potential of the communities of origin by generating a shortage of labor in certain sectors or specific industries and therefore tends to hinder economic growth. This issue is related to the conceptual framework in terms of cultural ecology and anthropology (Scudder & Colson, 1982; Scudder, 1973; Mongomery, Bennett, & Scudder, 1973).

Organizational ecological anthropology enhances sustainability and environmental resources management linked to production, service, quality, and managerial systems and promotes organizational learning on environmental and natural resources conservation (Cohen-Rosenthal, 2000; Ehrenfeld, 2000). The organizational abilities to respond to competitive environmental changes are dependent on the availability of ecological resources. Alignment of organizational ecological resources based on knowledge and learning systems development is related to sustainable and environmental management, resources conservation integrated into the organizational practices and activities aimed to improve operational performance.

Ecological anthropology addresses social and cultural forces that shape sustainability development. The ecological anthropological perspective on sustainability implies the joint ownership of resources claimed by stakeholders, performers, and community.

The application of social science approaches, sociology, and anthropology in organizational anthropological ecology. The anthropological ecology perspective of technology tends to solve specific problems in socio-ecological and organizational environments. Culture and ecology are concepts in organizational communication theory, and organizational anthropology theory and analysis, management sciences, and anthropological ecology (Everett, 1985).

Theoretical approaches, sustainable practices and belief systems of ecological populations lead to the generation of sustainability models with implicit ethics of the natural environmental care and particular ways of living. Sustainability accounting reporting systems are part of the ecological population systems that increase competition and survival to use as strategic resources for organizational performance measures associated with growth, survival, and death. Ecology and ethnoecology study the relationships between organisms and the physical, biological, and social factors they come into contact with, based on a political ecology principle that the contemporary human society and natural environment are not under control of exploitation and deterioration processes (Toledo, 1997).

Development of sustainable relationships with the natural environment focuses on sustainability models leading to some approaches in interdependence with human beings in their communities. The term Political ecology was used by Eric Wolf (1972), an anthropologist who described and analyzed the interactive relationship between human environment, territory, and space without an explicit connection to biological sciences.

The existing models of sustainability based on ethnoecology are already threatening to maintain the long-term sustainability of natural environments as the livelihoods of communities (Toledo, 1997). The ethno-ecology sustainable development at the community requires the natural and cultural endowments of the traditional ecological knowledge. The ecodevelopment has the objective defined as the strategy to produce and apply knowledge and techniques for the sustainable management of ecosystems and their struggles in communities for their social wealth and appropriation of natural resources (Leff, 1995).

The ethno-ecological knowledge kept by indigenous communities of people and increasingly used for natural history and anthropological studies (Heyd, 1995; Gragson & Blount, 1999). Ethno-ecology makes relevant the kind of knowledge in which humanity and their natural environments are integrated in sustainability. Many ecological populations are at the crossroads between the unsustainable and the traditional sustainable environments which resist change forces of sustainable development.

The ecosystem models the scientific and conventional logical habits among the ecological relations. The social ecological model is proposed by Golden & JoEarp, (2012), McLeroy et al., (1988), Stokols, (1992) and Stokols, (1996). Application of Eco semiotics considers humans and other species to co-constitute ecological phenomena (Maran & Kull 2014). Local technoscientific ecological models of the damage-centered environment with individual suffering related to the empirical ecological relations of human-environments (Mc-Clintock, 2014). The objective model of local ecology must make visible the spiritual connections to place, land, and the environment to generate the meaning. Thus, a comprehensive, systematic, and naturalistic vision of development is generated, understood as a complex process, which responds to the influence of multiple factors closely related to the ecological environment.

Technoscientific models of relationality, such as radioecology, enable us to visualize the cross-boundary movements. An ecological relationality framework brings relational arrangements by interrelated chance, supernatural force, and extra scientific logic, besides through blood, marriage, and fictive kinship (Hashimoto, 2013; Jensen et al. 2016; Goldfarb, 2016). The relational model proposed by Nozawa (2015) is different from the decontamination technoscientific policy and radioecology across the space and time linkages between humans, nonhumans, and the environment (Creager, 2013).

The radioecological framework isolates humans from contaminants forcing locals to coproduce nuclear waste. The radioecological across the multiple human relations to local accounts of connectivity. Locals consider that radioecology offers more than limited human relations and delimits an Eco semiotic or ineffable interrelated of living, decaying, and dying things.

The anthropocentric delimitation of human-environment rationality leads to socio ecological policy. The anthropocentric assumption about human separability from the environment to justify harming lives, values, historicity, relationships, things, and futurity (Morimoto, 2022) is complemented by a more functionalist approach to local forager culture guided by ecological anthropological principles and genetic determinism is adaptive to some animals, such as the elephants Rappaport, 2000; Reich, 2018).

A functionalist approach to local forager culture guided by ecological anthropological principles and genetic determinism is adaptive to the elephants Rappaport, 2000; Reich, 2018).

Wild boars are emblematic of the ecological upheaval (Walker, 2005) that residents experience in finding pre-disaster residence habituated by animals bided by anthropocentric division between wilderness and home (Cram, 2016). The dominance of biomedical and ecological orientation against the anthropocentric narratives illuminates the multiscale semiotic decontamination efforts.

The efforts of the state to remove and contain contaminants transform local ecology through decontamination presupposing the residents connected to the land, ancestors, nonhumans, and the environment. State and technoscientific decontamination rely on a radioecological model focusing on the environment to do entanglements of humans and contaminants (Masco, 2006).

Radioecology uses specific radioisotopes to investigate tracing and mapping connections of radiological contaminants and their transboundary movements between the earth and the atmosphere (Odum, 1959). The resultant production of accumulative nuclear waste brings ecology to collective recovery. Nuclearism is the ecological and technopolitical distribution of nuclear elements across time, space, human and non-human (Hecht, 2012).

Appearance and disappearance of radioisotopes and ecological diagrams among humans, nonhumans and abiotic factors leading to constitute models of ecological relationship in the post fallout environment. Liberation ecology is based on the idea that the time comes for the liberation of humanity and the natural environment (Marcuse, 1972; Gómez Hinojosa, 1990). Liberation ecology movements are more concerned for the environmental integrity for livelihood, entitlements, and social justice (Peet & Watts, 1996).

Traditional technology is part of the organizational space in relationship with organizational anthropology perspective in the articulation with modern technology in organizational productive developments for example, in rural organizations and business. The movements reveal the limits of the state to the post fallout ecology. Theoretical approach to fallout links human behaviors and actions, technological capabilities, atmospheres, environments, and ecologies in a new formation of contamination (Masco, 2015). The state`s policy approach to ecological redress post fallout impacts needs a local framework of relationality.

Achieving radiation safety implementing a strategy of decontamination produces ecological harm rendering wild boars, local food chains and land complicit in production of nuclear waste. The decontamination policies enable the gradual return of evacuees across the land with the opportunistic incursions of animals in asymmetrical redistribution of contaminants in the local ecology. Wild boars unconstrained move contaminants across the evacuation zones frustrating attempts at containment (Morimoto, 2022). Anthroponotic diseases are caused by global warming and changes in anthropological ecology (Desjeux, 2001) caused by the causative agent "L" tropica.

Scientific problematization of the post fallout environment and state usage of an objective and factual model of local ecology, ruptures, and redress. Post fallout ecology tends to threaten the local model of relationality among animate and inanimate things, animals and plants connected to the environment (Goldfarb, 2016; Nozawa, 2015). The ecological fallout in a planetary scope is the process of regional contamination manifested in local struggles with ecological upheaval. The ethnography of fallout ecology aims to localize the ecological relationality of the radiation-centered model framed by co-occurring scientific theorization of Eco semiotics and Eco semiotic concepts and dynamic representations of the environment Eco semiotics by various local actors and residents (Nöth, 1998; Kull, 1998; Maran, 2018; Maran & Kull, 2014).

Social ecology, natural ecosystems, cultural and anthropological ecology are relevant subjects to approach sustainable organizational, management and network ecology issues. Social ecology incorporates the biological orientation of human and anthropology ecologies to integrate the human agency socially embedded with natural environmental contexts. Social ecology embodies certain social movements with practical consequences for human wellbeing in natural sustainability movements such as the rescue of cultural diversity, relationships of human beings inserted in the environment and the social dimension of environmental problems.

SOCIO-ECOLOGICAL SUSTAINABLE DEVELOPMENT

In 1987, the Brundtland Report introduced the term sustainable development enabling humanity to continue the economic growth and expansion ensuring ecological viability in the commitment to growth in which sustainability is to degrowth (Hickel, 2020; Paulson, 2020). The Brundtland commission proposed Our Common Future (United Nations, 1987) concerned with a balance between ecological sustainability, global justice, and growth to propose the concept of sustainable development. The concept of sustainable development has been established in the United Nations report to reconcile this concept with economic growth.

According to the report, sustainable development meets the needs of the present without compromising the ability of future generations to meet their own needs based on alliance between present and future generations. (United Nations, 1987; The World Commission on Environment and Development, 1987; Mao & Robinson, 2019). In other words, sustainable development is the process to preserve, conserve and protect the planet's natural resources for the benefit of the benefit of current and future generations.

The sustainable development notion elaborated in sustainable development goals is a global vision of anthropology presupposing economic growth and cultural diversity. The concept of environment is essential to sustainable development. Later formulation of ecological sustainability leads to the sustainable development goals with an emphasis on global inequality but ignores cultural diversity. The seventeen sustainable development goals presented by the United Nations in 2015 guide the efforts of socio ecological challenges facing climate action, sustainable agriculture, sustainable industrialization, resilient and sustainable cities, sustainable energy, etc. It has been argued that fulfillment of the seventeen sustainable development goals.

The sustainability concept is divested on connection to the growth and development defined as the ability of the system to reproduce the conditions of existence as a societal challenge for humanity.

The concept of sustainable development is concerned with ecologically sustainable practices which are relevant to the study of societies framed by anthropology (Hendry, 2014). Sustainable development focuses on the continued generation of goods and services for their consumption by human beings, while sustainability is the condition for human and non-human natural beings, for a sustainable community. Community sustainable development brings about sustainability. The new sustainable development focuses on ecological solutions in natural, anthropological, social, cultural senses considered preservation values from both sustainability development and research investments (Zan & Lusiani, 2011).

Organizational and anthropological ecological practices in cultural adaptations, competition and strategy are the foundations for socio-ecological sustainable development. Anthropological and sociological ecologies provide the theoretical framework for the ecology of sustainability integration in accounting reporting (Hopwood, 2009, 2010) in specific conceptual areas such as the financial accounting, managerial accounting, and auditing.

Sustainability development includes corporate financial and accounting reporting as a differentiated competitive strategy among organizations. Sustainability reporting is a recent development in financial and accounting reporting. Sustainability accounting reporting integrates economic, social, and environmental development for community growth. Setting boards include guidelines for the conduct and development of sustainability reports assuring audit services co-evolving to secure legitimacy.

Socio-ecological sustainable development and environmental management are components based on ecological anthropology to establish anthropological, ecological, and organizational relationships. Social soundness analysis is related to socio-ecological sustainable development, culturally appropriate, ecologically sensitive, and self-regenerating change (Kottack, 1999). Creation of sustainable development implies a sustainable society with human beings and natural environments together, which are already suffering a severe crisis due to the failure to recognize human nature as the one to care (Boff, 1999).

The acts of socio ecological sustainable development are intentional, planned and designed to sustain ecological value considering the approaches and processes to the application of concepts and practices in sustainability and resilience. Sustainable development practices have implications in complex organizations. Sustainable development is not always tied to bureaucracy as the organizational form of modernity related to the human agency modes (du Gay, 1994, 2000; Kallinikos, 2001). There is a remarkable growth in the coverage of sustainable development practices in complex greening organizations. Complex societies are recent developments in human history with the tendency to fall to some chaos leaving the alternative the return to normal human condition of lower complexity and sustainable societies (Tainter, 2008).

Stable local communities and societies favored by anthropology with cosmologies and ethics dictating sustainable practices but in absence of indigenous concepts of societal development and change (Lévi-Strauss, 1962). Organizations involved in local communities practice the anthropological perspective of sustainability and ownership, sponsoring public culture, and public recreational events, etc., always involved in local sustainability and social development programs (Esty and Winston, 2006). Nongovernmental organizations (NGOs) including the social sector organizations address local needs in sustainable development.

Ecology promotes development sustainability management embedded in the natural selection of the Darwinian theory, individual actions based on economic choices that make stability and maintenance of social systems. Environmental and social data in development sustainability is related to co-existing ecological and natural resources management in geographical systems to ensure sustainable relationships and governance of individuals, groups and community behaviors, values, mores, and cultures of sustainability.

Cultural matrices are for sustainability, guiding everyday life, integrating the natural environment and human beings in feasible models for sustainable community development and appropriate ethics of care (Plumwood, 1995). Sustainable cultural matrices integrate nature and human beings in conditions of sustainable communities to guide the way of life in individuals which may point towards feasible models of sustainable development based on appropriate environmental ethics of care.

Historical, anthropological, and ethnographic sources support the analysis of the scale and the effects of human relationships in socio-ecology systems (Armstrong *et al.*, 2017; Jackson *et al.*, 2001; Kittinger *et al.*, 2015; Pandolfi *et al.*, 2003; Roberts, 2007; Tushingham *et al.*, 2020). Historical ecology fosters sustainable relationships and interactions between human communities with ecosystems to sustain the environment aiming to human wellbeing focused on international policy to improve socio-ecosystems and develop sustainable economies (European Commission, 2017; Food and Agriculture Organization, 2017; United Nations Environment Program, 2012; United Nations General Assembly, 2017).

The historical ecology sources of analysis used in human interactions and integrations with the socio-ecosystems dynamics to foster sustainable environmental relationships. Historical, ecological and environmental history analyzes the forms and functions of socio-ecosystems and socio-ecological changes aimed to foster the interactions of environmental sustainability and the good practices of sustainable communities (Caswell *et al.*, 2020; McClenachan & Kittinger, 2013). Environmental sustainability is a difficult social problem to find solutions that conflict with market capitalism producing ecological precarity.

The resource-based ecological view of an organization has heterogeneity in ecological physical, economic, financial, and natural resources. The resource-based ecological approaches of sociology and anthropology to integrate the co-evolution of organizational sustainability influencing corporate financial and accounting reporting. The resource-ecological view of organizational approaches to the development of sustainability financial and accounting reporting as the strategic management process. Internal legitimacy needs to be enforced and audited by external oversight to control the reporting quality of auditing including the organizational sustainability programs integrated into the strategic planning processes.

Urban political ecology is engaged in ecological imaginaries beyond the eco-criticism as the cultural articulation and counterpart to Anthropocene of alternative environmental sceneries (Gabriel, 2014; Millington, 2013). Urban political ecology contributes to critical analysis implicit on the Anthropocene and historical conception of urban space framework. The imaginary place related to the politics of the Anthropocene contribution by urban political ecology (Braun, 2015) Ecological imaginary is in the context of urban environmental sustainability. Urban political ecology is a concept gaining intellectual reach in the resilience field of the adaptive Anthropocene of social sciences. The urban political ecology takes a lead from history to distinguish between different forms of agency of nature in relation to anthropogenic environmental disturbance.

Sustainability planning evolves into ecological and natural resources providing ecosystem endowments to operate in competitive environmental resources conservation. Integration of sustainability as a by-product of organizational ecological and management of natural environmental resources embedded in development and conservation aimed to balance the economic, social, environmental, and ecological objectives of societies, communities, individuals, and organizations.

Based on the land development sustainability approach, contributions of urban ecological planning and design, sustainable urbanism, natural and green urban areas for protection and low impact development, green infrastructure, brownfield redevelopment, etc. (Arendt, 1999; Barnett, 2009; Benedict & McMahon, 2006; Farr, 2008; Palazzo & Steiner, 2011; Wang et al., 2011). Ecological planning and design apply bioecology and human ecology theoretical and conceptual frameworks to ecological urban development.

Sustainable urban land use and development policies must carry various capacities rather than concentrate on economic growth (Pfaffenbichler, Emberger, Shepherd, 2010; Jim, 1999; Postel, 1994). Urban sustainable and development policies for decision making processes are concerned with city planning and land use management based on system thinking of the phenomena resulting from the unlimited developable land and city sprawl. The system thinking of urban sustainability and city development identifies characteristics of dynamic interactions among the system structures and elements (Pfaffenbichler, Emberger, Shepherd, 2010; Department of Statistics, the Ministry of the Interior, 2014). Unlimited developable land leads to the loss of city growth with positive and negative effects for sustainability and development.

Research is exploring ecology of urban design and development and ecological urbanism (Lehmann, 2010; Mostafavi & Doherty, 2010; Spirn, 2011; Hill, 2007; Palazzo & Steiner, 2011) that are not using sound ecological principles in their practice and implementation.

Sustainable development and environmental management as components based on ecological anthropology to establish anthropological, ecological, and organizational relationships. Socio-ecological sustainability is integrated with natural resources and environmental management forming the core foundation of ecological ethics. The sustainability embedded in ecological and environmental ethics concerns natural resources conservation and ethics of land use. Transformations of land uses must consider past and present contributions of generations to shape the development without compromising the future needs of generations. Organizational sustainability programs support economic development programs more compatible with cultural practices (Bozzoli, 2000; Stone, 2003).

Long term socio-ecological sustainability interests require calculations for investments, production, and preservation of ecological and natural resources (Hopwood, 2009). The economic value-added measures increase the development of profitable investments with more responsible environmental policies (Epstein & Young, 1999). Sustainability enhances the value of ownership and sharing organizational resources used to formulate and implement strategic plans for long-term sustainable development promoting green innovation and organizational performance.

Structuring and mobilizing mechanisms are necessary to achieve the objectives of technological innovation capacities development to benefit the society. Indirect stimuli forms support creation of environments for innovation and competitiveness in regional development supported by venture capital.

CONCLUSION

This study aimed to analyze anthropological ecology and ecological anthropology which are then used as the framework approach for the analysis of socio-ecological sustainable development, from the assumption that sustainable development requires a socio-ecological approach sustained by anthropology and ecology. The study concludes that anthropology and ecology are sciences that combine conceptual, theoretical, and empirical elements to develop the framework to sustain the anthropological ecology and ecological anthropology approaches leading to the analysis of a socio-ecological sustainable development.

Ecological anthropology addresses social and cultural forces that shape sustainability development. The co-evolution of ecological anthropological and organizational ecological perspectives of sustainable development framed by the resource-based approach for the integration of sustainability accounting reporting. The adaptive Anthropocene supports the radical ecology critique combining neo-Marxian analysis of environmental change and critical race study. The historical ecology leverages the opportunities to foster sustainable relationships leading to reconnecting socio-ecological systems with human communities related to achieve sustainable human interactions. By the development of new sustainable economic, social, and environmental activities, political ecology is a powerful tool in cultivating a socially just and sustainable future.

The consumption theory is expanding leading to the development and economic anthropology behavior.

The threats of state-sponsored corporations to sustainable community development require taking control of the processes that arise. Long-term ecological, environmental, and sustainability concerns

in shaping new and emerging market strategies among organizations. Making it necessary for both individuals and companies to take into account aspects such as diversity (biological, genetic, ecological, landscape, productive, etc.), self-sufficiency, integration (of productive practices, natural cycles, etc.), equity (of resources, participation, productive, etc), economic justice, among other key aspects when it comes to operating in any economic environment. Regarding the sustainability reports within the ecological framework, they are internal and external reports used by management and the necessary parties, necessary to have a broader picture of the current situation and the challenges that organizations face.

REFERENCES

Alland, A. (1966). Jr. Medical Anthropology and the Study of Biological and Cultural Adaptation. *American Anthropologist, 68*, 40–51. doi:10.1525/aa.1966.68.1.02a00040

Arendt, R. (1999). *Growing greener: putting conservation into local plans and ordinances*. Island Press.

Armstrong, C. G., Shoemaker, A. C., McKechnie, I., Ekblom, A., Szabó, P., Lane, P. J., McAlvay, A. C., Boles, O. J., Walshaw, S., Petek, N., Gibbons, K. S., Quintana Morales, E., Anderson, E. N., Ibragimow, A., Podruczny, G., Vamosi, J. C., Marks-Block, T., LeCompte, J. K., Awâsis, S., & Crumley, C. L. (2017). Anthropological contributions to historical ecology: 50 questions, infinite prospects. *PLoS One, 12*(2), e0171883. doi:10.1371/journal.pone.0171883 PMID:28235093

Barnett, J. (2009). The way we were, the way we are the theory and practice of designing cities since 1956. A. Krieger, W.S. Saunders (Eds.), *Urban design*, University of Minnesota Press. https://www.jstor.org/stable/10.5749/j.ctttspsh.9View

Barnett, J., & Adger, W. N. (2003). Climate dangers and atoll countries. *Climatic Change, 61*(3), 321–337. doi:10.1023/B:CLIM.0000004559.08755.88

Belz, F.-M., & Peattie, K. (2010). Sustainability marketing: An innovative conception of marketing. *Marketing Review St. Gallen, 27*(5), 8–15. doi:10.100711621-010-0085-7

Benedict, M., & McMahon, E. T. (2006). *Green infrastructure: linking landscape and communities*. Island Press.

Berry, J. W. (2004). An Ecocultural Perspective on the Development of Competence. In R. Sternberg & E. Grigorenko (Eds.), *Culture and Competence: Contexts of Life Success* (pp. 3–22). American Psychological Association. doi:10.1037/10681-001

Bliege Bird, R., & Codding, B. F. (2022). Promise and peril of ecological and evolutionary modelling using cross-cultural datasets. *Nature Ecology & Evolution, 6*(1), 6–8. doi:10.103841559-021-01579-w PMID:34697456

Boff, L. (1999). *Saber Cuidar: Ética do humano – compaixão pela terra*. Editora Vozes.

Borland, H., & Lindgreen, A. (2013). Sustainability, epistemology, ecocentric business, and marketing strategy: Ideology, reality, and vision. *Journal of Business Ethics, 117*(September), 173–187. doi:10.100710551-012-1519-8

Bozzoli, M. E. (2000). A role for anthropology in sustainable development in Costa Rica. *Human Organization, 59*(3), 275–279. doi:10.17730/humo.59.3.eu5h42p81wk26578

Braun, B. (2014). A new urban dispositive? Governing life in an age of climate change. *Environment and Planning. D, Society & Space, 32*(1), 49–64. doi:10.1068/d4313

Cajee, L. (2014). Eco-Cultural Tourism: A Tool for Environmental, Cultural and Economic Sustainability (A Case Study of Darap Village, West Sikkim). *SHS Web Conf.*, 12, 01029. 10.1051hsconf/20141201029

Caswell, B., Klein, E. S., Alleway, H. K., Ball, J., Cardinale, M., Eero, M., & Thurstan, R. H. (2020). Something old, something new: 200 years of a blue growth agenda. *Fish and Fisheries, 21*, 774–796. doi:10.1111/faf.12460

Chen, Y.-W., Milstein, T., Anguiano, C., Sandoval, J., & Knudsen, L. (2012). Challenges and Benefits of Community-Based Participatory Research for Environmental Justice: A Case of Collaboratively Examining Ecocultural Struggles. *Environmental Communication, 6*(3), 403–421. doi:10.1080/17524032.2012.698291

Cohen-Rosenthal, E. (2000). A walk on the human side of industrial ecology. *The American Behavioral Scientist, 44*(4), 245–264. doi:10.1177/00027642000044002007

Cram, S. (2016). Wild and Scenic Wasteland: Conservation Politics in the Nuclear Wilderness. *Environmental Humanities, 7*(1), 89–105. doi:10.1215/22011919-3616344

Dahake, S. (2020). The making of fluid ecocultural identities in urban India. In T. Milstein & J. Castro-Sotomayor (Eds.), *Routledge Handbook of Ecocultural Identity* (pp. 333–350). Routledge. doi:10.4324/9781351068840-20

Daryl Slack, J. (2008). Resisting ecocultural studies. *Cultural Studies, 2008*(22), 477–497. doi:10.1080/09502380802012575

Davis, L. J., & Morris, D. B. (2007). Biocultures Manifesto. *New Literary History, 38*(3), 411–418. doi:10.1353/nlh.2007.0046

Dayton, P. K., Tenger, M. J., Edwards, P. B., & Riser, K. L. (1998). Sliding Baselines, Ghosts, and Reduced Expectations in Kelp Forest Communities. *Ecological Applications, 8*(2), 309–322. doi:10.1890/1051-0761(1998)008[0309:SBGARE]2.0.CO;2

Department of Statistics & the Ministry of the Interior (MOI), R.O.C. (Taiwan). (2014) *Statistical Yearbook of Interior.* Ministry of Interior. http://sowf.moi.gov.tw/stat/year/list.htm

Desjeux, P. (2001) Trans. of the Royal Society of *Trop. Med. and Hygiene, 95*(3) 239–43.

du Gay, P. (1994). Colossal immodesties and hopeful monsters. *Organization, 1*(1), 125–148. doi:10.1177/135050849400100114

du Gay, P. (2000) In praise of bureaucracy: Weber, organization, ethics. London: Sage. doi:10.4135/9781446217580

Bohm, S. (2015). Ecocultures: Towards sustainable ways of living. InBöhm, S., Bharucha, Z. P., & Pretty, J. (Eds.), *Ecocultures: Blueprints for Sustainable Communities* (pp. 3–26). Routledge.

Ehrenfeld, J. R. (2000). Industrial ecology: Paradigm shift or normal science? *The American Behavioral Scientist*, *44*(2), 229–244. doi:10.1177/00027642000044002006

Epstein, M. J., & Young, S. D. (1999). Greening' with EVA. *Management Accounting*, *80*(1), 45–49.

Winston, A. S. (2006). *Green to Gold*. Yale University Press.

European Commission. (2017). *Report on the blue growth strategy towards more sustainable growth and jobs in the blue economy*. European Commission.

Everett, J. L. (1985). *Culture and ecology in the analysis of organizations*. [PhD Thesis, Institute: University of Colorado at Boulder].

Farr, D. (2008). *Sustainable urbanism—urban design with nature*. John Wiley & Sons.

Food and Agriculture Organization. (2017). *Food and agriculture organization blue growth initiative: Partnering with countries to achieve the sustainable development goals*. United Nations.

Gabriel, N. (2014). Urban political ecology: Environmental imaginary, governance, and the non-human. *Geography Compass*, *8*(1), 38–48. doi:10.1111/gec3.12110

Ghatee, M. A., Taylor, W. R., & Karamian, M. (2020). The Geographical Distribution of Cutaneous Leishmaniasis Causative Agents in Iran and Its Neighboring Countries, A Review. *Frontiers in Public Health*, *8*, 111–113. doi:10.3389/fpubh.2020.00011 PMID:32391301

Giblett, R. (2012). Nature is ordinary too: Raymond Williams as the founder of ecocultural studies. *Cultural Studies*, *26*(6), 922–933. doi:10.1080/09502386.2012.707221

Golden, S. D., & Earp, J. A. L. (2012). Social Ecological Approaches to Individuals and Their Contexts: Twenty Years of Health Education & Behavior Health Promotion Interventions. *Health Education & Behavior*, *39*(3), 364–372. doi:10.1177/1090198111418634 PMID:22267868

Goldfarb, K. E. (2016). 'Coming to Look Alike': Materializing Affinity in Japanese Foster and Adoptive Care. *Social Analysis*, *60*(2), 47–64. doi:10.3167a.2016.600204

Gómez Hinojosa, J. F. (1990). De la ecología a la ecofilía: Apuntes para una ecología liberadora'. *Pasos (El Sauzal)*, *30*(July–August).

Gragson, T. L., & Blount Ben, G. (Eds.). (1999). *Introduction Ethnoecology: Knowledge, Resources, and Rights*. University of Georgia.

Hashimoto, K. (2013). *En no Shakaigaku: Fukushi Shakaigaku no Shitenkara*. Harvest-sha.

Hendry, J. (2014). *Science and Sustainability: Learning from Indigenous Wisdom*. Palgrave Macmillan. doi:10.1057/9781137430069

Heyd, T. (1995). Indigenous knowledge, emancipation, and alienation. *Knowledge in Society*, *8*(1), 63–73. doi:10.1007/BF02698557

Hickel, J. (2020). *Less is More: How Degrowth Will Save The World*. Windmill.

Hill, K. (2007). Urban ecological design and urban ecology: an assessment of the state of current knowledge and a suggested research agenda. V. Novotny, P. Brown (Eds.), Cities of the future: towards integrated sustainable water and landscape management. IWA Publishing, London.

Hoffmann, J. A. (2020). Memory, Waterways, and Ecocultural Identity. In T. Milstein & J. Castro-Sotomayor (Eds.), *The Routledge Handbook of Ecocultural Identity*. Routledge. doi:10.4324/9781351068840-9

Hoke, M. K., & Schell, L. M. (2020). Doing biocultural anthropology: Continuity and change. *American Journal of Human Biology*, *2020*(32), e23471. PMID:32681558

Hopwood, A. G. (2009). Accounting and the environment. *Accounting, Organizations and Society*, *34*(3-4), 433–439. doi:10.1016/j.aos.2009.03.002

Hopwood, A. G. (2010). In memoriam, Anthony G. Hopwood, 1944-2010. *Accounting, Organizations and Society*, *35*(4), 496–497. doi:10.1016/j.aos.2010.05.002

(2019). Introduction: Biocultures. InEhlers, N., & Krupar, S. (Eds.), *Deadly Biocultures: The Ethics of Life-Making*. The University of Minnesota Press.

Jackson, J. B. C., Kirby, M. X., Berger, W. H., Bjorndal, K. A., Botsford, L. W., Bourque, B. J., Bradbury, R. H., Cooke, R., Erlandson, J., Estes, J. A., Hughes, T. P., Kidwell, S., Lange, C. B., Lenihan, H. S., Pandolfi, J. M., Peterson, C. H., Steneck, R. S., Tegner, M. J., & Warner, R. R. (2001). Historical overfishing and the recent collapse of coastal ecosystems. *Science*, *293*(5530), 629–638. doi:10.1126cience.1059199 PMID:11474098

Jensen, C. B., & Miho, I. (2016). Attuning to the Webs of En: Ontography, Japanese Spirit Worlds, and the 'Tact' of Minakata Kumagusu. *HAU*, *6*(2), 149–172. doi:10.14318/hau6.2.012

Jim, C. Y. (1999). A planning strategy to augment the diversity and biomass of roadside trees in urban Hong Kong. *Landscape and Urban Planning*, *44*(1), 13–32. doi:10.1016/S0169-2046(98)00113-3

Kallinikos, J. (2001). 'Bureaucracy: Exploring organization forms in the age of contingency. Paper presented at the 17th EGOS Colloquium, The Odyssey of Organizing, 5–7 July, Lyon. (Organization, special issue on Bureaucracy, 2003). 616. *Organization Studies*, *24*(4).

Kempera, J. A., & Ballantine, P. W. (2019). What do we mean by sustainability marketing? *Journal of Marketing Management*, *35*(3–4), 277–309. doi:10.1080/0267257X.2019.1573845

Kilbourne, W., McDonagh, P., & Prothero, A. (1997). Sustainable consumption and the quality of life: A macromarketing challenge to the dominant social paradigm. *Journal of Macromarketing*, *17*(1), 4–24. doi:10.1177/027614679701700103

Kittinger, J. N., McClenachan, L., Gedan, K. B., & Blight, L. K. (2015). *Marine historical ecology in conservation: Applying the past to manage for the future*. University of California Press.

Kottack, C. P. (1999). The new ecological anthropology. *American Anthropologist*, *101*(1), 23–35. doi:10.1525/aa.1999.101.1.23

Krupar, S., & Ehlers, N. (2020). Biocultures: A critical approach to mundane biomedical governance. *Culture, Theory & Critique*, *61*(4), 440–456. doi:10.1080/14735784.2020.1857810

Kull, K. (1998). Semiotic Ecology: Different Natures in the Semiosphere. *Sign Systems Studies, 26*(1), 344–371. doi:10.12697/SSS.1998.26.15

Leff, E. (1995). *Green Production.* Guilford Press.

Lehmann, S. (2010). *The principles of green urbanism: transforming the city for sustainability.* Routledge.

Leonard, W. R. (2020). Re-examining Biocultural Approaches in Human Biology. *American Journal of Human Biology, 32*(4), e23475. doi:10.1002/ajhb.23475 PMID:32885558

Lévi-Strauss, C. (1962). *La Pensée Sauvage.* Plon.

Lyle, J. T. (1985). *Design for human ecosystems: landscape, land use, and natural resources.* Van Nostrand.

Mao, Y., & Robinson, D. (2019). Three pillars of sustainability: In search of conceptual origins. *Sustainability Science, 14*(3), 681–695. doi:10.100711625-018-0627-5

Maran, T. (2018). Two Decades of Ecosemiotics in Tartu. *Sign Systems Studies, 46*(4), 630–639. doi:10.12697/SSS.2018.46.4.11

Maran, T., & Kull, K. (2014). Ecosemiotics: Main Principles and Current Developments. *Geografiska Annaler. Series B, Human Geography, 96*(1), 41–50. doi:10.1111/geob.12035

Marcuse, H. (1972). *Counterrevolution and Revolt.* Boston Beacon Press.

Masco, J. (2006). *The Nuclear Borderlands: The Manhattan Project in Post–Cold War New Mexico.* Princeton University Press. doi:10.1515/9781400849680

McClenachan, L., & Kittinger, J. N. (2013). Multicentury trends and the sustainability of coral reef fisheries in Hawai'i and Florida. *Fish and Fisheries, 14*(3), 239–255. doi:10.1111/j.1467-2979.2012.00465.x

McClintock, P. (2014, March). *Box office milestone: 'Frozen' becomes No. 1 animated film of all time.* Hollywood Reporter. https://www.hollywoodreporter.com/news/box-office-milestone-frozen-becomes-692156

McCool, W. C., Vernon, K. B., Yaworsky, P. M., & Codding, B. F. (2022). Subsistence strategy mediates ecological drivers of human violence. *PLoS One, 17*(5), e0268257. doi:10.1371/journal.pone.0268257 PMID:35604917

McElroy, A. (1990). Biocultural Models in Studies of Human Health and Adaptation. *Medical Anthropology Quarterly, 4*(3), 243–265. doi:10.1525/maq.1990.4.3.02a00010

McHarg, I. (1996). *Design with nature.* The Natural History Press.

McLeroy, K. R., Bibeau, D., Steckler, A., & Glanz, K. (1988, Winter). An ecological perspective on health promotion programs. *Health Education Quarterly, 15*(4), 351–377. doi:10.1177/109019818801500401 PMID:3068205

Merçon, J., Vetter, S., Tengö, M., Cocks, M., Balvanera, P., Rosell, J. A., & Ayala-Orozco, B. (2019). From local landscapes to international policy: Contributions of the biocultural paradigm to global sustainability. *Global Sustainability, 2*(e7), 1–11. doi:10.1017us.2019.4

Millington, N. (2013). Post-industrial imaginaries: Nature, representation and ruin in Detroit, Michigan. *International Journal of Urban and Regional Research, 37*(1), 279–296. doi:10.1111/j.1468-2427.2012.01206.x

Milstein, T., & Castro-Sotomayor, J. (Eds.). (2020). *Routledge Handbook of Ecocultural Identity*. Routledge. doi:10.4324/9781351068840

Montgomery, E., Bennett, J. W., & Scudder, T. (1973). The impact of human activities on the physical and social environments: New directions in anthropological ecology. *Annual Review of Anthropology, 2*(1), 27–61. doi:10.1146/annurev.an.02.100173.000331

Morimoto, R. (2022). A wild boar chase: Ecology of Harm and Half-Life Politics in Coastal Fukushima. *Cultural Anthropology, 37*(1), 69–98. doi:10.14506/ca37.1.08

Mostafavi, M., & Doherty, G. (Eds.), *Ecological urbanism*. Lars Müller Publisher.

Ndubisi, F. (2002). *Ecological planning: a historical and comparative synthesis*. The Johns Hopkins University Press.

Ndubisi, F. (2014). *The ecological design and planning reader*. Island Press. doi:10.5822/978-1-61091-491-8

Netting, R. M. (1974). Agrarian ecology. *Annual Review of Anthropology, 3*(1), 21–56. doi:10.1146/annurev.an.03.100174.000321

Nöth, W. (1998). Ecosemiotics. *Sign Systems Studies, 26*(1), 332–343. doi:10.12697/SSS.1998.26.14

Nozawa, S. (2015). Phatic Traces: Sociality in Contemporary Japan. *Anthropological Quarterly, 88*(2), 373–400. https://muse.jhu.edu/article/582266. doi:10.1353/anq.2015.0014

Nsamenang, A. B. (2015). Ecocultural Theories of Development. In J. Wright (Ed.), *International Encyclopedia of the Social & Behavioral Sciences* (Vol. 6, pp. 838–844). Elsevier. doi:10.1016/B978-0-08-097086-8.23205-4

O'Neil, D. (2018, September 16). The EU needs a stability and wellbeing pact, not more growth. *The Guardian*. https://www.theguardian.com/politics/2018/sep/16/the-eu-needs-a-stability-and-wellbeing-pact-not-more-growth

Odum, E. P. (1959). *Fundamentals of Ecology* (2nd ed.). W. B. Saunders.

Palazzo, D. (1997). Sulle spalle di giganti. [On the Shoulders of Giants]. Franco Angeli, Milano.

Palazzo, D., & Steiner, F. R. (2011). *Urban ecological design—a process for regenerative places*. Island Press.

Pandolfi, J. M., Bradbury, R. H., Sala, E., Hughes, T. P., Bjorndal, K. A., Cooke, R. G., McArdle, D., McClenachan, L., Newman, M. J. H., Paredes, G., Warner, R. R., & Jackson, J. B. C. (2003). Global trajectories of the long-term decline of coral reef ecosystems. *Science, 301*(5635), 955–958. doi:10.1126cience.1085706 PMID:12920296

Paulson, S., D'Alisa, G., Demaria, F., & Kallis, G. (2020). *The Case for Degrowth*. Polity.

Pauly, D. (1995). Anecdotes and the Shifting Baseline Syndrome of Fisheries. *Trends in Ecology & Evolution, 10*(10), 430. doi:10.1016/S0169-5347(00)89171-5 PMID:21237093

Peet, R., & Watts, M. (1996) Liberation ecology: development, sustainability, and environment in an age of market triumphalism. in: R. Peet & M. Watts (eds.) Liberation Ecologies: Environment, Development, Social Movements. Routledge.

Petterson, M., Kensek, K., & Noble, D. (2017). Supple Skins: Considering the Relevance, Scalability, and Design Strategies for Façade System Resilience. *Journal of Architectural Education, 71*(1), 34–45. doi:10.1080/10464883.2017.1260919

Pfaffenbichler, P., Emberger, G., & Shepherd, S. A. (2010). System dynamics approach to land use transport interaction modelling: The strategic model MARS and its application. *System Dynamics Review, 26*(3), 262–282. doi:10.1002dr.451

Plumwood, V. (1995). Nature, self and gender: feminism, environmental philosophy, and the critique of rationalism. C. Pierce & D. VanDeVeer (Eds) People, Penguins and Plastic Trees. Wadsworth.

Postel, S. (1994). Carrying capacity: Earth's bottom line. In L. R. Brown, A. T. Durning, C. Flavin, H. F. French, N. Lenssen, M. D. Lowe, A. Misch, S. Postel, M. Renner, & P. Weber, (Eds.), *State of the World* (pp. 3–21). W.W. Norton.

Pretty, J. (2011). Interdisciplinary progress in approaches to address social-ecological and ecocultural *systems. Environmental Conservation, 38*(2), 127–139. doi:10.1017/S0376892910000937

Prothero, A., Dobscha, S., Freund, J., Kilbourne, W. E., Luchs, M. G., Ozanne, L. K., & Thøgersen, J. (2011). Sustainable consumption: Opportunities for consumer research and public policy. *Journal of Public Policy & Marketing, 30*(1), 31–38. doi:10.1509/jppm.30.1.31

Rappaport, R. A. (2000). *Pigs for the Ancestors: Ritual in the Ecology of a New Guinea People*. Waveland Press.

Raworth, K. (2012). A safe and just space for humanity: Can we live within the doughnut? Oxfam Policy and Practice. *Climate Change and Resilience, 8*, 1–26.

Reich, D. (2018). *Who We Are and How We Got Here: Ancient DNA and the New Science of the Human Past*. Oxford University Press.

Roberts, C. M. (2007). *The unnatural history of the sea*. Island Press.

Schnarch, B. (2004). Ownership, control, access, and possession (OCAP) or self-determination applied to research: A critical analysis of contemporary First Nations research and some options for First Nations communities. *Journal of Aboriginal Health, 1*, 80–97.

Scudder, T. (1973). The Human Ecology of Big Projects: River Basin Development and Resettlement. [Section 3 in Montgomery, Bennett, and Scudder (1973)] In B. J. Siegel, A. R. Beals, & S. A. Tyler (Eds.), *Annual Review of Anthropology*. Annual Reviews Inc.

Scudder, T., & Colson, E. (1982). From Welfare to Development: A Conceptual Framework for the Analysis of Dislocated People in Involuntary. In A. Hansen & A. Oliver-Smith (Eds.), *Migration and Resettlement: The Problems and Responses of Dislocated People* (pp. 267–287). Westview Press, Westview Special Study.

Simberloff, D. (2018). Nature, Culture, and Natureculture: The Role of Nonnative Species in Biocultures. In R. Rozzi, R. H. May Jr, F. S. Chapin III, F. Massardo, M. C. Gavin, I. J. Klaver, A. Pauchard, M. A. Nunez, & D. Simberloff (Eds.), *From Biocultural Homogenization to Biocultural Conservation* (pp. 207–218). Nature. doi:10.1007/978-3-319-99513-7_12

Slater, D., & Tonkiss, F. (2001). *Market society: Markets and modern social theory*. Polity Press.

Spirn, A. W. (2011). *Ecological urbanism: a framework for the design of resilient cities.* Anne Whistonspirn. http://annewhistonspirn.com/pdf/spirn_ecological_urbanism-2011.pdf

Steiner, F. R. (1991). *The living landscape: an ecological approach to landscape planning*. McGraw-Hill.

Steiner, F. R. (2001). *Human ecology: following nature's lead.* Island Press.

Steiner, F. R. (2016). Human ecology: how nature and culture shape our world. The Island Press: Washington, D.C.

Stiglitz, J. (2015). *Globalization and its discontents.* Penguin.

Stokols, D. (1992). Establishing and maintaining healthy environments: Toward a social ecology of health promotion. *The American Psychologist*, *47*(1), 6–22. doi:10.1037/0003-066X.47.1.6 PMID:1539925

Stokols, D. (1996). Translating social ecological theory into guidelines for community health promotion. *American Journal of Health Promotion, 10*(4), 282-98. doi:10.4278/0890-1171-10.4.282

Stone, M. (2003). Is sustainability for development anthropologists? *Human Organization*, *62*(2), 93–99. doi:10.17730/humo.62.2.9cfam6hd6e04yac8

Tainter, J. A. (2008). Collapse and Sustainability: Rome, the Maya, and the Modern World. *Archeological Papers of the American Anthropological Association*, *24*(1), 201–214. doi:10.1111/apaa.12038

Temesgen, A., Storsletten, V., & Jakobsen, O. (2021). Circular economy – reducing symptoms or radical change? *Philosophy of Management*, *20*(March), 37–56. doi:10.100740926-019-00112-1

Thompson, G. F., & Steiner, F. R. (Eds.). (1997). *Ecological design and planning*. John Wiley & Sons, Inc.

Toledo, V. M. (1997) Sustainable development at the village community level: a Third World perspective in: Fraser Smith (Ed.) Environmental Sustainability. St. Lucie Press. doi:10.1201/9781420048742.ch11

Tushingham, S., Hopt, J., Christiansen, C., Bommelyn, M. L., Green, J., Peterson, M. R., Steinruck, S., & Stewart, C. (2020). In the footsteps of Amelia Brown: Collaborative historical ecology at Shin-yvslh-sri~, a Tolowa village on the north coast of California. *Journal of Island & Coastal Archaeology*, *15*(1), 3–27. doi:10.1080/15564894.2018.1539789

United Nations. (1987). *Our Common Future. Report of the World Commission on Environment and Development.* Oxford University Press.

United Nations Environment Programme. (2012). *Green economy in a blue world: Synthesis report.* United Nations.

United Nations General Assembly. (2017). Resolution 71/312 Our ocean, our future: Call for action. In *Resolution adopted by the general assembly on 6 July 2017.* United Nations.

Van der Ryn, S., & Cowan, S. (2007). *Ecological design: tenth anniversary edition.* Island Press.

Walker, P. (2005). Political ecology: Where is the ecology? *Progress in Human Geography, 29*(1), 73–82. doi:10.1191/0309132505ph530pr

Wallace-Wells, D. (2019). *The uninhabitable earth: Life after warming.* Tim Duggan Books.

Wang, X., Shuster, W., Pal, C., Buchberger, S., Bonta, J., & Avadhanula, K. (2011). Low impact development design—Integrating suitability analysis and site planning for reduction of post-development stormwater quantity. *Sustainability (Basel), 2*(8), 2467–2482. doi:10.3390u2082467

Weick, K. E. (1969). *The Social Psychology of Organizing.* McGraw Hill.

Weston, A. (2009). *The incompleat eco-philosopher: essays from the edges of environmental ethics.* State University of New York Press.

Wiley, A. S., & Cullin, J. M. (2016). What Do Anthropologists Mean When They Use the Term Biocultural? *American Anthropologist, 2016*(118), 554–569. doi:10.1111/aman.12608

Willermet, C. (2016). Biological Anthropology in 2015: Open Access, Biocultural Interactions, and Social Change. *American Anthropologist, 118*(2), 118. doi:10.1111/aman.12529

Wolf, E. (1972). Ownership and political ecology. *Anthropological Quarterly, 45*(3), 201–205. doi:10.2307/3316532

World Commission on Environment and Development. (1987). *Our common future.* Oxford University Press.

Yongming, Z. (2010). How Our Village Becomes "Ecocultural." In Cultural Change and Persistence. Palgrave Macmillan US.

Zan, L., & Lusiani, M. (2011). Managing Change and Master Plans: Machu Picchu Between Conservation and Exploitation. *Archaeologies, 7*(2), 329–371. doi:10.100711759-011-9167-7

Zedler, J. B., & Stevens, M. L. (2018). Western and Traditional Ecological Knowledge in Ecocultural Restoration. *San Francisco Estuary and Watershed Science, 16*(3), 1–18. doi:10.15447fews.2018v16iss3art2

Chapter 3
Biodiversity Initiatives:
Theoretical and Practical Intersections Among TNFD, Biodiversity Credits, and ESG Reporting

Ruopiao Zhang
Macau University of Science and Technology, China

Carlos Noronha
University of Macau, China

ABSTRACT

The world is facing pressing environmental challenges, including climate change, biodiversity loss, and resource scarcity. These issues have led businesses and investors to concentrate on ESG factors, which are increasingly crucial for long-term sustainable growth. However, the existing ESG framework is not a panacea for urgent problems such as energy, climate, and biodiversity crises. This chapter explores the emerging role of the TNFD and the potential benefits of incorporating biodiversity credits in ESG reporting practices as a unified mechanism to combat biodiversity challenges and sustainability issues, highlighting the synergies among these concepts and the possible advantages of integrating them into business operations. By addressing nature-related risks and adopting market-based instruments like biodiversity credits, companies can better position themselves to achieve their ESG objectives and generate long-term value.

INTRODUCTION

In a bid to reduce the alarming rate of biodiversity loss, countries from around the globe gathered in Montreal, Canada, in 2022 for the 15th Conference of the Parties (COP15). As per a report by Briggs (2022), more than 200 countries have pledged to safeguard one-third of the planet's land and sea areas by 2030. The Living Planet Report 2022, which was recently released by the World Wildlife Fund (WWF), highlights some alarming statistics about the global wildlife population. According to the report, between

DOI: 10.4018/978-1-6684-8879-9.ch003

1970 and 2018, global wildlife abundance has declined by approximately 69%, with freshwater species being the most affected, with a decrease of up to 83%. This sharp decline in species puts them at a much higher risk of extinction (WWF, 2022). The United Nations Convention on Biological Diversity (COP 15), which was held in Montreal, Canada in December 2022, has established goals that are deemed to be the "best and last chance" to minimize and reverse natural decline.

Biodiversity refers to the variety of life in all its forms, from genetic diversity within individual species to the variety of ecosystems across the planet. It provides crucial services like pollination, pest control, and water purification that support human well-being and the global economy (Rands et al., 2010). The Kunming-Montreal Global Biodiversity Framework, which was established during COP15, lays out four global targets for 2050 and 23 specific objectives for 2030. Specific objective 15 states that large multinational corporations and financial institutions should regularly monitor, assess, and disclose their risks, dependencies, and impacts on biodiversity. Gradually reducing negative impacts, expanding positive ones, and promoting sustainable production patterns (CBD, 2022). As such, companies must consider and act upon the challenge of understanding how to measure the natural risks associated with their business activities and value chains (Jones and Solomon, 2013; Kurth et al., 2021).

From a business perspective, biodiversity loss poses various risks, such as disrupted supply chains, reduced availability of natural resources, and decreased resilience to climate change. Protecting biodiversity is, therefore, a moral and economic imperative that can enhance long-term corporate viability and value creation. The impact of corporate operations on global biodiversity is best understood within four primary categories: 1) changes in land and marine use, 2) direct consumption of living organisms, 3) pollution contributing to climate change, and 4) invasion of non-native species (IUCN, 2021). While the degree to which companies pressure biodiversity varies, loss of animals, plants, habitats and ecosystem services may occur as a result of various practices including land-use changes (e.g., construction, agriculture, and mining), direct utilization of living organisms (e.g., overfishing or deforestation), pollution (e.g., agricultural chemicals or waste disposal), climate change (e.g., greenhouse gas emissions or deforestation), the introduction of harmful invasive species (e.g., through shipping), the transformation of natural systems (e.g., dam building), and transportation corridors (e.g., roads and shipping channels).

ESG, which stands for Environmental, Social, and Governance, is a reporting framework used to measure a company's sustainability performance. It was introduced by the United Nations Global Compact in 2004 and evaluates a corporation's social responsibility by examining three significant dimensions: sustainable environment, social participation, and corporate governance (Halbritter and Dorfleitner, 2015). According to a sustainability audit conducted by KPMG, a renowned global accounting firm, 46% of the world's largest 250 companies by revenue have acknowledged the risks posed to their business operations by biodiversity loss. Similarly, 40% of the leading 100 companies by revenue in various regions have recognized the same risks (KPMG, 2022). However, the current environmental criteria within ESG mainly focus on climate change mitigation and natural resources management. Although these factors are critical, they do not fully address biodiversity loss, ecosystem resilience, and the interdependence between humans and nature. Consequently, there is an ongoing debate about the effectiveness and completeness of this approach in resolving the biodiversity crisis (Jones and Solomon, 2013; Lambooy et al., 2018). Moreover, most ESG frameworks, such as the MSCI ESG Ratings, tend to give limited emphasis to biodiversity risk compared to other environmental concerns, such as climate change, which could lead to an imbalance in issue prioritization. Additionally, ESG frameworks usually address the consequences of biodiversity loss rather than proactively promoting systemic changes necessary to prevent further biodiversity declines. For example, reporting on ESG indicators such as greenhouse gas emissions or water usage tends to be more prevalent than reporting on biodiversity metrics. Furthermore, ESG frameworks often address biodiversity in a piecemeal fashion, such as policies on wildlife protection in supply chains or initiatives to restore degraded lands. While these activities are commendable, they do not address the root causes of biodiversity loss. Biodiversity loss is driven by systemic factors, including land-use changes, habitat destruction, pollution, and climate change. Addressing these challenges requires systemic changes in the way that our economies operate and the way we lead our lives. ESG frameworks, which are generally geared towards incremental change, may not be sufficient to achieve the necessary systemic changes. The Taskforce on Nature-related Financial Disclosures (TNFD) is an initiative launched by Global Canopy and the United Nations Development Programme (UNDP) in collaboration with several other institutions (TNFD, 2023). TNFD aims to create a framework that would enable businesses and financial institutions to disclose their nature-related risks and opportunities (Cuckston, 2022; Nelson and Combe, 2022). In comparison, Biodiversity credits are a mechanism for protecting and restoring biodiversity. They operate similarly to carbon credits, which enable companies to offset their carbon emissions by supporting projects that reduce emissions elsewhere. Biodiversity credits, however, fund projects that support biodiversity protection or restoration, such as habitat conservation or reforestation, in close partnership with local communities (Conniff, 2018). The goal of biodiversity credits is to encourage companies to adopt sustainable practices and proactively reduce their impact on the environment (WEF, 2022a). The development of both TNFD and biodiversity credits initiatives is still in progress, with ongoing discussions and efforts aimed at establishing specific indicator criteria and standards. These initiatives are a response to a growing recognition of the importance of protecting the environment and biodiversity and the need for financial institutions and businesses to take an active role in these efforts (Kopnina et al., 2023).

This chapter aims to encourage companies to adopt a more comprehensive and ecologically informed ESG reporting framework that includes the recommendations put forth by the TNFD and the application of biodiversity credits. Though the specific disclosure practices of the two initiatives are still undergoing revision and enhancement, we propose that companies should take an accelerated and proactive approach and embrace a wider and all-encompassing ESG reporting philosophy to combat with the severe biodiversity crisis. By adopting a more comprehensive and ecologically informed ESG framework that incorporates TNFD recommendations and biodiversity credits in a forward-looking manner, companies can not only mitigate potential financial and reputational risks, but also actively contribute to global conservation efforts.

THEORETICAL UNDERPINNINGS

As mentioned in the introductory section above, this chapter aims to look into the intersections among the three elements of TNDF, biodiversity losses (biodiversity credits as well), and ESG reporting. It is firstly necessary to have a brief discussion of the source of such a combination of elements in a bio-economy (or even managerial, accounting or financial-wise) collection of this volume. We would like to arrange our theoretical discussions here based on both the micro/meso/macro (institutional/societal/international) as well as at the phenomenon *grande* (Donleavy and Noronha, 2023; Noronha et al., 2024) (Anthropocene) levels. And then we shall delve into the descriptive details of the three main elements so as to finally come up with a conclusion about how they can help mitigate (a word much used in the literature but not very desirable due to its firefighting connotation) issues of biodiversity, sustainability, and back to the basics (BTTB), corporate social responsibility (CSR).

THE MICRO/MESO/MACRO LEVELS

It is not new to the reader that businesses engage in CSR activities and reporting due to reasons of the social license to operate (Deegan, 2002) or the legitimacy theory (Dowling and Pfeffer, 1975; Suchman, 1995); the stakeholder theory (Freeman, 1984); as well as the resource-based view in terms of CSR (Branco and Rodrigues, 2006). That is to say, at the institutional level, firms need to gain recognition from the "social" to continue to operate through a "give and take" organizational mindset and to gain competitive advantages to stand out of the crowd. The instrumentalist's view of CSR is that disclosure or reporting, through sustainability accounting, or the topics under discussion now such as ESG reporting, is a major method to gain social acceptance for the capitalistic use of the factors of production. There has been an abundance of literature about how firms use disclosure strategies to attain genuine social acceptance or simply to achieve green-washing due to some normative aspects of institutionalism (see e.g., Lu, 2023 for an update discussion on window-dressing and the circular economy; and earlier literature such as Orange and Cohen, 2010). Due to the inconsistency among institutional and stakeholder levels of consensus on CSR and sustainability as evidenced by notorious industrial, labor, as well as customer-related incidents, normative acts (including the Taskforce on Climate-Related Financial Disclosure (TCFD), TNFD, as well as biodiversity guidelines) have been announced by the respective authorities. However, it is worth noting that the "authorities" referred to here can stem from what we call "compulsory" or "voluntary" sources. Both of them have demonstrated failures as well as success stories (which will be elaborated afterwards). At this point, all the makers and followers of the norms, as well as the norms themselves are brought onto the stage as actors, irrespective of their organic or inorganic nature, with their interactions among themselves called "translations", referring to relationships which can create changes (mutability) (Latour, 2005; Noronha and Zhang, 2023).

Pressures from international organizations originating from the 1987 Bruntland Report have stimulated the global focus on sustainability. The UN Sustainability Development Goals (UNSDGs) are probably the most comprehensive set of universal ideas which have become the "compulsory" source of businesses and societies to co-create shared-value in a sustainable way. Unfortunately, the UN Secretary General António Guterres reported in April 2023 that out of the 169 SDG targets, only 12% are on track, while progress on 50% is weak and inefficient, and there is an urgent need for the world to recommit to achieving the SDGs by the agreed 2030 deadline (United Nations, 2023). This is an example of the failing mechanism of the "compulsory" type of "authority" in pushing the SDGs through, sometimes, "pseudo-legislative" requirements. None of the SDGs can be fully achieved if the current state of the world remains with wars, aggressions, and political unrests, just to mention a few. From this perspective, the actors and the inorganic actants (Latour, 2005) are having functional difficulties not only on the macro multilateral translation level, but also at the hierarchical translation levels (Noronha and Zhang, 2023). "Black boxes" (Callon, 1991) or translations which are creating undesirable or unintended outcomes which cannot be corrected are all over the stage creating more and more uncertainties and obstacles. It is definitely not here to say that the UNSDGs are not effectively guiding where humans are heading toward; it is the fact that the process of the "exodus" has been interrupted by many black boxes, especially in the recent decade or so. Pereira et al. (2022) have assessed how the Russian-Ukrainian conflict impacts on the SDGs (biophysical, social, economic, partnership for the goals, in particular). They concluded their assessment with a slogan "no peace, no SDGs for 2030 and the future" (p. 286).

Our discussion at this point remains at the micro/meso/macro level. When the "pseudo-legislative" requirements are not working effectively at this level, the most likely way out is through the deployment of "private" ideas of creating institutional differentiation and eventually "isomorphism" (DiMaggio and Powell, 1983) among organizations and institutions. The idea is to work the way "up again" if the "top down" approach does not run smoothly. The importance of civil power and civil authorities is apparent here. Sjöstedt (1977, p. 21) lamented that civil power is a theoretical concept which rarely existed in reality. However, after 70 years of widespread democratic development, there have been many success stories of human and social advancement (e.g., the 1990 reunification of East and West Berlin and the end of the Cold War in 1991), but the over-representation of power or influence by military power noted by Sjöstedt is still prevalent today.

The idea of isomorphism has been discussed immensely, but the notion of "collective rationality" may have been ignored by many. In making rational decisions to reassemble the best in the "clan" (but sometimes rather difficult in the case when there is "uncodified" information diffused at the institutional level. See, e.g., Boisot and Child, 1996), rationality is often guided by the so-called "best practice". Now the question comes: Who sets the "best practice"? – the codified information or standard that is acceptable at the institutional level, and which eventually can be spread through isomorphism to the more macro and even global level?

Given the assumption of capitalism or even within the ideology of that such as a superpower like China where a market economy with socialist characteristics prevails, it is often the private sector that leads the so-called best practice or standard. In order to "build it up again" from the bottom to top, a host of private initiatives such as the TCFD and TNFD frameworks (together with the Global Reporting Initiative (GRI) and the Carbon Disclosure Project (CDP), for example) (Ardvisson, 2023) are guiding capitalistic activities to reflect on their responsibilities to the "social" and the "environmental" *-ities*. For example, Ricketts et al. (2023) have demonstrated how the "voluntary" disclosure guideline of BBFAW (Business Benchmark on Farm Animal Welfare), which is also TNFD-based, works in pulling the Australian beef procurement portfolios toward or against organizational or science-based sustainability objectives. Ricketts and colleagues' research represents an excellent example of how privately initiated "best practice" can help institutions, industries, and even societies to echo back the "pseudo-legislative" demands at the global level (such as the guiding principles of the UNSDGs). The authors underline that private standards shape and react to powerful external dynamics, including capital and consumer trends (Ricketts et al., 2023, p. 267). TNFD and BBFAW in their study are excellent examples of private initiatives to flex with vertical coordination of shifts in the entire supply chain.

The Anthropocene Level

The issue of the "Anthropocene" (Steffan et al., 2007) has always been discussed in the natural science arena, centering in areas including geo-science as well as in many other aspects of both natural and social sciences. In the accounting literature, Bebbington et al. (2019) and Bebbington and Rubin (2022) were among the first to introduce the idea to the related researchers. Given its multi- (inter-) disciplinary nature, the notion has been expanded by Noronha et al. (2024) and Donleavy and Noronha (2023) to depict an ontology of the "phenomenon *grande*". The phenomenon *grande* is different from the concept of "grand challenge" which mainly anchors on single disciplines (e.g., engineering, medicine, and so on). The phenomenon *grande* has a much wider connotation which embraces multidisciplinary and multi-method approaches born from Latour's (1987, 2005) train of thought, especially his actor-network theory. In other words, the topics of the current chapter on ESG reporting, biodiversity losses, as well as what was mentioned above as "best practices" initiated by private entities like TNFD, can all be viewed as actors laying on a flat plane, with inter-connections and "inter-meshings" so abundant while delicate. When viewed from a phenomenon *grande* perspective, the whole idea of TNFD is a response to the Anthropocenic issues of zoonosis, biodiversity losses and conservation; while ESG reporting in the related areas such as extinction accounting (Zhang and Noronha, 2023) are tools at the institutional (micro/meso) level for organizations and entities to create impactful translations. Figure 1 depicts the overall theoretical framework proposed by the authors.

TNFD as a Tool to Address Nature-Related Impacts

In recent years, efforts to address climate risks in the finance and corporate sectors have led to the development of trends which prioritize the disclosure, evaluation, and eventual action of such risks through the guidance provided by the TCFD. In a similar vein, on June 4, 2021, the TNFD announced its official establishment. This new initiative aims to aid financial institutions and companies from other sectors in assessing their impact on natural ecology, while encouraging them to allocate funds to promote biodiversity instead of causing harm.

The TNFD focuses on integrating nature into financial and business decision-making while redirecting global capital flows towards nature. The initiative combines risk management and disclosure, enabling

Figure 1. Theoretical framework of phenomenon grande applied to the Anthropocene and the macro/meso/micro levels

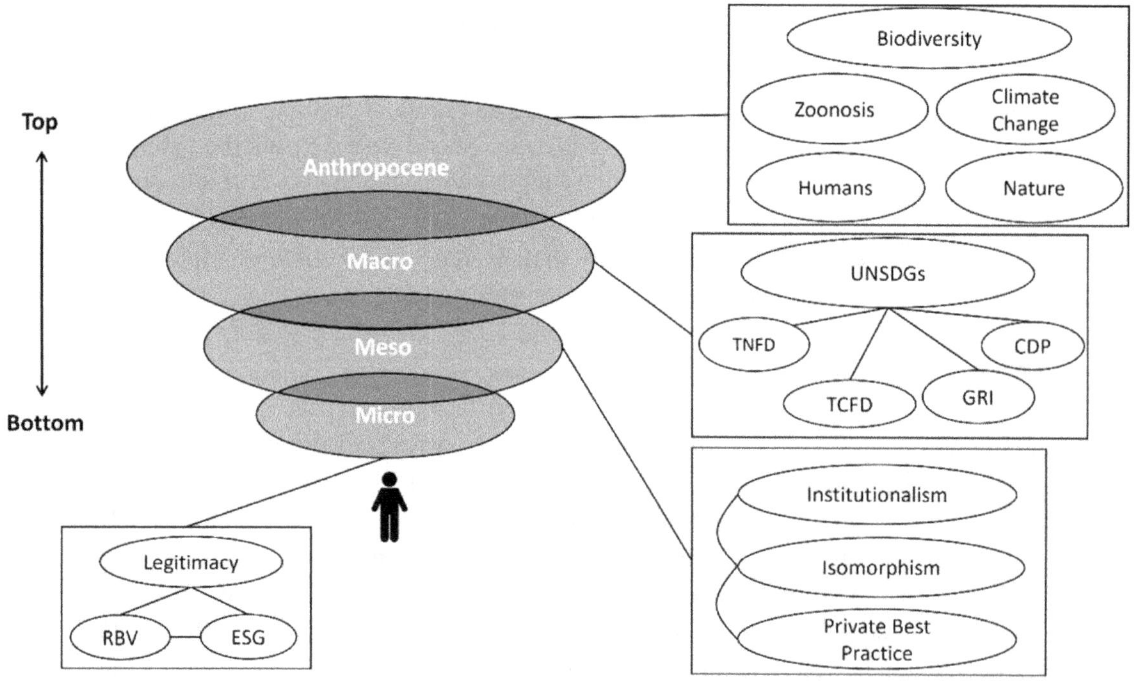

organizations to report on and respond to ever-changing natural risks, which include deforestation and habitat destruction, species loss, drought, and land use change. TNFD's beta framework provides a fundamental concept and definition for the term "nature", dividing it into four domains, namely land, oceans, freshwater, and atmosphere (TNFD, 2023). These domains differ significantly due to their organization, function, and geographic location, even for the same species.

Given the complexity of disclosing financing activities, such as loans, investments, insurance and advisory activities related to natural risks, TNFD proposes the LEAP guidelines: Locate, Evaluate, Assess, and Prepare. The guidelines emphasize a science-based and location-focused approach in incorporating enterprise and investment portfolio risk management (TNFD, 2023). Most importantly, the LEAP guidelines stress on location focus, where organizations should identify priority locations with high natural value during the locating process to avoid loopholes that may arise when assessing value differences between different regions and domains for natural risks and opportunities. The approach is designed to be flexible and scalable, allowing organizations of all sizes and sectors to use it. It involves engaging with stakeholders to identify nature-related risks and opportunities, assessing the organization's dependencies on nature, developing a plan to manage these risks and opportunities, and integrating this plan into the organization's overall strategy. For instance, organizations should carefully study the "biomes" where their operations and value chains are located before evaluating how they impact environmental assets and ecosystem services like water supply and flood control. This process ensures correct identification and quantification of risks and opportunities. Overall, TNFD provides a comprehensive framework that facilitates more informed business decision-making through risk management and disclosure, contributing to sustainable development (Nedopil, 2022; Petrie, 2021).

Figure 2 illustrate the most updated guidance for nature-related disclosure applicable for all sectors, especially for high nature-risk sectors. Specifically, TNFD seeks to create a coherent, consistent, and globally accepted set of metrics and disclosure recommendations, which would allow organizations to communicate information about their nature-related dependencies, impacts, and risks in a transparent manner. TNFD has four widely adoptable recommendations tied to governance, strategy, risk management, and metrics and targets. These recommendations yield 14 specific recommended disclosures that organizations should include in their financial filings to provide decision-useful information (Table 1). In this manner, TNFD's framework aims to enable financial institutions and businesses to identify investments that are aligned with global biodiversity and climate goals, thereby channeling resources towards initiatives that promote conservation, restoration, and sustainable use of ecosystems.

Biodiversity Credits

Aside from the goal of protecting biodiversity and promoting its sustainable use, COP15's key focus centers on implementing practical measures for biodiversity protection, including funding and technical capacity. Following that, a report published by the International Institute for Environment and Development (IIED) proposes the concept of biodiversity credits as a potential source of funding for biodiversity conservation (Ducros and Steele, 2022). Much like carbon credits, biodiversity credits can be purchased by companies to support the protection and management of biodiversity. When designed effectively, this mechanism for quantifying and tracking the conservation of biodiversity can provide direct funding toward persuasive biodiversity protection, while also directly supporting locally-led actions to guarantee the participation and benefit of indigenous peoples and local communities.

Figure 2. TNFD's approach to nature-related disclosure (adapted from TNFD Integrated Framework Version 0.4)

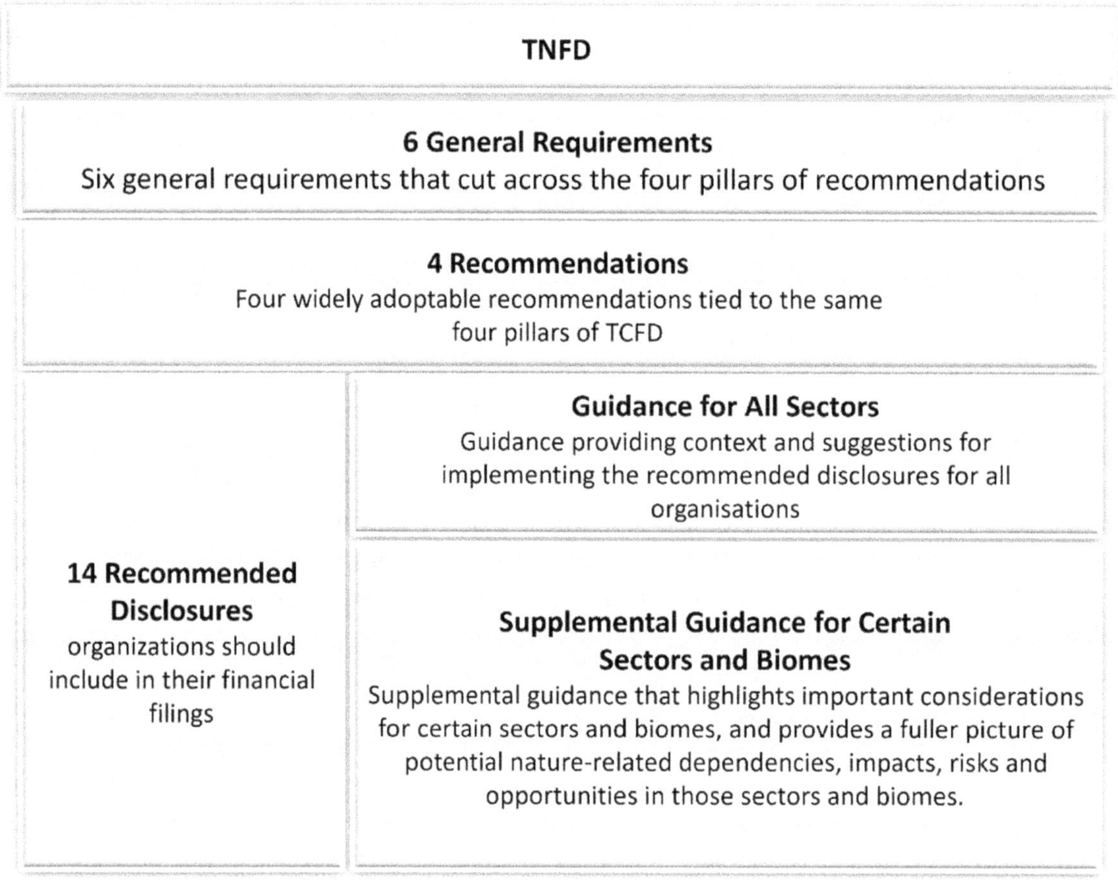

Biodiversity credits are divided into three categories: (1) protecting or preventing the loss of biological diversity, (2) ecosystem restoration, and (3) supporting existing conservation initiatives. A biodiversity plan has the potential to encompass all three categories (WEF, 2022a). The sellers of biodiversity credits can include government entities, private landowners, conservation organizations, project developers, and even corporations (NSW Environment Protection Authority, 2021). It is important to note that biodiversity credits differ from biodiversity offsets in that they are a proactive means of promoting effective biodiversity management through investment in projects, whereas offsets serve as compensation for corporations to mitigate potential biodiversity loss after all other measures have been taken (Simpson et al., 2022; WEF, 2022a).

Based on current regulations, individuals and entities are permitted to purchase biodiversity credits. This may include a variety of buyers such as developers seeking to offset their projects' impact on biodiversity, government agencies utilizing the market to achieve conservation objectives, and philanthropic organizations (WEF, 2022b). Several forward-thinking companies are already pioneering the practices of biodiversity credits according to the report by WEF (2022b). For instance, in May 2022, biodiversity credit product was introduced in Colombia. The product was developed by ClimateTrade, a blockchain-based

Table 1. TNFD's 14 disclosure recommendations (adapted from TNFD Integrated Framework Version 0.4)

TNFD's 14 Disclosure Recommendations (v0.4)			
Governance	**Strategy**	**Risk & Impact Management**	**Metrics & Targets**
A. Describe the board's oversight of nature-related dependencies, impacts, risks and opportunities. B. Describe management's role in assessing and managing nature-related dependencies, impacts, risks and opportunities.	A. Describe the nature-related dependencies, impacts, risks and opportunities the organization has identified over the short, medium, and long term. B. Describe the effect nature-related risks and opportunities have had and may have on the organization's businesses, strategy, and financial planning. C. Describe the resilience of the organization's strategy to nature-related risks and opportunities, taking into consideration different scenarios. D. Disclose the locations where there are assets and/or activities in the organization's direct operations, and upstream and/or downstream and/or financed where relevant, that are in priority areas.	A. (i) Describe the organization's processes for identifying and assessing nature related dependencies, impacts, risks and opportunities in its direct operations. A. (ii) Describe the organization's approach to identifying nature-related dependencies, impacts, risks and opportunities in its upstream and downstream value chain(s) and financed activities and assets. B. Describe the organization's processes for managing nature-related dependencies, impacts, risks and opportunities and actions taken in light of these processes. C. Describe how processes for identifying, assessing and managing nature-related risks are integrated into the organization's overall risk management. D. Describe how affected stakeholders are engaged by the organization in its assessment of, and response to, nature-related dependencies, impacts, risks and opportunities.	A. Disclose the metrics used by the organization to assess and manage material nature-related risks and opportunities in line with its strategy and risk management process. B. Disclose the metrics used by the organization to assess and manage dependencies and impacts on nature. C. Describe the targets and goals used by the organization to manage nature-related dependencies, impacts, risks and opportunities and its performance against these.

climate marketplace, and Terrasos, an organization dedicated to biodiversity conservation and habitat banking in Latin America. The first project to implement this initiative, known as voluntary biodiversity credits (VBCs), is the Bosque de Niebla-El Globo Habitat Bank, which is committed to preserving native species that are critically endangered within the High Andes. The Bosque de Niebla is a cloud forest that provides habitat for several threatened species, such as the spectacled bear, the yellow-eared parrot, and the black-and-chestnut eagle. A VBC from this project costs $30, and corresponds to 30 years of conservation or restoration of ten square meters of the Bosque de Niebla forest. This means that corporate entities may give back to nature and have a positive impact by purchasing VBCs, which in turn will align their business operations with biodiversity and ecosystem conservation (ClimateTrade, 2022).

In brief, there is substantial potential for biodiversity credits to become a valuable tool for alleviating environmental impacts. These credits enable organizations to offset any residual negative impacts by investing in conservation efforts, thereby promoting positive outcomes. Furthermore, the trading of biodiversity credits fosters partnerships between different sectors, opening up new avenues for collaboration on conservation initiatives and knowledge sharing.

Caveats for all Micro and Institutional Interventions

We have discussed a number of issues concerning the facilitation of trade through biodiversity credits as well as the rise of private-based norms to back up the so-called "pseudo legislations" at the more macro level. We take in particular Ferreira's (2017) alerting case study of UK's experience with "biodiversity offsets" which eventually went dead due to political pressures. The "mis-use" of carbon credits until today is a realistic reflection of a novel attempt to implement "biodiversity credits". Citing Trigedga (2013), Ferreira (2017, p. 1569) pointed out that a market-based "management" of biodiversity can result in the "trading" of different biodiversities. As we have in our minds the notion and the usage of the word "mitigation" has a post-hoc firefighting connotation, it is necessary to underline the consensus of "prevention" rather than "mitigation". Carbon or biodiversity offset schemes represent, to a certain extent, a contest of "who is not finally taking up all the mess". More than 10 years ago, Caney and Hepburn (2011, p. 5) have already discussed the moral and ethical aspects of carbon trading by citing Michael Sandle's writing that "turning pollution into a commodity to be bought and sold removes the moral stigma that is properly associated with it…[and] may undermine the sense of shared responsibility that increased global cooperations require".

Biodiversity credits involve all kinds of organic species, including soil and when at a more emancipated level, the "intra-matters" of human beings and all living things. The challenges posed in front of the TNFD, biodiversity credit schemes, and at the micro level of ESG reporting are not to be undermined. Therefore, we propose here an enhanced ESG framework to "work it from the bottom" and, BTTB (back to the basics), to the notion of CSR.

Furthermore, nowadays many transactions are conducted through a "decentralized" algorithm (e.g., "defi") which connects the related parties over blockchains and AI technologies including encryptions, smart contracts, and other advanced means to "avoid" intermediaries (including the government even). A further caveat here is that for nature-related credit transactions (including climate and biodiversity credits), the issue of AI must not be underestimated. The European Union's first Artificial Intelligence Act has already marked the emergence of the possible risks and the need to protect the public (WEF, 2023).

Towards an Enhanced ESG Framework

This chapter argues that the existing ESG framework should be expanded to include the concepts of TNFD and biodiversity credits by demonstrating how such enhancements can potentially drive deeper ecological values into an organization's philosophy. We believe that the commitment to increasing transparency, standardization, and robustness in nature-related disclosures should remain one of the cornerstones of the mission of ESG reporting.

First and foremost, the current ESG reporting frameworks address biodiversity risks rather minimally. In alignment with the TNFD philosophy, we recommend that under the environmental dimension of reporting, companies should disclose specific risks related to biodiversity loss, habitat degradation, and the ecosystem services they rely on (e.g., water provision, pollination, etc.). Moreover, detailed information on the biodiversity credits that they have engaged with should be included to quantify and track biodiversity-related actions and contributions. Regarding the social dimension, we advocate the identification of biodiversity custodians throughout the product life cycle. These custodians are critical stakeholders in the biodiversity value chain and ecosystems, having rights to or bearing responsibility for the natural resources within a project site (WEF, 2022b). Determining the relevant entities or individuals for a given land parcel is essential in ensuring that adequate incentives and compensation are in place for protecting and enhancing its ecological value. Consequently, community partnership programs tailored towards biodiversity conservation and restoration will incentivize companies to adopt best practices in biodiversity management and promote cross-sector collaboration in ecosystem protection and restoration. Lastly, the governance dimension should require companies to effectively manage and oversee nature-related risks and opportunities at the board and senior management levels. This component encompasses issues such as board oversight, management roles and responsibilities, internal control, and policies concerning nature-related risks and opportunities.

As TNFD incorporates specific metrics and targets for organizations from various sectors, both qualitative and quantitative information is necessary to enable users to understand their exposure to nature-related risks and opportunities. This information is crucial for meaningful comparisons between organizations. We argue that the critical aspect of enhancing ESG reporting is establishing transparent, sound, and coherent biodiversity conservation strategies to ensure that ecological impacts are genuine and sustainable for businesses. Moreover, it is important for organizations to establish corporate strategies that address nature-related risks and opportunities in the short, medium, and long term. The strategy should include elements such as scenario analysis, stress testing, resilience planning, and the integration of nature-related considerations in overall business strategy. We also reveal that collaboration between stakeholders, including investors, businesses, governments, and conservation organizations, is vital for an enhanced ESG reporting framework. Therefore, the above-discussed measures are intended to provide investors and biodiversity custodians with measurable ecological outcomes and long-term certainty. By incorporating the concepts of TNFD and biodiversity credits into the ESG framework, organizations can better demonstrate their commitment to preserving the environment and promoting sustainable business practices (See Figure 3). In short, to stave off the impending ecological collapse attributed to human activities, it is crucial for businesses to evolve and adapt their practices in a "work it from the bottom manner", bringing together accountability, transparency, and sustainability.

CONCLUSION

The Anthropocene era presents the world with unprecedented environmental challenges, such as climate change and biodiversity loss. These issues are not simply environmental concerns, but also encompass economic, developmental, security, social, moral, and ethical dimensions. Addressing these challenges in conjunction with the 17 UNSDGs is crucial to achieving objectives like food and water security, good health for all, poverty alleviation, and fostering a more equitable world (One Earth, 2021). In response, businesses and investors have increasingly prioritized ESG factors that underpin long-term sustainable growth. However, the existing ESG framework has limitations in addressing urgent concerns like energy, climate, and biodiversity crises. As such, adopting a stronger advocation for global peace and an Anthropocenic perspective is essential for the future of nature and humans.

The present chapter explores the advantages of incorporating TNFD and biodiversity credits as tools to enhance ESG performance. We assess the extent of environmental consideration in the ESG framework, the connection between TNFD and ESG, and the potential benefits of biodiversity credits as supplementary measures.

We contend that integrating TNFD guidelines into the ESG framework enables a more comprehensive evaluation of a company's ecological impact and related financial risks. This fusion can bolster ESG performance and facilitate the realization of global biodiversity conservation targets, such as those specified in the UNSDGs. Meanwhile, biodiversity credits offer a means to address conservation within the ESG framework. They incentivize companies to invest in ecosystem protection, restoration, and management, translating into financial benefits that can be traded on markets, thereby creating long-term value and promoting conservation. Incorporating TNFD recommendations and biodiversity credits into the ESG framework offers tremendous potential for enhancing nature conservation and financial sector resilience. By proactively recognizing and addressing nature-related financial risks, businesses can refine their decision-making processes to be more competitive in an increasingly sustainability-driven market.

Figure 3. The enhanced ESG reporting framework

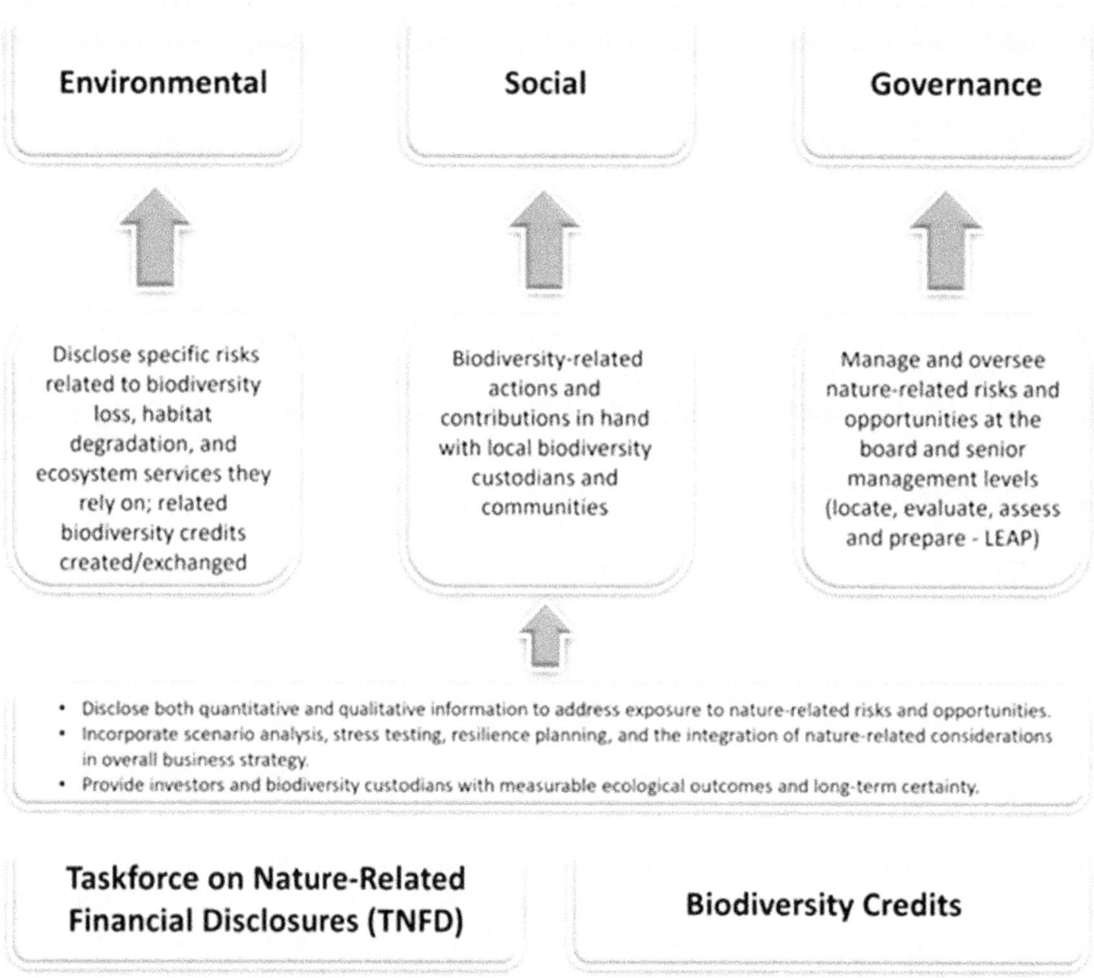

This chapter emphasizes the synergistic benefits of integrating TNFD, ESG reporting, and biodiversity credits. The inclusion of biodiversity credits into ESG reporting, accompanied by the establishment of TNFD, can catalyze transformative changes in how companies manage and disclose nature-related risks and opportunities. Furthermore, isomorphism and institutional theory highlight the potential drivers and constraints of enacting this transformation, stressing the importance of social norms, rules, and shared understandings that shape organizational behavior and response to external pressures. To fully capitalize on these benefits, organizations must proactively adapt to evolving social norms and regulatory pressures, re-evaluate their ESG reporting practices, and strategically employ market-based instruments like biodiversity credits. Ultimately, adopting a forward-thinking and proactive approach to TNFD and biodiversity credits can lay the foundation for a more sustainable and resilient global economy in the Anthropocene era.

REFERENCES

Ardvisson, S. (2023). CEO talk of sustainability in CEO letters: Towards the inclusion of a sustainability embeddedness and value-creation perspective. *Sustainability Accounting. Management and Policy Journal, 14*(7), 26–61.

Bebbington, J., Österblom, H., Crona, B., Jouffray, J. B., Larrinaga, C., Russell, S., & Scholtens, B. (2019). Accounting and accountability in the Anthropocene. *Accounting, Auditing & Accountability Journal, 33*(1), 152–171. doi:10.1108/AAAJ-11-2018-3745

Bebbington, J., & Rubin, A. (2022). Accounting in the Anthropocene: A roadmap for stewardship. *Accounting and Business Research, 52*(5), 582–596. doi:10.1080/00014788.2022.2079780

Boisot, M., & Child, J. (1996). From fiefs to clans and network capitalism: Explaining China's emerging economic order. *Administrative Science Quarterly, 41*(4), 600–628. doi:10.2307/2393869

Branco, M. C., & Rodrigues, L. L. (2006). Corporate social responsibility and resource-based perspective. *Journal of Business Ethics, 69*(2), 111–132. doi:10.100710551-006-9071-z

Briggs, H. (2022). COP15: Nations reach 'historic' deal to protect nature. *BBC News*. https://www.bbc.com/news/science-environment-64019324

Callon, M. (1991). Techno-economy networks and irreversibility. In L. Law (Ed.), *A Sociology of Monsters: Essay on Power, Technology and Domination*. Routledge.

Caney, S., & Hepburn, C. (2011). *Carbon trading: Unethical, unjust and ineffective?* Centre for Climate Change Economics and Policy working paper No. 59. & Grantham Research Institute on Climate Change and the Environment working paper No. 49.

ClimateTrade. (2022). *ClimateTrade and Terrasos jointly promote Voluntary Biodiversity Credits to support biodiversity conservation*. Climate Trade. https://climatetrade.com/climatetrade-and-terrasos-jointly-promote-voluntary-biodiversity-creditsto-support-biodiversity-conservation/#:~:text=The%20purchase%20of%20a%20VBC,to%2010%20m2%20of%20forest

Conniff, R. (2018). *Biodiversity Credits: How Policy Can Help Save Unprotected Species*. Yale Environment 360. https://e360.yale.edu/features/biodiversity-credits

Convention on Biological Diversity (CBD). (2022). *Kunming-Montreal Global Biodiversity Framework*. Montreal: CBD. https://www.cbd.int/gbf/

Cuckston, T. (2022). Book Review: Extinctions: Living and Dying in the Margin of Error. *Progress in Physical Geography, 46*(5), 819–821. doi:10.1177/03091333221109546

Deegan, C. (2002). Introduction – the legitimising effect of social and environmental disclosure – a theoretical foundation. *Accounting, Auditing & Accountability Journal, 15*(3), 283–311. doi:10.1108/09513570210435852

DiMaggio, P. J., & Powell, W. W. (1983). The iron cage revisited: Institutional isomorphism and collective rationality in organizational fields. *American Sociological Review, 48*(2), 147–160. doi:10.2307/2095101

Donleavy, G., & Noronha, C. (2023). Concluding remarks: Comparative CSR and sustainability research – An axial analysis. In G. Donleavy & C. Noronha (Eds.), *Comparative CSR and Sustainability: New Accounting for Social Consequences*. Routledge.

Dowling, J., & Pfeffer, J. (1975). Organizational legitimacy: Social values and organizational behavior. *Pacific Sociological Review*, *18*(1), 122–136. doi:10.2307/1388226

Ducros, A., & Steele, P. (2022). *Biodiversity Credits to Finance Nature and People: Emerging Lessons*. IIED.

Ferreira, C. (2017). The contested instruments of a new governance regime: Accounting for nature and building markets for biodiversity offsets. *Accounting, Auditing & Accountability Journal*, *30*(7), 1568–1591. doi:10.1108/AAAJ-12-2015-2336

Freeman, R. E. (1984). *Strategic Management: A Stakeholder Approach*. Pitman.

Halbritter, G., & Dorfleitner, G. (2015). The wages of social responsibility—Where are they? A critical review of ESG investing. *Review of Financial Economics*, *26*(1), 25–35. doi:10.1016/j.rfe.2015.03.004

International Union for Conservation of Nature (IUCN). (2021). *Guidelines for Planning and Monitoring Business Biodiversity Performance*. ICUN.

Jones, M. J., & Solomon, J. F. (2013). Problematising accounting for biodiversity. *Accounting, Auditing & Accountability Journal*, *26*(5), 668–687. doi:10.1108/AAAJ-03-2013-1255

Kopnina, H., Zhang, R., Anthony, S., Hassan, A., & Atkins, J. (2023). *How to account for extinction?: Discussion of Environmental, Social and Governance (ESG) in relation to biodiversity loss*. Working paper.

KPMG International. (2022). Big shifts, small steps. *Survey of Sustainability Reporting 2022*. KPMG International. https://kpmg.com/xx/en/home/insights/2022/09/survey-of-sustainability-reporting-2022.html

Kurth, T., Wübbels, G., Portafaix, A., Meyer zum Felde, A., & Zielcke, S. (2021). *The biodiversity crisis is a business crisis*. Boston Consulting Group.

Lambooy, T. E., Maas, K. E. H., van't Foort, S., & Van Tilburg, R. (2018). Biodiversity and natural capital: Investor influence on company reporting and performance. *Journal of Sustainable Finance & Investment*, *8*(2), 158–184. doi:10.1080/20430795.2017.1409524

Latour, B. (1987). *Science in Action: How to Follow Scientists and Engineers Through Society*. Harvard University Press.

Latour, B. (2005). *Reassembling the Social: An Introduction to Actor-Network Theory*. Oxford University Press.

Lu, F. (2023). Corporate green initiatives on real circular economy or window-dressing? In G. Donleavy & C. Noronha (Eds.), *Comparative CSR and sustainability: New Accounting for Social Consequences*. Routledge.

Nedopil, C. (2022). Integrating biodiversity into financial decision-making: Challenges and four principles. *Business Strategy and the Environment*. doi:10.1002/bse.2884

Nelson, F., & Combe, M. (2022). What is the taskforce on nature-related financial disclosures. *Equity*, *36*(8), 20–21.

Noronha, C., Zhang, R., & Donleavy, G. (2024). Developing the phenomenon *grande* in management research: The pandemic, lockdowns, and biodiversity extinction in the Anthropocene. In D. Crowther & S. Seifi (Eds.), *The Routledge Companion to the Future of Management Research*. Routledge.

NSW Environment Protection Authority. (2021). *What are biodiversity credits?* Biodiversity Offsets Scheme. https://www.environment.nsw.gov.au/topics/animals-and-plants/biodiversity-offsets-scheme/about-the-biodiversity-offsets-scheme/what-are-biodiversity-credits

One Earth. (2021). *Four Steps for Earth: Bringing the Post-2020 Global Biodiversity Framework into the Mainstream*. Elsevier. . doi:10.1016/j.oneear.2020.12.011

Orange, E., & Cohen, A. M. (2010). From eco-friendly to eco-intelligent. *The Futurist*, *44*(5), 24–32.

Pereira, P., Zhao, W., Symochko, L., Inacio, M., Bogunovic, I., & Barceo, D. (2022). The Russian-Ukrainian armed conflict will push back the sustainable development goals. *Geography and Sustainability*, *3*(3), 277–287. doi:10.1016/j.geosus.2022.09.003

Petrie, M. (2021). *Environmental Governance and Greening Fiscal Policy: Government Accountability for Environmental Stewardship*. Palgrave Macmillan. doi:10.1007/978-3-030-83796-9

Rands, M. R., Adams, W. M., Bennun, L., Butchart, S. H., Clements, A., Coomes, D., Entwistle, A., Hodge, I., Kapos, V., Scharlemann, J. P. W., Sutherland, W. J., & Vira, B. (2010). Biodiversity conservation: Challenges beyond 2010. *Science*, *329*(5997), 1298–1303. doi:10.1126cience.1189138 PMID:20829476

Ricketts, K. D., Palmer, J., Navarro-Garcia, J., Lee, C., Dominik, S., Barlow, R., Ridoutt, B., & Richards, A. (2023). Bridging organisational discourse and practice change: Exploring sustainable procurement for Australian beef. *Sustainability Accounting. Management and Policy Journal*, *14*(2), 265–288.

Simpson, K. H., de Vries, F. P., Dallimer, M., Armsworth, P. R., & Hanley, N. (2022). Ecological and economic implications of alternative metrics in biodiversity offset markets. *Conservation Biology*, *36*(5), e13906. doi:10.1111/cobi.13906 PMID:35288986

Sjöstedt, G. (1977). The exercise of international civil power: A framework for analysis. *Cooperation and Conflict*, *12*(1), 21–39. doi:10.1177/001083677701200102

Steffan, W., Crutzen, P., & McNeill, J. R. (2007). The Anthropocene: Are humans now overwhelming the great forces of nature? *Ambio*, *36*(8), 614–621. doi:10.1579/0044-7447(2007)36[614:TAAHNO]2.0.CO;2 PMID:18240674

Suchman, M. (1995). Managing legitimacy: Strategic and institutional approaches. *Academy of Management Review*, *20*(3), 571–610. doi:10.2307/258788

TNFD. (2023). *TNFD Integrated Framework Version 0.4*. TNFD. https://framework.tnfd.global/wp-content/uploads/2023/03/23-23882-TNFD_v0.4_Integrated_Framework_v7.pdf

Trigedga, H. (2013). Biodiversity offsetting: Problematisation of of an emerging governance regime. *Accounting, Auditing & Accountability Journal, 26*(5), 806–832. doi:10.1108/AAAJ-02-2013-1234

United Nations. (2023). Guterres urges countries to recommit to achieving SDGs by 2030 deadline. *UN New: Global perspective Human stories*. UN. https://news.un.org/en/story/2023/04/1136017

World Economic Forum (WEF). (2022a). *Biodiversity credits as an economic driver for nature can deliver net positive biodiversity gains*. WEF. https://www.weforum.org/agenda/2022/12/biodiversity-credits-nature-cop15/#:~:text=Biodiversity%20credits%20as%20an%20economic,%20deliver%20net%20positive%20biodiversity%20gains

World Economic Forum (WEF). (2022b). *Biodiversity Credits: Unlocking Financial Markets for Nature-Positive Outcomes*. Geneva, Switzerland: World Economic Forum.

World Economic Forum (WEF). (2023). *The European Union's Artificial Intelligence Act, explained*. WEF. https://www.weforum.org/agenda/2023/03/the-european-union-s-ai-act-explained/

World Wildlife Fund (WWF). (2022). Living. *The Planetary Report*. https://wwflpr.awsassets.panda.org/downloads/lpr_2022_full_report.pdf

Zhang, R., & Noronha, C. (2023). Assessing the nexus between cross-boarder infrastructure projects and extinction accounting - from the belt and road initiative. *Social and Environmental Accountability Journal, 43*(1), 30–55. doi:10.1080/0969160X.2022.2132969

Chapter 4
Boosting Local Economies:
The Impact of N-Contour Gesellian Money on Regional Growth

Rinat Galiautdinov
Independent Researcher, Italy

ABSTRACT

This article explores the practical use of Gesellian money in several local regions which allowed the authors to collect the data and research the influence of the Gesellian model on the local economies. The implementation of Gesellian money has resulted in a significant increase in the turnover of money within local economies, which has helped to save struggling businesses that were unable to pay their employees. By using Gesellian money, local businesses have been able to circulate more money within their communities, leading to a more sustainable and prosperous economy. The implementation of Gesellian money has proven to be an effective tool for stimulating local economies and supporting small businesses. The research also showed that introduction of Gesellian model allows to avoid crucial crises in economy. The article also provides the solution of introduction Gesellian model into big economies using n-contour which allows to consider different growth amplitude of local economies, and the instruments for interaction between local economies and local-to-international economies.

INTRODUCTION

Gesellian money, also known as complementary currencies, is a concept that has been gaining popularity in recent years as communities seek to promote local economic development. These currencies are used in addition to the national currency of a country and are designed to encourage people to spend money locally and support local businesses. The use of Gesellian money has been growing in popularity in recent years, with many communities implementing their own versions of these currencies. While the impact of Gesellian money on the overall economy of a region or country is still being studied, early results suggest that these currencies can have a positive effect on local businesses and help to stimulate economic growth.

DOI: 10.4018/978-1-6684-8879-9.ch004

What is Gesellian Money?

Gesellian money, named after the German economist Silvio Gesell, is a type of complementary currency that is designed to promote local economic development (Gesell, S., 1958). These currencies are used in addition to the national currency of a country and are typically used within a specific geographic area, such as a city or region. The purpose of Gesellian money is to encourage people to spend money locally and support local businesses, rather than spending their money on goods and services from outside the area.

Gesellian money can take many different forms, depending on the community that implements it. Some communities use physical currencies, such as paper bills or coins, while others use digital currencies that can be accessed through a mobile app or website. Some Gesellian money systems are backed by a local government or community organization, while others are entirely decentralized.

How Does Gesellian Money Work?

The basic idea behind Gesellian money is that it is designed to stimulate local economic activity by encouraging people to spend money locally. In order to do this, many Gesellian money systems are designed with certain features that make them more attractive to consumers than traditional currencies.

One of the most common features of Gesellian money is that it typically loses value over time. This is known as demurrage, and it is designed to encourage people to spend their money quickly, rather than hoarding it. By losing value over time, Gesellian money incentivizes people to spend their money locally and support local businesses, rather than holding onto it and spending it elsewhere.

Another feature of Gesellian money is that it is often designed to be used in specific ways. For example, some Gesellian money systems may only be accepted at certain businesses or for certain types of goods and services. This can help to promote local economic development by encouraging people to spend their money on goods and services that are produced locally.

Examples of Gesellian Money

One of the earliest examples of Gesellian money was in the town of Wörgl, Austria, in 1932. The town was suffering from high unemployment and economic stagnation, so the mayor, Michael Unterguggenberger, decided to introduce a local currency called the Wörgl Schilling. The currency had a stamp on it that needed to be updated each month to maintain its value. If the stamp was not updated, the currency lost its value by 1% per month.

The Wörgl Schilling was a huge success. It encouraged spending and investment in the local economy, and unemployment dropped dramatically. The town became a model for other communities looking to stimulate their local economies.

Another example of Gesellian money was in Brazil in the 1990s. The government introduced a currency called the Real, which had a built-in demurrage system. The currency lost its value over time, which encouraged spending and discouraged hoarding. The Real was a huge success, and it helped stabilize the Brazilian economy after years of hyperinflation.

In Switzerland, there is a currency called the WIR, which has been in use since 1934. The WIR is a complementary currency that is used by businesses to trade with each other. The currency has a demurrage system built into it, which encourages businesses to use it quickly and invest in their operations.

In the United States, there have been several experiments with Gesellian money. In the 1930s, during the Great Depression, a number of communities introduced local currencies with demurrage systems. These currencies were successful in promoting local economic growth, but they were eventually shut down by the government.

In recent years, there have been several experiments with digital currencies that use demurrage systems. These currencies are designed to encourage spending and investment, and they have been successful in promoting economic growth in local communities.

There are many different examples of Gesellian money systems in use around the world. One of the most well-known examples is the Chiemgauer, a local currency used in the Chiemgau region of Germany. The Chiemgauer is a physical currency that is designed to lose value over time, and it can only be used at businesses that are located within the Chiemgau region. The currency has been in use since 2003 and has been credited with helping to promote local economic development in the region.

Another example of Gesellian money is the Bristol Pound, a local currency used in the city of Bristol in the United Kingdom. The Bristol Pound is a digital currency that can be accessed through a mobile app or website, and it is designed to be used at businesses that are located within the city. The currency has been in use since 2012 and has been credited with helping to support local businesses and promote sustainable economic development in the city.

More modern example of Gesellian money are PrivateCurrency (privatecurrency.org), Turkic Reserve System and Shaimuratiki.

The Impact of Gesellian Money on the Economy

The impact of Gesellian money on the overall economy of a region or country is still being studied, but early results suggest that these currencies can have a positive effect on local businesses and help to stimulate economic growth.

One of the key benefits of Gesellian money is that it encourages people to spend money locally, which can help to support local businesses and create jobs. By keeping money within the local economy, Gesellian money can help to promote economic development and reduce the dependence of local communities on outside sources of income.

Another benefit of Gesellian money is that it can help to promote sustainable economic development. By encouraging people to spend money on locally produced goods and services, Gesellian money can help to reduce the environmental impact of economic activity and promote the use of renewable resources.

Finally, Gesellian money can help to promote social cohesion within communities. By encouraging people to spend money locally and support local businesses, Gesellian money can help to build stronger connections between individuals and promote a sense of community pride.

Gesellian money is a concept that has been gaining popularity in recent years as communities seek to promote local economic development. These currencies are used in addition to the national currency of a country and are designed to encourage people to spend money locally and support local businesses. While the impact of Gesellian money on the overall economy of a region or country is still being studied, early results suggest that these currencies can have a positive effect on local businesses and help to stimulate economic growth. By promoting local economic activity, reducing the environmental impact of economic activity, and building stronger connections between individuals, Gesellian money has the potential to promote sustainable economic development and create stronger, more resilient communities.

LITERATURE REVIEW

Gesellian money, also known as demurrage currency, is a type of currency that loses value over time. This means that the longer someone holds onto the currency, the less it is worth. This concept was first introduced by Silvio Gesell, a German economist, in the late 19th century. The idea behind Gesellian money is that it encourages people to spend their money rather than hoard it, which can stimulate economic activity.

There have been many studies and books written on the topic of Gesellian money and its influence on a region's or country's economy. The following is a review of some of the most significant works on the subject:

The Natural Economic Order by Silvio Gesell - This is the book that introduced the concept of Gesellian money. In it, Gesell argues that the current monetary system is flawed and that Gesellian money would create a more stable and equitable economy.

Money and the Mechanism of Exchange by William Stanley Jevons - Jevons was a British economist who wrote extensively on the theory of money. In this book, he discusses the concept of Gesellian money and its potential benefits.

The Theory of Money and Credit by Ludwig von Mises - Von Mises was an Austrian economist who is known for his work on the free market. In this book, he critiques the idea of Gesellian money and argues that it would be harmful to the economy.

The Economics of Inflation by Costantino Bresciani-Turroni - Bresciani-Turroni was an Italian economist who wrote about the effects of inflation on the economy. In this book, he discusses the potential benefits of Gesellian money in combating inflation.

The End of Money and the Future of Civilization by Thomas H. Greco Jr. - Greco is an American economist who has written extensively on alternative currencies. In this book, he discusses the potential benefits of Gesellian money and other alternative currencies.

A History of Money: From Ancient Times to the Present Day by Glyn Davies - Davies is a British economist who has written extensively on the history of money. In this book, he discusses the use of Gesellian money in various historical contexts.

The Future of Money by Bernard Lietaer - Lietaer is a Belgian economist who has written extensively on alternative currencies. In this book, he discusses the potential benefits of Gesellian money and other alternative currencies.

Money and Sustainability: The Missing Link by Bernard Lietaer and Christian Arnsperger - In this book, Lietaer and Arnsperger argue that Gesellian money and other alternative currencies can help create a more sustainable economy.

The Local Economy Solution: How Innovative, Self-Financing Pollinator Enterprises Can Grow Jobs and Prosperity by Michael H. Shuman - Shuman is an American economist who has written extensively on local economies. In this book, he discusses the potential benefits of Gesellian money in promoting local economic development.

The Ecology of Money by Richard Douthwaite - Douthwaite was an Irish economist who wrote extensively on alternative currencies. In this book, he discusses the potential benefits of Gesellian money and other alternative currencies in promoting sustainable development.

The Future of Money: Creating New Wealth, Work and a Wiser World by James Robertson - Robertson is a British economist who has written extensively on monetary reform. In this book, he discusses the potential benefits of Gesellian money in creating a more equitable and sustainable economy.

The End of Capitalism as We Knew It: A Feminist Critique of Political Economy by J. K. Gibson-Graham - Gibson-Graham is an Australian economist who has written extensively on feminist economics. In this book, she discusses the potential benefits of Gesellian money in promoting economic democracy.

The Wealth of Nature: Economics as if Survival Mattered by John Michael Greer - Greer is an American economist who has written extensively on sustainability. In this book, he discusses the potential benefits of Gesellian money and other alternative currencies in promoting a sustainable economy.

The Future of Money: From Financial Crisis to Public Resource by Mary Mellor - Mellor is a British economist who has written extensively on monetary reform. In this book, she discusses the potential benefits of Gesellian money and other alternative currencies in creating a more equitable and sustainable economy.

The Great Turning: From Empire to Earth Community by David C. Korten - Korten is an American economist who has written extensively on globalization and sustainability. In this book, he discusses the potential benefits of Gesellian money and other alternative currencies in promoting a more sustainable and equitable world.

The New Economics: A Bigger Picture by David Boyle - Boyle is a British economist who has written extensively on alternative economics. In this book, he discusses the potential benefits of Gesellian money and other alternative currencies in creating a more equitable and sustainable economy.

The Future of Money: How New Lending Technologies Will Transform the Global Economy by Neha Narula, Robleh Ali, and Alex Lipton - In this book, the authors discuss the potential benefits of Gesellian money and other alternative currencies in the context of new lending technologies.

Money and Sustainability: The Missing Link by Bernard Lietaer and Christian Arnsperger - In this book, Lietaer and Arnsperger argue that Gesellian money and other alternative currencies can help create a more sustainable economy.

The End of Banking: Money, Credit, and the Digital Revolution by Jonathan McMillan - In this book, McMillan discusses the potential benefits of Gesellian money and other alternative currencies in the context of digital banking.

The Future of Money: Bitcoin and Other Cryptocurrency Technologies Are a Way of Life in This Small Swiss Town by Gregor Schaefer - In this book, Schaefer discusses the potential benefits of Gesellian money and other alternative currencies in the context of cryptocurrency technologies.

METHODS

Study Design

This study was an observational study that aimed to evaluate the impact of Gesellian money on local economies. The study was conducted by introducing the Gesellian money in several local regions and comparing the economic outcomes in those regions to those in areas where Gesellian money was not used.

Sample Size and Characteristics

The study included multiple local regions where Gesellian money was introduced, and cooperatives in those regions where Gesellian money could be used to buy products and services. The sample size of the study was not explicitly stated, but it can be assumed that it included multiple cooperatives and local economies. The participants or subjects were not specifically identified or characterized by age, gender, or occupation.

Data Collection Methods

The data for this study was collected using a combination of surveys, interviews, and financial reports from the cooperatives. Surveys and interviews were used to gather information on the use of Gesellian money in the local economies, the impact of Gesellian money on the cooperatives' financial situations, and the opinions of local residents on Gesellian money. Financial reports from the cooperatives were also analyzed to evaluate their financial stability and performance before and after the introduction of Gesellian money.

Statistical Analyses

The data collected from this study was analyzed using statistical methods such as regression analyses and t-tests. Regression analyses were used to evaluate the relationship between the use of Gesellian money and the economic outcomes in the local economies. T-tests were used to compare the financial stability and performance of cooperatives before and after the introduction of Gesellian money. The significance level used for hypothesis testing was not explicitly stated in the study.

This observational study evaluated the impact of Gesellian money on local economies by introducing it in several regions and comparing the outcomes with those in areas where it was not used. The study collected data using surveys, interviews, and financial reports, and the data was analyzed using statistical methods such as regression analyses and t-tests.

It is necessary to note that the impact of Gesellian money on the economy may vary depending on many factors, such as:

1. Size of the economy: In small economies, where Gesellian money may be more significant, they can have a more noticeable effect on the economy than in larger economies.
2. Inflation level: If the inflation level is high, Gesellian money may have a smaller effect on the economy because they cannot prevent inflation.
3. Structure of the economy: In economies where small and medium-sized enterprises dominate, Gesellian money may be more effective than in economies dominated by large corporations.
4. Unemployment rate: If the unemployment rate is high, Gesellian money can stimulate the economy because they create new opportunities for work and entrepreneurship.
5. Political stability: In economies with political instability, Gesellian money may have a smaller effect on the economy because investors may not feel confident investing their money in such an economy.
6. Level of development of the financial system: If the financial system is developed and efficient, Gesellian money may have a smaller effect on the economy because existing financial instruments may be more attractive to investors.
7. Level of education and culture: In economies with a high level of education and culture, where people are more inclined to environmentally responsible behavior, Gesellian money may have a greater effect on the economy because they may be more attractive to such people.

Thus, the impact of Gesellian money on the economy may vary depending on specific conditions and context.

RESULTS

The Gesellian money experiment yielded several positive results, indicating its effectiveness in boosting local economies. The data collected on the impact of Gesellian money on the regional economy demonstrated a significant increase in turnover of the Gesellian money in the local economies where it was introduced, compared to regions where the local currency was used exclusively. The cooperatives in the regions where the Gesellian money was implemented experienced a significant improvement in their financial stability, indicating the effectiveness of the Gesellian money in supporting local businesses.

Hypothesis testing was conducted to determine if there was a significant difference in economic growth between areas where Gesellian money was used and areas where it was not. The results indicated that there was a statistically significant difference in economic growth between these regions. The regions where Gesellian money was introduced had a higher level of economic growth compared to regions where the local currency was used exclusively.

The magnitude of the impact of Gesellian money was substantial. The turnover of Gesellian money increased significantly, indicating that people were using it for transactions. The cooperatives that were close to bankruptcy saw their financial situation change for the better. Additionally, the local economies in the regions where the Gesellian money was introduced saw a rapid growth, indicating that the Gesellian money experiment had a positive impact on the overall economy of the region.

Overall, the results of the experiment indicate that the Gesellian money has the potential to significantly improve the economic situation in local regions. The increased turnover of the Gesellian money, the improvement in the financial situation of local businesses, and the growth of local economies demonstrate the positive impact of the Gesellian money experiment.

This model may have the following drawbacks:

Difficulties in conducting transactions: Gesellian money can only be used for transactions within a certain economy. This can create difficulties in international trade.

Difficulties in exchange: Gesellian money may not always be easily exchanged for other currencies, which can create problems for travel and international trade.

Difficulties in management: The Gesellian model may require the management of gold reserves or other stable assets to maintain price stability. This can create difficulties in managing the economy and require additional resources.

To improve the Gesellian model, the following methods can be considered:

Development of convenient tools for conducting transactions in international trade: The development of convenient tools for conducting transactions can make Gesellian money more attractive for international trade.

Creation of exchange means: The creation of exchange means for Gesellian money into other currencies can make them more convenient for use in travel and international trade.

Development of technologies for managing economic processes: The development of technologies for managing economic processes, such as automating the accounting of stable asset reserves and monitoring pricing, can simplify the management of the Gesellian economy.

In addition, mechanisms can be considered for managing the amount of Gesellian money in circulation and maintaining the stability of their prices to avoid the risk of inflation or deflation. It is also possible to consider increasing the diversification of stable assets used to support Gesellian money to reduce risks associated with a single asset.

Let's consider these difficulties in the context of solving the listed problems:

Difficulties with conducting transactions in international trade:

The difficulty with conducting transactions is solved through the use of an n-tier system. This means that for a particular country, it may be a 2 or 3 tier system depending on the size of the country. The outer tier operates with international currencies and is used in international settlements, the next inner tier operates with the official currency of the country and is used in settlements throughout the country, and the innermost tier represents Gesellian money of a particular region. It will be normal for each region to use its own regional Gesellian money, even with multiple regions.

Difficulties with exchange: Gesellian money may not always be easily exchanged for other currencies, which can create problems when traveling and in international trade.

The difficulty associated with currency exchange (outer tier), exchange for the official currency of the country (2nd tier), and exchange for Gesellian money of other regions is solved through the use of an exchange similar to Forex, where trading of Gesellian money takes place.

Difficulties in management: In the Gesellian model, there may be a need to manage reserves of gold or another stable asset to maintain price stability. This can create difficulties in managing the economy and require additional resources.

However, in reality, the Gesellian model is self-regulating, as money issuance is tied to the quantity of goods. Thus, with a large quantity of goods, we have a relatively large issuance, and with a small quantity of goods, the money supply automatically contracts due to the negative interest rate that is an integral part of Gesellian money. Based on this, it is evident that price stability is a self-regulating process in the Gesellian economy. Furthermore, the Gesellian model allows the economy to avoid destructive economic crises. In traditional economics, there are constant processes of imbalance between produced goods and the money supply, which ultimately leads to either uncontrolled inflation (when the money supply greatly exceeds produced goods) or deflation (when there is essentially a shortage of money in the economy). All of this leads to global economic crises, including the well-known economic depression of 1929-1932, the 2008 crisis, and the current crisis, which is rapidly gaining momentum.

In the Gesellian model, crises of such proportions are not possible, and the balance between goods and money is a self-regulating process.

Below are the graphics showing the result of introduction of Gesellian money in several local regions:

The result of implementing Gesellian money also showed that there is an explosive growth in the economy and as a result - the creation of many additional jobs.

In the regions where testing of the use of Gesellian money was carried out, it was also noted that access to borrowed funds for businesses is radically facilitated. Thus, if in the traditional economy the cost of a loan can vary up to 15-20% per year (depending on the country), then in the Gesellian model, where money is issued under goods, the cost of a loan is 0% per year and there was only a small commission related directly to account servicing.

Figure 1. Demonstrates dynamics of consumption (turnover) in Gesellian money (weekly breakdown)

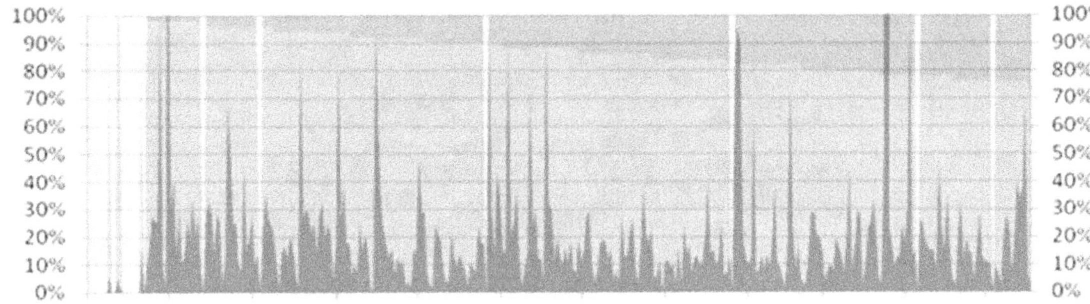

Figure 2. Demonstrates dynamics of consumption structure (turnover) using local currency and Gesellian money, where the orange-beige color indicates Gesellian money and magenta - local currency

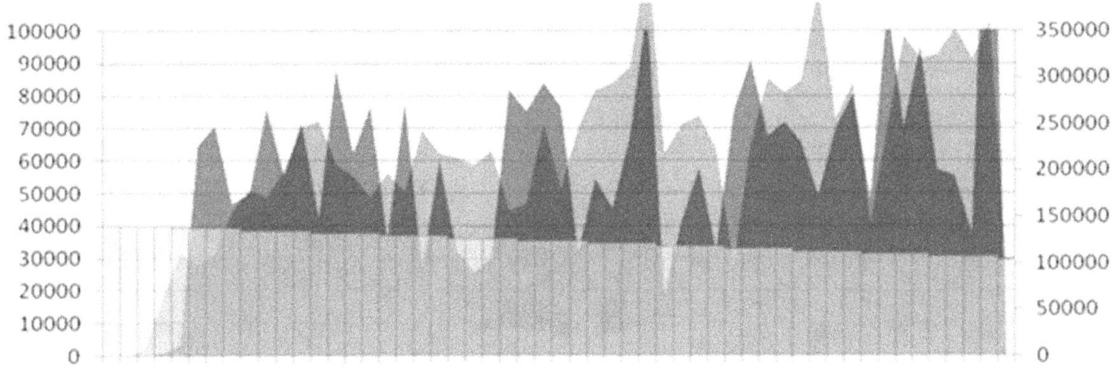

Figure 3. Demonstrates comparative dynamics of turnover in local currency and Gesellian money (weekly breakdown), where Gesellian money is represented in blue and local currency in pink

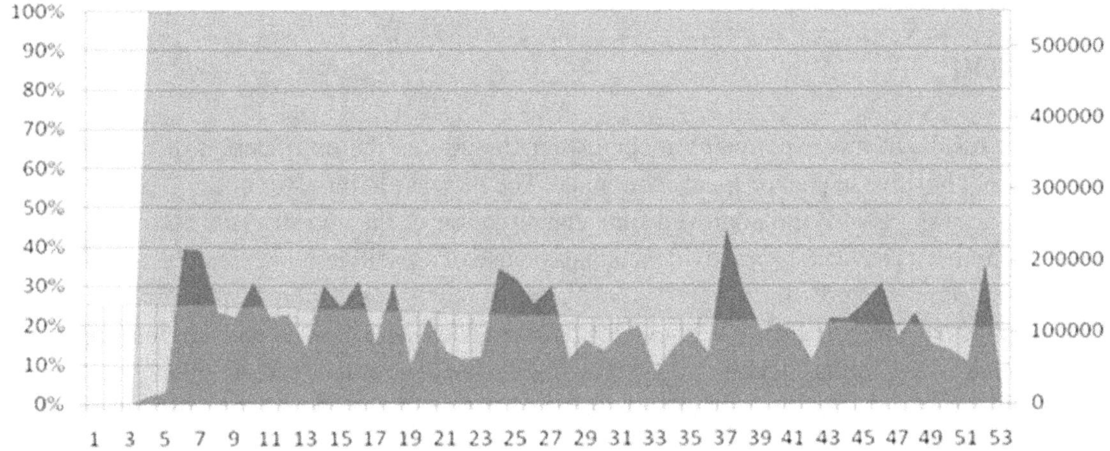

Figure 4. Demonstrates dynamics of consumption structure (turnover) in local currency and Gesellian money (weekly breakdown), where Gesellian money is represented in blue and local currency in pink

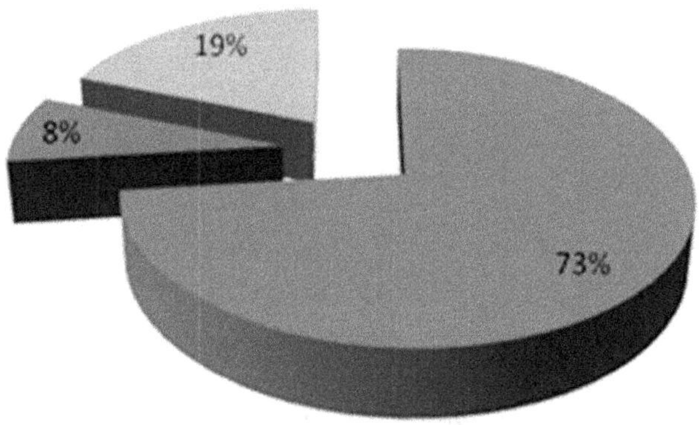

Figure 5. Demonstrates the structure of financial amortization distribution on average per year, where the blue color represents the cooperative through whose stores the use of Gesellian money was conducted, burgundy color - in trade, green color - among people

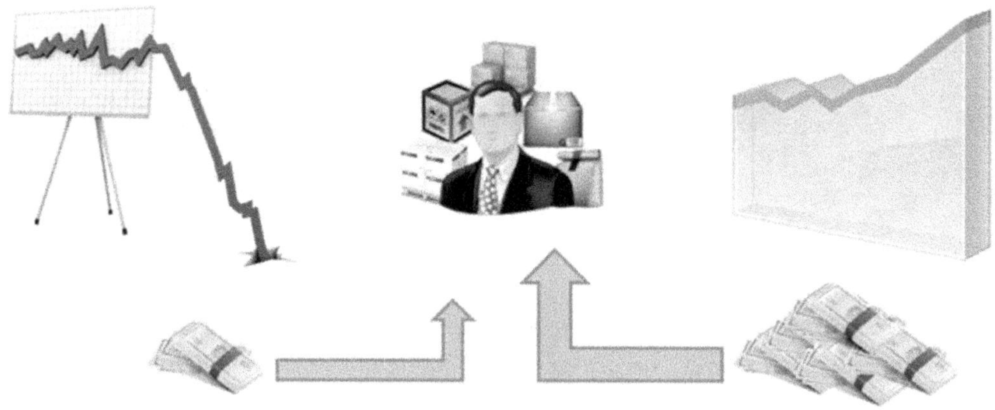

DISCUSSION

Based on the results of this experiment, it can be concluded that the introduction of Gesellian money has a significant positive impact on local economies. The increase in turnover of the Gesel money in the local economies, along with the improved financial situation of the cooperatives and rapid growth of the local economies, provides strong evidence that the use of Gesellian money can be an effective tool for stimulating economic growth.

The findings have important implications for policymakers who may be considering implementing Gesellian money. The results suggest that introducing a complementary currency like Gesellian money can help to stimulate economic growth in regions where the traditional currency is not functioning well. Policymakers can use these findings to inform the development of policies that encourage the use of complementary currencies to help improve economic conditions in regions that are struggling.

Figure 6. Demonstrates maintaining/increasing demand, where the left side of the image shows the aggregate demand of the traditional system, and the right side of the image shows the aggregate demand of the Free Settlement System based on Gesellian money

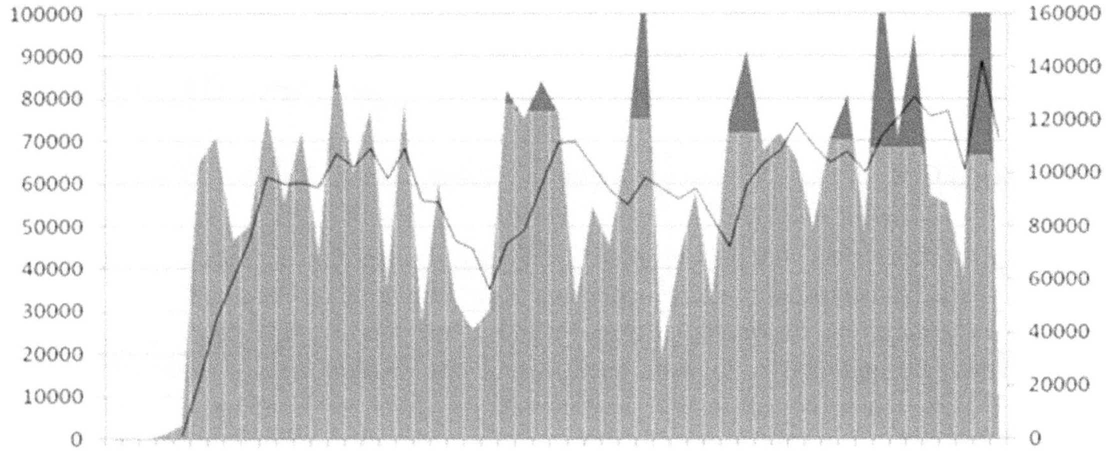

There are some limitations to this study that should be considered when interpreting the results. The study was conducted in a limited number of local regions and cooperatives, which may limit the generalizability of the findings. Furthermore, the study was not a randomized controlled trial, which may limit the ability to make causal inferences about the impact of Gesellian money on the regional economy. Future research in this area should focus on conducting randomized controlled trials in a wider range of regions and cooperatives to better understand the impact of Gesellian money on the regional economy.

In conclusion, the results of this experiment suggest that Gesellian money can be an effective tool for stimulating economic growth in local economies. The findings have important implications for policymakers who may be considering implementing complementary currencies to help improve economic conditions in regions that are struggling. While there are some limitations to this study, the results provide valuable insights into the potential benefits of using complementary currencies and highlight the need for further research in this area.

The results of the experiment with Gesell money in local regions showed promising results for the local economies. However, a big question remains on how to implement Gesellian money in larger economies with multiple regions and the need to trade with other countries. To address these challenges, the author proposes a solution of implementing unique Gesellian currencies for each region or local economy.

To enable trading between regions of the same country, a Gesellian currency exchange can be implemented, as developed by Mr. Rinat Galiautdinov in PrivateCurrency.org (a clearing payment system allowing to emit gesellian currencies, perform financial transactions using the system and trade the gesellian currencies). This would allow for the exchange of Gesellian currency between different regions and facilitate trade between local economies.

To address the issue of trading with other countries, the author proposes the implementation of a second contour. Within the inner contour, Gesellian currencies would operate within the regions, while on the outer contour, a Gesell-to-Fiat exchange market would be established to enable trading with other countries. This approach would allow for the integration of Gesellian money into a larger economy while still maintaining the benefits of a localized currency.

It is important to note that while this proposed solution addresses the challenges of implementing Gesellian money in larger economies, it has yet to be tested and implemented. Further research and experimentation are needed to determine the effectiveness and feasibility of this approach. However, we believe that this proposal offers a potential solution to the challenges of implementing Gesellian money in larger economies, and may offer a promising path towards a more sustainable and equitable economic system.

CONCLUSION

In conclusion, the experiment with the Gesellian money has shown promising results in improving local economies and the financial stability of cooperatives. The circulation of Gesellian money on an equal footing with the local currency has increased turnover and contributed to rapid growth in local economies.

However, there are limitations to this experiment that must be acknowledged. The study was conducted only in local regions and did not account for the complexities of a larger economy with multiple regions and international trade. To address these limitations, a proposed solution involving unique Gesell currencies for each region and a Gesell-To-Fiat exchange market for international trade has been presented.

The experiment with Gesellian money demonstrates the potential for alternative currencies to improve economic conditions and financial stability. Future research in this area can build upon the findings of this study and further explore the implementation of alternative currencies in larger economies.

In conclusion, the implementation of Gesellian money in various regions, including the USA, Austria, Belgium, and Switzerland, etc. has demonstrated the potential to significantly increase the turnover of money within local economies. This has been particularly beneficial for struggling businesses that were previously unable to pay their employees, as well as contributing to remarkable growth in the local economy. By increasing circulation of money within communities, local businesses have been able to create more sustainable and prosperous economies.

The success of Gesellian money, known as WIR in Switzerland and Torekes in Belgium, has also been observed in other countries such as Germany, Austria, and the United States, where it is known as Chiemgauer, Talent, and BerkShares, respectively. The use of Gesellian money has enabled local businesses to stimulate their economies and support small businesses.

Overall, the results of this article demonstrate that Gesellian money is an effective tool for supporting local economies and businesses. The implementation of Gesellian money has enabled businesses to circulate more money within their communities, which has contributed to more sustainable and prosperous economies. The use of Gesellian money has been successful in various regions, highlighting its potential for widespread adoption and impact.

REFERENCES

Boyle, D. (2001). *The New Economics: A Bigger Picture*. Earthscan Publications.

Bresciani-Turroni, C. (1937). *The Economics of Inflation*. George Allen & Unwin Ltd.

Davies, G. (2016). *A History of Money: From Ancient Times to the Present Day*. University of Wales Press.

Douthwaite, R. (1999). *The Growth Illusion: How Economic Growth Has Enriched the Few, Impoverished the Many and Endangered the Planet*. Green Books.

Douthwaite, R. (1999). *The Ecology of Money*. Green Books.

Gesell, S. (1919). *The Monetary and Economic Problem*. Peter Owen Publishers.

Gesell, S. (1932). *Free Money*. Ernst Wasmuth Verlag.

Gesell, S. (1958). *The Natural Economic Order*. Peter Owen Publishers.

Gibson-Graham, J. K. (1996). *The End of Capitalism as We Knew It: A Feminist Critique of Political Economy*. Blackwell Publishing.

Greco, T. H. (2009). *The End of Money and the Future of Civilization*. Chelsea Green Publishing.

Greer, J. M. (2012). *The Wealth of Nature: Economics as if Survival Mattered. Gabriola Island*. New Society Publishers.

Janssen, M. A., Jager, W., Bakker, M., & Weijer, M. (2009). Towards a Theory of Gesell Currencies. *Ecological Economics*, *68*(7), 1812–1824.

Jevons, W. S. (1875). *Money and the Mechanism of Exchange*. D. Appleton and Company.

Kennedy, M. (1995). *Interest and Inflation Free Money: Creating an Exchange Medium That Works for Everybody and Protects the Earth*. New Society Publishers.

Korten, D. C. (2006). *The Great Turning: From Empire to Earth Community*. Berrett-Koehler Publishers.

Lietaer, B., & Arnsperger, C. (2005). Money and Sustainability: The Missing Link. *Ecological Economics*, *54*(2-3), 164–168.

Lietaer, B., & Belgin, S. (2011). *New Money for a New World*. Qiterra Press.

Lietaer, B., & DeMeulenaere, S. (2005). Complementary Currencies: A Global Typology. *International Journal of Community Currency Research*, *9*, 1–23.

Lietaer, B., & Dunne, J. (2018a). *The Future of Money: Creating New Wealth, Work and a Wiser World*. Chelsea Green Publishing.

Lietaer, B., & Dunne, J. (2018b). *The Money Alternative: Creating Wealth in the 21st Century*. Watkins Publishing.

Lietaer, B., & Dunne, J. (2020a). Money and Sustainability: The Missing Link - A New Paradigm for the 21st Century. *Sustainability*, *12*(1), 1–24.

Lietaer, B., & Dunne, J. (2022a). *The Future of Money: Creating New Wealth, Work and a Wiser World*. Chelsea Green Publishing.

Lietaer, B., & Hallsmith, G. (2012). *Community Currency Guide: Guidelines for Community Currency Designers*. New Economics Institute.

Lietaer, B., & Kennedy, M. (1999). Complementary Currencies: A Beginner's Guide. *International Journal of Community Currency Research*, *3*, 1–10.

McMillan, J. (2014). *The End of Banking: Money, Credit, and the Digital Revolution*. Zero/One Economics.

Mellor, M. (2010). *The Future of Money: From Financial Crisis to Public Resource*. Pluto Press.

Narula, N., Ali, R., & Lipton, A. (2018). *The Future of Money: How New Lending Technologies Will Transform the Global Economy*. Harvard Business Review Press.

North, P. (2007). *Money and Liberation: The Micropolitics of Alternative Currency Movements*. University of Minnesota Press.

Orrell, D. (2014). *Money and Sustainability: The Missing Link*. Routledge.

Riegel, E. C. (1949). *Flight from Inflation: The Monetary Alternative*. Devin-Adair Company.

Robertson, J. (2012). *The Future of Money: Creating New Wealth, Work and a Wiser World*. Random House.

Schaefer, G. (2016). *The Future of Money: Bitcoin and Other Cryptocurrency Technologies Are a Way of Life in This Small Swiss Town*. St. Martin's Press.

Von Mises, L. (1912). *The Theory of Money and Credit*. Yale University Press.

Chapter 5
Environmental Regulation and Incentives in Socio-Ecological and Green Technological Innovation

José G Vargas-Hernandez
Tecnológico Superior de Jalisco, Mexico

Omar C. Vargas-González
Tecnológico Nacional de México, Ciudad Guzmán, Mexico

ABSTRACT

This study intends to analyze the impact of environmental regulation and incentives in socio-ecological and green technological innovation. It departs from the assumption that environmental regulation affects the performance of socio-ecological and green technological innovation in organizations. The method employed is critical analysis and reflective based on the theoretical, conceptual, and methodological literature. It is concluded that the analysis confirms that there is a direct relationship between the incentives of environmental regulation and the socio-ecological and green technological innovation in organizations.

INTRODUCTION

The rapid spread at global scope of innovative technologies has modernized the economic sectors that are contributors of environmental pollution and other health risk hazards with limiting findings in technological innovations and ecological solutions. The development of organizational socio-ecology is based on theoretical methodological and empirical research. The abundance of research in population socio-ecology is accompanied by criticisms of ecological research and organizational socio-ecology regarding the abandonment of organizations (Amburgey & Rao, 1996).

DOI: 10.4018/978-1-6684-8879-9.ch005

Research can be advanced at the intersection between sustainability and population socio-ecology (Salimath, & Jones, 2011). The theoretical proposal builds on the theory of organizational choice and the theory of resource dependency (Pfeffer and Salancik, 1978) and the ecological theory of organizations that allow concentrating on the specificities of the sector as exercises of elections.

The theoretical framework of organizational socio-ecology and sociological theory applies ecological principles to organizational analysis. The theoretical framework of organizational socio-ecology and sociological theory is a tool to formulate and implement socioecological principles to organizational strategies for turning around low-performance organizations, inducing the emergence of an economic and social environment and with an impact on non-linear development.

The organisational socio-ecology theory assumes that there is convergence between the paradigm of organizational socio-ecology and sociological perspectives leading to organizational research. The fundamental assumption of organizational socio-ecology is the differentiation of organizational populations. organizational socio-ecology emphasizes issues that are still under discussion, and they conclude with areas of progress and fertilization with other disciplines of organizational thought. Organizational socio-ecology is linked to other areas with the incorporation of methodological innovations such as strategic simulation models.

The forces of the organizational socio-ecology context give rise to criticism of the assumption that organizational destiny is a controllable variable. Organizational socio-ecology is dynamic and is based on the methodology of event history analysis in the analysis of temporal processes.

The theory of population socio-ecology is a tool for the theoretical and empirical analysis of organizational phenomena. The population of socio-ecology has several theoretical currents. Population socio-ecology is a perspective of convergent intersection with sustainability that integrates dimensions, levels of analysis and results. Routine socio-ecology deals at the organizational and inter-organizational analysis levels. The socio-ecology of routines at the organizational level, and the notion organizational character (Birnholtz, Cohen & Hoch, 2007) is defined as the ability to regenerate a coherent socio-ecology of patterns of action.

Research in organizational socio-ecology looks at the interactions of organizational populations. Organizational socio-ecology research intersects sustainability. The application of principles and assumptions of organizational socio-ecology to analyze the sustainability of organizations. The theory of organizational socio-ecology is based on a social Darwinism of organizational populations The organizational socio-ecology theory analyzes the alterations, difficulties, and restrictions of the organizational populations in application conditions (Dimaggio y Powell, 1983).

Research in the ecological theory of population is consolidated in the theory of population socio-ecology. The research focuses on the socio-ecology of populations in coordination and convergence with organizational sustainability considering that there are several areas related to the sustainability of populations that potentially contribute to population sustainability (Salimath, & Jones, 2011). The variables gas emissions, energy consumption, green technology innovation, institutional quality, and economic growth and urbanization are all interdependent and cointegrated.

There is a relationship between environmental regulation and green innovation. Values are assigned to environmental regulations leading to green innovation performance. When the environmental regulation system is weak, firms tend to emphasize maximization of profits, environmental taxes and engage in treatments to expand production scale and balance the regulation cost.

Green innovation research focuses on green product innovation and environmental management systems, the development of technological systems adopted by organizations, and the adoption of innovative technologies dealing with the integration of environmentally friendly practices. Environmental science and technology and strong normative and regulatory institutions support the transformation towards green innovation of firms.

This analysis on the impact of environmental regulation and incentives in socio-ecological and green technological innovation assumes that environmental regulation affects the performance of socio-ecological and green technological innovation in organizations. The study begins with some implications of the organizational socio-ecology to develop a framework for the analysis followed by the organizational green technological innovation leading to determine the impact of the environmental regulation policies and incentives. Finally, some conclusions are presented based on the analysis.

ORGANIZATIONAL SOCIO-ECOLOGY

Although the term ecology is not used Kremser, Pentland & Brunswicker (2017) they theorize the interdependence traced from the dynamics of social phenomena while sets of practices (Nicolini & Monteiro, 2017). The dynamic interplay of ecological and social systems leads to a socio-ecological systems (Berkes et al., 2003; Folke et al., 2002; Gunderson & Holling, 2002).

Organizational socio-ecology has implications for organizational theories. The emergence of models that value resources such as the resource-based theory, the stakeholder theory and theory of organizational socio-ecology that value external influences. Firms are restricted by internal factors of firms such as capital and technology and external factors such as inclusiveness, participants, and bureaucratic factors in the context of environmental regulation policies (Hu et al., 2020; Chen et al., 2021). Strict environmental policies lead small firms to be inclined to reduce expenditures to enhance pollution control rather than green technologies development. Organizational socio-ecology theory studies organizational size, sustaining that small organizations have more problems to capitalize, so they are prone to dissolve.

Organizational socio-ecology breaks the assumption of organization- environment by focusing the analysis not on the individual organization but on the aggregate propositions at the population level. In the classical approach to environment, as in socio-ecology theory and contingency theory, it is assessed with objective data to plan and adjust the organization to context. The contingency theory advocates organizational change based on the ability to adapt and adjust an alignment between the organization and the environment. The specific environment refers to the specific conditions of each organization, related to its sector, geography, stakeholders, etc. (Cadena-Roa, 2010). The intermediate environment in organizational socio-ecology theory refers to the conditions of organizations grouped by geographic area, theme, etc. (Hannan & Freeman, 1977, 1989).

Organizations compete for resources and resource positioning in organizational socio-ecology. The niche width theory is a category of organisational socio-ecology based on the survival opportunities of environmental changes of organizations (Hannan and Freeman, 1977). Ecological contingencies imposed by the environment require the fulfilment of components of organizational socio-ecology.

Organizational socio-ecology analyzes the characteristics of the vital manifestations of organizational populations, such as disappearance considering individual situations such as their size and age that are associated with survival (Aldrich & Auster, 1986). Population socio-ecology analyzes the sample as the entire population over time periods.

Organizational socio-ecology as a social demand on its organizations that exceed internal limits and affect the influence of human beings on the climate and natural environment. The internal social responsibility of organizations is introduced in the fields of organizational socio-ecology with the application of socio-ecological and environmental measures (Oller Alonso, 2021).

Building an organizational socio-ecology to learn as a management model of professional development must be based more on guiding than on control (Le Boterf, 1999). The organizational socio-ecology framework uses its knowledge to formulate and implement turn around organizational strategies. Organizational socio-ecology requires consensus for the cumulative generation of knowledge.

The socio-ecology of routines at the firm level, as well as the notion organizational character (Birnholtz, Cohen & Hoch, 2007) defined as the ability of participants to regenerate a coherent socio-ecology of patterns of action. Routine socio-ecology is analogous to the idea of a cluster of routines. Organizational routines are understood like ecologies (Pentland, 2004). Organizational socio-ecology of routines, which are often coupled and linked with actions, artifacts, and people. Routines in socio-ecology has the macro and micro levels of analysis of actors and activities.

The interdependence between routines challenges the scholars to meet the interactions in a socio-ecology between routines considering that routines never perform in a vacuum (Howard-Grenville et al. 2016) but are ubiquitous in organizations (Becker, 2008), configured in an interdependent network (Feldman et al., 2016; Pentland, 2011; Sele & Grand, 2016; Spee, Jarzabkowski, & Smets, 2016). Routines are intertwined by actions, artifacts and people linked to other routines. Ecological routines are co-located at a particular level of macro and micro analysis of actors and activities (Howard-Grenville et al., 2016).

An ecosystems adaptive governance and social-ecological systems (Dietz et al., 2003; Eckerberg & Joas, 2004; Folke et al., 2005; Ostrom, 2005). Space changes the stability properties of ecological interactions. The implications of chaos-induced dispersal in continuous-time models for socio-ecology are important (Sole & Bascompte, 1995). Elimination of species affected by habitual loss and fragmentation from spatial ecological patterns provides direct criteria (Hannan, 2005).

The practice of sustainability is related in convergence with the socio-ecology of the population. Population socio-ecology is a tool for the analysis of macro organizational phenomena that traces the use in theoretical and empirical research (Salimath, & Jones, 2011). Population socio-ecology supports organizational sustainability and explores the convergence that links the theory and practice of the organizational socio-ecology of populations. Population socio-ecology has areas of convergence with organizational sustainability for the analysis of population sustainability (Salimath, & Jones, 2011).

The industrial structure has effects on regional economic growth and ecological welfare (Zhao et al., 2020; Huang et al., 2021). Technological innovation infrastructure system improves the innovation efficiency by protecting the property rights. For rapid economic growth, most developing countries are dependent on non-renewable energy consumption, leading to significant CO 2 releases with adverse effects on environmental quality.

Improving the well-being via development increases output accompanied by the substantial consumption of fossil fuel with harmful influence on the environment. Energy consumption releasing more carbon is the major contributor to environmental footprint. The firms prefer to scale up production rather than use high-tech equipment exacerbating environmental pollution and weakening motivation (Bian et al., 2021). Many studies are used to evaluate environmental performance (Zhou et al., 2008; Sueyoshi & Goto 2012; Wu et al., 2019). The pollution emissions characteristics leading to the environmental performance affect the green development efficiency.

ORGANIZATIONAL GREEN TECHNOLOGICAL INNOVATION

The concept of green ergonomics is in the context of green technological innovation including the development of the products and services aimed at achieving eco-benefits and reducing the organizational environmental impacts (Bernroider, 2002; Damanpour, 1992). Green innovative ergonomics focuses on contributions to enhancing the human-nature connection and diminishing the environmental impact, by linking design of innovative products and services aimed to reduce emissions and pollution with other pro-nature benefits (Mandavilli, Rys, & Russell, 2008; Sanquist, 2008; Hilliard & Jamieson, 2008).

Organizations focus on technological innovation advances to implement green innovation characteristics that may affect the adoption and use (Sneideriene & Rugine, 2019; Kousar, Sabri, Zafar, & Akhtar, 2017; Weng & Lin, 2011; Lin & Ho, 2011). The government encourages organizations to adopt and use green technology innovation with technical assistance and provides financial support and training for adoption.

Organizations adopt technological innovation before use with elevated levels of environmental uncertainty that affects green technology innovation (Zailani, Iranmanesh, Nikbin, & Jumadi, 2014). The adoption of green technology innovation reduces negative organizational impacts of technological systems processing and of the internal administrative processing. Organizations may be able to try to adopt the green technology innovation several times before the final adoption (Baines, Brown, Benedettini, & Ball, 2012).

The green innovation of firms is related to the life cycles. Green organizational innovation adoption leads to offer ecological and reliable product and services environmentally friendly at competitive prices delivered on time (Thatte, 2007). Differences in the firm's ownership structure, life cycle, size, etc., affect the environmental regulations affecting green innovation. Environmental regulations have a positive direct relationship with green technological innovation. Some inconsistencies may lie on the measures of green innovation performance, such as using granted-patents and pollutant reduction (Amore & Bennedsen, 2016; Fujii & Managi 2016). Triability is the organizational innovation characteristic to adopt any trial of innovation before the use and implementation because the degree of uncertainty (Rogers, 2003).

Research finds that relative advantage is related to green innovation adoption The relative advantage is the perception of the characteristic of organizational technology innovation in terms of price, quality, satisfaction, use and life span over its costs. Perception of organizational relative advantage is positively related to green innovation. If the organizational perception of relative advantage of innovation is greater than the cost and better that the current technology, the organization may adopt the innovation. (Lin & Ho, 2011; Rogers, 2003; Kousar et al., 2017; Grandon & Pearson, 2004).

The innovation management is related to internal and external practices. Innovation externalities are being solved by means of government R&D subsidies (Bi et al., 2016; Li et al., 2018; Liu et al., 2020). Investments on research and development, human capital, and gross domestic product per capita have effects on the development of green innovation. Investments on research and development, has effects on human capital and gross domestic product and on the development of green innovation. High investments in research and development, human capital, and gross domestic product per capita, the green innovation performance improves.

Investments in research and development enhances the firm technological innovation capabilities. Green investment is the ratio between fiscal expenditure on energy conservation and environmental protection and total fiscal expenditure. Environmental protection is a social responsibility of governments (Li et al., 2015; Jiang et al., 2020; Wang & Jiang, 2021).

Regional economic development is a macroeconomic operation related to the investment level of innovation (Grossman & Krueger, 1995). Local governments may invest and encourage firms to develop innovative activities through technology spillovers from foreign investments and competition incentives.

Complexity of innovation is the degree of organizational perception of technology innovation as being difficult to use. If the complexity of innovation increases, the adoption decrease. The level of complexity increases with the difficulties of innovation diffusion and knowledge sharing. The perception of organizational complexity is negatively related to green innovation. Organization tends to adopt technology innovation when innovation is diffused, and knowledge shared. Innovation requires skills and efforts increasing the level of complexity while the level of adoption decreases (Kousar et al., 2017; Etzion, 2007; Rogers, 2003; Tornatzky & Klein, 1982; Deng & Ji, 2015; Bradford & Florin, 2003).

There is a lack of green technological innovations in developing countries compared to developed nations (Li, Pan, Kim, Linn, & Chiang, 2015). A systematic review by Taklo, Tooranloo, & Shahabaldini (2020) of multiple industrial sectors and regions on the emergence of green technologically innovative advances for enhancing technical systems and enrichment of the customer experience.

Technological innovation is adopted after observing the impacts (Shahrul & Normah, 2015; Tan & Eze, 2008). Use of innovation that is visible to potential adoption reflects the result of the characteristic observability (Rogers, 2003). The perception of observability is positively related to green innovation. Observability is a significant characteristic of technological innovation because enables organizations to identify the benefits of innovation with high possibilities to be adopted by organizations (Hatimtai & Hassan, 2018; Rogers, 2003).

Green innovation is categorized in green management, green process, and green product (Chen, Shih, Shyur, & Wu, 2012; Chen & Hsu, 2009). Green technology innovation capabilities of firms, green product innovation capabilities followed by green process innovation capabilities and governance technology capabilities are being promoted by firms. Green innovation capabilities, green product innovation, green process innovation, and governance technology capabilities must have to be implemented based on the green technology innovation capabilities of firms (Wang et al., 2019). Organizational technological innovations compatible with capabilities of the organization have positive effects on the environment (Etzion, 2007). The perception of organizational compatibility is positively related to green technological innovation.

The innovation of green process is related to updates in systems, methods and processes and process innovation is aimed to produce eco-friendly goods and services to meet the ecological concerns and environmental targets (Antonioli, Mancinelli, & Mazzanti, 2013). The indicators of a green process innovation capabilities are the strengthening transformation and enhancing protection design processes, pollution elimination and green process review. Actions of government environmental regulation has effects on the investments of pollution elimination. Pollution control benefits are greater than those of reduction of production promoting firms to invest in green innovation activities

The green technology innovation capabilities of firms are related to the elements of green product innovation, process innovation and technology governance. Organizations must intensify green technology innovation to meet the technical specifications and standards of green emissions set by governments, improve the technology governance capabilities through wastes and the innovation capabilities of green products, processes, and develop green technology innovation (Li, 2019; Wang et al., 2018; Saud et al., 2019).

Organizational learning and innovation thrive in informal settings (Gunderson,

1999). The internal fit of a learning organization has a positive relationship with organizational innovation, and this with organizational performance. The organizational learning intention has effects on innovation capability and this on organizational performance (Teece, 2016).

Organizational learning guides organizational innovation in knowledge-intensive organizations. Knowledge based view extends the resource-based view (Rahim, Eladwiah, & Rahman, 2013) and it is being used in various research of stakeholder theory, open innovation, big data, data management, outsourcing and offshoring (Hoskisson, Gambeta, Green, & Li, 2018; Santoro, Vrontis, Thrassou, & Dezi, 2018; Xu, Frankwick, & Ramirez, 2016 and Pereira, Munjal, & Ishizaka, 2019). Individual and organizational learning lead to innovation enhancing organizational performance. Organizational learning guides organizational innovation (Visser & Togt, 2016).

The ad-hocratic organization structure is one that is flexible, self-renewing, organic, innovative, and collaborative which can fuse different disciplines into a project team avoiding the traps of bureaucratic structures such as the labor division and coordination is the result of interactions (Toffler, 1970; Mintzberg, 1979).

From the perspective of motivation, green innovation should be separated from technological innovation ability. Organizational behaviors are classified based on their motivations as symbolic strategic environmental actions for cooperation and substantive environmental actions, or substantive strategies and cooperation (Delmas & Montes Sancho, 2010; Neumann, 2021; Truong et al., 2021).

ENVIRONMENTAL REGULATION POLICIES AND INCENTIVES

The mechanism of environmental regulation policy has influence on the motivation of green innovation firms to differentiate between the green innovation behaviors in regulation and regulated objects of government enterprise. Government environmental support increases the adoption of green innovation (Lin & Ho, 2010). Government environmental support refers to the systems that take the form of environmental regulations, policies, standards, and interventions that tend to reduce the environmental effect and encourage firms to engage in environmental technology innovation (Eiadat, Kelly, Roche, & Eyadat, 2008). Government environmental support motivates firms to adopt environmental methods of production (Gadenne, Kennedy, & McKeiver, 2009).

Mechanisms of environmental regulations policies affect organizational green innovation in different green innovations behaviors. State-owned enterprises assume responsibilities than non-state-owned do not assume in implementing the green innovation development under constraints of environmental regulations (Zhou et al., 2017; Bai et al., 2019). State-owned enterprises improve after the implementing of environmental regulations since constrained environmental regulations policy, firms respond faster to policies on green innovation due to the responsibility and use of scarce resources.

Environmental regulation policies cause small firms to enhance pollution control mora than implementing green technological innovation. Growing firms have access to abundant funds for R&D investments in green innovation to improve environment-friendly production and green technology (Cao & Cao, 2019). Environmental policy formulation must consider the specific measures for pollution control capacities of firms according to its firm´s ownership structure, size, life cycle, etc.

Policy on environmental regulations considering the value of carbon trading, it has a strong incentive effect in improving firm green innovation. Openness has no significant impact on green innovation because there is an indirect effect between openness and green innovation openness. That technological innovation in organizations reduces carbon emissions and enhances environmental quality (Dauda et al., 2021; Ko et al., 2021). Green innovation significantly lessens energy consumption and CO 2 gas emissions.

Carbon emission trading market must be liberalized positioning the emissions of firms in price mechanisms to stimulate green technological innovation activities. Carbon finance is the ratio between carbon dioxide emissions and gross domestic product. A higher level of CO2 emissions and environmental pollution as the causal links requires the establishment of effective institutional quality to green innovation. The causality link is established between the energy consumption and the institutional quality. Thus, there is a relationship between green innovation and institutional quality.

Research has found causal relationships among the variables are determined from CO2 emissions to economic growth, CO 2 emissions to green innovation and institutional

quality, urbanization and energy consumption to economic growth and institutional quality. From institutional quality to green innovation. Research also has found a bidirectional link between economic growth and gas emissions. Institutional quality and urbanization in developing countries may increase CO2 emissions.

The influence of environmental regulations on green technological innovation of growing and mature firms achieves the highest impact while the green technological innovation of startup firms has severe financial constraints, struggle to meet capital needs, face elevated level of risk-taking leading to achieve the lowest impact on investments in green innovation technologies (Shahzad et al., 2019).

The mechanism between capital misallocation, green development, and technological innovation leads to more efficient socio-ecology environment. Incorporating capital allocation into technological innovation in an analytical framework promotes regional green development efficiency. Regarding heterogeneity, the effect of capital allocation is significant. Capital allocation has a positive relationship with technological innovation enhancing the input of the firms with an impact on green development efficiency. Capital allocation enhances green development efficiency leading to reducing the level of technological innovation. The impact of capital allocation on technological innovation is positive (Schmidt et al., 2020; Gao et al., 2019).

Government intervention on capital misallocated may lead to deterioration of the environmental. Prices distortion of capital market, capital allocation cannot be optimized (Zhang & Lu., 2017). Firms have difficulties in obtaining the capital required for technological innovation of the firm leading to weaken the motivation to conduct research and development (Gao et al., 2019). If the price of capital controlled by government, firms tend to rent-seeking to obtain low-cost factors crowing research and development funds and reducing innovation efficiency (Murphy et al., 1993). The intervention of external factors may destroy the price of capital mechanism lowering costs of production factors for polluting industries. Technological innovation has a mediating role with capital allocation on green development efficiency.

Technological innovation increases the use of clean processes reducing the environmental pollution (Hua et al. 2018). Technological innovation stimulates firms to use efficient resources, adopt equipment and promote the upgrading of industries (Sun et al. 2020b). Technological innovation tends to facilitate green economy development having a mediating role on capital allocation on green economic efficiency while capital allocation improves the green development efficiency. Technological innovation has a mediating role with capital allocation and enhances green development efficiency.

Green finance addressing the financial gap is a relevant financial instrument that contributes to green innovation development of firms (Luo et al., 2021). Green finance motivates firms to invest in green technological innovation and is the driving force for green transformation of polluted firms through the engagement in green innovation activities. Environmental regulation improves the effects of green finance in green innovation. Environmental regulations strengthen the impact of green innovation projects and the effects of green finance.

Firms tend to adopt conservative strategies to invest in green technological projects with low-risk and intense environmental regulations (Boubakri et al. 2013; Chiu and Lee 2020). Green finance and strict environmental regulations have positive effects on green innovation. The moderation effects of environmental regulation advance the effects of green finance on green innovation. Environmental regulations at low levels leads to increasing of innovation performance as green finance performance increases.

An emphasis on green innovation outcomes by the firms requires more efficient green finance in producing green innovation. Green innovation output can be achieved by urging firms to use green finance and investments and avoiding time-consuming projects. Environmental regulation improves the effects of green finance in green innovation. Green finances are contingent to the implementation of environmental regulations and can improve with more intense regulations.

Large firms with more available resources tend to improve the quality of green technological innovation. Large firms have more mature development of organizational systems achieving high innovation management (Yu et al., 2022b). Large firms count on access to abundant innovation resources such as capital, R&D, patents, technological talents (Gupeng & Xiangdong, 2012; Hu et al., 2017). The green patent application is a measure that has effects of green finance in driving green technological innovation. The use of green patent application to measure innovative behaviors is not comprehensive for the different motivations on green innovation choices.

Large firms promote green technological innovation even if they are constrained by environmental regulations and small firms that are more restricted by capital, technology, talent, etc., have more difficulties in promoting green innovation. Firms' behaviors must be constrained by environmental regulations to develop green technological innovation. The implementation of an intense environmental regulations improves green innovation considering constraints of firms may not lead to transformation of technologically innovative production. Stringent environmental regulations tend to strength the motivation of firms to pursue technological innovation.

Green innovation as dependent variable can be measured by the number of patent applications associated to energy intensity (Wurlod & Noailly, 2018). Green patent applications indicate the advance of green innovation (Tolliver et al. 2021) and they are leading to innovation activities without institutional preferences and time-lag, used as indicators to evaluate green innovation (Fujii & Managi 2016; Cho & Sohn, 2018; Fujii 2016).

Environmental regulation has a positive impact on green innovation patents and green utility model patents, although from the perspective of influence, the impact of environmental regulation on green innovation may decrease if firms have not eliminated strategic green innovation. Green technology innovation strategies must be formulated and implemented with sensitivity considering cultural differences that have an impact on innovation processes (Westwood & Low, 2003).

Environmental regulations influence government subsidies to eliminate compliance costs improving green innovation with minimal impact on green innovation behaviors. Government subsidies on green innovation must be classified in the diverse types to determine the effect on the choices of innovative behaviors with different motivations. Government innovation subsidies policies must be more targeted and flexible according to the qualifications of firms in need and the appropriateness of subsidies, matching the development stage at the life cycle to enhance core competitiveness and realize positive effects on green technological innovation.

Regulatory capture and government subsidies are moderating variables. Government subsidies supplying funds provide firms to conduct green innovation aligned to the government goals of technological innovation (Boeing, 2016; Bai et al., 2019). Government subsidies promote green innovation; however, regulatory capture has a negative role in the regulatory process of environmental regulation promoting green innovation of firms. Increasing green subsidies lead to firms to have more resources and funds to invest in green technology development enhancing green product, process innovation and technology governance capabilities (Saud et al., 2019; Lin et al., 2020). The regulatory capture behavior requires more accurate indicators to provide verification of results on the impact of green innovation effects.

The regulatory capture leads to different green innovation behaviors. The regulatory capture has a distorting effect on the high quality substantive green innovation incentives leading to economic consequences of the environmental regulations. Regulatory capture has a negative effect on environmental regulation in promoting firms as the explanation for the different green innovation behaviors. Regulatory captures may lapse for regulatory stringency as the leading cause for environmental accidents intentionally balancing profits (Graham et al., 2011; Steinzor, 2012). Firms taking a profit maximization orientation may have more potential benefits from regulatory capture behavior than those having intense motivational behavior for regulatory capture (Li et al., 2021; Zhang et al., 2021).

Environmental regulations tend to influence green innovation of growing and mature firms at the highest level, while the startup stage has the lowest impact due to the available resources in capital, technology, and talent.

CONCLUSION

In general terms, this analysis concludes that there is a direct relationship between the incentives of environmental regulation and the socio-ecological and green technological innovation in organizations. The socio-ecology of the organizational population has implications for sustainability. Institutional reforms are essential to reduce ecological emissions and achieve climate change goals.

Firms realize high quality green innovation behavior to promote green technology to obtain a competitive advantage. Green innovation firms are higher during the growth and maturity periods when they improve green technological innovation and develop a keen sense of social responsibility for more active innovation activities. The role of technological innovation needs to be strengthened to improve green development efficiency. The organization must implement and scrutinize green product innovation capabilities implementation, green process innovation capabilities, and technology governance capabilities.

Environmental regulation affects the firm´s green innovation through the mechanisms of innovation compensation and compliance costs. The scientific design of environmental regulation policy tools must be strengthened to monitor the emission reduction and conduct green innovation. Environmental regulation intensity must be enhanced according to the specific situations of firms. Firms that bear excessive environmental costs must be inclined to regulatory capture rather than improve green technological environmental protection which may destroy the incentive mechanism with extended time and higher costs.

Governments in policymaking and stakeholders must facilitate organizational green strategies to invest in technological innovation projects aimed to reduce environmental pollution. Policymaking need to strengthen institutional quality and enact more effective policies to address the environmental challenges. The different impacts of institutional environments must take the research model. Institutional environment is relevant in factor allocation and environmental performance

Firm differences in ownership, life cycle, size, etc., have a significant impact on the environmental regulation green innovation incentives. Environmental regulations and green finance have effects on green technological innovation. The green finance levels on green innovation have heterogeneous impacts with a moderating effect on environmental regulations. Green finance, green innovation and environmental regulations tend to be stable with fixed effects.

Green finance and environmental regulations have effects on green innovation. Environmental regulation is a moderator is used to examine the effects. Regional green innovation can be measured by the level of contribution of green finance and the green patent application. Governments must support with fiscal incentives and financial subsidies the development of innovative firms encouraging more flows of capital into the green technology innovations and clean processes.

Research findings suggest that subsidies do not explain the gap between the green technological innovation behaviors.

Future research in applying population socio-ecology principles and assumptions to examine organizational sustainability in organizations.

Future research for measurement methods must separate the individual effects of the mechanisms of government subsidies and regulatory capture on green innovation behavior with different motivations.

REFERENCES

Aldrich, H., & Auster, E. R. (1986). Even dwarfs started small: Liabilities of age and size and their strategic implications. In B. M. Staw & L. L. Cummings (Eds.), *Research in Organizational Behavior* (Vol. 8, pp. 165–198). JAI Press.

Amburgey, T., & Rao, H. (1996). Organizational ecology: Past, present, and future directions. *Academy of Management Journal*, *39*(5), 1265–1286. doi:10.2307/256999

Amore, M. D., & Bennedsen, M. (2016). Corporate governance and green innovation. *Journal of Environmental Economics and Management*, *75*, 54–72. doi:10.1016/j.jeem.2015.11.003

Antonioli, D., Mancinelli, S., & Mazzanti, M. (2013). Is environmental innovation embedded within high- performance organizational changes? The role of human resource management and complementarity in green business strategies. *Research Policy*, *42*(4), 975–988. doi:10.1016/j.respol.2012.12.005

Bai, Y., Song, S., Jiao, J., & Yang, R. (2019). The impacts of government R&D subsidies on green innovation: Evidence from Chinese energy-intensive firms. *Journal of Cleaner Production*, *233*(9), 819–829. Advance online publication. doi:10.1016/j.jclepro.2019.06.107

Baines, T., Brown, S., Benedettini, O., & Ball, P. (2012). Examining green production and its role within the competitive strategy of manufacturers. *Journal of Industrial Engineering and Management*, *5*(1), 53–87. doi:10.3926/jiem.405

Becker, M. C. (Ed.). (2008). Handbook of organizational routines. Cheltenham, UK: Edward Elgar Publishing. doi:10.4337/9781848442702

Berkes, F., Folke, C., & Colding, J. (Eds.). (2003). *Navigating Social-Ecological Systems: Building Resilience for Complexity and Change*. Cambridge University Press.

Bernroider, E. (2002). Factors in SWOT analysis applied to micro, small-to-medium, and large software enterprises: An Austrian study. *European Management Journal, 20*(5), 562–573. doi:10.1016/S0263-2373(02)00095-6

Bi, K., Huang, P., & Wang, X. (2016). Innovation performance and influencing factors of low-carbon technological innovation under the global value chain: A case of Chinese manufacturing industry. *Technological Forecasting and Social Change, 111*, 275–284. doi:10.1016/j.techfore.2016.07.024

Bian, Y., Wu, L., Bai, J., & Yang, Y. (2021). Does factor market distortions inhibit green economic growth? *World Econ Papers, 2*, 105–119.

Birnholtz, J. P., Cohen, M. D., & Hoch, S. V. (2007). Organizational character: On the regeneration of Camp Poplar Grove. *Organization Science, 18*(2), 315–332. doi:10.1287/orsc.1070.0248

Boeing, P. (2016). The allocation and effectiveness of China's R&D subsidies—Evidence from listed firms. *Research Policy, 45*(9), 1774–1789. doi:10.1016/j.respol.2016.05.007

Boubakri, N., Cosset, J. C., & Saffar, W. (2013). The role of state and foreign owners in corporate risk-taking: Evidence from privatization. *Journal of Financial Economics, 108*(3), 641–658. doi:10.1016/j.jfineco.2012.12.007

Bradford, M., & Florin, J. (2003). Examining the role of innovation diffusion factors on the implementation success of enterprise resource planning systems. *International Journal of Accounting Information Systems, 4*(3), 205–225. doi:10.1016/S1467-0895(03)00026-5

Cadena-roa, J. (Coordinador) (2022). *Contribuciones para el estudio del desempeño asociativo resultados de la puesta a prueba del protocolo para la evaluación de asociaciones.* Universidad Nacional Autónoma de México Centro de Investigaciones Interdisciplinarias en Ciencias y Humanidades México.

Cao, X., Deng, M., Song, F., Zhong, S., & Zhu, J. (2019). Direct and moderating effects of environmental regulation intensity on enterprise technological innovation: The case of China. *PLoS One, 14*(10), e0223175. doi:10.1371/journal.pone.0223175 PMID:31589643

Chen, C. C., Shih, H. S., Shyur, H. J., & Wu, K. S. (2012). A business strategy selection of green supply chain management via an analytic network process. *Computers & Mathematics with Applications (Oxford, England), 64*(8), 2544–2557. doi:10.1016/j.camwa.2012.06.013

Chen, H. L., & Hsu, W. T. (2009). Family ownership, board independence, and R&D investment. *Family Business Review, 22*(4), 347–362. doi:10.1177/0894486509341062

Chen, Z., Zhang, X., & Chen, F. (2021). Do carbon emission trading schemes stimulate green innovation in enterprises? Evidence from China. *Technological Forecasting and Social Change, 168*, 120744. doi:10.1016/j.techfore.2021.120744

Chiu, Y. B., & Lee, C. C. (2020). Effects of financial development on energy consumption: The role of country risks. *Energy Economics, 90*, 104833. doi:10.1016/j.eneco.2020.104833

Cho, J. H., & Sohn, S. Y. (2018). A novel decomposition analysis of green patent applications for the evaluation of R&D efforts to reduce CO2 emissions from fossil fuel energy consumption. *Journal of Cleaner Production, 193*, 290–299. doi:10.1016/j.jclepro.2018.05.060

Damanpour, F. (1992). Organizational size and innovation. *Organization Studies*, *13*(3), 375–402. doi:10.1177/017084069201300304

Dauda, L., Long, X., Mensah, C. N., Salman, M., Boamah, K. B., Ampon-Wireko, S., & Dogbe, C. S. K. (2021). Innovation, trade openness and CO2 emissions in selected countries in Africa. *Journal of Cleaner Production*, *281*, 125143. doi:10.1016/j.jclepro.2020.125143

Delmas, M. A., & Montes-Sancho, M. J. (2010). Voluntary agreements to improve environmental quality: Symbolic and substantive cooperation. *Strategic Management Journal*, *31*(6), 575–601.

Deng, Q., & Ji, S. (2015). Organizational green IT adoption: *Concept and evidence. Sustainability (Basel)*, *7*(12), 16737–16755. doi:10.3390u71215843

Dietz, T., Ostrom, E., & Stern, P. L. (2003). The Struggle to Govern the Commons. *Science*, *302*(5652), 1907–1912. doi:10.1126cience.1091015 PMID:14671286

DiMaggio, P., & Powell, W. W. (1983). The iron cage revisited: Collective rationality and institutional isomorphism in organizational fields. *American Sociological Review*, *48*(2), 147–160. doi:10.2307/2095101

Eckerberg, K., & Joas, M. (2004). Multi-level Environmental Governance: A Concept under Stress? *Local Environment*, *9*(5), 405–412. doi:10.1080/1354983042000255315

Eiadat, Y., Kelly, A., Roche, F., & Eyadat, H. (2008). Green and competitive? An empirical test of the mediating role of environmental innovation strategy. *Journal of World Business*, *43*(2), 131–145. doi:. jwb.2007.11.012 doi:10.1016/j

Etzion, D. (2007). Research on organizations and the natural environment, 1992-present: A review. *Journal of Management*, *33*(4), 637–664. doi:10.1177/0149206307302553

Feldman, M. S. (2016). Past, Present, and future. In J. Howard-Grenville, C., Rerup, A., Langley, & H., Tsoukas (Eds.). Organizational Routines: How They are Created, Maintained, and Changed (pp. 23-46) Oxford: Oxford University Press

Folke, C., Carpenter, S., Elmqvist, T., Gunderson, L., Holling, C. S., & Walker, B. (2002). Resilience and Sustainable Development: Building Adaptive Capacity in a World of Transformations. *Ambio*, *31*(5), 437–440. doi:10.1579/0044-7447-31.5.437 PMID:12374053

Folke, C., Hahn, T., Olsson, P., and Norberg, J. (2005). Adaptive Governance of Social–ecological Systems. *Annual Review of Environment and Resources*.

Fujii, H. (2016). Decomposition analysis of green chemical technology inventions from 1971 to 2010 in Japan. *Journal of Cleaner Production*, *112*, 4835–4843. doi:10.1016/j.jclepro.2015.07.123

Fujii, H., & Managi, S. (2016). Research and development strategy for environmental technology in Japan: A comparative study of the private and public sectors. *Technological Forecasting and Social Change*, *112*, 293–302. doi:10.1016/j.techfore.2016.02.012

Gadenne, D. L., Kennedy, J., & McKeiver, C. (2009). An empirical study of environmental awareness and practices in SMEs. *Journal of Business Ethics*, *84*(1), 45–63. doi:10.100710551-008-9672-9

Gao, X., Lyu, Y., Shi, F., Zeng, J., & Liu, C. (2019). The impact of financial fac-tor market distortion on green innovation efficiency of high-tech industry. *Ekoloji*, *28*(107), 3449–3461.

Graham, B., Reilly, W., Beinecke, F., Boesch, D., Garcia, T., Murray, C., & Ulmer, F. (2011). *Deep water: The Gulf oil disaster and the future of offshore drilling. Report to the President*. National Commission on the BP Deepwater Horizon Oil Spill and Offshore Drilling.

Grossman, G. M., & Krueger, A. B. (1995). Economic growth and the environment. *The Quarterly Journal of Economics*, *110*(2), 353–377. doi:10.2307/2118443

Gunderson, L. (1999). Resilience, Flexibility and Adaptive Management: Antidotes for Spurious Certitude? *Conservation Ecology*, *3*(1), 7. http://www.consecol.org/vol3/iss1/art7. doi:10.5751/ES-00089-030107

Gunderson, L. H., & Holling, C. S. (Eds.). (2002). Panarchy; Understanding Transformations in Human and Natural Systems, Island, Washington, DC.

Gupeng, Z., & Xiangdong, C. (2012). The value of invention patents in China: Country origin and technology field differences. *China Economic Review*, *23*(2), 357–370. doi:10.1016/j.chieco.2012.02.002

Hannan, M. (2005). Ecologies of Organizations: Diversity and Identity. *The Journal of Economic Perspectives*, *19*(1), 51–70. doi:10.1257/0895330053147985

Hannan, M. T., & Freeman, J. (1989). *Organizational Ecology*. Harvard University Press. doi:10.4159/9780674038288

Hannan, M. T., & Freeman, J. H. (1977). The Population Ecology of Organizations. *American Journal of Sociology*, *82*(5), 929–963. doi:10.1086/226424

Hatimtai, M. H., & Hassan, H. (2018). The relationship between the characteristics of innovation towards the effectiveness of ICT in Malaysia productivity corporation. *Malaysian Journal of Communication*, *34*(1), 253–269. doi:10.17576/JKMJC-2018-3401-15

Hilliard, A., & Jamieson, G. A. (2008). Winning solar races with interface design. *Ergonomics in Design*, *16*(2), 6–11. doi:10.1518/106480407X312374

Hoskisson, R. E., Gambeta, E., Green, C. D., & Li, T. X. (2018). Is my firm-specific investment protected? Overcoming the stakeholder investment dilemma in the resource-based view. *Academy of Management Review*, *43*(2), 284–306. doi:10.5465/amr.2015.0411

Howard-Grenville, J., Rerup, C., Langley, A., & Tsoukas, H. (2016). Introduction: Advancing a process perspective on routines by zooming out and zooming in. In J. Howard-Grenville, C., Rerup, A., Langley, & H., Tsoukas (Eds.). Organizational routines: how they are created, maintained, and changed (pp.1-22.) Oxford: Oxford University Press.

Hu, A. G. Z., Zhang, P., & Zhao, L. (2017). China as number one? Evidence from China's most recent patenting surge. *Journal of Development Economics*, *124*, 107–119. doi:10.1016/j.jdeveco.2016.09.004

Hua, Y., Xie, R., & Su, Y. (2018). Fiscal spending and air pollution in Chinese cities: Identifying composition and technique effects. *China Economic Review*, *47*, 156–169. doi:10.1016/j.chieco.2017.09.007

Huang, M., Ding, R., & Xin, C. (2021). Impact of technological innovation and industrial-structure upgrades on ecological efficiency in China in terms of spatial spillover and the threshold effect. *Integrated Environmental Assessment and Management*, *17*(4), 852–865. doi:10.1002/ieam.4381 PMID:33325155

Jiang, Z., Wang, Z., and Lan, X. (2021). How Environmental Regulations Affect Corporate Innovation? the Coupling Mechanism of Mandatory Rules and Voluntary Management. *Technol. Soc.*, *65*, 101575. doi:. 2021.101575 doi:10.1016/j.techsoc

Ko, Y.-C., Zigan, K., & Liu, Y.-L. (2021). Carbon capture and storage in South Africa: A technological innovation system with a political economy focus. *Technological Forecasting and Social Change*, *166*, 120633. . techfore. 2021. 120633 doi:10.1016/j

Kousar, S., Sabri, P. S. U., Zafar, M., & Akhtar, A. (2017). Technological factors and adoption of green innovation- moderating role of government intervention: A case of SMEs in Pakistan. *Pakistan Journal of Commerce and Social Sciences*, *11*(3), 833–861.

Kremser, W., Pentland, B. T., & Brunswicker, S. (2017). *The Continuous Transformation of Interdependence in Networks of Routines*. Presented at European Group for Organizational.

Li, J., Pan, S. Y., Kim, H., Linn, J. H., & Chiang, P. C. (2015). Building green supply chains in eco-industrial parks towards a green economy: Barriers and strategies. *Journal of Environmental Management*, *162*, 158–170. doi:10.1016/j.jenvman.2015.07.030 PMID:26241931

Li, N. B. (2019). Environmental Regulation and Corporate Green Technology Innovation - a Conditional Process Analysis. *Inn. Mong. Soc. Sci. Chin. Ed*, *40*(6), 109–115. doi:10.14137/j.cnki.issn1003-5281.2019.06.016

Li, S., Song, X., & Wu, H. (2015). Political connection, ownership structure, and corporate philanthropy in China: A strategic-political perspective. *Journal of Business Ethics*, *129*(2), 399–411. doi:10.100710551-014-2167-y

Li, Z., Liao, G., Wang, Z., & Huang, Z. (2018). Green loan and subsidy for promoting clean production innovation. *Journal of Cleaner Production*, *187*, 421–431. doi:10.1016/j.jclepro.2018.03.066

Lin, C. Y., & Ho, Y. H. (2010). The influences of environmental uncertainty on corporate green behavior: An empirical study with small and medium-size enterprises. *Social Behavior and Personality*, *38*(5), 691–696. doi:10.2224bp.2010.38.5.691

Lin, Y. X., Sha, K. C., & Wang, J. (2020). Practical Experience and Inspiration of Foreign Emission Permit System. *Environ. Impact. Assess.*, *42*(1), 14–16.

Liu, J., Zhao, M., & Wang, Y. (2020). Impacts of government subsidies and environmental regulations on green process innovation: A nonlinear approach. *Technology in Society*, *63*, 101417. doi:10.1016/j.techsoc.2020.101417

Liu, Y., Wang, A., & Wu, Y. (2021). Environmental regulation and green innovation: Evi-dence from China's new environmental protection law. *Journal of Cleaner Production*, *297*, 126698. doi:10.1016/j.jclepro.2021.126698

Luo, Y., Salman, M., & Lu, Z. (2021). Heterogeneous impacts of environmental regulations and foreign direct investment on green innovation across different regions in China. *The Science of the Total Environment, 759*, 143744. doi:10.1016/j.scitotenv.2020.143744 PMID:33341514

Mandavilli, S., Rys, M. J., & Russell, E. R. (2008). Environmental impact of modern roundabouts. *International Journal of Industrial Ergonomics, 38*(2), 135–142. doi:10.1016/j.ergon.2006.11.003

Mintzberg, H. (1979). *The structuring of organizations: A synthesis of the research*. Prentice Hall.

Murphy, K. M., Shleifer, A., & Vishny, R. W. (1993). Why is rent-seeking so costly to growth? *The American Economic Review, 83*(2), 409–414. https://www.jstor.org/stable/2117699

Neumann, T. (2021). Does it pay for new firms to be green? An empirical analysis of when and how different greening strategies affect the performance of new firms. *Journal of Cleaner Production, 317*, 128403. doi:10.1016/j.jclepro.2021.128403

Nicolini, D., & Monteiro, P. (2017). The practice approach: For a praxeology of organizational and management studies. In A. Langley & H. Tsoukas (Eds.), *The Sage Handbook of Process Organization Studies*. Sage.

Oller Alonso, M. (2021). *La Responsabilidad Social Corporativa de las empresas del mármol en la comarca del Almanzora, Almería (2019-2021): análisis de sus estrategias de comunicación integral.* [Doctorado en Gestión de la Información y de la Comunicación en las Organizaciones Universidad de Murcia Escuela Internacional de Doctorado].

Ostrom, E. (2005). Understanding Institutional Diversity, Princeton University Press, Princeton New Jersey, USA.

Pentland, A. (2004, October). *Social Dynamics: Signals and Behavior. International Conference on Developmental Learning*, Salk Institute, San Diego, CA

Pentland, B. T., Hærem, T., & Hillison, D. (2011). The (N)Ever-Changing World: Stability and Change in Organizational Routines. *Organization Science, 22*(6), 1369–1383. doi:10.1287/orsc.1110.0624

Pereira, V., Munjal, S., & Ishizaka, A. (2019, October). Outsourcing and offshoring decision making and its implications for businesses- A synthesis of research pursuing five pertinent questions. *Journal of Business Research, 103*, 348–355. doi:10.1016/j.jbusres.2019.07.009

Pfeffer, J., & Salancik, G. (1978). *The External Control of Organizations: A Resource Dependence Perspective*. Harper & Row.

Rahim, A., Eladwiah, R., & Abdul Rahman, A. (2013). Resource-based framework of green IT capability toward firms 'competitive advantage. *Resource (Saint Joseph, Mich.), 6*, 18–2013.

Rogers, E. M. (2003). Elements of diffusion. *Diffusion of Innovations, 5*, 1–38.

Salimath, M., & Jones, R. III. (2011). Population ecology theory: Implications for sustainability. *Management Decision, 49*(6), 874–910. doi:10.1108/00251741111143595

Sanquist, T. F. (2008). Human factors and energy use. *HFES Bulletin, 51*(11), 1–3.

Santoro, G., Vrontis, D., Thrassou, A., & Dezi, L. (2018). The Internet of Things: Building a knowledge management system for open innovation and knowledge management capacity. *Technological Forecasting and Social Change*, *136*, 347–354. doi:10.1016/j.techfore.2017.02.034

Saud, S., Chen, S., Haseeb, A., Khan, K., & Imran, M. (2019). The Nexus between Financial Development, Income Level, and Environment in Central and Eastern European Countries: A Perspective on Belt and Road Initiative. *Environmental Science and Pollution Research International*, *26*(16), 16053–16075. doi:10.100711356-019-05004-5 PMID:30968296

SchmidtC.SchneiderY.SteffenS.StreitzD. (2020). Capital misal-location and innovation. SSRN 3489801. doi:10.2139/ssrn.3489801

Sele, K., & Grand, S. (2016). Unpacking the dynamics of ecologies of routines: Mediators and their generative effects in routine interactions. *Organization Science*, *27*(3), 722–738. doi:10.1287/orsc.2015.1031

Shahrul, N. S., & Normah, M. (2015). Digital version newspaper: Implication towards printed newspaper circulation in Malaysia. *Malaysian Journal of Communication*, *31*(2), 687–701.

Shahzad, F., Lu, J., & Fareed, Z. (2019). Does firm life cycle impact corporate risk taking and performance? *Journal of Multinational Financial Management*, *51*, 23–44. doi:10.1016/j.mulfin.2019.05.001

Sneideriene, A., & Rugine, H. (2019). Theoretical approach on the green technologies development. *Regional Formation and Development Studies*, *2*(28), 124–134.

Sole, R. V., & Bascompte, J. (1995). Spatial Self Organization Self-Organization in complex Ecosystems Princeton University Press.

Spee, P., Jarzabkowski, P., & Smets, M. (2016). The influence of routine interdependence and skillful accomplishment on the coordination of standardizing and customizing. *Organization Science*, *27*(3), 759–781. doi:10.1287/orsc.2016.1050

Steinzor, R. (2012). The case for abolishing centralized white house regulatory review. *Michigan Journal of Environmental & Administrative Law*, *1*(1.1), 209. doi:10.36640/mjeal.1.1.case

Studies: EGOS Le Boterf, G. (1999). *L'ingénierie des compétences.* París: Éditions d'Organisation.

Sueyoshi, T., & Goto, M. (2012). Efficiency-based rank assessment for electric power industry: A combined use of data envelopment analysis (DEA) and DEA-discriminant analysis (DA). *Energy Economics*, *34*(3), 634–644. doi:10.1016/j.eneco.2011.04.001

Sun, X., Zhou, X., Chen, Z., & Yang, Y. (2020b). Environmental efficiency of electric power industry, market segmentation and technologi-cal innovation: Empirical evidence from China. *The Science of the Total Environment*, *706*, 135749. doi:10.1016/j.scitotenv.2019.135749 PMID:31940733

Taklo, S. K., & Tooranloo, H. S. & Shahabaldini parizi, Z. (. (2020). Green Innovation: A Systematic Literature Review. *Journal of Cleaner Production*, *2020*(7). doi:10.1016/j.jclepro.2020.122474

Tan, K. S., & Eze, U. C. (2008). An empirical study of internet-based ICT adoption among Malaysian SMEs. *Communications of the IBIMA*, *1*, 1–12.

Teece, D. J. (2016). Dynamic capabilities and entrepreneurial management in large organizations: Toward a theory of the (entrepreneurial) firm. *European Economic Review*, *86*, 202–216. doi:10.1016/j.euroecorev.2015.11.006

Thatte, A. A. (2007). *Competitive advantage of a firm through supply chain responsiveness and SCM practices* [Doctoral dissertation, The University of Toledo].

Toffler, A. (1970). *Future shock*. Random House.

Tolliver, C., Fujii, H., Keeley, A. R., & Managi, S. (2021). Green innovation and finance in Asia. *Asian Economic Policy Review*, *16*(1), 67–87. doi:10.1111/aepr.12320

Tornatzky, L. G., & Klein, K. J. (1982). Innovation characteristics and innovation adoption-implementation: A meta-analysis of findings. *IEEE Transactions on Engineering Management*, *29*(1), 28–45. doi:10.1109/TEM.1982.6447463

Truong, Y., Mazloomi, H., & Berrone, P. (2021). Understanding the impact of symbolic and substantive environmental actions on organizational reputation. *Industrial Marketing Management*, *92*, 307–320. doi:10.1016/j.indmarman.2020.05.006

Visser, M., & Togt, K. (2016). Learning in Public Sector Organizations: A Theory of Action, Approach. *Public Organization Review*, *16*(2), 235–249. doi:10.100711115-015-0303-5

Wang, F. Z., Jiang, T., & Guo, X. C. (2018). Government Quality, Environmental Regulation and Corporate Green Technology Innovation. *Sci. Res. Manage*, *39*(1), 26–33.

Wang, K., & Jiang, W. (2021). State ownership and green innovation in China: The con-tingent roles of environmental and organizational factors. *Journal of Cleaner Production*, *314*, 128029. doi:10.1016/j.jclepro.2021.128029

Wang, Y., Sun, X., & Guo, X. (2019). Environmental regulation and green productivity growth: Empirical evidence on the Porter Hypothesis from OECD industrial sectors. *Energy Policy*, *132*, 611–619. doi:10.1016/j.enpol.2019.06.016

Weng, M. H., & Lin, C. Y. (2011). Determinants of green innovation adoption for small and medium-size enterprises (SMES). *African Journal of Business Management*, *5*(22), 9154–9163.

Westwood, R., & Low, D. R. (2003). The multicultural muse: Culture, creativity, and innovation. *International Journal of Cross Cultural Management*, *3*(2), 235–259. doi:10.1177/14705958030032006

Wu, H. Y., Tsai, A., & Wu, H. S. (2019). A HYBRID MULTI-CRITERIA DECISION ANALYSIS APPROACH FOR ENVIRONMENTAL PERFORMANCE EVALUATION: AN EXAMPLE OF THE TFT-LCD MANUFACTURERS IN TAIWAN. [EEMJ]. *Environmental Engineering and Management Journal*, *18*(3), 597–616. doi:10.30638/eemj.2019.056

Wurlod, J. D., & Noailly, J. (2018). The impact of green innovation on energy intensity: An empirical analysis for 14 industrial sectors in OECD countries. *Energy Economics*, *71*, 47–61. doi:10.1016/j.eneco.2017.12.012

Xu, Z., Frankwick, G. L., & Ramirez, E. (2016). Effects of big data analytics and traditional marketing analytics on new product success: A knowledge fusion perspective. *Journal of Business Research*, *69*(5), 1562–1566. doi:10.1016/j.jbusres.2015.10.017

Yu, Z., Khan, S. A. R., Ponce, P., & Jabbour, A. B. L. (2022). Factors affecting carbon emissions in emerging economies in the context of a green recovery: Implications for sustainable development goals. *Technological Forecasting and Social Change*, *176*, 121417. doi:10.1016/j.techfore.2021.121417

Zailani, S., Iranmanesh, M., Nikbin, D., & Jumadi, H. B. (2014). Determinants and environmental outcome of green technology innovation adoption in the transportation industry in Malaysia. *Asian Journal of Technology Innovation*, *22*(2), 286–301. doi:10.1080/19761597.2014.973167

Zhang, D. G., & Lu, Y. Q. (2017). Impact of market segmentation on energy efficiency. *China Popul Resour Environ*, *27*(1), 65–72.

Zhang, W., He, L., & Yuan, H. (2021). Enterprises' decisions on adopting low-carbon technology by considering consumer perception disparity. *Technovation*, *102238*. Advance online publication. doi:10.1016/j.technovation.2021.102238

Zhao, X., Shang, Y., & Song, M. (2020). Industrial structure distortion and urban ecological efficiency from the perspective of green entre-preneurial ecosystems. *Socio-Economic Planning Sciences*, *72*, 100757. doi:10.1016/j.seps.2019.100757

Zhou, K. Z., Gao, G. Y., & Zhao, H. (2017). State Ownership and firm innovation in China: An integrated view of institutional and efficiency logics. *Administrative Science Quarterly*, *62*(2), 375–404. doi:10.1177/0001839216674457

Zhou, P., Ang, B. W., & Poh, K. L. (2008). Measuring environmental performance under different environmental DEA technologies. *Energy Economics*, *30*(1), 1–14. doi:10.1016/j.eneco.2006.05.001

Chapter 6
Estimating the Impact of the Sustainable Development Goals on the Lebanese Society:
Effects on Bio-Economic Indicators

Geoges Bellos
Lebanese International University, Lebanon

Mazen Muhieddine Kotob
Lebanese International University, Lebanon

ABSTRACT

The current project thesis aims to examine the influence of the sustainable development goals on Lebanese society and whether or not they enhance the Lebanese people's quality of life. For this purpose, via sequential exploratory mixed research methods, the researcher intended to target one expert interviewer in the field of sustainability followed by a survey sent to over 40 respondents. Our findings demonstrate that the shift from service economies to production economies is needed in order to build up a strong economy, and sustainability will indeed ultimately help us lead healthier and better lives. Thus, the link with the aforesaid variables was determined and indeed, a positive relationship was determined between the implementation of SDGs and the improved quality of life for the Lebanese people. Thus, demonstrating the proportional relationship with both variables,

INTRODUCION

This study aspires to discover how can people wishing to act as responsible quality citizens ought to behave if they want to live a sustainable, through considering adopting a life surrounded by sustainable development goals (henceforth referred to as SDGs), which both individuals and their respective communities can consider if they want to have a dignified life, and consider Patrick Geddes's concept of "thinking globally for acting locally" (see: Kaufman & Herman, 1991).

DOI: 10.4018/978-1-6684-8879-9.ch006

The Concept that Geddes (1915) coined a century ago is similar to the reality that Kauffmann et al; (1996) disused along the lines that people ought to live sustainably and manage all resources wisely so that they can be safeguarded for future generations. Thus, the following inquiry will be addressed as the researcher sets to find out *how applying SDGs in our daily lives will enhance the Lebanese daily life quality*.

The study at hand examines how, we as Lebanese people need to be masters of our own future, and as the SDGs can help us reach the stability we want, as we know neither the governments not the central bank will get us out of the economic trough we were in since 1993, it is our job to be quality and responsible citizens as Bellos (2021b) argumented, through the data pressnted in Table A.1 in the Appendix. A concept that in the last year seemed to fail the Authorities as Lebanon faced drastic economic collapses, a dysfunctional system crippled n debt, with an estimated debt to GDP ratio nearing 454%, fed by the corrupt elites, more concerned than the quick gain than they are concerned for economic growth among other impediments. A more recent paper addresses the economics issue further, as per recent figures (Noaman et al., 2022). Research evidenced that from the summary economic indicators table presented by Hobeika (2019) detailed the situation of all industrial sectors in Lebanon, pre-pandemic, with extrapolated Debt to GDP values in Lebanon ranging from 165% to 454% by Noaman et al. (2022), as shown in Table A.2 in the Appendix.

Some of the aforesaid impediments were locally induced, which recent researches notified that the most two major obstacles to economic development were terrorism and political corruption which also were linked to money laundry among other illicit transactions (Massoud, 2021; Zreik, 2021). Yet, among the other external factors which impeded on the economic growth, several sources pointed the finger to the incompetent ruling class since die to the fact that the same leaders roamed around since the last three decades, and ruled in a way that was defined by some as an oligarchic mafia, there was no governance or they were not held accountable for their actions. Worldwide impacts on Covid 19 were raised by Rogoff (2020).

Thus, the local outcomes of the Covid 19 disaster, which visibly the state mismanaged the pandemic prevention, as it mishandled its assets (which recently ex. CEO of Renault and Nissan, Mr: Carlos Ghosn, said their net worth at raw state were 40 bn USD, so had they utilized them wisely, we would not have be n in that mess to start with) and were dissimulating the fact the harbor had stocked ammonium nitrate, what caused the Beirut Port, to explode, and with it half of our Nation's capital to be blasted to oblivion. The mismanagement was summarized by El-Bacha (2020), Gerges (2020) and Chatterjee (In Press). Yet when HE Hassan Diab was blamed, the prosecutors should have pressed charges against all parties, aware since 2013, which equates to the entire political oligarchy.

PROBLEM STATEMENT

Our direct concern ought not to be in the next government, in the parliament or in the Presidential Palace, as they are all equally to blame for the socioeconomic nightmare we are currently facing: The researcher is convinced the entire corrupt class is responsible for the geopolitical tensions, hence the mentioned impediments as such as "terrorism" and "corruption" which were mentioned by El-Hajj (2020), Abdalla (2021), Massoud (2021) and Zreik (2021) as they mentioned some of the issues linked to money laundering/were among the obstacles that impinged on the implementation of the SDGs in Lebanon.

Why conducting this study? The researcher firmly believes that entrepreneurship is the key to get us out of this endless nightmare Lebanon faced since 1993, which the elite was unable to hide since

Summer 2019. The people need to be proactive thinkers and social entrepreneurs is they want to build a sustainable solution. That is all there is to it.

PURPOSE OF THE STUDY

As discussed above, the study aims to discuss what needs to be done at the micro level (individual) in order to influence the mega level (societal), so we all ought to consider thinking and planning strategically, and thus become proactive thinkers: It is our only way out of this mess, and maybe the only way the country can be prosperous again (Kauffmann et al., 1996). The best way to do that is to "think global, and act local" as Geddes (1915) and Kauffmann et al. (1996) proposed regarding doing the right thing in terms of actions, which we call ought to be held accountable for; once our actions today will harm the sanity of future generations.

RESEARCH QUESTION AND HYPOTHESES

The research will be guided by the following questions: RQ1: *How will the application of the sustainable development goals enhance the Lebanese daily life quality?* RQ2: *Will engaging in the SDGs help the Lebanese society work for a better future?* The researcher has identified more research (sub-) questions, or related ones that could be explored throughout this research, leading to design a research methodology that is relying on mixed methods, but effectively involving action research, through collecting both hard and soft data, as part of the research may involve a strategic plan, and thus leading to a proposed course of action pertaining to Agenda 2030. As the intended study involves solving organizational problems (pragmatism), utilizing both a priori (confirmative) and a posteriori (explorative) approaches

The researcher firmly believes in adopting the mixed methods approach and thus will go for a research design to encompass action research, utilizing qualitative methods for collecting expert opinions through interviews (using purposive sampling) and a structured questionnaire disseminated to study subjects at random. According to Creswell (2012), action research is a mixed methods approach, however other methodologists may consider it qualitative, like Sekaran (2000). Since a quantitative component is included, in the proposed research design as well, a priori hypotheses will be included (a posteriori hypotheses will be considered for the qualitative part as that don't need to confirm anything). *The numeric hypotheses are:*

H0: No Correlation between implementing the SDGs on the Lebanese Society
H1: Positive Correlation between implementing the SDGs on the Lebanese Society
H2: Negative Correlation between implementing the SDGs on the Lebanese Society

RESEARCH AIM AND OBJECTIVES

The current research sets out to investigate the impact of the Sustainable Development Goals on the Lebanese Society through sequential exploratory mixed research methods.

SIGNIFICANCE OF THE STUDY

This study is extremely significant inasmuch as it proposes an action type research in which we will be the masters of our own future as we need to rebuild our country which the mafia elite destroyed since like forever. We are not blaming them for being corrupt, as it is in their nature, but we should be ashamed of ourselves for accepting this elite to rob us for decades and to keep endorsing them for elections. This study ought to show that leaders show the ways things are done correctly and thus lead by example so; if we show positive changes through proactive thinking the elites may be realizing and regaining conscience of their wrongdoings.

STUDY OVERVIEW

This study's synopsis focuses on studying how applying SDGs affects the Lebanese society. Thus, keeping in mind the hypotheses we came up with in our current section, together with our study objectives, will be developed in details in the following section. Those are the points we need to look into prior to working on our methodology, findings and concluding parts. This is why Bellos (2008) considered a potential source for offshore hydrocarbon, onshore. Yet, we need to find ways to valorize our natural and all our resources, as we no longer have the luxury of time to waste with futile political or sectarian conflicts whilst 75% of the population became poor now, and need to lead dignified lives.

LITERATURE REVIEW

A recent research discusses how studying which SDG(s) work(s) in synergy with others, and which one(s) work(s) against each other (Pradhan et al., 2017). As a result, it was discovered that working with SDGs 12 and 15 will have tradeoffs in comparison with the other SDGS if they are working towards the others. Several experts studying SDGs noticed that in the past three years implementing SDGs in the ways we do business will improve the way in which business transactions are done (Youness, 2017). This is evident as Mathers & Deonandan (2018) mention that implementing SDGs will bring more peace, more equity and more justice to the world. This will add impetus to the research in *Estimating the impact of sustainable development goals on the Lebanese people's everyday life*. Once we do not work towards SDGs, the following overview includes major issues, which will be bestowed upon us. So, potential readers could get the feel what will happen once societies chose to ignore working towards aiming for sustainability. As such, Pradhan et al. (2017), Youness (2017) and Mathers, & Deonandan et al. (2018) discuss some of the issues this literature review will go over.

WHAT ARE THE SDGS AND HOW DO THEY AFFECT PEOPLES' LIVES

The SDGs, have been derived in 2015 from the Millennium Development Goals (MDGs) and serve as a transformational accomplishment for humanity and were one of the most striving global health initiatives that has were ever been undertaken on a global scale level (Mathers, & Deonandan, 2018). Their main aims were to build "*a more peaceful, prosperous and just world*". Since they were meant to work on the roadway to reach the 2030 agenda, Mathers, & Deonandan (2018) discussed that this 15-year landmark, which became the successor of the MDGs, will fulfill the 2030 milestone.

Therefore, and as specified by Mathers, & Deonandan (2018) they encompassed 17 distinct goals and 169 targets, but were more aspiring and less inflexible than their MDG counterparts were (Mathers, & Deonandan, 2018). Once the simplistic results of the MDGs were met, the initial valuation opportunity arose to assess the following SDGs, which had a potential for being based upon its predecessor framework's accomplishments while avoiding the setbacks of its failures. So, in their study, Mathers & Deonandan (2018) focused on expert opinions to better understand the precise role that these SDGs would've played. The current researcher will consider the same approach in its qualitative study to sample key expert opinions through purposive sampling.

IMPEDIMENTS PREVENTING ATTAINING THE SDGS

Among the impediments Mathers, & Deonandan (2018) discussed in their qualitative approach pinpointed the fact that the researchers agreed that there won't be any true sustainable development without an effort to encourage good governance. They agree that Political Corruption is a major impediment in countries like Lebanon (Mathers & Deonandan, 2018). According to them; definitely, when good governance lacks, the addition of resources will encourage and prolong exploitation and poor financial systems (Mathers & Deonandan, 2018). So, in this case, the SDGs ought to be both more inclusive and targeted. Mathers & Deonandan (2018) concede that the urge to satisfy conflicting political agendas brought about a list consisting of aims and objectives that lacked both a unified viewpoint and, in some situations, quantifiable results.

Their multiplicity was too wide and dilute to be optimally beneficial and their aspirational nature can be advantageous in its inclusiveness, but problematic when it comes to serving as a tool to assess programmatic success (Mathers & Deonandan, 2018). The experts that Mathers & Deonandan (2018) interviewed believed that SDGs should approach health inequalities from a deterministic and an intersectional outlook. Presently, the health biomedical model succeeds, inasmuch as objectives lean toward traditional quantitative population health indicators (Mathers & Deonandan, 2018). As Mathers, & Deonandan (2018) proposed, reaching sustainable development in the health sector, one must imperatively address issues, such corruption, governance and taxation, while remembering to target health dissimilarities through an intersectional lens.

WHAT ARE THE SDGS AND HOW DO THEY AFFECT PEOPLES' LIVES

The SDGs, have been derived in 2015 from the Millennium Development Goals (MDGs) and serve as a transformational accomplishment for humanity and were one of the most striving global health initiatives that has were ever been undertaken on a global scale level (Mathers, & Deonandan, 2018). Their main aims were to build "*a more peaceful, prosperous and just world*". Since they were meant to work on the roadway to reach the 2030 agenda, Mathers, & Deonandan (2018) discussed that this 15-year landmark, which became the successor of the MDGs, will fulfill the 2030 milestone.

Therefore, and as specified by Mathers, & Deonandan (2018) they encompassed 17 distinct goals and 169 targets, but were more aspiring and less inflexible than their MDG counterparts were (Mathers,

& Deonandan, 2018). Once the simplistic results of the MDGs were met, the initial valuation opportunity arose to assess the following SDGs, which had a potential for being based upon its predecessor framework's accomplishments while avoiding the setbacks of its failures. So, in their study, Mathers & Deonandan (2018) focused on expert opinions to better understand the precise role that these SDGs would've played. The current researcher will consider the same approach in its qualitative study to sample key expert opinions through purposive sampling.

IMPEDIMENTS PREVENTING ATTAINING THE SDGS

Among the impediments Mathers, & Deonandan (2018) discussed in their qualitative approach pinpointed the fact that the researchers agreed that there won't be any true sustainable development without an effort to encourage good governance. They agree that Political Corruption is a major impediment in countries like Lebanon (Mathers & Deonandan, 2018). According to them; definitely, when good governance lacks, the addition of resources will encourage and prolong exploitation and poor financial systems (Mathers & Deonandan, 2018). So, in this case, the SDGs ought to be both more inclusive and targeted. Mathers & Deonandan (2018) concede that the urge to satisfy conflicting political agendas brought about a list consisting of aims and objectives that lacked both a unified viewpoint and, in some situations, quantifiable results.

Their multiplicity were too wide and dilute to be optimally beneficial and their aspirational nature can be advantageous in its inclusiveness, but problematic when it comes to serving as a tool to assess programmatic success (Mathers & Deonandan, 2018).

The experts that Mathers & Deonandan (2018) interviewed believed that SDGs should approach health inequalities from a deterministic and an intersectional outlook. Presently, the health biomedical model succeeds, inasmuch as objectives lean toward traditional quantitative population health indicators (Mathers & Deonandan, 2018). As Mathers, & Deonandan (2018) proposed, reaching sustainable development in the health sector, one must imperatively address issues, such corruption, governance and taxation, while remembering to target health dissimilarities through an intersectional lens. A more recent paper addresses the economics issue further, as per recent figures.

PROBLEMS ARISING WHEN NOT AIMING FOR SDGS

Mathers, & Deonandan (2018) pointed out that ignoring corruption or other impediments that other researchers believe is related to illicit trafficking, terrorism, and among others, money laundering activities leave us with the incapacity to aim towards SDGs. So; taking in to account Massoud (2021) and Zreik (2021) who discussed these impediments in details, here the arguments advanced by Mathers & Deonandan (2018) concerning dealing with conflicting agendas make sense. This point was raised by Chaaban (2019) and Hobeika (2019) as well.

The researcher, knowing that dwelling on this issue is unnecessary, as by now in Lebanon we know who the corrupt leaders are and what illegal action they try to mask, agrees to the above and invites for caution when considering working for SDGs in concerned, to not take the daunting task lightly and only world on the SDG most important to us, or paired with those that will help us reach our goals through synergy, as Pradhan et al. (2017) advises us.

So, estimating the impact of sustainable development goals on the Lebanese people's everyday life should show a link. Not working towards the 17 SDGs will be catastrophic for Lebanon as shown in Zreik's (2021) survey results, if one refers to the aforementioned impediments, or if the state does not take immediate action to fight corruption. The data presented here from Zreik's (2021) research help us understand the deep problems with the currency crisis, where all impediments are listed. (see: Figure 1). These indicators worsened from 2 years prior. Salame (2022) presented arguments on the severity of this crisis, now evidencing the terrible conditions and how worse off Lebanon will be if nothing is done at local, regional, or global scale.

Figure 1. Major economic indicators to look for
(Farnham, 2015, Zreik, 2021)

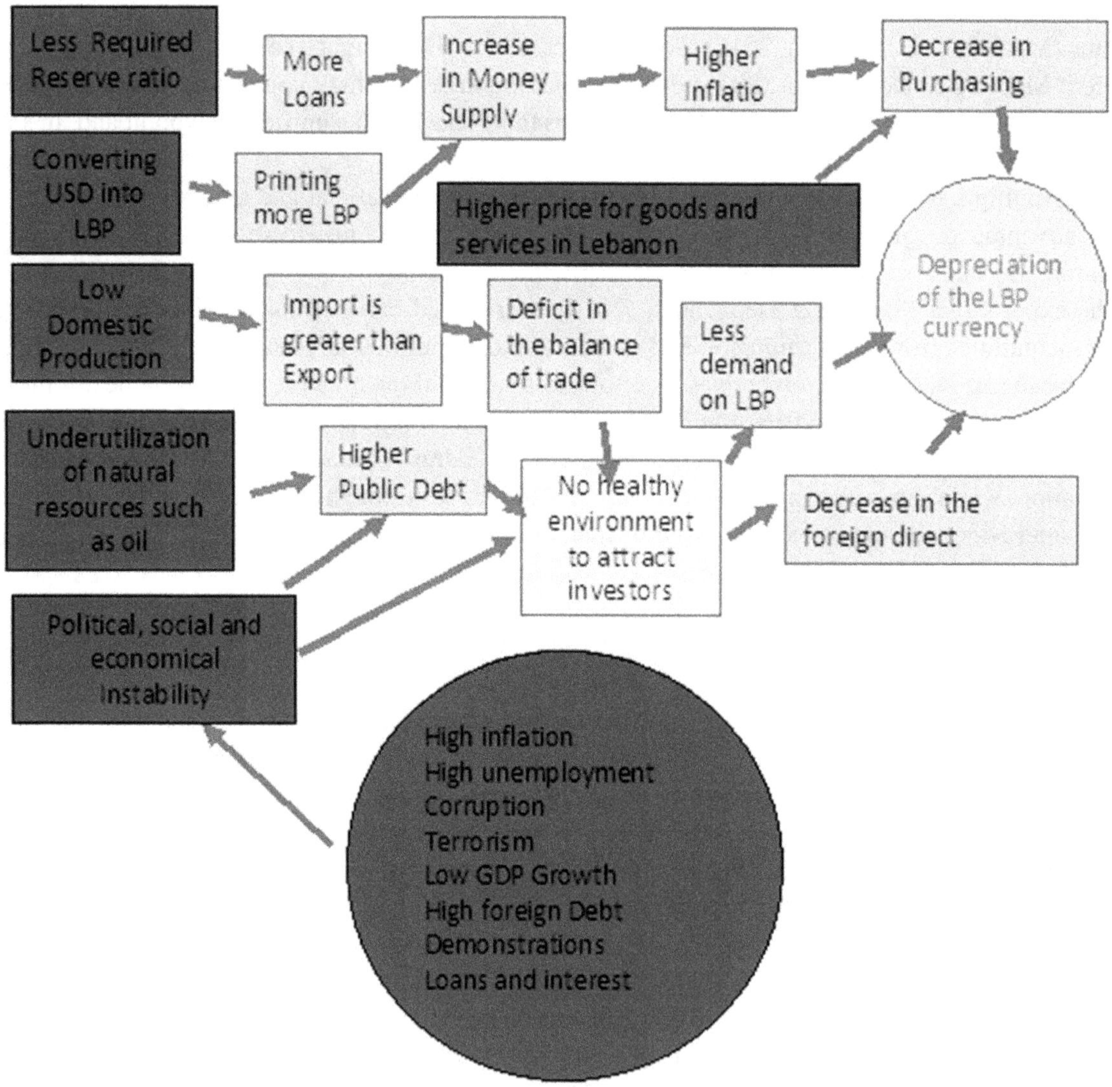

SOLUTIONS THAT SDGS CAN BRING COMMUNITIES

Considering to work on the right SDGs as Pradhan et al. (2017) commend us to do, like aiming for SDGs 1, 2, 3, and 4 simultaneously will help us attain all four goals as they related to each other. And, as such according to Mathers & Deonandan (2018) working on rendering health and well-being services attainable to all is a way to start since working towards the four SDGs together will allow us to reach synergy, like working on the 2020-2021 Hult Prize challenge, if we chose among the seven proposed SDGs wisely, we could target the ones which will guarantee synergy (see: www.hultprize.org). Thus, and in accordance to both Pradhan et al. (2017) and Mathers & Deonandan (2018), working on SDGs 1, 2, and 3 seems to be the best way to go about this, since working with SDGs will allow us to reach synergy.

Taking the Hult Prize challenge as an example, SDGs 5, 6, 7, and 8 show no effect of synergy or tradeoff, as Pradhan et al. (2017) discuss. So, since SDG 4, 8, 13 and 17 are a set of SDGs of concern to many institutions operating currently in Lebanon, there may be a chance to work on these goals together. However, Pradhan et al. (2017) cautions to avoid working with SDGs 12 and 15 together with the others since these were reporting to show the most tradeoffs.

Currently, the Hult Prize Challenge theme is on redesigning fashion, hence, with the 6 SDGs to select from, competitors need to find solutions to make the Fashion industry more sustainable by at least selecting one problem in order to propose a solution, using at least one of the proposed SDGs, such as SDGs 5, 9, 8, 9, 12, 13, all of which are related to the Fashion industry and in effect could affect countries' economies, which in turn will affect Global GDP values

SUMMARY OF LITERATURE REVIEW

Briefly this section introduced us about the SDGs, what happens if we don't aim to reaching Agenda 2030 and if we do not detect the causes (corruption, terrorism, ...) why in certain countries one cannot implement them and in others there are part of the citizens' everyday lives (Mathers & Deonandan, 2018). Pradhan et al. (2017) discussed that working on one of the SDGs from the set of SDGs from SDG 1 – 4 together will help reaching another SDG in that set. And aiming towards SDGs 5 – 8 together show no effect. Yet, some SDGs show tradeoffs. Thus enhancing the purpose of this research, in proposing a course of action (Creswell, 2012).

METHODOLOGY

The sampling used for the qualitative component, involving interviews, was a purposive sampling, whereas the surveys utilized a random sampling technique (Creswell, 2012). We also need to explain how we intend to process the data, through triangulation, and explain why we chose the methodology we selected, before, proposing our full research design, as using both Cooper & Schindler (2014) and Edson et al., (2017) allowed us to select our final design, which was executed for our data collection as raised in our aims and objectives, this study encompasses solving organizational problems while adopting a pragmatist's viewpoint, applying both confirmative and explorative methods (see: Figure 2). So, based Sekaran (2000), Creswell (2012), Olivia (2012), Sekaran & Bougie (2013, 2016), the research design will utilize the mixed methods approach and collect data through action research. It utilized qualitative methods for collecting expert opinions through interviews (utilizing judgmental sampling) and a structured questionnaire disseminated to randomly selected study subjects. Therefore, in the next section, we will present our findings. Figure 2, below, shows the Saunders Research onion which helped us in setting up our research design.

FINDINGS AND ANALYSIS

According to Geddes (1915), a Scottish urban planner and social activist, the "think globally, act locally" statement (or "think global, act local") clearly was a pioneering statement, as this reality has been utilized in many sectors, including development, education, science, and management. Recently, with the 17 UN Sustainable development goals (SDGs), this has added more impetus. Kauffman et al (1996) define this statement via a societal need on a global (mega) scale and translates it to a local (or micro) scale, for any institution's direct applications, one firms consider reaching the 17 UN SDGs. As Pradhan et al. (2017) mentioned that not all SDGS can work together, as the impediments of one SDG can impinge the development of another, so care must be taken whilst aiming towards all 17 SDGs.

Figure 2. Saunders' research onion
(Saunders et al., 2019: 130)

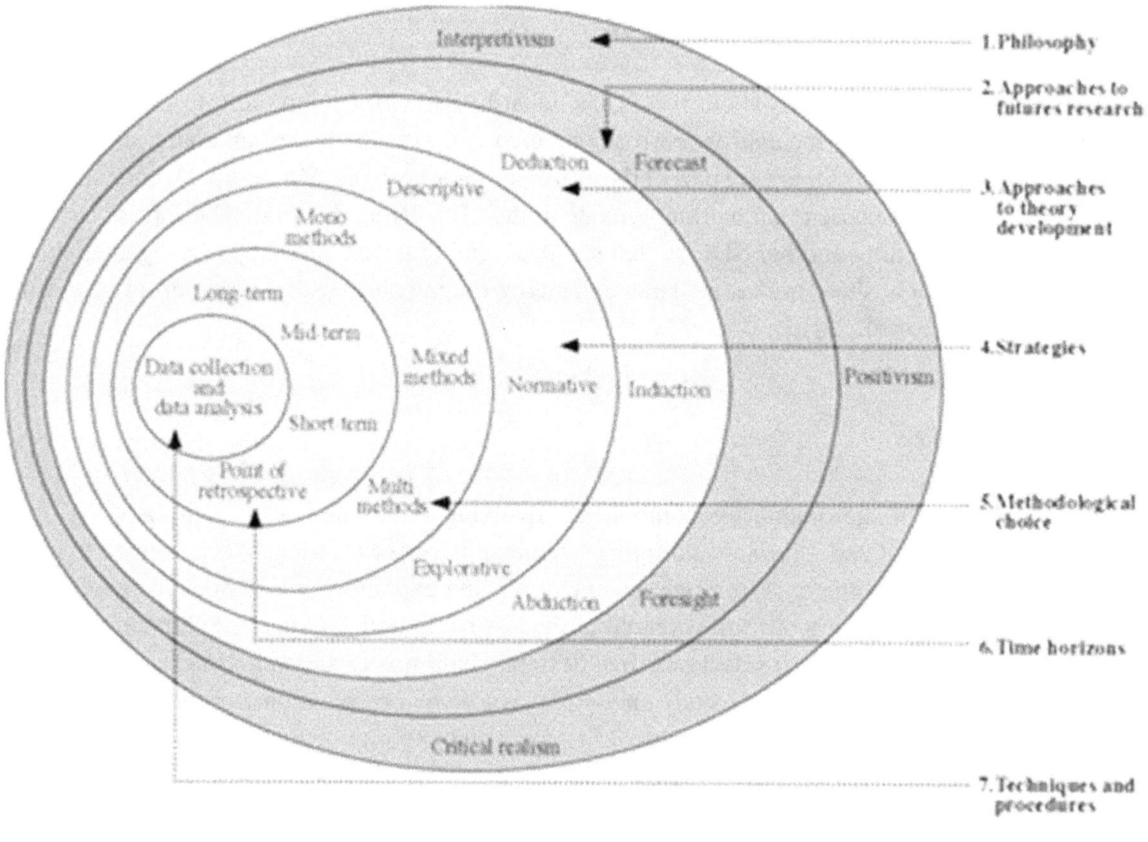

Thus, both Pradhan et al. (2017) and Mathers, & Deonandan, (2018) denote that .since individuals possessing these social responsibilities can work towards these SDGs (on a local, or individual scale) through thinking of the "International" or "global" scale and acting at a "local" scale would benefit any institution wanting to invest in SDGs. Youness (2017, 2020) provides the mindset some modern businessmen need to be in in in order to accomplish successful careers both interns of personal achievements as in company achievements through the trends of business. Whereas Pradhan et al. (2017) and Mathers, & Deonandan, (2018) mention that the SDGs that ant institution in any third world country like Lebanon ought to invest in need to be synergetic, as working on either SDGs 4, 8, 13 and 17 will allow them to aim towards the others as well, and commend us to take care in terms of what impediments affect most implementing the SDGs.

So, as Both Pradhan et al. (2017) and Mathers, & Deonandan, (2018) commended institutions to work on SDGs 1 to 4 will definitely give any company a chance to correctly estimate which SDGs to work on, although it ultimately on all 17 SDGs, to not just jump into action hastily, and aim towards their achievements wisely. This is why Creswell (2012) and Saunders et al (2019) recommend the institution to use mixed methods and undergo action research. The next two subsections will respectively over the proposed strategic plan the institution ought to consider in order to conduct the proposed operational action as an outcome of action research as discussed by Creswell (2012) as both hard data (quantitative) and soft data (qualitative) are conducted. Although Saunders et al (2019) believe it is qualitative, we believe like Creswell (2012) that these types apply to mixed methods as it is used to solve organizational problems through adopting a pragmatist view and using adductive approaches (see: Olivia, 2012).

RESULTS

The following findings were noticed from our pilot study which was carried out in April 2021. The results were well correlated as the Cronbach alpha results (see: Table 1), from a similar survey using similar variables, were over 0.95 since while we conducted it (Cronbach, 1951). Results of those are shown in the Appendix, while our Cronbach Alpha value is shown in Table 1, for the pilot study, which is highly correlated as it is above 0.99. The point of our research is to see whether "*Implementing SDGs in Lebanon*" and the "*Improved Access to basic health care services and encouraging a healthy lifestyle*" are important indicators for the well-being of the Lebanese people's daily lives (cf. Pozzato, 2017-2018). There is an almost unanimous agreement, since 94% of our survey subjects agree with these statements, whereas 6% don't necessarily consider it as being important (Bosch et al., 2017). These variables are discussed first as they relate to my research question and will help me in determining if out to respond to the hypotheses I for the study, which indicates a link between SDGs and improving the Lebanese people's daily lives. Moreover, the same applies to the "*Improved access to educational resources*" (SDG 4) and findings agree with Bosch et al. (2017) as well as ongoing research.

These define the variables we are attempting to investigate in the current research, yet, we will display all survey results in the Appendix. Since we went for exploratory sequential research designs, we still decided to show the pilot study first, only because the elements of the research were derived from the literature, so we wanted to be sure of the validity of the survey before we collect the actual data, while the findings agree with Bosch et al. (2017) as well as ongoing research. Therefore, part of our research process entails triangulating the data. For this purpose, I will compare the results from the pilot study to the literature review, and with the survey results, data gathered through exploratory sequential data

Table 1. Survey reliability results

Cronbach's Alpha	0.99
Split-Half (odd-even) Correlation	0.99
Split-Half with Spearman-Brown Adjustment	0.99
Mean for Test	3337
Standard Deviation for Test	2403.33
KR21 (use only 0 and 1 to enter data for this)	1.03
KR20 (use only 0 and 1 to enter data for this)	1.03

collection tools I used to gather my primary data, to compare them back to the Literature Review. Findings agree with Bosch et al. (2017) as well as ongoing research.

ANALYSIS OF THE PRIMARY DATA COLLECTION AND INTERPRETATION

"*Implementing SDGs in Lebanon*" is considered as being extremely important by 37.50% from our respondents, and very important by 31.25% of our subjects, somewhat important by 25.00% of the respondents. While 6.25% of the respondents think it is not so important (Bosch et al., 2017). Rephrasing this, we notice that this points out to 94% think it is important and 6% do not (Bellos, 2021b). So, in terms of "***Improved access to basic health care services, and encouraging a healthy lifestyle***" indeed, existing data, as well as our collect primary data confirm the importance of this indicator as a measure of the effectiveness of SDGs in improving our daily lives, thus validating H1.

This agrees with the query about the level of importance "***in implementing SGGs for enhancing Lebanese people's daily lives***". This evidences the importance in "***engaging in the SDGs to help the Lebanese society work for a better future***" as sustainability means living smartly and providing a safe and prosperous future for generations to come without massively depleting our scarce resources (Pozzato, 2017-2018). The "***Connection to the existing cultural heritage while designing a sense of place***" is extremely important for us citizens to feel at home and to want pro preserve our country and way of life as a species (Knio & Bellos, 2018; Pozzato, 2017-2018; Bellos, 2019).

One way to improve our lives while implementing SDGs clearly is to "***attain economic efficiency in Lebanon***" since this ought to "***permanently solve the hyperinflation, the catastrophic unemployment, the elitist corruption, terrorism, low GDP growth, skyrocketing foreign debt, and high loans***". Zreik (2021) agrees since she estimated that we need to attain economic efficiency in Lebanon, and that our expert interview demonstrated that this will solve the hyperinflation problem we are facing since the oligarchy took office. Marouni (2021) also agrees. Figure 1 summarizes which factor(s) impede(s) the Lebanese Lira. A more recent paper addresses the economics issue further, as per recent figures.

We can say that it is somewhat agreed that we indeed are in a n unprecedented economic crunch, never seen before, although Zreik (2021) found out that some think Lebanon is currently in an economic boom, which is of course unlikely as all indicators point to dramatic economic collapse. Zreik (2021) also agreed that economic efficiency is a must to attain in Lebanon. She also agrees with that Bellos (2021a, b) presented.

Through sustainability we could "***limit the Lebanese Lira devaluation through implementing SDGs***" through developing "***economic rational plans to allow us to implement SDGs and improve our daily lives***" as, indeed we must "***find a solution to make Lebanon a healthy environment to attract investors once we consider implementing SDGs to improve our daily lives***".

Pozzato (2017-2018) shows similar findings when discussing the issues pertaining to mart cities and health care. It was noticed that "***Rentier or service economies are to blame for our economic meltdown. Yet cyclical economies are what we need to reach in order to attain sustainability***" so, attaining them is a must in order to proceed towards a production or cyclical economy, as implementing SDGs will help countries develop their economies, create jobs, prepare more stable futures, all of which are much needed in Lebanon now.

Our expert interview agrees with this too. Indeed, and as Marouni (2021) and Zreik (2021) discuss, service and rentier economies have had their share of the blame in the failure of Lebanon's economic system, yet many other factors also contributed to its demise. Among the impediments mentioned both Massoud (2020) and Zreik (2021) agreed that terrorism and corruption are among them most important obstacles, that caused the failure of the current economy, some scholars pointed that the anti-corruption riots were also impediments, but both Bellos (2021a, b) and Marouni (2021) disagreed.

Therefore, "***How important, in our context, is it to move from a service economy to a productive one for SDG implementation and/or improving our daily lives*?**" was answered in the following way: 50.00% of the surveyed subjects believe that it is extremely important, and 37.50% of them think that it is very important, while 12.50% of the respondents believe it is somewhat important (Bosch et al., 2017). There is an agreement that this is important, yet in terms of statistics, 87% think it is important and 13% think it is somewhat important (Bellos, 2021b). This again shows to validate H1.

The results of Question 11, concur that in general it is agreed that as Zreik (2021), and Marouni (2021) discuss, we are in need to shift to a production economy, and to do o fast. This is since the current economy is weak and the various crises give it occasional shocks, and as the last and current one gave it its final blow. This indicator "***Improved cybersecurity, digital literacy and data privacy***" also evidences the effectiveness of implementing SDGs in improving our daily lives, and again, validating H1. Likewise as "***Improved access to vehicle sharing solutions***" helps solving traffic issues, will also benefit our lives. So, in the end, we should seriously consider sustainability (Pozzato, 2017-2018).

As the "***Reduction in annual final energy consumption and in life cycle energy use***", "***Share of recycled input materials and of renewable materials***", ***Decreased emissions of Particulate matter (PM2,5)***", "***CO2 reduction cost efficiency***", as the "***Increase in green and blue space***", among other indicators preventing air pollution, already seen in some Smart cities in Eurasia again show lots of benefits for our daily lives, thus validating H1 (Brozovsky et al., 2021).

"***Affordability of housing***", "***Internal rate of return (IRR)***", "***Net Present Value (NPV)***" and "***New forms of financing***", among other financial asset measurement variables, and all of the indicators listed by Bosch et al. (2017) and Pozzato (2017-2018), which we present out data pertaining to them in the Appendix confirm the positive relationship between implementing SDGs and improving our daily lives (see: Table 2). Indeed, the link among the variables are clearly proportional, since implementing SDGs will positively improve our daily lives. Finally, our Hypothesis validation table is shown in Table 2. Therefore, we accepted HI, and correspondingly reject H0 and H2.

Table 2. Hypothesis verification

Hypothesis	Decision
H0: No correlation between implementing the SDGs on the Lebanese Society	Rejected
H1: Positive correlation between implementing the SDGs on the Lebanese Society	Accepted
H2: Negative correlation between implementing the SDGs on the Lebeanee Society	Rejected

CONCLUSION, LIMITATIONS, AND RECOMMENDATIONS

Conclusion

As we noticed, the shift from service economies to production economies is what we need in order to build up a strong economy, and sustainability will eventually help us leading healthier and better lives. (Pozzato, 2017-2018). The link we discussed from the accepted hypothesis (H1) confirms the positive link of increasingly working towards SDGs will greatly mope our daily lives. The first set of research queries were purposed to addressing *how applying SDGs in our daily lives will enhance the Lebanese daily life quality* and addressing through our study rationale, w*hy were we conducting this study* allowed us to clearly and accurately respond to the main (two) research questions. Recall that the following two questions "*How will the application of the sustainable development goals enhance the Lebanese daily life quality?*" (RQ1) and "*Will engaging in the SDGs help the Lebanese society work for a better future?*" (RQ2) were also clearly addressed.

Our 1st research question, which was responded through the quantitative analysis of our study demonstrated how engaging in SDGs will improve the Lebanese People's daily lives, as on many accounts, from the data collection, this was evidenced. Thus, this helped us validate H1, while rejecting H0 and H2. Our 2nd research question was positively responded, as stating that without a doubt, engaging in SDGs will improve our lives tremendously. Thus, the aforementioned research questions in *Estimating the impact of sustainable development goals on the Lebanese people's everyday life*, agree that applying the SDGs ultimately will lead to "*a more peaceful, prosperous and just world*".

Therefore, while we were focusing on "*Implementing SDGs in Lebanon*", naturally the impact of them on the "*Improved access to basic health care services and encouraging a healthy lifestyle*" was proven to be proportional, according to Bosch et. al. (2017) and Pozzato (2017-2018). Most of these indicators were studied by Pozzato (2017-2018) in a different context.

The following questions, which were responded pertaining to our research will be listed in terms of significance. So, "*As per context, how important is it to attain economic efficiency in Lebanon?*" was shown to be highly significant. As well as "*How important, in our context, is it to move from a service economy to a productive one for SDG implementation and/or improving our daily lives*?"

The reasons being that sustainability enhances economic development through SDG 8, while the few variables we selected to discuss enhance SDG 11. So, "*How important is it, within our context, to permanently solve the hyperinflation, the catastrophic unemployment, the elitist corruption, terrorism, low GDP growth, skyrocketing foreign debt, and high loans (since they are all impediments to implementing SDGs in Lebanon) for finding a durable solution to this never ending issue*?" is also shown to be quite significant as well. A more recent paper addresses the economics issue further, as per recent figures.

While, "*Connection to the existing cultural heritage, while designing a sense of place*" was shown to be, based on the importance of valorizing cultural resources also somewhat significant (Bellos, 2017-2018, 2019). For us, this is considered also extremely significant, -especially in terms of revamping the sustainable tourism sector, post Covid 19. The following query "*How important is it to implement SGGs in enhancing Lebanese people's daily lives?*" was majoritarily deemed as significant by our findings. This is also the case in "*engaging in the SDGs to help the Lebanese society work for a better future*" as the SDGs are proven to offer tremendous benefits to all nations choosing to follow Agenda 2030.

Therefore, within our case, it was deemed significant to "*limit the Lebanese Lira devaluation through implementing SDGs*" as working towards productivity will enhance employment and eventually curb inflation. More research is needed in order to study the link between Inflation and unemployment, if one considers the Philips curves among other ways to estimate these through economic indicators (Bellos, 2021b; Salame, 2022; Noaman et al., 2022).

It is likewise, significant to "*implement economic rational plans to allow us to implement SDGs and improve our daily lives*" as these will ultimately get us out of recession and propose stable programs in revamping our economy. It is as well, as the above, also of significant impact, to "*find a solution to make Lebanon a healthy environment to attract investors once we consider implementing SDGs to improve our daily lives*" as once we become sustainable, revamp our economy, install all the needed corporate governance measures, and present clear solutions, will investors be willing to invest in Lebanon again. At the same time, the IMF will not pour in a penny once the State doesn't deliver its dues, and undergoes the International Community's imposed reforms (see: Pozzato, 2017-2018). The discussion that was recurring since decades pointed the finger to the Rentier economies in Lebanon, and how they were implemented. Indeed "*rentier or service economies are to blame for our economic meltdown. Yet cyclical economies are what we need to reach in order to attain sustainability*". This was deemed necessary to reach that, as it was a significant step towards "*implementing SDGs to improve our daily lives*".

Last, but not least, this joins with this statement to also be a significant step to accomplish once we consider moving "*from a service economy to a productive one for SDG implementation and/or improving our daily lives*". Thus, this ends our research. The best section goes over our limitations, while the last one presents various recommendations to follow so that Lebanon choses sustainability as a mode of survival, as we will not survive another devaluation as we are the 5th worst currency in the world (cf. Hjeij, 2021). Lastly, Figure 3 shows reconstructed interpretations estimated from secondary data triangulated with ours in order to estimate the significance of implementing SDGs as a way to improve our daily lives. The same data was used to estimate the results in response to the corresponding research questions (see: Figure 3).

Rearranging the above responses, we obtain a 96% overall rate of responses through interpreting all our primary and secondary data which agree to the significance of implementing SDGs in improving our daily lives, against a 4% rate of responses showing disagreement. In terms of bulk Likert scale values for the research, overall, there is an agreement in the significance in working towards the SDGs.

Limitations

Unfortunately, due to limited allotted time to conduct the study, we needed to gather some of our data through secondary means, and compare and contrast them with our data. The main setback remains Covid 19, as the survey subjects or the expert interviewees were reached out through online modes of

Figure 3. Interpreted results

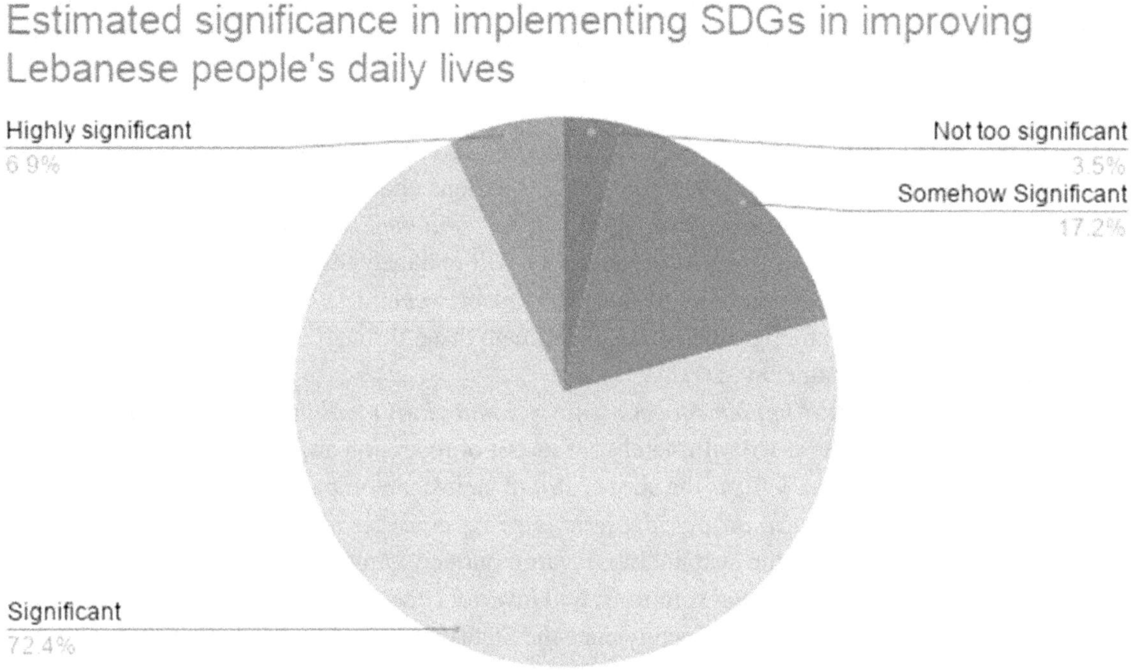

communication. Yet, the information we gathered were mostly from secondary data, some of which were graciously obtained from ongoing researches and others obtained from consulting the available literature.

Thus the data and methodological triangulations helped us tremendously in ensuring that the data we collected aligns with the literature review as well as the collected secondary data while ensuring the data are valid, through cross checking them from various sources; our data has been very well correlated due to the high Cronbach Alpha value results, confirming the very high internal correlation (see: Cronbach, 1951).

Recommendations

It is highly recommended to conduct the IMF imposed reforms, eliminate all corruption, nepotism and terrorism, while working on limiting the effects of illegal and fraudulent acts as well as increasing corporate governance measures to render any acts embezzlement of fraud harder, and to take immediate punitive actions, and as per the stipulation of the Sarbanes Oaxley 2002 Act, to forbid CEO duality, impose an expert, independent, unbiased and professional auditing committee capable to impose check and balances on all state mechanisms, or sectors susceptible to generate illicit activities (Farhoud, 2021).

The MOF and BDL are welcome to adopt expansionary policies; it is time they get us out of recessions as their constructional policies did us more harm than benefit, as their sole aims were to slow economies down (Farnham, 2015; Zreik, 2021) and only once this tackled can we effectively implement SDGs and as Pradhan et al. (2017) and as Mathers & Deonandan (2018) denote, to work on them smartly. Already implementing consensus spaces and imposing the adoption of Circular 551 will be much needed as since Atallah (2019) discussed them nothing was done yet, while Circular 551 was introduced since 2013.

Since the SDGs can help in transferring the useless rentier (or service) economy we had to a cyclical (or production) one, Atallah (2019), Marouni (2021) and Zreik (2021) agree on this as well, via different methods, to explain the urge in executing this transformation. Thus, by doing so can we become transformative and socially responsible citizens, as we ought to be the custodians of our country, not this corrupt oligarchy.

REFERENCES

Abdalla, M. (2021). *The impact of money laundering in foreign exchange firms in the Lebanese financial market.*. [MBA Project submitted to the Lebanese International University].

Al-Sawi, M. (2018). *The Effective Valorization of Resources for the Economic Development of Tyre.* [MBA Project presented at the Lebanese International University].

Bellos, G. S. G. (2008). *Sedimentology and Diagenesis of Some Neocomian-Barremian Rocks (Chouf Formation), Southern Lebanon* [M.Sc. Thesis Presented to de The Department of Geology at the American University of Beirut].

Bellos, G. S. G. (2017-2018). *Évolution culturelle du peuplement de l'âge du Fer dans la Phénicie du Nord.* [MA Thesis Presented and Defended at the Lebanese University].

Bellos, G. S. G. (2019). *Estimating the Impact of the Effective Valorization of Cultural and Natural Resources for the Economic Development of Beirut.* [MBA Project presented at the Lebanese International University].

Bellos, G. S. G. (2020). Enquête sur les effets de la récente dépression économique sur le licenciement en masse des employés: Les huit derniers indicateurs économiques. *Libna News.* https://libnanews.com/enquete-sur-les-effets-de-la-recente-depression-economique-sur-le-licenciement-en-masse-des-employes-les-huit-derniers-indicateurs-economiques/

Bellos, G. S. G. (2021a). Historique de la Thawra depuis le 17 Octobre 2019 à nos jours. *Libna News.* https://libnanews.com/historique-de-la-thawra-depuis-le-17-octobre-2019-a-nos-jours/

Bellos, G. S. G. (2021b). *Estimating the impact of the Sustainable Development Goals on the Lebanese Society.* Thesis, Lebanese International University, Beirut Campus, Beirut, Lebanon].

Bosch, P., Jongeneel, S., Rovers, V., Neumann, H. M., Airaksinen, M., & Huovila, A. (2017). CITYkeys indicators for smart city projects and smart cities. *Eurocities.* http://nws.eurocities.eu/MediaShell/media/CITYkeystheindicators.pdf

Brozovsky, J., Gustavsen, A., & Gaitani, N. (2021). Sustainable Cities and Society Zero emission neighbourhoods and positive energy districts – A state-of-the-art review. *Sustainable Cities and Society*, 72.

Bryman, A. (2012). *Social Research Methods* (4th ed.). Oxford University Press.

Chaaban, J. (2019). Lebanon Needs a Budget for Its People, Not is Ruling Class. *Beirut Today.*

Chatterjee, S. (in press). COVID 19. *Tackling Global Pandemics Through Scientific and Social Tools, Elsevier.*

Cooper, D., & Schindler, P. S. (2006). *Business Research Methods*. McGraw-Hill.

Cooper, D., & Schindler, P. S. (2014). *Business Research Methods* (14th ed.). McGraw-Hill.

Creswell, J.-W. (1998). *Qualitative Inquiry and Research Design: Choosing among Five Traditions*. Sage Publications.

Creswell, J.-W. (2002). *Educational Research: Planning, Conducting and Evaluating Quantitative and Qualitative Research*. Prentice Hall.

Creswell, J.-W. (2003). *Research Design: Qualitative, Quantitative and Mixed Methods Approach* (2nd ed.). Sage Publications.

Creswell, J.-W. (2012). *Educational Research: Planning, Conducting, and Evaluating Quantitative and Qualitative Research* (4th ed.). Pearson.

Cronbach, L. J. (1951). Coefficient alpha and the internal structure of tests. *Psychometrika, 16*(3), 297–334. doi:10.1007/BF02310555

Denzin, N. K., & Lincoln, Y. S. (2005). *The Sage handbook of qualitative research* (3rd ed.). Sage.

Edson, M. C., Henning, P. B., & Sankaran, S. (2017). *A Guide to System Research* (2nd ed.). Springer. doi:10.1007/978-981-10-0263-2

El-Bacha, F. (2020a). Coronavirus: L'erreur libanaise. *Libna News*. https://libnanews.com/coronavirus-lerreur-libanaise/

El Hajj, M. G. (2020). *Fighting Money Laundering in Foreign Exchange Firms in Lebanon*. [MBA Thesis, Middle East University].

Farhoud, N. (2021). Estimating the impact of financial accounting on risk management. [MBA Thesis, Lebanese International University].

Fleddermann, C.-B. (2003). *Engineering Ethics* (2nd ed.). Pearson.

Geddes, P. (1915). *Cities in Evolution*. Williams.

Gerges, F.-A. (2020). Will Beirut's blast be a catalyst for change? *Dev Discourse*. https://www.devdiscourse.com/article/politics/1159968-will-beiruts-blast-be-a-catalyst-for-change

Hjeij, Z. (2021). *Estimating the impact of the Lebanese Currency Depreciation on Small investors*. [MBA Thesis, Lebanese International University].

Hobeika, L. (2019). *High-interest rates mean a weaker currency*. Anna Har. https://en.annahar.com/article/1039542-highinterest-rates-mean-a-weaker-currency

Jabbour, P. (2018). *The challenges facing e-commerce businesses in Lebanon*. [MBA Thesis, Lebanese International University].

Kaufman, R., & Herman, J. (1991). *Strategic planning in education: Rethinking, Restructuring, Revitalizing*. Technomic.

Kaufmann, R., Hermann, J., & Watters, K. (1996). *Educational Planning: Strategic, Tactical, Operational*. Technomic Publishing Inc.

KnioM. S. (2021). *Personal Communication*.

Knio, M. S., & Bellos, G. S. G. (2018). The Effective Valorization of Resources for the Economic Development of Tyre. *AJFM*, *11*(11), 1794–1800. doi:10.31142/afmj/v3i11.04

Marouni, H. (2021). *Estimating the impact of entrepreneurial ventures and startups on the Lebanese economy*. [MBA Thesis, Lebanese International University].

Massoud, F. (2021). *The Effects of the Consecutive Crises in Lebanon on Financial Markets Efficiency and Money Laundering*. [MBA Thesis, Lebanese International University].

Mathers, A., & Deonandan, R. (2018). Are the Sustainable Development Goals Realistic and Effective: A Qualitative Analysis of Key Informant Opinions. *OIDA International Journal of Sustainable Development*, *11*(3), 11–16.

Mazjoub, Z. (2018). Business Ethics (1st ed.). Open Stax.

Noaman, S. B., Harguem, S., & Bellos, G. S. G. (2022). Estimating the Impact of the Lebanese Economic Crisis on Employee Turnover through Motivation among other mediating variables as impeded by COVID-19 [Retrived from: http://www.resmilitaris.net.]. *Res Militaris*, *12*(6), 1351–1374.

Olivia, P. F. (2012). *Developing the Curriculum* (6th ed.).

Pozzato, A. (2017-2018). Smart cities: an innovative solution for the management of global risks. [MA thesis, Ca' Foscari University of Venice, Italy].

Pradhan, P., Pradhan, L., Rybski, D., Lucht, W., & Kropp, J. P. (2017). A Systematic Study of Sustainable Development Goal (SDG) Interactions. *Earth's Future*, *5*(11), 1169–1179. doi:10.1002/2017EF000632

Riemer, F. J., Lapan, S. D., & Quartaroli, M. T. (2012). *Qualitative research: an introduction to methods and designs*. Jossey-Bass.

Rmeity, R. (2021). *Estimating the impact of the Banking Industry on the Financial Crisis*. [MBA Thesis, Lebanese International University].

Rogoff, K. (2020). *Mapping the COVID-19 Recession*. Project Syndicate. https://www.project-syndicate.org/commentary/mapping-covid19-global-recession-worst-in-150-years-by-kenneth-rogoff-2020-04

Salame, R. (2022). Lebanese lira lost 82 percent of its purchasing power in two-year period, UN data shows. *L'Orientle Jour.* https://today.lorientlejour.com/article/1289780/lebanese-lira-lost-82-percent-of-its-purchasing-power-in-two-year-period-un-data-shows.html

Saule, M. (2004) *Rural Tourism as a tool for Local Development, Diagnosis of the French area of Haute-Corrèze*. [Master's thesis, the Bournemouth University, UK].

Saunders, M. (2019). *Research Methods for Business Students* (8th ed.). Pearson.

Sekaran, U. (2000). *Research Methods for Business: A skill-building approach* (4th ed.). Wiley.

Sekaran, U., & Bougie, R. (2013). *Research Methods for Business: A skill-building approach* (6th ed.). John Wiley & Sons.

Sekaran, U., & Bougie, R. (2016). *Research Methods for Business: A skill-building approach* (7th ed.). John Wiley & Sons.

Youness, H. (2017). *Trends of Business* (1st ed.).

Zreik, Z. (2021). *The transition of the Lebanese Economy from a service economy to a productive economy,* [MBA Thesis, Lebanese International University].

APPENDIX

Table A.1: The below secondary data were collected from Bellos's (2021b) study and placed here following the below indicators, based on Bosch et al. (2017). Values, in Likert Scale format, show overall agreement of the importance in implementing SDGs for positively affecting the Lebanese people's daily lives, through bio-economic indicators like those shown in Table A.2.

Table A.1.

Selected Variables, filled as per secondary data collection

Improved access to vehicle sharing solutions	4
Improved cybersecurity, digital literacy and data privacy	4
Reduction in annual final energy consumption	4
Decreased emissions of Particulate matter (PM2,5)	4
Increase in green and blue space	4
CO2 reduction cost efficiency	4
Internal rate of return (IRR)	4
Quality of open data	4
Leadership	4
Smart City Policy	4
Market demand	4

Source: Bosch et al. (2017)

Table A.2: Summary Economics indicators table discussed by Hobeika (2019) and adapted by Noaman et al. (2022).

Table A.2.

	2016 Act.	2017	2018	Proj. 2019	2020	2021	2022	2023
Output and prices	(Annual percentage change)							
Real GDP (market prices)	1.0	1.2	1.5	1.8	2.2	2.6	2.9	2.9
GDP deflator	-0.7	2.5	2.7	3.3	2.4	2.3	2.1	2.1
Consumer prices (end-of-period)	3.1	5.0	3.5	2.5	2.5	2.5	2.5	2.5
Consumer prices (period average)	-0.8	4.5	4.3	3.0	2.5	2.5	2.5	2.5
Investment and saving	(In percent of GDP)							
Gross capital formation	20.4	22.7	21.4	20.4	19.4	18.4	18.4	17.8
Government	1.4	1.4	1.4	1.4	1.4	1.4	1.4	1.4
Nongovernment	19.0	21.3	20.0	18.9	18.0	17.0	17.0	16.4
Gross national savings	-2.1	-2.2	-4.3	-4.8	-5.2	-5.8	-5.3	-5.5
Government	-7.7	-5.9	-9.1	-9.6	-9.9	-10.3	-11.0	-11.6
Nongovernment	5.6	3.6	4.8	4.8	4.6	4.5	5.7	6.1
Central government finances (cash basis)	(In percent of GDP)							
Revenue (including grants)	20.0	22.1	23.0	23.2	23.3	23.1	23.0	23.0
Expenditure	29.2	29.4	33.6	34.2	34.6	34.9	35.4	36.0
Budget balance (including grants)	-9.2	-7.3	-10.6	-11.0	-11.3	-11.7	-12.4	-13.0
Primary balance (including grants)	0.1	2.5	0.2	0.3	0.5	0.4	0.3	0.3
Total government debt	151	153	157	161[4]	165[5]	169	173	178
Monetary sector	(Annual percentage change, unless otherwise indicated)							
Credit to the private sector	6.1	6.1	4.2	5.1	4.7	4.9	5.0	5.0
Reserve money	27.7	32.0	-11.5	2.2	2.2	2.2	2.1	2.0
Broad money[2]	7.6	4.1	5.0	5.0	5.0	5.0	5.0	5.0
Deposit dollarization (level)	65.9	68.8	69.0	69.0	69.0	69.0	69.0	69.0
	Interest rates (period average, in percent)							
Three-month treasury bill yield	4.4	5.1	5.8	6.1	6.2	6.3	6.4	6.4
Five-year treasury bill yield	6.8	7.9	8.2	8.4	8.5	8.6	8.7	n.a.
External sector	(In percent of GDP, unless otherwise indicated)							
Exports of goods and services (in US$, percentage change)	-4.5	2.6	6.0	5.2	5.3	5.2	5.5	5.6
Imports of goods and services (in US$, percentage change)	0.4	6.9	7.0	3.0	3.1	4.0	4.3	4.4
Balance of goods and services	-24.0	-26.3	-27.4	-26.1	-24.9	-24.2	-23.6	-23.0
Current account	-22.5	-25.0	-25.8	-25.2	-24.7	-24.2	-23.7	-23.3
Foreign direct investment	4.0	4.6	4.0	4.0	4.0	4.0	4.0	4.0
Total external debt[3]	190	198	202	205	208	211	213	216
Gross reserves (in billions of U.S. dollars)[4]	40.2	40.6	37.5	33.7	29.9	26.1	22.4	18.6
In months of next year imports of goods and services	14.7	13.9	12.4	10.8	9.2	7.7	6.3	5.1
In percent of short-term external debt[5]	48.0	44.8	38.8	32.7	27.3	22.4	18.1	14.1
In percent of banking system foreign currency deposits	37.6	35.1	30.7	26.2	22.2	18.4	15.0	11.9
In percent of total banking system deposits	24.8	24.1	21.2	18.1	15.3	12.7	10.4	8.2
Memorandum items:								
Nominal GDP (in billions of U.S. dollars)	49.6	51.5	53.6	56.4	59.0	61.9	65.0	68.3
Non-resident deposits (staff estimate, percent change)	11.7	3.8	5.0	5.0	5.0	5.0	5.0	5.0
Commercial bank total assets (percent of GDP)	394	396	394	394	392	389	385	n.a.

[2]Defined as currency in circulation plus resident and nonresident deposits.
[3]Includes nonresident deposits.
[4] Excluding gold and encumbered assets.
[5] Short-term debt on a remaining maturity basis, including short-term nonresident deposits.
Note. Actual figures for 2019 were close to the projected culminating between 160 to 170%. Under COVID-19, this value nearly doubled from its initial estimate in March 2020; by December, the value nearly tripled.
(Source: Hobeika, 2019).

Chapter 7
Experiential Marketing:
A Tool to Maximize the Contribution of Natural History Museums to the Bioeconomy

Hasan Hüseyin Erdoğan
https://orcid.org/0000-0002-9838-0421
Pamukkale University, Turkey

ABSTRACT

Natural history museums are key stakeholders of the bioeconomy. They contribute to the bioeconomy by increasing the bioecological knowledge of the public. This chapter suggests that their contribution can be maximized if they apply the right marketing practices. In this sense, experiential marketing, which aims to provide individual, enjoyable, and engaging educational experiences to customers, seems to be the best approach. Accordingly, the three most visited natural history museums were examined in order to highlight experiential marketing practices that can help to maximize the contribution of natural history museums to the bioeconomy. The findings revealed six marketing practices: holistic experience design, storytelling boosters, sensory interactive elements, participatory interactive elements, co-creation, and online interactive elements. In the last section, detailed practical suggestions on these practices are presented for natural history museum professionals.

INTRODUCTION

The mission of natural history museums is to increase the public's knowledge of the natural world in order to help preserve it (Mclean, 1997). Therefore, they are key stakeholders of the bioeconomy, which seeks to raise public awareness of bioecology conservation (Bugge et al., 2016). Considering that three of the twenty most visited museums are natural history museums (Palicki, 2021), their role is vital to the bioeconomy.

However, not every natural history museum's contribution to the bioeconomy is equal. While some museums are successful at conveying messages about conservation by offering enjoyable and memorable educational experiences to visitors, others fail. So, marketing applications seem to be a significant differentiator among these museums.

DOI: 10.4018/978-1-6684-8879-9.ch007

In general, museums have recognized the importance of the marketing field and have been implementing marketing strategies for years. They expect to raise their income through marketing in order to survive the competition in cultural markets (Cole, 2008). However, if they aim to maximize their contribution to the bioeconomy, they must not only rely on traditional marketing principles but also benefit from the experiential marketing.

Experiential marketing is a more powerful differentiator than traditional marketing (Shaw & Ivens, 2002). It seeks to ensure that every touchpoint between the customer and company is meticulously designed (Lemon & Verhoef, 2016) in order to provide the customer with an individual and memorable experience (Pine & Gilmore, 1998). The customer experience is staged by making use of its sensory, affective, cognitive, social, and symbolic dimensions (Schmitt, 1999). Also, the optimal mix of education, entertainment, escape and esthetics experiences is sought (Pine & Gilmore, 1998).

Museums are considered both formal and informal education institutions (Ambrose & Paine, 2006). Therefore, specifically education experience matters for museums. However, in most cases, the reason why people visit museums is not just education but recreation or social interaction (Falk & Dierking, 2016). So, if the experiential marketing perspective is to be transferred to the museum context, providing the visitor with an individual, memorable, and enjoyable education experience should be the main consideration of museum managers (Kirezli, 2011). Education experience should be enriched with entertainment, escapist, esthetic, social and symbolic experiences. Such visitor-oriented marketing applications help to provide existing visitors with more satisfactory enjoyment and education experiences while attracting more visitors to the museum (Cole, 2008).

Hence, natural history museums employing experiential marketing strategies are expected to enhance existing visitors' knowledge of the natural world and attract more visitors to educate. It raises public awareness of the conservation of the natural world. Thus, they can contribute to the bioeconomy more than museums that do not employ experiential marketing. Hence, they should employ experiential marketing if they want to maximize their contribution to the bioeconomy.

Accordingly, this study aims to offer practical implications about experiential marketing for natural history museum professionals. For this, the marketing practices of the three most visited natural history museums (i.e., Natural History Museum, London (NHM London), American Museum of Natural History (American MNH), Smithsonian National Museum of Natural History (Smithsonian NMNH)) (Palicki, 2021)) were examined.

In the following section, how experiential marketing can enhance natural history museums' contribution to the bioeconomy is discussed. Then, the research methodology and findings are reported. After that, discussion and practical implications are presented. Next, limitations of the present study and future research agenda are argued. Finally, conclusion of the study is presented.

NATURAL HISTORY MUSEUMS AS KEY STAKEHOLDERS OF THE BIOECONOMY

A "participatory governance that engages the general public and key stakeholders in an open and informed dialogue" (McCormick & Kautto, 2013, p. 2589) is required to maintain a powerful and sustainable bioeconomy. Bugge et al.'s (2016) literature review suggests three visions of the bioeconomy: biotechnology, bioresource and bioecology. Of these, the mission of natural history museums is related to the bioecology vision, which "highlights sustainability and ecological processes that optimize the use of energy and nutrients, promote biodiversity, and avoid monocultures and soil degradation" (Bugge et al., 2016, p. 1). A mission statement derived from the mission statements of the three most visited natural history museums in the world would sound like "to discover, interpret, and disseminate—through scientific research and education—knowledge about" and "to promote understanding of the natural world" in order "to tackle the biggest challenges facing the world today" (Table 1). Consequently, it is clear that natural history museums are key stakeholders of the bioeconomy.

EDUCATION EXPERIENCE AT NATURAL HISTORY MUSEUMS

Since education is one of the fundamental practices of museums, many studies have analyzed it (Siu et al., 2022; Huang et al., 2012; Mehmetoglu & Engen, 2011). Although some studies have reported that education experience leads to visitor satisfaction (Siu et al., 2022; Mehmetoglu & Engen, 2011) others have found that information overload at the museum dissatisfies visitors (Huang et al., 2012). Accordingly, some academics argued that education experience should be presented with some fun elements. They researched concepts such as edutainment (Nowacki & Kruczek, 2021; Bideci & Albayrak, 2018; Radder & Han, 2015) and edumotion (Vesci et al., 2021), which express entertaining educational experience. Also, they reported that these experiences result in visitor satisfaction.

Ambrose & Paine (2006) reported that visitors to a natural history museum found specimens exhibited in an area such as their natural habitat more interesting than those exhibited individually. Hence, it seems that a natural history museum could engage and create a more memorable experience for visitors by designing the exhibition in a delightful way and avoiding information overload. In such settings, visitors are expected to be better educated about the natural environment than in museums with traditional exhibition design. Therefore, it is vital for museums to design visitor experiences in order to improve the quality of education and enjoyment experiences for existing visitors. Cole (2008) suggested that this is possible only when the right marketing practices are employed. But the critical question is: what are the right marketing practices?

EXPERIENTIAL MARKETING AT NATURAL HISTORY MUSEUMS

Museums, which once did not attach sufficient importance to marketing, have recently understood the vital role of marketing (Kawashima, 1998) and have begun to apply basic marketing principles. These

Table 1. Mission statements of the most visited natural history museums in the world

Rank	Museum	Mission
1	Natural History Museum, London (https://www.nhm.ac.uk)	To use our unique collections and unrivalled expertise to tackle the biggest challenges facing the world today.
2	American Museum of Natural History (http://www.amnh.org)	To discover, interpret, and disseminate—through scientific research and education—knowledge about human cultures, the natural world, and the universe.
3	Smithsonian National Museum of Natural History (https://naturalhistory.si.edu)	To promote understanding of the natural world and our place in it.

Source: Adapted from Palicki (2021) and the museums' websites stated in the table

principles basically help to raise their revenues and profits (Cole, 2008). However, traditional marketing perspective is no longer enough to satisfy visitors. Therefore, a number of museums have begun to leverage experiential marketing principles although some still adhere to the traditional marketing (Kirezli, 2011). Dirsehan (2012) identifies the former as dynamic museums and the latter static museums. Similarly, Falk & Dierking (2016) classify museums as hands-on and hands-off museums according to their exhibition styles. Accordingly, hands-on museums allow visitors to actively interact with and touch the exhibits whereas hands-off museums expect visitors to act passively and silently during the visit. In this chapter, they are called experiential and static, respectively.

Experiential marketing aims to provide customers with a satisfying experience that is individual and memorable (Pine & Gilmore, 1998). For this, it seeks to activate the sensory, affective, cognitive, social, and symbolic dimensions of the consumption experience (Schmitt, 1999). Also, it seeks to create an optimal mix of education, entertainment, aesthetics, and escape experiences (Pine & Gilmore, 1998). To this end, it is important to design experiences using the tools of theming, multi-sensory elements, memorabilia, and touchpoints supporting the theme (Pine & Gilmore, 1998).

Åstrøm (2018, p. 20) described theming as "a highly context-specific process meant to instill a place with a voice and a personality". The author also identified ten dimensions of theming applied in the tourism industry. These are authenticity, chronotope, cohesion, digital technology, immersion, interaction/co-creation, multi-sensory, novelty, relatability, and storytelling. All these dimensions have been applied to stage museum experiences. For example, Ponsignon et al. (2017) identified immersive and participative touchpoints, social interactions, personalization, variety, cognitive-emotional engagement, architecture, and atmosphere as experiential design characteristics at a museum context. Of these, the authors particularly stressed the importance of immersive and participative touchpoints in staging enjoyable education experiences. Nowacki & Kruczek (2021, p.65) suggested that co-creation at the museum can occur "through participation in organized sightseeing with a guide, or through participation in programs and activities offered by the visitor attractions, such as arts and crafts workshops for children, use of touchscreens, or interaction with animals at the zoo". Kempiak et al. (2017) found that engagement experience, which consists of interactive and "hands on" workshops, educational games, dressing up, kids' zones, re-enactment and living history, and special events, is one the main factors shaping the overall museum experience.

As an experiential museum, Pera (https://www.peramuzesi.org.tr) offers visitors a virtual reality (VR) experience called "a journey into the world of Osman Hamdi Bey". Integrating elements such as digital technology, immersion, interaction, multi-sensory features, novelty, and storytelling, this experience allows visitors to immerse themselves in Osman Hamdi Bey's atelier and his most well-known painting, the Tortoise Trainer. With this interactive experience, Pera provides an engaging and enjoyable education about its most admired art piece, The Tortoise Trainer. The Louvre (https://www.louvre.fr) does the same for Leonardo da Vinci's Mona Lisa with its "Mona Lisa Beyond the Glass" VR experience. Moreover, Lighthouse Artspace enterprise (https://www.lighthouseartspace.com) creates immersive experiences on themes ranging from Tutankhamun to Vincent van Gogh, using cutting-edge projections.

According to Lukas (2012) natural history museums are interpretive environments where natural aspects of the world are displayed. The author argued that these environments have begun to benefit from the design, architecture and theme strategies implemented by theme parks to create a combination of education and entertainment experiences for their visitors. For example, Falk & Dierking (2016) noted that the "Hall of Human Evolution" in NHM London demonstrates a very good example of storytelling practice as it effectively explains human evolution with minimum labels but a successful design.

Moreover, NHM London launched its augmented reality (AR) experience where visitors could watch the award-winning movie "Who Do You Think You Really Are?" and interact with dinosaurs and other extinct creatures (Barry et al., 2012). Barry et al. (2012) reported that most visitors who used the AR experience stated that it was more engaging and interactive than a usual museum experience.

With such interactive and engaging experiences, museums are likely to communicate their corporate message more effectively. Hence, natural history museums using experiential marketing practices are likely to engage visitors more and convey the message of bioecology conservation more effectively. Accordingly, experiential marketing at natural history museums seems to be an important determinant of the contribution to the bioeconomy. Therefore, it can be posited that experiential natural history museums provide a more engaging, entertaining, and satisfying education experience than static museums, which employ classic exhibition methods that provide only simple information. Moreover, they are expected to attract more new visitors than static museums to educate about the natural world. Hence, they are more effective than static museums in raising public awareness of the bioecology. Thus, experiential natural history museums contribute more to the bioeconomy than static museums (Figure 1).

METHODOLOGY

The case study is a preferred research method by business research because it allows to explore "a contemporary phenomenon within some real-life context" (Yin, 2003, p. 1). Accordingly, the author chose the case study as research strategy because the aim of this chapter is to offer practical implications about experiential marketing for museum managers.

Luck et al. (2006) suggested that the case study should be considered as a flexible method that allows researchers to use mixed techniques during data collection and analysis depending on what is being analyzed. The research problem of this study requires highlighting successful experiential marketing applications for natural history museums. Therefore, the author has focused on the three most visited natural history museums in the world and their visitors to collect data. NHM London is the world's most visited natural history museum located in the United Kingdom. American MNH in New York and Smithsonian NMNH in Washington are the second and third most visited museums, respectively and both are located in the United States. (Palicki, 2021). Data were collected by analyzing these museums' official online media such as websites and social media. In addition, their visitors' comments on TripAdvisor, the most well-known review site of the tourism industry, were examined.

The experiential marketing practices of these three museums were carefully analyzed. Especially, the role of these practices in enhancing the effect of the "natural conservation" message that the museums want to convey was examined in depth. The practices were categorized according to their characteristics. As a result, six types of marketing applications were detected that have the potential to create engaging education experiences. In addition, practical implications were presented drawing on these applications.

FINDINGS

The case study identified six types of experiential marketing applications: holistic experience design, storytelling boosters, sensory interactive elements, participatory interactive elements, value co-creation, and online interactive elements (Figure 2).

Figure 1. The effect of experiential natural history museum marketing on the bioeconomy
Source: Author's own work

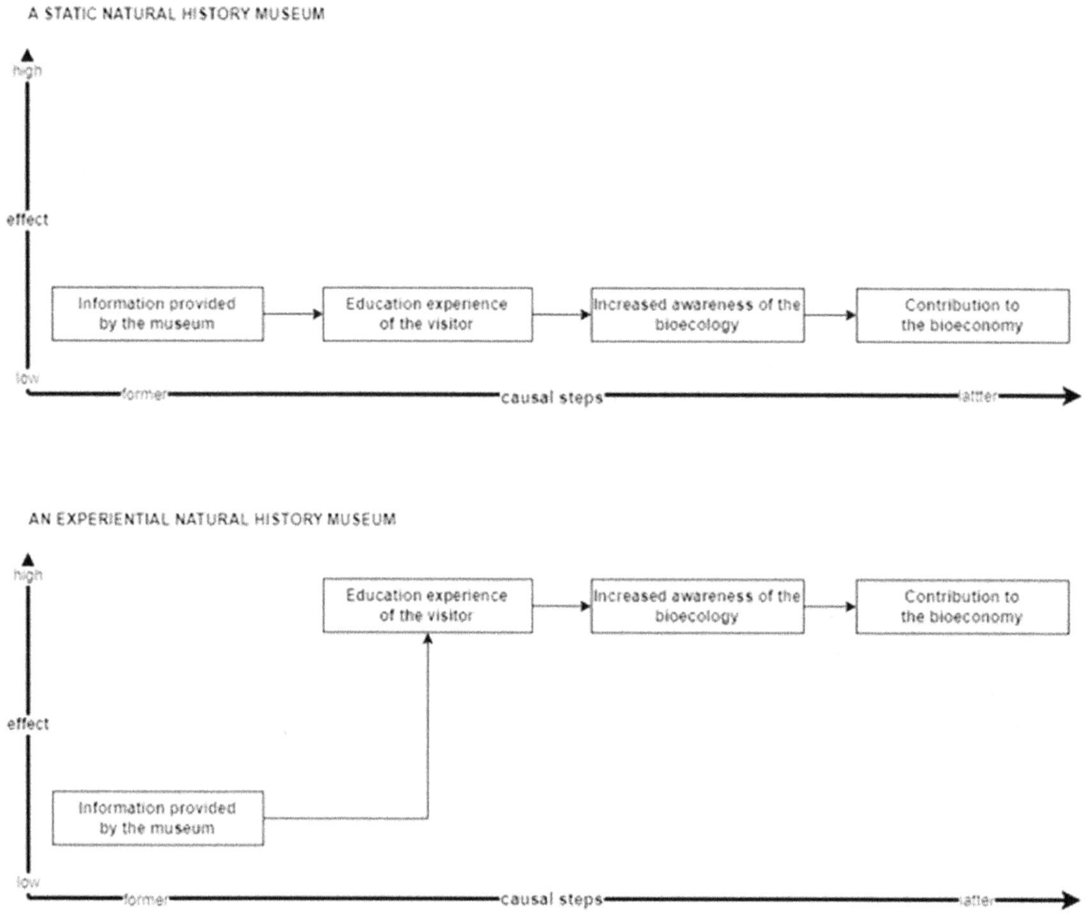

Holistic Experience Design

Previous research suggested that every touchpoint needs to be carefully designed to create a holistic customer experience (Lemon & Verhoef, 2016). Holistic experience design is the practice in which the museum considers every touchpoint to stage a holistic visitation experience. Natural history museums' main objective is to educate the society about the natural world. Therefore, they should provide a holistic visitation experience, which tells a story about the beauty of the natural world. The findings indicate that all the three museums stage a holistic visitation experience that conveys messages about the beauty of the natural world. This is reflected in the visitor reviews on TripAdvisor. Specifically, this has been repeatedly emphasized by visitors to American MNH:

Figure 2. Six marketing practices that emerged from the findings
Source: Derived from the case analysis of this study

"This museum simply has everything. The Hayden Planetarium alone is a full day visit, the walk through the halls of pre-history, dinosaurs, early humans, the evolution and walk through our world is told on an unending and massive storyboard. Simply the best museum of its kind probably anywhere…"

"Considered a gem of an Institution by New Yorkers and tourist alike, is the American Museum of Natural History. It possesses an innate ability to take things we have only seen in books and movies and brings them to life.... a virtual to reality transformer. It boasts jaw-smashing collections from meteorites to minerals, from prehistoric animals to modern alligators and everything 'nature' in between.

I will not forget the gigantic 90-foot-long blue whale (a blue Moby Dick) hanging above the Hall of Ocean Life or the awesome view portrayed in the Hall of Dinosaurs featuring the skeleton of a T Rex and a herd of other dinos. My eyes glittered in the sparkling collection of more than 5000 gems housed in the Hall of Gems and Minerals…"

"What an amazing museum! Sooo much history and done beautifully with obvious consideration of detail in each exhibit and diorama. Each area makes one truly reflect on how amazing our planet is and how precious each aspect of the planet is…"

"…Exhibits were fantastic. Well planned and beautifully executed. Seemed like you were actually there. Has been many years since I went to any museum and am glad to be reintroduced. It's now a new source of inspiration."

"We all loved the exhibits and the way the items are presented and the descriptions. This surely won't be my last visit. I highly recommend a trip to the museum."

"…We prioritized the animal-based exhibits over the peoples and cultures exhibits, just because we knew we couldn't see it all. We started on the second floor with the African Mammals, Asian Mammals, and Birds of the World exhibits. It's basically like a taxidermy zoo, and the animals and accompanying displays are incredibly done. We next viewed the American Mammals and Marine Life exhibits, then moved on to Reptiles and Amphibians, Primates, and finally the spectacular Dinosaur and Primitive Mammals exhibits. The museum spaces are all bright and spaced to move about easily and enjoy the sometimes huge displays…"

"Beautiful museum full of stunning displays of animals. Loved the large halls with the blue whale and the dioramas. So much to see!"

"Fantastic!

Huge and stunningly presented collection of fauna & culture from around the world.

Yes, it's sad to see stuffed animals, but it needs to be recognized that these were collected many years ago.

Dioramas are superb, and great if you loved Night at the Museum!

Dinosaur collection is vast, and has all the favorites.

Wonderful!"

"All the exhibits are well presented and interesting. While the dioramas and dinosaurs are certainly cool, I particularly enjoyed the Hall of Planet Earth and Hayden Planetarium."

"What's more iconic then the AMNH in NYC? Such a beautiful building full of info, history and cool exhibits! Many floors to explore around. Clean and plentiful restrooms."

These comments indicate that the overall layout of American MNH was successfully designed considering every detail, presentation principle and harmony between multiple elements. By making use of dioramas and stuffed animals, the museum has succeeded in creating an atmosphere of natural habitat in the museum. With this, the museum can tell the story of and educate about the natural world successfully. Accordingly, some comments about visitors' education experiences are:

"We have been to natural history museums all over the world. It's a passion of one of my daughters. This is better than most. The gallery about evolution and dinosaurs upstairs is truly educational, fascinating and well organized. Downstairs is the standard taxidermized animal dioramas. But they were recently reimagined, and even those of us who have really seen plenty of stuffed bears could enjoy the staging and artwork."

"This vast museum close to central park is situated in a beautiful, castle-like building. Inside, one will find spacious halls filled with dioramas with stuffed animals from around the world, a large selection of artifacts, clothing, masks etc. representing cultures of the people of Africa, Asia etc. There is a pretty good collection of dinosaurs and prehistoric mammals' skeletons. We saw the skeletons of a saber-toothed tiger, wooly mammoth, T-rex, triceratops, and duckbill dinosaurs to name a few. There is also a collection of gemstones etc. With your ticket you get a chance to see a large screen show. For us we had a choice between the Universe or Oceans. For an extra charge you could see a show about T-rex. There is a planetarium as well. It would definitely take you more than a day to see all of the displays. I suggest you pick up a map, decide on what you are most interested in seeing, and go there. I liked the hall of African and Asian peoples and the dioramas of the animals found on these 2 continents, as I like learning about exotic places. I have also been a dinosaur fan since childhood, so I loved the dinosaur hall. The skeleton of the T rex was the most impressive! I can see how this museum would appeal to families with children. Its very informative and can keep young learners entertained..."

"Very big museum. Very informative. The planetarium exhibit was fantastic. Would recommend to everyone who wants to spend a good day learning something new."

"I was really impressed. Not only by the quality of the items, this museum has it all (from an incredible amount of gems to Lucy, a 32 million-year-old skeleton and the famous skull from Flores), but also by the scale and the presentation. I can see why this museum is such a success with families and (small) children. It is learning without knowing it is...

Some visitors to Smithsonian NMNH also mentioned the holistic visitation experience, often stressing the museum's good layout:

"The museum was so interesting, beautifully laid out, at times interactive & full of knowledge as one would expect."

"This museum is fantastic. It's packed with varied exhibits and it's a great way to spend the day. They have it all from insects to dinosaurs, and everything in between. It's all well maintained and well-curated."

"This museum is absolutely fantastic. It's absolutely packed full of different exhibits, so many that it's almost impossible to see everything in a single day. They have everything from butterflies to dinosaurs, and everything in between. Everything is well-curated..."

A comment from a visitor to Smithsonian NMNH highlights the importance of exhibiting specimens in their natural environment:

"Great to see the updated exhibits...beautifully done and worth the visit, we were in awe of the changes made to the "dinosaurs" by creating a complete environment versus the previous exhibit of individual dinosaurs not in any environment or relationship to the rest of the flora and fauna."

Moreover, many of the reviews on the holistic experience design providing education by Smithsonian NMNH visitors emphasize the role of the modern and up-to-date features and technology:

"This is a great museum in DC, it has so much areas to learn, from the sea world, to the pandemics, mineral and animals. It is so rich of knowledge with so much detail, I love the way they use technology to recreate and explain all information, is so easy to grasp despite your age.

My son and wife enjoy the visit again, we always take away things from the visit, this was our second one."

"I used to come here over the years at different stages of my life... This is truly an amazing museum on so many levels. I still saw displays and artifacts I recognized from Jr. High, but I also saw modern, interactive exhibits that my daughter loved. I also really liked the updated exhibits while also really enjoying the old bones of the building. You can spend one day here or multiple days going slowly through. It is a must see."

"...The halls were also typically well-lit and felt inviting. We didn't manage to see everything in the museum, but everything we did see felt modern and quite up to date...."

On TripAdvisor, the "beautiful building" is one of the popular mentions of NHM London visitors. Many visitors mentioned the museum's building while describing their holistic visitation experience:

"If the collection didn't impress you, the building sure will. This is where you can combine your love for nature, science, history and architecture. This magnificent building that looks like a cathedral/palace, has millions of objects that are million-year-old. From most famous T-Rex, rare animals/birds to some exotic plants whose tiles decorate the museum's ceilings, this is one place that I can never get tired of.

We should all be thankful to the father of dinosaurs Sir Richard Owen who wanted the building large enough to display the skeleton of the largest animal on earth, The Blue Whale."

"Other than the objects displayed, the architecture of a building itself is a fruitful site to visit."

"This is a truly magnificent museum. A must visit when in London. The building alone and whale at the entrance are beautiful."

"Very beautiful old building with amazing statues and decor, enjoyed the open space and clean surroundings."

"The architecture of the building alone is worth the visit, and this is noticed all throughout the museum.

The authenticity of the objects was emphasized by some visitors. Most visitors to American MNH in particular commented on how real the exhibits looked:

"...The exhibitions are excellent in quality. One really thinks the dolphins might be real, as well as the skeletons of the dinosaurs. Which they are not. If only for one hipbone of a huge dinosaur weighs about two tons."

"...There is too much to see in one visit and I think you should take your time but seeing a real life T. rex and the blue whale are awesome..."

NHM London visitors also mentioned the authenticity of the exhibits. However, some visitors commented positively on the originality of the exhibits whereas others commented negatively on the fake appearance of the exhibits:

"We then went to the Human Evolution section, which was really interesting. There were so many real-life skeletons and reconstructions of our distant relatives."

"If you choose to go you should know that the T. rex isn't real, it's animatronic. Most of the whale skeletons are authentic, but dino's are replicas. There is so much more to see and do in London and you can see real dino's in the US."

"A wonderful building, housing a wide range of fascinating objects, but overall in my view there is an overly modern focus on digital displays, animatronic dinosaurs and dramatic framing and staging of certain exhibits."

"...do not enter the dinosaur chamber, unless you want to spend at leats 40 minutes in the queue just to see 2 dino fossils a few plastic models and a plastic model t-rex roaring..."

"Was hoping to see an actual T-Rex skeleton and not an animatronics T-Rex, although it was amusing..."

Storytelling Boosters: Iconic Exhibits, Memorabilia, and Guides

The findings reveal that besides the holistic experience design, there are also other tools to support the storytelling in the museum. These are iconic exhibits, memorabilia, and guides. All three museums have iconic exhibits. The findings show that with iconic exhibits the museums enhance the memorability of the museum visit and the impact of the message they want to convey. NHM London has the blue whale skeleton suspended symmetrically through its historical building. "Blue whale" is one of the popular mentions in the reviews on TripAdvisor. Here are some examples of comments about the blue whale:

"...The building itself is so impressive, and the entrance hall blows you away with its suspended skeleton of a blue whale..."

"The Museum itself was incredibly interesting; the best parts were the massive Blue Whale skeleton and the dinosaur area."

American MNH has two iconic exhibits, specifically the blue whale and Tyrannosaurus rex (T. rex). The museum has focused specifically on the blue whale as a storytelling tool. Oceanographer Sylvia Earle, who made the first scientific studies on blue whales, draws attention to the key role of the American MNH's blue whale in raising awareness about sustainable oceans, in the video posted on the museum's Youtube channel:

"...The ocean is not infinite. It is not too big to fail. It is failing. The big question—how do we create a real awareness?

This museum, the American Museum of Natural History and fellow institutions around the world—it's our collective knowledge, our history.

As a kid, I remember museum experiences that inspired me to want to go see the creatures that I could not see in my backyard, but I knew they existed.

I know that the kids who come through this institution have experiences like that. And years later, they will look back and say, "It's when I stood under the great blue whale that I knew I had to go see one for myself."

I know of no faster way to make people aware and create a feeling of humility and respect and to realize we're a part of it, not apart from it. To put ourselves in perspective—that is so worth doing."

The findings show that American MNH uses the blue whale as a memorable icon to raise awareness. Both blue whale and T. rex are among the popular mentions in the visitor reviews on TripAdvisor. Here are some examples of the reviews about blue whale and T. rex:

"I will not forget the gigantic 90-foot-long blue whale (a blue Moby Dick) hanging above the Hall of Ocean Life or the awesome view portrayed in the Hall of Dinosaurs featuring the skeleton of a T Rex and a herd of other dinos."

"This is a world class museum.....seeing the blue whale is mind boggling and I waited to last to gaze at the T-Rex."

Smithsonian NMNH's iconic exhibit is the Hope Diamond which is also a popular mention on TripAdvisor. Here are some examples of the comments on the Hope Diamond:

"Important site to visit, especially seeing the Hope Diamond and learning about the story of the gem. Fabulous collection of fossils and gems / minerals. One of the best in the world!"

"We had researched this particular Smithsonian in order to see the Hope diamond. Well that was fantastic but there was so much more! The whole family enjoyed learning the history of that diamond and all of the other gemstones and rock formations housed at the museum."

All three museums supply memorabilia associated with their iconic exhibits. A Smithsonian NMNH visitor's review reveals the memorabilia's potential to enhance the memorability of the iconic exhibits:

"...You could spend all days, several days or a week here and not see everything. I think we probably spent 4-5 hours and while we didn't see everything, we were satisfied with the amount we saw. My favorite was seeing the "Hope Diamond" and they also have some nice Hope Diamond replica gifts (i got keychains for my staff and a necklace for my mother in law and a bracelet for my bff)..."

All three museums have gift stores in their museum building. NHM London has three different stores, including a Jurassic Park themed store. American MNH and NHM London have special shop sections on their websites. Smithsonian NMNH's online shopping site is reached through the general Smithsonian Institute's website. American MNH, even, offers a high-quality 360-degree virtual tour of its giftshop on its website. Furthermore, NHM London deliver messages on sustainability and conservation issues, even in activities involving these stores. For example, the museum states on its website:

"In our shops we strive to source products made from organic, recycled or sustainable materials. All sales support the work of the Museum."

"Shop with us and help protect our planet's future - find out how"

"Our values and sustainability: why buy from us?

The Museum's strategy to 2031 is to work towards a future where both people and planet thrive. At the heart of this is sustainability and a mission to create advocates for the planet.

The online shop supports this commitment by offering customers a choice of products that are as sustainably and ethically made as possible. For example:

We only work with companies that enforce CITES, prohibit child labour, modern slavery and testing on animals.

Many of our suppliers are members of Sedex, which works to protect people and the environment in the global supply chain.

Orders sent from the Museum are delivered in boxes made from FSC-certified material, printed with plant-based recycled inks and manufactured in the UK.

All Museum-branded soft toys contain 100% recycled filling.

Since the end of 2022, all our t-shirts are made with organic or recycled cotton and printed with vegetable-based inks.

We know there is a long way to go and are always looking for better solutions..."

It seems that guides also play an important role in creating enjoyable education experiences at the museum. Visitors to all three museums expressed this fact on TripAdvisor. For example, a visitor to Smithsonian NMNH wrote:

"We want to thank our tour guide Amanda. There were only a few of us in the party, which was great. It was truly a private tour. Amanda was extremely knowledgeable. Her anecdotes, in addition to description of the exhibits made this experience very enjoyable and educational. The museum is so huge. It was good to have someone so familiar with the exhibit highlights."

Similarly, some visitors commented on their experience of the "exclusive guided tour" to American MNH:

"Did a private tour with Mark and this was one of the best tours we have ever done. He was so knowledgeable, enthusiastic, and courteous throughout. He made sure all our interests and questions were answered making the whole experience unique and definitely worthwhile. His passion for the museum made us want to see and learn more. I couldn't highly recommend it more!"

"I find that when you go yourself, you miss so much, Jeff was excellent and gave us a great overview and understanding of the museum, particularly how the exhibition was created and the history behind it."

Furthermore, the experience of the "private guided museum tour" to the NHM London was explained by a visitor as follows:

"I am so glad I booked this tour. Becky our tour guide was incredibly engaging and knowledgeable. She led us through the museum from artifact to artifact with fascinating stories and details about the importance and background story of each display. We learned an incredible amount of information about natural history. Becky brought the museum to life and allowed us to experience the collections in a way that we would not have on our own."

Sensory Interactive Elements

Astrøm (2018) stressed the importance of multi-sensory elements to theme tourism experiences. Similarly, sensory experience was identified as a factor determining the overall museum experience (Ober-Heilig et al., 2014; Dirsehan, 2012). Accordingly, findings of this study indicate that sensory experiences are an important determinant of visitor experience in natural history museums as well. Many reviewers mentioned sensory experiences when explaining their overall visitation experience. In this chapter, the tools that museums use to engage visitors' visual and auditory senses are called sensory interactive elements. For example, NHM London's Dinosaurs gallery is a good example of sensory interactive elements. The gallery is themed on the life of dinosaurs, with dinosaur skeletons and fossils and a roaring T. rex. Most visitors on TripAdvisor mentioned this gallery. Here are some examples:

"I visited here with a friend a couple of weeks ago and must say we had a really informative and enjoyable time. There are so many things to look at and read about, definitely allow a good few hours to walk around here... My favourite part however (as I'm sure is the same for everyone) was the dinosaur exhibition which was brilliant..."

"I went with my 4yr old daughter... We headed for the dinosaurs first and went round twice to see the T. rex and then Dippy which was pretty good but I was probably more interested than she was. She liked the moving dinos."

"The dinosaur room was our favourite, especially the real life, moving t-rex."

Moreover, NHM London also has volcano and earthquake galleries. These were also mentioned by visitors. Some examples of comments about these sections are as follows:

"...there were some really breathtaking moments... The third was the incredible escalator through the centre of the planet. Dippy was also brilliant and that exhibition is well worth seeing!

We also loved the animatronic t-rex in the dinosaur section (extremely busy) and the earthquake room! These were really cool interactive experiences and the museum could do with more of these!"

"We then went up to the Volcano and Earthquake section - best part was probably the earthquake simulator."

"The earthquake stimulator was cool, soo much information and videos.

The escalator where you end up in the moon/ egg was really cool, we all enjoyed this."

Smithsonian NMNH visitors especially mentioned the interactive videos in the Hall of Human Origins that appeal to the visual and auditory senses:

"...We also very much liked the section on Human Origins...

There were many interactive and audiovisual elements to complement the items on display, and written information was brief and mostly at a level and height suitable for children to read also."

"The sleeper hit was the Human Origins exhibit - lots of interactive video that explains how we evolved - my 6-year-old spent the most time here (surprisingly!)"

"...We especially enjoyed the interactive exhibit encompassing human evolution. This exhibit alone was a great and quick lesson in our human evolution."

Moreover, Smithsonian NMNH's fossil lab, where scientists at work could be watched, is an interesting sensory interactive practice. This was mentioned in a review as follows:

"We browsed the Fossil Hall quickly (my husband was a bit tired of looking at dinosaurs, but I was interested to see how it compared to the AMNH), and enjoyed the Fossil Lab, where we could observe a scientist at work; there were also signs at various stations describing some of the work that is done in the Fossil Lab..."

American MNH has facilities such as the Hayden Planetarium and LeFrak Theater to stage high-quality multi-sensory experiences. The planetarium currently screens the "Worlds Beyond Earth" exhibit, which the museum promotes on its website as follows:

"Immersive visualizations of distant worlds. Groundbreaking space missions. Breathtaking scenes depicting the evolution of our solar system".

Some American MNH visitors commented positively on the Hayden Planetarium:

"I would recommend the Worlds Beyond Earth planetarium exhibit as an extra, this was very informative."

"The planetarium is world class. The program was outstanding. The museum has wonderful presentations. Their movies regarding volcanoes were very good."

"While visiting NYC for the holidays we spent one day at the Museum of Natural History which includes the famous Hayden Planetarium. With 423 seats this is the largest in the USA. The staff and show were world class. For anyone interested in the stars or science this show about Dark Matter is excellent."

American MNH's LeFrak Theater currently screens the Serengeti movie on its 40-foot-high, 60-foot-wide screen. The promotion of the movie on the museum's website demonstrates the potential of movie screening events to convey the nature conservation message to visitors:

"Experience one of nature's most spectacular events—the annual migration in the Serengeti—on a 40-foot-high, 60-foot-wide screen.

Nestled in northern Tanzania and spanning into the southern edge of Kenya, the vast plains of the Serengeti host one of the largest migrations of mammals on the planet.

More than one million wildebeest traverse this ecosystem every year, encountering lions, zebra, cheetahs, elephants, crocodiles, and many other remarkable animals along the way.

Audiences will experience this amazing journey through the eyes of the youngest animals."

Moreover, American MNH also have some special exhibits with high-quality digital features. For example, some visitors commented positively on the Sharks exhibit:

"Great experience! You could spend multiple days exploring the different exhibits. The gem / rock exhibit was beautiful! Paid for the new shark exhibit and I loved how interactive it was."

"...Our tickets included the Shark exhibition and the Space show in the planetarium, and we had to book separate timeslots for those.

They were both great and the rest of the museum was extremely interesting and informative..."

Participatory Interactive Elements

In this chapter, sensory elements that appeal to the touch sense and enable visitors to move physically are identified as participatory interactive elements. Since these elements include the active participation of the visitor, they are expected to affect the visitors more than other sensory elements.

Also popularly mentioned in Smithsonian NMNH's reviews on TripAdvisor, the butterfly exhibition is a good example of participatory interactive experiences. This exhibition allows the visitor to physically get in touch with real butterflies. Most visitors commented positively about this exhibition. Here are some examples of such reviews:

"Loved this visit to the museum. Highlight of the day was a venture into the Butterfly Pavilion (pre-booked). I was totally blown away with the gorgeous butterflies. Had three land on me at one time. Their beauty was extraordinary. "

"We love this museum and used to go here all the time. There is so much to do, so much to learn, and the perfect place for science lovers. Lots of interactive exhibits for children. Make sure to check out the new Insect Zoo and the Butterfly Garden (where you can walk among a ton of butterflies!). The prehistoric exhibits are wonderful too. I can't even explain it all. Just go!!"

"...There was a live butterfly exhibit which we loved even though I'm not a big fan of insects. Very entertaining and interesting to see and learn about them."

"The Butterfly Pavilion was GREAT. Had two butterflies land on me; it is not very wide but the amount of butterflies and flowers make up for it."

Although temporarily closed at the time of this study, Smithsonian NMNH's the Q?rius initiative also provides another good example of participatory interactive experiences. This initiative allows visitors to have an interactive learning experience by touching original specimens and asking scientists questions. The promotion of the Q?rius on the museum's website is as follows:

"Explore Your World

Q?rius (pronounced "curious") is an interactive and experimental learning space that brings the unique assets of the Smithsonian's National Museum of Natural History — the science, researchers, and collections — out from behind the scenes. Visitors can use microscopes, handle some of the 6,000 collection objects, solve science puzzles, do touchscreen activities, and even meet a scientist. In Q?rius and the Q?rius jr. area, visitors can unleash their curiosity with surprising results."

A visitor's review of Q?rius Jr, the kids' version of Q?rius, demonstrates how interactive this program is:

"Q?rius Jr.: fantastic for the preschool set

I was very impressed with this children's museum-within-a-museum-- they had a wide range of hands-on, active learning materials and spaces for young children to explore in a secure environment... the museum is so kid friendly it's not like it's the only thing to see there. They had a microscope, puzzles, coloring materials, dress-up stuff, and lots of interesting objects all over the walls, with an "if you can reach if, you can touch it" motto. Great place to retreat from the crowds and hang out for a while during a busy day of sightseeing."

At the time of this study, NHM London announced on the home page of its website the interactive exhibition "Titanosaur: Life as the Biggest Dinosaur" that will run from March 2023 to January 2024. The experience described on this announcement provides a good example of an interactive education experience. It aims to stimulate visitors' multiple senses, including smell and touch, to create a strong bond with an extinct creature. Aroused visitors are then subjected to a storytelling about the challenges existing animals face and how they can be protected. The transcription of the announcement of this exhibition is as follows:

"As you weave your way around the gallery, you'll handle specimens, explore interactives and learn how a creature of this colossal size could have survived, and thrived, on Earth. You'll discover how Patagotitan stayed safe, found food and grew to tower head and shoulders above its fellow Cretaceous critters. Get ready to size yourself up against its huge thigh bone, stare into its gigantic sauropod skull and even smell its poo!

After discovering what life was like as the biggest dinosaur, get to know the giants we share our planet with today, learn about their lives, the challenges they face and how we can protect them."

Similarly American MNH announced the "Invisible Worlds" experience, which will be available in the Hayden Planetarium starting Spring of 2023. The introduction of this experience on the museum's Youtube channel shows that it will be an interesting interactive experience that enhances the relationship between human and nature:

"Be transported into nature's hidden realms—from the depths of the ocean to the DNA strands in all living organisms. Invisible Worlds is a new 360-degree immersive experience opening Spring 2023, as part of the Richard Gilder Center for Science, Education, and innovation at the American Museum of Natural History. Housed in a custom venue, Invisible Worlds is the latest presentation innovation, part of the Museum's long tradition of transporting visitors across the world via its iconic habitat dioramas and throughout the universe in its science-visualization-driven Hayden Planetarium Space Shows. Just as the 180-degree planetarium dome creates the experience of looking up at the night sky, Invisible Worlds is designed to evoke the relationship humans have within nature—vast and ever-changing—in

a wide, oval space with 23-foot-high walls and a mirrored ceiling that surrounds visitors with projections at all scales. At key moments, visitors are drawn into the story as their own movements affect the images around them."

Value Co-Creation

Co-creation is a "process, by which products, services, and experiences are developed jointly by companies and their stakeholders, opening up a whole new world of value" (Ramaswamy, 2009, p.11). Accordingly, in this study, value co-creation is identified as the practice of integrating visitors into the museum's value creation process. In this context, NHM London's membership program is a good example of integrating visitors into the museum's mission. Clicking the "Become a Member" section on the homepage of the museum's website, people are confronted by a direct storytelling of belonging to a community that protects the environment:

"... Our scientists are working hard to combat the biggest issues facing the natural world, while our Museum team care for the collection which inspires millions of visitors worldwide to treasure and protect our planet. Together, with your support, we can do even more."

Furthermore, every year, NHM London arranges the "Wildlife Photographer of the Year" (WPY) competition. Anyone from amateur to professional can participate in the competition by submitting photographs on the museum's website. The photographs are expected to highlight the impact of humans on the natural world. Thus, the main purpose of this competition is to raise public awareness for natural conservation. The photographs are exhibited on the website and at the museum. Moreover, they are shared on the WildlifePhotographerOfTheYear (nhm_wpy) Instagram account, which has 230000 followers. The museum awards 100 photographers annually. Additionally, 25 photographs are voted by the public to be selected as the winner of the People's Choice Award. It is a successful example of integrating visitors into co-creation. It allows visitors to create and rate photos. In addition, the exhibition of these photographs significantly affects the visitors. A visitor's comment about the wildlife photographer exhibition shows how this co-created exhibition create satisfaction:

"The Wildlife Photographer exhibition needs to be booked in advance and there is a charge to get in (unlike the museum, which is free). It was mind-blowing, like stepping into another world - the quiet, dark surroundings with the illuminated, otherworldly photos is so special and moving. We just took our time, read every word of every caption and marveled at the absolute miracles of nature captured on film.

Spent about an hour and a half there. Simply marvelous."

Smithsonian NMNH's Do-It-Yourself (DiY) exhibits campaign is another good example of co-creation. With this campaign, the museum allows the public to use some digital materials free of charge to set up their own exhibitions in places such as libraries, hospitals, shopping centers, offices, schools, and universities. The mission of this campaign is to increase public awareness of nature conservation by integrating the public into the co-creation process. This is clearly stated on the museum's website:

"The mission of Do-It-Yourself exhibits is to provide information and tools to engage global communities and help the public understand the natural world and our place in it."

Moreover, Smithsonian NMNH has a temporary exhibition called "Our Places: Connecting People and Nature". With this exhibit the museum engages the public in process of co-creation that produces stories about the connection between people and nature. The exhibit is introduced on the website as follows:

"People + Nature = Inspiration

Nature is all around us, and we are connected to it in countless ways. Our experiences with nature can inspire us to connect, care, and act.

Learn the personal stories of connection and inspiration from scientists and community leaders. Participate in activities exploring your own personal connections. And, in a first for the museum, we have left parts of this exhibit unfinished waiting for audience contribution. By sharing YOUR STORY, you can co-create an unforgettable exhibition experience for everyone!"

"...Nature is all around us, even in the heart of DC. Listen to the sounds of nature and tell us about the animals that live near you.

Life depends on water, and communities across the nation have different relationships with it. What does water mean to you? Leave a note, a poem, or a drawing behind exploring your connection.

Not everyone has equal access to nature and natural spaces. Learn about the barriers that different communities face and plan a park that addresses these issues."

Online Interactive Elements

Online interactive elements are the tools that the museums use to engage visitors virtually. NHM London offers an interactive online platform in partnership with Google Arts & Culture (https://artsandculture.google.com/partner/natural-history-museum). This platform contains 35 online exhibitions with different themes, and many collections. In addition, it includes an interactive experience consisting of 10 different themed-educational episodes voiced by scientists. These episodes have the theme of discovering biological diversity to secure a sustainable future. They move chronologically from 4.6 billion years ago, "the origins of the solar system" to the present, "still discovering".

The platform also allows visitors to take a 360-degree virtual tour. More importantly, on the platform, there is a good example of an interactive VR experience: Rhomaleosaurus: Back to Life in Virtual Reality. This VR application was formed on the idea of bringing an extinct creature back to life in a 3D environment. With this 3D experience, NHM London aims for virtual visitors to establish an emotional bond with the Rhomaleosaurus and to become aware of humanity's responsibility in the extinction of this creature. Thus, the experience concludes with a good example of a storytelling on the negative effects of humanity on the possible extinction of existing animals. The transcription of this experience's storytelling is as follows:

"Welcome to the world-famous Natural History Museum in London. You're now standing in the fossil marine, reptile gallery. While dinosaurs ruled on land these strange creatures dominated the oceans. These are all plesiosaurs. Their sharp teeth and long necks earned them the nickname, sea dragons. But a far more dangerous predator roamed the Jurassic Sea, the Rhomaleosaurus, this extraordinary creature was like nothing that exists today. So, it used its keen eyesight and sensitive snout to track down prey.

The Rhomaleosaurus gradually disappeared from our waters and became extinct over 170 million years ago. Today, sharks often play the same vital role in the ecosystem that Rhomaleosaurus once did. But they're in danger of disappearing much more quickly. We kill an estimated 100 million of them a year and it doesn't end there, human activity affects the whole marine ecosystem. Some scientists believe that species are becoming extinct a thousand times faster because of us. If we don't change our lifestyles and respect our ecosystem one day our grandchildren might only be able to see many of these animals in museums. But if each and every one of us chooses to learn from the past we can help to preserve the amazing diversity of life on our planet for the future."

This experience was also published on the YouTube channel of Google Arts & Culture (https://youtu.be/BH1AvqYXwHQ). Up to this research, the video has been viewed 1.788.951 times and received more than 11000 likes and 143 comments. Some comments on YouTube show that this experience created satisfaction:

"It was amazing, being able to go back in history, see one of the most amazing animals that almost seem impossible to have existed and bringing it to life in just a video is awesome almost as incredible as the animal itself."

"This is a great use of 3d video technology! Incredibly immersive."

Moreover, a viewer's comment supports that this experience has raised public awareness of humanity's negative impact on nature:

"Please be respectful to the nature, human."

In 2009, NHM London established Attenborough Studio, a media facility with digital technologies such as hand-held computers, projection screens and AR to screen interactive movies (Barry et al., 2012). Barry et al. (2012) explained how the museum used this studio and interactive movies to stage engaging education experiences:

"…A key goal of the project was to combine the strengths of traditional film narrative and cutting-edge interactive technology into a new and innovative user experience more powerful and informative than what either medium could achieve on its own. Personalization was essential to engaging an adolescent educational audience.

As the film's theme—an exploration of the tree of life back through time—emerged, bringing the museum's specimens "to life" in the studio environment seemed the most evocative way to allow the audience to travel back to the distant past and see extinct creatures and our human ancestors. To ensure that visitors would experience a real-time 3D experience that was appropriate to their location within the facility, the production team considered numerous technologies, including stereoscopic 3D projections and Pepper's ghost presentations that give the illusion of floating objects, but ultimately deemed AR to be the best fit with the proposed content, existing layout, and budget..."

In the following years, NHM London embarked on creating VR experiences, mostly in partnership with renowned biologist and natural historian David Attenborough. Some examples of these experiences are First Life, Great Barrier Reef Dive, and Hold the World. Sir David Attenborough summarizes the potential of VR to convey messages on the natural world in an article on the museum's website:

"Sharing my passion for the natural world is something which I have done for many years through different technologies, from the days of black-and-white TV to colour, HD, 3D, 4K and now virtual reality.

Hold the World is an extraordinary next step in how we can communicate and educate people about experiences they wouldn't usually have access to in the real world, and I am delighted about what users can learn and discover from the Museum's treasures in this new VR experience.

It really is one of the most convincing and bewitching experiences that the world of technology has yet produced."

American MNH also offers an online exhibition platform in partnership with Google Arts & Culture (https://artsandculture.google.com/partner/natural-history-museum). This platform contains 5 online exhibits and thousands of photos of collections. Also, a 360-degree virtual tour is available. There is also a science-themed online platform "ology" for kids on the museum's website. On this platform, kids can play games, read stories and watch videos on many subjects such as artifact, bacteria, bird, coral reef, dinosaur, DNA, ecosystem, endangered, extinct, fish, fossil, gravity, greenhouse effect, ice, insect, mammals, Mars, myth, neuron, rock, star, and volcano.

Smithsonian NMNH also collaborates with Google Arts & Culture to offer visitors 3 online exhibitions, many collections, and a 360-degree tour (https://artsandculture.google.com/partner/ smithsonian-national-museum-of-natural-history). However, the museum additionally offers another virtual platform on its webpage. The virtual tours on this platform have higher graphical and visual quality than the other two museums reviewed. In addition, permanent and temporary exhibitions, both current and former, can be visited online. Moreover, the virtual tours on this platform offer a variety of options (i.e., stops, fossils, text panels, murals) for visitors to navigate, as well as detailed information and extra images on the exhibits. Some reviewers on TripAdvisor commented in detail about the virtual tour of the museum:

"I would like users to note that this is a review on the virtual tour of the museum.

Firstly the UI part of the website was pretty straightforward and easy to use. The 360 degree images were of good quality and most of the things were visible clearly. Some spots were at an odd angle and looked weird but other than that everything was fine. Navigation was pretty smooth and with good internet one would not face any problems. I liked the fact that there were separate images of each exhibition, which can be viewed by tapping on them. They also provided a map on the top right corner, which helped me navigate incase I lost my way. They also had a VR mode for VR users which I think would be a more immersive experience."

Of course a Virtual Experience comes nowhere near the real thing, but for those who are not able to visit in person due to one reason or another can definitely opt for this virtual tour."

"Today I am going to share my review on a virtual tour of the Smithsonian National Museum of Natural History.

At first, I started it and thought that this is going to be a boring tour because we do not expect much from a virtual tour. At least I was thinking like that but, I was wrong. Yes! you hear me right. This is an amazing virtual tour. During the whole tour, it feels like I am walking in the museum itself. And finally, I just wanted to tell you that from the virtual tour, you can experience the phenomenal and astonishing beauty of our planet Earth. I loved it."

"I went on a virtual trip. The place had a lot to offer. The enormous fossils and the wall cravings left me speechless. If anyone has time constricts or difficulties being present physically, I would recommend a virtual tour of the space.

They offer it on their official page. Try it once to make sure that it is your cup of tea."

"My wonderful Virtual tour of the Smithsonian NMNH

I recently visited the Smithsonian National Museum of Natural History, based in Washington DC from the comfort of my bedroom. How was this made possible? With the technological advancements of humankind and great initiatives, the NMNH has created a 360 virtual tour of their entire museum.

The controls for the tour are remarkably simple. You use your mouse to drag the screen to look anywhere, even behind you. There are arrows to guide you towards the various sections of the museum. There is also a map if you want to go straight to a specific section. There are many artifacts and models so that you can visualize that period.

The high definition of the images is a wonderful feature as you can zoom in on the artifacts and this adds a realistic aspect to it.

The only downside is that there are a few spots where the lighting is bright and thereby blurs out certain written texts.

Overall, it is truly an enchanting experience to have a deep dive into time, whilst sitting on my bed!"

In addition, on this virtual platform there are also narrated virtual tours called the David H. Koch Hall of Fossils – Deep Time, the David H. Koch Hall of Human Origins, the Janet Annenberg Hooker Hall of Geology, Gems, and Minerals, the Objects of Wonder Exhibit, the Sant Ocean Hall". These narrated virtual tours can be accessed from the museum's YouTube channel and other sections of its website. In the introduction video of these narrated tours, the museum's Sant Director, Kirk Johnson, explains how these videos enable visitors to virtually experience the exhibitions. Then, he introduces exhibitions in the videos and concludes with a sentence on natural conservation. The transcription of this introduction video is as follows:

"...In this virtual age, the Museum provides a special opportunity to see and interact with real things. You can examine stardust as old as the Solar System, get nose-to-nose with some of the largest creatures to walk the Earth, and compare fossils of early humans that came and went long ago. These treasures belong to the National Collections-an unsurpassed trove of more than 147 million natural and cultural objects that enables both scientists and the public to better understand our world. The Museum is also an important center for scientific discovery across many fields, and our scientific research is the foundation of our exhibitions.

The diversity of living animals is on parade in our biology halls, including the award-winning Kenneth E. Behring Family Hall of Mammals, where you can get to know more than 60 of your mammal relatives. The Sant Ocean Hall features unparalleled collections, state-of-the-art technology, and dramatic displays that work together to communicate a critical message: the ocean is a global ecosystem essential to all life on our planet. In the Janet Annenberg Hooker Hall of Geology, Gems, and Minerals, home of the Hope Diamond, you will see how tiny atoms came together to create our vast Solar System. You can walk through 6 million years of humans' evolutionary history in the David H. Koch Hall of Human Origins and find out how dramatic climate change drove the evolution of the characteristics that make us human. And the newly opened David H. Koch Hall of Fossils-Deep Time will take you on a journey like no other through the 3.7-billion-year history of life on Earth, as told through interactive exhibits and over 700 extraordinary fossils-including the Nation's T. rex. Along the way, you'll discover how Earth's past is connected to the present, and how our actions are shaping our future..."

On Smithsonian NMNH's webpage, there is also an interactively designed exhibits section. The colorful brochure-like design in this section helps visitors to learn about exhibits in a pleasant way. Moreover, the museum organizes online webinars with a variety of topics about the natural world. However, the weak side of Smithsonian NMNH's online offerings is that all these offerings are scattered across different parts of the museum's web page.

DISCUSSION OF FINDINGS AND PRACTICAL IMPLICATIONS

The findings of this study revealed six types of marketing applications: holistic experience design, storytelling boosters, sensory interactive elements, participatory interactive elements, co-creation, online interactive elements. Of them, the holistic experience design is the most important. In this context, particularly American MNH provides a good example. According to the findings, most American MNH visitors stated that the holistic experience that the museum staged resulted in better knowledge and awareness of the natural world. They mostly noted that the museum designed the entire experience down to the last detail. They especially emphasized that the museum effectively uses dioramas and stuffed animals to reflect the beauties of the natural world. Therefore, natural history museums should build a natural atmosphere by meticulously utilizing dioramas and stuffed animals to provide a satisfying visitor experience.

Using modern elements and technology to tell stories about natural beauty and man's place in nature is another good strategy. For example, most visitors to Smithsonian NMNH mentioned the up-to-date tools that the museum uses when describing their holistic visitation experience in the museum. On the other hand, some visitors to NHM London stated that some exhibits looked too modern and fake. Therefore, natural history museums should also consider the authenticity of the exhibits. While benefitting from the technology, it is critical to maintain the balance between the authenticity and modernity. Moreover, most visitors to NHM London described their holistic visitation experience mentioning the museum's historical building. Thus, a historical building could also add to the quality of the holistic visitation experience.

As the mission of natural history museums is to raise public awareness of the natural world, they aim to tell a variety of stories to draw visitors' attention to natural conservation issues. The findings indicate that museums can leverage iconic exhibits, memorabilia and guides to increase the impact of the storytelling. All the three museums have at least one iconic exhibit. In particular, American MNH, which stage the biggest animal in nature, the giant blue whale suspended in the middle of the main hall of the museum, sets a good example in this context. The museum uses the blue whale at the center of most communication activities and creates a variety of stories about it. In addition, the museum is highly focused on commercializing the memorabilia related to the blue whale and other nature elements on display at the museum, both in the museum and on its website. Thus, the museum can increase the memorability of the messages conveyed during the visit.

Also, findings from all three museums show that private guided tours can play a vital role in communicating the museums' message to visitors. Guides who are well-informed about natural issues and aware of the museum's mission are likely to communicate museums' conservation messages effectively. However, the guided tours in all the three museums are quite costly and are offered by independent tour agencies. Most museums may find hiring a guide costly. However, if they want to capitalize on the remarkable value-added impact of social contact, they need to figure out how to offer guided tours at a reasonable price.

The findings indicate that all three museums stage interactive sensory experiences. NHM London's roaring t-rex is a good application that engages the visual and auditory senses in an interactive way. Kids in particular are expected to be greatly influenced by animated and roaring animals. Thus, natural history museums could create animated and roaring models that are compatible with their specific themes or messages. In this way, they could provide more engaging and memorable experiences. NHM London's other good practices in this context are to create mobile platforms and simulators. Other natural history museums should strive to create such engaging experiences, that align with their mission. They should also examine how Smithsonian NMNH's interactive videos are produced and staged. Additionally, Smithsonian NMNH's working scientist practice set a good example for other museums.

American MNH uses the facilities of a planetarium and a theater very effectively. The museum screens high quality and interactive movies with these facilities. In this way, the museum can better engage visitors and communicate messages more effectively. Therefore, natural history museums should build such facilities if they have enough space. Moreover, American MNH uses digital technology to create special exhibitions. This also could engage visitors more effectively. Therefore, it also sets a good example for other natural history museums.

The author identifies more sensory interactive experiences, which also awaken the senses of touch and smell, as participatory interactive experiences. These experiences are expected to be more influential than the other sensory interactive experiences. A very good example of these experiences is Smithsonian NMNH's butterfly exhibition, which allow visitors to physically contact butterflies in their real-life environment. With this exhibition, the museum offers its visitors an environment to empathize with butterflies more deeply. Thus, this leads visitors to establish a closer connection with the natural environment and to develop a higher awareness of natural conservation.

Providing an environment where visitors can engage with scientists is also a successful practice of Smithsonian NMNH. Museums can strengthen their bond with visitors through such experiences. In this context, other museums can examine Smithsonian NMNH's q?rius practice. Moreover, the findings indicate that the museums seek to offer more interactive experiences that trigger the senses of touch and smell using digital technology. Such experiences can engage visitors more effectively. In this context, the upcoming experiences of NHM London's "Titanosaur: Life as the Biggest Dinosaur" and American MNH's "Invisible Worlds" may serve as good examples for natural history museums.

Co-creation makes the experience more valuable because the value is jointly created. Accordingly, Ramaswamy (2009: 11) suggests that "firms must seek to engage people as active co-creators of value everywhere in the system". The value that natural history museums want to create is about natural conservation. Therefore, natural history museums should integrate the public into the museums' value creation about natural conservation with accurate practices. NHM London's WPY competition sets a good example for other museums in this context. With this competition, the museum integrates the public in co-creation in every process from the creation to voting of the photographs. Besides, the museum displays publicly produced content in the museum and on its social media and website.

Moreover, NHM London uses its membership program as a means of creating a sense of belonging to a nature-conserving community. Thus, it also sets a good example for other museums. Natural history museums must also supply exhibition materials for the public to install self-exhibitions in order to integrate the public in co-creation. For this, they can check out Smithsonian NMNH's DiY practice. Moreover, they can also check out Smithsonian NMNH's "Our Places: Connecting People and Nature" exhibit to integrate visitors in co-creation of exhibition contents.

Natural history museums should also use online platforms effectively to reach visitors without being dependent to the physical place. They should create a single platform which combines various amenities such as 360-degree tours, narrated tours, gaming and video watching platforms. Although American MNH offers all these facilities, they are scattered across different sections on the museum's website. Therefore, other museums can review American MNH's practices in this context. But they should provide a particular platform combining all these amenities for visitors to navigate easily. Also, collaborating with a celebrity to deliver stories on online tools will add to the quality of online experiences. For example, NHM London's collaboration with David Attenborough is a very good example in this context. All these practical suggestions are summarized in the Table 2.

LIMITATIONS AND FUTURE RESEARCH

This study is the first attempt to highlight the importance of implementing successful experiential marketing practices for natural history museums to increase public interest in the conservation of bioecology, one of the visions of bioeconomy. Therefore, the three most visited museums were examined because they are expected to be the most successful in this context. It is likely that the suggestions presented by drawing attention to the successful marketing practices of these museums will offer useful implications

Table 2. Practical suggestions for natural history museum professionals to maximize the museum's contribution to the bioeconomy

Main Suggestions	Detailed Suggestions	Good Practices
Stage a holistic experience	Use dioramas and stuffed animals to create the atmosphere of a natural habitat	American MNH
	Maintain a modern look by utilizing modern-up-to-date elements and technology	Smithsonian NMNH
	Having a historical building, as an historical element itself, would add to the story of the natural history	NHM London
	Construct the authenticity by avoiding over-modern and fake looking of exhibits.	American MNH
Boost storytelling with iconic exhibits, memorabilia, and guides	Create some iconic exhibits to enhance memorability of the visit	American MNH
	Supply memorabilia which help to convey the museum's natural message	American MNH
	Collaborate with or employ guides who could comply with the message that the museum wants to convey	n/a
Create sensory interactive experiences	Construct moving and roaring animals	NHM London (Roaring T. rex)
	Offer moving platforms and simulators	NHM London (Earthquake simulator, escalator)
	Use interactive videos to convey messages effectively	Smithsonian NMNH (Human Origins Hall)
	Stage working scientists along with the exhibits	Smithsonian NMNH (Fossil lab)
	Screen high-quality movies in Planetarium and Theaters	American MNH (Hayden, LeFrak)
	Create high-quality digital exhibits	American MNH (Shark exhibit)
Offer participatory interactive experiences	Bring visitors together with real-life animals	Smithsonian NMNH (Butterfly exhibit)
	Bring visitors together with scientists and let them touch the real specimens and ask questions	Smithsonian NMNH (Q?rius)
	Provide highly interactive and participatory experiences by stimulating touch and smell senses	NHM London (Titanosaur)
Integrate visitors into value co-creation	Integrate visitors as members with a proper storytelling	NHM London
	Initiate natural life-themed photograph competitions	NHM London (WPY)
	Encourage self-exhibitions by supplying material	Smithsonian NMNH (DiY)
	Integrate visitors into the co-creation of exhibition contents	Smithsonian NMNH (Our Places)
Offer interactive online platforms	Create high quality virtual tours and narrated videos with detailed information	American MNH
	When creating online material, collaborate with a famous person who is congruent with the museum's mission	NHM London (David Attenborough)
	Create a science-themed game and video platform for the public	American MNH (Ology)
	Gather all online amenities in one, specific place	n/a

Source: Author's own work

for natural history museums. However, some examples may not be suitable for every museum because museums vary in size and budget. Therefore, future studies can examine small-size and low-budget museums to reveal more suitable practices for them.

CONCLUSION

This chapter aims to highlight the role of the experiential marketing in maximizing the contribution of the natural history museums to the bioeconomy. Some useful implications for natural history museum professionals were reported drawing on the good applications of three most visited natural history museums. By implementing the suggested experiential marketing strategies, natural history museums can improve the quality and effectiveness of their education experiences. Accordingly, they can communicate messages on nature conservation more effectively. This will lead to increased public awareness of the conservation of bioecology, one of the visions of bioeconomy. Thus, their contribution to the bioeconomy will increase.

REFERENCES

Ambrose, T., & Paine, C. (2006). *Museum basics*. Routledge. doi:10.4324/9780203018989

Åstrøm, J. (2018). Exploring theming dimensions in a tourism context. *European Journal of Tourism Research*, *20*, 5–27. doi:10.54055/ejtr.v20i.337

Barry, A., Thomas, G., Debenham, P., & Trout, J. (2012). Augmented reality in a public space: The Natural History Museum, London. *Computer*, *45*(7), 42–47. doi:10.1109/MC.2012.106

Bideci, M., & Albayrak, T. (2018). An investigation of the domestic and foreign tourists' museum visit experiences. *International Journal of Culture, Tourism and Hospitality Research*, *12*(3), 366–377. doi:10.1108/IJCTHR-02-2018-0029

Bugge, M., Hansen, T., & Klitkou, A. (2016). What is the bioeconomy? A review of the literature. *Sustainability (Basel)*, *8*(7), 691. doi:10.3390u8070691

Cole, D. (2008). Museum marketing as a tool for survival and creativity: The Mining Museum Perspective. *Museum Management and Curatorship*, *23*(2), 177–192. doi:10.1080/09647770701865576

Dirsehan, T. (2012). Analyzing museum visitor experiences and post experience dimensions using sem. *Bogazici Journal*, *26*(1), 103–125. doi:10.21773/boun.26.1.6

Falk, J. H., & Dierking, L. D. (2016). *The museum experience*. Whalesback books.

Huang, Y., Liu, C., Lee, C., & Huang, Y. (2012). Designing a Personalized Guide Recommendation System to Mitigate Information Overload in Museum Learning. *Journal of Educational Technology & Society*, *15*(4), 150–166.

Kawashima, N. (1998). Knowing the public. A review of museum marketing literature and research1. *Museum Management and Curatorship*, *17*(1), 21–39. doi:10.1080/09647779800301701

Kempiak, J., Hollywood, L., Bolan, P., & McMahon-Beattie, U. (2017). The Heritage Tourist: An understanding of the visitor experience at Heritage Attractions. *International Journal of Heritage Studies*, *23*(4), 375–392. doi:10.1080/13527258.2016.1277776

Kirezli, Ö. (2011). Museum marketing: Shift from traditional to experiential marketing. *International Journal of Management Cases*, *13*(4), 173–184. doi:10.5848/APBJ.2011.00125

Lemon, K. N., & Verhoef, P. C. (2016). Understanding customer experience throughout the customer journey. *Journal of Marketing*, *80*(6), 69–96. doi:10.1509/jm.15.0420

Luck, L., Jackson, D., & Usher, K. (2006). Case study: A bridge across the paradigms. *Nursing Inquiry*, *13*(2), 103–109. doi:10.1111/j.1440-1800.2006.00309.x PMID:16700753

McCormick, K., & Kautto, N. (2013). The Bioeconomy in Europe: An overview. *Sustainability (Basel)*, *5*(6), 2589–2608. doi:10.3390u5062589

Mclean, F. (1997). *Marketing the museum*. Routledge. doi:10.4324/9780203277744

Mehmetoglu, M., & Engen, M. (2011). Pine and Gilmore's concept of experience economy and its dimensions: An empirical examination in tourism. *Journal of Quality Assurance in Hospitality & Tourism*, *12*(4), 237–255. doi:10.1080/1528008X.2011.541847

Nowacki, M., & Kruczek, Z. (2020). Experience marketing at Polish museums and visitor attractions: The co-creation of visitor experiences, emotions and satisfaction. *Museum Management and Curatorship*, *36*(1), 62–81. doi:10.1080/09647775.2020.1730228

Ober-Heilig, N., Bekmeier-Feuerhahn, S., & Sikkenga, J. (2014). Enhancing museum brands with experiential design to attract low-involvement visitors. *Arts Marketing: An International Journal*, *4*(1/2), 67–86. doi:10.1108/AM-01-2014-0006

Palicki, M. (Ed.). (2021). (rep.). TEA/AECOM 2021 Index and Museum Index: The Global Attractions Attendance Report (pp. 1–89). Themed Entertainment Association (TEA).

Pine, B. J. I. I., & Gilmore, J. H. (1998). Welcome to the Experience Economy. *Harvard Business Review*, *76*(4), 97–105. PMID:10181589

Ponsignon, F., Durrieu, F., & Bouzdine-Chameeva, T. (2017). Customer experience design: A case study in the cultural sector. *Journal of Service Management*, *28*(4), 763–787. doi:10.1108/JOSM-01-2017-0016

Radder, L., & Han, X. (2015). An examination of the museum experience based on pine and Gilmore's experience economy realms. *Journal of Applied Business Research*, *31*(2), 455–470. doi:10.19030/jabr.v31i2.9129

Ramaswamy, V. (2009). Co-Creation of Value: Towards an Expanded Paradigm of Value Creation. *Marketing Review St. Gallen*, *6*(6), 11–17. doi:10.100711621-009-0085-7

Schmitt, B. (1999). Experiential Marketing. *Journal of Marketing Management*, *15*(1-3), 53–67. doi:10.1362/026725799784870496

Shaw, C., & Ivens, J. (2002). *Building Great Customer experiences*. Palgrave Macmillan. doi:10.1057/9780230554719

Siu, N. Y., Zhang, T. J., & Kwan, H. Y. (2021). Reference effects and customer engagement in a museum visit. *International Journal of Contemporary Hospitality Management*, *34*(2), 482–508. doi:10.1108/IJCHM-02-2021-0208

Vesci, M., Conti, E., Rossato, C., & Castellani, P. (2021). The mediating role of visitor satisfaction in the relationship between museum experience and word of mouth: Evidence from Italy. *The TQM Journal, 33*(1), 141–162. doi:10.1108/TQM-02-2020-0022

Yin, R. K. (2003). *Case study research: Design and methods.* SAGE Publications.

KEY TERMS AND DEFINITIONS

Biocology: The vision of bioeconomy that emphasizes the importance of using bioresources properly and avoiding harmful practices such as monocultures and soil degradation that threaten bioresources and sustainability.

Bioeconomy: A type of economy based on biotechnology, bioenergy, and bioecology. It uses bioresources to create economic value. Also, it aims to protect bioecology and biodiversity for a sustainable growth.

Co-creation: The practice of integrating customers into a business's value creation process.

Experiential marketing: A type of marketing that aims to provide an individual, engaging, and enjoyable customer experience. In this type of marketing, the sensory, affective, cognitive, social, and symbolic dimensions of the customer experience are tried to be triggered.

Natural history museum: It is a type of museum that aims to increase the public's knowledge of natural world. The collections of natural history museums often include specimens, objects, artifacts related to animals, plants, minerals, and humans.

Storytelling: A practice used by marketers to enhance meaningfulness and memorability of the customer experience using specific symbols, narratives, and communications.

Theming: A practice used by marketers to add meaningful and memorable soul to a particular space using specific objects, colors, and decorations.

Touchpoint: Any encounter where customers interact with a business or brand. They play a key role in shaping the customer experience.

Chapter 8
Green Knowledge in Green Roofs and Organizational Green Innovation

José G. Vargas-Hernandez
Tecnológico Superior de Jalisco, Mexico

Carlos Alberto A. Rodriguez Maillard
Universidad Cristóbal Colón, Mexico

Omar C. Vargas-González
Tecnológico Nacional de México, Ciudad Guzmán, Mexico

ABSTRACT

This study aims to analyze the implications of green knowledge and technology in organizational green innovation, urban green innovation, and green roofs. Green roofs can be an effective tool for cities to improve the thermal environment, save energy, combat climate change, and are an appropriate method of saving energy. The analysis is supported by the assumption that green technology is basic to organizational green innovation and urban green innovation areas practices, operations, and activities. Making the balance between urban developments and environmental issues, in consideration to sustainable development principle and its innovative green solutions. The methods employed are based on the analytical-reflective and descriptive supported with the review of theoretical and empirical literature. The analysis concludes that green knowledge sharing is relevant to create and develop green technology with positive implications for organizational green innovation, urban green innovation areas and green roofs.

INTRODUCTION

The COVID-19 pandemic is a sanitary crisis that questions many activities to become greener and more sustainable. Sustainable and green are two concepts increasingly used to mean the same, however, sustainability refers to the persistence and indefinite future of necessary and desired characteristics of

DOI: 10.4018/978-1-6684-8879-9.ch008

the human subsystem within and the ecosystem (Hodge, 1997), at the same time, megacities recently are experiencing a shortage of green spaces due to the rapid growth of urbanization in the cities and the increase in the construction of different buildings. That is why it is important to consider that the sustainable urban development is essential since the expansion of city facilities need to be in line with the various aspects (social, economic, and environmental) (Gohari et al., 2023).

Organizational environmental and green knowledge learning is linked to green technology for the environmental protection to stimulate organizational green innovation, urban green innovation areas and green roofs. Sustainable development is related to decision making in the economic and social effects (Wiering et al., 2020). Sustainable development meets the needs of the present without compromising the ability of future generations to meet their own needs (Brundtland, 1987). Urban sustainability including social and economic has been highlighted as one of the leading features of cities (Dwift Consulting, 2017). Some green trends are proposed as smart city, sustainable city, and so forth. Use of green and ecofriendly technology offers more sustainability environment with zero gas emissions to the environment and some other opportunities and challenges.

People living in dense communities tend to use public urban green innovation spaces and public parks more frequently to have more relaxation time and may travel more to the countryside for leisure, experimenting and testing new greener ways of conducting organizations concentrated on environmental sustainability, enable to introduce green innovations (Anderson et al., 2010; Stubbs & Cocklin, 2008). Organizations embracing the concept of saving money, by creating recycling programs and monitoring thermostats, focusing on environmental sustainability may contribute to support nongovernmental agencies involved in environmental troublesome areas.

The frequency of use of a green space living environment support individuals to be satisfied with public spaces and improves the social space and the mental health (Hadavi, 2017). People using public and private urban green innovation spaces attach meanings, identity, and psychological experiences to diverse types of green spaces, as described in the Place identity integrated model and environmental representation (Bernardini & Irvine, 2007). Research on green organization identity has focused on individual level (Chen & Chang, 2013). Green organizational identity supports individual tasks of organizational members related to the organizational environmental activities and strengthens the ability to cope with organizational green oriented conditions. Landscape connectivity has some differences between urban green innovation roofs and urban open space management.

Organizations are facing challenges regarding the compliance with green sustainability strategy. The deployment of a green strategy to face the negative effects that industrialization has on the environment. Organizational business sustainability considers the green growth, green branding, and green sustainable reporting. Branding sustainability enhances the ability of the organization to appeal customers concerned about the environment.

Green sustainable development combined with economic growth, social justice and progress and environmental security concepts are relevant issues in research such as green entrepreneurship in organic farming (Gupta & Vegelin, 2016; Ihnatenko & Novak, 2018; Kucher, 2019; Mohd & Norhidayah, 2016; Savickiene & Miceikiene, 2018; Shevchenko & Petrenko, 2020, Skydan et al., 2020). To make the balance between urban developments and environmental issues, and at the same time to combat climate change and global warming it is important to consider a sustainable development principle and an innovative solution (Gohari et al., 2023).

Organizations engage in green initiatives to develop sustainable competitive advantages and competitiveness (Galdeano-Gómez et al., 2008; López-Gamero et al., 2008; Wysocki, 2021). These sustainable components give organizations a competitive advantage (Bansal, 2005; Barnet, 2007; Chang & Fong, 2010; Namkung & Jang, 20139. Organizational business systems are the foundation for any kind of sustainable business, and they are formulating and implementing green initiatives to attain sustainability and competitiveness. The ecological partnership between organizations and society provides sustainable competitiveness.

The organizational ecological and sustainable procedures follow a strategic orientation (Fernandes & Solimun, 2017). Sustainability had previously been considered an additional cost for organizations, but nowadays it is viewed as a sustainability strategy and tool to derive value and lead to innovation (Van Holt et al., 2020). An organizational shared vision enables development of green behaviors, making meaningful contributions (Aragon-Correa et al., 2008; Renwick et al., 2008).

BACKGROUND

Numerous studies, including Castro & Orellana (2019), deal with green roofs. They look at how adding green roofs could improve the sustainability and inventiveness of cities. According to the authors, green roofs can help create urban green innovation zones, which are intended to encourage sustainable urban growth by fusing green infrastructure and innovation. The paper includes a review of the literature on the advantages of green roofs, including how they can be incorporated into urban green innovation spaces and how they can reduce the effects of urban heat islands, improve stormwater management, and enhance biodiversity.

Green roofs have proven to be energy efficient, environment friendly, and economical in a long run. It has become necessary to implement green roofs as an effective and sustainable method that help in the achievement of urban sustainability (Alqahtany, 2023). The design of a green roof system is a complex procedure that depends on several factors that include social, cultural, climatic, and environmental considerations (Williams et al., 2010).

There are diverse types of green roofs. The green roof system that is an extensive green roof system, The semi-intensive living roof, and it is planted with horizontal cover plants, and finally the intensive green roof, which includes growth medium, plantation, and irrigation and must be implemented above a concrete floor base (Karachaliou et al., 2016). The innovation in the creation of the roofs has been important in the las years. Klemm et al. (2019), investigate the role of green roofs as part of urban green infrastructure and their potential to contribute to sustainable and innovative cities. The authors argue the challenges and barriers to the implementation of green roofs, such as lack of knowledge and expertise, and the need for supportive policies and regulations. The authors conclude that green roofs are a valuable component of urban green infrastructure for innovative cities, and that their full potential should be realized through supportive policies and regulations.

In addition, Hemsley-Brown & Laing (2019) focused their inquiry on potential roadblocks to the adoption of green roofs as a method for sustainable urban development. The lack of understanding and knowledge of green roofs, the high initial installation costs, the absence of encouraging policies and regulations, and the lack of incentives for property owners to install green roofs are just a few of the potential barriers that the authors identified after conducting a literature review. The study also looks at social and cultural variables including aesthetic preferences and notions of the worth of green space that could influence the adoption of green roofs. According to the authors, overcoming these obstacles would necessitate a multifaceted strategy involving cooperation amongst several stakeholders, including policymakers, researchers, practitioners, and the public.

A thorough analysis of the study on the influence of green roofs on the urban heat island effect is also provided by Palomino & Durán (2020). The urban heat island effect—a phenomenon in which urban regions experience hotter temperatures than their rural neighbors because of the built environment and human activity—and its causes are covered by the writers. For policymakers, practitioners, and academics interested in the conception and application of green roofs in urban environments, it emphasizes the potential of green roofs to reduce the urban heat island effect and enhance thermal comfort.

MAIN FOCUS

A thorough overview of the literature on the functions of green roofs in sustainable urban development is provided by Wu & Jim (2021). The study addresses the advantages of green roofs, including how they lower the urban heat island effect, enhance biodiversity, improve stormwater management, and provide social benefits, as well as how these advantages support the broader objectives of sustainable urbanization. The authors conclude that green roofs are a significant part of urban green infrastructure, which is a network of green areas and green technologies that offer numerous social and environmental benefits, and that they have a huge potential to support sustainable urban growth, in the same page Jamei et al. (2023), conclude that green roof installation reduces the heat exchange between outdoor and indoor environments this help to reduce the consumption of building energy. This is why the knowledge, technology, and green innovation it is important to considered to consider the impact of the green roofs.

Green roofs have been a promising alternative to conventional rooftop materials for realigning urbanized areas with the natural environment, in details, a green roof typically consists of a roofing membrane, drainage layer, geotextile filter, growth substrate, and vegetation that is selected for being use in the regional climate, and could help in the stormwater retention, urban heat island mitigation and building energy reduction (Gunn et al., 2022). Understand the importance of the green roofs to alleviate environmental problems in cities and to mitigate the urban heat island effect is important to help to decrease air pollution and enhance urban species biodiversity.

Green Knowledge and Technology

New insights into green organizational and business models are still uncertain and volatile in today's business ecosystem conditions, challenging corporate risk management and decision making. Technological knowledge of economic, social, and environmental benefits of constructing green roofs enhances the good construction of green roofs without any problems of maintaining the roof vegetation. Organizational green knowledge shares issues adopting a multilevel framework to make contributions to green technology. Organizational management must formulate green knowledge and green technology. Knowledge management of environmental issues is a source for learning, solving problems and creating competitiveness (Liao et al., 2004). Green knowledge management is an effective tool for organizational improvement in transformation environment (Raudeli ¯ unien˙ et al., 2018). The dynamic complexity of organizational green innovation requires environmental knowledge learning (Li et al., 2021) and green technology. Organizational environmental and green knowledge learning drives green innovation.

Organizational environmental knowledge learning is linked to green technology for the environmental protection to stimulate green innovation (Redman, 2014).

Knowledge sharing is the basis for knowledge management (Zhang & Sundaresan, 2010) in terms of sustainable development goals such as eliminating pollution, environmental protection, and SDGs, etc. with other organizational members is good (Bock et al, 2005). Green knowledge sharing is the behavior of organizational members who pass information, data and knowledge about green concerns and issues to other members and create new knowledge and learning opportunities to encourage others (Chen & Chang, 2013; Norton et al., 2015). Green Knowledge Sharing (GKS) is the organizational members' behavior, which means to pass on personal information and new knowledge learning opportunities about green issues to other members (Bock et al., 2005). Green property psychological ownership and green knowledge sharing are influenced by green organizational identity. Knowledge sharing of people under pressure of time is reduced (Chang et al., 2020).

Organizational knowledge sharing focuses on creating an atmosphere of environmental protection to create green knowledge sharing related to green issues. Green technology is a socio-environmental healing technology capable to reduce environmental damages and degradation while conserving natural resources. Sustainable green technologies used in processes and applications do not create footprint (Aithal & Shubhrajyotsna, 2016).

Urban green innovation technology can be transferred by public government and confidential business pursuing inclusive green growth and ensuring social integration by eradicating poverty of vulnerable groups and maintaining the footprint of humanity within the ecological boundaries. Building an inclusive and green organization of available and retained talent is an arduous task (Brands & Fernandez-Mateo, 2017). Green roof technology is used to tackle social and environmental concerns to enhance climate resilience. Green plant based green chemistry principles are used for green synthesis, nanomaterials, and nanoparticles (Maas & Hox, 2005). Green and eco-friendly nanotechnology based on green chemistry techniques tends to decrease the risks associated to industrial applications.

Green eco-friendly nanotechnology solutions used in sustainable development goals reduces the technification threat. Environmental degradation and climate pose a threat to global sustainable development and the Sustainable Development Goals. The concept of achieving sustainable development goals by means of using and developing proven green ecofriendly nanotechnology processes further accelerates the spread and growth of many other systems and devices. Green nanotechnology is based in miniature sized communication and computational devices and nano sensor devices, high density, and memory chips. Ensuring ecological communication among the stakeholders disclose the relevance of environmental concerns (Abimbola et al., 2010). Sustainable green nanotechnology principles and processes in the primary sector of the economy is related to the extraction of raw materials from nature.

Long-term effects of human food are required by green manufactured nanomaterials designed according to principles of green nanotechnology complemented with current regulations aimed to address sustainable development of green technology. Manufactured and processed inorganic food has increased the cases of illnesses leading to healthy food safety. Green design and development of health and environment using nanoscale materials is a concern to improve sustainable solutions. Green synthesis methods can be used to take care of adverse effects on nanomaterials on the environment and user health. Complementarity of mixed methods to detect contradictions, improve the results of one method using the results of the other method and enhance the scope and breath for different components of research (Greene et al. 1989).

Green nanotechnology in the service industry is intangible in nature and affected directly and indirectly. Green nanotechnology innovates in educational technology through higher quality and low-cost and ubiquitous online education. ICCT technologies working with green nanotechnology provide intelligent services in artificial intelligence, internet of things, cloud and quantum computing, 3D printing, etc. Green nanotechnology supports ICCT technologies to develop super-intelligent machines and human beings. Green nanotechnology improves the durability, speed, and reachability of digital entertainment instruments (Aithal & Shubhrajyotsna, 2016).

Organizational Green Innovation

Organizational green business innovations follow the principles, values and norms of economic concern, societal parity, and ecological accountability (Alawattage & Fernando, 2017). New green organizational innovations are created either in-house or external. Organizational green innovation activities are stimulated by green contexts-oriented factors. Organizations have attached more relevance to environmental issues based on green organizational identity that enhances organizational green innovation and competitiveness and ensures organizational competitive advantages (Geraie 2015). Organizational identity theory and psychological ownership theory are combined to integrate a conceptual model.

Organizational green process innovation is friendly to the environment through developing processes of manufacturing goods according to their cycle (Kam-Sing 2012). Organizational green innovation and actions are designed to reduce adverse environmental consequences. There is a relationship between organizational green innovation strategy and actions because the green innovation behaviors carried out under a strategy. Motivational mechanisms stimulate innovation to facilitate organizational green innovation. Motivated organizational stakeholders to pursue collective efforts towards green innovative outcomes with their teammates lead with values of organizational green innovation (Norton et al., 2015). According to meta-analytical research, prosocial motivation is a mechanism to link contextual factors and innovative and creative outcomes (Liu et al., 2016).

Organizational green innovation addresses the potential risks and challenges of implementing innovative resources and inputs supported by green efficacy (Liu, et al. 2015). Green organizational innovation is related to individual tasks of organizational environmental activities to strengthen the organization. Society gives relevance to environmental issues and concerns where green organizational identity enhancing organizational creativity and innovation and ensuring sustainable competitive advantages (Geraie, 2015). Social acceptance enables the organization to attain and develop sustainable organizational green competitiveness capability.

The green competitive advantage is the concept describing the conditions to hold a position in sustainable and ecological management and innovation of the firm which cannot be imitated by others (Lin & Chen, 2017). Resources and capabilities that are not imitable provide sustainable competitive advantage (Barney, 1991). The resource-based view of organizational research focuses on the characterization of organizations (Pereira & Vence, 2012) and organizational resources and capabilities (Fong & Chang, 2012; Marchi, 2012) such as organizational redundancy, innovation capabilities, network, embedding, strategy orientation and green innovation strategy as the drivers of green innovation practices. Organizations depend on intangible resources to enhance environmental sustainability (Singh et al., 2020). In a resource intensive sector, a green pioneer organization must have an organizational model (Elkington,1994) with a social and environmental impact.

Human resources green management deals with valuable organizational resources and assets considered based on environmental sustainability. There is evidence that the workforce engages strongly about the environmental sustainability development and become more organizational committed and satisfied. Human resource green management is engaged in organizational environmental sustainability management (Shoeb & Nisar 2015; Wirtenberg et al., 2007). Green dynamic capability is in the nature within the organization processes. The green dynamic capabilities are permanently reliable in the organizations (Chen, 2008).

Urban Green Innovation Spaces and Green Roofs

The rapid growth of the urban population and the low rate of economic development due to the pandemic, are leading to serious civilization problems, among others the lack of urban greener innovations spaces, pollution, and lack of health. Smart growth is supported by sustainable growth based on adoption of greener, efficient technologies, inclusive growth, fostering employment and social cohesion (Gunn & Mintrom, 2016). Organizations must take an approach to stay in inclusive green organizations embracing elements such as equity and social justice needed to create goals, formulate, and implement strategies and develop public commitment (Johnson, 2017). Organizational green innovation refers to the innovation that emphasizes the implementation of organizational green culture, environmental prevention of pollution and waste reduction to enhance organizational sustainability. Urban green innovation areas are increasing in green metropolis driven by the preservation and expansion of sustainable environmental development (Kalandides, 2021).

Urban green innovation spaces must be defined by the limits of construction and protection. The interactions of people and urban green innovation spaces are divided in public and private (Vijayaraghavan, 2016). Public and private urban green innovation spaces co-exist and can be incorporated green roofs as open green spaces aimed to use multi-functionality framed in an effective landscape management. these spaces have differences in perception (Coolen & Meesters 2012) although they should co-exist (Mesimaki, et al. 2017).

People have different perceptions on private and green urban spaces with social identity and environmental sustainability. Green behaviors are defined as the actions and behaviors on which employees engage linked with and contribute to environmental sustainability (Ones & Dilchert, 2012). This concept includes voluntary green behaviors beyond the job tasks referred as organizational citizenship, also considered as task-related environmental behavior (Bissing-Olson et al., 2013; Boiral & Paillé, 2012). The disappearance of green urban areas due to residential and commercial purposes, enhanced by the urban sprawl, leads to loss of habitats, biodiversity, and recreation areas, and decrease of natural resources and elements (Burszta-Adamiak &, Fiałkiewicza, 2019). Urban green spaces with high social and environmental vulnerability are the most urbanized areas with high land temperature might be prioritized for the implementation of urban green innovation areas and green roofs.

Traditional streets can be transformed into green urban connecting areas considered as an open space system and park lands. An urban green innovation belt prevents urban expansion and urban sprawl by merging with nearby settlements. Greenbelt conserves accessibility of open spaces along roadways and connect to a regional trail. Nevertheless, greenbelt was not successful to prevent urban expansion and sprawl but facilitates the access to open spaces (Maruani & Amit-Cohen 2007). Intensification of investments to develop accessible green spaces conceived to cover economic, social and leisure needs, and activities contribute to create the sense of community and respond to the pressures of construction.

Urban green innovation areas are open to enhance urban green innovation space networks and extend the green infrastructure with the functions of green roofs (Fung, 2018). Green infrastructure is the urban physical environment in a network of green open spaces that brings economic, social, and environmental benefits to residents of communities in cities, towns, and villages.

Green infrastructure is a way to create an attractive environment, increase the urban resilience benefits by better managing stormwater, mitigating the urban heat island effect, purifying air. Green roofs are considered passive open green spaces that have environmental benefits and provide open spaces for more urban livability (Mesimaki, et al. 2017). Eco roofs are required for new city infrastructures and facilities. Multi-functional urban green innovation infrastructure should allocate funding for green initiatives of development.

Scholars have applied quantitative and qualitative analysis to address the needs and perceptions of people on green roofs. The environmental effectiveness of green roofs should improve public health and connect the social needs of vulnerable people (Vijayaraghavan 2016). Green roofs and solar and living roofs considered a multi-benefit asset bringing significant economic benefits beyond stormwater management, energy efficiency for cooling, improving air quality, reducing urban heat island. Loss of green spaces intensify heat island effect and storm water runoffs. The development of urban green innovation spaces and roofs are required to comply with regulations on stormwater control. A motivation program for the installation of eco roofs provides incentives to better management of the stormwater.

Green roofs have historically evolved from the hanging gardens in Babylon embracing the aesthetic values and facilitating the human interactions and increase the insulation. Le Corbusier applied green roofs to modern architecture. (Berardi et al., 2014). In Germany followed by France and Switzerland occurred the research and implementation of green roofs, in USA, in Portland in 2005 the authority requires the city-owned buildings to build green roof, in Canada in Toronto the authority passed a bylaw requiring all the new developed buildings meanwhile in Japan in the city of Tokyo, the new constructed buildings must have green roofs (Vijayaraghavan 2016).

Urban green innovation areas should provide spaces for green infrastructure solutions such as the building green roofs, which have beneficial influence on the urban environment. The stage of operation of green infrastructure is the maintenance solutions. The green infrastructure is single function derived from the ability to effectively manage stormwater and focusing on ecological landscape conservation than social and economic benefits (Mell, 2010). Urban green innovation roofs integrated into open space management have multifunctional opportunities to develop green infrastructure and benefits in urban communities and neighborhoods. Private and public green roofs as open spaces a multi-functional green infrastructure and should accommodate the urban open space needs of residents (Meerow & Newell, 2017)

Well-developed urban green innovation infrastructure needs to be supported by a multi-faceted approach. The increasing awareness of green roofs to alleviate global and local environmental impacts. Vegetated green roofs reduce carbon dioxide in the atmosphere naturally sequestered through the photosynthesis processes (Rowe, 2011). It has been confirmed the capacity of green roofs to retain rainwater, reduce air pollutants and improve microclimate that have a positive influence in urban heat island effect (Burszta-Adamiak et al., 2019; Shafi que et al., 2018). There are diverse types of intensive and extensive green roofs which include vegetation with different soil depth and weight, irrigation, plant species and maintenance. Urban green innovation projects include urban gardening, permaculture, guerrilla gardening, smart city and building architecture. Productive urban green innovation combines urban agriculture, allotment gardens and self-sufficiency culture ecologically motivated (Chen et al., 2020).

Mitigation of greenhouse gas emission technique to improve the environmental impact of agriculture needs to consider animal welfare the largest anthropogenic contributor (Kucher et al., 2018; Llonch et al., 2017; Reisinger & Clark, 2018). Wasteland spaces can be converted into a place of productive greenery and developed into natural and green innovation areas that preserve and increase the ecological value and be used also as place for joint creation and learning, (Green, 2020; Nachhaltiges, 2020).

Rehabilitation of buildings using energy optimization should consider the creation of new alternative green spaces transformed into green innovation areas for free recreational purposes. Green roofs provide leisure spaces and enhance aesthetical values of buildings (Sutton, 2014) reduce carbon footprint in urban areas (Ugai, 2016) and direct water footprint (Fialkiewicz et al., 2018). Matching urban places needing urban green innovation areas and green roofs in suitable buildings to determine opportunities to implement in communities and neighborhood. Administrative and technical support are necessary in designing, constructing, developing, and maintaining urban green innovation areas and green roofs. This task recommendation enhances the allocation of urban green innovation areas based on the structural design of buildings and the surrounding environmental characteristics in the community. The design and construction of green roofs need to consider the plant species and the development of small fauna.

The construction and application of green roofs can be achieved in large urban areas are located near another urban constructed areas and should not be single investments. Stovin et al. (2012) calculate that the surface of roofs accounts for almost 50% of the sealed urban areas with the potential for urban green innovation roofs. Green roofs decrease the costs for discharging snowmelt and stormwater. The infrastructure of green storm water initiative reduces greenhouse gas emissions and improves the air quality. The development initiatives of green roofs are linked with storm water management to meet the flood management control required. Local building codes must allow green roofs exempted from the building floor area and related to open spaces. The incentives for zoning to encourage the implementation of the urban green innovation roof should include the development of parking landscapes as one measure adopted in a stormwater management plan that may provide solutions (Murray, 2017).

The development of green roofs value investments as an additional incentive for potential customers. Meeting the ecological needs of customers (Jain & Kaur, 2004). There is a trend of organizations to shift operations towards the direction of environment-friendly processes despite the challenges of customer cynicism regarding organizational green actions (Kumar & Christodoulopoulou, 2014). Co-financing and local legal regulations for the construction of green roofs are most often used incentives. Green development finance has been assigned to Multilateral Development Banks, alongside public-private partnerships (PPPs) and new forms of finance, blended finance, bond instruments, green bonds, social impact bonds and development impact bonds. Urban green innovation roofs are the result of the implementation of motivational tools and other incentives for the different stakeholders.

The existence of types of barriers encourages local authorities to improve the use of incentives for the creation and functioning of urban green innovation areas and urban green innovation roofs and the most commonly use of incentives are the direct ones and legal regulations for the construction of regulations (Mentens et al., 2006). Incentives must be translated from sufficient interest into motivation for the excessively strict requirements for the construction of green roofs such as the minimum surface area, to become eligible for co-financing and tax allowance. Market demand-related incentives such as promotional instruments, are tools to determine demand for buildings with greenery in comparison to traditional. The implementation of financial and non-financial incentives is a crucial factor for the green roofs as a relevant element of urban environment (Brudermann & Sangkakool, 2017). Financial incentives include subsidies or donations granted to reimburse costs of investments that are aimed to support potential investors.

The development of rural and urban green innovation tourism close to nature and based on settlements is becoming more relevant for the impact on green economy and ecology, with the self-employment of residents to provide environmentally accommodation and food to the visitors (Kolomiets & Tomashuk, 2021; Mazur & Tomashuk, 2019; Tomashuk & Baldynyuk, 2021). The green economy development and the pursuit of environmental protection is bringing profits to the organizations (Chen et al., 2006).

STUDIES OF CASE

While green roof application has long been popular in Germany and Norway (Köhler and poll, 2010), There have recently been reports that addressed different advantages of green roofs in tropical areas (hot and humid climatic conditions) such as declining urban heat islands and reducing energy consumption for cooling, now there are currently some studies that are beginning to list the impacts generated in green roofs, some of them are presented below.

Most green roof studies have focused on reducing energy consumption. In developed countries, air conditioning in buildings consumes a large amount of electricity, these studies show that green roofs are an appropriate method of saving energy in hot climates, to improve the energy efficiency of indoor areas, the air quality of indoor buildings, and the air temperature of outdoor buildings and decrease the urban heat island effect, presents technological developments and how to use in urban green innovations spaces.

This topic takes greater relevance in these days, as it presents the challenge that countries must face to ensure that climate change and the heat island effect is increasing, as well as the obstacles that green roofs can confront, while the countries redesigning the laws in this new economic, political, and social moment they are facing.

CONCLUSION

This study analyzes the organizational environmental and green knowledge learning is linked to green technology for the environmental protection to stimulate organizational green innovation, urban green innovation areas and green roofs. Green roof technology can be considered an effective solution for sustainable urban developments in megacities due to its various economic, environmental, and social benefits. However, these benefits are not similarly important for diverse types of buildings, and their relative importance according to the building's types need to be investigated.

Organizations transform to mitigate and neutralize the environmental impact and to adapt environmental sustainability. Green transformation is a multifaceted process incorporating interconnected and overlapping processes posing several managerial challenges. Environmental community involvement promotes and improve the natural environment aiming to sustainable growth of society, inspiring for voluntary participation in social activities. The green government should make and secure access to affordable sustainable housing in urban green innovation spaces and open buildings to create an urban social cohesion process consolidated in urban social movements, The design of an effective green roof system is a complex procedure that depends on several factors of sustainability, including social, cultural, climatic, and environmental considerations.

Table 1. Main articles on green roofs innovations and urban spaces

Article	Author(s)	Main topic
Evaluation of cool roof and vegetations in mitigating urban heat island in a tropical city, Singapore	(Li & Norford, 2016)	"The results reveal that the city-scale deployment of cool roofs can greatly reduce the near-surface air temperature and surface skin temperature during the daytime (especially at noon) but has little effect during nighttime in Singapore."
Green Roof: Its Awareness Among Professionals and Potential in Malaysian Market	(Rahman et al., 2013)	"The awareness of green roof technology among Malaysian professionals in built environment industry and its potential in the country's market. A total of thirty projects of local green roofs has also been identified. There is a growing tendency to deploy green roof technology in Malaysia in the construction discipline, the landscape architect, architect, developer, government officer, urban planner, project executive and academicians."
Using Green Roofs for Social Housing to Improve Energy Consumption in New Cities. (An Applied Study of Social Housing in Egypt's New Cairo City)	(Goda et al., 2023)	"The study shows the improve in energy consumption of cooling and heating energy in social housing in new urban cities in Egypt, reduce high temperatures inside and outside building, and decrease the urban heat island effect, the study suggests a method that is based on modifying the traditional roof of residential construction in Egypt by converting it to a green roof. An integrated analysis model including a strategy shows the environmental and climatic benefits of using green roofs to provide the external outdoor temperature of social housing buildings. The result revealed that green roofs, which are used by 50% of green roofs in traditional buildings, were shown to be efficient in reducing annual energy consumption by 12% and decreasing the average temperature inside buildings by 2.44°C, whereas the average temperature outside was reduced by 3°C."
Green Roofs as an Approach to Enhance Urban Sustainability: A Study of Public Perception in Riyadh, Saudi Arabia	(Alqahtany, 2023)	Most of the population of Riyadh understands the benefits related to green roofs, and they have shown a positive response to a few factors. Almost 94% of people agreed with the fact that green roofs enhance the aesthetics of the building, and the same number of people agreed that they play a role in controlling the air quality. Similarly, 93% of people that green roofs provide additional recreational space in the building. On the other hand, 91% of the respondents identified the climate of the area as the biggest challenge in implementing green roofs in the buildings
The use of green roofs in urban green innovation spaces: A case study in Barcelona.	(Nogue & Salas, 2018)	offers a broader perspective on the motivations, advantages, and challenges of implementing green roofs in urban green innovation spaces, The study investigates the reasons for installing green roofs in urban green innovation areas, the advantages, and difficulties of doing so, and the part stakeholders play in the planning and maintenance of green roofs.
How to design green roofs to enhance urban green innovation space? Evidence from Beijing	(Gong et al., 2020)	provides a more in-depth analysis of the design features and effectiveness of green roofs. who also examine their design elements and efficiency in increasing urban green innovation spaces. They offer suggestions on how to design and deploy green roofs to support sustainable urban development and innovation for policymakers and practitioners.
The application of green roofs in urban green innovation spaces: A case study in Jinan City, China	(Li et al., 2021)	A case study on the use of green roofs in urban green innovation areas in Jinan City, China provides a more detailed analysis of the design and implementation of green roofs in specific urban green innovation spaces. They evaluate the effectiveness of green roofs in providing a variety of environmental and social benefits, including lowering the urban heat island effect, enhancing air quality, and supplying space for community activities
Green roof innovation in Singapore: Policies, performance, and potential	(Tan et al., 2019)	a case study of green roof innovation in Singapore offers a broader perspective on the policies and regulations that have facilitated the implementation of green roofs in an entire city look at the laws and policies that have made it easier for green roofs to be installed in Singapore, how well they work at offering a variety of environmental and social advantages, and what obstacles stand in the way of their use. They also offer suggestions on how to design and execute green roofs in Singapore and other urban contexts for policymakers and practitioners.

Source: Own elaboration based on references included.

Organizational networks are vital to connecting each other in the space of a green movement. The environmental movements along the nongovernment organizations and environmental foundations lack of racial diversity despite the efforts to increase diversity and inclusion. Residents also express their intentions to be involved in discussions and debates to develop a green urban framework. It is recommended that the local authorities initiate strong campaigns in favor of green technologies at least for the new buildings or the buildings under construction. They must educate people about the benefits these green roofs can provide. The authorities must facilitate the residents and house owners in designing, executing, and installing such systems.

Adding green components to technology can become sustainable green technologies able to avoid degradation and provide clean environment. Green roof technology is used to tackle social and environmental concerns to enhance climate resilience. Green roof is related to be linked to resilience, restorative, and accessible. Green nanotechnology represents opportunities and challenges in industry to encourage growth by supporting nanotechnology usage. Green roof technology has been recommended as an effective solution for the growth of green spaces per capita and improving sustainability means of urban development's due to its diverse advantages.

Gain competitive advantage through the implementation of a sustainable organizational strategy is relevant for the survival. The organizational green shared vision should involve sustainable business goals. Long-term organizational green innovation strategy must stimulate followers to implement green operations. Organizational socio ecological strategies are formulated an implemented to achieve green competitiveness. Despite of significant benefits, there are other factors including high construction costs and maintenance issues, which are required to be considered for future applications globally.

REFERENCES

Abimbola, T., Lim, M., Hillestad, T., Xie, C., & Haugland, S. A. (2010). Innovative corporate social responsibility: The founder's role in creating a trustworthy corporate brand through "green innovation.". *Journal of Product and Brand Management*, *19*(6), 440–451. doi:10.1108/10610421011085758

Aithal, P., & Shubhrajyotsna, A. (2016). Nanotechnology Innovations and Commercialization – Opportunities, Challenges & Reasons for Delay. In *Proceedings of National Conference on Changing Perspectives of Management, IT, and Social Sciences in Contemporary Environment*, (pp. 1-12). IJEM. 10.5815/ijem.2016.06.02

Aithal, P., & Shubhrajyotsna, A. (2018). Nanotechnology based Innovations and Human Life Comfortability –Are we Marching towards Immortality? *International Journal of Applied Engineering and Management Letters*, *2*(1), 1–8. doi:10.5281/zenodo.1485048

Alawattage, C., & Fernando, S. (2017). Postcoloniality in corporate social and environmental accountability. *Accounting, Organizations and Society*, *60*, 1–20. doi:10.1016/j.aos.2017.07.002

Alqahtany, A. (2022). Green Roofs as an Approach to Enhance Urban Sustainability: A Study of Public Perception in Riyadh, Saudi Arabia. *Buildings*, *12*(12), 2202. doi:10.3390/buildings12122202

Aragon-Correa, J. A., Hurtado-Torres, N. E., Sharma, S. & García-Morales, V. J. (2008). Environmental strategy and performance in small firms: A resource-based perspective. *Journal of Environmental Management*, *86*(1), 88–103. doi:10.1016/j.jenvman.2006.12.032

Bansal, P. (2005). Evolving sustainably: A longitudinal study of corporate sustainable development. *Strategic Management Journal*, *26*(3), 197–218. doi:10.1002mj.441

Barnet, M. L. (2007). Stakeholder influence capacity and the variability of financial returns to corporate social responsibility. *Academy of Management Review*, *33*(3), 794–816. doi:10.5465/amr.2007.25275520

Barney, J. (1991). Firm Resources and Sustained Competitive Advantage. *Journal of Management*, *17*(1), 99–120. doi:10.1177/014920639101700108

Berardi, U., GhaffarianHoseini, A. H., & GhaffarianHoseini, A. (2014). State-of-the-art analysis of the environmental benefits of green roofs. *Applied Energy*, *115*, 411–428. doi:10.1016/j.apenergy.2013.10.047

Berlin government.. (2020). *Capital of Green Trends: How Berlin leads the Way in Urban Sustainability*. Visit Berlin. https://about.visitberlin.de/en/green berlin

Berlin, N. (2020). *Stadt der grünen Trends-Wie Berlin den Weg der urbanen Nachhaltigkeit geht.* Visit Berlin. https://about.visitberlin.de/nachhaltiges-berlin

Bernardini, C. & K. N. Irvine, K. N. (2007). The 'nature' of urban sustainability: Private or public green spaces. *Sustainable Development and Planning*, *2*, 661-674. doi:10.2495/SDP070642

Bissing-Olson, M. J., Iyer, A., Fielding, K. S., & Zacher, H. (2013). Relationships between daily affect and pro-environmental behavior at work: The moderating role of pro-environmental attitude. *Journal of Organizational Behavior*, *175*(2), 156–175. doi:10.1002/job.1788

Bock, G. W., Zmud, R. W., Kim, Y. G., & Lee, J. N. (2005). Behavioral intention formation in knowledge sharing: Examining the roles of extrinsic motivators, social-psychological forces, and organizational climate. *Management Information Systems Quarterly*, *29*(1), 87–112. doi:10.2307/25148669

Boiral, O., & Paillé, P. (2012). Organizational citizenship behaviour for the environment: Measurement and validation. *Journal of Business Ethics*, *109*(4), 431–445. doi:10.100710551-011-1138-9

Brands, R. A., & Fernandez-Mateo, I. (2017). Leaning out: How negative recruitment experiences shape women's decisions to compete for executive roles. *Administrative Science Quarterly*, *62*(3), 405–442. doi:10.1177/0001839216682728

Brudermann, T., & Sangkakool, T. (2017). Green roofs in temperate climate cities in Europe - An analysis of key decision factors. *Urban Forestry & Urban Greening*, *21*, 224–234. doi:10.1016/j.ufug.2016.12.008

Brundtland, G. (1987). *Our Common Future: The World Commission on Environment and Development*. Oxford University Press.

Burszta-Adamiak, E., & Fiałkiewicza, W. (2019). Review of green roof incentives as motivators for the expansion of green infrastructure in European cities. *Scientific Review – Engineering and Environmental Sciences*, *28* (4), 641–652.

Burszta-Adamiak, E., Stańczyk, J., & Łomotowski, J. (2019). Hydrological performance of green roofs in the context of the meteorological factors during the 5-year monitoring period. *Water and Environment Journal : the Journal / the Chartered Institution of Water and Environmental Management*, *33*(1), 144–154. doi:10.1111/wej.12385

Castro, A. J. L., & Orellana, F. (2019). Green roofs as a strategy for urban green innovation spaces. *Sustainable Cities and Society*, *44*, 697–705. doi:10.1016/j.scs.2018.10.034

Chang, N. J., & Fong, C. M. (2010). Green product quality, green corporate image, green customer satisfaction, and green customer loyalty. *African Journal of Business Management, 4*(13), 2336–2344.

Chang, T. W., Chen, Y. S., Yeh, Y. L., & Li, H. X. (2020). Sustainable consumption models for customers: Investigating the significant antecedents of green purchase behavior from the perspective of information asymmetry. *Journal of Environmental Planning and Management, 64*(9), 1668–1688. doi:10.1080/09 640568.2020.1837087

Chen, Y. S. (2008). The driver of green innovation and green image–green core competence. *Journal of Business Ethics, 81*(3), 531–543. doi:10.100710551-007-9522-1

Chen, Y. S., & Chang, C. H. (2013). The determinants of green product development performance: Green dynamic capabilities, green transformational leadership, and green creativity. *Journal of Business Ethics, 116*(1), 107–119. doi:10.100710551-012-1452-x

Chen, Y. S., Lai, S. B., & Wen, C. T. (2006). The influence of green innovation performance on corporate advantage in Taiwan. *Journal of Business Ethics, 67*(4), 331–339. doi:10.100710551-006-9025-5

Chen, Y. S., Lin, S. H., Lin, C. Y., Hung, S. T., Chang, C. W., & Huang, C. W. (2020). Improving green product development performance from green vision and organizational culture perspectives. *Corporate Social Responsibility and Environmental Management, 27*(1), 222–231. doi:10.1002/csr.1794

Consulting, D. (2017) *12 mal Berliner Leben, 12 mal Berlin Erleben. Konzept für einen stadtverträglichen Berlin-Tourismus 2018+;* VisitBerlin https://about.visitberlin.de/sites/default/files/2018 02/Tou rismuskonzept_Berlin_Studie_2017.pdf

Coolen, H., & Meesters, J. (2012). Private and public green spaces: Meaningful but different settings. *Journal of Housing and the Built Environment, 27*(1), 49–67. doi:10.100710901-011-9246-5

Elkington, J. (1994). Towards the sustainable corporation: Win-win-win business strategies for sustainable development. *Cal. Management Review, 36* (3), 90e100.

Fernandes, A. A. R., & Solimun. (2017). The mediating effect of strategic orientation and innovations on the effect of environmental uncertainties on the performance of business in the Indonesian aviation industry. *International Journal of Law and Management, 59*(6), 1269–1278. doi:10.1108/IJLMA-10-2016-0087

Fialkiewicz, W., Burszta-Adamiak, E., Kolonko-Wiercik, A., Manzardo, A., Loss, A., Mikovits, C., & Scipioni, A. (2018). Simplified direct water footprint model to support urban water management. *Water (Basel), 10*(5), 630. doi:10.3390/w10050630

Fong, C. M., & Chang, N. J. (2012). The impact of green learning orientation on proactive environmental innovation capability and firm performance. *African Journal of Business Management, 6*(32), 727–735.

Fung, K. L. (2018). *Expanding the green network on rooftops: A study of integrating green roofs as a part of urban green innovation space planning* (Order No. 10932095). ProQuest One Academic. (2124999351). http://wdg.biblio.udg.mx:2048/login?url=https://www.proquest .com/dissertations-theses/expanding-green-network-on-rooftop s-study/docview/2124999351/se-2?accountid=28915

Galdeano-Gómez, E., Céspedes-Lorente, J., & Martínez-del-Río, J. (2008). Environmental performance and spillover effects on productivity: Evidence from horticultural firms. *Journal of Environmental Management*, *88*(4), 1552–1561. doi:10.1016/j.jenvman.2007.07.028 PMID:17825476

Geraie, M. S., & Rad, F. M. (2015). Mediator role of the organizational identity green in relationship between total quality management and perceived innovation with sustainable competitive advantage. *International Journal of Biology, Pharmacy and Allied Sciences*, *4*, 266–276.

Goda, Z., Foda, M. & Elsayyad, N. (2023). Using Green Roofs for Social Housing to Improve Energy Consumption in New Cities. (An Applied Study of Social Housing in Egypt's New Cairo City). *Future Cities and Environment*, *9*(1), 2, 1–11. DOI: doi:10.5334/fce.165

Gohari, A., Gohari, A., & Bin, A. (2023). Importance of green roof criteria for residential and governmental buildings: A multi-criteria decision analysis. *Environmental Science and Pollution Research International*, *30*(2), 3707–3725. doi:10.100711356-022-22472-4 PMID:35953748

Gong, Y., Li, X., & Li, Z. (2020). How to design green roofs to enhance urban green innovation space? Evidence from Beijing. *Habitat International*, *107*, 102–113. doi:10.1016/j.habitatint.2020.102113

Greene, J., Caracelli, V., & Graham, W. (1989). Toward a conceptual framework for mixed-method evaluation designs. *Educational Evaluation and Policy Analysis*, *11*(3), 255–274. doi:10.3102/01623737011003255

Gunn, A., & Mintrom, M. (2016). Higher Education Policy Change in Europe: Academic Research Funding and the Impact Agenda. *European Education*, *48*(4), 241–257. doi:10.1080/10564934.2016.1237703

Gunn, P., Burak, H., Van Geel, P., & Baldwin, C. (2022). Development of resistor-capacitor and finite difference models to evaluate green roof thermal performance. *Building and Environment*, *226*, 1–15. doi:10.1016/j.buildenv.2022.109700

Gupta, J., & Vegelin, C. (2016). Sustainable development goals and inclusive development. *International Environmental Agreement: Politics, Law and Economics*, *16*(3), 433–448. doi:10.100710784-016-9323-z

Hadavi, S. (2017). Direct and indirect effects of the physical aspects of the environment on mental well-being. *Environment and Behavior*, *49*(10), 1071–1104. doi:10.1177/0013916516679876

Hemsley-Brown, T., & Laing, A. (2019). Green roofs and sustainable urban development: An exploratory review of potential barriers to adoption. *Sustainable Cities and Society*, *50*, 101648. doi:10.1016/j.scs.2019.101648

Hodge, T. (1997). Toward a conceptual framework for assessing progress toward sustainability. *Social Indicators Research*, *40*(1/2), 5–98. doi:10.1023/A:1006847209030

Huang, J., Kong, F., Yin, H., Middel, A., Liu, H., & Meadows, M. E. (2023). Green roof effects on urban building surface processes and energy budgets. *Energy Conversion and Management*, *287*, 1–13. doi:10.1016/j.enconman.2023.117100

Ihnatenko, M., & Novak, N. (2018). Development of regional programs for the development of agrarian enterprises with organic production based on the European and international experience. *Baltic Journal of Economic Studies*, *4*(4), 126–133. doi:10.30525/2256-0742/2018-4-4-126-133

Jain, S. K., & Kaur, G. (2004). Green marketing: An Indian perspective. *Decision (Washington, D.C.), 31*(2), 161–209.

Jamei, E., Chau, H., Seyedmahmoudian, M., Mekhilef, S., & Alzahraa, F. (2023). Green roof and energy – role of climate and design elements in hot and temperate climates. *Heliyon, 9*(5), 1–16. doi:10.1016/j.heliyon.2023.e15917 PMID:37215798

Johnson, S. K. (2017, August 17th). What 11 CEOs have learned about championing diversity. *Harvard Business Review.*

Johnson, S. K. (2019) *Leaking Talent How People of Color are Pushed Out of Environmental Organizations How People of Color are Pushed Out of Environmental Organizations.* Diverse Green. diversegreen.org/research/leaking-talent/

Kalandides, A., & Grésillon, B. (2021). The Ambiguities of "Sustainable" Berlin. *Sustainability (Basel), 2021*(13), 1666. doi:10.3390u13041666

Kam-Sing Wong, S. (2012). The influence of green product competitiveness on the success of green product innovation: Empirical evidence from the Chinese electrical and electronics industry. *European Journal of Innovation Management, 15*(4), 468–490. doi:10.1108/14601061211272385

Karachaliou, P., Santamouris, M., & Pangalou, H. (2016). Experimental and numerical analysis of the energy performance of a large scale intensive green roof system installed on an office building in Athens. *Energy and Buildings, 114*, 256–264. https://doi.org/. enbuild.2015.04.055 doi:10.1016/j

Klemm, W., Zinngrebe, Y., & Heidenreich, S. (2019). Green roofs as part of urban green infrastructure for innovative cities. *Environmental Innovation and Societal Transitions, 31*, 50–56. doi:10.1016/j.eist.2018.11.003

Köhler, M., & Poll, P. H. (2010). Long-term performance of selected old Berlin greenroofs in comparison to younger extensive green roofs in Berlin. *Ecological Engineering, 36*(5), 722–729. doi:10.1016/j.ecoleng.2009.12.019

Kolomiets, T. V., & Tomashuk, I. V. (2021). Entrepreneurship and development of rural areas in Ukraine. *Colloquium-journal, 9*(96), 29–42.

KucherA. (2019). *Sustainable soil management in the formation of competitiveness of agricultural enterprises.* Academic Publishing House «Talent». https://doi.org/ doi:10.13140/RG.2.2.19554.07366

KucherL.HeldakM.OrlenkoA. (2018). Project management in organic agricultural production. *Agricultural and Resource Economics, 4*(3), 104–128. doi:10.22004/ag.econ.281753

Kumar, V., & Christodoulopoulou, A. (2014). Sustainability and branding: An integrated perspective. *Industrial Marketing Management, 43*(1), 6–15. doi:10.1016/j.indmarman.2013.06.008

Li, J., Yin, Z., Li, Y., Sun, S., Li, M., & Li, Y. (2021). The application of green roofs in urban green innovation spaces: A case study in Jinan City, China. *Environmental Science and Pollution Research International, 28*(3), 3525–3537. doi:10.100711356-020-11544-2 PMID:32892283

Li, X., & Norford, L. (2016). Evaluation of cool roof and vegetations in mitigating urban heat island in a tropical city, Singapore. *Environmental Science and Pollution Research*, 30, 3707–3725. *Urban Climate*, *16*, 59–74. doi:10.1016/j.uclim.2015.12.002

Liao, S. H., Chang, J. C., Cheng, S. C., & Kuo, C. M. (2004). Employee relationship and knowledge sharing: A case study of a Taiwanese finance and securities firm. *Knowledge Management Research and Practice*, *2*(1), 24–34. doi:10.1057/palgrave.kmrp.8500016

Lin, Y. H., & Chen, Y. S. (2017). Determinants of green competitive advantage: The roles of green knowledge sharing, green dynamic capabilities, and green service innovation. *Quality & Quantity*, *51*(4), 1663–1685. doi:10.100711135-016-0358-6

Liu, D., Jiang, K., Shalley, C. E., Keem, S., & Zhou, J. (2016). Motivational mechanisms of employee creativity: A meta-analytic examination and theoretical extension of the creativity literature. *Organizational Behavior and Human Decision Processes*, *137*, 236–263. doi:10.1016/j.obhdp.2016.08.001

Liu, J., Chen, J., & Tao, Y. (2015). Innovation Performance in New Product Development Teams in China's Technology Ventures: The Role of Behavioral Integration Dimensions and Collective Efficacy. *Journal of Product Innovation Management*, *32*(1), 29–44. doi:10.1111/jpim.12177

Llonch, P., Haskell, M. J., Dewhurst, R. J., & Turner, S. P. (2017). Current available strategies to mitigate greenhouse gas emissions in livestock systems: An animal welfare perspective. *Animal*, *11*(2), 274–284. doi:10.1017/S1751731116001440 PMID:27406001

López-Gamero, M. D., Claver-Cortés, E., & Molina-Azorín, J. F. (2008). Complementary resources and capabilities for an ethical and environmental management: A qual/quan study. *Journal of Business Ethics*, *82*(3), 701–732. doi:10.100710551-007-9587-x

Maas, C. J., & Hox, J. J. (2005). Sufficient sample sizes for multilevel modeling. *Methodology: European Journal of Research Methods for the Behavioral and Social Sciences*, *1*(3), 86–92. doi:10.1027/1614-2241.1.3.86

Marchi, V. D. (2012). Environmental innovation and R&D cooperation: Empirical evidence from Spanish manufacturing firms. *Research Policy*, *41*(3), 614–623. doi:10.1016/j.respol.2011.10.002

Maruani, T., & Amit-Cohen, I. (2007). Open Space planning models: A review of approaches and methods. *Landscape and Urban Planning*, *81*(1–2), 1–13. doi:10.1016/j.landurbplan.2007.01.003

Mazur, K. V., & Tomashuk, I. V. (2019). Governance, and regulation as an indispensable condition for developing the potential of rural areas. *Baltic Journal of Economic Studies*, *5*(5), 67–78. doi:10.30525/2256-0742/2019-5-5-67-78

Meerow, S., & Newell, J. P. (2017). Detroit, Spatial planning for multifunctional green infrastructure: Growing resilience in. *Landscape and Urban Planning*, *72*, 62–75. doi:10.1016/j.landurbplan.2016.10.005

Mell, I. C. (2010). *Green infrastructure: concepts, perceptions, and its use in spatial planning*. [Doctoral Thesis, School of Architecture, Planning and Landscape Newcastle University].

Mentens, J., Raes, D., & Hermy, M. (2006). Green roofs as a tool for solving the rainwater runoff problem in the urbanized 21st century? *Landscape and Urban Planning*, *77*(3), 217–226. doi:10.1016/j.landurbplan.2005.02.010

Mesimaki, M., Hauru, K., Kotze, D. J., & Lehvavirta, S. (2017). Neo-spaces for urban livibility? Urbanities' versatile mental image of green roofs in the Helsinki metropolitan area, Finland. *Land Use Policy*, *61*, 587–600. doi:10.1016/j.landusepol.2016.11.021

Mohd, H. A. & Norhidayah, S. (2016), Sustainable food production: insights of Malaysian halal small and medium sized enterprises. *International Journal of Production Economics, 181* (part B), 303–314. . doi:10.1016/j.ijpe.2016.06.003

Murray, N. (2017). Urban disaster risk governance. A systemic review, PPI-Centre. UCL Institute of Education.

Namkung, Y., & Jang, S. (2013). Effects of restaurant green practices on brand equity formation: Do green practices really matter? *International Journal of Hospitality Management*, *33*(2), 85–95. doi:10.1016/j.ijhm.2012.06.006

Nogué, S., & Sala, M. (2018). The use of green roofs in urban green innovation spaces: A case study in Barcelona. *Urban Forestry & Urban Greening*, *33*, 19–29. doi:10.1016/j.ufug.2018.02.008

Norton, T. A., Parker, S. L., Zacher, H., & Ashkanasy, N. M. (2015). Employee green behavior: A theoretical framework, multilevel review, and future research agenda. *Organization & Environment*, *28*(1), 103–125. doi:10.1177/1086026615575773

Ones, D. S., & Dilchert, S. (2012). Employee green behaviors. In D. S. S. E. Jackson (Ed.), *Managing human resource for environmental sustainability*. Jossey-Bass.

Palomino, J. C., & Durán, P. (2020). The influence of green roofs on the urban heat island effect: A review of the current state of knowledge. *Urban Climate*, *34*, 100696. doi:10.1016/j.uclim.2020.100696

Pereira, Á., & Vence, X. (2012). Key business factors for eco-innovation: An overview of recent firm-level empirical studies. *Cuadernos de Gestión*, *12*, 73–103. doi:10.5295/cdg.110308ap

Rahman S., Ahmad, H., & Rosley, M. (2013) Green roof: its awareness among professionals and potential in Malaysian market. *Procedia Social and Behavioral Science, 85*, 443–453. https://doi.org/.sbspro.2013.08.373 doi:10.1016/j

Redman, C. (2014). Should sustainability and resilience be combined or remain distinct pursuits? *Ecology and Society*, *19*(2), 190–202. doi:10.5751/ES-06390-190237

Reisinger, A., & Clark, H. (2018). How much do direct livestock emissions contribute to global warming? *Global Change Biology*, *24*(4), 1749–1761. doi:10.1111/gcb.13975 PMID:29105912

Renwick, D. W. S., Redman, T., & Maguire, S. (2008). Green human resource management: A review and research agenda. *International Journal of Management Reviews*, *10*(1), 1–18. doi:10.1111/j.1468-2370.2011.00328.x

Rowe, D. B. (2011). Green Roofs as a means of pollution abatement. *Environmental Pollution, 159*(8-9), 2100–2110. doi:10.1016/j.envpol.2010.10.029 PMID:21074914

Savickiene, J. & Miceikiene, A. (2018), Sustainable economic development assessment model for family farms. *AgriculturalEconomics – Czech, 64*, 527–535. . doi:10.17221/310/2017-AGRICECON

Shafi Que, M., Kim, R. & Rafi Q, M. (2018). Green roof benefits, opportunities, and challenges– A review. *Renewable and Sustainable Energy Reviews, 90*, 757-773.

Shevchenko, A., & Petrenko, O. (2020). Current state of micro and small agribusiness in Ukraine. *Agricultural and Resource Economics, 6*(1), 146–160. doi:10.51599/are.2020.06.01.10

Shoeb, A., & Nisar, T. (2015) Green Human Resource Management: Policies and practices, *Cogent Business & Management, 2*(1), 1030817-103. https://www.tandfonline.com/doi/full/10.1080/23311975.2015.1030817 doi:10.1080/23311975.2015.1030817

Singh, S. K., Del Giudice, M., Chierici, R., & Graziano, D. (2020). Green innovation and environmental performance: The role of green transformational leadership and green human resource management. *Technological Forecasting and Social Change, 150*, 119762. doi:10.1016/j.techfore.2019.119762

Skydan, O., Nykolyuk, O., Pyvovar, P., & Martynchuk, I. (2020). Methodological approach to the evaluation of agricultural business system flexibility. *Management Theory and Studies for Rural Business and Infrastructure Development, 41*(4), 444–462. doi:10.15544/mts.2019.36

Stovin, V., Vesuviano, G., & Kasmin, H. (2012). The hydrological performance of a green roof test bed under UK climatic conditions. *Journal of Hydrology (Amsterdam), 414*, 148–161. doi:10.1016/j.jhydrol.2011.10.022

Stubbs, W., & Cocklin, C., (2008). Conceptualizing a sustainability business model. *Organizational Environment, 21* (2), 103e127.

Sutton, R. (2014). Aesthetics for green roofs and green walls. *The Journal of Living Architecture, 1*(2), 1–20. doi:10.46534/jliv.2014.01.02.001

Tan, P. Y., Wong, N. H., Tan, T. K., & Wong, K. W. (2019). Green roof innovation in Singapore: Policies, performance, and potential. *Journal of Environmental Management, 233*, 128–135. doi:10.1016/j.jenvman.2018.12.060

TomashukI. V.BaldynyukV. M. (2021). Identification of problems and prospects of rural infrastructure development of Ukraine *Economic sciences, 13* (100), 58-70. doi:10.24412/2520-6990-2021-13100-58-70

Ugai, T. (2016). Evaluation of sustainable roof from various aspects and benefits of agriculture roofing in urban core. *Procedia: Social and Behavioral Sciences, 216*, 850–860. doi:10.1016/j.sbspro.2015.12.082

Van Holt, T., Statler, M., Atz, U., Whelan, T., van Loggerenberg, M., & Cebulla, J. (2020). The cultural consensus of sustainability-driven innovation: Strategies for success. *Business Strategy and the Environment, 29*(2), 3399–3409. doi:10.1002/bse.2584

Vijayaraghavan, K. (2016). Green roofs: A critical review on the role of components, benefits, limitations, and trends. *Renewable & Sustainable Energy Reviews, 57*, 740–752. doi:10.1016/j.rser.2015.12.119

Wang, Y., Hu, H., Dai, W., & Burns, K. (2021). Evaluation of industrial green development and industrial green competitiveness: Evidence from Chinese urban agglomerations. *Ecological Indicators*, *124*, 107371. doi:10.1016/j.ecolind.2021.107371

Wiering, M., Liefferink, D., Boezeman, D., Kaufmann, M., Crabbé, A., & Kurstjens, N. (2020). The Wicked Problem the Water Framework Directive Cannot Solve. The Governance Approach in Dealing with Pollution of Nutrients in Surface Water in the Netherlands, Flanders, Lower Saxony, Denmark, and Ireland. *Water (Basel)*, *12*(5), 1240. doi:10.3390/w12051240

Williams, N., Rayner, J. & Raynor, K. (2010). Green roofs for a wide brown land: Opportunities and barriers for rooftop greening in Australia. *Urban Forestry & Urban Greening*, *9*(3), 245–251. https://doi.org/. ufug.2010.01.005 doi:10.1016/j

Wirtenberg, J., Harmon, K. D., Russell, W. G., & Fairfield, K. D. (2007). HR's role in building a sustainable enterprise. *Human Resource Planning*, *30*, 10–20.

Wu, W., & Jim, C. Y. (2021). Urban green infrastructure and sustainable urbanization: A comprehensive review of the roles of green roofs. *Landscape and Urban Planning*, *212*, 104097. doi:10.1016/j.landurbplan.2021.104097

Wysocki, J. (2021). Innovative green initiatives in the manufacturing SME sector in Poland. *Sustainability (Basel)*, *13*(4), 2386. doi:10.3390u13042386

Zhang, Z., & Sundaresan, S. (2010). Knowledge markets in firms: Knowledge sharing with trust and signalling. *Knowledge Management Research and Practice*, *2010*(8), 322–339. doi:10.1057/kmrp.2010.22

APPENDIX

Questions

1. Why and how technological developments can minimize the occurrence of environmental problems through the adoption of green innovation and green roofs in urban areas?
2. What are the main technological factors that this study aims to identify in the use of urban green innovation spaces?
3. What are the main factors that this study aims to identify in the use of technology in green roofs?
4. What are some of the technological factors that drive the adoption of green roofs in horizontal property by means of a technology acceptance of the place integrated identity model.
5. Under what assumptions are sustained the technological dynamics of the urban green innovation?
6. How can be described the perceived use of a technology in urban green innovation areas when addressing concerns such as attitude toward use, intention to use, perceived ease of use, and perceived usefulness?
7. Which methodology you may suggest being used to determine the perceived use of technology in green roofs?
8. For an efficient use of technology in urban green innovation areas and green roofs, which criteria suggest helping to minimize the chaos, the occurrence of environmental breakdowns, such as air pollution, increased temperature, urban heat islands, etc.
9. How could technology help to address such socio-ecological problems through the implementation of simple strategies?

Chapter 9
Impact-Driven Productivity and Innovation With Unified Communication and Collaboration Technologies:
Drivers to Be Addressed Across Global Regions

Anthony D. Bolton
https://orcid.org/0000-0002-1259-7479
University of South Africa, South Africa

Leila Goosen
https://orcid.org/0000-0003-4948-2699
University of South Africa, South Africa

Elmarie Kritzinger
https://orcid.org/0000-0002-5141-4348
University of South Africa, South Africa

ABSTRACT

The purpose of this chapter is to introduce research on the impact-driven productivity and innovation gains achieved through the adoption of unified communication and collaboration (UC&C) technologies as drivers of digital transformation across a global automotive organization. Against the background of research on the bioeconomy and economic ecosystems, indicators of cultural change in the workplace and other transformation factors shown to have an impact included the ease of use and adoption, legacy reduction, and elimination, increased organizational efficiency and generating savings.

INTRODUCTION

Although a press release by the United Nations Environment Program (2023) indicated that the "ozone layer is on track to recover within four decades", the **bioeconomy** as a new approach emphasizing "re-

DOI: 10.4018/978-1-6684-8879-9.ch009

using, recycling, and recovering materials" (Satya, et al., 2023, p. 1) is one of the essentials as part of the "strong and functional couplings among ecological, **economic**, social, and technological processes" explaining "the complexification of human-made systems, and phenomena", such as the use of artificial intelligence for sustainable complex socio-technical-**economic ecosystems** (Martínez-García, 2022, p. 1). Such **ecosystems** are acutely threatened by serious degradation. This state of affairs not only leads to acute **economic** losses, but also entails catastrophic ecological, social, and cultural *damage*.

This is because after decades of intensely *damaging* use, in "recent years, the growing trend of energy consumption from fossil fuels in the world" had presented humankind with major **crises** related to environmental **pollution**, hydrological stress and "increasing acceleration in the depletion of energy resources" (Wang, et al., 2023, p. 1). In this scenario, some of the **challenges** related to biogas generation from biomass as a cleaner alternative towards a circular **bioeconomy**, which is considered to be a natural investment in citizen well-being and *societal* prosperity, can be solved using artificial intelligence.

The **bioeconomy** is one of the priorities of the European Commission (EC), and the topics of the **bioeconomy** and **economic** and social **ecosystems** have been included as related to **emerging trends** with regard to **challenges** in e.g., the proceedings of the international conference on *Sustainable Research and **Innovation*** (SRI) (Mwongera & Langat, 2014) and when building the hyperconnected society in terms of Internet of Things (IoT) *research and innovation* value chains, **ecosystems** and markets (Vermesan & Friess, 2015), as well as considered in the 2030 **Sustainable Development Goals** (SDGs).

This specific chapter will address especially SDG 9: Industry, **Innovation** and Infrastructure, and to a lesser extent, SDG 12: Responsible Consumption and Production, as related to **productivity**.

Societal Impacts

The importance of this theme lies primarily in the conservation of **ecosystems** and their biodiversity, but also in discussing potential new **challenges,** "namely the **economic**, technological, and *societal" impacts* and dimensions of entrepreneurial **ecosystems** (Audretsch, Cunningham, Kuratko, Lehmann, & Menter, 2019, p. 313).

These topics further aim to prepare for the implementation of clinical research, which "evolved side-by-side with technology, leading to exponential data generation contributing to social", **economic**, technological and environmental "development. Nevertheless, data storage, integrity," *blockchain* and clinical data economics had led to the tokenization of clinical research in the European Union (EU) (Pego, Raposo, & Loureiro, 2021, p. 269).

Against the background of the *political and **economic** implications of blockchain technology in business and healthcare,* the latter authors indicated that the "single critical output of a *blockchain* is creating trust where previously impossible. While this feature delivers compelling value for many use cases", such as bitcoin for money, the real *blockchain* game changer is related to protocols and Decentralized Autonomous Organizations (DAOs) for coordinating work to provide goods and services (Loureiro, Pêgo, & Raposo, 2021, p. 160).

The motivation for the thesis discussed in this chapter was the increased priority and drive towards digital transformation across global industry, specifically at enterprises engaging in large-scale manufacturing and product development. The inspiration for the research study was found in the unique **challenges** encountered in planning, designing and affecting the digital change of communication and collaboration in rapidly transforming industries. The goal of this research was the creation of a technical and services framework facilitating the digital transformation of Unified Communication and Collaboration (UC&C) **across a global** enterprise, based on the principles and technologies associated with UC&C services.

Klewes, Popp and Rost-Hein (2017a) presented an introduction on **challenges** with regard to **unified communication and collaboration technologies** and digital transformation. The research portrayed in this chapter examined the **challenges** that a global automotive enterprise experienced in establishing and executing the transformation of digital communication and collaboration technologies in the face of rapidly changing global markets and enterprise landscapes. The MAIN FOCUS will be on how digital communication and collaboration technologies are opening up new opportunities in response to **challenges** in many fields, in particular in socio-cultural milieus. The convergence of data and new media allows for the generation of new experiences using the Internet of Things, augmented reality, virtual reality and mixed realities, as well as providing people with the opportunity to have increasingly in-depth experiences.

"Be it the first corporate websites in the early 1990s or, ten years later" (Klewes, et al., 2017a, p. 1), transformation via digital communication and collaboration technologies has existed in the lexicon of human thought for over fifty years. The potential of such transformation to enhance digital communication and collaboration in society and across industries is beginning to be realized.

The research reported on in this chapter built on developments and industry-shaping **drivers** as discussed and presented in a review of scholarly literature within the scope of the study. The empirical research presented in this study aimed to further contribute to the body of academic knowledge associated with processes associated with the transformation of digital communication and collaboration technologies, while also advancing the development of practical methods and digital communication and collaboration technologies that realize the benefits of transformation in industry.

In assessing the effect and potential significance of digital communication and collaboration technologies on transformation in the automotive industry, an empirical study, focused on a single subject case study, was conducted into the **impact-driven productivity and innovation** *achieved via* **UC&C technologies** *(Bolton, Goosen, & Kritzinger, 2021b).*

At a commercial level, the automotive industry represents a diverse, fast moving and global business segment. General Motors (GM), for example, sells a vehicle every eight seconds across a business spanning over 125 countries and manufactures over 9 million vehicles globally through its plants. There is an increasing demand for communication and collaboration across automotive multi-nationals, and a framework for partnerships promoting the transformation of digital communication and collaboration technologies is a key component of continued support towards success in the marketplace and innovation in terms of technologies and new vehicles.

In addition to digital communication and collaboration technologies, transformation *impacts* people and social interactions. With increasing digital opportunities around the world, an institutional perspective on the exploitation of Information and Communication Technologies (ICTs) is critical to boosting the development of communication and collaboration among stakeholders (Huerta-Carvajal & Luna-Reyes, 2009).

Professional digital communication and collaboration can also be an influencing factor in the process of transformation through digital communication and collaboration technologies. Literature suggests that Industry 4.0 and the Internet of Things transformation will only succeed with the aid of digital communication and collaboration technologies (Klewes, Popp, & Rost-Hein, 2017b). Klewes, et al.

(2017b) further suggested neglect and a lack of explicit analysis of the role of digital communication and collaboration technologies in acceptance of Industry 4.0 or IoT transformation, from the perspectives of *societal*, organizational, market and corporate communication.

"Seldom have there been phases in which the development of the communications profession was so confusing as it is today" (Klewes, Popp, & Rost-Hein, 2017c, p. 187). "Depending on the type of company, its size and the size of the team—and subject above all, of course, to the market in question—the paths" towards managing transformation via digital communication and collaboration technologies can be followed by implementing criteria and guidelines for the selection of communication professionals (Goosen, 2004).

Target Audience

Like that of the book that it proposes to form part of, the **target audience** of this chapter includes students, academics, researchers and others interested in the bioeconomy and economic ecosystems.

Recommended Topics

Based on the **recommended topics** suggested for this book, the chapter will specifically pay attention to:

- Bioeconomy
- Economic Ecosystems
- Societal Impacts
- Bioeconomy Sectors
- The Circular Economy
- Sustainable Development Goals

Objectives

Based on the **objectives** set out for this book, that of the chapter is to:

- present **research on** creating an **impact-driven** cross-regional **productivity and innovation ecosystem** equipped **with unified communication and collaboration technologies** for tapping organizational resources and leading to significant economic outcomes,

as well as to demonstrate:

- linkages between **drivers to be addressed across global regions**, and
- economic and social models related to the **bioeconomy and economic ecosystems**.

BACKGROUND

"Building on a clarification of the IoT, the Industrial Internet and other key concepts", as did Klewes, et al. (2017b, p. 7), this section of the chapter will provide broad **definitions** and discussions on the topic of **Impact-Driven Productivity and Innovation with Unified Communication and Collaboration Technologies**. It will also incorporate the views of others, in the form of a *literature review*, into the discussion to support, refute, or demonstrate the authors' position on this topic.

Definitions: Productivity

The research presented within this study concentrated on how the process of transformation through digital communication and collaboration technologies **impacted** the innovation and productivity of workers within the case study of General Motors. According to Attar, Gupta and Desai (2012), productivity can be **defined** in many ways; however, productivity is usually taken to mean labor productivity. In a study of various factors affecting labor productivity and methods to improve it, against the background of a journal article on mechanical and civil engineering, Attar, et al. (2012) **defined** productivity as the ratio of output to all or some of the resources leveraged to produce the desired output.

Against the **background** of Information Systems research, Aral, Brynjolfsson and Van Alstyne (2012) highlighted that groups of Information Technology (IT) and information employees are working to gather, process and analyze information through technology-supported research methods, tools and process towards productivity. Relative to the focus on the influence of digital communication and collaboration technologies on productivity within this study, Dedrick, Kraemer and Shih (2013) found strong evidence of productivity growth being associated with the introduction of information technology systems and IT investment. The journal article on management information systems reporting on the latter authors' study included a mix of 45 developed and developing countries, suggesting that the effects of IT on productivity depended on both the level of IT use and the availability of technology resources.

Definitions: Internet of Things

In terms of the Internet of Things and its relevance, Kangra and Singh (2021, p. 19) indicated that the Internet of Things "model connects physical devices to the virtual world and enables" these to interact. It also "enables smart devices to communicate with other devices to exchange information" and to link via "a wireless network or cloud".

According to Singh, Singh and Bedi (2021, p. 217), the internet of things is a device composed of actuators that "should be used for enhancing agricultural efficiency in precision agriculture", which is adding to the "bedrock of the Indian economy".

The "IoT and cloud computing are … novel fields that are rapidly progressing in the world of internet" technologies (Guria & Bhattacharyya, 2021, p. 190). A massive amount of data is being communicated via the IoT and cloud devices, along with highly configured devices.

Providing a flexible forum for applications of the cloud-based internet of things (Navadia, et al., 2021, p. 65), cloud "storage is a great way for companies to fulfill more of their data-driven needs and excellent" technologies that allow "the company to evolve and grow at a faster pace", as well as accelerating growth.

The mission of the chapter by Alasti (2021, p. 288) was "to review and investigate the requirements and applications of using cloud-based internet of things (CIoT) for shifting … legacy robotic manufacturing towards Industry 4.0" through sensing and communication.

Bioeconomy

Vogelpohl (2023, p. 2) did "not aim to subscribe to or endorse one specific understanding or **definition** of the **bioeconomy**". However, the latter author somewhat involuntarily focused "on those bioeconomy visions that have become politically relevant and manifest in actual bioeconomy strategies" towards understanding the **bioeconomy** through its instruments in terms of standardizing *sustainability science*.

According to the journal article on CO_2 utilization by Raza, et al. (2023, p. 3), the **bioeconomy** "is a diversified field involving various production activities". Raza, et al. (2023, p. 2) also pointed out that the "circular **bioeconomy** depends on the capturing and purity of the CO_2". Thus, the bioeconomy affects ecosystems "because the economy needs raw materials". Engineering approaches for converting CO_2 to biomass coupled with nano-biomaterials as bio-mediated towards the circular **bioeconomy** are therefore needed.

"Microalgae are the ideal source contributing to" the circular bioeconomy. More importantly, progress on microalgae cultivation in wastewater for bioremediation and the circular **bioeconomy** and *environmental research* into these will "play a role in the circular bioeconomy" (Satya, et al., 2023, p. 1).

Against the background of renewable and sustainable energy reviews, Sharma and Malaviya (2023, p. 1) highlighted links between the "circular **bioeconomy** and internationally agreed SDG targets", as the "circular **bioeconomy** provides significant opportunities to achieve" the 17 SDGs. In the review by the latter authors, a critical synthesis on ecosystem services and climate action from a circular **bioeconomy** perspective was presented "to assure the *sustainability* of bioeconomy" programs.

In the context of *sustainability*, a "review of the current practices of **bioeconomy** *education* and training in the EU" by Paris, et al. (2023, p. 2) will contribute to an "understanding of how **bioeconomy** *education* and training is conducted, evaluate the different approaches, and provide the **background** for" the chapter by the latter authors.

The paper by Goosen (2018a) in the proceedings of the 26th Conference of the Southern African Association for Research in Mathematics, Science and Technology Education (SAARMSTE) showed how *sustainable* and inclusive quality *education* could be achieved through research-informed practice on information and communication technologies.

Various proceedings from the South Africa International Conference on *Educational Technologies* (SAICET) showed how such educational technologies can be used to increase students' access to an ICTs for Development (ICT4D) Massive Open Online Course (MOOC) (Goosen, 2018c) or for such an ICT4D MOOC in the 21st century (Goosen, 2015). Goosen and Mukasa-Lwanga (2017) later indicated how *educational technologies* were being used in distance education, taken beyond the horizon with qualitative perspectives, while Van Heerden and Goosen (2012) presented a brief history on choosing to use vodcasts to teach programming in an Open Distance Learning (ODL) environment (Goosen, 2008).

Bioeconomy Sectors

The article by Ferreira, Pié, Mainar-Causapé and Terceño (2023, p. 19) presented "a structural analysis of the **bioeconomy sectors** in Spain" as a new economic paradigm, and the role of key **sectors** with different approaches to "several products within the bioeconomy. Considering" **bioeconomy sectors**, the article focused on the environment, development and *sustainability*.

Economic Ecosystems

Against the background of *sustainability,* Escalona-Orcao, et al. (2020, p. 3) showed "how the cultural dynamics of the **ecosystems** in a post-pandemic scenario will continue ... according to the nature of their cultural and creative **ecosystems**" in medium-sized cities with regard to evolution in times of **economic** crisis and the pandemic. Peter (2021, p. 1) was of the opinion that social innovation can be used for sustainable urban developmental transitions in Sub-Saharan Africa and across the continent by leveraging '**economic ecosystems**' and the entrepreneurial state. Local scales can be implemented "by using **economic ecosystems** as the framework" and "innovation-based **economic ecosystems** for catalyzing" multi-scalar transitions to *sustainability.*

In the contexts of regional **ecosystems** and research policy, Auerswald and Dani (2022, p. 6) focused "on the high positive interdependencies between industries." The latter authors "expect that **economic ecosystems** with a higher density of **economic** complementarities" will lead to entrepreneurial opportunity and related specialization in **economic ecosystems.**

A journal article on open innovation in terms of technology, the markets and complexity presented the role of **collaboration** "between ecosystem actors and between **ecosystems** themselves", as well as "studying the symbiosis of socio-**economic ecosystems** in the megapolis" being proposed (Tolstykh, Gamidullaeva, Shmeleva, Gromov, & Ermolenko, 2022, p. 1).

Regarding *artificiality and sustainability in entrepreneurship*, Saha, Sáha and Sáha (2022) asked whether entrepreneurial university and social innovation **ecosystems** support and/or help with the knowledge-based **economic** growth and development of Higher Education Institutions (HEIs).

The Circular Economy

"Recent studies on *renewable* energy and its links to the **circular economy** point to the importance of creating a business model that supports this relationship. For example, the" Technium Social Sciences journal article by Pego (2021, p. 499) suggested a business model for the **circular economy** based on a Portuguese offshore energy cluster.

An earlier journal article in the Physics Conference Series, also by Pego (2019, p. 1), on a Portuguese offshore energy Strengths, Weaknesses, Opportunities, and Threats (SWOT) analysis, indicated that *renewable* energy had "gained importance in recent years because of its cooperation in **ecosystems**, welfare and the **circular economy**. The new business model" described therein comprised new values.

With regard to progress in *renewable* energy offshore, the paper by Pego, Marques, Salvador, Soares and Monteiro (2016, p. 867) studied "the economic features and territorial *implications* of the potential offshore energy cluster in Portugal," which originated from "investments in wave and wind energy prototypes".

MAIN FOCUS OF THE CHAPTER

Issues, Problems, Challenges

This section of the chapter will present the authors' perspective on the **issues**, **problems**, **challenges**, etc., as these relate to the main theme of the **Handbook of Research on Bioeconomy and Economic Ecosystems** and arguments supporting the authors' position. It will also compare and contrast with what has been, or is currently being, done as it relates to the specific topic of the chapter, on **Impact-Driven Productivity and Innovation with Unified Communication and Collaboration Technologies**.

In the same context with which the previous section of this chapter concluded, the influence of *renewable* energy in *smart cities* from a European perspective on new business was presented by Pego (2018, p. 63), pointing out that one "of the main **issues** of the" Horizon 2020 Research and Innovation funding program of the European Union was the *smart cities*' "*sustainability* and the relevance of implementing the green economy. The" cities' functionality was associated with positive externalities.

Urban living laboratories "(ULL) are a new concept which involves users in **innovation** and development and are regarded as a way of meeting the **innovation challenges** faced by information and communication technology" service providers. Describing research on entrepreneurship and marketing for global reach in the digital economy, Pego and Bernardo (2019, p. 203) discussed the role of such urban living labs in entrepreneurship, energy, and the governance of *smart cities*.

Bioeconomy

"In order to create a" low-carbon **bioeconomy**, the review by Karimi-Maleh, et al. (2023, p. 3) looked at integrated approaches for waste-to-biohydrogen using nano-biomediated *advanced composites and hybrid materials*. Addressing the **challenges** related to "the low-carbon **bioeconomy**, which could be overcome" through the use of such *advanced composites and hybrid materials*, Karimi-Maleh, et al. (2023, p. 2) were convinced that is possible to move "from a fossil economy to a **bioeconomy** with net-zero" greenhouse gas (GHG) emissions.

Economic Ecosystems

In the context of **ecosystems**, the journal article on technology transfer "and the outcome measures of the respective **ecosystems**" by Audretsch, et al. (2019, p. 313) in a "special **issue** was a first attempt".

The Circular Economy

According to Pego (2020b, p. 224), **the** "**circular economy** (CE) is a new concept" that evolved "out of a new management assumption in organizations based on recycling, reuse, and" the repurposing of products. "The use of renewable energy and recycling products" are some "of the main achievements of organizations." In the context of *networked business models in **the circular economy***, the latter chapter looked at the features of new business **challenges** against the background of circularity in Portugal.

New **challenges** for the tourism sector in the Algarve region based on an evaluation of the **circular economy** was the topic of another chapter by Pego (2020a, p. 185) in the context of strategic business models to support demand, supply, and destination management in the tourism and hospitality industry. The latter chapter indicated that during "recent years, the increase in tourism in the Algarve" region had "shown an increase in changes in supply for consumers. This means that the tourism industry" had started to change.

The multi-disciplinary e-journal article by Pego and de Matos (2017, p. 39) aimed "to present the results of a study that investigated the use" and **impact** "of information systems and technologies (IST) in tourism in" a "rural space (TRS) in the Algarve region." In the *Handbook of Research on Entrepreneurship and Marketing for Global Reach in the Digital Economy,* the chapter by Pego and Bernardo

(2018, p. 274) analyzed "the importance of information systems (IS) on the decision-making process at rural organizations" in a case study of Algarve, as decision making in rural tourism management "is an important role performed by managers."

Sustainable Development Goals

With regard to globalization, societies and education, Leite (2022, p. 2) indicated that the "global indicator framework is a call for increased education in three key areas", including global citizenship. Since there seemed to be "no common agreement on the *definitions* of global citizenship education and education for sustainable development", the latter author was using the **sustainable development goals** for global citizenship education in terms of definitions, **challenges**, and opportunities.

Study Structure

The study reported on in this chapter had a structure consisting of four parts. This structure started with a focus on the development of an understanding of the *primary research question*. Relevant scholarly opinions in literature were subsequently reviewed to establish the context of the **problem domain** and *primary research question*. The process of *empirical research*, focusing on the transformation of digital communication and collaboration technologies at a global automotive enterprise was presented. Finally, an analysis of the observed outcomes resulting from the study was reviewed and discussed, leading to the outcomes reported on in the **conclusion**.

Part 1: Challenges to Transforming Digital Communication Collaboration Technologies

Part 1 established the context that supported the development of the **problem domain of the study** associated with the research and primary question. This part of the study introduced the concepts with regard to the transformation of digital communication and collaboration technologies and discussed the primary ***drivers*** that are influencing ***emerging trends*** toward the digitization and transformation of global enterprises. A review of the literature associated with generic ***drivers*** of enterprise transformation via digital communication and collaboration technologies was presented in the chapter by Bolton, et al. (2021b).

Part 2: Emerging Trends Related to and Drivers of Innovative Technologies

Part 2 focused on identifying and developing an understanding in terms of the context of the primary question with regard to the critical *societal* and technological ***drivers*** that motivate progress towards the digitization of communication and collaboration technologies. In a chapter on the integration and implementation of the Internet of Things through the transformation of digital communication and collaboration technologies, and the **impact** on productivity and innovation (Bolton, Goosen, & Kritzinger, 2021c), the origins of the Internet were explored through a review of the existing scholarly literature. The incremental, progressive development of Internet technologies was presented in conjunction with the combinatorial process of development toward the modern-day concept of the Internet of Things. In the latter chapter, the concept of people as an integrated part (thing) within the context of a connected Internet was also established.

In the chapter by Bolton, Goosen and Kritzinger (2020b), the future **impact** of digital communication and collaboration technologies on productivity and innovation towards promotion for the Fourth Industrial Revolution (4IR) was introduced, building an understanding of digital communication and collaboration technologies that promise to integrate people into the digital world of connected people, processes and things with a vision to ease the burden and transition of people and the establishment of their digital persona. The concepts of Industry 4.0 were further reviewed from literature, providing insight into the ***challenges*** that people and organizations face within the broader context and progression of the digitization process (Bolton, Goosen, & Kritzinger, 2022c). An additional understanding of the influence that the IoT is having on the evolution of industry towards a connected, digital age was also established.

More details regarding enterprise digitization enablement through digital communication and collaboration technologies can be found in the conference paper by Bolton, Goosen and Kritzinger (2016).

Part 3: Empirical Investigation

With the *primary research question* identified and the foundational context and understanding of the question, as well as associated ***drivers*** and related scholarly opinion and research, established, Part 3 focused on the process of the *empirical investigation*.

In terms of digital communication and collaboration technologies at a global automotive organization, chapters by Bolton, Goosen and Kritzinger (2021a; 2022b) outlined the framework, research design, methodology and structured approach adopted to facilitate the development of research data. The primary theoretical constructs developed to frame the investigation process were presented, along with the instruments for data collection.

The chapters by Bolton, et al. (2022b, 2022c) also discussed ***emerging trends*** *related to technologies* for innovation and productivity *management* in the automotive and software industries with regard to the **impact** of digital transformation on communication and collaboration, as well as ***challenges, emerging trends*** and *strategies* for a global automotive enterprise and global networking towards a **post-Covid-19 era** in terms of the **impact** of digital transformation.

Another chapter discussed the actual empirical research process undertaken within the study and its original artefacts (Bolton, Goosen, & Kritzinger, 2022a). The latter chapter identified and discussed the research case study subject, the General Motors automotive corporation and the framework for the digital transformation via unified communication and collaboration technologies specified within the research process. That chapter also conveyed the structure and processes associated with the data collection instruments that were used to gather research data, which were later used in analysis and the investigation of observed outcomes.

Part 4: Discussion of Findings and Conclusion

Part 4 explored the established research data and analysis, provided a *discussion of* the *findings* in terms of observations and new opportunities identified with regard to **future research directions**, and empirical investigations were presented and discussed, building further on this unique body of research. Having established research data through the empirical research process, attention and focus shifted to a *discussion of findings and* outcomes.

The chapter on the results and interpretation discussed the findings and observations identified through review and analysis of the established qualitative and quantitative research data (Bolton, et al., 2022c). A summary of the research data was presented, along with interpretive discussion and evidence of outcomes associated with the primary question and established research hypotheses. The final chapter summarized **recommendations**, before the last section provided a discussion of the overall coverage of the chapter and concluding remarks about the **conclusion** developed via the process of empirical research and results interpretation.

Further particulars regarding specifically the security aspects of the empirical study into the **impact** of transformation via digital communication and collaboration technologies on the productivity and innovation of a global automotive enterprise, that cuts across different aspects of the study, can be obtained from the journal article by Bolton, Goosen and Kritzinger (2020a).

Problem Domain of the Study

Organizational Culture

The **challenges** discussed in the chapter by Bolton, et al. (2021b) focused on the **transformation** of digital communication and collaboration technologies in an enterprise and highlighted the breadth and depth of the **future impact** that digitization has on all levels of the enterprise value chain and its business model. These **challenges** included fundamental shifts from traditional business models to the development of new business models and consumer experiences (Bolton, et al., 2022a), the threat of digital commoditization, increased global competition and changes in *organizational culture* and structures. **Challenges** also encompassed broad and constant change in digital C&C technologies, requiring planning for adoption, as well as rapid cycles of innovation management of technology change **impacting** on the industry.

Challenges should be considered at a human level, especially for those employees, who manage and maintain the business operations and product innovation. The landscape of digital **challenges** is extensive and will differ in elements between organizations; however, common themes emerge. Due to the potential for broad variance regarding enterprise business needs and required change, the scope of this research was focused on one business segment (automotive), therefore allowing for consideration within a single study.

The following **issues** were considered and formed the foundation of the research focus:

Issue 1: Integrating Employees Through Digital Communication and Collaboration

Issues associated with *integrating employees through digital communication and collaboration* technologies within the transforming enterprise were considered.

A subsection on "Technology **Challenges** – Scope, Speed, Complexity, Risk" demonstrated that due to shifts towards the development of new business models and consumer experiences (Bolton, et al., 2022a), as enterprises shift towards digitization, the **impact** of communication and collaboration extends beyond that of people-to-people C&C with the addition of real-time people-to-system C&C to facilitate decision making (Bolton, et al., 2021b, p. 137).

Some of the fundamental **challenges** that digitization introduced to the digital communication and collaboration landscape within an enterprise, because of transformation, was presented in subsections on "People **Challenges** – Culture, Skills, Communication" (Bolton, et al., 2021b, p. 141) and "Common **Challenges** – Rapid, Continuous Change and Communication" (Bolton, et al., 2021b, p. 142).

A subsection on *combinatorial innovation differentiation opportunities* established how IoT technologies' development, fueled by combinatorial innovation (Bolton, et al., 2021b, p. 148), resulted in the development of new business *models* and innovation (Bolton, et al., 2022a).

Subsections on "Analytics Across the Value Chain" (Bolton, et al., 2021b, p. 149) and *digital customer engagement* (Bolton, et al., 2021b, p. 150) highlighted current **emerging trends** in the shift towards, and development of, intelligent analytics and digital customer collaboration, as well as the associated **future impact** on communication and collaboration between the enterprise and its employees, customers and partners.

Issue 2: Managing Constant Change

Issues associated with *managing* communication and collaboration, as well as employee integration with the digital industry through permanent and consistent *change*, were also considered.

The **challenges** and emerging strategies associated with communication and collaboration within the digitally transforming global automotive enterprise are compounded by the expected continuous development in digital technologies, new business models and the feature of permanent change (Bolton, et al., 2022c).

The literature presented in the subsection on "Technology Challenges – Scope, Speed, Complexity, Risk" showed how enterprises are at various levels of digital maturity (Bolton, et al., 2021b, p. 137), with the subsection on "People **Challenges** – Culture, Skills, Communication" providing a contrast between readiness and capability at the human level maturity (Bolton, et al., 2021b, p. 141).

A subsection on "Business **Challenges** – Model, Vision, Culture, Security" supported the position that enterprises and employees are likely to experience rapid and broad change (Bolton, et al., 2021b, p. 139), because of combinatorial innovation, and coupled with the literature reviewed in the latter subsection, as well as the one on technology **challenges**, that this change is likely to be persistent to compete with new digital business entrants and the risk of industrial commoditization. This disparity between capability and **indicators of cultural change** for employees **in the** digital **workplace**, processes and communication and collaboration will require planning and close monitoring to ensure successful outcomes.

Another subsection determined how *digital disruption* develops through the process of combinatorial innovation, and literature highlighted the disparity in levels of preparedness across enterprises to handle digital transformation maturity (Bolton, et al., 2021b, p. 150).

Limitations of the Study

The empirical study was carried out within the context of the General Motors corporation. Starting in 2013, General Motors embarked on a transformation of its ICT environment. This transformation included a significant shift from an outsourced ICT environment to insourced, with a focus on establishing a robust ICT capability that could support the digital transformation of its business. Prior to the empirical research study, employee communication and collaboration were limited to legacy methods of voice C&C, physical meetings and travel. Limited use of instant messaging technologies was also evident within the GM enterprise.

The research case study focused on the process of **transformation** as General Motors shifted its ICT infrastructure and services to enable integrated digital communication and collaboration. The Enhanced Unified Communication and Communication (E-UC&C) model developed within the scope of the research focused on the specific needs and business environment of General Motors. The design was tailored to the changing business environment within General Motors and facilitated the transition from its existing ICT systems and *changing* employee *work practices*. While the study focused on a specific global automotive original equipment manufacturer, the **challenges** that GM faces as it digitally transforms its systems, service and *work practices* are likely to be familiar to many large global enterprises.

The resulting E-UC&C **ecosystem** of technologies, services and end-user functionalities were not designed or developed to individually meet automotive work practices and are suitable for deployment within any business environment. The research study was executed over a five-year duration, culminating in the development of the strategy framework. Implementation in practice and subsequent **impact** was observed and measured.

Primary Research Question

The research and literature presented as described earlier in *Part 1* of this section of the chapter established the *primary research question*: What was the extent of the **impact-driven productivity and innovation** gains achieved via the use of **unified communication and collaboration technologies** for digital transformation **across** a **global** automotive organization?

Secondary Research Questions

To facilitate evaluation, the *primary research question* was further expressed through the following secondary questions:

1. Which *indicators of cultural change* in the workplace as transformation factor could be identified during the digital transformation of **unified communication and collaboration technologies across** a **global** automotive organization?
2. To what extent was transformation factors related to *ease of use and adoption* of **unified communication and collaboration technologies**, *legacy reduction and elimination, increased organizational efficiency* and *generating savings* observed during digital transformation **across** a **global** automotive organization?
3. To what extent was an **impact-driven** cross-regional **productivity and innovation ecosystem** equipped with **unified communication and collaboration technologies** created during the digital transformation **across** a **global** automotive organization?

Human Resource Management and Diversity

The methodology designed and implemented in this research study facilitated the analysis of a process undertaken to achieve digitization and the resulting **future impact** of the digital integration of what scholars identify as 'human capital' (Popescu, Comanescu, & Sabie, 2016). The latter authors (Popescu, et al., 2016) investigated the role of human capital in the knowledge-networked economy against the background of psycho-sociological **issues** in *human resource management*.

Aligned with a theme of digitization, the research design and methodology of this study encompassed a combination of qualitative and quantitative methods to generate a rich set of data, variables and observations.

This information provided a solid foundation for analysis and the exploration of the human and technical factors associated with the *primary research question* and associated *research secondary questions.*

SOLUTIONS AND RECOMMENDATIONS

This section of the chapter will discuss solutions and recommendations in dealing with the **issues, problems** and/or **challenges** presented in the preceding section.

Solutions Through Digital Communication and Collaboration Technologies

Research Approach and Methodology Implemented Towards Obtaining Solutions

Scholarly literature relating to the emergence of the digital economy, such as the research presented by Sebastian, et al. (2017), drew the focus to how the process of digitization provides an opportunity for increased measurement and observation. Goldfarb, Greenstein and Tucker (2015) highlighted how digitization opens new opportunities to empirically analyze questions, such as those central to this research study, by leveraging new data sources. The central **issues** addressed in this study were the relationships between the transformation of productivity and innovation through digital communication and collaboration technologies, the Internet of Things, and the resulting **future impact** on a global automotive enterprise.

Exploration of the research questions required observation of the process and effects of transformation in a working enterprise. Transforming an existing enterprise and observing the process helps to establish a deeper understanding of the resulting outcomes associated with employee productivity and innovation. Observing the process of transformation enabled the process of analysis through gathering data and provided insight into the underlying digitization processes. Kompf (2012) pointed out that the effective leverage of knowledge and information is a crucial element in the emerging digital economy. Kompf (2012, p. 42) also noted that digitization is the "move from atoms (physical) to bits (information)". It is important to consider the **impact** of digital communication and collaboration technologies on employees within the enterprise, because they will have to contend with a higher volume of richer digital communication and collaboration technologies within the transformed enterprise.

Epistemological Perspective

Characterized by the approach of observing an object or **issue** from multiple standpoints, a pluralist posture was adopted when approaching this research and seeking alignment with the *epistemological perspective*. Scholarly perspectives on cross- and trans-disciplinary approaches to action research and action learning (Goosen, 2018b) for managing the growing complexity of information in the 'Age of the IoT' (Koverman, 2016), such as the meta-praxis of integral metatheory suggested by Esbjorn-Hargens (2016), have influenced this pluralist stance.

The amalgamated theoretical approach of the integral realist arguably facilitated the integration of broad sets of growing digital and non-digital information and the development of these into knowledge.

Philosophies associated with the integral pluralist metatheory influenced the research approach. From a realist theory of science, the critical realist *epistemological perspective* advocated by scholars, such as Bhaskar (2008), was adopted during the analysis phase of the research study. Employing critical realism and its integration of transcendental realism (philosophy of science) and critical naturalism (philosophy of social science) facilitated the process of information abstraction and generalization in the analytical process of empirical evaluation.

The Next Generation Towards Solutions for the Post-Covid-19 Era

Towards the post-Covid-19 era, Ngugi and Goosen (2021) investigated innovation, entrepreneurship and *sustainability* for ICT students. In the same context, Bolton, et al. (2021b) conducted an empirical study into the **impact** on innovation and productivity with regard to the digital transformation of a global automotive enterprise, while Van Heerden and Goosen (2021) reported on students' perceptions of e-assessment in the context of **Covid-19** in a case study of the University of South Africa (UNISA).

As part of the book, this chapter focuses on **research on** creating an **impact-driven** cross-regional **productivity and innovation** ecosystem equipped **with unified communication and collaboration technologies** for tapping organizational resources and leading to significant economic outcomes, as well as on demonstrating linkages between **drivers to be addressed across global regions**, particularly **towards the post-Covid-19 era** (Bolton, et al., 2021b).

FUTURE RESEARCH DIRECTIONS

This section of the chapter will discuss **future** and **emerging trends** and provide insight about the **future** of the theme of the **Handbook of Research on Bioeconomy and Economic Ecosystems**, from the perspective of the chapter focus on **Impact-Driven Productivity and Innovation with Unified Communication and Collaboration Technologies**. The viability of a **paradigm**, model, implementation **issues** of proposed programs, etc., may also be included in this section. **Future research directions** within the domain of the topic will finally be suggested.

Emerging Trends

In the contexts of *resources, the environment and sustainability,* Aworunse, Olorunsola, Ahuekwe and Obembe (2023, p. 2) saw the "bioeconomy as an **emerging paradigm**, with biotechnology contributing a major role." The latter authors also highlighted the contributions of an African bioeconomy, with a special focus towards a sustainable bioeconomy in a post-oil era for Nigeria.

Global mega**trends**, such as personalization and technology developments, including machine-to-machine communication and collaboration and the Internet of Things, are reshaping the global market and ushering in the next revolution in terms of the concept of Industry 4.0 (Bartodziej, 2017), and the latter author therefore conducted an empirical analysis of technologies and applications in production logistics.

According to Klewes, et al. (2017a, p. 1), communications "professionals, especially those with an Anglo-American orientation, are often seen as pioneers in identifying" **emerging "trends** in business and society and putting" these to use for their organizations. In the following chapter, Klewes, et al. (2017b, p. 7) addressed "fundamental **challenges** with which, in the context of" the **impact** of "digital transformation, professional communication for companies and organizations must cope" and how key **emerging trends** will transform the way companies communicate and are out-thinking organizational communications.

The chapter by Bolton, et al. (2022b) discussed **emerging trends** with regard to technologies for innovation and productivity management in the automotive and software industries in terms of the **impact** of digital transformation on communication and collaboration. **Challenges** and **emerging trends** related to strategies for a global automotive enterprise and networking **towards** a **post-Covid-19 era**, together with the **impact** of digital transformation, were discussed in Bolton, et al. (2022c).

In terms of public health and surveillance, Basch and Basch (2022) agreed that because Covid-19 and other **emerging trends** related to "infectious diseases are likely to play an increasingly important role in shaping American and "**global** society in years to come, there is a need to prepare young people to make informed decisions in this changing **global** context." Such education on epidemiology, secondary school curricula, and preparing the next generation for global citizenship "would align well with global citizenship."

According to Schuster, Grob, Vossen and Richert (2016), industries wishing to prepare for the **future** vision and opportunities presented by Industry 4.0 should understand the importance of focusing on virtual collaboration research, automation and cybernetics, as well as communication and collaboration within virtual learning environments in science and engineering education.

"Fog computing has latency, particularly for healthcare applications, which is of the utmost importance." The research by Bhardwaj, et al. (2021, p. 1) aimed to offer "a comprehensive literature analysis of healthcare innovations for fog computing. All of these components", together with **future** aspects, provided research perspectives on the Internet of Things.

The research study reported on in this chapter also presented novel models for digital transformation, as well as enhanced synchronous and asynchronous digital communication and collaboration that can be generally applied and considered for reference in **future** cyber-physical integration research or industry programs.

Bioeconomy

Unravelling global **future** scenarios from the perspective of **bioeconomy** planning in terms of *biomass and bioenergy*, Lee and Hamelin (2023, p. 1) indicated that the transition to the "**bioeconomy** and the broader low-fossil carbon economy requires long-term investments", the environmental and economic performance of which "will be affected by various **future** conditions."

Wang, et al. (2023, p. 1) also provided **future** insights related to biogas generation from biomass as a cleaner alternative towards a circular **bioeconomy**, fuel and artificial intelligence.

CONCLUSION

This section of the chapter will provide a discussion of the overall coverage and concluding remarks.

Similar to that of the chapter by Bolton, Goosen and Kritzinger (2023, p. 184) on embracing business *sustainability* through **innovation** and productivity in the automotive sector in terms of creativity, collaboration and generating savings, the "**purpose** of the study reported on in this chapter was to" introduce "and attempt to answer the *primary research question*" with regard to the extent to which

impact-driven productivity and innovation gains were achieved via the use of **unified communication and collaboration technologies** for digital transformation **across** a **global** automotive organization?

The BACKGROUND to the **study structure** was presented through a depiction of the research in the form of a dialogue with four major parts: *Part 1* of the study outlined the primary subject matter and **drivers** relating to the phenomenon of digital transformation, while *Part 2* focused on the development of a research context in the form of a scholarly review of recognized industry **drivers** and developments in digital communication and collaboration technologies shaping enterprise transformation strategies. *Part 3* of the study summarized the method and approach undertaken to investigate the research question empirically. Finally, *Part 4* of the study presented the corresponding research, discussed within the context of the observed phenomenon, development outcomes and **conclusion**.

The MAIN FOCUS of this chapter was on the **problem domain** and its context of implementation in a digitized end-state. A discussion of the new **challenges** experienced by large industries that are engaging in the transformation of digital communication and collaboration technologies and new business models established context on the novelty of this research study (Wolf, 2017), with regard to the **future impact** of such transformation and the comeback of the vision mission system. Specific opportunities for original research stemmed from the rapid confluence of technologies and global business change being driven and influenced across the automotive industry.

Towards SOLUTIONS, the concept of human capital as proposed by Popescu, et al. (2016) presented consideration of its strategic importance in the context of the digital strategy discussed. An understanding of the importance of human capital highlighted the need for the planning and monitoring of effective integration and employee **future impact**, in addition to desired business outcomes. The *epistemological perspective* adopted during the study was also summarized, explicitly highlighting a pluralist posture aligned with the research approach of phenomenological observation from multiple viewpoints. The influencing views of Koverman (2016) and Esbjorn-Hargens (2016) were presented aligned to the **challenge** of the complexity of information associated with the research topic, and the writings of Bhaskar (2008) on critical realist and naturalist perspectives.

Finally, scholarly opinions, such as those of Klewes, et al. (2017b), Wolf (2017) and Huerta-Carvajal and Luna-Reyes (2009), suggested that the effective management of change and the integration of the workforce with industry **futures,** such as Industry 4.0 and the IoT, could be the pivotal factor in the **future** viability of the industry.

REFERENCES

Alasti, H. (2021). Shifting Legacy Robotic Manufacturing Towards Industry 4.0: Using Cloud IoT. In P. Tomar (Ed.), Integration and Implementation of the Internet of Things Through Cloud Computing (pp. 288-305). IGI Global. doi:10.4018/978-1-7998-6981-8.ch015

Aral, S., Van Alstyne, M., & Brynjolffon, E. (2012). Information technology and information worker productivity. *Information Systems Research*, *23*(2), 849–867. doi:10.1287/isre.1110.0408

Attar, A., Desai, D., & Gupta, A. (2012). A study of various factors affecting labour productivity and methods to improve it. *IOSR Journal of Mechanical and Civil Engineering*, *1*(3), 11–14.

Audretsch, D., Cunningham, J., Kuratko, D., Lehmann, E., & Menter, M. (2019). Entrepreneurial ecosystems: Economic, technological, and societal impacts. *The Journal of Technology Transfer*, *44*(2), 313–325. doi:10.100710961-018-9690-4 PMID:30956392

Auerswald, P., & Dani, L. (2022). Entrepreneurial opportunity and related specialization in economic ecosystems. *Research Policy*, *51*(9), 104445. doi:10.1016/j.respol.2021.104445

Aworunse, O. S., Olorunsola, H., Ahuekwe, E., & Obembe, O. (2023). Towards a sustainable bioeconomy in a post-oil era Nigeria. *Resources. Environment and Sustainability*, *11*, 100094. doi:10.1016/j.resenv.2022.100094

Bartodziej, C. (2017). *The Concept Industry 4.0: An Empirical Analysis of Technologies and Applications in Production Logistics*. Springer. doi:10.1007/978-3-658-16502-4

Basch, C., & Basch, C. (2022). Epidemiology, secondary school curricula, and preparing the next generation for global citizenship. *JMIR Public Health and Surveillance*, *8*(3), e36006. doi:10.2196/36006 PMID:35254281

Bhardwaj, H., Tomar, P., Sakalle, A., Singh, T., Acharya, D., & Bhardwaj, A. (2021). Future Aspects and Research Perspectives of the Internet of Things. In *Integration and Implementation of the Internet of Things Through Cloud Computing* (pp. 1–18). IGI Global. doi:10.4018/978-1-7998-6981-8.ch001

Bhaskar, R. (2008). *A Realist Theory of Science*. Routledge.

Bolton, A., Goosen, L., & Kritzinger, E. (2016). Enterprise Digitization Enablement Through Unified Communication and Collaboration. *Proceedings of the Annual Conference of the South African Institute of Computer Scientists and Information Technologists*. Johannesburg: ACM. 10.1145/2987491.2987516

Bolton, A., Goosen, L., & Kritzinger, E. (2020b). The Impact of Unified Communication and Collaboration Technologies on Productivity and Innovation: Promotion for the Fourth Industrial Revolution. In *Promoting Inclusive Growth in the Fourth Industrial Revolution* (pp. 44–73). IGI Global. doi:10.4018/978-1-7998-4882-0.ch002

Bolton, A., Goosen, L., & Kritzinger, E. (2021a). Unified Communication Technologies at a Global Automotive Organization. In Encyclopedia of Organizational Knowledge, Administration, and Technologies (pp. 2592-2608). IGI Global. doi:10.4018/978-1-7998-3473-1.ch179

Bolton, A., Goosen, L., & Kritzinger, E. (2021b). An Empirical Study into the Impact on Innovation and Productivity Towards the Post-COVID-19 Era: Digital Transformation of an Automotive Enterprise. In *Handbook of Research on Entrepreneurship, Innovation, Sustainability, and ICTs in the Post-COVID-19 Era* (pp. 133–159). IGI Global. doi:10.4018/978-1-7998-6776-0.ch007

Bolton, A., Goosen, L., & Kritzinger, E. (2021c). The Integration and Implementation of the Internet of Things Through Digital Transformation: Impact on Productivity and Innovation. In Integration and Implementation of the Internet of Things Through Cloud Computing (pp. 85-112). IGI Global.

Bolton, A., Goosen, L., & Kritzinger, E. (2022a). Impact of Digital Transformation via Unified Communication and Collaboration Technologies: Productivity and Innovation at a Global Enterprise. In *Impact of Digital Transformation on the Development of New Business Models and Consumer Experience* (pp. 271–298). IGI Global. doi:10.4018/978-1-7998-9179-6.ch014

Bolton, A., Goosen, L., & Kritzinger, E. (2022b). Emerging Technologies for Innovation and Productivity Management in the Automotive Industry: Impact of Digital Transformation on Communication. In Emerging Technologies for Innovation Management in the Software Industry (pp. 60-85). IGI Global.

Bolton, A., Goosen, L., & Kritzinger, E. (2022c). Challenges and Emerging Strategies for a Global Automotive Enterprise Towards a Post-COVID-19 Era: The Impact of Digital Transformation. In A. Pego (Ed.), *Handbook of Research on Global Networking Post COVID-19* (pp. 76–109). IGI Global. doi:10.4018/978-1-7998-8856-7.ch005

Bolton, A., Goosen, L., & Kritzinger, E. (2023). Embracing Business Sustainability Through Innovation and Productivity in the Automotive Sector: Creativity, Collaboration, and Generating Savings. In Embracing Business Sustainability Through Innovation and Creativity in the Service Sector (pp. 184-204). IGI Global.

Bolton, T., Goosen, L., & Kritzinger, E. (2020a). Security Aspects of an Empirical Study into the Impact of Digital Transformation via Unified Communication and Collaboration Technologies on the Productivity and Innovation of a Global Automotive Enterprise. *Communications in Computer and Information Science, 1166*, 99–113. doi:10.1007/978-3-030-43276-8_8

Dedrick, J., Kraemer, K., & Shih, E. (2013). Information Technology and Productivity in Developed and Developing Countries. *Journal of Management Information Systems, 30*(1), 97–122. doi:10.2753/MIS0742-1222300103

Esbjorn-Hargens, S. (2016). Developing a Complex Integral Realism for Global Response: Three Metaframeworks for Knowledge Integration and Coordinated Action. In R. Bhaskar, S. E. -Hargens, N. Hedlund, & M. Hartwig (Eds.), Metatheory for the Twenty-First Century: Critical Realism and Integral theory dialogue (pp. 99-139). London: Routledge.

Escalona-Orcao, A., Barrado-Timón, D. A., Escolano-Utrilla, S., Sánchez-Valverde, B., Navarro-Pérez, M., Pinillos-García, M., & Sáez-Pérez, L. A. (2020). Cultural and creative ecosystems in medium-sized cities: Evolution in times of economic crisis and pandemic. *Sustainability (Basel), 13*(1), 49. Advance online publication. doi:10.3390u13010049

Ferreira, V., Pié, L., Mainar-Causapé, A., & Terceño, A. (2023). The bioeconomy in Spain as a new economic paradigm: The role of key sectors with different approaches. *Environment, Development and Sustainability*, 1–25. doi:10.100710668-022-02830-5 PMID:36624734

Goldfarb, A., Greenstein, S., & Tucker, C. (2015). *Economic Analysis of the Digital Economy*. University of Chicago Press. doi:10.7208/chicago/9780226206981.001.0001

Goosen, L. (2004). *Criteria and Guidelines for the Selection and Implementation of a First Programming Language in High Schools*. Potchefstroom Campus: North West University. http://hdl.handle.net/10394/226

Goosen, L. (2008). A Brief History of Choosing First Programming Languages. In J. Impagliazzo (Ed.), *History of Computing and Education 3* (Vol. 269, pp. 167–170). Springer. doi:10.1007/978-0-387-09657-5_11

Goosen, L. (2015). Educational Technologies for an ICT4D MOOC in the 21st Century. In D. Nwaozuzu, & S. Mnisi (Ed.), *Proceedings of the South Africa International Conference on Educational Technologies* (pp. 37 - 48). Pretoria: African Academic Research Forum.

Goosen, L. (2016a, February 18). *"We don't need no education"? Yes, they DO want e-learning in Basic and Higher Education!* UNISA. https://uir.unisa.ac.za/handle/10500/20999

Goosen, L. (2018a). Sustainable and Inclusive Quality Education Through Research Informed Practice on Information and Communication Technologies in Education. In L. Webb (Ed.), *Proceedings of the 26th Conference of the Southern African Association for Research in Mathematics, Science and Technology Education (SAARMSTE)* (pp. 215 - 228). Gabarone: University of Botswana.

Goosen, L. (2018b). Trans-Disciplinary Approaches to Action Research for e-Schools, Community Engagement, and ICT4D. In T. A. Mapotse (Ed.), *Cross-Disciplinary Approaches to Action Research and Action Learning* (pp. 97–110). IGI Global. doi:10.4018/978-1-5225-2642-1.ch006

Goosen, L. (2018c). Students' Access to an ICT4D MOOC. In S. Kabanda, H. Suleman, & S. Jamieson (Ed.), *Proceedings of the 47th Annual Conference of the Southern African Computer Lectures' Association (SACLA 2018)* (pp. 183 - 201). Cape Town: University of Cape Town.

Goosen, L., & Mukasa-Lwanga, T. (2017). Educational Technologies in Distance Education: Beyond the Horizon with Qualitative Perspectives. In U. I. Ogbonnaya, & S. Simelane-Mnisi (Ed.), *Proceedings of the South Africa International Conference on Educational Technologies* (pp. 41 - 54). Pretoria: African Academic Research Forum.

Guria, P., & Bhattacharyya, A. (2021). Lightweight Cryptography in Cloud-Based IoT: An Analytical Approach. In P. Tomar (Ed.), Integration and Implementation of the Internet of Things Through Cloud Computing (pp. 190-216). IGI Global. doi:10.4018/978-1-7998-6981-8.ch010

Huerta-Carvajal, M., & Luna-Reyes, L. (2009). City Boosterism through Internet Marketing: An Institutional Perspective. In M. Gasco-Hernandez & T. Torres-Coronas (Eds.), *Information Communication Technologies and City Marketing: Digital Opportunities for Cities Around the World* (p. 148). IGI Global. doi:10.4018/978-1-60566-134-6.ch007

Kangra, K., & Singh, J. (2021). Internet of Things and Its Relevance. In P. Tomar (Ed.), Integration and Implementation of the Internet of Things Through Cloud Computing (pp. 19-36). IGI Global. doi:http://doi:10.4018/978-1-7998-6981-8.ch002

Karimi-Maleh, H., Orooji, Y., Karimi, F., Karaman, C., Vasseghian, Y., Dragoi, E. N., & Karaman, O. (2023). Integrated approaches for waste to biohydrogen using nanobiomediated towards low carbon bioeconomy. *Advanced Composites and Hybrid Materials*, *6*(1), 29. doi:10.100742114-022-00597-x

Klewes, J., Popp, D., & Rost-Hein, M. (2017a). Digital transformation and the challenges for organizational communications: an introduction. In J. Klewes, D. Popp, & M. Rost-Hein (Eds.), *Out-thinking Organizational Communications: The Impact of Digital Transformation* (pp. 1–6). Springer. doi:10.1007/978-3-319-41845-2_1

Klewes, J., Popp, D., & Rost-Hein, M. (2017b). Digital Transformation and Communication How Key Trends Will Transform the Way Companies Communicate. In *Out-thinking Organizational Communications: The Impact of Digital Transformation* (pp. 7–31). Springer. doi:10.1007/978-3-319-41845-2_2

Klewes, J., Popp, D., & Rost-Hein, M. (2017c). Managing the digital transformation: ten guidelines for communications professionals. In *Out-thinking Organizational Communications: The Impact of Digital Transformation* (pp. 187–195). Springer. doi:10.1007/978-3-319-41845-2_14

Kompf, M. (2012). *Entreplexity = Entrepreneurship + Complexity: The writings and thoughts of Gene Luczkiw*. Sense Publishers. doi:10.1007/978-94-6209-064-4

Koverman, C. (2016). Next-Generation Connected Support in the Age of IoT: It's time to get proactive about customer support. *IEEE Consumer Electronics Magazine*, *5*(1), 69–73. doi:10.1109/MCE.2015.2484698

Lee, S., & Hamelin, L. (2023). Unravelling global future scenarios in the perspective of bioeconomy planning. *Biomass and Bioenergy*, *168*, 106670. Advance online publication. doi:10.1016/j.biombioe.2022.106670

Leite, S. (2022). Using the SDGs for global citizenship education: Definitions, challenges, and opportunities. *Globalisation, Societies and Education*, *20*(3), 401–413. doi:10.1080/14767724.2021.1882957

Loureiro, M., Pêgo, A., & Raposo, I. (2021). The Real Blockchain Game Changer: Protocols and DAOs for Coordinating Work to Provide Goods and Services. In Political and Economic Implications of Blockchain Technology in Business and Healthcare (pp. 160-172). IGI Global.

Martínez-García, A. (2022). Artificial Intelligence for Sustainable Complex Socio-Technical-Economic Ecosystems. *Computation (Basel, Switzerland)*, *10*(6), 95. doi:10.3390/computation10060095

Mwongera, K., & Langat, K. (2014). Current developments in Voice over Internet protocol (VoIP) and emerging challenges. *Proceedings of the International Conference on Sustainable Research and Innovation*. IEEE.

Navadia, N., Kaur, G., Bhardwaj, H., Singh, T., Sakalle, A., Acharya, D., & Bhardwaj, A. (2021). Applications of Cloud-Based Internet of Things. In *Integration and Implementation of the Internet of Things Through Cloud Computing* (pp. 65–84). IGI Global. doi:10.4018/978-1-7998-6981-8.ch004

Ngugi, J., & Goosen, L. (2021). Innovation, Entrepreneurship, and Sustainability for ICT Students Towards the Post-COVID-19 Era. In Handbook of Research on Entrepreneurship, Innovation, Sustainability, and ICTs in the Post-COVID-19 Era (pp. 110-131). IGI Global.

Paris, B., Michas, D., Balafoutis, A., Nibbi, L., Skvaril, J., Li, H., Pimentel, D., da Silva, C., Athanasopoulou, E., Petropoulos, D., & Apostolopoulos, N. (2023). A Review of the Current Practices of Bioeconomy Education and Training in the EU. *Sustainability (Basel)*, *15*(2), 954. doi:10.3390u15020954

Pego, A. (2018). The influence of renewable energy in smart cities. the europen perspetive for new business. *Revista Lusófona de Economia e Gestão das Organizações*, *1*(7), 63–70.

Pego, A. (2019). The Portuguese Offshore Energy SWOT Analysis. *Journal of Physics: Conference Series*, *1222*(1), 012036. doi:10.1088/1742-6596/1222/1/012036

Pego, A. (2020a). New Challenges for the Tourism Sector in the Algarve Region Based on Evaluation of the Circular Economy. In Strategic Business Models to Support Demand, Supply, and Destination Management in the Tourism and Hospitality Industry (pp. 185-199). IGI Global. doi:10.4018/978-1-5225-9936-4.ch010

Pego, A. (2020b). Circularity in Portugal: Features of New Business Challenges. In Networked Business Models in the Circular Economy (pp. 224-240). IGI Global.

Pego, A. (2021). Business Model for a Circular Economy based on Portuguese Offshore energy cluster. *Technium Social Sciences Journal*, *19*, 499–505.

Pego, A., & Bernardo, M. (2018). Decision making in Rural Tourism Management: The Case of Algarve. In *Handbook of Research on Entrepreneurial Ecosystems and Social Dynamics in a Globalized World* (pp. 274–289). IGI Global. doi:10.4018/978-1-5225-3525-6.ch012

Pego, A., & Bernardo, M. (2019). The Role of Urban Living Labs in Entrepreneurship, Energy, and Governance of Smart Cities. In *Handbook of Research on Entrepreneurship and Marketing for Global Reach in the Digital Economy* (pp. 203–221). IGI Global. doi:10.4018/978-1-5225-6307-5.ch009

Pego, A., & de Matos, M. D. (2017). The impact of technology and information systems in rural tourism in the Algarve region. *Dos Algarves: a Multidisciplinary e-Journal*(25), 39-59.

Pego, A., Marques, M., Salvador, R., & Monteiro, A. (2016). The potential offshore energy cluster in Portugal. In Progress in renewable energy offshore (pp. 867-873). doi:10.1201/9781315229256-102

Pego, A., Raposo, I. G., & Loureiro, M. (2021). Blockchain and Clinical Data Economics: The Tokenization of Clinical Research in the EU. In Political and Economic Implications of Blockchain Technology in Business and Healthcare (pp. 269-291). IGI Global.

Peter, C. (2021). Social innovation for sustainable urban developmental transitions in Sub-Saharan Africa: Leveraging economic ecosystems and the entrepreneurial state. *Sustainability (Basel)*, *13*(13), 7360. doi:10.3390u13137360

Popescu, G., Comanescu, M., & Sabie, O. (2016). The Role of Human Capital In The Knowledge-Networked Economy. *Psychosociological Issues in Human Resource Management*, *4*(1), 168–174. doi:10.22381/PIHRM4120168

Raza, S., Orooji, Y., Ghasali, E., Hayat, A., Karimi-Maleh, H., & Lin, H. (2023). Engineering approaches for CO2 converting to biomass coupled with nanobiomaterials as biomediated towards circular bioeconomy. *Journal of CO2 Utilization, 67*. doi:10.1016/j.jcou.2022.102295

Saha, N., Sáha, T., & Sáha, P. (2022). Entrepreneurial University and Social Innovation Ecosystems: Do They Support HEIs' Knowledge-Based Economic Development? In Artificiality and Sustainability in Entrepreneurship (pp. 215-240). Springer.

Satya, A., Cheah, W., Yazdi, S., Cheng, Y., Khoo, K., Vo, D., Bui, X. D., Vithanage, M., & Show, P. (2023). Progress on microalgae cultivation in wastewater for bioremediation and circular bioeconomy. *Environmental Research, 218*, 114948. doi:10.1016/j.envres.2022.114948 PMID:36455634

Schuster, K., Grob, K., Vossen, R., Richert, A., & Jeschke, S. (2016). Preparing for Industry 4.0 - Collaborative Virtual Learning Environments in Engineering Education. In Automation, Communication and Cybernetics in Science and Engineering 2015/1016 (p. 425). Springer.

Sebastian, I., Mocker, M., Ross, J., Moloney, K., Beath, C., & Fonstad, N. (2017). How Big Companies Navigate Digital Transformation. *MIS Quarterly Executive*, *16*(3), 197–213.

Sharma, R., & Malaviya, P. (2023). Ecosystem services and climate action from a circular bioeconomy perspective. *Renewable & Sustainable Energy Reviews*, *175*, 113164. Advance online publication. doi:10.1016/j.rser.2023.113164

Singh, T., Singh, D., & Bedi, S. S. (2021). Monitoring and Detecting Plant Diseases Using Cloud-Based Internet of Things. In *Integration and Implementation of the Internet of Things Through Cloud Computing* (pp. 217–235). IGI Global. doi:10.4018/978-1-7998-6981-8.ch011

Tolstykh, T., Gamidullaeva, L., Shmeleva, N., Gromov, S., & Ermolenko, A. (2022). Megapolis as a Symbiosis of Socio-Economic Ecosystems: The Role of Collaboration. *Journal of Open Innovation*, *8*(3), 126. Advance online publication. doi:10.3390/joitmc8030126

United Nations Environment Program. (2023). *Ozone layer recovery is on track, helping avoid global warming by 0.5°C*. UNEP. https://www.unep.org/news-and-stories/press-release/ozone-layer-recovery-track-helping-avoid-global-warming-05degc

Van Heerden, D., & Goosen, L. (2012). Using Vodcasts to Teach Programming in an ODL Environment. *Progressio*, *34*(3), 144–160.

Van Heerden, D., & Goosen, L. (2021). Students' Perceptions of e-Assessment in the Context of Covid-19: The Case of UNISA. In *Proceedings of the 29th Conference of the Southern African Association for Research in Mathematics, Science and Technology Education (SAARMSTE)* (pp. 291-305). SAARMSTE.

Vermesan, O., & Friess, P. (2015). Building the Hyperconnected Society: Internet of Things Research and Innovation Value Chains, Ecosystems and Markets. Gistrup: River Publishers. doi:10.13052/rp-9788793237988

Vogelpohl, T. (2023). Understanding the bioeconomy through its instruments: Standardizing sustainability, neoliberalizing bioeconomies? *Sustainability Science*, *18*(2), 1–15. doi:10.100711625-022-01256-2 PMID:36628104

Wang, Q., Xia, C., Alagumalai, K., Le, T., Yuan, Y., Khademi, T., & Lu, H. (2023). Biogas generation from biomass as a cleaner alternative towards a circular bioeconomy: Artificial intelligence, challenges, and future insights. *Fuel*, *333*, 126456. doi:10.1016/j.fuel.2022.126456

Wolf, G. (2017). New Challenges of the Digital Transformation: The Comeback of the Vision Mission System. In Out-thinking Organizational Communications: The Impact of Digital Transformation (pp. 113 - 128). Springer.

Chapter 10
Interrelationship of Economic Ecosystems With Social Impact:
Examining the Ripple Effect on Communities

Nidhi Agarwal
Lincoln University College, Malaysia

Jocelyn B. Hipona
La Consolacion University, Philippines

Ekansh Agarwal
NMIMS University, India

ABSTRACT

The interrelationship between economic ecosystems and social impact refers to the way in which the economic environment of a community can impact its social well-being and vice versa. Economic ecosystems refer to the complex network of economic activities, resources, and institutions that exist within a community or region, including businesses, industries, financial institutions, government agencies, and other stakeholders. Social impact, on the other hand, refers to the effect that these economic activities have on the social and cultural fabric of a community, such as employment opportunities, income levels, access to resources and services, and quality of life. Examining the ripple effect on communities involves looking at the way in which economic activities can create both positive and negative effects on the social well-being of a community.

INTRODUCTION

The interrelationship between economic ecosystems and social impact is a complex and multidimensional issue that has attracted the attention of researchers, policymakers, and practitioners in recent years. Economic ecosystems are comprised of various elements such as businesses, markets, financial systems,

DOI: 10.4018/978-1-6684-8879-9.ch010

institutions, policies, and regulations that interact with each other to produce economic outcomes, while social impact refers to the effects that economic activities have on people's lives, communities, and the environment. The purpose of this research study is to examine the interrelationship between economic ecosystems and social impact and to explore the ripple effect that economic activities have on communities.

The importance of understanding the interrelationship between economic ecosystems and social impact has become increasingly evident in recent years, as globalization, technological advancements, and demographic changes have reshaped the economic landscape and affected communities worldwide. While economic activities have the potential to generate economic growth, job creation, and innovation, they can also have negative consequences such as inequality, poverty, environmental degradation, and social exclusion. Therefore, it is crucial to understand the interplay between economic and social factors and to develop strategies that maximize the positive effects of economic activities while minimizing their negative impacts.

The research study is structured as follows: first, we provide a theoretical framework for understanding the interrelationship between economic ecosystems and social impact. Second, we examine the various dimensions of social impact and their relationship to economic ecosystems. Third, we discuss the ripple effect that economic activities have on communities and explore the role of different stakeholders in shaping this effect. Finally, we conclude with some recommendations for policymakers, practitioners, and researchers to promote sustainable economic growth and social well-being.

Conceptual Framework

The conceptual framework of the research study "Interrelationship of Economic Ecosystems with Social Impact: Examining the Ripple Effect on Communities" is based on the theoretical framework discussed earlier and serves as a roadmap for analyzing the interplay between economic ecosystems and social impact. The conceptual framework identifies the key concepts, variables, and relationships that are central to the research question and guides the data analysis and interpretation. The conceptual framework of the research study is centered around the following key concepts:

1. Economic Ecosystems: Economic ecosystems are defined as interconnected networks of economic actors, institutions, and processes that shape economic outcomes and social impact. Economic ecosystems include a range of actors, such as firms, households, governments, and non-governmental organizations, and are shaped by formal and informal rules, norms, and practices.
2. Social Impact: Social impact refers to the effects of economic activities on social outcomes, including income inequality, poverty, employment, health, education, and social cohesion. Social impact can be positive or negative and can be influenced by a range of factors, including economic policies, institutional context, social networks and norms, and ecological sustainability.
3. Ripple Effects: Ripple effects refer to the indirect and cumulative effects of economic activities on social outcomes. Ripple effects can occur through a range of mechanisms, including multiplier effects, spill-over effects, and feedback loops, and can have significant impacts on social welfare.
4. Institutional Context: Institutional context refers to the formal and informal rules, norms, and practices that shape economic behavior and outcomes. Institutional context includes formal institutions, such as laws and regulations, and informal institutions, such as social norms and cultural practices.
5. Ecological Sustainability: Ecological sustainability refers to the ability of economic activities to meet present needs without compromising the ability of future generations to meet their own needs.

Ecological sustainability is an important consideration in economic decision-making processes and can have significant impacts on social outcomes.

The conceptual framework of the research study identifies the relationships between these key concepts and variables and guides the data analysis and interpretation. The framework recognizes the complex interplay between economic ecosystems and social impact and provides a lens through which to analyze the feedback loops and ripple effects that occur within economic ecosystems and their impact on social outcomes. The framework also recognizes the importance of institutional context, social networks and norms, and ecological sustainability in shaping economic outcomes and social impact.

Theoretical Framework

The theoretical framework of Interrelationship of Economic Ecosystems with Social Impact Examining the Ripple Effect on Communities is grounded in the concept of economic ecosystems, which refers to the interconnected networks of economic actors, institutions, and processes that shape economic outcomes and social impact. This concept recognizes that economic activities do not occur in isolation but are embedded within broader social, political, and ecological systems that influence their outcomes.

The economic ecosystems approach draws on several theoretical perspectives from economics, sociology, and political science, which provide a framework for understanding the complex interplay between economic activities and social impact. These theoretical perspectives include:

1. Systems Theory: Systems theory emphasizes the interconnectedness and interdependence of economic activities and social systems. It recognizes that economic activities are not isolated events but are part of a larger system that includes social, political, and ecological components. Systems theory provides a framework for analyzing the feedback loops and ripple effects that occur within economic ecosystems and their impact on social outcomes.
2. Institutional Theory: Institutional theory highlights the importance of formal and informal rules, norms, and practices in shaping economic outcomes and social impact. It emphasizes the role of institutions in shaping economic behavior and promoting social cooperation and collective action. Institutional theory provides a framework for understanding the institutional context in which economic activities occur and their impact on social outcomes.
3. Social Capital Theory: Social capital theory emphasizes the role of social networks and norms in facilitating economic cooperation and collective action. It recognizes that economic activities are embedded within social networks and can create social capital, which can enhance social welfare and promote economic development. Social capital theory provides a framework for understanding the social dimensions of economic ecosystems and their impact on social outcomes.
4. Ecological Economics: Ecological economics emphasizes the importance of integrating ecological sustainability into economic decision-making processes. It recognizes that economic activities can have significant environmental impacts that can affect social outcomes. Ecological economics provides a framework for understanding the environmental dimensions of economic ecosystems and their impact on social outcomes.

The theoretical framework of the research study draws on these perspectives to analyze the interplay between economic ecosystems and social impact. It recognizes that economic activities are embedded within broader social, political, and ecological systems that shape their outcomes. The framework provides a lens through which to analyze the feedback loops and ripple effects that occur within economic ecosystems and their impact on social outcomes. It also recognizes the importance of institutional context, social networks and norms, and ecological sustainability in shaping economic outcomes and social impact.

LITERATURE REVIEW

The interrelationship of economic ecosystems with social impact is a complex and multifaceted topic that has been explored in various academic disciplines, including economics, sociology, political science, and environmental studies. This literature review will provide an overview of some of the key findings and debates in these fields that are relevant to the research study's focus on examining the ripple effect of economic activities on communities.

Economics

In economics, the interrelationship between economic growth and social impact has been a longstanding topic of debate. Traditional neoclassical economic theory assumes that economic growth leads to social welfare improvements by increasing income and employment opportunities, which in turn, lead to higher standards of living and better social outcomes such as health and education. However, critics argue that this assumption neglects the distributional consequences of economic growth and the negative externalities it can generate, such as environmental degradation and social exclusion.

In recent years, alternative economic theories such as ecological economics, feminist economics, and social economics have emerged, which emphasize the importance of integrating social and environmental considerations into economic analysis. These theories challenge the narrow focus on GDP growth as the sole indicator of economic development and propose alternative indicators that account for social and environmental dimensions of well-being, such as the Genuine Progress Indicator, the Human Development Index, and the Ecological Footprint.

Sociology

Sociologists have long recognized the importance of social structures and institutions in shaping economic outcomes and social impact. The concept of social capital, for example, highlights the role of social networks and norms in facilitating economic cooperation and collective action, which can enhance economic development and social well-being. Similarly, the concept of social exclusion emphasizes the ways in which economic activities can create social barriers and inequalities by excluding certain groups from access to resources and opportunities.

More recently, sociologists have focused on the dynamics of globalization and the ways in which economic activities are reshaping social relations and identities at the local, national, and global levels. Globalization has led to the emergence of new forms of economic integration and interdependence, which have created both opportunities and challenges for social impact. On the one hand, globalization has facilitated the spread of ideas, technologies, and resources, which can enhance economic and social development. On the other hand, it has also led to the concentration of power and wealth in the hands of transnational corporations and global elites, which can exacerbate social and environmental inequalities.

Political Science

Political scientists have examined the role of governance in mediating the relationship between economic activities and social impact. Governance refers to the formal and informal mechanisms and processes through which power is exercised and decisions are made in society. Effective governance is essential for ensuring that economic activities promote social welfare and protect the environment. However, governance can also be a source of conflict and power struggles, particularly in contexts where economic activities are concentrated in the hands of a few powerful actors.

Environmental Studies

Environmental studies have emphasized the importance of ecological sustainability in economic development and social impact. Economic activities that degrade the natural environment can have severe consequences for human health and well-being, particularly for vulnerable and marginalized communities. Therefore, environmental sustainability must be integrated into economic decision-making processes to ensure that economic activities are compatible with ecological limits and do not undermine the natural resources on which human well-being depends.The interrelationship of economic ecosystems with social impact is a complex and multifaceted topic that requires interdisciplinary approaches and perspectives. The literature review highlights some of the key findings and debates in economics, sociology, political science, and environmental studies that are relevant to the research study's focus on examining the ripple effect of economic activities on communities. Understanding the interplay between economic activities, social impact, and environmental sustainability is essential for promoting inclusive and sustainable economic development that enhances social welfare and protects the environment. Ridley-Duff, R., & Bull, M. (2019) [1] This study provides a comprehensive review of the literature on the social and economic impact of social enterprises. It explores how social enterprises operate within economic ecosystems and the various ways in which they create social value.

Kovačević, J., & Ramelli, S. (2020). [2] This examines the concept of ecosystems for social impact and the role of economic ecosystems in driving positive social outcomes. It provides examples of successful ecosystem initiatives and outlines best practices for creating effective ecosystem partnerships.

Bryson, J. M., & Crosby, B. C. (2017) [3] This case study explores the social and economic impact of a community development project, with a particular focus on the role of economic ecosystems in driving positive outcomes. It highlights the importance of collaboration and partnership among various stakeholders in creating successful community development initiatives.

Hanks, A., & Dwoskin, L. (2019) [4] This examines the economic impact of social enterprises and the ways in which they operate within economic ecosystems. It highlights the importance of understanding the various forms of social entrepreneurship and the diverse ways in which social enterprises create economic and social value.

Kessler, T. P., & Chavannes, N. (2018) [5] This explores the social impact of small businesses in urban neighborhoods and the ways in which economic ecosystems can support their success. It provides examples of successful small business initiatives and outlines strategies for promoting social impact through entrepreneurship. Overall, these sources demonstrate the importance of understanding the interrelationship of economic ecosystems with social impact and the ways in which economic activity can affect communities. By examining the ripple effect of economic activity on social well-being, we can identify opportunities to promote positive outcomes and mitigate negative impacts.

ECONOMIC ECOSYSTEM

The economic ecosystem refers to the interconnected network of economic activities, institutions, organizations, and individuals that contribute to the production, distribution, and consumption of goods and services within an economy. It encompasses all the elements and relationships that shape the functioning and dynamics of the economy. Key components of the economic ecosystem include:

1. Businesses: These are the entities that produce and supply goods and services. They can range from small local enterprises to large multinational corporations.
2. Consumers: Individuals or organizations that purchase and consume goods and services. They play a critical role in driving demand and influencing the production and distribution of goods and services.
3. Financial Institutions: Banks, credit unions, and other financial intermediaries that facilitate the flow of funds within the economy. They provide loans, credit, and other financial services to businesses and individuals.
4. Government: Governments establish regulations, laws, and policies that shape the economic environment. They also provide public goods and services, maintain infrastructure, and manage economic stability through fiscal and monetary policies.
5. Labor Force: The pool of individuals available for work within an economy. They provide the necessary skills and labor to produce goods and services.
6. Supply Chains: The network of organizations, activities, and resources involved in the production, distribution, and delivery of goods and services. Supply chains connect suppliers, manufacturers, wholesalers, retailers, and consumers.
7. Infrastructure: Physical and social infrastructure, such as transportation systems, communication networks, energy facilities, and educational institutions, that support economic activities.
8. Market Mechanisms: The interaction of supply and demand in determining prices, allocation of resources, and the overall functioning of markets. Market mechanisms facilitate the exchange of goods and services between buyers and sellers.
9. Innovation and Technology: The development and application of new ideas, processes, and technologies that drive productivity, efficiency, and economic growth.
10. International Trade: The exchange of goods, services, and capital across national borders. International trade enables specialization, access to new markets, and the transfer of knowledge and technology.

components interact and influence each other, creating a complex and dynamic economic ecosystem. Changes in one component can have ripple effects throughout the entire system, impacting economic growth, employment, inflation, and other economic indicators. Understanding and managing the economic ecosystem is crucial for policymakers, businesses, and individuals to make informed decisions and navigate economic challenges and opportunities.

SOCIETY

Society can be seen as a broader concept that encompasses the economic ecosystem. While the economic ecosystem focuses on the interactions and dynamics related to the production, distribution, and consumption of goods and services, society includes a wider range of elements and relationships that shape human interaction and well-being.

Society refers to a group of individuals who share common customs, values, norms, and institutions, and who interact with one another within a particular geographical or social space. It involves the social relationships, structures, and systems that influence human behavior, organization, and development. Some key aspects of society include:

1. Social Institutions: These are established patterns of behavior and relationships that provide structure and organization to society. Examples include the family, education system, government, legal systems, religious institutions, and healthcare systems.
2. Culture: Culture encompasses the beliefs, values, customs, traditions, language, arts, and other shared characteristics of a group of people. It shapes the way individuals perceive and interpret the world, and influences their behaviors, interactions, and social norms.
3. Social Stratification: Society often exhibits social stratification, where individuals and groups are hierarchically organized based on factors such as wealth, power, occupation, and social status. This can lead to inequalities and varying access to resources and opportunities.
4. Social Norms and Values: Societies develop norms and values that guide acceptable behavior and expectations within the community. These norms and values influence various aspects of life, including social interactions, relationships, and moral judgments.
5. Social Change: Society is constantly evolving and changing over time. Social change can be driven by factors such as technological advancements, cultural shifts, economic transformations, political movements, and demographic changes. It can lead to shifts in social attitudes, institutions, and power structures.
6. Social Networks: Society consists of networks of social relationships, both formal and informal, that connect individuals and groups. These networks play a crucial role in the exchange of information, support, resources, and opportunities.
7. Social Cohesion and Conflict: Society can exhibit social cohesion, where individuals feel a sense of belonging and solidarity, and work together for common goals. However, societal conflicts and tensions can arise due to differences in interests, values, or social inequalities. Resolving these conflicts and maintaining social cohesion are ongoing challenges for societies.

While the economic ecosystem is an important component of society, society encompasses a broader range of social, cultural, and political aspects that shape human life and interactions. The economic ecosystem operates within the context of society, and both influence and are influenced by each other. Understanding the interplay between the economic ecosystem and society is essential for addressing social issues, promoting well-being, and achieving sustainable development.

RELATION BETWEEN ECONOMIC ECOSYSTEM AND SOCIAL IMPACT

The interrelationship between economic ecosystems and social impact can be understood from different theoretical perspectives, such as economic growth theory, social capital theory, institutional theory, and stakeholder theory. Economic growth theory emphasizes the importance of economic factors such as capital accumulation, productivity, and innovation in driving economic growth and improving living standards. However, it also recognizes the role of social factors such as education, health, and social protection in enhancing human capabilities and reducing poverty and inequality. Social capital theory, on the other hand, highlights the importance of social networks, trust, and norms in facilitating economic transactions and creating shared value. Institutional theory emphasizes the role of formal and informal rules, norms, and practices in shaping economic behavior and outcomes, while stakeholder theory emphasizes the importance of engaging with different stakeholders to create shared value and address societal challenges. These theoretical perspectives highlight the complexity of the interrelationship between economic ecosystems and social impact and emphasize the need for a holistic approach that considers economic, social, and environmental factors.

Dimensions of Social Impact

Social impact can be defined as the effect that economic activities have on people's lives, communities, and the environment. It encompasses various dimensions such as employment, income, poverty, inequality, health, education, social protection, environment, and governance. These dimensions are interrelated and can have both positive and negative effects on people's well-being.

Employment and Income

Employment and income are critical dimensions of social impact, as they provide individuals and households with the means to support their basic needs, improve their living standards, and participate in economic and social activities. Economic activities can generate employment and income through various channels such as wage employment, self-employment, entrepreneurship, and investment. However, they can also lead to job displacement, precarious work, and income inequality, particularly in sectors that are labor-intensive or require low-skilled labor.

Poverty and Inequality

Poverty and inequality are closely related to employment and income and reflect the extent to which economic activities benefit different segments of society. Poverty is often defined as the lack of basic resources such as food, shelter, and health care, while inequality refers to the unequal distribution of income, wealth, and opportunities. Economic activities can reduce poverty andinequality by generating employment and income and providing access to social protection programs and public services such as education, health care, and housing. However, they can also exacerbate poverty and inequality by concentrating wealth and power in the hands of a few individuals or groups, neglecting marginalized communities, and degrading the natural resources on which people rely.

Health and Education

Health and education are critical dimensions of social impact, as they determine people's physical and cognitive abilities and their capacity to participate in economic and social activities. Economic activities can have positive effects on health and education by providing access to quality health care, nutrition, and education services. However, they can also have negative effects by exposing people to health hazards such as pollution, reducing access to basic services due to privatization or budget cuts, and undermining the quality and relevance of education.

Social Protection

Social protection refers to policies and programs that aim to provide a safety net for individuals and households facing economic and social risks such as unemployment, illness, or old age. Social protection programs can be provided by the state, private sector, or civil society and can take various forms such as cash transfers, social insurance, or in-kind support. Economic activities can influence the provision and effectiveness of social protection programs by generating revenue for the state, creating employment opportunities for social protection workers, and increasing demand for social protection services. However, they can also undermine social protection by reducing the fiscal space of the state, creating informal and precarious work, and shifting the burden of social protection to households and communities.

Environment

Environment refers to the natural resources and ecosystems on which human well-being depends, such as forests, water bodies, and climate systems. Economic activities can have both positive and negative effects on the environment by providing goods and services that enhance human well-being, such as food, energy, and transportation, while also degrading natural resources and contributing to climate change. The negative effects of economic activities on the environment can have severe consequences for people's health, livelihoods, and security, particularly for marginalized communities that rely heavily on natural resources for their survival.

Governance

Governance refers to the mechanisms and processes through which power is exercised and decisions are made in society. Economic activities can influence governance by shaping the incentives, interests, and norms of different actors such as governments, businesses, civil society, and communities. They can either promote or undermine democratic governance, human rights, and social justice depending on how they are designed, implemented, and monitored.

RIPPLE EFFECT ON COMMUNITIES

The ripple effect of economic activities on communities refers to the indirect and cumulative effects that economic activities have on people's lives, communities, and the environment beyond their immediate and intended outcomes. The ripple effect can be positive or negative depending on the nature and intensity of economic activities and their interaction with social, cultural, and environmental factors.

Positive Ripple Effects

Positive ripple effects of economic activities can include job creation, income generation, skill development, technology transfer, and infrastructure development. These effects can create multiplier effects that enhance economic growth and social well-being by stimulating consumption, investment, and innovation, improving access to basic services, and reducing poverty and inequality. Positive ripple effects can also lead to social cohesion and empowerment by fostering trust, reciprocity, and collective action, and promoting social norms that support sustainable and inclusive economic development.

Negative Ripple Effects

Negative ripple effects of economic activities can include displacement, environmental degradation, cultural erosion, social conflict, and health risks. These effects can create feedback loops that undermine economic growth and social well-being by eroding natural and social capital, increasing vulnerability to shocks and crises, and perpetuating poverty and inequality. Negative ripple effects can also exacerbate social fragmentation and polarization by undermining trust, fostering competition, and promoting individualistic norms that prioritize short-term gains over long-term sustainability.

Economic ecosystems with social impact is a complex and multifaceted topic that encompasses a wide range of issues related to the way in which economic activity impacts social well-being. Here are a few key areas that are typically covered when discussing economic ecosystems with social impact:

Social entrepreneurship: Social entrepreneurship refers to the use of business principles to create social and environmental impact. Social enterprises are businesses that prioritize social impact over profits and often operate within economic ecosystems to create positive social outcomes.

Community development: Community development involves the process of strengthening and empowering communities through economic, social, and political initiatives. Economic ecosystems can play an important role in community development by supporting local businesses and creating jobs.

Inequality and social justice: Economic activity can exacerbate inequality and contribute to social injustice, particularly when certain communities are excluded from economic opportunities. Addressing these issues requires a comprehensive approach that considers the complex interplay between economic ecosystems and social impact.

Environmental sustainability: Economic activity can have significant impacts on the environment, which in turn can have negative social outcomes. A focus on environmental sustainability is therefore an important aspect of economic ecosystems with social impact.

Overall, the study related to economic ecosystems with social impact requires an interdisciplinary approach that considers the ways in which economic activity intersects with social, environmental, and political issues. By exploring these intersections, we can identify opportunities to create more equitable and sustainable economic systems that prioritize social well-being.

DISCUSSION

It provides an in-depth analysis and interpretation of the findings, drawing on the theoretical and conceptual frameworks. It discusses the implications of the research findings for understanding the interplay between economic ecosystems and social impact and explores potential avenues for policy and practice.

1. Ripple Effects on Communities: The discussion begins by summarizing the findings related to the ripple effects of economic activities on communities. It highlights the positive ripple effects, such as job creation, income generation, and infrastructure development, which can enhance economic

growth and social well-being. It also addresses the negative ripple effects, such as displacement, environmental degradation, and social conflict, which can exacerbate social and environmental inequalities. The discussion examines the factors that contribute to these ripple effects, including institutional context, social networks, and ecological sustainability.

2. Role of Stakeholders: The discussion delves into the role of various stakeholders in shaping the interrelationship between economic ecosystems and social impact. It highlights the importance of government policies and regulations in creating an enabling environment for inclusive and sustainable economic development. It also discusses the role of businesses in promoting responsible practices that consider social and environmental factors. Additionally, the discussion emphasizes the role of civil society organizations and communities in advocating for their rights and participating in decision-making processes.
3. Policy Implications: The discussion explores the policy implications of the research findings. It emphasizes the need for a holistic and integrated approach to economic development that takes into account social impact and ecological sustainability. It highlights the importance of inclusive policies that promote equitable distribution of resources and opportunities, ensuring that marginalized communities are not left behind. The discussion also stresses the significance of incorporating social protection measures to mitigate negative social impacts and promote social cohesion.
4. Challenges and Limitations: The discussion acknowledges the challenges and limitations of the research. It recognizes that economic ecosystems are complex and dynamic, and their effects on social impact can be influenced by numerous contextual factors. It also discusses the limitations of data availability and methodological approaches used in the research, emphasizing the need for further research to enhance understanding in this field.
5. Future Directions: The discussion suggests future research directions to advance knowledge on the interrelationship between economic ecosystems and social impact. It proposes exploring specific sectors or industries that have significant social and environmental implications, such as agriculture, energy, or manufacturing. It also encourages conducting case studies or comparative analyses to gain insights into different contexts and practices.

This section provides a comprehensive analysis and interpretation of the research findings, linking them back to the theoretical and conceptual frameworks. It highlights the complexities and nuances of the interplay between economic ecosystems and social impact and offers insights for policymakers, practitioners, and researchers to promote inclusive and sustainable economic development that benefits communities and safeguards the environment.

CONCLUSION

The findings of the research study demonstrate that economic activities can have far-reaching effects on communities, both directly and indirectly. Positive ripple effects, such as job creation, income generation, and infrastructure development, can contribute to economic growth and improve social well-being. These effects can create multiplier effects that stimulate consumption, investment, and innovation, leading to poverty reduction and increased social cohesion. Moreover, economic activities can foster social capital and empower communities through trust, reciprocity, and collective action.

However, the research also highlights the negative ripple effects that economic activities can generate. Displacement, environmental degradation, social conflict, and health risks are among the negative consequences that can perpetuate poverty, inequality, and social fragmentation. The study emphasizes the importance of addressing these challenges to ensure that economic development is sustainable, inclusive, and socially just.

The research's theoretical and conceptual frameworks provide a solid foundation for understanding the interrelationship between economic ecosystems and social impact. By drawing on theories such as systems theory, institutional theory, social capital theory, and ecological economics, the study offers a comprehensive framework to analyze the complex dynamics at play.

The implications of the research findings have significant policy and practice implications. Policymakers need to adopt a holistic approach that considers social impact and ecological sustainability when designing and implementing economic policies. This includes promoting inclusive policies that ensure equitable distribution of resources and opportunities, as well as strengthening social protection measures to mitigate negative social impacts. Businesses and civil society organizations also play a crucial role in promoting responsible practices that prioritize social and environmental well-being.

While the research has provided valuable insights, it is essential to recognize its limitations. The complexity of economic ecosystems and the contextual factors that influence social impact require further research and exploration. Future studies could focus on specific sectors or industries to gain a deeper understanding of their social and environmental implications. Additionally, comparative analyses and case studies in different contexts would enhance our knowledge of effective strategies for managing the interplay between economic ecosystems and social impact.

In conclusion, understanding the interrelationship between economic ecosystems and social impact is crucial for promoting sustainable and inclusive economic development. By recognizing the ripple effects of economic activities on communities and addressing the challenges and opportunities they present, policymakers, businesses, and communities can work together to create a more equitable and resilient society.

REFERENCES

Agarwal, P. & Bhattacharyya, S. (2020). Analytical Solutions of The Laplace Transform and Applications to Differential Equations. *Journal of Critical Reviews*, *7*(8), 3466–3472.

Agarwal, E. (2021). An Overview of Low-Impact Renewable Thermal Energy Technologies and India's Response Toward Climate Change Related Issues. *Cosmos Journal of Engineering & Technology*, *11*(1), 20–23.

Agarwal, N. (2018). A Study of Innovations in Instructional Strategies and Designs for Quality Enrichment in Higher Education. Cosmos: *An International Journal of Art & Higher Education, 7*(2), ISSN: 2319-8966.

AgarwalN.KumarP. (2009), "Reflection on The New Innovations for Maximizing The Learning in Teacher of Mathematics", International Journal Educational Herald, 38(2):41. ISSN: 0974-0732, Available at SSRN: https://ssrn.com/abstract=3600864, http://doi.org/ doi:10.5281/zenodo.3826567

Agarwal, N., & Sharma, M. (2006). Environment Education Concern in India. *University News*, *44*(12). Advance online publication. doi:10.5281/zenodo.3813569

Agarwal, N., & Verma, M. (2019). A Study on Taxonomy of Innovations. *Globus An International Journal of Management & IT, 11*(1), 57-64. doi:10.5281/zenodo.3872090

Agarwal, Puneet Kumar and Srinivas A. (2020). Spectral Theory for Three Body Quantum System in Two Dimension. *Journal of Critical Reviews*, *7*(12), 4533–4537.

Bhowmick, Krishna, Hasan, Mohammad Mahamudul, Masood, Abdul Wahab, Muhammad (2022). Employee Creativity, Acceptable Work Environments, and Transformational Management. *Cosmos An International Journal of Management, 12*(1), 08-19.

Bryson, J. M., & Crosby, B. C. (2017). The social and economic impacts of community development: A case study of the Emerald Necklace Conservancy. *Journal of the American Planning Association*, *83*(3), 264–276.

Chauhan, Priyanka and Singh, N.P. (2023). An Immaculate Study on The Challenges, Opportunities and Innovation in Waste Management Industry. *Cosmos An International Journal Art & Higher Education*, *12*(2), 69–72.

Clemente, P. A. (2023). Operations Strategy of Air Bed and Breakfast (Airbnb): Basis for An Enhanced Business Model. *Globus An International Journal of Management & IT, 14*(2), 06-25.

Deswal, A. & Kumar, S. (2022). Technological Aspects in Cloud Computing. *Cosmos Journal of Engineering & Technology, 12*(2), 05-07.

Dwivedi, D. M. C. (2022). A Study on Advanced Innovation Managing Risk and Reward. *Cosmos Journal of Engineering & Technology*, *12*(1), 45–47.

Fatima, S. (2022). Impact of Covid-19 Pandemic on The Microfinance Sector in India. *Globus An International Journal of Management & IT, 14*(1), 25-27.

Fenequito, A. C., Robles, C. F., & Santos, A. F. & Valencia, K. C. (2023). Effects of Caregiver Burden to The Psychosocial Adjustment and Quality of Life of Caregivers of Older People with Debilitating Diseases. Globus An International Journal of Medical Science. *Engineering and Technology*, *12*(1), 41–53.

Galvez, R. S. (2022). Learners' Psychological Well-Being and Academic Performance in The Midst of Health Crisis: A Correlational Study. *Cosmos An International Journal Art & Higher Education*, *11*(1), 45–52.

Guiamalon, T. S. (2022). Stakeholders' Awareness and Acceptance of The Institution's Vision, Mission and Goals and Information Technology Program Objectives in A State University in The Philippines. *Globus Journal of Progressive Education*, *12*(1), 51–60.

Gupta, T., & Khan, S. (2023). Education and International Collaboration in The Digital Age. *Globus Journal of Progressive Education*, *13*(1), 79–87.

Hanks, A., & Dwoskin, L. (2019). The economic impact of social enterprises in the United States. *Journal of Entrepreneurship and Public Policy*, *8*(1), 1–17.

Karamillo, R. N, Liezl Mae G. M., Bucao, T., Bueno, B., Bulatao, L., Cabrera, S., & Cacho, L. (2022). Projected Action Plan and Timeline in Pilot Testing of Simulation-Based Learning in The College of Nursing at Phinma-University of Pangasinan. Cosmos An International Journal Art & Higher Education, 11(2), 04-15.

Kessler, T. P., & Chavannes, N. (2018). Leveraging the social impact of small business in urban neighborhoods. *Journal of Small Business Management*, *56*(4), 602–616.

Kovačević, J., & Ramelli, S. (2020). Ecosystems for social impact: A review of literature and practice. *Journal of Social Entrepreneurship*, *11*(2), 161–182.

Punla, N., Regalado, D., & Navarro, R. M. (2022). Perceived Social Support as Moderator Between Self-Stigma and Help-Seeking Attitudes. *Globus An International Journal of Medical Science, Engineering and Technology, 11*(1), 87-98.

Ridley-Duff, R., & Bull, M. (2019). The social and economic impact of social enterprises: A systematic review of the evidence. *Social Enterprise Journal*, *15*(1), 1–25.

Shukla, D. A. (2022). Urgency of Current Times – Peace and Non-Violence. *Globus Journal of Progressive Education*, *12*(1), 61–63.

Chapter 11
Systematic Literature Review of Maritime Clusters and Competitiveness

Maria Manuela Santos Natário
https://orcid.org/0000-0002-5983-8399
Polytechnics of Guarda, Portugal

Elisabeth T. Pereira
https://orcid.org/0000-0001-8997-1249
University of Aveiro, Portugal

João Pedro A. Couto
University of the Azores, Portugal

ABSTRACT

The aim of this chapter is to provide a systematic literature review, using bibliometric methods, on the topics of Maritime Cluster and Competitiveness, complemented with the combination of innovation, networks, and co-petition. Data were collected from the web of science and quantitative and qualitative analyses were applied to understand the relevance of the topics, the years and countries of production and the types of articles published. It was possible to identify three approaches: one that relates the theme from a strategic management perspective, another that associates the theme with regional development policies, and the third associated with the development of networks. It was also found that although co-petition does not appear explicitly, it appears as a fundamental element for the promotion of maritime clusters of the simultaneous combination of competition and collaboration of different actors within the cluster and as a mobilizing element of innovation networks.

1. INTRODUCTION

Although maritime clusters have existed for a long time and have a strong contribution to the development of regions and nations, only in recent decades have their importance been recognized in studies

DOI: 10.4018/978-1-6684-8879-9.ch011

on the agglomeration effects (Li & Luo, 2021) of economic activities.

The importance and popularity of this theme among academics and politicians arise associated with the competitiveness policies of an economic sector (Doloreux, 2017), but also at regional and national level (Stavroulakis et al., 2020a), as a key strategic tool to support economic development and business. For Paramio et al. (2013) it emerges as a smart mechanism that allows monitoring the dynamics derived from the European and international agenda on sea affairs, particularly in the context of economic growth and the blue economy.

Effectively, governments recognize the important role that maritime clusters play, not only in promoting economic development, but also in supporting potential innovation and technological development (Doloreux, 2017). As in the case of other types of clusters, the relational and dynamic characteristics of maritime clusters promote knowledge and innovation creation and thus regional and national competitiveness (Koliousis et al., 2018b). The effectiveness of a cluster is found in the network of trust between competitors that acts as a catalyst for knowledge and innovation creation (Stavroulakis et al., 2020b). For Li et al. (2021) the expansion of institutions' maritime science and/or maritime scientific research and the promotion of the maritime industry has a positive impact on efficient technological innovation with repercussions at the level of regions.

However, researchers contributing to cluster research have mainly focused on describing the actors and institutions relevant to maritime clusters, analyzing the potential benefits of clustering for maritime industrial and innovation activities, and investigating policy practices related to maritime clusters (Doloreux, 2017). Or they have sought to identify the main characteristics and critical success factors that can inform the proposal of a differentiation framework for maritime clusters (Monteiro et al., 2013). Shi et al. (2020), for example, explore the development trends of traditional maritime clusters originating in ports from two perspectives: the development of a cluster and its development path.

This study is part of a broader analysis to explore the most relevant issues associated with the Sea cluster in Portugal to see whether it foreshadows and presents a structure for functioning as a cluster, and to understand whether it satisfies the conditions and functions as a cluster.

Thus, to understand the maritime cluster and its combination with competitiveness, as well as with innovation, network relations and co-opetition, the bibliometric analysis method was chosen. This analysis is intended to contribute to the study of the topic, in particular to the study of the fields of networks, co-opetition and innovation in a context of maritime cluster competitiveness, and seeks to highlight the importance of these concepts and the benefits that can result from the combination of the four concepts in the promotion of the maritime cluster.

In terms of data collection and analysis, data were collected from the Web of Science, a very complete and reliable research database. After presenting the methodology used, the results achieved and the development prospects will be presented.

2. METHODOLOGY

In this study, a systematic literature review is conducted using bibliometric methods. According to Carver et al. (2013), a systematic literature review can be used to establish the foundations of new research or to summarize what is currently known or unknown about a topic. Bibliometric analysis is a review method used to draw the structure and development of certain research areas and to assess their evolution

(Boyack et al., 2005; Zupic & Čater, 2015). Bibliometric methods are used to examine a specific field of research by linking individual articles with each other (Zupic & Čater, 2015).

In this sense, it was tried to understand how the theme of the maritime cluster, its combination with competitiveness and innovation, networks and co-opetition have been studied. The terms defined for the selection of articles were: maritime cluster AND competitiveness AND innovation AND networks AND co-opetition.

The database selected was the Web of Science. This database contains numerous academic publications, including bibliographic details about their authors, affiliations and number of citations. The Web of Science database has been used by several researchers (e.g., Olczyk, 2016; Zhu & Liu, 2020), so the same database was also chosen in this review. The search was performed by "topic", during March 2022 and without a filter.

In a first phase, the topic "maritime cluster" was explored, obtaining a total of 3524 articles with publication years ranging from 1990 (4 articles) to 2022 (49 articles), distributed by the different areas and types of documents. Simultaneously the combination (maritime cluster AND competitiveness AND innovation AND networks AND co-opetition) was explored and no results were obtained in the Web of Science database.

Based on these results, a second phase was to explore the combination (maritime cluster AND competitiveness). A total of 121 documents were obtained, with publication dates varying between 2005 (1 article) and 2022 (1 article) (2021- 19 articles) and distributed by the different scientific categories of the Web of Science database.

Subsequently, some filters were applied to limit and refine the selection in terms of document types, excluding *Review Articles* (3), *Book Reviews* (2), *Data Papers* (1) and categories or areas of research, with these being limited to: *Transportation- 30, Economics – 25, Environmental Studies - 12, Transportation Science Technology - 10, Business – 8, International Relations – 8, Management – 8, Regional Urban Planning – 5, Geography- 4- Multidisciplinary Sciences- 3, Oceanography- 3, Operations Research Management Science – 3, Business Finance – 2, Engineering Marine - 2, Environmental Sciences – 2, Development Studies – 1, Engineering Industrial – 1, Engineering Ocean – 1, Green Sustainable Science Technology – 1, Hospitality Leisure Sport Tourism – 1, Marine Freshwater Biology – 1, Urban Studies -1*. The application of these filters resulted in 72 publications with publication dates ranging from 2005 (1 article) to 2022 (1 article) (2021- 12 articles).

Finally, in a next phase (2nd selection) the following combination was chosen (maritime cluster AND competitiveness AND innovation AND networks), obtaining a total of ten articles.

The articles were extracted from the Web of Science and a descriptive analysis of the articles was carried out, mainly using graphic methods, frequency tables and descriptive measures. All articles of the 2nd selection (ten documents) were read and manually retrieved from the Web of Science. The steps for article identification and retrieval are schematized in Figure 1.

3. BIBLIOMETRIC ANALYSIS

In order to understand the maritime cluster and the combination with competitiveness, innovation, network relations and co-opetition, in this study the bibliometric analysis method was chosen. This analysis aims to contribute to the study of the theme, in particular to the study of the fields of networks, co-competition and innovation in the context of competitiveness of the maritime cluster, and seeks to

Figure 1. Summary of the article identification and collection steps
Source: Own elaboration

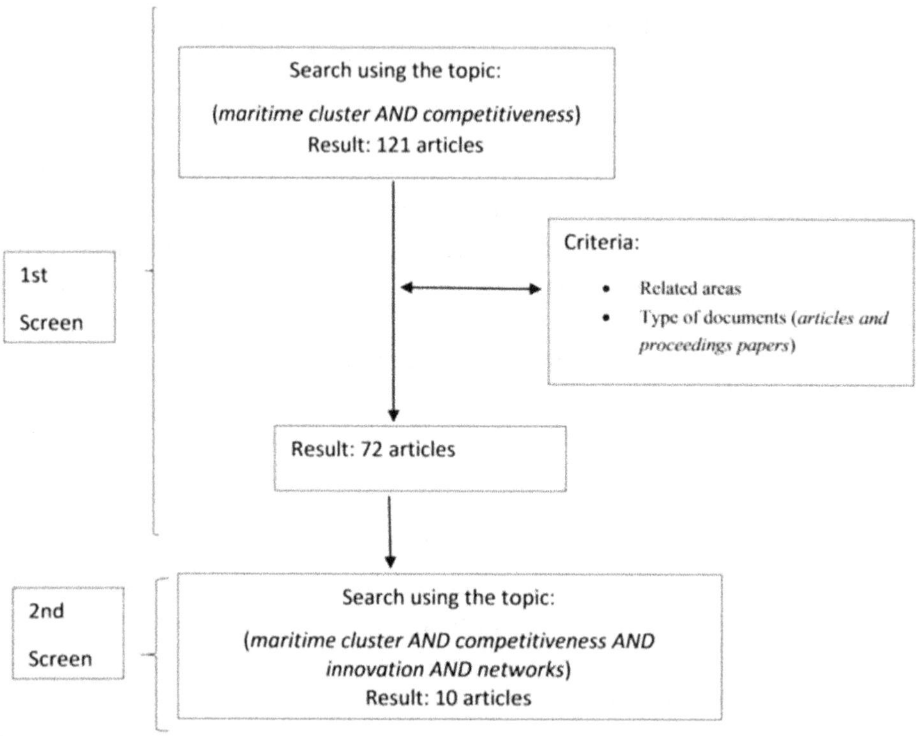

highlight the importance of these concepts and the benefits that can result from the combination of the four concepts in the promotion of the maritime cluster

The data were collected from the Web of Science, a very comprehensive and reliable research database. Automatically extracted, the database covers a period from 2005 to 2022 and covers all types of articles published in the period. The search was carried out on a first level for "maritime cluster", being refined to a second level, "competitiveness", and, on the last level, "innovation and networks". This search resulted in a total of 10 articles that combine the maritime cluster with issues of innovation and networks and competitiveness.

Quantitative and qualitative analyses were applied to the treatment and analysis of the data to understand the relevance of the themes, the years and countries of production, and the types of articles published.

The process begins with the analysis of the combination of "maritime cluster" and "competitiveness" (72 articles), and a brief presentation of the main results is made for better framing of the theme and to understand how the intensification of studies in these areas has been.

Figure 2 presents the annual evolution of the number of published articles and the number of citations under analysis, by year, on the maritime cluster and competitiveness.

It can be seen that until the year 2015, the number of articles on this topic remained below 5 articles published per year, however, from 2016, there is a growth until reaching a maximum of 12 articles in 2021.

Figure 2 also shows that the year with the highest number of citations is 2021 and that in 2005, 2006 and 2008 the number of citations is zero.

Figure 2. Annual evolution of the number of articles and of citations of the sample (2005-2022)

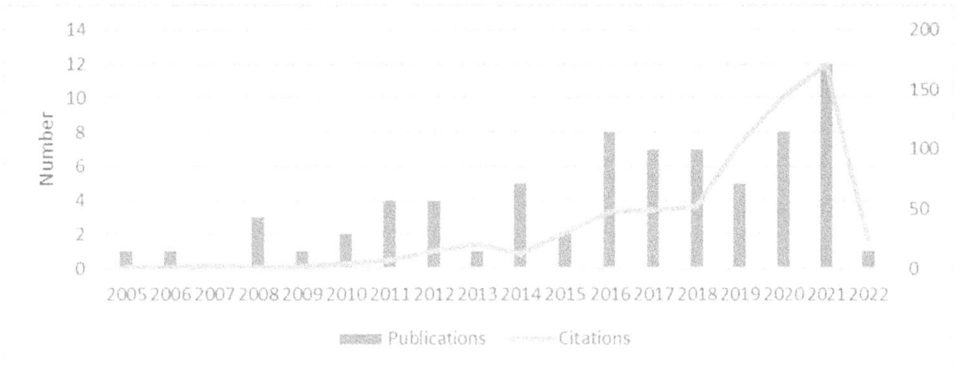

Figure 3 shows the authors with the most articles related to the theme: "maritime cluster" and "competitiveness," highlighting 5 authors who have more than 5 articles in the area: Papadimitriou (8), Stavroulakis (8), Riza (6) Koliousis (5) and Tsioumas (5).).

Table 1 shows the six most cited articles related to the topic under analysis.

As shown in Table 1, the most cited article, with 65 citations, is the study "*Typology and distribution of small farms in Europe: Towards a better picture*" in 2018, in the context of the European policy for the structuring of the Nomenclature of Territorial Units for Statistics (NUTS 3) for the agri-food sector. This is followed by the article "*Cluster Life Cycles, Path Dependency and Regional Economic Development: Insights from a Meta-Study on Basque Clusters*", from 2012, with 41 citations that studies the historical trajectory of industrial clusters in the Basque Country (paper manufacturing, electronics and information and communication technology, maritime industries and aeronautics) and the factors that explain

Figure 3. The authors with most articles related to maritime cluster and competitiveness

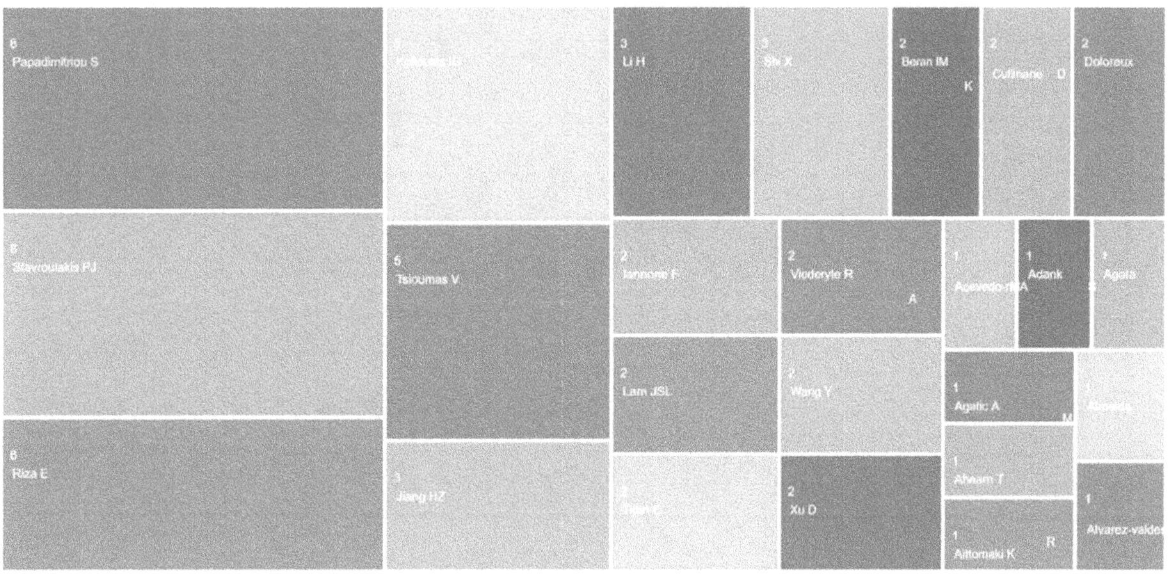

Table 1. Six top articles related to maritime cluster and competitiveness

Authors	Title of Publication	Methodology	Local	CWoS
Guiomar *et al.* (2018)	*Typology and distribution of small farms in Europe: Towards a better picture*	Quantitative	Europe	65
Elola *et al.* (2012)	*Cluster Life Cycles, Path Dependency and Regional Economic Development: Insights from a Meta-Study on Basque Clusters*	Qualitative	Spain	41
Doloreux eShearmur (2009)	*Maritime clusters in diverse regional contexts: The case of Canada*	Qualitative	Canadá	40
Cabral e Ramos (2014)	*Cluster analysis of the competitiveness of container ports in Brazil*	Quantitative	Brasil	39
Lee *et al.* (2014)	*A cross-country study of competitiveness of the shipping industry*	Quantitative	Coreia do Sul	36
Doloreux (2017)	*What is a maritime cluster?*	Qualitative	Canadá	36

Source: Own elaboration from the number of citations in the Web of Science
Note: CWoS - Number of citations in the Web of Science; Local - Place where the study was conducted.

the origin, development, maturity and, if so, decline of these clusters, considering factors related both to cluster dynamics and the regional environment. The third article with the highest number of citations corresponds to the article "*Maritime clusters in diverse regional contexts: The case of Canada*", in the year 2009, with 40 citations. This article uses a quantitative and qualitative methodology.

Figure 4 shows the origin of the articles by countries/regions. China stands out with 18 publications, followed by Spain with 12 publications.

In the top countries with the most articles related to maritime cluster and competitiveness, Portugal presents three articles. The first most cited article, already mentioned above, is the study by Guiomar et al. (2018) on "*Typology and distribution of small farms in Europe: Towards a better picture*", the second by Monteiro et al. (2013) with the study on "*A Differentiation Framework for Maritime Clusters:*

Figure 4. Top countries with the most articles related to maritime cluster and competitiveness

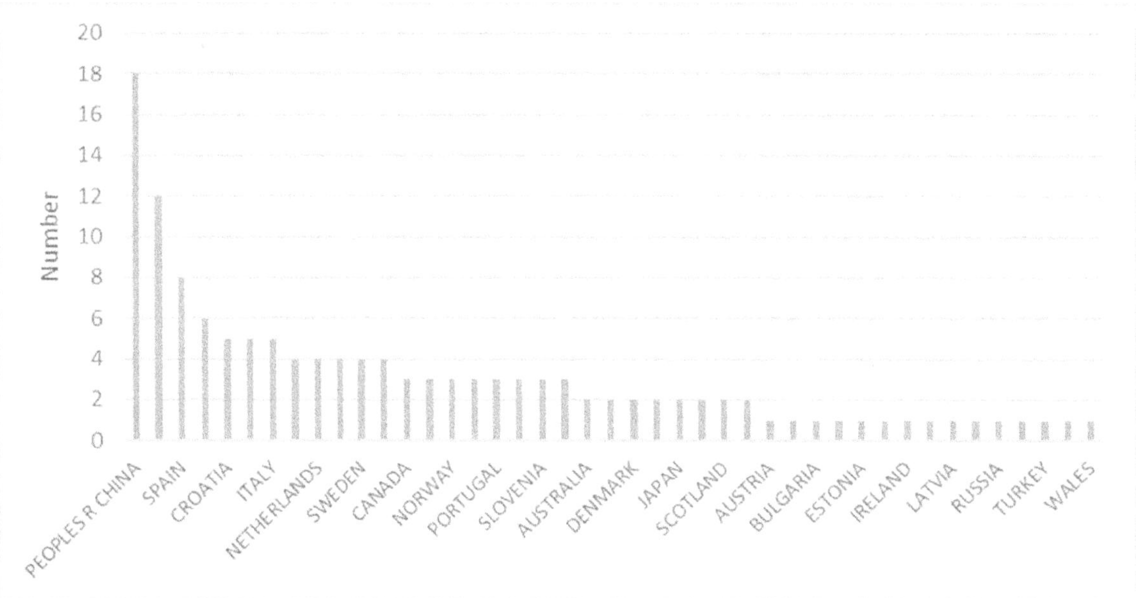

Comparisons across Europe", and the third by Vale et al. (2011) on "*Intellectual Capital in a Seaport Context*".

Figure 5 presents the journals with the highest number of articles published on the theme Maritime Cluster and Competitiveness. *Marine Policy* and the *Journal of Transport Geography* stand out with 8 articles each.

The analysis of the articles that combine the terms "maritime cluster" and "competitiveness" with issues of "innovation" and "cooperation networks" (10 articles) is now presented, in order to understand the relevance of the themes, the years and countries of production, and the types of articles published. Quantitative and qualitative analyses were used to treat the data.

Table 2 summarizes the list of the ten articles analyzed that relate maritime clusters, innovation, networks and competitiveness. The majority of the articles were published in journals (9 articles), particularly *Marine Policy* (with 4 articles). In relation to the country of origin of the studies, the production is as follows: Greece stands out with 3 articles, followed by Portugal and Canada with 2 articles (each).

Referring to the document titles (Table 2) seven documents have *Maritime cluster* in the title, one has *seaport*, one has *cluster* and one has *network*.

In the time period under analysis 2009-2021, it can be seen that the highest production occurred in 2017, 2018 and 2021 (respectively three, two and two articles) and that the level of citations had its highest number in 2021 (Figure 6) revealing that these topics combined in the studies are very recent

In Figure 7 is showed the authors with the most articles that relate *maritime cluster, innovation, networks and competitiveness*, standing out are Papadimitriou and Stavroulakis (with three publications each) and Doloreux, Koliousis and Tsioumas (with two publications each).

The main key words identified in the documents were those presented in Figure 8, highlighting strategic management, industrial cluster, and network. It is also highlighted that all articles explicitly

Figure 5. The main scientific publications related to maritime cluster and competitiveness

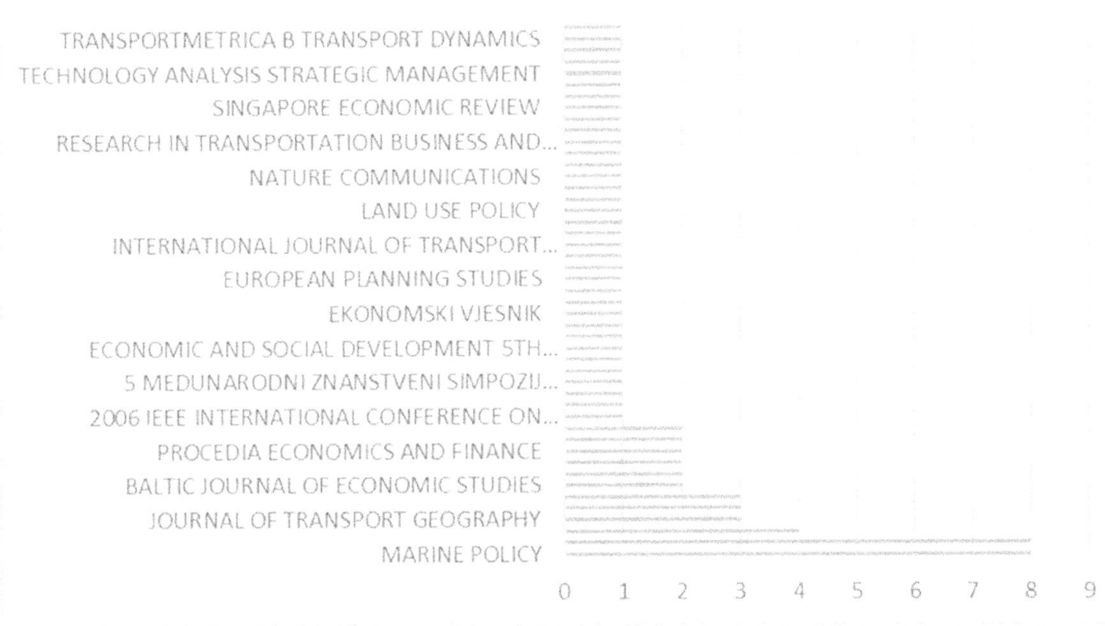

Table 2. List of the ten articles analyzed that relate maritime cluster, innovation, networks and competitiveness

Authors	Title	Publication	Year	Country	Citations CWoS
Doloreux (2017)	*What is a maritime cluster?*	MARINE POLICY	2017	Canadá	36
Stavroulakis and Papadimitriou (2017)	*Situation analysis forecasting: the case of European maritime clusters*	MARITIME POLICY & MANAGEMENT	2017	Greece	12
Koliousis *et al.* (2018b).	*The Management of Change within Maritime Clusters*	FME TRANSACTIONS	2018	Greece	8
Monteiro *et al.* (2013)	*A Differentiation Framework for Maritime Clusters: Comparisons across Europe*	SUSTAINABILITY	2013	Portugal	18
Li *et al.* (2021)	*Invulnerability in innovation networks: the importance of technology cluster coupling and network centralization*	TECHNOLOGY ANALYSIS & STRATEGIC MANAGEMENT	2021	China	1
Doloreux and Shearmur (2009)	*Maritime clusters in diverse regional contexts: The case of Canada*	MARINE POLICY	2009	Canada	40
Zhou *et al.* (2021a)	*Maritime knowledge clusters: A conceptual model and empirical evidence*	MARINE POLICY	2021	Singapura	3
Koliousis *et al.* (2018a)	*Scarcity theory and maritime clusters: From paradox to modelling*	MARINE POLICY	2018	Greece	12
Palyvoda and Seliverstova (2017).	*Management of The Industry's Innovative Development in The European Union Based on Cluster Infrastructure Formation*	SCIENTIFIC BULLETIN OF POLISSIA	2017	Ukraine	
Vale *et al.* (2011)	*Intellectual Capital in a Seaport Context*	PROCEEDINGS OF THE 3RD EUROPEAN CONFERENCE ON INTELLECTUAL CAPITAL	2011	Portugal	

Source: Own elaboration

Figure 6. Annual evolution of the number of articles and the number of citations of the sample (2009-2021)

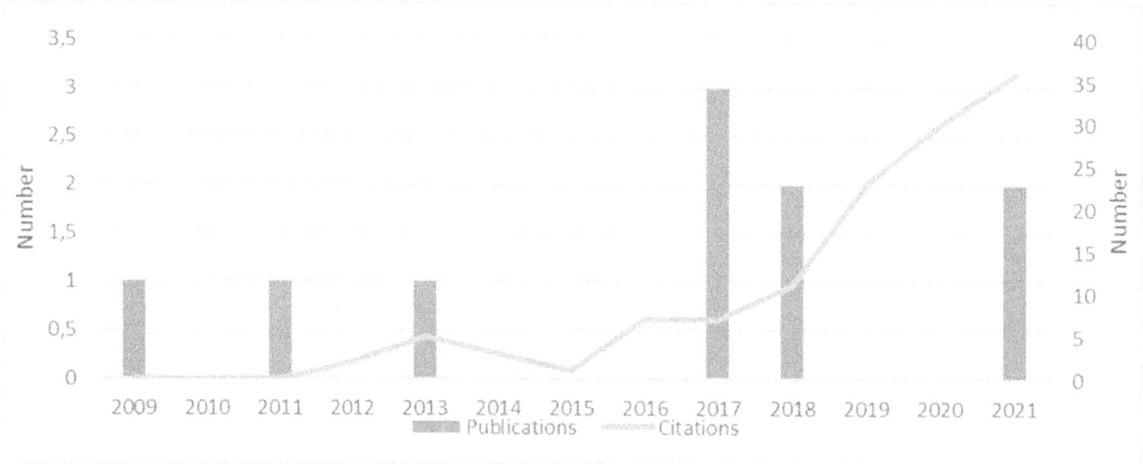

Figure 7. The authors with the most articles relating maritime cluster, innovation, networks, and competitiveness

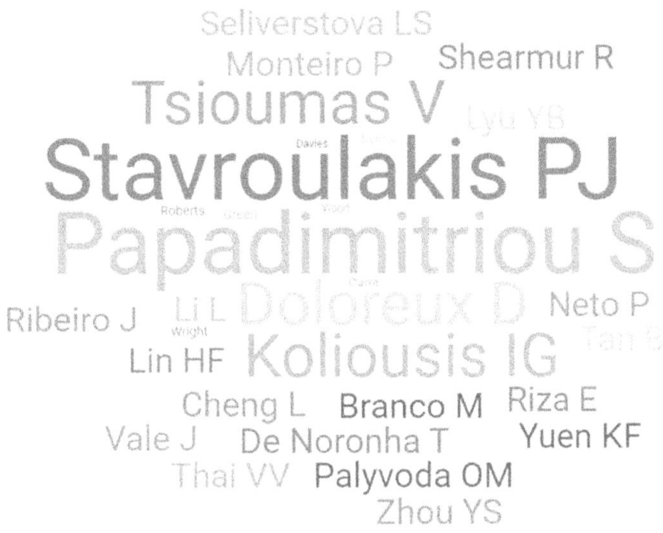

highlight *collaboration/cooperation* and that six documents highlight *collaboration/cooperation* together with *competition* simultaneously.

From the analysis of the ten documents previously identified that combine the study of the maritime cluster with competitiveness, networks and innovation, it was possible to identify three different perspectives of analysis:

a) one that relates the theme from a strategic management perspective,
b) another that associates the theme with regional development policies, and
c) the third associated with the development of networks (knowledge, innovation, relational capital).

The first group of studies fall within the domain of strategic management: Monteiro et al. (2013), Stavroulakis and Papadimitriou (2017), Koliousis et al. (2018a). Through the use of Porter's Diamond

Figure 8. Key words of the documents relating maritime cluster, innovation, networks, and competitiveness

Model, the study by Monteiro et al. (2013) seeks to analyze the competitiveness of maritime clusters by identifying the main characteristics and critical success factors that can substantiate the proposal of a differentiation framework for maritime clusters. The study carries out a comparative analysis aimed at distinguishing the most relevant aspects that can or should be observed in these types of clusters, applied in the following countries: Spain (Basque Country), Germany (Lander of Schleswig-Holstein), the Netherlands and Norway.

The main differentiating factors identified concern agglomeration economies and endogenous conditions derived from geographical proximity, essential for reducing transaction costs, enhancing the leverage of public/private cooperation through centers of maritime excellence, while providing an appropriate local environment that favors positive interactions between different maritime industries and actors.

In this view, the study by Stavroulakis and Papadimitriou (2017) considers SWOT analysis to design a forecasting and foresight model, and this was materialized through research on the maritime cluster in Europe. The emphasis is thus on the strategic management of the maritime cluster and on the strategic foresight models indicating the strategies to be adopted and policies to be defined considering the SWOT dimensions and the promotion of the cluster.

The strategic management of maritime clusters influences their competitiveness (Koliousis et al., 2018a). Thus, the study by Koliousis et al. (2018a) falls within the management domain with emphasis on strategic management and policy formulation for the maritime cluster and seeks to create tools for change management within maritime clusters.

The second group of studies is in the field of regional development policies: Doloreux and Shearmur (2009), Doloreux (2017), Palyvoda and Seliverstova (2017).

Within the fields of regional development strategies and policies for the promotion of clusters, in particular for the promotion of maritime clusters, the study by Doloreux and Shearmur (2009) seeks to analyze and evaluate three clusters in the different provinces of Canada: *Quebec, Newfoundland and Labrador, and British Columbia.* Thus, it aims to describe the driving forces and development processes of the different maritime clusters in Canada and discuss the role of cluster policy in promoting maritime competitiveness in Canada as well as the main institutional and geographic barriers affecting their growth and development.

In this context, Doloreux and Shearmur (2009) identify the main actors of the maritime cluster, their roles and the interactions that occur between them; but also, the main institutional and economic challenges that hinder their development and growth. Policy settings that promote interactions between local actors to benefit from participation in an innovation network, for example between companies and universities or research institutes, or between small start-ups and larger client companies, are important for the success of clusters. These interactions can be localized, but can also reach out to the wider business community and governance structure. Consequently, policy strategies can be oriented towards promoting accessibility both among local actors themselves and between these actors and external actors, and towards developing local comparative advantages linked to existing local resources.

In this domain and considering economic performance, national policies and regional development, actors and institutions, innovation and networks are important (Doloreux, 2017). Thus, Doloreux (2017) sought to find definitions for the maritime cluster and identified three different perspectives*: Maritime cluster as an industrial complex, Maritime cluster as an agglomeration of interlinked industries and Maritime cluster as a community-based network.*

In a broader perspective, but not directly related to the maritime cluster, the study by Palyvoda and Seliverstova (2017) seeks to analyze approaches to design and implement different types of service clusters provided by service clusters in the European Union. The study focused on the practical experience of European service clusters in an organization through market research, mediation in partner search, project support, organization of technology transfer and support in obtaining funding. They demonstrate the role of these services in strengthening the participants of the cross-sectoral innovation cooperation network. For the authors cluster organizations provide the right environment for the development of advanced technologies and new business models, acting as "kickers" to attract companies and research institutions from different sectors and countries to jointly develop innovative solutions. They also facilitate SMEs' access to the global value chain and make it easier for them to engage in long-term cooperation with strategic partners in other countries.

The third group of studies falls within the domain of network development (knowledge, innovation, relational capital): Vale et al. (2011), Li et al. (2020) e Zhou et al. (2021a).

Conceptualizing a cluster as formed by interrelated organizations in a complex and specific network system, Vale et al. (2011), in their study, seek to illustrate how and why the concept of Intellectual Capital (IC) can be applied to a seaport. In this perspective, also aiming at conceptualizing the maritime cluster, Zhou et al. (2021a) introduce the concept of Maritime Knowledge Clusters (MKCs), understood as a stage of evolution of such a cluster and empirically examine the interrelationships between its components. The objectives of the study are anchored in relational theory, knowledge-based theory and resource-based theory, and are to (1) propose and examine an integrated model underlying the components and dimensions of MKCs, and (2) empirically explore the relationships between the dimensions. MKCs are embodied in three critical components: local knowledge pooling, global systemic mechanisms, and global linkages. In this regard, policymakers should create a favorable local cluster environment with maritime firms located in geographical proximity, having local networks, knowledge base, and knowledge flows.

The study of Zhou et al. (2021a) draws on relational theory and knowledge theory and from the perspective of identifying the components and dimensions of a new stage of MKC maritime cluster development and the interactions between them to serve as guidance and best practices for policymakers to manage these clusters effectively. It has to do with cluster governance.

MKCs are substantiated by three critical components, in which the local knowledge cluster can develop global systemic mechanisms, thereby further strengthening global linkages. Policymakers should strive to create a favorable local cluster environment, with maritime firms located in geographical proximity, and with local networks, knowledge base and knowledge flows. They form the fundamental basis for the growth of MKCs. In addition, maritime firms should optimize MKCs' local knowledge cluster to facilitate their global expansion, which in turn will drive MKC internationalization (Zhou et al., 2021a).

Also, the study by Li et al. (2020) offers a network perspective on how technology clusters enhance innovation network invulnerability seeking to provide implications for policy makers seeking to govern technology clusters aimed at enhancing the stable development of innovation networks. However, it is not explicitly targeted at the maritime cluster and for this reason it has been excluded from the final analysis.

The study of Koliousis et al. (2018a) does not fall within only one of the three aforementioned perspectives and may refer to all three at the same time. Applying the scarcity theory to clusters and analyzing the maritime cluster, highlighting the different paradoxes, the study of Koliousis et al. (2018a), aims to formulate a model that deconstructs the paradox of scarcity within clusters (within maritime clusters), recognizing a new frontier for empirical and theoretical research, with distinct implications for management, governance and policy. Of the different paradoxes identified in cluster characteristics centralization and agglomeration, competition and cooperation, globalization and localization, specialization and diversification, and creative destruction concerning to innovation stand out.

3. Co-opetition: collaboration versus competition a challenge for the maritime cluster

To understand how the theme of the maritime cluster is being studied, its combination with competitiveness, innovation, networks and co-opetition the combination (*maritime cluster AND competitiveness AND innovation AND networks AND co-opetition*) was explored, but no results were obtained in the Web of Science database.

However, from the analysis of the ten documents identified above, it was found that although co-opetition does not appear explicitly, it appears as a fundamental element for the promotion of maritime clusters through the simultaneous combination of competition and collaboration of the different actors within the cluster and as a mobilizing element of innovation networks.

Co-opetition is highlighted in the study by Koliousis et al. (2018a) which highlights that firms compete and cooperate with each other to innovate and create dynamics and synergies of a cluster. One can also see that coo-petition (cooperating and collaborating in a network and competition/competition) is present in Doloreux's (2017) approach. This author highlights the importance of collaborative networks and the fact that clusters are groups of interconnected firms that are geographically close and whose spatial proximity creates an economic environment conducive to competition and innovation.

Co-competition in the maritime cluster, associated with the dynamics of cooperation and competition is very important for strategic management with various applications. These dynamics of cooperation and competition, within the maritime cluster influence its innovation, competitiveness and value creation performance and can affect national and regional competitiveness (Koliousis et al., 2018a).

However, this concept of co-opetition is more evident in the study of Monteiro et al. (2013), highlighted by the association of the maritime cluster as a driver of the dynamics of the innovation system in this sector. In addition to geographical concentration, specialization, the presence of actors and critical mass, the relations and dynamics of cluster relations (the links and interrelationships between cluster actors) are fundamental: cooperation and competition. Typically, firms and individuals in competing with each other generate pressures to improve. Depending on the characteristics of the market, actors may strive to gain advantages by reducing costs or prices, increasing quality, acquiring new customers, or entering new markets. At the same time, actors in a cluster may cooperate around a core activity, using their key competencies to complement each other. By operating together, firms may also be able to attract resources and services that are not available to them when working in isolation.

CONCLUSION

With the aim of understanding the maritime cluster and the combination of its paradigms with competitiveness, innovation, network relations and co-opetition, a bibliometric analysis was carried out. This analysis is intended to contribute to the study of the topic, in particular to the study of the fields of networks, co-opetition and innovation in a context of maritime cluster competitiveness, and seeks to highlight the importance of these concepts and the benefits that can result from the combination of the four concepts in the promotion of the maritime cluster. Li and Luo (2021) also used a similar methodology to understand current developments in the study of maritime clusters, but from a broader perspective.

The study began with a more comprehensive analysis of the combination of "maritime cluster" and "competitiveness", which allowed a better framing of the theme and an understanding of the intensification of studies in these areas. The more restricted analysis of the articles that combine the study of the maritime cluster with competitiveness, networks and innovation, allowed the identification of three different perspectives of analysis in the maritime clusters: a) one that relates the theme in a strategic management perspective b) another that associates the theme to regional development policies and c) the third associated with the development of networks (knowledge, innovation, relational capital).

The combination of the theme of maritime cluster and competitiveness, with innovation, networks and co-opetition was also carried out. Although co-opetition does not appear explicitly in the documents of the Web of Science database, it appears as a fundamental element for the promotion of maritime clusters of the simultaneous combination of competition and collaboration of the different actors within the cluster and as a mobilizing element of innovation networks. It is a trend that leads to the development of maritime clusters.

This study contributes to a better understanding of the most relevant issues associated with the Sea cluster, particularly in Portugal, to understand if this cluster is prefigured as a cluster and if it fulfils the conditions of a cluster and also if it has a structure to function as such.

The main limitation of the present study is the fact that no explicit evidence on the theme and co-opetition was found and therefore it is necessary, as future research clues, to complete the analysis with a field study of the relationships of the stakeholders involved to assess the importance of this theme.

REFERENCES

Boyack, K. W., Klavans, R., & Börner, K. (2005). Mapping the backbone of science. *Scientometrics*, *64*(3), 351–374. doi:10.100711192-005-0255-6

Cabral, A. M. R., & de Sousa Ramos, F. (2014). Cluster analysis of the competitiveness of container ports in Brazil. *Transportation Research Part A, Policy and Practice*, *69*, 423–431. doi:10.1016/j.tra.2014.09.005

Doloreux, D. (2017). What is a maritime cluster? *Marine Policy*, *83*, 215–220. doi:10.1016/j.marpol.2017.06.006

Doloreux, D., Gaviria de la Puerta, J., Pastor-López, I., Gómez, I. P., Sanz, B., & Zabala-Iturriagagoitia, J. M. (2019). Territorial innovation models: To be or not to be, that's the question. *Scientometrics*, *120*(3), 1163–1191. doi:10.100711192-019-03181-1

Doloreux, D., & Shearmur, R. (2009). Maritime clusters in diverse regional contexts: The case of Canada. *Marine Policy*, *33*(3), 520–527. doi:10.1016/j.marpol.2008.12.001

Doloreux, D., Shearmur, R., & Figueiredo, D. (2016). Québec'coastal maritime cluster: Its impact on regional economic development, 2001–2011. *Marine Policy*, *71*, 201–209. doi:10.1016/j.marpol.2016.05.028

Elola, A., Valdaliso, J. M., López, S. M., & Aranguren, M. J. (2012). Cluster life cycles, path dependency and regional economic development: Insights from a meta-study on Basque clusters. *European Planning Studies*, *20*(2), 257–279. doi:10.1080/09654313.2012.650902

Guiomar, N., Godinho, S., Pinto-Correia, T., Almeida, M., Bartolini, F., Bezak, P., Biró, M., Bjørkhaug, H., Bojnec, Š., Brunori, G., Corazzin, M., Czekaj, M., Davidova, S., Kania, J., Kristensen, S., Marraccini, E., Molnár, Z., Niedermayr, J., O'Rourke, E., & Wästfelt, A. (2018). Typology and distribution of small farms in Europe: Towards a better picture. *Land Use Policy*, *75*, 784–798. doi:10.1016/j.landusepol.2018.04.012

Koliousis, I. G., Papadimitriou, S., Riza, E., Stavroulakis, P. J., & Tsioumas, V. (2018a). Scarcity theory and maritime clusters: From paradox to modelling. *Marine Policy*, *93*, 40–46. doi:10.1016/j.marpol.2018.03.029

Koliousis, I. G., Papadimitriou, S., Stavroulakis, P. J., & Tsioumas, V. (2018b). The management of change within maritime clusters. *FME Transactions*, *46*(3), 360–366. doi:10.5937/fmet1803360K

Lee, C. B., Wan, J., Shi, W., & Li, K. (2014). A cross-country study of competitiveness of the shipping industry. *Transport Policy*, *35*, 366–376. doi:10.1016/j.tranpol.2014.04.010

Li, G., Wang, X., Su, S., & Su, Y. (2019). How green technological innovation ability influences enterprise competitiveness. *Technology in Society*, *59*, 101136. doi:10.1016/j.techsoc.2019.04.012

Li, G., Zhou, Y., Liu, F., & Tian, A. (2021). Regional difference and convergence analysis of marine science and technology innovation efficiency in China. *Ocean and Coastal Management*, *205*, 105581. doi:10.1016/j.ocecoaman.2021.105581

Li, L., Lin, H., Lyu, Y., & Cheng, L. (2021). Invulnerability in innovation networks: The importance of technology cluster coupling and network centralization. *Technology Analysis and Strategic Management*, 1–13. doi:10.1080/09537325.2021.1979202

Li, M., & Luo, M. (2021). Review of existing studies on maritime clusters. *Maritime Policy & Management*, *48*(6), 795–810. doi:10.1080/03088839.2020.1802786

Monteiro, P. V., Noronha, T. D., & Neto, P. A. (2013). A Differentiation Framework for Maritime Clusters: Comparisons across Europe. *Sustainability (Basel)*, *5*(9), 4076–4105. doi:10.3390u5094076

Olczyk, M. (2016). A systematic retrieval of international competitiveness literature: A bibliometric study. *Eurasian Economic Review*, *6*(3), 429–457. doi:10.100740822-016-0054-9

Palyvoda, O. M., & Seliverstova, L. S. (2017). Management of the Industry's Innovative Development in the European Union Based on Cluster Infrastructure Formation. *Scientific Bulletin of Polissia*, *1*(1), 185–191. doi:10.25140/2410-9576-2017-1-1(9)-185-191

Paramio, L., Tiago, M. T. B., Tiago, F. B., & Vieira, J. C. (2013). Contributions of Marine Cluster for Regional Sustainability: Economic Prospective for Azores. *Maria Scientia-Scientific e-Journal*, *6*, 48-67.

Shi, X., Jiang, H., Li, H., & Wang, Y. (2020). Upgrading port-originated maritime clusters: Insights from Shanghai's experience. *Transport Policy*, *87*, 19–32. doi:10.1016/j.tranpol.2019.11.002

Stavroulakis, P. J., & Papadimitriou, S. (2017). Situation analysis forecasting: The case of European maritime clusters. *Maritime Policy & Management*, *44*(6), 779–789. doi:10.1080/03088839.2017.1330560

Stavroulakis, P. J., Papadimitriou, S., Tsioumas, V., Koliousis, I., Riza, E., & Kontolatou, E. O. (2020a). Strategic competitiveness in maritime clusters. *Case Studies on Transport Policy*, *8*(2), 341–348. doi:10.1016/j.cstp.2019.10.008

Stavroulakis, P. J., Papadimitriou, S., Tsioumas, V., Koliousis, I. G., Rizad, E., & Tsirikou, F. (2020b). Exploratory spatial analysis of maritime clusters. *Marine Policy*, *120*, 104125. Advance online publication. doi:10.1016/j.marpol.2020.104125

Stavroulakis, P. J., Papadimitriou, S., & Tsirikou, F. (2021). Perceptions of competitiveness for maritime clusters. *Ocean and Coastal Management*, *205*(1), 105546. Advance online publication. doi:10.1016/j.ocecoaman.2021.105546

Vale, J., Branco, M., & Ribeiro, J. (2011). *Intellectual Capital in a Seaport Context*. ECIC.

Zhou, Y., Yuen, K. F., Tan, B., & Thai, V. V. (2021a). Maritime knowledge clusters: A conceptual model and empirical evidence. *Marine Policy*, *123*, 104299. doi:10.1016/j.marpol.2020.104299

Zhou, Y., Yuen, K. F., Tan, B., & Thai, V. V. (2021b). The effect of maritime knowledge clusters on maritime firms' performance: An organizational learning perspective. *Marine Policy*, *128*, 104472. doi:10.1016/j.marpol.2021.104472

Zhu, J., & Liu, W. (2020). A tale of two databases: The use of Web of Science and Scopus in academic papers. *Scientometrics*, *123*(1), 321–335. doi:10.100711192-020-03387-8

Zupic, I., & Čater, T. (2015). Bibliometric methods in management and organization. *Organizational Research Methods*, *18*(3), 429–472. doi:10.1177/1094428114562629

Chapter 12
Climate Change and the Ocean Economy:
A Review on OECD's Ocean Economy

Seda Yildirim
https://orcid.org/0000-0003-4367-6652
Tekirdag Namık Kemal University, Turkey

Merve Kaplan
https://orcid.org/0000-0001-5096-5546
Canakkale Onsekiz Mart University, Turkey

Durmus Cagri Yildirim
Tekirdag Namık Kemal University, Turkey

ABSTRACT

The ocean economy is directly linked to the biodiversity of the ocean. Biodiversity is under great threat as a result of the effects of climate change and global warming. Therefore, the ocean economy and the marine economy are among the economies that will be most affected by the effects of climate change. Each change in ocean and sea levels causes changes in biodiversity. Considering its importance in terms of food health, economic and social development, the sustainability of the marine economy is an important issue. The aim of this study is to present indicators of ocean economy among OECD countries and show the importance of ocean and marine economy in the long term.

INTRODUCTION

Climate change and global warming change the ecological structure of our planet and affect the continuity of natural life. With the acceptance of the reality of climate change, discussions on the continuity of humanity have increased. World leaders have begun to feel the effects of climate change and global warming in recent years, Although climate change is essentially a natural weather phenomenon, it has

DOI: 10.4018/978-1-6684-8879-9.ch012

become a major problem in the years that started with the industrial revolution due to human-induced effects (NASA Global Climate Change, n.d.; European Commission, n.d.; Turrentine, 2022; Yıldırım and Yıldırım, 2022). As a result of fossil fuels used for industry and production, excessive use of natural resources, inefficient use of water resources and neglecting environmental pollution for a long time, world leaders have been faced with the problem of sustainability (Bostancı ve Yıldırım, 2021). The reality accepted today can be said to be the danger of the rapid end of life as a result of an increase in the temperature of the world by more than one and a half degrees. The physical and chemical properties of the oceans have changed as a result of climate change. In particular, warming caused by CO2 and other greenhouse gases has also caused an increase in water temperatures in the oceans (Jiao et al., 2015). While scientists and researchers emphasize that climate change is a global problem, they also state that there should be global solutions to combat its effects (Koyuncu, 2021; Shukman, 2021).

The first affected factors as a result of climate change and global warming are ecosystem and biodiversity. Every change that occurs in the ecosystem also creates a change in the population and living conditions of the living things in the relevant environment (Lake et al., 2012). At this point, the most observable evidence can be considered changes in the oceans and seas. As a result of temperature differences in sea water, extinction problem arises in local species, and some invasive alien marine species are displaced, threatening local marine species and endangering the marine economy (Yildirim and Kaplan, 2022; Kaplan et al., 2022; Kaplan and Yildirim, 2023).

It is known that climate change and global warming affect economic activities based on natural resources such as "agriculture, livestock, fisheries, forestry etc." therefore, the arrangements and measures to be taken regarding the continuity of these economies will also be of vital importance for the continuity of human life (Hoeden et al., 2007; Reid et al., 2008; Hossain et al., 2012; FAO, 2015; Ladrera and Cagasan, 2022;). Every country in the world has different sources of wealth. Although the countries with natural gas and oil seem lucky in terms of energy, the food and water crisis has also been among the prominent issues after the Ukraine-Russia war and the Covid-19 pandemic (Acer, 2022; Murphy, 2022; Bozkuş, 2022). With the year 2023, the food crisis is estimated to be the first issue that world leaders focus on. In this context, developed and developing countries will start to develop new prevention programs for food security. Fisheries, on the other hand, have an important position in many countries for the continuity of food security. In the studies, it is emphasized that the continuity of the marine economy is important for both food security, economic development and social welfare (Yildirim and Kaplan, 2020; Yıldırım and Yıldırım, 2021).

In recent years, the world has been faced with many risks, especially the global pandemic, the Russia-Ukraine war, and then the energy crisis, and the blue economy issue has begun to be reconsidered globally. Blue economy, which is perceived as the sustainable use and protection of ocean and marine resources at the point of establishment of economic growth; It is gaining increasing attention in the context of poverty reduction, food security and tackling the many risks facing our planet, including climate change. The blue economy is defined by the World Bank as the sustainable use of ocean resources for economic growth, better livelihoods and employment creation while maintaining the health of the ocean ecosystem. The European Commission has stated that all economic activities related to the oceans, seas and coasts can be considered within the scope of the blue economy (NTV, 2023).

"Blue economy" is a concept that also foresees sustainability within the framework of "blue" that gives life to the oceans. The blue economy creates economic opportunities and positively supports the revenues of the states (The Commonwealth, n.d.). The oceans should not be thought of only as a source of jobs and wealth. The continuity of the sea areas is essential for the continuity of the functioning of the life of the planet. Therefore, the blue economy should be considered as an important force with the potential to achieve sustainable development (Sustainability for all, n.d.). Blue Economy sees ocean and sea resources as leverage for economic growth. However, the exploitation of seas and oceans by many states has also revealed the problem of sustainability. The blue economy, which has an important place in the development of island countries, is an important sector with a promising future globally (Zogopoulos, 2020).

This study focuses on ocean economy in the context of Bioeconomy and Economic Ecosystems. The impacts of climate change on bioeconomy are investigated by many scholars since decades. Especially, the numbers of scientific papers increased in the last decade. There is a close relationship between climate change and marine economy that most of studies examined this relationship for different countries in the literature. For example, Haward et al (2013) investigated climate change and marine biodiversity by developing some scenarios in Australia. There are many factors affecting marine biodiversity and climate change is one of them. In terms of protecting biodiversity, it is an important issue to combat climate change and examine what can be done against other factors. The authors examined what can be done against threatening factors for the protection of marine biodiversity with the scenarios they developed (Haward et al., 2013). According to Dey et al. (2016a) examined the relationships between climate change, economy and fisheries in Fiji. Fishing is an important industry for Fiji and the negative effects of climate change are taking a toll on the marine economy. The Republic of Fiji has developed a National Climate Change Policy to combat climate change (Dey et al., 2016a). In another study, Dey et al. (2016b) conducted a food security and climate change review for the Solomon Islands. The primary sources of food for the Solomon Islands are agriculture, fishing and hunting. The potential impacts of climate change on food production in the Solomon Islands should be considered (Dey et al., 2016b). Dinga et al. (2017) studied on the impacts of climate change on fishery and food security in 109 countries. In the study, the contribution of fisheries to national food and food security and vulnerabilities in the impact of climate change were scored for each country. Accordingly, Africa, Asia, Oceania and Latin America have been identified as the most vulnerable countries. New policies should be developed to combat climate change, especially in the most vulnerable countries in terms of the continuity of food security (Dinga et al., 2017).

This chapter aims to review marine and ocean economy indicators in OECD countries by investigating the link between blue economy and climate change. By using the data from OECD.Stat, this study will review the profile of ocean blue economy in OECD countries.

OCEAN AND MARINE ECONOMY IN OECD

The climate change and global warming has changed ecosystem since decades. The world leaders aim to prevent potential higher temperature in the future to save natural ecosystem. Marines and oceans are both directly affected by climate change and many countries launched sustainable programs to save biodiversity of marines and oceans in the long term. In this part, there are two main parts as "marine economy and climate change" for OECD countries. Firstly, blue economy in OECD countries is presented. Then, the potential impact of climate change on ocean economy in OCED countries is presented.

Table 1 shows indicators as "employment in fisheries, aquaculture and processing" for OECD countries. It can be said that some of the OECD countries have an economic structure based on the marine economy more than others. Because it cannot be said that high levels of fishing are carried out in every

Table 1. Employment in fisheries, aquaculture, and processing

Country	Working domain	Year						
		2010	2015	2016	2017	2018	2019	2020
Australia	Aquaculture	7000	7000	6000	6000	6000	6000	7000
	Fishing sector	13000	11000	10000	10000	11000	10000	10000
Austria	Aquaculture	2240	2300	2300	2300	2400	2500	2526
	Fishing sector	190	180	180	180	172	171	169
	Processing	125	..	153	153	153	153	153
Belgium	Aquaculture	143	143	143	143	143	143	143
	Fishing sector	614	563	491	483	521	490	504
Canada	Aquaculture	3375	3280	3355	3464	3505	3785	3934
	Fishing sector	49681	42507	44783	46011	45933	51381	49074
	Processing	21067	23136	23129	21826	21602	21433	19716
Chile	Aquaculture	26070	18341	25703	19266	18315	8881	11135
	Fishing sector	35147	37008	37651	37687	37249	34973	32978
	Processing	42752	49984	51324	55754	61794	41501	40537
Colombia	Aquaculture	54837	72801	76811	85318	91985	106125	109364
	Fishing sector	72294	53295	46548	44895	48830	61998	56381
Costa rica	Aquaculture	1031	1358	1298	1235	1704	891	1776
	Fishing sector	5747	3695	3888	3338	5763	6315	5348
	Processing	1660	1332	1105	1251	2055	756	1563
Czech republic	Aquaculture	1330	1581	1537	1533	1463	1433	1433
	Processing	135	150	170	170	190	200	203
Denmark	Aquaculture	442	437	489	486	516	487	493
	Fishing sector	2120	1836	1828	1842	1824	1801	1607
	Processing	3661	3641	3728	3606	3622	3418	3133
Estonia	Aquaculture	89	97	87	100	107	109	98
	Fishing sector	4604	3962	3962	3937	3901	3905	4043
	Processing	1772	1931	1583	1641	1381	1313	1400
Finland	Aquaculture	1256	1085	1122	1015	1093	960	1019
	Fishing sector	2534	2489	4126	4257	4308	3829	3913
	Processing	1796	1873	1873	1717	1717	1704	1704
France	Aquaculture	20694	15546	16520	17494	20475	19475	18687
	Fishing sector	15052	13955	13935	14022	13732	13566	13549
	Processing	15633	17523	38785	13996	14767	14767	14767
Germany	Aquaculture	1684	5290	4949	4746	4631	4535	2052
	Fishing sector	3205	3023	2824	3660	3690	2112	2141
	Processing	7206	7091	7173	7579	7675	7542	7393
Greece	Aquaculture	4253	4073	4291	4404	4260	4159	4237
	Fishing sector	12169	10819	21600	21153	20565	19889	18535
Hungary	Aquaculture	1252	1663	2092	1714	1449	1393	1415
	Fishing sector	1860	201	..	..	..	..	..
Iceland	Aquaculture	240	340	560	580	580	600	600
	Fishing sector	4416	4046	4363	4157	3944	3730	3460

contiued on following page

Table 1. Continued

		Year						
Country	**Working domain**	**2010**	**2015**	**2016**	**2017**	**2018**	**2019**	**2020**
Ireland	Aquaculture	1719	1835	1950	1918	1945	1980	1852
	Fishing sector	4422	3451	3461	3062	2944	3297	3188
	Processing	2867	3797	3949	4076	4210	4323	5116
Israel	Aquaculture	400	400	400	400	400	400	350
	Fishing sector	1652	1250	1090	1050	1117	1185	1151
	Processing	..	..	..	..	..	..	250
Italy	Aquaculture	5836	4917	4546	4488	4761	4765	4765
	Fishing sector	29414	26004	26125	25716	26060	24435	21669
	Processing	5950	5926	5905	5968	6101	6037	6037
Japan	Aquaculture	49260	42069	41355	40642	39928	39928	39928
	Fishing sector	219916	183646	177056	170526	168737	161786	152696
Korea	Aquaculture	29712	35103	35160	35297	37995	37034	36118
	Fishing sector	138288	105311	100296	98228	90728	88535	85434
	Processing	37455	37455	37455	36998	38064	37921	43167
Latvia	Aquaculture	333	376	363	356	340	305	330
	Fishing sector	730	617	1111	549	598	520	509
Lithuania	Aquaculture	348	475	512	434	410	430	446
	Fishing sector	1009	1011	927	811	758	774	720
	Processing	4379	5668	5362	5052	5186	5273	5199
Mexico	Aquaculture	43123	56250	56250	56250	56250	56250	56250
	Fishing sector	250680	238590	237995	238783	242195	240346	238950
Netherlands	Aquaculture	369	200	200	200	235	234	231
	Fishing sector	2165	1779	1881	1981	2333	2109	1997
	Processing	2506	2800	2800	2800	2800	2600	2470
New zealand	Aquaculture	650	650	750	780	840	840	900
	Fishing sector	1874	2117	2010	2507	2602	2601	2688
	Processing	5650	4960	5300	5100	5150	5150	5150
Norway	Aquaculture	5525	6875	7826	8173	8552	9501	9996
	Fishing sector	12993	11130	11236	11340	11219	11048	10978
	Processing	10591	11209	11165	11382	11600	12135	12416
Poland	Aquaculture	5400	7162	6348	6262	6255	6172	6131
	Fishing sector	4124	4342	4344	4787	4385	4374	3948
	Processing	16971	19156	20146	21293	20972	18673	18234
Portugal	Aquaculture	2314	2691	2754	1576	1653	1927	1936
	Fishing sector	16920	17190	17285	17642	16208	15309	15055
	Processing	14752	14296	14904	15356	15997	16618	16722
Slovak republic	Aquaculture	812	1079	992	1027	615	579	943
	Processing	849	715	695	709	614	596	578
Slovenia	Aquaculture	218	215	212	198	199	198	181
	Fishing sector	124	124	101	96	89	80	86
	Processing	266	209	209	209	209	209	209

contiued on following page

Table 1. Continued

Country	Working domain	Year 2010	2015	2016	2017	2018	2019	2020
Spain	Aquaculture	6377	6938	6655	8081	8334	8267	8015
	Fishing sector	41062	25863	24744	22823	22647	21867	21270
Sweden	Aquaculture	399	411	469	364	412	436	543
	Fishing sector	1974	1696	1432	1628	1638	1558	1489
	Processing	2007	2171	2113	2022	2015	1984	1955
Switzerland	Aquaculture	150	150	150	150	150	350	350
	Fishing sector	310	274	274	270	270	249	248
Türkiye	Aquaculture	6600	9850	9950	10500	10600	10750	10800
	Fishing sector	54172	35821	36907	36664	35937	34177	41365
	Processing	5833	6200	6250	6500	6400	6450	6500
United kingdom	Aquaculture	3514	3264	3285	3240	3302	3375	3375
	Fishing sector	12405	11757	11692	11692	11961	12043	11298
United states	Aquaculture	6457	6952	6973	7154	7334	7543	7344
	Fishing sector	159739	164047	166952	168746	158811	164522	164616
	Processing	36469	36624	36440	35579	34597	35406	32298

Source: OECD.Stat (2022a)

OECD country. When countries are examined in terms of marine economy, “Australia; Canada; Chile; Colombia; France; Greece; Italy; Japan; Korea; Mexico; Norway; Portugal; Spain; Turkey; United kingdom; United states” are the prominent countries.

Table 2 shows the revenues from international trade in fisheries products (US dollars) in OECD countries. In the 10-year period covering the years 2010-2020, it is seen that there is an increase in income in some countries and a decrease in others. For example; While there is a decrease in international fishery products trade in 2020 compared to previous years in Australia; It is observed that there is an increase in the international fishery trade in Austria in 2020 compared to previous years.

Table 3 shows indicators of inland fisheries for OECD countries. Inland waters fishing activities are quite high in some OECD countries. Therefore, it is important for countries to take precautions against the effects of climate change in terms of the sustainability of inland fisheries and the sustainability of the marine economy. It cannot be said that inland fisheries are very active in every OECD country.

Table 4 gives the total number of fishing fleets in OECD countries. The number of fishing fleets compiled according to September 2022 data also provides important data for the marine economy.

In table 5, the total aquaculture production results (in tons) compiled by all seafood species are shown separately for each OECD country. While there is an increase in production from 2010 to 2020 in some of the countries; some appear to be in decline. Data refer to OECD’s september-2022 period.

Table 6 shows indicators including marine landings and Table 7 shows indicators including the Vessel registration of in the context of data of fisheries and aquaculture in OECD countries.

Figure 1 shows aquaculture production (tonnes) for OECD countries. Looking at the data between 1960-2020; The amount of aquaculture production in OECD countries is lower in total compared to European Union countries.

Table 2. International trade of fisheries commodities (US Dollar)

	YEAR						
Country	**2010**	**2015**	**2016**	**2017**	**2018**	**2019**	**2020**
Australia	952.236.000	1.078.156.000	1.049.644.000	1.112.029.000	1.120.939.000	1.089.841.429	880.246.879
Austria	52.584.000	76.443.000	80.600.000	81.597.000	93.310.000	96.332.877	102.232.600
Belgium	1.140.616.000	1.069.004.000	1.148.541.000	1.285.343.000	1.181.491.000	1.090.367.201	1.117.593.453
Canada	3.875.003.000	4.659.554.000	4.947.013.000	5.295.436.000	5.345.335.000	5.612.642.013	4.797.027.858
Chile	3.510.784.000	4.963.976.000	5.292.426.000	6.161.316.000	6.932.195.000	6.675.177.103	6.057.604.745
Colombia	179.903.000	167.515.000	161.744.000	135.535.000	149.465.000	153.639.762	160.778.775
Costa Rica	106.679.000	139.069.000	118.207.000	131.442.000	134.888.000	125.281.561	110.605.889
Czech Republic	101.569.000	159.589.000	187.886.000	208.547.000	191.975.000	195.825.266	207.419.685
Denmark	4.208.403.000	4.283.325.000	4.710.353.000	4.894.491.000	5.066.814.000	4.860.505.956	4.784.828.777
Estonia	184.690.000	190.874.000	151.406.000	159.412.000	175.883.000	224.396.679	200.287.998
Finland	46.790.000	42.041.000	63.699.000	167.399.000	214.008.000	183.993.139	203.944.504
France	1.639.140.000	1.642.327.000	1.723.648.000	1.792.703.000	1.912.379.000	1.798.570.017	1.655.003.984
Germany	2.507.327.000	2.807.259.400	3.002.877.400	2.919.692.100	3.025.102.000	2.768.325.299	2.704.058.461
Greece	710.621.000	661.958.000	738.906.000	777.505.000	821.540.000	773.459.189	841.760.653
Hungary	9.104.000	21.680.000	24.106.000	22.816.000	22.195.000	16.420.609	13.819.450
Iceland	1.849.340.000	2.075.235.000	2.025.863.000	2.018.878.000	2.379.402.000	2.370.709.933	2.256.250.405
Ireland	505.577.000	682.087.000	669.156.000	766.790.000	770.703.000	717.327.811	686.700.921
Israel	31.100.000	28.791.000	25.895.000	22.923.000	21.379.000	24.296.000	23.445.000
Italy	714.973.000	765.120.000	791.940.000	845.968.000	903.642.000	838.828.062	872.789.222
Japan	2.014.055.000	1.986.418.000	2.113.853.000	2.112.314.000	2.392.716.000	2.294.063.522	2.041.126.905
Korea	1.710.983.000	1.681.150.000	1.869.301.000	2.006.989.000	2.024.606.000	2.104.744.132	1.852.932.033
Latvia	173.477.000	189.160.000	207.964.000	231.220.000	252.924.000	237.824.701	262.490.596
Lithuania	380.178.000	520.918.000	583.701.000	647.165.000	682.843.000	706.370.274	687.260.999
Luxembourg	19.379.000	13.978.000	14.762.000	16.849.000	18.944.000	17.228.849	14.101.156
Mexico	773.770.000	1.054.073.000	1.035.726.000	1.298.053.000	1.468.076.000	1.394.570.449	1.225.125.177
Netherlands	3.231.463.000	3.647.941.000	4.219.151.000	5.306.274.000	5.669.730.000	5.723.789.229	5.577.824.688
New Zealand	1.078.938.000	1.112.486.000	1.214.043.000	1.240.166.000	1.230.786.000	1.299.766.707	1.122.816.786
Norway	8.852.961.000	9.210.984.000	10.798.031.000	11.311.852.000	12.014.212.000	12.022.774.894	11.153.465.985
Poland	1.380.165.000	1.736.433.000	1.958.034.000	2.203.749.000	2.540.709.000	2.542.471.841	2.724.116.184
Portugal	918.119.000	1.117.586.000	1.062.377.000	1.190.020.000	1.301.215.000	1.196.743.550	1.030.824.823
Slovak Republic	10.054.000	11.648.000	10.106.000	11.942.000	19.395.000	18.015.975	15.657.562
Slovenia	24.878.000	28.547.000	39.539.000	39.327.000	44.510.000	45.413.741	46.986.993
Spain	3.339.477.000	3.807.001.000	4.155.816.000	4.711.052.000	5.093.480.000	4.753.455.599	4.566.398.030
Sweden	2.670.699.000	3.680.508.000	4.428.092.000	4.143.061.000	4.839.445.000	4.500.620.369	4.364.749.583
Switzerland	21.451.000	20.863.000	19.647.000	20.928.000	21.646.000	24.438.872	28.017.064
Türkiye	361.239.000	698.910.000	806.592.000	862.127.000	976.347.000	1.047.065.078	1.108.380.769
United Kingdom	2.433.461.000	2.515.271.000	2.674.742.000	2.930.105.000	2.863.615.000	3.013.916.339	2.544.599.067
United States	4.774.879.000	5.670.390.000	5.588.117.000	6.246.034.000	5.787.552.000	5.423.038.910	4.505.723.694

Source: OECD.Stat (2022b)

Table 3. Inland fisheries

	YEAR						
Country	**2010**	**2015**	**2016**	**2017**	**2018**	**2019**	**2020**
Australia	1376	1039	982,84	1161	1006,62	884,303	1014,69
Austria	350	350	350	350	350	350	350
Belgium	511	283	283	283	283	283	283
Canada	27385	32203,049	34948,604	32556,51	43124,177	26543,612	21863,9
Chile	0	0	0	0	0	0	..
Colombia	21049	18554	18830	27125	21757	22495,168	20757,859
Costa rica	100	50	50	50	50	50	50
Czech republic	3990	3841	3507	3557	3650	3721	3729
Denmark	114	121	121	122	110	238	224,494
Estonia	2869	2656	3109	3036	3243	3014	3715,272
Finland	28874	29469	28045	28709	23428	24653	24652
France	1950	1187	1087	1134	1107	1107	1112,336
Germany	15010	21349	20673	18773	17997	17169,8	16945,5
Greece	940	940	940	940	940	940	940
Hungary	6216	5926	5048	5607	5234	4816	5406,442
Iceland	260	201	176	137	124	119	0,012
Ireland	116	78	73	64	81	85	85
Israel	426	503	331	932	926	926	906
Italy	3852	3800	3800	3800	3800	3800	3800
Japan	39844	32920	27937	25215	26956	21769	21759
Korea	10256	9133	9103	5726	5755	5755	5755
Latvia	329	226	243	226	302	298	229
Lithuania	1310	1437	1210	1091	926	1278,545	1340,842
Luxembourg	0	0	0	0	0	0	0
Mexico	116410	151416	199665	168072	223625	155714	148551,16
Netherlands	2051	1904	1915	1256	1321	1325	1315
New zealand	892	832	807	818	817	440,042	298,186
Norway	437	408	393	424	309	330	354
Poland	18508	18376	18393	18086	18201	18162	18389
Portugal	1	1	2	1	3	1	1
Slovak republic	1608	1971	1866	1870	1937	1853	1736
Slovenia	169	141	145	142	129	146	155,228
Spain	6000	6000	6000	6000	6000	6000	6000
Sweden	1368	9017	7493	6904	6668	6200	8520
Switzerland	2101	2023	1851	1932	1553	1382,313	1485,92
Türkiye	40259	34176	33856	32145	30139	31596	33119
United kingdom	2473	747	656	581	1157	1150	1066,166
United states	21151	18249	21734	18290	20412	13730	11678

Source: OECD.Stat (2022c)

Table 4. Fishing fleet

Country	Year						
	2010	2015	2016	2017	2018	2019	2020
Australia	318	298	290	315	285	294	302
Austria	190	180	180	180	..	..	..
Belgium	89	76	72	70	68	68	64
Canada	19.906	17.856	17.703	17.522	18.430	16.911	16.891
Chile	7.815	9.391	9.540	9.622	9.563	8.872	8.350
Colombia	158	133	175	124	120	118	46
Costa Rica	5.035	1.912	1.574	2.644	3.155	2.715	2.803
Denmark	2.820	2.356	2.261	2.198	2.123	2.077	2.033
Estonia	934	1.538	1.557	1.595	1.665	1.742	1.828
Finland	3.366	2.723	3.093	3.224	3.245	3.207	3.137
France	7.216	6.904	6.833	6.509	6.373	6.237	6.224
Germany	1.673	1.443	1.414	1.377	1.331	1.309	1.292
Greece	16.913	15.351	15.176	14.982	14.745	14.694	14.625
Iceland	1.375	1.312	1.279	1.216	1.148	..	..
Ireland	2.144	2.141	2.114	2.022	2.035	2.039	2.033
Italy	13.431	12.300	12.260	12.252	12.102	12.125	12.152
Japan	292.822	250.817	244.569	237.503	230.504	225.276	219.485
Korea	76.974	67.226	66.970	66.736	65.906	65.835	65.744
Latvia	786	686	686	675	671	663	663
Lithuania	171	145	142	144	143	138	139
Mexico	94.111	76.285	75.997	76.306	77.483	76.879	76.842
Netherlands	846	829	844	849	832	836	834
New Zealand	1.401	1.324	1.254	1.221	1.168	1.140	1.060
Norway	6.310	5.884	5.947	6.134	6.018	5.982	5.839
Poland	793	875	843	834	827	826	822
Portugal	8.425	8.035	7.955	7.907	7.834	7.757	7.714
Slovenia	182	168	170	170	134	137	136
Spain	10.855	9.397	9.244	9.145	8.973	8.883	8.839
Sweden	1.360	1.318	1.277	1.230	1.199	1.136	1.136
Türkiye	17.440	15.680	15.663	15.406	15.352	15.315	15.302
United Kingdom	6.460	6.232	6.197	6.151	6.046	5.964	5.824
United States	77.695	75.231	75.231	75.231	75.231	75.231	75.231

Source: OECD.Stat (2022d)

Figure 2 shows total fisheries production for OECD countries. Looking at the data between 1960-2020; Fish production in OECD countries has declined over the last decade.

Table 5. Aquaculture production

Year		2010	2015	2016	2017	2018	2019	2020
Country	**Unit**							
Australia	Tonnes	75.679	89.318	97.222	93.818	97.672	94.086	106.088
Austria	Tonnes	2.167	3.503	3.483	3.862	3.991	4.250	4.527
Belgium	Tonnes	539	82	44	75	111	86	209
Canada	Tonnes	162.341	187.374	200.765	191.616	191.373	187.226	171.007
Chile	Tonnes	713.241	1.057.742	1.050.117	1.219.747	1.287.233	1.407.289	1.505.486
Colombia	Tonnes	80.367	95.857	106.685	118.485	132.756	171.025	179.351
Costa rica	Tonnes	26.839	23.560	22.212	21.124	21.955	22.046	16.269
Czech republic	Tonnes	20.420	20.200	20.952	21.685	21.751	20.986	20.401
Denmark	Tonnes	36.175	35.867	36.337	37.498	36.453	40.221	42.629
Estonia	Tonnes	573	799	868	870	944	1.062	1.090
Finland	Tonnes	11.772	14.877	14.412	14.587	14.164	15.296	15.053
France	Tonnes	203.409	163.353	180.611	188.606	187.465	196.151	191.350
Germany	Tonnes	40.694	29.909	41.801	35.979	34.196	40.756	32.258
Greece	Tonnes	121.244	107.162	123.707	125.525	132.365	128.822	131.737
Hungary	Tonnes	14.245	17.337	16.248	18.258	17.852	17.283	18.373
Iceland	Tonnes	5.050	8.383	15.129	20.859	19.185	33.959	40.595
Ireland	Tonnes	46.490	39.650	40.244	45.433	36.896	37.914	37.751
Israel	Tonnes	19.895	19.640	18.914	17.997	17.025	16.950	14.700
Italy	Tonnes	153.494	148.763	156.460	156.625	145.000	132.378	122.778
Japan	Tonnes	1.151.101	1.105.652	1.068.034	1.023.192	1.034.920	946.643	996.297
Korea	Tonnes	1.377.233	1.676.485	1.859.220	2.334.719	2.278.850	2.399.720	2.327.903
Latvia	Tonnes	549	863	788	808	830	609	717
Lithuania	Tonnes	3.191	4.450	4.393	3.749	3.744	4.215	4.477
Mexico	Tonnes	126.238	211.622	221.327	243.307	247.222	251.242	278.694
Netherlands	Tonnes	67.040	62.795	62.185	52.185	53.485	46.350	39.940
New zealand	Tonnes	110.592	91.275	109.016	116.530	104.549	114.582	118.582
Norway	Tonnes	1.019.802	1.380.890	1.326.216	1.308.634	1.355.117	1.453.042	1.490.412
Poland	Tonnes	30.751	36.971	38.293	38.800	43.361	44.719	47.700
Portugal	Tonnes	8.225	9.563	9.787	12.509	13.991	13.691	14.552
Slovak republic	Tonnes	702	1.309	2.169	2.646	2.224	2.688	2.296
Slovenia	Tonnes	778	1.617	1.859	1.750	1.944	2.144	1.673
Spain	Tonnes	256.786	296.572	290.483	318.142	356.897	311.033	276.571
Sweden	Tonnes	10.644	12.020	15.747	14.793	13.095	11.579	12.090
Switzerland	Tonnes	1.261	1.593	1.733	1.733	1.843	2.048	2.048
Türkiye	Tonnes	167.721	240.334	253.395	276.502	314.537	373.356	421.411
United kingdom	Tonnes	201.364	223.026	204.564	227.672	189.951	233.238	221.000
United states	Tonnes	496.699	426.015	444.698	439.698	466.174	490.305	448.535

Source: OECD.Stat (2022e)

Table 6. Marine landings

Year		2010	2015	2016	2017	2018	2019	2020
Country	Unit							
Australia	Tonnes	173.553	152.433	174.246	166.016	173.434	167.022	179.080
Belgium	Tonnes	19.773	22.489	24.583	22.142	20.633	19.309	18.306
Canada	Tonnes	951.529	838.411	853.006	827.028	803.172	743.233	719.249
Chile	Tonnes	3.048.322	2.098.360	1.921.691	2.312.603	2.357.237	2.371.212	2.603.715
Colombia	Tonnes	..	41.325	12.143	47.438	47.441	71.199	71.114
Costa rica	Tonnes	13.649	13.382	10.002	11.833	14.084	17.612	25.633
Denmark	Tonnes	821.378	861.701	662.252	897.041	782.130	632.258	728.266
Estonia	Tonnes	104.325	83.186	84.331	90.287	94.387	83.924	70.811
Finland	Tonnes	121.850	148.008	157.253	154.427	147.580	135.059	112.315
France	Tonnes	254.876	457.585	449.270	419.707	482.205	503.959	436.190
Germany	Tonnes	214.067	170.146	250.596	256.317	270.488	192.937	181.134
Greece	Tonnes	70.089	64.431	61.359	62.554	53.580	73.214	50.280
Iceland	Tonnes	1.093.034	1.313.201	1.079.258	1.158.718	1.255.080	..	..
Ireland	Tonnes	318.055	234.028	241.846	245.479	213.339	..	..
Italy	Tonnes	224.758	192.213	192.356	191.916	199.825	180.060	136.392
Japan	Tonnes	4.122.104	3.492.430	3.263.566	3.258.018	3.415.243	3.268.357	3.164.413
Korea	Tonnes	1.119.493	1.050.496	898.718	956.345	1.036.169	943.857	945.346
Latvia	Tonnes	164.489	81.303	114.655	117.913	135.235	110.589	103.265
Lithuania	Tonnes	33.106	68.559	95.381	88.675	68.955	103.947	87.154
Mexico	Tonnes	1.318.828	1.367.753	1.400.409	1.454.298	1.772.336	1.541.933	1.662.203
Netherlands	Tonnes	386.750	330.464	366.795	411.766	394.746	309.893	297.866
New zealand	Tonnes	429.950	439.981	422.102	422.112	400.333	420.873	366.453
Norway	Tonnes	2.549.381	2.160.172	1.905.277	2.247.997	2.340.437	2.140.105	2.265.307
Poland	Tonnes	108.395	132.856	137.462	135.734	153.451	121.317	113.967
Portugal	Tonnes	217.337	183.336	187.590	175.405	172.135	190.613	161.296
Slovenia	Tonnes	763	194	152	128	126	120	156
Spain	Tonnes	755.465	838.977	799.599	851.353	920.834	..	662.480
Sweden	Tonnes	205.310	202.880	198.002	221.769	214.968	178.226	171.100
Türkiye	Tonnes	445.680	397.731	301.582	322.173	283.955	431.572	331.281
United kingdom	Tonnes	608.807	709.122	700.846	728.182	700.173	619.548	624.808
United states	Tonnes	4.540.234	5.132.215	4.984.654	5.180.631	4.934.883	4.951.855	4.351.990

Source: OECD.stat (2022f)

Figure 3 shows captured fisheries production for OECD countries, European Union and the World. Looking at the data between 1960 and 2020; In general, it is seen that there has been a decrease in OECD countries, European Union countries and the world in the last ten years.

Table 7. Fisheries and aquaculture indicators

Indicator		IUU policy indicators, A - Vessel registration (Indicator)*	
Year		2016	2018
Country			
Australia		0,88	0,83
Belgium		1,00	0,90
Canada		0,97	0,97
Chile		0,68	0,68
Colombia		0,82	0,82
Costa Rica		0,47	0,47
Denmark		0,97	1,00
Estonia		0,91	0,91
France		0,45	0,45
Germany		0,93	0,97
Greece		0,74	0,82
Iceland		0,85	..
Ireland		0,88	0,88
Italy		0,72	0,72
Japan		0,92	0,92
Korea		0,95	0,90
Latvia		0,63	0,70
Lithuania		0,75	0,75
Mexico		1,00	1,00
Netherlands		0,72	0,72
New Zealand		0,90	0,90
Norway		0,98	0,98
Poland		0,96	0,96
Slovenia		1,00	1,00
Sweden		0,77	0,87
Türkiye		0,53	0,55
United Kingdom		0,74	0,79
United States		0,83	0,83
OECD - Total		0,82	0,83
Non-OECD Economies	Argentina	0,85	0,85
	China (People's Republic of)	1,00	1,00
	Indonesia	0,57	0,87
	Chinese Taipei	0,85	0,85
	Thailand	0,75	0,75
	Viet Nam	0,23	0,23
*The IUU policy indicator 'a - vessel registration' describes the progress made by countries and economies in collecting and publicising information on vessels or flying their flag.			

Source: OECD.Stat (2022g)

Figure 1. Aquaculture production (metric tons): OECD members
Source: The World Bank, (2020a)

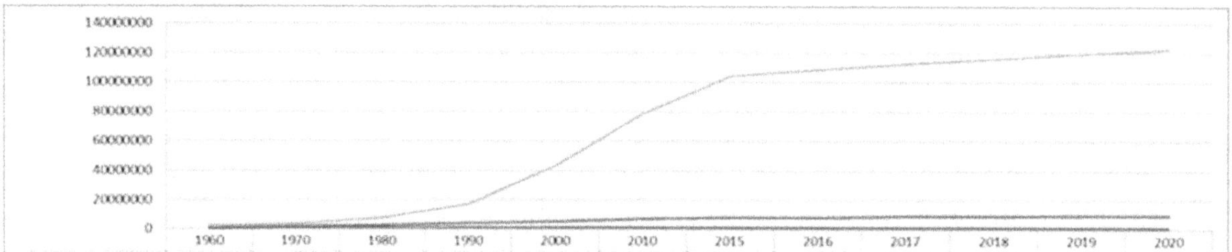

	1960	1970	1980	1990	2000	2010	2015	2016	2017	2018	2019	2020
European Union	259939	470838	717360	988944	1255292	1073503,23	1065372,18	1126759,23	1163044,83	1195717,6	1143939,55	1094315,15
OECD members	699362	1363396	2520861	3831063,04	5191383,09	6765310,95	7846425,2	8060119,43	8750017,93	8881120,95	9268989,43	9356551,15
World	2047815	3607571	7604596	17244868,04	43012663,75	77993683,31	104003294,6	108215372,4	112236485,8	115911560,8	119799127,7	122580059,7

Figure 2. Total fisheries production (metric tons): OECD members
Source: The World Bank, (2020b)

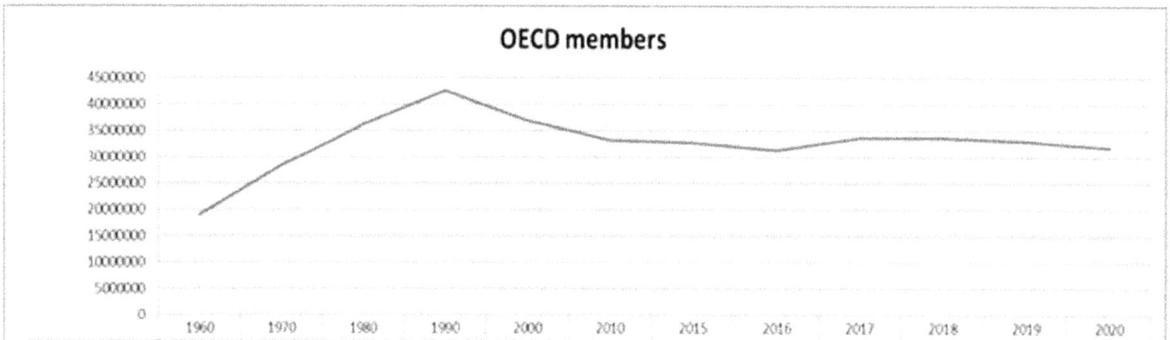

	1960	1970	1980	1990	2000	2010	2015	2016	2017	2018	2019	2020
OECD members	18929324	28420347	36125856,9	42549282,55	37015122,69	33192383,49	32689883,44	31276432,41	33647767,29	33701133,6	33028550,73	31835366,41

Figure 3. Captured fisheries production (metric tons)
Source: The World Bank, (2020c)

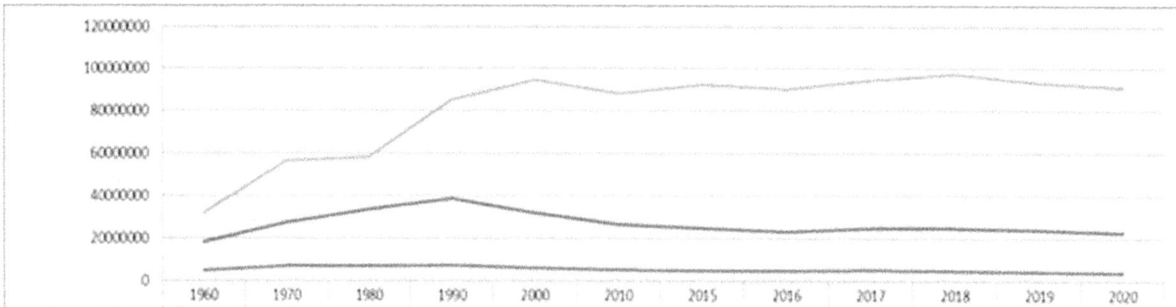

	1960	1970	1980	1990	2000	2010	2015	2016	2017	2018	2019	2020
European Union	4549642	6576532	6915035,9	7302139,17	6092709,39	4883769,7	4693889,41	4560089,69	4952053,26	4786802,13	4265072,78	4018657,68
OECD members	18229962	27056951	33604995,9	38718219,51	31823739,6	26427072,54	24843458,24	23216312,98	24897749,36	24820012,65	23759561,3	22478815,26
World	31681220,01	56302178	58362135	85383567,52	94442193,42	88082278,5	92511839,15	90529210,88	94392329,86	97328274,9	93187798,29	91336074,76

As of 2021, the OECD has 37 members. 20 countries within the organization are founding members. Turkey, USA, France, Canada, Austria, England, Belgium, Germany, Italy, Netherlands, Sweden, Spain, Portugal, Norway, Iceland, Switzerland, Denmark, Greece, Ireland and Luxembourg are founding members of the OECD. The founding countries were later joined by Japan, South Korea, Finland, Israel, Latvia, Colombia, Lithuania, Australia, Estonia, New Zealand, Czechia, Hungary, Slovenia, Poland, Costa Rica, Chile, Slovakia and Mexico. Within the scope of the OECD, there are non-members as well as member countries of the organization (KuveytTürk, 2022).

Due to the functions of the oceans, the temperature is balanced and events such as precipitation, drought and flood can be arranged. If there is a change in the temperature of the oceans, natural disasters will increase even more. In particular, global warming accelerates climate change as a result of the inability to absorb large amounts of greenhouse gases by the oceans. Oceanic and coastal communities are directly exposed to the effects of greenhouse gases. (OECD, n.d.). As the accelerating effect of many factors, the environmental quality of coastal areas is deteriorating. Due to rapid urban development, unsustainable consumption of coastal resources and inefficient land use, marine biodiversity and its ecosystem are in grave danger (OECD, 2021). In this context, it can be said that coastal areas and oceans are more vulnerable as a result of climate change. In the long run, coastal communities will go into an economically, socially and environmentally unsustainable cycle (OECD, 2021).

The possible effects of climate change on seas and oceans can be summarized as follows (OECD, 2021): "*Sea level change: Sea level rise; Ocean acidification*; *Coastal storm surge*; *Ocean warming*; *Broader alterations in the hydrological cycle*"

Table 8 shows Sustainable Ocean Economy (total marine landings, all species) for OECD countries . These data is belonged to 2022 data of OECD countries.

Table 8. Sustainable marine economy indicators in OECD

Country	Value (thousand tonnes)
Canada	799,56
Chile	2357,24
Denmark	782,13
France	538,23
Germany	270,49
Iceland	1255,08
Ireland	213,34
Japan	3415,33
Korea	1036,83
Mexico	1772,34
Netherlands	404,11
New Zealand	395,57
Norway	2330,99
Spain	920,83
Türkiye	283,95
United Kingdom	698,2
United States	4890,79

Source: OECD, (2022).

SOLUTIONS AND RECOMMENDATIONS

OECD's aim is to help member states in the struggle of global problems and crises. Climate change, which is the biggest problem of today, is a problem that concerns all countries globally. OECD creates various road maps against climate change, increased carbon emissions and global warming. The higher economic concerns of developing countries in OECD can lead to the background. However, data on climate change show that the continuation of temperature changes will create unsustainable results for the future. Regional common work of countries close to each other in the protection of seas will be a more sustainable way for the continuity of marine creatures. In this case, common cooperation between the same sea or neighboring countries will benefit countries for the sustainable maritime economy.

FUTURE RESEARCH DIRECTIONS

This study was carried out in review format. The OECD data is taken from the OECD.stat website and the most up -to -date data existing are turned into tables and shapes. Future studies can calculate the marine economy indicators of the future by riding various statistics.

CONCLUSION

Climate change and global warming affect the natural environment and ecosystem, resulting in some problems such as depletion and reduction in food resources, clean water resources, and biodiversity. Therefore, it can be said that bio-economy is one of the areas where the effects of climate change are most evident. In the report titled "Principles of a Sustainable Blue Economy" by the World Wide Fund for Nature (WWF), it is stated that oceans, seas and coastal zone ecosystems are vital for the livelihood of billions of people around the world, as well as for food security and the economic well-being of countries. In the report, which states that more than 24 trillion dollars of assets are obtained from the oceans, which is described as the "seventh largest economy in the world", it is underlined that the long-term capacity of sea and ocean environments to provide jobs and nutrition (Kahveci and Biçer, 2019).

In this study, the basic indicators about Ocean and Marine Economy were examined for OECD countries. During this examination, climate change was also emphasized. It can be said that especially among the OECD countries, especially developing countries, are more threatened by crooked problems due to climatic change. Food and water scarcity in the economies of large population and development will be the biggest problem of future periods. Within the scope of Blue Economy, activities of fishing and fish growth and other aquaculture market have a vital value for many societies both economically and socially. While economically creates an important value in employment and income, Ocean Economy sustainability is an important issue for the sustainability of societies in terms of health and food. In terms of food security, the sustainability of biodiversity and ecosystem is the fore. According to the OECD (2016) policy booklet, the following recommendations have been made for the sustainability of the ocean economy in OECD countries:

1. To support a sustainable blue economy by promoting greater international cooperation in maritime science and maritime technology,

2. Concentrating on the field of ocean management, making use of scientific techniques and long-term economic forecasting,
3. Development of the OECD's Database on Ocean Economics and benefiting all OECD countries.
4. Working within the framework of national and international cooperation

The value of ocean resources also plays a vital role in establishing a sustainable future. Ocean resources as food, employment and income are of primary value to approximately 3 billion people today. On the other hand, the oceans and seas are most under threat from the effects of global warming and climate change. Ensuring the security of the oceans and seas at the global level, especially in coastal countries, should be among the long-term policies (UNCTAD, 2022).

The usage of natural sources has a limit and the economy should adopt new approaches to keep economic growth as using natural sources efficiently in the long term (Yildirim, 2023). When there is a disruption of the ecosystem in sea and oceans, a significant portion of food resources will be destroyed. On the other hand, the decrease in aquaculture and the amount of fish causes the loss of employment and income in the marine economy. For both economic and social sustainability, the struggle against climate change should be discussed in terms of marine economy and regional solutions should be produced.

REFERENCES

Acer, Y. (2022). Rusya-Ukrayna Savaşinin Gida Arzina Etkileri Ve Tahil Koridoru Anlaşmasi [The Food Supply Effects of the Russia-Ukraine War and the Grain Corridor Agreement]. *Perspektif*, 343. https://setav.org/assets/uploads/2022/07/P343.pdf

Bostancı, S. H., & Yıldırım, S. (2021). Sustainable Communities vs. Climate Refugees: Two Opposite Results of Climate Change. In C. R. G. Popescu (Ed.), *Handbook of Research on Novel Practices and Current Successes in Achieving the Sustainable Development Goals* (pp. 298–319). IGI Global. doi:10.4018/978-1-7998-8426-2.ch015

Bozkuş, Y. D. (2022). *Rusya-Ukrayna savaşıyla başlayan tahıl sorunu küresel bir gıda krizine dönüşebilir mi?* (*Could the grain problem that started with the Russia-Ukraine war turn into a global food crisis?*), Indy Turk. https://www.indyturk.com/node/520321/t%C3%BCrki%CC%87yeden-sesler/rusya-ukrayna-sava%C5%9F%C4%B1yla-ba%C5%9Flayan-tah%C4%B1l-sorunu-k%C3%BCresel-bir-g%C4%B1da

Dey, M. M., Gosh, K., Valmonte-Santos, R., Rosegrant, M. W., & Chen, O. L. (2016a). Economic impact of climate change and climate change adaptation strategies for fisheries sector in Fiji. *Marine Policy*, *67*, 164–170. doi:10.1016/j.marpol.2015.12.023

Dey, M. M., Gosh, K., Valmonte-Santos, R., Rosegrant, M. W., & Chen, O. L. (2016b). Economic impact of climate change and climate change adaptation strategies for fisheries sector in Solomon Islands:Implication for food security. *Marine Policy*, *67*, 171–178. doi:10.1016/j.marpol.2016.01.004

Dinga, Q., Chena, X., Hilbornd, R., & Chene, Y. (2017). Vulnerability to impacts of climate change on marine fisheries and food security. *Marine Policy*, *83*, 55–61. doi:10.1016/j.marpol.2017.05.011

European Commission. (n.d.). *Causes of climate change*. EC. https://climate.ec.europa.eu/climate-change/causes-climate-change_en

FAO. (2015). *Climate change and food security: risks and responses*. FAO. https://www.fao.org/3/i5188e/I5188E.pdf

Haward, M., Davidson, J., Lockwood, M., Hockings, M., Kriwoken, L., & Allchin, R. (2013). Climate change,scenarios and marine biodiversity conservation. *Marine Policy*, *38*, 438–446. doi:10.1016/j.marpol.2012.07.004

Hoeden, S. M., Soussana, J. F., & Tubiello, F. N. (2007). Adapting agriculture to climate change. *Agricultural Sciences*, *104*(50), 19691–19696. PMID:18077402

Hossain, A., Reza, I., Rahman, S., & Kayes, I. (2012). Climate Change and its Impacts on the Livelihoods of the Vulnerable People in the Southwestern Coastal Zone in Bangladesh. In W. Leal Filho (Ed.), *Climate Change and the Sustainable Use of Water Resources. Climate Change Management*. Springer. doi:10.1007/978-3-642-22266-5_15

Jiao, N.Z., Chen, D.K., Luo, Y.M., Huang, X.P., Zhang, R., Zhang, H.B., Jıang, Z.J. & Zhang, F. (2015). Climate change and anthropogenic impacts on marine ecosystems and countermeasures in China. *Advances in Climate Change Research 6* (2015) 118e125

Kahveci, Z., & Biçer, A. (2019). *Mavi dünya' için mavi ekonomi (Blue economy for the blue world)*, AA. https://www.aa.com.tr/tr/temiz-cevre-temiz-su/mavi-dunya-icin-mavi-ekonomi/1508543

Kaplan, M., & Yildirim, S. (2023). The Sustainability of Seafood Market Under the Attack of Invasive Alien Species: A Review of Turkish Marines. In A. Ben Salem, L. Rhazi, & A. Karmaoui (Eds.), *Climatic and Environmental Significance of Wetlands: Case Studies from Eurasia and North Africa* (pp. 74–95). IGI Global.

Kaplan, M., Yildirim, S., & Yildirim, D. C. (2022). Pufferfish versus lionfish: Comparing risks for Turkish marine economics. *Marine Economics and Management*, *5*(2), 173–187. doi:10.1108/MAEM-09-2022-0008

Koyuncu, H. (2021). İklim krizi: 1,5 ve 2 derecelik küresel ısınma arasındaki fark nedir, nasıl sonuçlara yol açabilir? *Euronews*. https://tr.euronews.com/2021/11/08/iklim-krizi-1-5-ve-2-derecelik-kuresel-s-nma-aras-ndaki-fark-nedir-nas-l-sonuclara-yol-aca

KuveytTürk. (2022). *OECD (Ekonomik Kalkınma ve İşbirliği Örgütü) Nedir? Hangi Ülkeler OECD Kapsamında Yer Alır?* OECD. https://www.kuveytturk.com.tr/blog/finans/oecd-ekonomik-kalkinma-ve-isbirligi-orgutu-nedir-hangi-ulkeler-oecd-kapsaminda-yer-alir#:~:text=T%C3%BCrkiye%2C%20ABD%2C%20Fransa%2C%20Kanada,ve%20L%C3%BCksemburg%20OECD%20kurucu%20%C3%BCyeleridir

Ladrera, R.C. & Cagasan, U.A. (2022). A Review on the Impact of Climate Change on Agriculture Sector:Its Implication to Crop Production and Management. *Eurasian Journal of Agricultural Research*, *6*(1), 52-61.

Lake, I. R. (2012, November). Hooper, L., Abdelhamid, A., Bentham, G., Boxall, A.B.A., Draper, A., Fairweather-Tait, S., Hulme, M., Paul R. Hunter, P.R., Gordon, N., & Waldron, K.W. (2012). Climate Change and Food Security: Health Impacts in Developed Countries. *Environmental Health Perspectives, 120*(11). PMID:23124134

Murphy, M. (2022). *Gıda krizi: BM, Ukrayna'daki savaşın küresel gıda krizine yol açabileceği uyarısında bulundu*, BBC. https://www.bbc.com/turkce/haberler-dunya-61503829

NASA Global Climate Change. (n.d.). *The Causes of Climate Change.* NASA. https://climate.nasa.gov/causes/

NTV. (2023). *Mavi ekonominin yeşil kalkınmadaki rolü (The role of the blue economy in green development)*,. NTV. https://www.ntv.com.tr/turkiye/mavi-ekonominin-yesil-kalkinmadaki-rolu,5qPFiFcgh0WAdX-g_yj9RQ

OECD. (2016), *The Ocean Economy in 2030, Directorate for Science, Technology and Innovation Policy Note. OECD.* https://www.oecd.org/sti/futures/Policy-Note-Ocean-Economy.pdf

OECD. (2021), *Adapting to a changing climate in the management of coastal zones.* OECD. https://www.oecd-ilibrary.org/docserver/b21083c5-en.pdf?expires=1679401663&id=id&accname=guest&checksum=3C774E8FE5923025E42023931179DE73

OECD. (2022). *Sustainable Ocean Economy.* OECD. https://stats.oecd.org/Index.aspx?DataSetCode=OCEAN

OECD. Stat, (2022a). Employment in fisheries, aquaculture and processing. *Agriculture and Fisheries, Fisheries and Aquaculture, Fisheries and Aquaculture statistics.* OECD. https://stats.oecd.org/

OECD. Stat, (2022b). International trade of fisheries commodities. *Agriculture and Fisheries, Fisheries and Aquaculture, Fisheries and Aquaculture statistics.* OECD. https://stats.oecd.org/

OECD. Stat, (2022c). Inland Fisheries. *Agriculture and Fisheries, Fisheries and Aquaculture, Fisheries and Aquaculture statistics.* OECD. https://stats.oecd.org/

OECD. Stat, (2022d). Fishing fleet. *Agriculture and Fisheries, Fisheries and Aquaculture, Fisheries and Aquaculture statistics.* OECD. https://stats.oecd.org/

OECD. Stat, (2022e). Aquaculture production. *Agriculture and Fisheries, Fisheries and Aquaculture, Fisheries and Aquaculture statistics.* OECD. https://stats.oecd.org/

OECD. Stat, (2022f). *Marine Landings. Agriculture and Fisheries, Fisheries and Aquaculture, Fisheries and Aquaculture statistics.* OECD. https://stats.oecd.org/

OECD. Stat, (2022g). Fisheries And Aquaculture Indicators. *Agriculture and Fisheries, Fisheries and Aquaculture, Fisheries and Aquaculture statistics.* OECD. https://stats.oecd.org/

OECD. (n.d.). Ocean and climate change. *Agriculture and Fisheries, Fisheries and Aquaculture, Fisheries and Aquaculture statistics.* OECD. https://www.oecd.org/ocean/topics/climate-change-ocean/

Reid, H., Sahlén, L., Stage, J., & Macgregor, J. (2008). Climate change impacts on Namibia's natural resources and economy. *Climate Policy, 8*(5), 452–466. doi:10.1080/14693062.2008.9685709

Shukman, D. (2021). *Dünya küresel ısınma eşiği olan 1.5 C dereceye 'gelecek beş yıl içinde ulaşabilir [The world may reach 1.5°C, the global warming threshold, within the next five years].* BBC. https://www.bbc.com/turkce/haberler-dunya-57264316

Sustainability for all, (n.d.).*The Blue Economy: Benefits Of Investing In Ocean Conservation*, Active Sustainability. https://www.activesustainability.com/water/blue-economy/?_adin=02021864894

The Commonwealth. (n.d.). *Action Group on Sustainable Blue Economy*. The Commonwealth. https://thecommonwealth.org/bluecharter/sustainable-blue-economy#:~:text=The%20blue%20economy%20embraces%20economic,of%20poverty%20and%20climate%20change

World Bank. (2020c). *Captured Fisheries Production (Metric Tons)*. World Bank. https://data.worldbank.org/indicator/ER.FSH.CAPT.MT?locations=OE

World Bank. (2020b). *Total fisheries production (metric tons) - OECD members*. World Bank. https://data.worldbank.org/indicator/ER.FSH.PROD.MT?locations

Turrentine, J. (2022). *What Are the Causes of Climate Change?* NRDC. https://www.nrdc.org/stories/what-are-causes-climate-change

UNCTAD. (2022). *5 global actions needed to build a sustainable ocean economy.* UNCTAD. https://unctad.org/news/5-global-actions-needed-build-sustainable-ocean-economy

Yildirim, S. (2023), Greenwashing: a rapid escape from sustainability or a slow transition?, *LBS Journal of Management & Research*. https://doi.org/ doi:10.1108/LBSJMR-11-2022-0077

Yildirim, S., & Kaplan, M. (2020). Seafood security and sustainability through sustainable development: a review of Turkish seafood market. In A. Obayelu & O. Obayelu (Eds.), *Developing Sustainable Food Systems, Policies, and Securities* (pp. 16–36). IGI Global. doi:10.4018/978-1-7998-2599-9.ch002

Yıldırım, S., & Yıldırım, D. C. (2021). Achieving seafood security in the Mediterranean region: a case of Turkey. In R. Castanho & J. Martın Gallardo (Eds.), *Management and Conservation of Mediterranean Environments* (pp. 175–195). IGI Global. doi:10.4018/978-1-7998-7391-4.ch011

Yıldırım, S., & Yıldırım, D. Ç. (2022). İklim Değişikliği ve Gıda Güvenliği İlişkisine Kısa Bir Bakış [A Brief Overview of the Relationship between Climate Change and Food Security]. Dirençli Şehir Yolunda Toplum Politikaları, Ed. Lehimler, H.M., Nobel yayıncılık.

Yildirim, S., & Kaplan, M. (2022). The threat of invasive alien marine species to the blue economy: the mediterranean case. In L. Raimi & J. Kah (Eds.), *Implications for Entrepreneurship and Enterprise Development in the Blue Economy* (pp. 50–80). IGI Global.

Zogopoulos, E. (2020). Why Blue Economy Is So Important. *Energy Industry Review*. https://energy-industryreview.com/analysis/why-blue-economy-is-so-important/

ADDITIONAL READING

Kaplan, M., & Yildirim, S. (2023). The Sustainability of Seafood Market Under the Attack of Invasive Alien Species: A Review of Turkish Marines. In A. Ben Salem, L. Rhazi, & A. Karmaoui (Eds.), *Climatic and Environmental Significance of Wetlands: Case Studies from Eurasia and North Africa* (pp. 74–95). IGI Global.

Kaplan, M., Yildirim, S., & Yildirim, D. C. (2022). Pufferfish versus lionfish: Comparing risks for Turkish marine economics. *Marine Economics and Management*, *5*(2), 173–187. doi:10.1108/MAEM-09-2022-0008

Yildirim, S. (2020). The consumer role for sustainable development: how consumers contribute sustainable development goals. In V. Chkoniya, A. O. Madsen, & P. Bukhrashvili (Eds.), *Anthropological Approaches to Understanding Consumption Patterns and Consumer Behavior* (pp. 325–341). IGI Global. doi:10.4018/978-1-7998-3115-0.ch018

Yildirim, S. (2021). 3R Principle in Household Energy Consumption Patterns: A Case of Turkish Households. In M. Danish & T. Senjyu (Eds.), *Eco-Friendly Energy Processes and Technologies for Achieving Sustainable Development* (pp. 88–107). IGI Global. doi:10.4018/978-1-7998-4915-5.ch005

Yildirim, S. (2023), Greenwashing: a rapid escape from sustainability or a slow transition?, *LBS Journal of Management & Research*. https://doi.org/ doi:10.1108/LBSJMR-11-2022-0077

Yıldırım, S., & Bostancı, S. H. (2021). The efficiency of e-government portal management from a citizen perspective: Evidences from Turkey, *World Journal of Science. Technology and Sustainable Development*, *18*(3), 259–273. doi:10.1108/WJSTSD-04-2021-0049

Yildirim, S., Demirtas, I., & Yildirim, D. C. (2022). A Review of Alternative Economic Approaches to Achieve Sustainable Development: The Rising Digitalization and Degrowth Post COVID-19. In R. Castanho (Ed.), *Handbook of Research on Sustainable Development Goals, Climate Change, and Digitalization* (pp. 288–307). IGI Global. doi:10.4018/978-1-7998-8482-8.ch018

Yıldırım, S., & Kantarcı, T. (2022). A Review on Sustainability Policies of Businesses: Recycling and Waste Reduction. *Journal of Recycling Economy & Sustainability Policy*, *1*(1), 1–9.

Yildirim, S., & Kaplan, M. (2020). Seafood security and sustainability through sustainable development: a review of Turkish seafood market. In A. Obayelu & O. Obayelu (Eds.), *Developing Sustainable Food Systems, Policies, and Securities* (pp. 16–36). IGI Global. doi:10.4018/978-1-7998-2599-9.ch002

Yildirim, S., & Kaplan, M. (2022). The threat of invasive alien marine species to the blue economy: the mediterranean case. In L. Raimi & J. Kah (Eds.), *Implications for Entrepreneurship and Enterprise Development in the Blue Economy* (pp. 50–80). IGI Global. doi:10.4018/978-1-6684-3393-5.ch004

Yıldırım, S., & Yıldırım, D. Ç. (2020). Achieving Sustainable Development Through a Green Economy Approach. In S. Patti & G. Trizzino (Eds.), *Advanced Integrated Approaches to Environmental Economics and Policy: Emerging Research and Opportunities* (pp. 1–22). IGI Global. doi:10.4018/978-1-5225-9562-5.ch001

Yıldırım, S., & Yıldırım, D. C. (2021). Achieving seafood security in the Mediterranean region: a case of Turkey. In R. Castanho & J. Martın Gallardo (Eds.), *Management and Conservation of Mediterranean Environments* (pp. 175–195). IGI Global. doi:10.4018/978-1-7998-7391-4.ch011

Yıldırım, S., Yıldırım, D. Ç., & Gedikli, A. (2016). Sustainable Consumption Trends in the World in the Context of Green Economy and Sustainability. In M. Erdoğdu, T. Arun, & I. Ahmad (Eds.), *Handbook of Research on Green Economic Development Initiatives and Strategies* (pp. 65–84). IGI Global. doi:10.4018/978-1-5225-0440-5.ch004

KEY TERMS AND DEFINITIONS

Blue Economy: It refers to sustainable ocean and marine economy in the long term.

Climate Change: It is an climate event that refers to changing temperatures of the Earth.

Ocean Economy: A type of bio-economy that using sources of oceans and marines to provide food, employment and income.

OECD: Organization for Economic Co-operation and Development.

Chapter 13
The Global Waste Trade:
A Decade in Review

Ahmed Yusuf Sarihan
https://orcid.org/0000-0001-7119-9852
Bandırma Onyedi Eylül University, Turkey

ABSTRACT

The global waste trade refers to the movement of waste materials across international borders for disposal or recycling. This trade has grown rapidly in recent years as countries seek to capitalize on the economic benefits of exporting waste while also addressing the challenges of managing increasing volumes of waste at home. This chapter will provide an overview of the global waste trade over the past decade, including key trends and developments, major players and destinations, and this rapidly growing industry's environmental and economic impacts. The chapter will also explore the challenges and opportunities presented by the global waste trade and offer insights into the future of this complex and dynamic field.

INTRODUCTION

The past decade has seen unprecedented growth in the global waste trade, driven by increased global consumption and the need for effective waste management. The world population is constantly growing, leading to a corresponding increase in consumption and related waste. The non-stop consumption activities, where people crave more and more products, regardless of their origin, lead to more production and hence more waste. This endless cycle has resulted in the global circulation of products and waste, with globalization making the circulation of waste just as essential as the circulation of domestically produced products.

The international trade of waste materials has emerged as a rapidly growing industry over the past decade, allowing countries to manage their waste effectively, either by processing it within their own borders or by exporting it to other countries equipped to do so. This trade has become a vital component of the global economy. Its growth is driven by the increasing demand for waste management solutions and the need to regulate waste flows, protect the environment, and ensure social equity in distributing waste management responsibilities. Since the commercial circulation of waste is a valuable research

DOI: 10.4018/978-1-6684-8879-9.ch013

opportunity, especially in terms of bioeconomics and international trade, this book chapter is written on the basis of the following research questions:

- What theoretical foundations explain the global flow of waste?
- What conclusions have been reached in the empirical studies conducted regarding the theory?
- What topics do the studies conducted in the last ten years and indexed in respected journals focus on?
- From which aspects will it be more beneficial to look at the waste trade in the coming years?
- Which countries are the largest global exporters of waste banned or restricted by China?
- Which countries are the biggest importers of these wastes?

The following objectives have been achieved to answer these questions.

1. The pollution haven theory and its opposing views are explained.
2. In particular, empirical studies testing the pollution haven are cited.
3. Articles written in the last ten years and indexed in the Web of science database and containing the words "waste trade" in the title were examined.
4. The results of the research and empirical studies were interpreted and the issues that are important in the future were emphasized.
5. With data from important statistical databases, major players and major destinations have been revealed.

In conclusion, this chapter will provide a comprehensive overview of the past decade of global waste trade, examining the academic trends, challenges faced by importers, and opportunities that have emerged for exporters in this rapidly evolving field. By providing valuable insights into the current state of international waste trade, this chapter will serve as a valuable resource for policymakers, researchers, and practitioners working in waste management and international trade.

BACKGROUND

Pollution Haven Theory

The reality of waste trade, which researchers have also observed, has become an area of study that has been theorized on. Of course, researchers are not the only side of the issue. Governments have started to set new rules for waste in their policies. In addition, non-governmental organizations are also drawing attention to this issue and seeking to raise awareness. Whichever way one looks at it, waste becomes a problem at some point. Firstly, from a scientific approach, thanks to Pethig's (1976) work which brought forward pollution haven theory, and Walter and Ugelow's (1979) tracking study of ecology and economy sciences paid attention to this topic. Antweiler et al. (2001) draw attention to two theories regarding the direction of the flow of waste in the world. According to the first of these theories, the "pollution haven" theory, wastes move toward low-income countries.

On the other hand, the second theory emphasizes the opposite. In other words, it argues that countries with abundant capital also have significant waste. Empirical research on both approaches will be presented in the literature review section of this chapter. However, before addressing this, it is worth mentioning the previously mentioned governments' points of view. Starting in 1992, China, which imported 45% of the world's plastic waste, made a policy change in 2018 and imposed restrictions on plastic waste imports (Brooks et al., 2018). As a final perspective before the details of a conceptual framework, it is worth mentioning the approach of Green Peace, one of the world's leading non-governmental organizations. Green Peace (2019) emphasized that there are failures in recycling rather than the existence of waste trade and pointed out that plastic waste will increase by 40% in the future, especially if plastic production continues.

Waste trade, which occupies an important place in global trade in terms of meaning, if not amount, has brought the issue of waste management to the agenda. At this point, many types of waste come to mind. Hazardous wastes and plastic wastes are at the top of these. Kaya (2012) stated that since the 1970s, hazardous wastes have been sold from developed to developing countries. Baggs (2009) drew attention to the flow of waste in global trade, emphasizing the excessive pollution observed in African countries with low populations. O'Neill (2001) attributed the global movement of waste to the high cost of transportation, disposal, minimization, and storage. From this point of view, the picture clearly shows that developed countries ignore the rest of the world for less cost and, therefore, more profitability. Moreover, this picture confirms the "pollution haven" theory emphasized in research (see Antweiler, et al., 2011; Baggs, 2009). Therefore, developing countries must position themselves, especially in this regard. China's ban on plastic waste imports in 2018 is an essential political step taken in this area and can serve as an example for developing countries. At this point, the question of how developing countries will bear the costs avoided by developed countries comes to mind. The answer is quite simple. It is stated that developing countries do not store these wastes properly, do not incur high costs to dispose of them, and keep them in landfills. Hamilton and Viscusi (1999) confirm this belief by stating that processes such as cleaning and disposal of hazardous wastes are more common in high-income countries. On the other hand, Sigman (1996) emphasized that developed countries impose high taxes on waste disposal, and therefore waste trade is shifting towards developing countries. Kaya (2012) stated that strict environmental policies are another reason developed countries want to eliminate these wastes. As it is known, one of the problems frequently encountered in underdeveloped and developing countries is the need for more environmental awareness and the failure to implement policies to protect the environment. This theoretical approach also suggests that waste trade is shifting towards developing countries.

Regarding the second theory, that countries with abundant capital also attract waste, Baggs (2009) argues that countries with capital-intensive industries are the largest pollution emitters, while countries with high levels of prosperity are also capital-abundant. Some studies have argued that waste flows are related to capital intensity and production inputs. For example, Grossman and Krueger (1991) argue that since the factors available to low-capital countries to participate in global trade allow them to offer energy- and pollution-free products, the waste of the products they sell will also go to developed countries. This finding, which has a limited approach, can be attributed to the research conducted in the form of a comparison between the United States and Mexico. Jaffe et al. (1995) argue that environmental policies do not impose high costs and therefore do not affect international competitiveness, so there is no reason for waste trade to be directed to developing countries. Tobey (1990) similarly found no effect on trade flows in many of his tests where environmental policies were included as a variable. Therefore, it has been argued that developed countries should adopt more stringent environmental policies. It is also emphasized that most empirical studies do not consider transportation costs, but these costs can be a negative barrier to waste diversion.

The fact that the theoretical framework consists of a pollution haven theory and an approach that rejects it in contrast to it has aroused a scientific curiosity for the writing of this book chapter. Therefore, the following section of the paper provides a literature review and discusses the similarities and differences between the empirical studies on the trajectory of waste trade and the theoretical approaches.

Empirical Studies on the Theory

As the world's climate changes, human health becomes more fragile, and different ecological situations arise, the focus of scientific research is shifting. Looking at global economic studies, all scientists working in this field can state that their interest in sustainability, carbon emissions, and energy has increased significantly. On the other hand, it can be determined by simple internet searches that marketing professionals also examine consumers' attitudes towards healthier, recyclable, and environmentally friendly products. However, the studies conducted within the scope of international trade and economics are limited at this point concerning the circulation of waste because they focus more on the issues mentioned above. In this part of the book chapter, a general overview of research on global waste trade is given. Then, as a result of the research conducted in the "web of science," database at the level of titles such as "waste trade," the studies conducted between 2011-2022 are discussed separately, and some conclusions are presented for these studies.

International trade is a complex economic phenomenon affected by inter-country relations, regulated by strict legal practices, restricted by countries' protectionist measures, and shaped by competitiveness. While even the buying and selling of commodities between countries are such a complex process, the waste movement between countries can be assumed to be even more complex. Strohm (1993) quoted that it is emphasized that the import of hazardous wastes, which is argued was more damaging than the generation of hazardous wastes; the study emphasized that the political infrastructure of the period contained great dangers in terms of waste trade. According to the study, in the context of waste trade policy, countries are caught between various dilemmas, and this situation causes damages in the countries where waste import takes place. Hence the point was made that waste should only be traded with equality and ethically sound policies. It is known that there are many negative economic consequences when the political infrastructure is of low quality, especially in less developed countries. However, Strohm (1993) brought necessary research to the world of science by emphasizing that these effects are pregnant with great disasters in terms of environment and health. Similarly, Clapp (1994) draws attention to two important details in his research on hazardous waste exports to less industrialized countries. The first is that the world economy and trade are inextricably linked to the world ecology, and the second is that non-governmental organizations have great power in the waste trade. According to Clapp's (1994) study, the global economic system causes the wastes of producing and developed countries to flow into poor and underdeveloped countries. In addition, underdeveloped countries, unfortunately, allow wastes and even hazardous wastes to import into their countries for both recycling and storage due to the search for foreign exchange. In contrast to this situation, non-governmental organizations such as Greenpeace have a preventive role in the connections they establish with local non-governmental organizations in undeveloped or developing countries. It is noteworthy that going back thirty years; some scientists have examined the role of waste in their research on the economy and the environment. On the other hand, these studies are much more critical, considering that they focus on less developed countries and examine the sources of environmental problems that those countries may face.

Given the pollution haven theory proposed in the background section of this chapter, some researchers have tried to test the theory empirically. Balsalobre-Lorente et al. (2019) used a series of econometric methods in a study aiming to test the theory on Mexico, India, Nigeria, and Turkey. In the related study, FDI was added to the model in addition to the economic growth variable, and the pollution level was measured through the ecological footprint. According to the findings of this originally modeled study on MINT countries, the ecological footprint also increases as the level of economic development and FDI in these countries increases. On the other hand, this increase peaks after a certain point and then tends to decrease. With its results, the study provides empirical evidence for the pollution haven theory because it shows that countries do indeed attract pollution until they reach a certain point of economic growth. Another issue that should be emphasized here is foreign direct investments. This is because these investments bring not only cash but also employment (Santiago, 1987; Athukorala & Menon, 1996), market diversity through export (Caves, 1981; Scherer, 1980), technology (Maskus, 2004) and sub-industries to countries. From this point of view, it will become a regular target to utilize the products obtained as exports and earn more foreign currency for the country. Therefore, the impact of FDIs on economic growth has not been ignored in this study. In another pollution haven study by Gill et al. (2018: 172), a review of the studies conducted on the theory was conducted. The study concluded that it is impossible to prove this theory definitively because different studies have reached different conclusions. It was emphasized that it does not make sense for developed countries to be clean because the world continues to remain globally polluted in one way or another. It was stated that the vital issue here is to reduce polluting production in developed countries.

Cole (2004) tried to explain the inverted U-shape of the environmental Kuznet curves due to the shift of polluting industries from developed to developing countries. Not only that, but he also stated that trade also has effects in the context of environmental pollution. In doing so, as expected, he started with the pollution haven theory. The findings of the study show that the pollution haven theory is valid. Nevertheless, according to the study, the possibility that the effects may not be as significant as this theory suggests draws attention as a reality that should be considered. Speaking of the environmental Kuznet curve, it would also be fitting to clarify this issue. According to the environmental Kuznet curve, environmental pollution initially increases with economic development. However, after income reaches a certain level, environmental pollution starts to decline (Dam & Karakaya, 2013). This hypothesis determines the relationship between economic growth and the concentration of greenhouse gases, a component of environmental pollution. Different methods can be used to measure environmental pollution. Sulfur dioxide, nitrogen oxides, chlorofluorocarbons, wastewater, and other chemicals released directly into the air or water cause many environmental pollutants (Brajer et al., 2011). Grether, et al. (2012) attempted to test the pollution haven theory of pollution-causing products in global trade worldwide in their research. They believe this study differs from single-country and small-product group studies considering it had a large sample with different product groups. Although the researchers claimed that the study was enormous in terms of scope, they also admitted that it was similar to previous narrow-scope studies in terms of its limited findings. According to the research results, the pollution haven theory is partially supported, i.e., pollution-causing products flow to developing countries. On the other hand, according to the research, sometimes pollution-generating substances remain in developed countries, and transportation costs are also important at this point. The research draws attention to a different point and brings up another variable that modern studies should include in their research models. This variable is transportation costs. As a result, since the traded products are wastes, the cost increase may push exporters to search for different options. Sometimes it is cheaper to set up a local recycling or disposal facility than to ship these wastes to other countries continually. As Manunen (2000) mentioned, the logistic cost might rise by 10% in a company's total production costs. Because of that, the flow of savings starting from businesses can reach the macro level and affect the global waste trade.

In a study examining the relationship between the environment and trade policies implemented and whether the pollution haven theory will be realized or not, Kellenberg (2009) found that external affiliates established through the US, i.e., FDIs of the US to different countries, cause effects that prove the pollution haven theory in the destination countries. Moreover, according to the related study, this pollution is not necessarily seen in the sectors grouped as "dirty," and other sectors are also seen to cause this pollution. Of course, when we look at the relationship between the study and policies, it is seen that strict environmental policies prevent pollution in some sectors. The different perspectives of scientific research continue to surprise the human brain and illuminate the points where knowledge is insufficient. As can be seen, when it comes to the pollution caused by trade and the effects of the economy on the environment, the originality of the models not only allows for a better understanding of the nature of the phenomena but also provides empirical evidence to humanity. It would not be wrong to expect the policies implemented to have different effects. For example, a strict environmental policy may reduce pollution in a country, but it is a big mystery how it will affect other countries with which they have trade relations. Levinson and Taylor (2008) revealed exciting findings in their study that show that the pollution haven theory is invalid. Moreover, this study is a study that includes policies and laws in its model, similar to the ones discussed earlier. According to the researchers, sectors in which companies increase their costs due to strict rules introduced to reduce environmental impacts tend to import more. It is argued that pollution increases even more with imported products. As a result of these original findings, it is emphasized that the pollution haven theory needs to be revised to explain this complexity. The fact that many studies reach different findings with different perspectives actually expresses the best of science. It explains why waste trade will remain a unique issue and why research on this subject can be valuable. Millimet and List (2004), another study that tests the pollution haven theory through policies and their outcomes, emphasizes that the policies implemented do not always reduce pollution. After reviewing these studies in detail, it would be helpful to discuss the findings of some empirical studies quickly. In this way, it will be possible to have a general understanding of the subject before moving on to the research conducted in journals indexed in scientifically reputable indexes in the last decade.

Many studies empirically test the pollution haven theory across countries. In this respect, we should first emphasize the studies that prove the theory's validity. For instance, some studies on Turkey prove that the pollution haven theory is valid for this country (Mutafoglu, 2012; Gokmenoglu & Taspinar, 2016; Seker et al., 2015; Kaya, et al., 2017; Kilicarslan & Dumrul, 2017; Kocak & Sarkgunesi, 2017; Yildirim, et al. 2017; Kurt et al., 2019; Mike, 2020). Although the standard features of these studies are that the period they use consists of similar intervals, the methods used in the studies differ from time to time. Nevertheless, it would not be correct to state that all studies use econometric methods. When we look at the research on China, some studies (He, 2006; Liang, 2008; Lan et al., 2012; Jiang, 2015; Liu et al., 2018) prove that the pollution haven theory is valid, while some studies (Bao et al., 2010) argue the opposite. Apart from these studies, it is also possible to find empirical evidence that the pollution haven theory is valid in countries such as India (Acharyya, 2009), Malaysia (Lee, 2009), Viet Nam (Tang & Tan, 2015). On the other hand, many studies on country groups also prove that the theory is realized (Sapkota & Bastola, 2017; Asghari, 2013; Al-Mulali & Tang, 2013). It should be re-emphasized that these studies have also been conducted through econometric models.

Last Decade of Respected Research

When we look at the waste trade articles indexed in the "web of science" database of the last decade, the following studies and their results draw attention. Notably, the studies conducted in the last decade have started to be presented in the "web of science" database since 2015. It should be reminded again that these studies include studies with the word group "waste trade" in the title. In this respect, the studies focus on the plastic waste trade, hazardous waste trade, China's policy, electronic waste trade, and polyethylene waste trade.

As everyone knows, plastic waste has many negative impacts on public health and the environment. These issues are being addressed in more depth by health and environmental scientists. Regarding commercial and economic impacts, some research has focused on this issue in the last decade. Starting in 2018 Brooks et al. (2018) argue that China's ban on waste imports necessitates different countries to build new local waste management facilities. Balkevicius et al. (2019) stated that the policy implemented by China has reduced plastic waste from developed countries to China and that this waste has shifted to other developing countries. Similarly, Wang et al. (2020) emphasized that China's ban on the import of plastic products shifted these imports to other countries. Liang et al. (2021) examined plastic waste trade and management in Asia and found that changes in environmental policies implemented by countries impact waste imports. Liu et al. (2021) emphasized that urgent steps should be taken at critical levels for the recycling of plastic waste because the global plastic waste trade is distributed across different countries. Wen et al., (2021) found that due to China's policy, plastic waste from developed countries to China decreased by 92%, but this increased waste imports from less developed and developing countries. Zhao et al. (2021) emphasized that after China's policies, the center of the global plastic waste trade network shifted from China to other South East Asian countries. Liu et al., (2022) pointed out that a network was formed in the plastic waste trade category, and a stable plastic waste trade flow was formed within this network. Zhao et al., (2022) emphasized that the international community should bring up tariff-based preventive measures against the global plastic waste trade that will exist in the future. As can be seen, the plastic waste trade is a continuum, and China, which imports most of this waste, has given up on this policy. Therefore, the new importer gap is filled by less developed and developing countries, which creates the basis for the increase in health and environmental problems in these countries. Although we have mentioned polyethylene as a different title above, it is worth mentioning the research conducted by Xu et al., (2020) since this subject is technically included in the subject of plastics. Here, it is essential to note that the researchers have also investigated the local recycling of these wastes.

Another issue that has attracted attention in the last decade of research is the global trade of electronic waste. Among the studies on this subject, it is seen that the studies indexed in the "web of science" database date back to 2016. Efthymiou et al. (2016) observed illegal e-waste trade from economically developed to underdeveloped countries. It was emphasized that this situation includes risks in terms of public health. Petridis et al. (2020) pointed out that the e-waste trade is affected by many different factors (including the economic and geographical locations of the countries) while drawing attention to human health in the management of these wastes. Cotta (2020) found that used electronic product wastes in northern countries are traded to southern countries, causing significant health and environmental problems in these countries. Considering that electronic waste may contain radioactive and toxic materials, it would be correct to state that these studies are more important. These issues will be emphasized more in the future. As a result, the production of electronic products is increasing day by day, and the life cycle of the products is getting shorter day by day.

Finally, the trade of hazardous waste draws attention in the literature. It is seen that the history of the articles scanned on the same database dates back to 2015. Lucier and Gareau (2015) argue that although there are still regulations and laws regarding the hazardous waste trade, many countries are moving toward rock bottom. This indicates inadequate laws and careless waste management. Moore et al., (2018) provided valuable research to science in terms of tracking waste in North America by demonstrating the flow of hazardous waste, the detection of waste havens, and the visualization methods from which institutional networks the waste originates. Vincent, et al., (2019) stated that interface complexity in the circulation of hazardous waste greatly influences decisions about waste management.

As can be seen, the articles indexed in reputable scientific indexes in recent years, although limited in diversity, have critical research topics. The information presented by these studies is also remarkable for future studies. This part of the book proves the unique value of the approach to be presented in the next stage and the abundance of waste categories addressed.

APPROACH, DATA AND FINDINGS

Approach

The research aims to reveal the world waste trade situation and offer ecological suggestions by observing the leading waste exporters and the countries where the waste is piled up. An exploratory and descriptive analysis method based on secondary data was used to achieve this aim. From this point of view, the study interprets quantitative data with a qualitative approach. Three methods are generally recommended in qualitative data analysis. First, it is essential to stick to the original form of the collected data as much as possible. This basis is invalid in this study as the data were obtained from secondary sources. In the second method, although the data is presented with a descriptive approach, it is possible to determine some codes and go from codes to categories and from categories to themes. In the third method, the researcher analyzes the data using his own interpretations, description, and thematic analysis. (Baltaci, 2017; Creswell, 2002; Eysenbach & Köhler, 2002; Miles & Huberman, 1994). It is possible to say that the second and third methods are carried out together in this book chapter.

Data

The data used in the research were obtained from the "trademap," which is the database of the International Trade Center (I.T.C.). The data obtained were evaluated based on 2012 - 2021 for each waste item, and the largest exporters and importers were reported. The data used in the research process and the clarity of the standards for waste classification could not be determined. For this reason, the classification of waste as waste was made according to the "Imported Waste Management List" published by the Ministry of Ecology and Environment (2018) of the People's Republic of China (P.R.C.). However, some of these products were not included in the scope of the research since they were not emphasized much in the literature review, especially in the last ten years. Therefore, it was decided to conduct the research for nine product groups. The global classification of the waste in this list as a product has been standardized to follow the worldwide trade according to the "H.S. Code," the harmonized system.

Table 1 above contains the harmonized system codes and product groups of the wastes whose 10-year increase and decrease values were examined and interpreted during the research process.

Findings

Major Players

The last decade's leading exporters of the 261800 HS Code product group are shown in the table below.

Table 1. Examined waste's commercial codes and explanations

#	HS Code	Explanation
1	261800	Granulated slag (slag sand) from the manufacture of iron or steel
2	261900	Slag, dross, scalings and other waste from the manufacture of iron or steel (excluding granulated slag)
3	391510	Waste, parings and scrap, of polymers of ethylene
4	391520	Waste, parings and scrap, of polymers of styrene
5	391530	Waste, parings and scrap, of polymers of vinyl chloride
6	391590	Waste, parings and scrap of plastics (excluding that of polymers of ethylene, styrene and vinyl chloride)
7	720449	Waste and scrap of iron or steel (excluding slag, scale and other waste of the production of iron and steel; radioactive waste and scrap; fragments of pigs, blocks or other primary forms of pig iron or spiegeleisen; waste and scrap of cast iron, alloy steel or tinned iron or steel; turnings, shavings, chips, milling waste, sawdust, filings, trimmings and stampings; waste and scrap of primary cells, primary batteries and electric accumulators)
8	740400	Waste and scrap, of copper (excluding ingots or other similar unwrought shapes, of remelted copper waste and scrap, ashes and residues containing copper, and waste and scrap of primary cells, primary batteries and electric accumulators)
9	760200	Waste and scrap, of aluminium (excluding slags, scale and the like from iron and steel production, containing recoverable aluminium in the form of silicates, ingots or other similar unwrought shapes, of remelted waste and scrap, of aluminium, ashes and residues from aluminium production)

Source: ITC, 2023(1), Trademap, trademap.org

Table 2. Leading exporters for 261800 HS code between 2012-2021 (thousand US dollars)

Countries	2012	2013	2014	2015	2016	2017	2018	2019	2020	2021
South Africa	264.958	338.007	232.327	230.144	152.506	235.058	248.626	243.992	216.632	266.619
Japan	33.703	57.141	79.992	82.342	80.508	73.559	92.254	112.838	84.091	94.343
Germany	28.963	26.314	27.216	22.485	18.279	30.534	33.846	34.649	31.833	55.013
India	5.009	10.226	20.015	15.852	12.084	17.007	52.646	70.674	48.170	50.252
Netherlands	22.531	20.533	20.891	25.875	31.524	32.708	35.957	31.921	34.114	42.971
Spain	24.493	21.579	17.949	11.144	22.006	30.562	35.511	29.671	28.170	39.825
Malaysia	0	20	0	631	358	6.094	13.210	23.609	29.520	34.598
Viet Nam	303	203	2.070	1.376	472	5.715	17.006	17.117	16.721	23.056
Türkiye	980	2.819	2.235	998	1.551	1.935	3.226	5.151	13.135	19.839
France	9.539	10.568	15.489	10.749	21.763	22.900	27.394	33.078	28.376	18.669

Source: ITC, 2023(2), Trademap, trademap.org

As shown in Figure 1. below, South Africa is the leading country for 261800 product group wastes, which can be summarized as granulated slag over the years.

Although the slope towards exporters appears in different colors for each annual period, the place where the slope is concentrated is sufficient to understand the leading country. South Africa has been the biggest exporter of this product group for the last decade, and when looking at the slope, it is understood that Japan and Germany are the other exporters in this waste group.

The last decade's leading exporters of the 291900 HS Code product group are shown in the table below.

As shown in Figure 2. below, Russian Federation emerges as the top country for 261900 product group wastes, which can be summarized as other slags over the years.

Figure 1. 261800 HS coded waste group exporter countries density chart

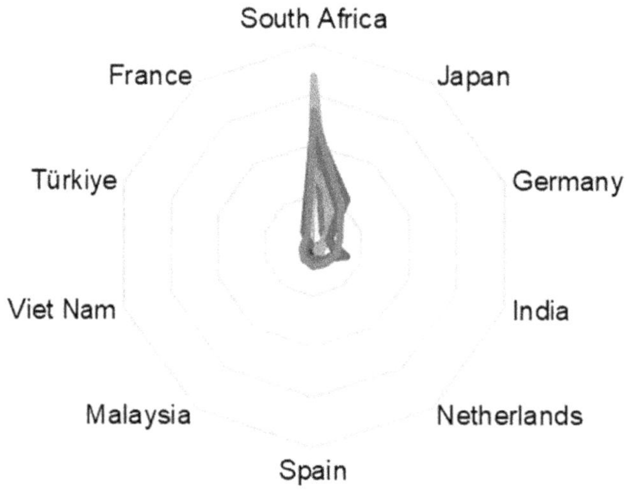

Table 3. Leading exporters for 261900 HS code between 2012-2021 (US dollars thousand)

Countries	2012	2013	2014	2015	2016	2017	2018	2019	2020	2021
Russian Federation	83.024	104.732	138.122	16.679	32.492	77.518	227.113	104.158	60.380	92.330
Belgium	13.939	6.915	12.023	5.823	4.563	8.261	8.137	10.037	23.382	41.742
India	152.353	153.386	70.575	22.387	13.558	19.574	8.680	8.521	13.611	39.271
Egypt	6.631	18.507	6.589	9.191	12.449	11.706	12.344	17.729	12.540	28.988
Germany	12.505	23.721	25.700	23.184	39.168	46.536	45.282	41.490	26.543	28.571
Türkiye	35.303	29.647	20.991	4.284	7.700	10.055	12.621	14.059	9.503	27.341
Taipei, Chinese	11.117	22.680	18.502	12.381	7.828	6.983	13.539	13.159	14.938	24.853
United States of America	15.507	27.125	17.278	10.368	4.225	4.288	6.016	11.077	9.297	22.297
Korea, Republic of	42.186	32.386	14.912	13.483	9.178	5.237	6.912	6.670	4.648	20.527
New Zealand	5.608	6.223	5.068	2.680	3.705	15.872	1.218	2.351	6.038	17.475

Source: ITC, 2023(3), Trademap, trademap.org

Figure 2. 261900 HS coded waste group exporter countries density chart

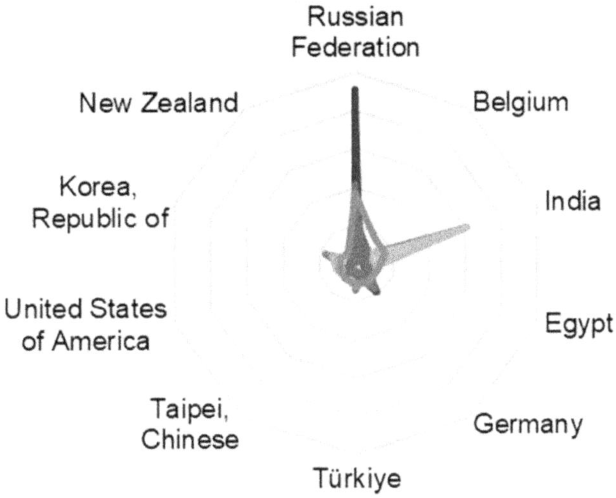

As recalled from the previous graph interpretation, the slopes of the lines are of higher importance. Russia has been the biggest exporter of this product group for the last few years, and when looking at the slope, it is understood that India was the old leader for this waste, which remains third place in 2021.

The last decade's leading exporters of the 391510 HS Code product group are shown in the table below.

As can be seen from the Figure 3. below, Germany emerges as the leading country for 391510 product group wastes, which can be summarized as waste of polymers of ethylene over the years.

Although Germany is seen as the leader of the last decade, it can be said that the USA follows it.

The last decade's leading exporters of the 391520 HS Code product group are shown in the table below.

As seen in Figure 4. below, Japan emerges as the top country for 391520 product group wastes, which can be summarized as a waste of polymers of styrene over the years.

Table 4. Leading exporters for 391510 HS code between 2012-2021 (US dollars thousand)

Countries	2012	2013	2014	2015	2016	2017	2018	2019	2020	2021
Germany	424.565	413.501	470.098	369.613	341.847	291.350	269.130	248.444	192.651	232.027
Netherlands	81.075	95.003	106.723	81.336	80.762	69.277	58.785	61.514	72.375	142.988
USA	284.318	287.331	352.083	290.027	284.976	246.442	173.033	106.927	83.570	126.694
Belgium	96.735	98.815	116.292	100.965	91.609	99.989	97.607	85.862	76.231	101.825
United Kingdom	185.556	122.717	164.436	183.208	161.969	123.312	81.235	65.525	56.070	97.919
Philippines	6.302	8.328	44.165	16.560	6.785	2.131	17.784	55.604	43.988	74.249
France	145.987	140.178	124.310	100.572	71.665	64.332	68.593	60.752	38.851	71.509
Poland	14.310	21.329	30.865	36.229	34.069	27.573	30.565	32.206	29.646	48.023
Japan	190.876	153.213	157.095	137.744	128.513	113.857	69.135	54.421	42.521	43.862
Austria	26.587	21.101	25.301	25.528	18.915	16.229	51.882	36.849	33.237	29.218

Source: ITC, 2023(4), Trademap, trademap.org

Figure 3. 391510 HS coded waste group exporter countries density chart

Table 5. Leading exporters for 391520 HS code between 2012-2021 (US dollars thousand)

Countries	2012	2013	2014	2015	2016	2017	2018	2019	2020	2021
Japan	130.956	118.646	121.056	88.217	72.267	72.662	62.124	56.493	50.000	59.668
France	21.805	20.219	17.871	11.671	10.820	10.740	12.406	14.504	10.699	14.178
United Kingdom	11.389	9.647	9.948	3.906	5.470	5.428	8.361	5.420	3.297	13.694
USA	32.044	22.449	17.943	19.135	19.435	23.222	18.800	23.966	14.798	10.654
Belgium	5.735	4.753	3.294	3.221	4.026	4.102	2.970	5.376	6.680	7.969
Czech Republic	1.779	2.073	6.116	5.381	4.696	5.763	6.942	6.352	6.002	6.739
Germany	27.556	47.396	37.053	18.524	13.230	12.605	11.642	16.427	15.385	6.523
Mexico	3.133	3.823	5.013	4.169	4.566	3.336	6.623	6.555	6.413	6.406
Poland	3.538	3.665	4.790	5.523	5.861	5.382	5.610	5.817	5.436	5.878
Netherlands	3.316	5.836	5.410	4.559	4.642	3.168	3.246	2.009	2.254	5.026

Source: ITC, 2023(5), Trademap, trademap.org

Japan has been the biggest exporter of this product group for the last few years, and when looking at the slope, it is understood that the USA and Germany follow the leader.

The last decade's leading exporters of the 391530 HS Code product group are shown in the table below.

As can be seen in Figure 5. below, China emerges as the top country for 391530 product group wastes, which can be summarized as a waste of polymers of vinyl chloride over the years.

China has been the biggest exporter of this product group for the last few years, but one can understand that the USA was the leader before. Also, Japan has a significant potential for this waste export.

The last decade's leading exporters of the 391590 HS Code product group are shown in the table below.

Figure 4. 391520 HS coded waste group exporter countries density chart

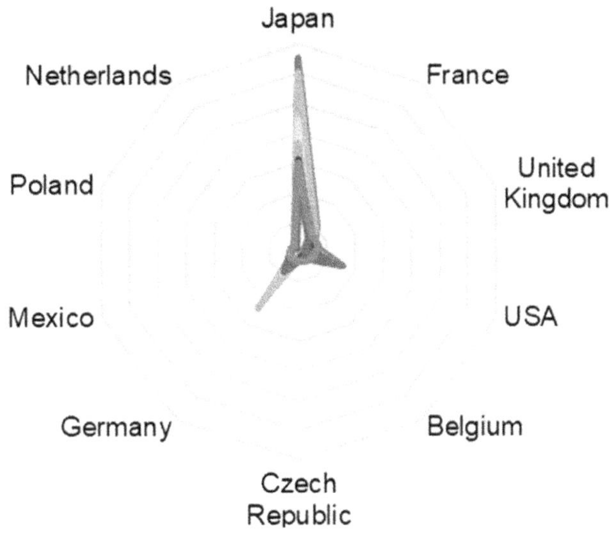

Table 6. Leading exporters for 391530 HS code between 2012-2021 (US dollars thousand)

Exporters	2012	2013	2014	2015	2016	2017	2018	2019	2020	2021
China	3.702	1.417	7.987	7.649	5.863	13.640	4.849	5.788	25.772	38.752
Japan	34.043	29.746	25.450	20.973	22.694	21.010	19.473	20.334	23.171	19.395
Poland	4.387	5.249	3.539	5.560	5.274	6.130	10.132	8.495	9.654	14.281
Netherlands	1.829	1.405	1.422	1.014	1.436	1.359	1.748	6.094	4.667	12.989
Thailand	437	172	37	75	102	9	26	337	4.318	10.876
USA	83.284	97.146	102.383	76.172	72.149	48.956	42.114	15.362	12.418	10.090
Canada	5.235	4.721	4.584	5.386	4.658	4.371	3.597	3.666	3.791	6.621
UK	12.172	20.358	22.435	19.036	16.792	14.832	17.302	19.898	13.574	6.433
France	11.965	11.895	10.427	8.912	5.765	6.746	6.902	4.321	3.124	5.173
Spain	1.687	306	545	913	723	924	1.101	1.562	2.538	4.708

Source: ITC, 2023(6), Trademap, trademap.org

As seen from Figure 6. below, Japan emerges as the top country for 391590 product group wastes, which can be summarized as a waste of other polymers over the years.

Japan has been the biggest exporter of this product group for the last few years, and the USA looks distinctively competitive after Japan.

The last decade's leading exporters of the 720449 HS Code product group are shown in the table below.

As seen from Figure 7. below, the USA emerges as the top country for 720449 product group wastes, which can be summarized as waste and scrap of iron or steel other than slag over the years.

The USA has been the biggest exporter of this product group for the last few years, and the United Kingdom and Japan are the followers of this waste group.

The last decade's leading exporters of the 740400 HS Code product group are shown in the table below

Figure 5. 391530 HS coded waste group exporter countries density chart

Table 7. Leading exporters for 391590 HS code between 2012-2021 (US dollars thousand)

Countries	2012	2013	2014	2015	2016	2017	2018	2019	2020	2021
Japan	569.040	509.096	458.826	381.230	325.917	324.787	242.876	221.694	172.312	168.139
USA	547.854	462.459	480.360	431.444	357.304	318.246	212.504	131.986	114.759	153.459
Netherlands	151.892	169.360	133.662	124.723	104.881	98.357	98.435	87.788	70.798	119.559
Mexico	259.719	237.812	229.155	159.088	124.882	114.628	114.064	89.225	63.424	100.434
China	44.742	56.703	50.616	58.265	59.258	97.972	59.783	71.126	108.374	90.170
Germany	238.915	222.361	211.072	148.202	132.308	123.900	132.603	114.562	87.637	79.742
Italy	51.940	57.359	46.345	35.668	36.059	50.642	58.825	52.803	44.290	76.468
Belgium	106.578	122.495	92.192	75.115	76.986	76.278	72.359	57.580	54.807	69.562
France	109.019	99.601	117.672	92.079	90.079	90.821	84.098	78.408	58.837	61.055
Canada	92.420	77.129	82.047	74.935	65.692	63.419	45.684	39.416	34.338	57.915

Source: ITC, 2023(7), Trademap, trademap.org

As seen from Figure 8. below, the USA emerges as the top country for 740400 product group wastes, which can be summarized as waste and scrap of copper over the years.

The USA has been the biggest exporter of this product group for the last few years, and Germany and Japan are the followers of it this waste group.

The last decade's leading exporters of the 760200 HS Code product group are shown in the table below.

As seen in Figure 9. below, the USA has emerged as the leading country for 760200 product group wastes, which can be summarized as waste and scrap of aluminum over the years.

The USA has been the biggest exporter of this product group for the last few years, and Germany and Canada are the followers of this waste group.

Figure 6. 391590 HS coded waste group exporter countries density chart

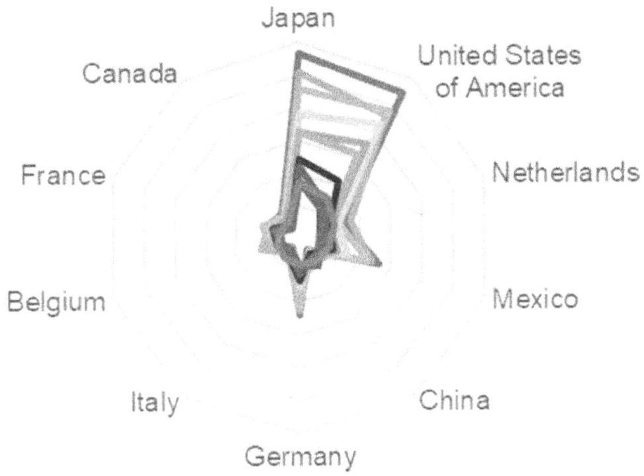

Table 8. Leading exporters for 720449 HS code between 2012-2021 (US dollars thousand)

Countries	2012	2013	2014	2015	2016	2017	2018	2019	2020	2021
USA	7.461.989	5.980.647	4.778.570	2.905.548	2.632.621	3.762.471	4.871.135	3.956.157	3.681.491	6.209.698
UK	2.327.682	2.049.464	2.047.042	1.392.410	1.314.614	1.970.700	2.174.772	1.800.950	1.523.035	3.133.307
Japan	3.497.083	3.157.732	2.688.540	1.916.036	1.934.304	2.403.099	2.369.709	1.731.332	1.836.892	2.498.227
Netherlands	1.225.665	1.070.374	948.369	743.342	879.063	1.290.002	1.578.942	1.421.340	1.371.489	2.482.797
Germany	2.113.740	1.946.784	1.902.921	1.110.228	1.022.334	1.395.276	1.649.903	1.358.744	1.352.567	2.215.264
France	1.394.659	1.269.203	1.183.575	698.607	654.029	949.632	1.119.577	917.700	847.259	1.653.343
Belgium	1.119.520	950.820	1.041.164	531.652	541.382	783.718	974.361	829.549	901.597	1.519.879
Russia	1.133.620	1.147.539	1.463.047	1.041.593	936.646	1.123.480	1.411.775	980.019	1.055.031	1.362.481
Czech Republic	573.982	484.104	497.410	291.738	289.347	487.116	580.604	473.705	442.446	873.748
Denmark	522.507	401.539	604.532	325.653	350.539	419.668	446.698	376.154	375.492	840.894

Source: ITC, 2023(8), Trademap, trademap.org

Since the largest exporters of waste products are identified based on their export figures, understanding the economic development of these leading countries is also of great importance for pollution haven theory. Leader countries' most current economic situation can be seen in Table 11. below.

As can be seen from the table, out of these countries, which represent more than 50% of the world economy, all except South Africa are classified as high-income countries. This is consistent with the pollution haven theory as it shows the outflow of waste from developed countries.

Major Destinations

The last decade's leading importers of the 261810 HS Code product group are shown in the table below.

Figure 7. 720449 HS coded waste group exporter countries density chart

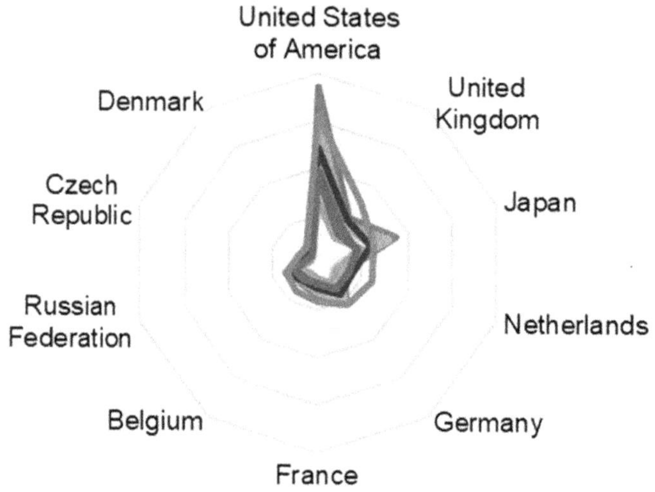

Table 9. Leading exporters for 740400 HS code between 2012-2021 (US dollars thousand)

Countries	2012	2013	2014	2015	2016	2017	2018	2019	2020	2021
USA	4.471.697	4.129.993	3.516.027	2.789.215	2.232.812	2.761.231	3.166.813	2.825.123	2.675.659	4.497.184
Germany	2.531.753	2.094.531	2.025.523	1.605.298	1.311.561	1.687.725	1.589.233	1.502.108	1.510.280	2.525.175
Japan	1.011.291	984.617	839.448	625.891	544.038	801.176	811.752	954.012	1.260.605	2.341.477
France	1.407.141	1.211.332	1.199.986	854.233	770.344	1.014.751	1.032.329	900.146	902.784	1.666.453
UK	1.680.721	1.273.648	1.205.491	906.544	918.479	1.011.770	1.012.725	1.009.303	892.315	1.655.249
Netherlands	1.198.354	976.885	1.030.057	969.793	744.008	864.809	925.171	791.377	850.447	1.385.028
Thailand	624.030	433.851	372.383	268.268	263.066	393.119	483.521	401.593	527.102	1.008.221
Italy	756.891	602.880	578.016	510.943	535.544	611.177	616.068	611.405	604.435	1.001.852
Spain	559.059	530.943	481.880	431.526	356.148	545.787	639.373	646.453	541.485	968.285
Hong Kong, China	356.534	290.782	304.898	175.968	131.741	178.102	467.961	782.226	415.050	911.242

Source: ITC, 2023(9), Trademap, trademap.org

As can be seen in Figure 10. below, the USA emerges as the leading major destination for 261800 product group wastes.

Like in the major player's section, although the slope toward exporters appears in different colors for each annual period, the place where the slope is concentrated is sufficient to understand the leading country. The USA has been the biggest importer of this product group for the last decade, and when looking at the slope, it is understood that Singapore, UAE, and Taipei are the other exporters in this waste group.

The last decade's leading importers of the 261900 HS Code product group are shown in the table below.

As can be seen in Figure 11. below, Austria emerges as the leading major destination for 261900 product group wastes.

Figure 8. 740400 HS coded waste group exporter countries density chart

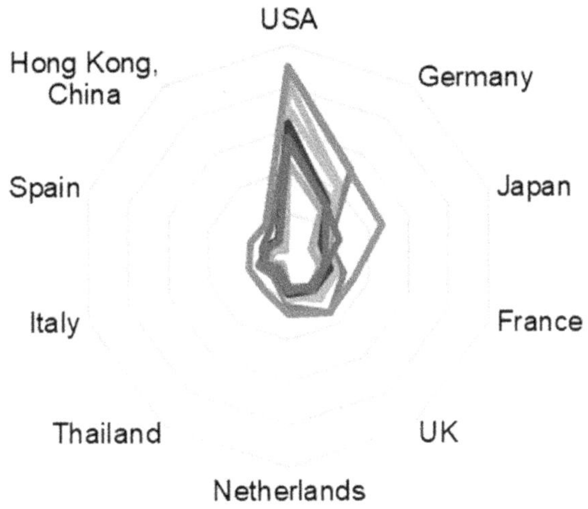

Table 10. Leading exporters for 760200 HS code between 2012-2021 (US dollars thousand)

Countries	2012	2013	2014	2015	2016	2017	2018	2019	2020	2021
USA	3.337.485	3.155.135	2.767.571	2.342.239	1.811.970	2.246.126	2.493.178	2.180.570	2.169.262	3.720.722
Germany	1.584.581	1.529.921	1.713.948	1.521.624	1.338.504	1.530.001	1.630.522	1.291.004	1.298.202	2.114.346
Canada	770.897	702.909	825.316	740.155	664.794	815.281	895.050	746.358	626.628	1.131.659
Hong Kong, China	66.185	72.782	63.736	68.411	110.215	124.242	173.587	232.940	169.070	1.123.204
UK	908.825	734.880	796.718	619.849	576.077	567.558	668.141	590.851	488.630	956.919
France	600.297	621.439	658.904	531.726	474.175	541.666	590.910	495.437	407.471	765.710
Netherlands	561.690	448.530	469.129	534.254	434.431	483.463	498.910	409.964	395.765	674.002
Australia	381.842	362.918	370.661	382.407	341.947	397.956	448.478	383.745	358.873	651.186
Mexico	301.046	258.350	229.168	202.544	224.304	310.549	383.481	380.408	365.060	605.505
Belgium	451.452	423.229	413.057	353.306	312.873	409.015	402.157	392.629	315.740	582.515

Source: ITC, 2023(10), Trademap, trademap.org

Japan has been the largest importer in the last decade, although Austria has emerged as the leader in recent years.

The last decade's leading importers of the 391510 HS Code product group are shown in the table below.

As can be seen in Figure 12. below, Malaysia emerges as the leading major destination for 391510 product group wastes.

Although Malaysia has emerged as the leader in recent years, it is noteworthy that Türkiye, Germany, and the Netherlands are also importing these wastes, especially in 2021. On the other hand, it is possible to say that Germany is the overall leader when it is talked about the last decade.

The last decade's leading importers of the 391520 HS Code product group are shown in the table below.

Figure 9. 760200 HS coded waste group exporter countries density chart

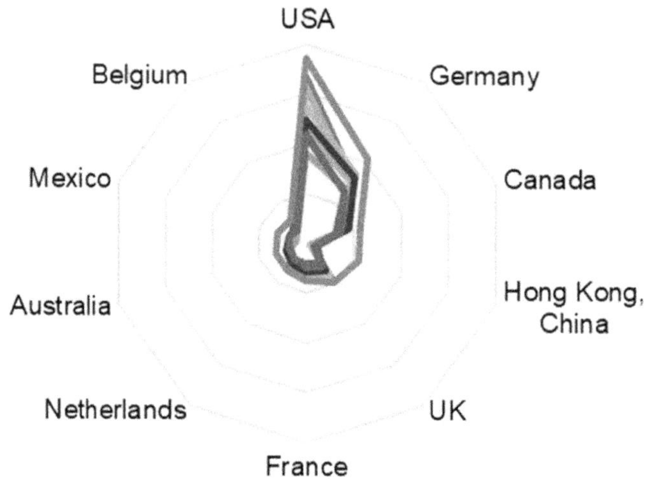

Table 11. Leading exporters' economic conditions

Country	GDP (US Dollar millions)	GDP per capita (US Dollars)
World	96.510.000	(average) 12.235
USA	23.315.080	70.248
Germany	4.259.934	51.203
Canada	1.988.336	51.987
Japan	4.940.877	39.312
UK	3.131.377	46.510
Netherlands	1.012.846	57.767
China	17.734.062	12.556
France	2.957.879	43.659
Russia	1.778.782	12.194
South Africa	419.015	7.055

Source: WB, 2023(1), data.worldbank.org

As can be seen in Figure 13. below, Spain emerges as the leading major destination for 391520 product group wastes.

Spain has emerged as the leader in the last year. On the other hand, it is possible to say that Germany is the overall leader when it is talked about the last decade.

The last decade's leading importers of the 391530 HS Code product group are shown in the table below.

As can be seen in Figure 14. below, the USA emerges as the leading major destination for 391530 product group wastes.

The USA has emerged as the leader in the last year. On the other hand, it is possible to say that Germany is the overall leader when it is talked about the last decade.

The last decade's leading exporters of the 391590 HS Code product group are shown in the table below.

Table 12. Leading importers for 261800 HS code between 2012-2021 (thousand US dollars)

Countries	2012	2013	2014	2015	2016	2017	2018	2019	2020	2021
USA	52.394	35.686	33.674	32.457	39.341	62.053	79.442	70.083	85.976	113.475
Singapore	45.268	54.591	54.070	81.300	55.294	51.896	51.289	65.614	42.304	77.658
Taipei, Chinese	37.623	50.508	57.474	46.541	28.608	23.486	34.659	51.028	60.290	54.348
Bangladesh	46.073	47.003		60.750	21.272	23.998	53.438	54.006	52.972	45.563
UK	2.785	6.870	11.910	12.735	28.085	48.656	45.146	49.852	19.331	44.014
Belgium	19.016	18.610	14.961	15.806	13.859	20.785	22.050	40.881	33.524	40.834
UAE	44.172	48.663	62.382	76.730	96.589	87.974	124.626	84.651	45.794	32.110
Ireland	6.627	9.443	6.864	5.093	11.910	14.170	12.773	16.158	14.558	29.608
Netherlands	6.927	4.956	5.472	4.894	4.889	7.861	13.081	11.452	20.050	26.025
Canada	2.093	1.890	2.408	1.835	3.095	1.541	1.819	1.838	1.015	18.095

Source: ITC, 2023(11), Trademap, trademap.org

Figure 10. 261800 HS coded waste group importer countries density chart

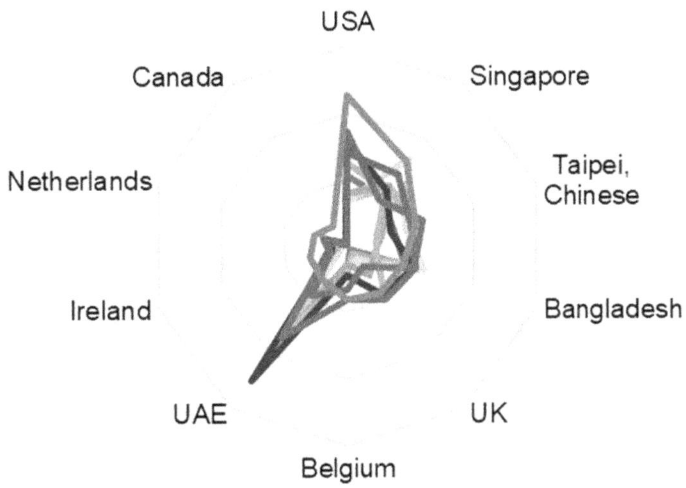

As can be seen in Figure 15. below, the USA emerges as the leading major destination for 391590 product group wastes.

The USA has emerged as the leader in the decade. The USA is one of the most important import destinations for this waste group by a large margin. The USA alone imports significantly more than even the Netherlands, which is the closest follower.

The last decade's leading importers of the 720449 HS Code product group are shown in the table below.

As can be seen in Figure 16. below, Türkiye emerges as the leading major destination for 720449 product group wastes.

Türkiye is clearly the largest importer in the last decade for this waste group. Since 2012, it has been the largest importer of these wastes every year. Moreover, this is not common when looking at previous waste statistics.

Table 13. Leading importers for 261900 HS code between 2012-2021 (US dollars thousand)

Countries	2012	2013	2014	2015	2016	2017	2018	2019	2020	2021
Austria	0	0	0	0	0	80.172	213.731	150.769	69.615	97.857
Malaysia	3.780	2.782	1.854	774	1.205	4.134	4.415	19.425	59.721	89.412
Japan	148.565	153.645	137.648	55.290	50.344	46.645	44.599	40.414	21.225	64.922
Indonesia	0	0	0	0	0	0	957	2.319	1.072	29.065
France	18.016	12.367	11.957	11.561	9.945	13.706	20.835	29.234	21.568	27.913
South Africa	1	0	10	2.148	9.144	9.404	24.402	18.360	7.630	20.640
Belgium	18.775	13.208	15.050	6.385	6.234	13.748	7.069	6.246	6.855	19.313
South Korea	8.426	9.168	7.573	6.785	9.810	13.210	28.969	27.624	13.468	19.015
Norway	16.638	12.472	13.574	10.743	9.772	7.671	19.407	16.477	9.512	15.242
USA	15.499	9.132	12.824	5.964	3.290	5.690	7.265	13.431	7.377	11.168

Source: ITC, 2023(12), Trademap, trademap.org

Figure 11. 261900 HS coded waste group importer countries density chart

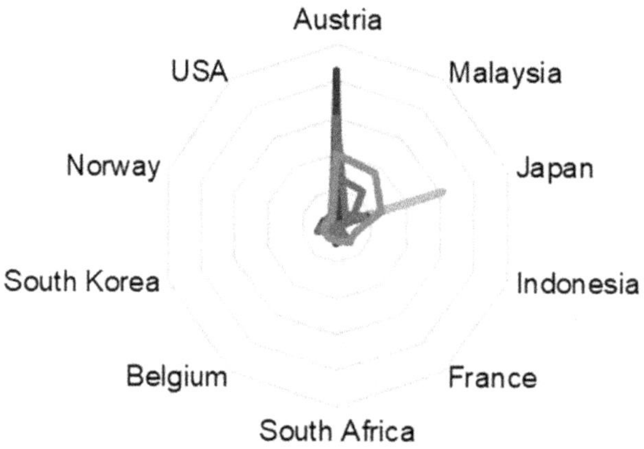

The last decade's leading importers of the 740400 HS Code product group are shown in the table below As can be seen in Figure 17. below, China emerges as the leading major destination for 740400 product group wastes.

Despite China's waste import bans and restrictions, this waste group has been left out of the relevant situation. China has been the most frequent port of call for these wastes in the last decade.

The last decade's leading importers of the 760200 HS Code product group are shown in the table below As can be seen in Figure 18. below, India emerges as the leading major destination for 760200 product group wastes.

Despite China's waste import bans and restrictions, this waste group has been left out of the relevant situation, just like the previous waste group. In 2021, India imported the most, but China has been the largest importer of this waste in the last decade.

Table 14. Leading importers for 391510 HS code between 2012-2021 (US dollars thousand)

Countries	2012	2013	2014	2015	2016	2017	2018	2019	2020	2021
Malaysia	19.121	46.967	41.918	38.691	44.978	73.324	122.590	70.222	120.344	130.981
Türkiye	4.244	7.969	11.056	17.922	24.161	28.557	47.217	48.902	72.259	114.836
Germany	76.688	82.193	106.843	92.092	95.845	76.405	64.338	73.825	70.440	110.044
Netherlands	82.544	98.901	118.191	92.772	79.238	88.828	77.940	66.392	76.169	108.911
Indonesia	5.625	8.331	9.375	10.802	9.672	10.470	43.969	44.553	55.058	70.182
Viet Nam	1.691	2.587	5.352	6.321	10.081	21.192	16.722	38.474	39.993	65.474
USA	24.883	27.896	37.789	36.071	44.963	40.862	35.833	28.736	29.288	64.259
Belgium	91.671	83.852	58.430	65.299	78.739	65.048	56.364	46.480	38.540	63.265
Italy	25.637	26.540	32.565	36.379	50.160	52.629	41.912	37.247	29.844	53.896
Poland	3.978	10.987	11.542	15.534	17.509	21.300	22.583	15.458	21.762	32.131

Source: ITC, 2023(13), Trademap, trademap.org

Figure 12. 391510 HS coded waste group importer countries density chart

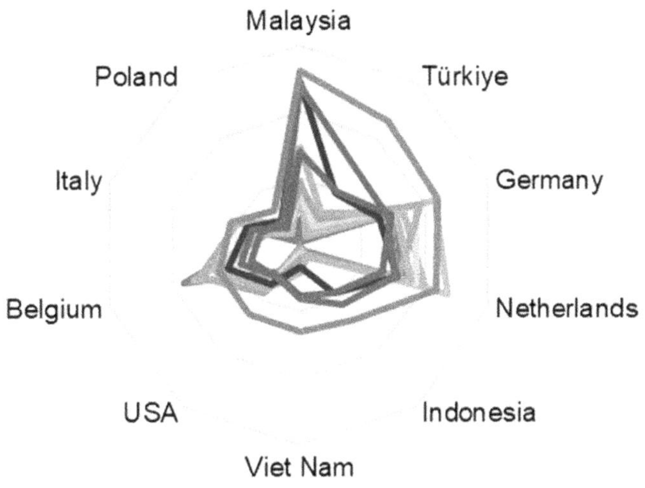

Since the largest importers of waste products are identified based on their import figures, understanding the economic development of these leading countries is also of great importance for pollution haven theory. Destination countries' most current economic situation can be seen in Table 21. Below.

As seen from the table above, it is understood that economically developing or underdeveloped countries also have a share in the import of waste products. Considering the pollution haven theory, for some wastes, there is a tendency from developed countries to developing or underdeveloped countries. However, it is impossible to claim that large countries do not attract waste in light of the last decade's data.

Table 15. Leading importers for 391520 HS code between 2012-2021 (US dollars thousand)

Countries	2012	2013	2014	2015	2016	2017	2018	2019	2020	2021
Spain	4.500	8.970	9.354	6.814	6.948	13.735	17.410	23.954	11.336	26.409
Germany	36.737	16.163	18.698	13.255	12.047	12.779	14.687	14.908	10.070	16.504
Netherlands	2.632	4.007	1.935	1.460	1.202	1.788	1.145	4.109	7.460	14.933
Austria	4.245	5.544	4.984	2.936	3.115	3.487	5.478	7.455	6.594	9.869
USA	11.743	9.677	7.012	4.219	5.228	4.920	6.961	8.154	6.583	8.456
Türkiye	241	888	1.334	424	527	1.778	4.437	2.779	3.191	8.440
Canada	4.264	3.075	5.368	6.674	6.644	8.248	10.056	8.266	6.895	5.907
Czech Republic	608	649	2.988	2.798	3.548	3.804	4.880	4.500	3.233	5.378
Poland	130	417	2.329	2.463	2.206	1.966	4.559	4.169	2.862	4.890
Portugal	207	205	1.086	95	875	1.547	2.233	1.274	1.249	4.790

Source: ITC, 2023(14), Trademap, trademap.org

Figure 13. 391520 HS coded waste group importer countries density chart

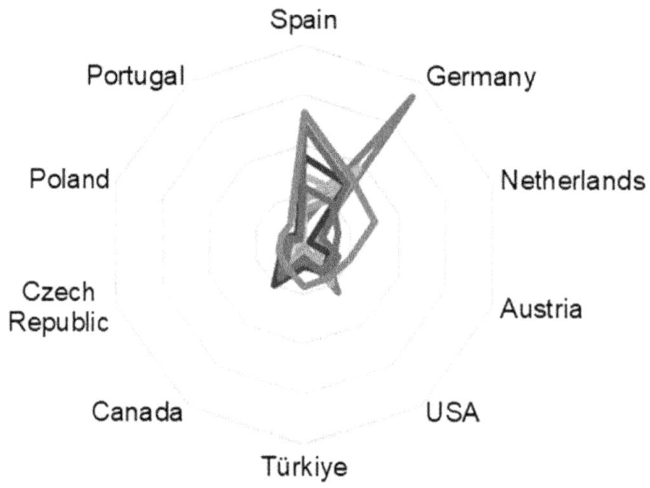

FUTURE RESEARCH DIRECTIONS

This research revealed that there will be debates in near future about waste trade. Past studies and current data show that theories and empirical findings still can not explain the waste trade stream worldwide. Therefore, some future research may be useful to light the road. It is recommended to:

- Conduct empirical research on waste trade by using waste types and examining countries' commercial and economic time series data,
- Conduct research that tries to explain the conditions under which imported waste is reused or what impact it has on the environment,

Table 16. Leading importers for 391530 HS code between 2012-2021 (US dollars thousand)

Countries	2012	2013	2014	2015	2016	2017	2018	2019	2020	2021
USA	6.333	7.530	8.521	9.705	11.027	8.999	8.775	9.811	13.434	26.480
France	6.485	5.882	6.766	4.698	5.041	6.978	9.699	12.266	8.551	11.804
Germany	28.945	22.176	16.566	15.957	14.068	15.631	16.322	18.372	11.496	11.656
Pakistan	11.574	9.615	9.956	4.628	2.973	8.328	11.520	6.971	7.869	8.959
Thailand	5	19	74	244	350	322	119	713	2.851	7.989
Taipei, Chinese	7.676	9.548	8.106	6.318	5.223	4.990	7.676	10.578	12.605	6.974
Poland	3.130	6.080	4.191	2.552	4.921	5.337	6.578	5.510	5.889	6.410
Czech Republic	664	1.729	929	1.099	2.986	4.694	7.007	6.872	5.554	5.886
Netherlands	2.567	2.607	3.509	2.480	3.007	2.579	3.393	4.570	3.841	4.641
Mexico	5.652	6.335	4.928	5.732	4.941	5.041	4.627	3.912	2.388	4.253

Source: ITC, 2023(15), Trademap, trademap.org

Figure 14. 391530 HS coded waste group importer countries density chart

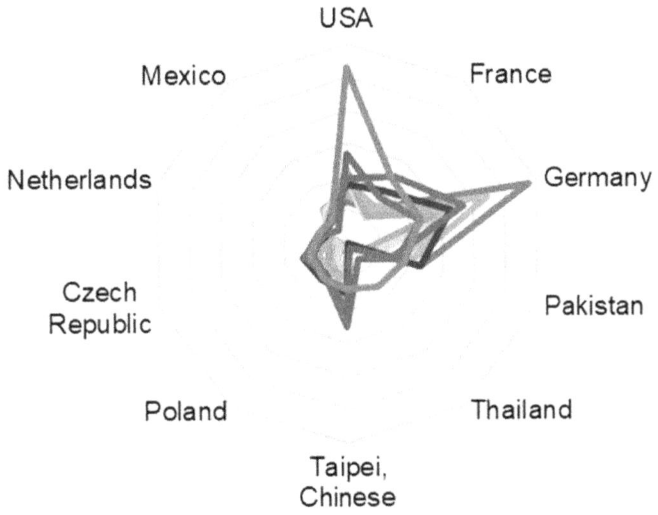

- Develop different theoretical approaches to overcome world waste trade and environmental problems.

CONCLUSION

In this book chapter, the most influential leaders of the global waste trade in terms of importers and exporters and their favorite destinations are revealed through a descriptive interpretation of secondary data. With the theoretical framework discussed before in background and past empirical research, it has been stated why this examination is necessary and vital. Findings from both theory, empirical literature,

Table 17. Leading importers for 391590 HS code between 2012-2021 (US dollars thousand)

Countries	2012	2013	2014	2015	2016	2017	2018	2019	2020	2021
USA	180.673	217.145	234.012	184.810	172.570	189.389	217.286	200.275	183.653	269.535
Netherlands	120.268	143.286	146.467	120.784	111.292	115.699	111.681	113.699	87.626	175.223
Viet Nam	28.436	41.643	46.351	44.332	32.283	48.757	89.282	109.133	98.430	138.337
Taipei, Chinese	56.919	56.101	57.538	46.786	41.231	55.468	141.688	130.350	96.142	130.598
Canada	65.857	52.107	58.453	54.418	52.104	44.823	47.397	53.799	52.464	77.624
Türkiye	26.270	21.088	31.707	17.025	24.509	44.929	63.960	60.703	63.682	69.899
Austria	23.945	22.578	54.071	45.505	38.185	37.655	44.131	42.339	39.458	69.816
Belgium	54.108	61.051	70.681	60.484	50.056	54.885	53.001	45.609	36.013	59.204
Germany	60.266	71.334	92.828	74.331	65.833	63.737	63.549	60.685	49.919	55.203
Italy	69.902	62.718	70.630	49.117	47.589	54.695	60.822	45.148	39.368	51.115

Source: ITC, 2023(16), Trademap, trademap.org

Figure 15. 391590 HS coded waste group importer countries density chart

and this research show that it is highly probable that the debate about waste circulation will continue. Because there is a need to increase the effort required to make the world a more sustainable place, it should be discussed which countries will have a more significant role in spending these efforts.

Research findings show different commercial flows from product group to product group. It is possible to say that results in parallel with the research on the distribution of plastics, whose importation is prohibited by China, are observed in different countries. For example, Turkey has become a significant pollution haven in recent years in the waste group with the HS codes 391510 and 391520. Similarly, it can be said that Malaysia has made a giant leap in this respect

When looking at iron and steel waste and scrap, it is understood that China is a big importer, as well as Turkey and India. Iron and steel wastes and scraps are quite easily recyclable for the foundry industry, and this can be considered normal, as these sectors are strong in these countries. However, when

Table 18. Leading importers for 720449 HS code between 2012-2021 (US dollars thousand)

Countries	2012	2013	2014	2015	2016	2017	2018	2019	2020	2021
Türkiye	9.259.900	7.431.631	6.892.992	4.258.008	3.904.386	6.056.670	7.061.333	5.555.510	6.061.417	10.777.646
Viet Nam	1.401.678	1.234.390	1.233.460	801.767	863.671	1.381.740	1.866.132	1.603.331	1.581.285	2.607.847
Korea, Republic of	4.061.232	3.339.632	2.828.226	1.313.016	1.236.048	1.700.149	2.238.712	2.013.531	1.108.587	2.145.777
Italy	1.508.850	1.356.817	1.279.671	755.326	567.221	918.841	1.142.755	869.843	849.700	1.962.737
Egypt	841.304	852.985	849.926	301.008	275.716	538.466	867.255	778.028	754.528	1.611.784
India	3.189.924	1.948.027	1.838.163	1.872.279	1.277.859	1.263.180	1.867.815	1.820.601	1.230.915	1.568.813
Germany	1.402.422	1.330.967	1.171.798	751.201	620.631	898.986	849.400	700.333	652.615	1.413.281
Bangladesh	93.094	102.682		231.622	305.168	447.919	619.316	838.713	793.201	1.301.733
Taipei, Chinese	1.935.893	1.535.189	1.391.430	742.595	593.286	682.172	1.146.204	886.035	853.510	1.251.239
Belgium	945.660	886.705	888.305	506.320	459.976	649.595	719.150	662.436	619.811	1.237.252

Source: ITC, 2023(17), Trademap, trademap.org

Figure 16. 720449 HS coded waste group importer countries density chart

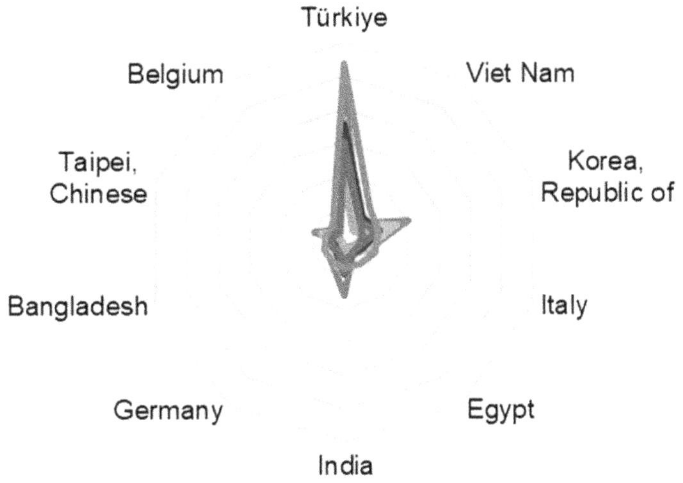

considered in the context of harmful gases that these sectors will leave to the world, it can be stated that it is possible to cause severe problems in these countries regarding the health of people close to these sectors and air pollution.

It is impossible to say that the world will produce less waste due to increasing consumption and shortening product life. On the other hand, these wastes must absolutely be reprocessed or disposed of in a way that does not harm other resources of the world in order to use limited resources accurately. The data used in the study show that these problems are not unique to underdeveloped or developing countries but that the whole world is ready to face them. Especially when looking at the countries where low-added products are imported, question marks arise about the reprocessing of these wastes. Non-governmental organizations emphasize the heaps that are not recycled and become waste islands with each passing day and argue that developing countries should not be seen as garbage dumps of developed countries.

Table 19. Leading importers for 740400 HS code between 2012-2021 (US dollars thousand)

Countries	2012	2013	2014	2015	2016	2017	2018	2019	2020	2021
China	14.840.883	13.732.988	10.913.497	8.450.144	6.182.737	9.152.484	9.370.722	7.366.793	4.568.402	12.664.080
Germany	3.749.614	3.215.934	3.329.652	2.481.638	2.263.790	2.737.813	2.949.704	2.777.190	2.598.999	3.926.054
Korea, Republic of	2.134.047	1.950.533	1.920.442	1.536.958	1.190.313	1.700.827	1.989.196	1.892.855	1.711.549	2.679.466
Belgium	1.759.425	1.767.677	1.416.911	898.299	1.042.757	1.140.760	1.238.486	1.275.056	1.475.561	2.086.503
Japan	1.031.446	840.972	1.000.352	889.361	997.484	1.150.657	1.332.221	1.195.647	955.509	1.575.810
India	1.173.900	819.814	899.632	780.264	668.963	792.495	902.141	966.353	854.300	1.524.490
Italy	807.755	853.247	940.069	730.275	553.698	749.258	979.541	820.313	725.151	1.355.129
Poland	363.327	315.207	368.930	317.374	287.583	326.821	421.051	462.122	591.751	980.442
Taipei, Chinese	632.697	646.690	591.060	493.063	495.070	653.439	823.173	669.883	613.105	977.130
USA	532.928	525.450	566.641	460.010	462.342	765.562	769.992	605.768	495.768	830.493

Source: ITC, 2023(18), Trademap, trademap.org

Figure 17. 740400 HS coded waste group importer countries density chart

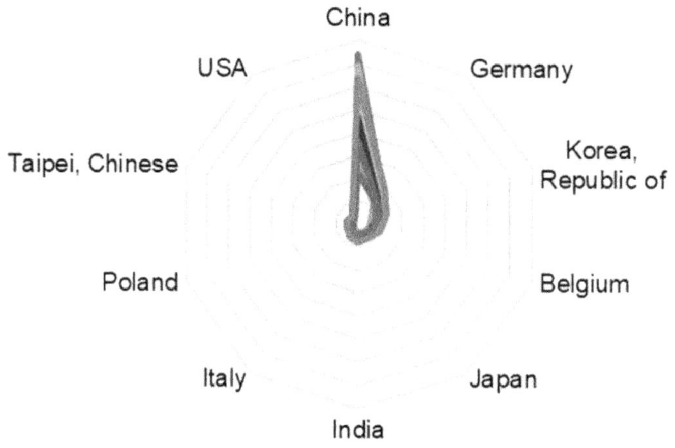

Table 20. Leading importers for 760200 HS code between 2012-2021 (US dollars thousand)

Countries	2012	2013	2014	2015	2016	2017	2018	2019	2020	2021
India	1.311.365	1.284.201	1.490.517	1.437.274	1.330.577	1.833.006	2.486.510	2.046.364	1.810.250	3.141.290
China	4.127.918	3.915.962	3.461.849	2.875.164	2.201.272	2.828.853	2.509.357	1.927.214	1.032.871	2.029.438
Korea, Republic of	1.108.548	1.179.374	1.333.397	1.107.455	894.714	1.132.557	1.165.436	1.091.552	1.099.558	1.789.799
Germany	871.509	844.876	1.011.942	1.081.220	951.558	1.163.010	1.407.411	1.307.246	1.113.408	1.673.250
USA	883.358	814.785	902.227	781.295	783.307	1.021.284	1.108.319	805.266	668.995	1.224.520
Malaysia	116.402	89.763	110.304	107.083	150.651	194.377	361.613	406.125	566.008	1.137.969
Italy	687.332	723.190	792.506	709.442	644.234	758.276	874.520	754.574	654.894	1.081.706
Hong Kong, China	46.780	27.665	22.749	30.382	33.845	39.800	78.287	196.036	112.624	818.964
Thailand	112.447	125.674	157.337	99.617	96.576	151.842	190.242	223.139	294.013	622.913
Austria	556.337	488.875	480.681	427.238	416.066	489.005	537.264	429.714	351.755	604.882

Source: ITC, 2023(19), Trademap, trademap.org

Figure 18. 760200 HS coded waste group importer countries density chart

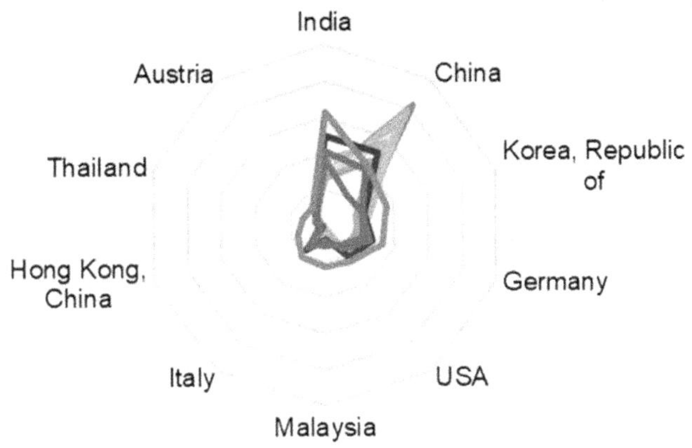

Table 21. Leading importers' economic conditions

Country	GDP (US Dollar millions)	GDP per capita (US Dollars)
World	96.510.000	(average) 12.235
India	3.176.295	2.256
China	17.734.062	12.556
Korea Republic	1.810.955	34.997
Germany	4.259.934	51.203
Türkiye	819.035	9.661
Viet Nam	366.137	3.756
USA	23.315.080	70.248
Netherlands	1.012.846	57.767
France	2.957.879	43.659
Spain	1.427.380	30.103
Malaysia	372.980	11.109
Austria	480.368	53.637
Japan	4.940.877	39.312
Singapore	396.986	72.794

Source: WB, 2023(1), data.worldbank.org

The fact that developed countries import waste may also attract attention as misleading data. At this point, when it is assumed that it will be easier to recycle these wastes for the economy with the high and new technologies they have in developed countries, the issue of destroying the non-recyclable wastes after recycling comes to mind. If the facilities that ensure the healthy disposal of these wastes are available in these countries, using them at full capacity is happily recommended. If it is not enough, the number of these facilities should be increased. On the other hand, developed countries and the world public should take action regarding the management of these wastes sent to underdeveloped or underdeveloped countries. If necessary, relevant facilities should be established within the scope of joint projects in those countries to dispose of waste sent to these countries.

Under the assumption that there are many alternatives for managing iron and steel wastes and scraps, the circulation of plastic waste remains the most critical point. Restricted recycling or bad reuse of plastic waste can cause cancer etc., in terms of human health. Although it contains risks that may result in diseases, it is also risky for nature. With the mixing of some materials, which are almost impossible to dissolve in nature, into the soil, water, and air, the war waged by the world in the climate crisis becomes meaningless. On the other hand, the life of creatures living in garbage-like areas where these wastes are accumulated only attracts the attention of some non-governmental organizations. In today's world, where human hands have changed even the route of global bird migrations, it may be possible for millions of different creatures to suffer due to the mismanagement of these wastes and even for some species to disappear. In particular, the effects of wastes spilled into the oceans and seas on fish life are frequently encountered examples.

REFERENCES

Acharyya, J. (2009). FDI, growth and the environment: Evidence from India on CO2 emission during the last two decades. *Journal of Economic Development*, *34*(1), 43–58. doi:10.35866/caujed.2009.34.1.003

Al-Mulali, U., & Foon Tang, C. (2013). Investigating the validity of pollution haven hypothesis in the Gulf Cooperation Council (GCC) countries. *Energy Policy*, *60*, 813–819. doi:10.1016/j.enpol.2013.05.055

Antweiler, W., Copeland, B. R., & Taylor, M. S. (2001). Is free trade good for the environment? *The American Economic Review*, *91*(4), 877–908. doi:10.1257/aer.91.4.877

Asghari, M. (2013). Does FDI promote MENA Region's environment quality? Pollution halo or pollution haven hypothesis. [IJSRES]. *International Journal of Scientific Research in Environmental Sciences*, *1*(6), 92–100. doi:10.12983/ijsres-2013-p092-100

Athukorala, P., & Menon, J. (1996). Foreign investment and industrialization in Malaysia: Exports, employment and spillovers. *Asian Economic Journal*, *10*(1), 29–44. doi:10.1111/j.1467-8381.1996.tb00156.x

Baggs, J. (2009). International trade in hazardous waste. *Review of International Economics*, *17*(1), 1–16. doi:10.1111/j.1467-9396.2008.00778.x

Balkevicius, A., Sanctuary, M., & Zvirblyte, S. (2020). Fending off waste from the west: The impact of China's Operation Green Fence on the international waste trade. *World Economy*, *43*(10), 2742–2761. doi:10.1111/twec.12949

Balsalobre-Lorente, D., Gokmenoglu, K. K., Taspinar, N., & Cantos-Cantos, J. M. (2019). An approach to the pollution haven and pollution halo hypotheses in MINT countries. *Environmental Science and Pollution Research International*, *26*(22), 23010–23026. . doi:10.100711356-019-05446-x PMID:31183758

Baltacı, A. (2018). Nitel araştırmalarda örnekleme yöntemleri ve örnek hacmi sorunsalı üzerine kavramsal bir inceleme. *Bitlis Eren Üniversitesi Sosyal Bilimler Enstitüsü Dergisi, 7(1),* 231–274. Retrieved from https://dergipark.org.tr/en/pub/bitlissos/issue/38061/399955

Bao, Q., Chen, Y., & Song, L. (2010). Foreign direct investment and environmental pollution in China: A simultaneous equations estimation. *Environment and Development Economics*, *16*(1), 71–92. doi:10.1017/S1355770X10000380

Brajer, V., Mead, R. W., & Xiao, F. (2011). Searching for an environmental kuznets curve in China's air pollution. *China Economic Review*, *22*(3), 383–397. doi:10.1016/j.chieco.2011.05.001

Brooks, A. L., Wang, S., & Jambeck, J. R. (2018). The Chinese import ban and its impact on global plastic waste trade. *Science Advances*, *4*(6), eaat0131. . doi:10.1126ciadv.aat0131 PMID:29938223

Caves, R. E. (1981). Diversification and seller concentration: Evidence from change. *The Review of Economics and Statistics*, *63*(2), 289–293. doi:10.2307/1924100

Clapp, J. (1994). The toxic waste trade with less-industrialised countries: Economic linkages and political alliances. *Third World Quarterly*, *15*(3), 505–518. doi:10.1080/01436599408420393

Cole, M. A. (2004). Trade, the pollution haven hypothesis and the environmental Kuznets curve: Examining the linkages. *Ecological Economics*, *48*(1), 71–81. doi:10.1016/j.ecolecon.2003.09.007

Cotta, B. (2020). What goes around, comes around? Access and allocation problems in Global North–South waste trade. *International Environmental Agreement: Politics, Law and Economics*, *20*(2), 255–269. doi:10.100710784-020-09479-3

Creswell, J. W. (2002). *Educational research: Planning, conducting, and evaluating quantitative*. Prentice Hall Upper Saddle River.

Dal, M. M., Karakaya, E., & Bulut, Ş. (2014). Çevresel kuznets eğrisi ve Türkiye: Ampirik bir analiz. *Dumlupınar Üniversitesi Sosyal Bilimler Dergisi, Special Issue of XIV. International Symposium on Econometrics, Operations Research and Statistics*, 85-96. https://dergipark.org.tr/en/pub/dpusbe/issue/31805/34898

Efthymiou, L., Mavragani, A., & Tsagarakis, K. P. (2016). Quantifying the effect of macroeconomic and social factors on illegal e-waste trade. *International Journal of Environmental Research and Public Health*, *13*(8), 789. . doi:10.3390/ijerph13080789 PMID:27527200

Eysenbach, G., & Köhler, C. (2002). How do consumers search for and appraise health information on the world wide web? Qualitative study using focus groups, usability tests, and in-depth interviews. *BMJ (Clinical Research Ed.)*, *324*(7337), 573–577. . doi:10.1136/bmj.324.7337.573 PMID:11884321

Gill, F. L., Viswanathan, K. K., & Karim, M. Z. A. (2018). The critical review of the pollution haven hypothesis. *International Journal of Energy Economics and Policy*, *8*(1), 167–174. http://hdl.handle.net/11159/1929

Gökmenoğlu, K., & Taspinar, N. (2016). The relationship between CO2 emissions, energy consumption, economic growth and FDI: The case of Turkey. *The Journal of International Trade & Economic Development*, *25*(5), 706–723. doi:10.1080/09638199.2015.1119876

Grether, J. M., Mathys, N. A., & De Melo, J. (2012). Unravelling the worldwide pollution haven effect. *The Journal of International Trade & Economic Development*, *21*(1), 131–162. doi:10.1080/09638190903552040

Grossman, G. M., & Krueger, A. B. (1991). *Environmental impacts of a North American free trade agreement (No. w3914)*. National Bureau of Economic Research., doi:10.3386/w3914

Hamilton, J. T., & Kip Viscusi, W. (1999). Regulation of economic activity. In *Calculating risks? The spatial and political dimensions of hazardous waste policy* (Vol. 21). MIT Press.

He, J. (2006). Pollution haven hypothesis and environmental impacts of foreign direct investment: The case of industrial emission of sulfur dioxide (SO2) in Chinese provinces. *Ecological Economics*, *60*(1), 228–245. doi:10.1016/j.ecolecon.2005.12.008

International Trade Center. (2023). World trade statistics. *Trade Map*. ITC. https://www.trademap.org/Index.aspx?nvpm=1%7c%7c%7c%7c%7c%7c%7c%7c%7c%7c%7c%7c%7c%7c%7c%7c%7c

Jaffe, A. B., Peterson, S. R., Portney, P. R., & Stavins, R. N. (1995). Environmental regulation and the competitiveness of US manufacturing: What does the evidence tell us? *Journal of Economic Literature*, *33*(1), 132–163. https://www.jstor.org/stable/2728912

Jiang, Y. (2015). Foreign direct investment, pollution, and the environmental quality: A model with empirical evidence from the Chinese regions. *The International Trade Journal*, *29*(3), 1–16. doi:10.1080/08853908.2014.1001538

Kaya, G., Kayalica, M. O., Kumas, M., & Ulengin, B. (2017). The role of foreign direct investment and trade on carbon emissions in Turkey. *Environment and Ecology*, *8*(1), 8–17. doi:10.21511/ee.08(1).2017.01 PMID:28241803

Kaya, Y. (2012). Waste trade and the effectiviness of the legal arrengements for tranfortier movement of hazardous wastes. *Paradoks: The Journal of Economics. Social Politics*, *8*(2), 63–82.

Kellenberg, D. K. (2009). An empirical investigation of the pollution haven effect with strategic environment and trade policy. *Journal of International Economics*, *78*(2), 242–255. doi:10.1016/j.jinteco.2009.04.004

Kılıçarslan, Z., & Dumrul, Y. (2017). Foreign direct investments and CO2 emissions relationship: The case of Turkey. *Business and Economics Research Journal,* 8(4), 647-660. https://www.ceeol.com/search/article-detail?id=586046

Koçak, E., & Şarkgüneşi, A. (2017). The impact of foreign direct investment on CO2 emissions in Turkey: New evidence from cointegration and bootstrap causality analysis. *Environmental Science and Pollution Research International*, *25*(1), 790–804. . doi:10.100711356-017-0468-2 PMID:29063405

Kurt, Ü., Kılıç, C., & Özekicioğlu, H. (2019). Doğrudan yabancı yatırımların co2 emisyonu üzerindeki etkisi: Türkiye için ardl sınır testi yaklaşımı. *Selçuk Üniversitesi Sosyal Bilimler Meslek Yüksekokulu Dergisi*, *22*(1), 213–224. doi:10.29249elcuksbmyd.487139

Lan, J., Kakinaka, M., & Huang, X. (2012). Foreign direct investment, human capital and environmental pollution in China. *Environmental and Resource Economics*, *51*(2), 255–275. doi:10.100710640-011-9498-2

Lee, C. G. (2009). Foreign direct investment, pollution and economic growth: Evidence from Malaysia. *Applied Economics*, *41*(13), 1709–1716. doi:10.1080/00036840701564376

Levinson, A., & Taylor, M. S. (2008). Unmasking the pollution haven effect. *International Economic Review*, *49*(1), 223–254. doi:10.1111/j.1468-2354.2008.00478.x

Liang, F. (2008). *Does foreign direct investment harm the host country's environment? Evidence from China*. Rutgers University.

Liang, Y., Tan, Q., Song, Q., & Li, J. (2021). An analysis of the plastic waste trade and management in Asia. *Waste Management (New York, N.Y.)*, *119*, 242–253. . doi:10.1016/j.wasman.2020.09.049 PMID:33091837

Liu, Q., Wang, S., Zhang, W., Zhan, D., & Li, J. (2018). Does foreign direct investment affect environmental pollution in China's cities? A spatial econometric perspective. *The Science of the Total Environment*, *613*, 521–529. . doi:10.1016/j.scitotenv.2017.09.110 PMID:28923755

Liu, X., Lei, T., Boré, A., Lou, Z., Abdouraman, B., & Ma, W. (2022). Evolution of global plastic waste trade flows from 2000 to 2020 and its predicted trade sinks in 2030. *Journal of Cleaner Production*, *376*, 134373. doi:10.1016/j.jclepro.2022.134373

Liu, Z., Liu, W., Walker, T. R., Adams, M., & Zhao, J. (2021). How does the global plastic waste trade contribute to environmental benefits: Implication for reductions of greenhouse gas emissions? *Journal of Environmental Management*, *287*, 112283. . doi:10.1016/j.jenvman.2021.112283 PMID:33706087

Lucier, C. A., & Gareau, B. J. (2015). From waste to resources? Interrogating 'race to the bottom'in the global environmental governance of the hazardous waste trade. *Journal of World-systems Research*, *21*(2), 495–520. doi:10.5195/jwsr.2015.11

Manunen, O. (2000). An activity-based costing model for logistics operations of manufacturers and wholesalers. *International Journal of Logistics*, *3*(1), 53–65. doi:10.1080/13675560050006673

Maskus, K. E. (2004). Encouraging international technology transfer. In UNCTAD-ICTSD Project on IPRs and Sustainable Development, 7. https://unctad.org/system/files/official-document/ictsd2004ipd7_en.pdf

Mike, F. (2020). Kirlilik sığınağı hipotezi Türkiye için geçerli mi? ardl sınır testi yaklaşımından bulgular. *Dogus University Journal, 21(2)*. 107-121. Retrieved from https://dergipark.org.tr/en/pub/doujournal/issue/66683/1043221

Miles, M. B., & Huberman, A. M. (1994). *Qualitative data analysis: An expanded sourcebook*. Sage Publications, Inc.

Millimet, D. L., & List, J. A. (2004). The case of the missing pollution haven hypothesis. *Journal of Regulatory Economics*, *26*(3), 239–262. doi:10.100711149-004-7550-7

Ministry Of Ecology And Environment of The People's Republic Of China. (2023). *Announcement on the Regulation of the Imported Waste Management Catalogue*. https://www.mee.gov.cn/xxgk2018/xxgk/xxgk01/201804/t20180419_629582.html

Moore, S. A., Rosenfeld, H., Nost, E., Vincent, K., & Roth, R. E. (2018). Undermining methodological nationalism: Cosmopolitan analysis and visualization of the North American hazardous waste trade. *Environment and Planning A. Environment & Planning A*, *50*(8), 1558–1579. doi:10.1177/0308518X18784023

Mutafoglu, T. H. (2012). Foreign direct investment, pollution, and economic growth: Evidence from Turkey. *Journal of Developing Societies*, *28*(3), 281–297. doi:10.1177/0169796X12453780

O'Neill, K. (2001). The changing nature of global waste management for the 21st century: A mixed blessing? *Global Environmental Politics*, *1*(1), 77–98. doi:10.1162/152638001570642

Peace, G. (2023, February, 4). *Data from, the global plastic waste trade 2016-2018 and the offshore impact of China's foreign waste import ban*. Greeen Peace. https://www.greenpeace.org/eastasia/publication/5907/data-from-the-global-plastics-waste-trade-2016-2018-and-the-offshore-impact-of-chinas-foreign-waste-import-ban/

Pethig, R. (1976). Pollution, welfare and environmental policy in the theory of comparative advantage. *Journal of Environmental Economics and Management*, *2*(3), 160–169. doi:10.1016/0095-0696(76)90031-0

Petridis, N. E., Petridis, K., & Stiakakis, E. (2020). Global e-waste trade network analysis. *Resources, Conservation and Recycling*, *158*, 104742. doi:10.1016/j.resconrec.2020.104742

Santiago, C. E. (1987). The impact of foreign direct investment on export structure and employment generation. *World Development*, *15*(3), 317–328. doi:10.1016/0305-750X(87)90016-7

Sapkota, P., & Bastola, U. (2017). Foreign direct investment, income, and environmental pollution in developing countries: Panel data analysis of Latin America. *Energy Economics*, *64*, 206–212. doi:10.1016/j.eneco.2017.04.001

Scherer, F. M. (1980). *Industrial Market Structure and Economic Performance*. Rand McNally.

Seker, F., Ertugrul, H. M., & Cetin, M. (2015). The impact of foreign direct investment on environmental quality: A bounds testing and causality analysis for Turkey. *Renewable & Sustainable Energy Reviews*, *52*, 347–356. doi:10.1016/j.rser.2015.07.118

Sigman, H. (1996). The effects of hazardous waste taxes on waste generation and disposal. *Journal of Environmental Economics and Management*, *30*(2), 199–217. doi:10.1006/jeem.1996.0014

Strohm, L. A. (1993). The environmental politics of the international waste trade. *Journal of Environment & Development*, *2*(2), 129–153. doi:10.1177/107049659300200209

Tang, C. F., & Tan, B. W. (2015). The impact of energy consumption, income and foreign direct investment on carbon dioxide emissions in Vietnam. *Energy*, *79*, 447–454. doi:10.1016/j.energy.2014.11.033

Tobey, J. A. (1990). The Effects of Domestic Environmental Policies on Patterns of World Trade: An Empirical Test. *Kyklos*, *43*(2), 191–209. doi:10.1111/j.1467-6435.1990.tb00207.x

Tran, T., Goto, H., & Matsuda, T. (2021). The impact of China's tightening environmental regulations on international waste trade and logistics. *Sustainability (Basel)*, *13*(2), 987. doi:10.3390u13020987

Vincent, K., Roth, R. E., Moore, S. A., Huang, Q., Lally, N., Sack, C. M., Nost, E., & Rosenfeld, H. (2019). Improving spatial decision making using interactive maps: An empirical study on interface complexity and decision complexity in the North American hazardous waste trade. *Environment and Planning. B, Urban Analytics and City Science*, *46*(9), 1706–1723. doi:10.1177/2399808318764122

Walter, I., & Ugelow, J. L. (1979) Environmental policies in developing countries. *Technology, Development and Environmental Impact 8(2)*:102–109. https://www.jstor.org/stable/4312437

Wang, C., Zhao, L., Lim, M. K., Chen, W. Q., & Sutherland, J. W. (2020). Structure of the global plastic waste trade network and the impact of China's import Ban. *Resources, Conservation and Recycling*, *153*, 104591. doi:10.1016/j.resconrec.2019.104591

Wen, Z., Xie, Y., Chen, M., & Dinga, C. D. (2021). China's plastic import ban increases prospects of environmental impact mitigation of plastic waste trade flow worldwide. *Nature Communications*, *12*(1), 425. . doi:10.103841467-020-20741-9 PMID:33462243

World Bank. (2023). *World GDP statistics,* https://data.worldbank.org/

Xu, W., Chen, W. Q., Jiang, D., Zhang, C., Ma, Z., Ren, Y., & Shi, L. (2020). Evolution of the global polyethylene waste trade system. *Ecosystem Health and Sustainability*, *6*(1), 1756925. doi:10.1080/20964129.2020.1756925

Yıldırım, M., & Destek, M. A. ve Özsoy F. N. (2017). Doğrudan yabancı yatırımlar ve kirlilik sığınağı hipotezi. *C. Ü. İktisadi ve İdari Bilimler Dergisi, 18(2),* 99-111. Retrieved from http://esjournal.cumhuriyet.edu.tr/en/download/article-file/370376

Zhao, C., Liu, M., Du, H., & Gong, Y. (2021). The evolutionary trend and impact of global plastic waste trade network. *Sustainability (Basel)*, *13*(7), 3662. doi:10.3390u13073662

Zhao, C., Qi, X., Wang, J., Du, F., & Shi, X. (2022). Predicting possible new links to future global plastic waste trade networks. *Sustainability (Basel)*, *14*(8), 4692. doi:10.3390u14084692

KEY TERMS AND DEFINITIONS

Environmental Kuznets Curve: The Environmental Kuznets Curve (EKC) explains the hypothetic relationship between deterioration of environmental conditions and income per capita level. In the relationship between environmental pollution and income per capita, the quality of life is initially deteriorating and then healing.

Foreign Direct Investment (FDI): FDI is an investment in the form of a controlling ownership in a business, in real estate or in productive assets such as factories in one country by an entity based in another country.

Pollution Haven: The countries or regions that attract industries that cause pollution due to flexible environmental policies are called as a pollution haven.

Chapter 14
The Pension System in the Ecuadorian Social Security Institute (IESS):
Analysis and Perspectives in the Economic and Legal Fields

Arturo Luque
https://orcid.org/0000-0002-4872-891X
Universidad Técnica de Manabí, Ecuador & Observatorio Euromediterráneo de Espacio Público y Democracia URJC, Spain

Marcelo Centeno Ponce
Universidad Técnica de Manabí, Ecuador

Ana Corrales Mendoza
Universidad Técnica de Manabí, Ecuador

ABSTRACT

The financial stability of millions of older adults around the world rests on the shoulders of well-functioning pension systems. For this reason, the sustainability of such systems is one of the most important challenges facing states, especially in the face of low birth rates and increased life expectancy. Hence the growing importance of the debate over taxation and economic policy. This study presents the historical backgrounds of the pension system in Ecuador to give an all-encompassing view of its present configuration, and examines the relevant data of the factors that affect the pension system with the purpose of evaluating the sustainability of the Ecuadorian pension system in the long term. Both the historical and evaluative aspects of the study lead to an analysis of fiscal policies, the political relationships of the system and numerous exogenous processes that reveal aspects of the system that lack the expected solidarity.

INTRODUCTION

The pension system has become an indispensable worldwide institution due to its catalytic function as a guarantor of a dignified future for older people who have completed their working life in active

DOI: 10.4018/978-1-6684-8879-9.ch014

employment (Tuesta, 2017). Currently, this system is facing great challenges at the global level, despite the fact that pension systems differ considerably from country to country; the systems in the Nordic countries are generally considered the most advanced and secure (Aidukaite, Hort & Ainsaar, 2022). The differences are caused by various circumstances, such as varying social and cultural developments faced by contemporary society, together with policies of privatizations and unequal levels of spending (Development Bank of Latin America, 2020). Ecuador is no exception and is one of the countries in which this problem can be most clearly seen (Knell, 2018; Barrientos & Malerba, 2019). In fact, the current pension system in Ecuador harbors dysfunctions that compromise financial stability (Hyde & Shand, 2017). Therefore, there is a clear need to carry out an economic and legal analysis of the Ecuadorian Social Security Institute with special emphasis on pension insurance.

It is first necessary to define what this institute is and what function it has. The Ecuadorian Institute of Social Security, better known by its acronym as IESS, is an independent public body that is responsible for offering services and managing economic resources. These are mostly derived from contributing taxpayers, but also, in exceptional cases, from public financing from the state coffers in order to maintain the services it offers to the population (Durán Valverde, 2013). The main function of the IESS is to manage mandatory national insurance, which in turn is part of the national social security system. Its mission is to provide the public with health coverage in case of maternity, disability, illness and death. Nonetheless, it should be noted that not all members of the public who receive pensions are necessarily affiliated, as in the cases of the national police and the armed forces, which are part of the national social security system but not of the IESS (Casalí, et al., 2021). Pension insurance aims to ensure a dignified future for retired contributors through a scheme of monthly remunerations in proportion to the contributions that the pensioner has made throughout their working life (Actuarial Directorate, Research and Statistics of the IESS, 2020). In fact, in Ecuador "the average amount paid for retirement is $590 dollars, which represents a high pension with respect to the wage contribution, and which currently does not correspond to the demographic evolution that the country is experiencing" (Ibarra, 2021). In other words, Pension insurance is a process for providing a regular monetary payout for retirees (International Labour Organization, 2020). Most commonly this pension is administered and provided by the state, but private, non-governmental institution can also be commissioned (Coordinating Ministry of Social Development, 2016). Consequently, pension insurance can be regulated by three classes of entities—public, private and mixed—which can be administered in four different ways: capitalization pensions, national accounts pensions, self-enrollment pensions and pay-as-you-earn pensions. (Central Bank of Ecuador, 2004) This last case is the most common way to pay and administer pensions worldwide as it requires workers, from the beginning of their working life, to contribute monthly to the pension system (all deducted directly from the employee salary) (Contreras-Jaramillo, 2018).

It should be noted that the pension distribution system managed by the Ecuadorian Social Security Institute is raising doubts as to its stability due to numerous exogenous factors, chief of which is the increase in the life expectancy beyond retirement age (Economic Commission for Latin America and the Caribbean, 2009). In light of this, there is a clear need to propose certain corrections to the current system in order to secure the financial sustainability of retirees (Marotta, 2019). This study undertakes to: 1) make a critical and proactive analysis of the Ecuadorian pension system; 2) identify the risks to which the pension system is subjected in the current economic climate and the context of globalization; 3) compile a syllabus of the various inconsistencies of the contemporary pension system that will serve as a preamble to the understanding of public policies that affect the sustainability of pensions.

DESCRIPTION OF THE CONSTITUTION OF ECUADOR

The body that governs the pension system in Ecuador is a public entity, whose privatization is forbidden by the constitution. Nonetheless, it interacts continually with the private sector because it offers its services to workers in both the public and private sphere (Villasmil et al., 2018). The Ecuadorian Social Security Institute is under the protection and jurisdiction of the constitution of Ecuador, specifically in part VII, chapter 1, section 3, articles 367 to 374, in which the obligations, commitments and guarantees that the IESS must undertake with respect to taxpayers are stipulated in order to prevent illegal and unconstitutional acts (International Labor Organization, 2022a). Ecuador is constituted as a state that guarantees rights and follows the rule of law, as set out in the Constitution of the Republic of 2008, which establishes the basis of social coexistence as the endeavor to afford security and equality to all citizens (Nikken, 2010). According to the Constitution of the Republic of Ecuador (2008), in article 34:

The right to social security is an inalienable right of all persons, and is a primary duty and responsibility of the state. Social security shall be governed by the principles of solidarity, compulsion, universality, equity, efficiency, subsidiarity, sufficiency, transparency and participation, for the fulfillment of individual and collective needs.

This right, as established by the constitution, must extend even to unpaid domestic workers, unemployed persons or for any form of productive independent work (CARE, 2018). At the global level, social security is recognized as a universal right in the Declaration of Human Rights of the General Assembly of the United Nations (UN) in articles 22 and 25, in which the importance of social security for citizens is established as an equal and inalienable right (United Nations General Assembly, 1948). Labor is both a right and a social duty and is protected by the state, which is why laws and programs are promoted that seek to regulate working conditions (Arroyo & Montoya, 2017; International Labour Organization, 2022b). At the national level, based on the information provided through the National Survey of Employment, Unemployment and Underemployment (ENEMDU), the percentage of the economically active population actively contributing to the social security system (Ecuadorian Institute of Social Security (IESS), Social Security Institute of the Armed Forces of Ecuador (ISSFA) and the Social Security Institute of the National Police (ISSPOL)) increased between 2007 (26.3%) and 2016 (42.1%). However, there are gaps in certain population groups, which in 2016 were in the first quintile (18%), the indigenous population (21%) and in the age group of 65 years and over (34%)[1] (National Secretariat of Bakery and Development, 2017). The human right to social security in essence acts as a regulatory concept that guarantees its contributors access to economic and health benefits through which they can deal with different situations arising in daily life and thus enjoy a better quality of life. (Carranza & Cisneros, 2014). The recent economic crisis caused by the COVID-19 pandemic, together with the effects of the war between NATO and Ukraine and Russia, have further highlighted the importance of social protections to offset the effects of such crises. Policies must be implemented that protect those who require assistance, especially those in a condition of social exclusion, by ensuring access to progressive non-contributory social security mechanisms (Ministry of Economic and Social Inclusion - MIES, 2013). Consequently, the sustainability processes of pension systems can be seen as one of the most important challenges currently in the region. According to Arenas de Mesa, (2020a, p. 12):

Pension systems are part of a fiscal-social pact, in which it is agreed what type of protection the elderly and other groups will have, under what requirements and with what contributions, the state being an explicit or implicit guarantor of both pension undertakings and the sustainability of pension systems, ensuring that this can be repeated for the coming generations.

The main task of social security and, in particular, of pension insurance, is to collectively combat the financial insecurities that arise throughout people's lives. As such, it is clear that there is a need to give greater importance to these guarantees in the face of greater economic and social uncertainty, such as the danger of privatization or the lack of economic resources (Fitzpatrick, 2019). While governments of all political persuasions are tempted to cut public spending and prioritize their electability, the quality of life of pensioners must always be maintained. In the case of Spain, the increase in pensions was set according to the average of the year-on-year CPI from December 2021 to November 2022, as established by the pension reform; the government expects that pensions will rise by around "8.5% in 2023" (Olías, 2022). The commitment to sustainable and fair public spending is paramount, so that future generations can enjoy a dignified retirement. (Chowdhury, Ortiz, Capaldo, Yamabana, & Urban, 2016).

Currently, there is an international pact regarding the importance of social security and its place as an essential human right, which is the result of a historical process of the work of multiple international bodies such as the International Labour Organization (ILO), the United Nations (UN), and supranational bodies such as the International Social Security Association (ISSA), the Inter-American Conference on Social Security (CISS) and the Ibero-American Social Security Organization (Autenne, 2017). However, during crises, when public spending is threatened with reductions, pension systems are often endangered (Cheneval, 2021). The Constitution of Ecuador, in article 371, states that social security cash benefits are not to be susceptible to diversion, seizure or retention, with the exception of certain cases (Betancur, 2020). This was put to the test in the recent health crisis of COVID-19, in which there was less economic activity and the financing of pensions was greatly reduced due to a fall in employment and widespread reductions in salaries (Rodriguez, 2020).

At this point, it is necessary to consider certain challenges of common interest, such as the increase in life expectancy after retirement and the decrease in the fertility rate (Jensen & Luchsinger, 2019; Pessino & TerMinassian, 2021). To solve these problems, a paradigm shift will be needed in the Ecuadorian pension system. However, first it is necessary to know the history of the pension system generally and the history of national insurance in Ecuador.

LITERATURE REVIEW

The pension insurance that currently prevails in Ecuador is the result of a historical evolution internationally and, later, domestically (Lekander, 2017) that has its origins in other countries before being implanted in Ecuador (Schieber, 2017) The direct antecedent of the current pension system is the collection system called "tontine"; this system was created in the seventeenth century by the Italian banker, Lorenzo de Tonti, as an impulse to economic recovery after the 30 years' war (Weinert & Gründl, 2021). The tontine functioned as an investment system and consisted of a group of people contributing a certain amount of money until a given maturity date; upon fulfillment, the investors who were still alive collected the final amount (Milevsky & Salisbury, 2015). This means that every time one of the contributors died, their stake was distributed among the other investors (Sabin, 2010). This happened successively until there was only one person left alive, who kept all the capital (Chen, Rach, & Sehner, 2020). This led to the question of what would happen if all investors died before the payment date. In this case, the money of the deceased went to the government, which usually used it for public works (Bernhardt & Donnelly, 2019). To ensure that the system functioned fairly, contributors had to be of the same age (Forman & Sabin, 2015). Initially, the system seemed to work well, but this changed when the amounts of money that were handled began to increase exponentially. Unfortunately, this led to many of the members of the tontines suffering violent deaths (Schmidt, 2016). The tontine was also susceptible to corruption: with the slogan of "the longer you live, the more you get", many investors consulted doctors in search of families with a history of longevity to select as members of the system (Fullmer & Sabin, 2019). Such cases like these were responsible for the ban on tontines (Milevsky, 2015). However, the system gave rise to some of the first sources of mortality data, since it was essential for the tontines to keep accurate records of deaths (Hieber, Klein, & Chen, 2019), and its most important contribution to modern finance was the creation of mechanisms for the state to finance itself. Ultimately, it led to the current pension systems used around the world (Salisbury & Milevsky, 2016).

Later, the nineteenth century witnessed a series of cultural and social phenomena, among which was the organization of the working classes with the objective of demanding an improvement in working conditions and other fields related to protections (Ungolo, Christiansen, Kleinow, & Macdonald, 2019). This era was also characterized by revolutionary events such as the unification of Germany, the revolutions in Poland and France, the independence of Belgium and the development of the Industrial Revolution (Schludi, 2005). The social changes brought by the Industrial Revolution left workers helpless and adrift in terms of social security (de Luxán, 2017). Moreover, increasingly common occupational diseases and accidents were not covered by employers, as they were considered to be expenses that made production processes less efficient; associations between wage earners were prohibited and wages remained very low (Cammeraat, 2020).

Regarding the development of the pension system, late 19th-century Germany saw the first approach to what we know today as "social security" (Kaminska, 2022). During this period, Germany was newly unified and a federal political system was established with William I as king and Otto von Bismarck as federal chancellor (Hipp, 2020). Bismarck, who had previously been ambassador of Prussia, faced various new conflicts at that time, which were to have important repercussions on the future of the country (International Labour Organization, 2009). The country was suffering a great economic crisis, with significant housing shortages in urban areas and a series of protests by sectors of the workforce that was increasingly aggravated by the spread of communism and other revolutionary forces (Betthäuser, 2017). Although Bismarck's first reaction was to impose laws to curb unrest, as he lacked sufficient support in parliament or from the Kaiser, a more inclusive solution was adopted instead (Stamm, 2017). Three projects were presented to parliament; the first, in 1883, was a project of "compulsory insurance against accidents or diseases", which was approved in the middle of the same year, and consisted of protecting employees on a mandatory basis against events such as illness while also providing for workers' wives during maternity (Hicks, 2019). Subsequently, in 1884, the law on "insurance against accidents at work" for workers and employees of industrial companies was approved, which protected employees from accidents occurring during the working day (Bigo, 2017). Thirdly, in 1889, the "disability and old age insurance" law was approved (Carney, 2019). These measures ushered in a period of significant advances in the area of social security throughout the globe, including the creation of the International Labour Organization (ILO) in 1919 and the enactment of the social security law in the United States (Kilgour, 2017). More was to follow until, by the end of the Second World War, all legislative bodies around the world began to propose, analyze and pass laws on welfare and the pension system that today is known as social security policy. (Diamond, 2018). The final elements in this process are the "Declaration of Philadelphia" of 1944, which set out the intention to extend social security to many more countries and, in 1948, the "Universal Declaration of Human Rights" which to this day protects pensions as an inalienable right for all and an integral part of the social security system that society has provided to protect its members (Conde-Ruiz & González, 2014).

THE PENSION SYSTEM IN ECUADOR

The first pension system implemented in Ecuador was during the government of president Isidro Ayora Cueva. Decree No. 018 of March 8, 1928 established three pension funds (Gachet, Grijalva, Ponce, & Rodríguez, 2019) through three savings banks: Caja de Jubilaciones y Montepío Civil, Caja de Retiro y Montepío Militares and Caja de Ahorro y Cooperativa (Bonilla, Cabrera, & López, 2019). The purpose of these funds was to grant military, private and public employees the benefits of a mortuary trust, mutual assistance and retirement funds (Social Security Administration, 2020). The Pension Fund was established as an institution with assets separate from state power and control in both the public or private labor sectors (Duran Valverde, 2008). From this point, greater importance was given to the issue of pensions, with the next step being taken in Supreme Decree No. 12 of 1935 with the enactment of the Compulsory Social Security Law and the establishment of the National Pension Institute as the highest organ of Social Security System from May 1, 1936 (Bistoletti, 2019). The purpose of the institute was to specify the activities of Compulsory Social Security, oversee the "Board of Trustees of the Indian and the Montubio" and promote Voluntary Insurance (Ceni, 2017). In March of that year, the executive approved the Statutes of the Insurance Fund for Private Employees and National Workers as proposed by the Institute of Welfare. Decree No. 1179 of July 14, 1942 enacted these measures as the Law of Compulsory Social Security and promulgated the statutes of the Insurance Fund in January 1944, as the last piece of the social security system in Ecuador (Granda & Jimenez, 2019).

In July 1958, the various amendments to the Compulsory Social Insurance Act brought the insurance fund into financial balance and placed it on an equal footing with the pension fund in terms of the amounts of allowances and benefits. In September 1963, Supreme Decree No. 517 merged both the Pension Fund and the Insurance Fund into a single fund called the "National Social Security Fund", under the jurisdiction of the former National Welfare Institute (Mason, 2018). Artisan Insurance, Occupational Risk Insurance, Domestic Workers Insurance, and Professional Insurance were all established in 1965, followed by Secular Clergy Insurance in 1966 (Razavi, 2022). In 1968, several studies that were carried out under the supervision of foreign and domestic observers concluded the need to reconsider the principles and guidelines adopted thirty years before in the fields of administration and services. This resulted in the Social Security Code as an apparatus for the application and development of the principles of social justice based on the philosophical bases of social security systems; these principles were universality, solidarity and compulsion, although it must also be said that the code had only a brief existence (Apella, 2020). Following Supreme Decree No. 40 of July 25, 1970 and its publication in the Official Gazette No. 15 on July 10, 1970, the National Social Security Fund was transformed into the Ecuadorian Social Security Institute (IESS), which is its present form (Pontarollo, Orellana, & Segovia, 2020). Other reforms followed, such as the Voluntary Insurance, the Compulsory Agricultural Worker Insurance, and the Marginal Social Security Fund, all established in 1986 to protect populations with incomes below the minimum living wage (Santoro, 2021). In 1987, the National Congress created the Higher Council on an equal and tripartite basis and, together with representation from the Executive, also established the obligation to register the items corresponding to the payment of state obligations in the state annual budget (Naranjo Bonilla, 2014). In 1991, The Inter-American Development Bank produced a special report on social security in which it proposed the permanent separation of health and pension insurance and the private administration of these funds (Rofman, Apella, & Vezza, 2015). This proposal was adopted in Ecuador in 1995 following a referendum, which also recommended the exclusion of the private sector in matters of social security, or indeed any institution other than the IESS from the administration of its resources (Espinosa, 2021).

In 1998, the National Assembly was convened to make changes to the constitution; one of these changes was the confirmation of the IESS as the only institution to be independently responsible for the management of Compulsory General Insurance. To the present, the IESS, as stated in the Compulsory Social Security Law, remains an autonomous body with its own resources, separate from those of the government (Valenzuela, 2021). Official Gazette No. 465 of November 30, 2001 published the Social Security Act, a law that remains in force to this day. However, it has been subject to varying interpretations and reforms in accordance with the changing social reality (McKinnon, 2020). The social security law consists of 23 transitory provisions, 308 articles, a general provision and a single special provision and is covered by the 2008 Constitution of the Republic of Ecuador (Porras, 2015). More recently, social protection mechanisms have been substantially hybridized, and the Bismarck model for the management of the social insurance system continues to predominate (Guinnane, Jopp, & Streb, 2021).

FINANCING SOCIAL SECURITY IN ECUADOR

Marked inequality and elevated poverty rates are a reality in Ecuador. Such indicators place inexorable demands on public policy to improve the comprehensive social protection system (ECLAC, 2020). In this analysis of the processes of sustainability of pensions, it is crucial to describe exactly how they are financed in order to identify their weaknesses. It should be borne in mind that "a decade ago, Ecuador had eight contributors for each retiree while today the ratio is 5 to 1 and in the future, it is estimated to be 3 to 1" (de la Torre, 2021). In Ecuador, pensions are financed in two different ways: privately, in the form of the IESS as an autonomous body, and by the state acting as guarantor of the fulfillment of the functions of the IESS itself. In the latter case, in order to shoulder the enormous burden of serving as guarantor for the pensions of the entire country, the state has to secure its own financial resources. The Ecuadorian state is financed in various ways, including taxes and duties, external and domestic debt, the sale of real estate and financial assets, donations and transactions, state ventures and investments; all the income derived from these means is used to fulfill the obligations contracted by the state, among which is funding the IESS in case of a shortfall (Ministry of Economy and Finance, 2022). It is worth examining these sources of finance in more detail.

Taxes and duties are traditionally the most effective way for states to raise funds, based on a clear intentional component that has evolved over time, moving from private and individual to public and general interests. In the case of Ecuador, "its citizens contribute with their taxes so that a small percentage of the population (33% of the workforce) may receive a pension from the IESS in the future" (Torres, 2021). When taxation is insufficient to prevent a fiscal deficit, external and domestic public debt are common levers used by governments to raise funds for public services (Ministry of Economy and Finance, 2021). The sale of real estate and financial assets is an infrequent method of balancing public finances since it tends to be unpopular with the public and seen as tantamount to selling off the "public property of the government" and running counter to the ethical functioning of the state (Martin, 2012; Luque & Herrero-García, 2019). In the case of donations and transactions, as the terms suggest, these are non-reciprocal economic transactions initiated by agents external to the state that effectively give it economic resources without expecting anything in return. Therefore, in this type of transaction, the state is under no obligation to return the economic aid offered. State ventures and investments are possibly the most studied and commonly used forms of finance since they allow the state to generate income independently, that is, without the need to depend on external agents (Coordinating Ministry of Social Development, 2015).

Income from these activities is divided into two categories, namely, "permanent income" and "non-permanent income". Permanent income remains stable for a period of time and is reusable and predictable. This includes the various types of taxes such as income tax, VAT, customs and excise duties and, in the case of Ecuador, the special levy on outflows of foreign currency. Non-permanent income is not predictable over time and cannot be continually reused over the long and short term. These include revenues from petroleum extraction, the disbursement of credit and the sale of assets (Ministry of Finance, 2015). In understanding the way in which the state is financed, it is important to emphasize that only a part of the income obtained is directed toward meeting the needs of the IESS. In any event, the Ecuadorian social security system is an autonomous agent of the government, with its own means of finance.

The IESS is financed by the continuous income from contributions of the people who are affiliated to the system. The system of operation is simple and tends to be somewhat inefficient: when contributors retire, they are paid with the money of active contributors who pay their dues from their economic activity. It is also worth comprehending the details of these contributions. Public or private workers pay 0.35% of their taxable income, as do all employees affiliated under the Compulsory General Insurance. Similarly, the state makes a compulsory financial contribution of 0.30% on the taxable income of affiliated workers under the Compulsory General Insurance of the IESS. The compulsory contribution of Private and Public Insurance for members of the Social Security System is set at a minimum of 0.5% (Social Security Law, 2001).

METHODOLOGY

This study takes a critical approach to the pension system in Ecuador, and is aimed at provoking debate into whether the current pay-as-you-earn system is sustainable in the long term (Inter-American Development Bank, 2016). It is based on the perspective of a world whose main political leaders no longer want people to take charge of their lives, but rather to remain subservient to an inert pension system (Thompson, 1998). A content analysis was used, based upon a synthetic-analytical methodology, which sought to analyze theories, doctrines, norms and legislation within a complex object of study that comprises a large volume of different public policies, as well as the operations of national and supranational bodies that affect the pension system itself (López-Roldán & Fachelli, 2021). Specifically, the following steps were taken:

1. A search of the relevant literature was carried out between 26/04/2022 and 11/10/2022 in various databases such as Scopus, Web of Science and Latindex, in addition to reports related to pri-

mary sources such as The International Social Security Association (ISSA), The Social Security Administration (SSA), the Economic Commission for Latin America and the Caribbean (ECLAC), The United Nations (UN), The International Labour Organization (ILO), as well as official documents provided by the state. This had the aim of compiling all of the details of the operation of the system to achieve a comprehensive understanding of it, and in turn to check whether the results of the evaluation adjust to the social, political, economic and cultural reality of the country.

2. A full description of social security in Ecuador was narrated, based on historical and background analysis in order to explain the current state of the pension system and how its respective functions evolved to their present form. This had the aim of providing a global understanding of the object of study in order to validate the critical analysis and explore viable alternatives.
3. An analysis was carried out of the government proposals to reform the pension system in order to guarantee its long-term sustainability. Specifically, a five-item plan of action was compiled to guide the state toward a solution that is capable of facing the contemporary challenges of globalization (Wolf, 2021).

RESULTS AND DISCUSSION

The Ecuadorian pension system currently faces the serious dilemma of a low birth rate at the same time as an increase in life expectancy, compounded by the little interest that governments have shown in this issue in recent years (Amaglobeli, et al., 2019). As can be seen in Table No.1, State Contribution, the money invested to maintain pensions during 2016, 2017 and 2018 decreased compared to previous years. That is why, as of 2019, the Constitutional Court made it compulsory for the state to pay 40% of the pension bill (Medina, 2017). Subsequently, the Organic Law for Labor Justice and Recognition of Work in the Home[2] exempted the government from this obligation, which was established in article 237 of the Social Security Law. However, the Constitutional Court responded again on March 21, 2021 with judgment 002-18-SIN-CC, which declared the unconstitutionality of the third paragraph of article 68.1 of the new law, thus changing article 237 of the Social Security Law again and re-imposing the obligation of the state to finance 40% of the IESS (International Labor Organization, 2022c).

Table 1. State contribution

Year	Contribution	Pensions paid for old age, disability and montepío	Participation of the State in the payment of pensions
2010	534.380.038,98	1.207.116.134,16	0,44
2011	603.474.213,61	1.383.552.598,40	0,44
2012	697.436.878,32	1.625.076.833,73	0,43
2013	849.489.938,28	1.625.076.833,73	0,42
2014	1.034.098.035,10	2.453.453.749,05	0,42
2015	349.443.929,15	2.872.976.482,10	0,12
2016	94.225.805,28	3.276.903.632,51	0,03
2017	90.650.359,71	3.650.398.400,97	0,02
2018	135.918.826,89	3.904.868.756,30	0,03
Source: Ecuadorian Social Security Institute, 2019, p. 105.			

Due to this uncertainty of state funding of pensions, finance must be almost exclusively obtained from the contributions of active IESS affiliates, which affects the long-term solvency of insurance by failing to provide an adequate economic cushion in cases of emergency (Cueva, 2021). Another factor affecting the sustainability of pensions is the level of unemployment since this prevents a significant part of the population from joining the insurance system: few individuals out of work have the necessary economic resources to keep up their contributions (Toledo, 2016). In Table No.2, Unemployment Indicators, it appears that unemployment and informal work have decreased in the last year with respect to the previous one, but this is largely due to the post-lockdown recovery; to know the real current employment situation, it is necessary to examine the pre-pandemic situation and the trends in the unemployment rate at that time (Pinto, 2021). When this longer-term picture is analyzed, it is clear that the unemployment situation in the country is gradually returning to the situation present in 2019 (International Labor Organization, 2022d).

The increase in life expectancy, while certainly a positive development for the country in terms of the health of the population, causes serious issues for the sustainability of the pension system (Coppola, Russolillo, & Simone, 2022). When people live longer after retirement age, the period over which their pensions must be maintained increases, causing an imbalance between the contributions paid in during the working life of the population and the amount to be paid out in retirement (Forttes, 2020). Figure Nº.1, Population of Ecuador 1960-2020, shows how the demographic index of Ecuador has decreased over time, putting further strain on the proper functioning of the pension system. When the contributions of one population sector (young affiliates) is used to finance another population sector (retired affiliates), the decrease of one of these will have consequences for the other. In this case, decline of the birth rate means fewer active contributors to support retirees, making it likely that they be forced to pay more in order to support the ageing sector of the population (Paredes, 2022).

The political and demographic phenomenon of an ageing population is the result of the increase in the survival of individuals in relation to the decrease in the average number of children born in a given location (Miller & Mejía, 2020). Life expectancy in Ecuador is defined as the life-span of an individual under the mortality conditions of a specific territory (Ortiz-Crespo, 2022). The decrease in the rate of population increase may be due to several factors, among which is the reduction in the number of households and the individual choice of families not to have children (ECLAC, 2022). The reduction in the birth rate means fewer contributors in the future, and the level of their contribution must be measured in relation to an increasingly elderly population. This inevitably impacts on the different social demands regarding pension and insurance systems (Fárez, 2020).

All of the challenges facing the social security system are problems for the state since the national finances are the only mechanism strong enough to find a solution to these problems, either by changing

Table 2. Unemployment indicators

National Indicators (in % with respect to the economically active population)	sep-19	nov-20	may-21	oct-21	apr-22
Global Participation Rate	67,8	65,0	67,2	66,0	66,5
Adequate employment rate	38,5	28,5	30,2	32,0	32,5
Underemployment rate	19,7	24,1	23,2	23,9	23,9
Unemployment rate	4,9	5,6	6,3	4,6	4,7
Source: National Institute of Statistics and Censuses, 2022.					

Figure 1. Population of Ecuador 1960-2020
Source: own elaboration based on United Nations data. Department of Economic and Social Affairs, 2020.

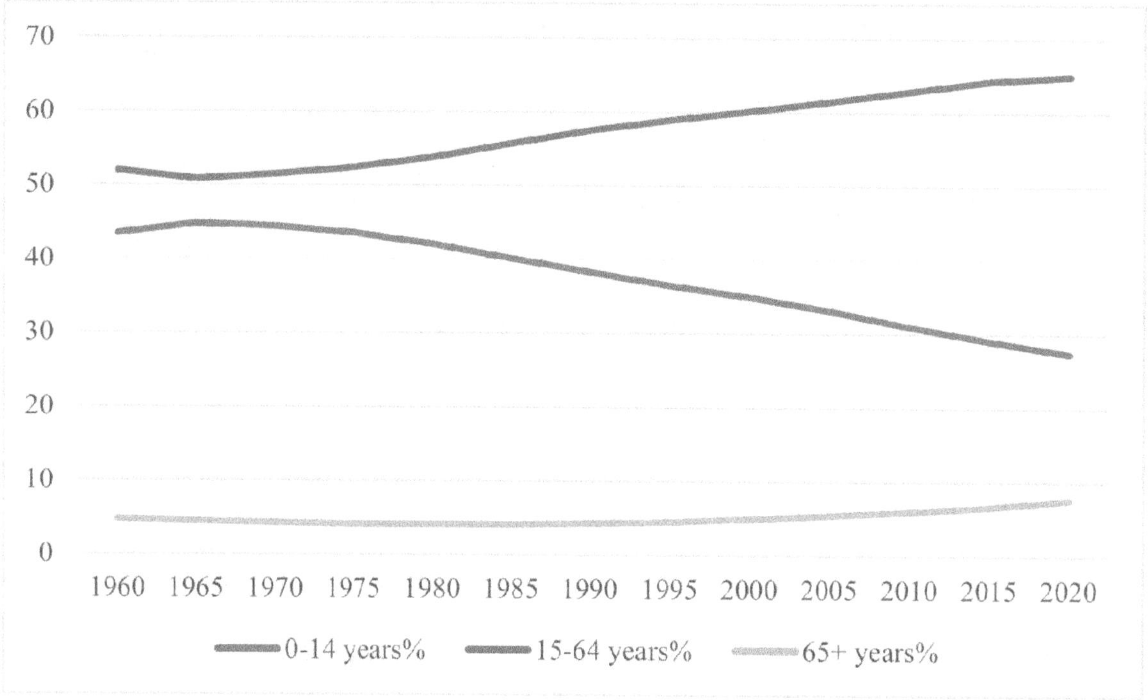

the current pension model to a universal one, or by modifying the voluntary contribution system for those most vulnerable (Ibarra-Jarrín, & Corral, 2017). The coverage of the current social security system is exclusively for those that are able to make contributions while those who lack the resources to pay in are excluded. This greatly increases processes of inequality and informal economy (Luque, 2022), making it crucial to support the insurance system with state-financed social protection policies. In a universal insurance model, all people, with or without formal employment, are obliged to keep up payments while taking into account personal differences between individuals such as old age or disability. The various types of insurance managed by the IESS must continue to be financed by both employees and employers, but with the support of the state to supplement payments by minimizing the existing asymmetries. Compulsory affiliation must be reformed to allow certain individuals with different economic or residence circumstances to join the insurance system freely in return for certain benefits. This is especially true in today's world due to the processes of globalization and the needs of the growing class of the working poor, that is, those with jobs (one or several) who have their basic needs covered but are exhausted by the prevailing precariousness. Public policies in the social protection sector must take a problem-solving approach to the labor issues that push thousands of people into poverty and instead seek to foster a more ethical, fairer and more supportive society.

CONCLUSION AND RECOMMENDATION

Based on the sources of information provided by both the Ecuadorian state and the IESS, it is clear that the Ecuadorian social security system is functioning ineffectively. The analysis of the data confirms the central thesis of this article that the pension system is unsustainable in the long term and that the social reality requires a structural change so as to avoid a national pension crisis. The issue will have a great impact on the Ecuadorian public in the future, despite its being under-reported in the media. Indeed, the mainstream media have in effect ignored the crisis, presumably because it is seen as less pressing and lacking in sensationalism, but possibly also due to information control by the government. Similarly, the Ecuadorian Social Security Institute has so far remained passive in the face of the growing problem; it is not actively hiding the disquieting figures, but it is certainly not disseminating information in an ethical and accessible manner, which makes this a social problem as well as a financial one. It is true that the crisis is not exclusive to Ecuador; indeed, most of Latin America is experiencing similar circumstances to those described in this study, such as the decrease in the birth rate, lower levels of paying in and poor state financing. Some of these problems are explained by the negative influence of political and governmental interests on the structure of social security policies. However, they are also explained by the fact that the current system is based on outmoded models that were designed for other demographics and historical contexts. From this perspective, the reason for making changes to the pension system is not only about incorporating current paradigms such as the "social protection floor", but is also about strengthening it to become a solidarity system that provides effective social protection to the elderly. In this sense, a solidarity system would be a state policy aimed at taking a greater contribution from those who have more economic resources through direct taxation to enlarge the state budget and redistribute it through social protection mechanisms to the population that does not have an income. In view of these observations, the conclusions of this study are that:

- The declining economic situation in which the IESS finds itself is mainly caused by the low level of contributions it receives from both its members and the state, which leads to a smaller field of action for the development of the Institute's financial activities and lower remunerations for retirees and other affiliates.
- There is an urgent need to inform the general population about financial issues and to disseminate relevant information that in turn will generate support for certain reforms of the social security system. This should be achieved through educational and social programs focused on raising citizen awareness.
- The drive to encourage a culture of citizen responsibility is justified in order to combat the challenges facing the country. This is especially true of economic issues such as tax evasion because such behavior depletes the national coffers, while citizen responsibility leads to greater state resources
- The need for public policies aimed at pension sustainability and the prevention of economic mismanagement by the IESS.

These circumstances suggest an action plan aimed at solving all of the problems set out here, such as these five points to be followed by both the public and the state.

Action Plan

1. Make compulsory contributions to a private savings account for the entire working life (Ortiz, et al., 2018).
2. Establish a constitutional guarantee of the right to a minimum subsidy by the state for those who do not have sufficient balance in their private savings account (Van Regenmortel, et al., 2016).
3. State financial aid to pay for pensions in the case of financial vulnerability of retirees (López-Arranz & Picatoste, 2022).
4. In case of insolvency of the Ecuadorian Institute of Social Security, the obligation of the state to stand in and assume its liabilities (Mesa-Lago & Valero, 2020).
5. Severe criminal consequences for cessation of payments from a private insurance company and the obligation of the state to act as guarantor for the fulfillment of the debts (Wolf, 2022).

ACKNOWLEDGMENT

This research received no specific grant from any funding agency in the public, commercial, or not-for-profit sectors. Author Contributions: Conceptualization, A.L.G. (Arturo Luque González), M.C.P. (Marcelo Centeno Ponce) and A.C.M. (Ana Corrales Mendoza); methodology, A.L.G., M.C.P. and A.C.M.; software, A.L.G., M.C.P. and A.C.M.; validation, A.L.G., M.C.P. and A.C.M.; formal analysis, A.L.G., M.C.P. and A.C.M.; investigation, A.L.G., M.C.P. and A.C.M.; resources, A.L.G., M.C.P. and A.C.M.; data curation, A.L.G., M.C.P. and A.C.M.; writing—original draft preparation, A.L.G., M.C.P. and A.C.M.; writing—review and editing, A.L.G., M.C.P. and A.C.M.; visualization, A.L.G., M.C.P. and A.C.M.; supervision, A.L.G., M.C.P. and A.C.M.; project administration, A.L.G., M.C.P. and A.C.M.; funding acquisition, A.L.G., M.C.P. and A.C.M. All authors have equally contributed to the manuscript. All authors have read and agreed to the published version of the manuscript.

REFERENCES

Actuarial Directorate, Research and Statistics of the IESS. (2020). *Boletín Estadístico del IESS*. Quito: Instituto Ecuatoriano de Seguridad Social, IEES. https://bit.ly/3fKBN1b

Aidukaite, J., Hort, S., & Ainsaar, M. (2022). Current trends in social welfare policies toward the older people in the Baltic and Nordic countries: An explorative study. *Journal of Baltic Studies*, *53*(2), 147–167. doi:10.1080/01629778.2021.1998166

Amaglobeli, D., Chai, H., Dabla-Norris, E., Dybczak, K., Soto, M., & Tieman, A. (2019). *The Future of Saving: The Role of Pension System Design in an Aging World. IMF Staff Discussion Note*. International Monetary Fund. doi:10.5089/9781484388990.006

Apella, I. (2020). The performance of the income protection system for older adults in Ecuador and future challenges. *International Social Security Review*, *73*(2), 51–74. doi:10.1111/issr.12234

Arenas de Mesa, A. (2020a). Los sistemas de pensiones en la encrucijada: desafíos para la sostenibilidad en América Latina (Vols. Libros de la CEPAL, N° 159). Santiago: Comisión Económica para América Latina y el Caribe (CEPAL). https://doi.org/ doi:10.18356/64c72454-es

Arenas De Mesa, A. (2020b). *Los sistemas de pensiones en América Latina: institucionalidad, gasto público y sostenibilidad financiera en tiempos del COVID-19*. Serie Macroeconomía del Desarrollo, N° 212, 9-52. https://bit.ly/3yot3V0

Autenne, A. (2017). Occupational pension funds. *European Journal of Social Security*, *19*(2), 158–171. doi:10.1177/1388262717712152

Banco de Desarrollo de América Latina. (2020). *Los sistemas de pensiones y salud en América Latina: los desafíos del envejecimiento, el cambio tecnológico y la informalidad*. Buenos Aires: Corporación Andina de Fomento. https://bit.ly/3REtAcl

Barrientos, A., & Malerba, D. (2019). Social protection in Latin America: One region, two systems. In *Welfare and social protection in contemporary Latin America* (pp. 59–71). Routledge. doi:10.1111/issr.12244

Bernhardt, T., & Donnelly, C. (2019). Modern tontine with bequest: Innovation in pooled annuity products. *Insurance, Mathematics & Economics*, *86*(1), 168–188. doi:10.1016/j.insmatheco.2019.03.002

Betancur, A. (2020). *Efectos Fiscales del Sistema de Pensiones en Chile: institucionalidad, gasto público y sostenibilidad financiera*. Santiago: Comisión Económica para América Latina y el Caribe (CEPAL). https://bit.ly/3ygQccl

Betthäuser, B. (2017). Protecting outsiders? Corporatism and the dualisation of unemployment protection in Germany and Austria. *European Journal of Social Security*, *19*(3), 209–224. doi:10.1177/1388262717724892

Bigo, D. (2017). International flows, political order and social change: (in)security, by-product of the will of order over change. *Global Crime*, 303–321. doi:10.1080/17440572.2017.1350428

Bistoletti, E. (2019). *The Power Struggles over the Post-neoliberal Social Security System Reforms in Venezuela and Ecuador*. Springer. doi:10.1007/978-3-319-98168-0

Bonilla, L., Cabrera, J., & López, M. (2019). La jubilación: un derecho del trabajador o una formalidad de las prestaciones de la seguridad social en Ecuador. Pro Sciences: Revista De Producción. *Ciencia e Investigacion*, *3*(29), 74–83. doi:10.29018/issn.2588-1000vol3iss29.2019pp74-83

Cammeraat, E. (2020). The relationship between different social expenditure schemes and poverty, inequality and economic growth. *International Social Security Review*, *73*(1), 101- 123. doi:. . doi:123

CARE. (2018). *Situación del trabajo remunerado del hogar en ecuador (Primera ed.)*. Quito: Publiasesores Cía. Ltda. https://bit.ly/3Vdp1sg

Carney, T. (2019). Automation in Social Security: Implications for Merits Review. *The Australian Journal of Social Issues*, *55*(3), 260–274. doi:10.1002/ajs4.95

Carranza, C., & Cisneros, M. (2014). Hacia un sistema de protección social más inclusivo en el Ecuador Seguimiento y desenlace de un proceso de construcción de consensos en la búsqueda del Buen Vivir. *Serie Políticas Sociales, 205*, 8-84. https://bit.ly/3Ve3kZl

Casalí, P., Vallejo, F., Jaramillo, D., Lucio, R., & Velásquez, M. (2021). *Ecuador. Panorama de la protección social: gobernanza, cobertura y sostenibilidad. (Primera ed.).* Lima: Organización Internacional del Trabajo. https://bit.ly/3T33aBV

Ceni, R. (2017). Pension schemes and labor supply in the formal and informal sector. *IZA Journal of Labor Policy*, *6*(8), 1–29. doi:10.118640173-017-0085-1

Central Bank of Ecuador. (2004). La Seguridad social en el Ecuador. Apuntes de Economía (47), 17-25. https://bit.ly/3ylWYNK

Chen, A., Rach, M., & Sehner, T. (2020). On the Optimal Combination of Annuities and Tontines. Astin Bulletin. *The Journal of the IAA*, *50*(1), 95–129. doi:10.1017/asb.2019.37

Cheneval, F. (2021). Property rights of personal data and the financing of pensions. *Critical Review of International Social and Political Philosophy*, *24*(2), 253–275. doi:10.1080/13698230.2018.1528521

Chowdhury, A., Ortiz, I., Capaldo, J., Yamabana, H., & Urban, S. (2016). *Ecuador: Financing Social Protection through Debt Restructuring*. Ginebra: International Labour Organization. from https://bit.ly/3fVctG4

Conde-Ruiz, J., & González, C. (2014). *From Bismarck to Beveridge: The other Pension Reform in Spain*. Madrid: Banco de España. https://bit.ly/3T3A9WN

Constitution of the Republic of Ecuador. (2008). *Asamblea Nacional Constituyente*. Defensa. https://www.defensa.gob.ec/wp-content/uploads/downloads/2021/02/Constitucion-de-la-Republica-del-Ecuador_act_ene-2021.pdf

Contreras-Jaramillo, M. A. (2018). Análisis de la sostenibilidad del sistema de pensiones ecuatoriano, periodo 2013-2025. *Papeles de población, 24*(96), 29-62. doi:http://dx.doi.org/ doi:10.22185/24487147.2018.96.14

Coordinating Ministry of Social Development. (2015). *La Articulación de un Sistema de Seguridad Social (Primera ed.).* Quito: Ministerio Coordinador de Desarrollo Social. https://www.social-protection.org/gimi/RessourcePDF.action?id=53636

Coordinating Ministry of Social Development. (2016). *La institucionalidad normativa y funcional de los sistemas de seguridad social (Primera ed.).* Quito: MCDS. https://www.social-protection.org/gimi/RessourcePDF.action?id=53638

Coppola, M., Russolillo, M., & Simone, R. (2022). On the evolution of the gender gap in life expectancy at normal retirement age for OECD countries. *Genus*, *78*(27). doi:10.118641118-022-00175-5

Cueva, C. (2021). Una mirada interpretativa y constitucional de los derechos sociales: La reducción de las pensiones jubilares. Foro. *Revista de Derecho*, (35), 106–125. doi:10.32719/26312484.2021.35.6

de la Torre, A. (2021). *Retos de los sistemas de pensiones y salud*. CAF. https://www.caf.com/es/actualidad/noticias/2021/09/retos-de-los-sistemas-de-pensiones-y-salud-en-ecuador-y-america-latina/

de Luxán, J. (2017). ¿Cobraremos la pensión? Cómo sostener el sistema público de pensiones. *Cuadernos de Gobierno y Administración Pública*, *4*(1), 107–110. doi:10.5209/CGAP.56025

Desai, N. (2017). Universal Basic Income or Social Security. *Indian Journal of Human Development.*, *11*(2), 203–205. doi:10.1177/0973703017742759

Diamond, P. (2018). Questions and Answers: The Future of Social Security. *Economic Inquiry*, *56*(2), 682–685. doi:10.1111/ecin.12552

Duran Valverde, F. (2008). *Diagnóstico del sistema de seguridad social del Ecuador. Oficina Subregional para los Países, Departamento de Seguridad Social*. Lima: Organización Internacional del Trabajo. https://bit.ly/3V9f3Iy

Durán Valverde, F. (2013). *Innovaciones en la extensión de la cobertura del seguro social a los trabajadores independientes: Departamento de Protección Social de la OIT*. Ginebra: Organización Internacional del Trabajo- OIT. https://bit.ly/3CGHjLn

Economic Commission for Latin America and the Caribbean (ECLAC). (2009). *Envejecimiento, derechos humanos y políticas públicas* (S. Huenchuan, Ed.). Naciones Unidas. https://bit.ly/3VmxYzM

Economic Commission for Latin America and the Caribbean (ECLAC). (2020). El desafío social en tiempos del COVID-19. *Informe Especial COVID-19, Nº3*. Santiago: Naciones Unidas. https://bit.ly/3V8QCuJ

Economic Commission for Latin America and the Caribbean (ECLAC). (2022). *Panorama Social de América Latina, 2021*. Santiago: Naciones Unidas. https://repositorio.cepal.org/bitstream/handle/11362/47718/1/S2100655_es.pdf

Ecuadorian Institute of Social Security. (2019). *Valuación Actuarial del Seguro de Invalidez, Vejez y Muerte del Seguro General Obligatorio*. EISS. https://www.iess.gob.ec/documents/10162/14444609/IESS_IVM_estudio_actuarial_011.pdf

Ecuadorian Institute of Social Security. (2022). *Estadisticas laborales de Mayo 2022*. EISS. https://bit.ly/3Ci4euY

Espinosa, S. (2021). The Evolution of the System of Foreign Investment Protection in Ecuador. In *Handbook of International Investment Law and Policy* (pp. 2645–2676). Springer. doi:10.1007/978-981-13-3615-7_85

Fárez, M. (2020). *Jubilación en Ecuador: informalidad, desigualdad e insostenibilidad*. https://bit.ly/3fLfXe1

Fitzpatrick, M. (2019). Pension reform and return-to-work policies. *Journal of Pension Economics and Finance*, *18*(4), 500–514. doi:10.1017/S1474747218000367

Forman, J., & Sabin, M. (2015). Tontine Pensions. *University of Pennsylvania Law Review*, *163*(3), 755–831. doi:10.2139srn.2393152

Forttes, P. (2020). Envejecimiento y Atención a la Dependencia en Ecuador. https://bit.ly/3CjrV6k

Fullmer, R., & Sabin, M. (2019). Individual Tontine Accounts. *Journal of Accounting and Finance, 8*(19), 2–46. doi:10.2139srn.3217551

Gachet, I., Grijalva, D., Ponce, P., & Rodríguez, D. (2019). Vertical and Horizontal Inequality in Ecuador: The Lack of Sustainability. *Social Indicators Research, 145*, 861–900. doi:10.100711205-017-1810-2

Granda, M., & Jimenez, W. (2019). The evolution of socioeconomic health inequalities in Ecuador during a public health system reform (2006–2014). *International Journal for Equity in Health, 18*(31), 1–12. doi:10.118612939-018-0905-y PMID:30736808

Guinnane, T., Jopp, T., & Streb, J. (2021). Bismarck's social security and its influence on Germany's demographic change. *Journal for Economic Policy, 10*(1), 262–265. doi:10.100710273-021-2893-6

Hicks, J. (2019). Campaigning for social security rights: Women in the informal economy and maternity benefits. *Agenda (Durban, South Africa), 33*(2), 32–41. doi:10.1080/10130950.2019.1609809

Hipp, L. (2020). Feeling secure vs. being secure? Qualitative evidence on the relationship between labour market institutions and employees' perceived job security from Germany and the U.S. *Contemporary Social Science, 15*(4), 416–429. doi:10.1080/21582041.2019.1656816

Hyde, M., & Shand, R. (2017). *Retirement, Pensions and Justice: A Philosophical Analysis*. Palgrave Macmillan. doi:10.1057/978-1-137-60066-0

Ibarra, R. (2021). *Retos de los sistemas de pensiones y salud*. https://bit.ly/3EvnZlL

Ibarra-Jarrín, R., & Corral, C. (2017). IESS: riesgo de descapitalización y posibles medidas. *Coyuntura. 11*, 1-4. https://bit.ly/3fESHyb

Inter-American Development Bank. (2016). *Saving for Development*. Palgrave Macmillan. doi:10.1057/978-1-349-94929-8_7

International Labour Organization. (2009). De Bismarck a Beveridge: seguridad social para todos. *Trabajo, 67*, 2-31. https://bit.ly/3rJwHVy

International Labour Organization. (2020). *Valuación actuarial del régimen de invalidez, vejez y muerte del Instituto Ecuatoriano de Seguridad (Primera ed.)*. Quito: OIT. https://bit.ly/3fJX1fz

International Labour Organization. (2022a). *Análisis de la legislación ecuatoriana de seguridad social a la luz del Convenio sobre la seguridad social (Primera ed.)*. Quito: OIT. https://bit.ly/3T3nL9s

International Labour Organization. (2022b). *Declaración de la OIT relativa a los principios y derechos fundamentales en el Trabajo*. ILO. https://www.ilo.org/declaration/lang--es/index.htm

International Labour Organization. (2022c). *Informe de análisis de la coherencia normativa del sistema de seguridad social administrado por el IESS*. Quito: OIT / Oficina de la OIT para los Países Andinos. https://bit.ly/3MgwvH9

International Labour Organization. (2022d). *Análisis de la afectación de la pandemia de la COVID-19 en el riesgo de informalidad laboral y pobreza en el Ecuador*. Quito: OIT /Oficina de la OIT para los Países Andinos. https://bit.ly/3SZMfQE

Jensen, J., & Luchsinger, G. (2019). *Estado de la población mundial 2019*. New York: Fondo de población de las Naciones Unidas- Unfpa. https://bit.ly/3fQ6gLr

Kaminska, M. (2022). Variations on Bismarck: Translations of Social Health Insurance in Post-Communist Healthcare Reforms in Central and Eastern Europe. In International Impacts on Social Policy (pp 449–462). Palgrave Macmillan. doi.org/ doi:10.1007/978-3-030-86645-7_35

Kilgour, J. (2017). Social Security Retirement Income Replacement Rates. *Compensation and Benefits Review*, *48*(3), 1–9. doi:10.1177/0886368717736666

Knell, M. (2018). Increasing life expectancy and NDC pension systems. *Journal of Pension Economics and Finance*, *17*(2), 170–199. doi:10.1017/S1474747216000226

Lekander, J. (2017). How do institutional pension managers consider real estate. *Journal of Property Investment & Finance*, *35*(1), 26–43. doi:10.1108/JPIF-05-2016-0033

Ley para la Justicia Laboral y el Reconocimiento del Trabajo en el Hogar. (2015). *Asamblea Nacional Constituyente*. https://bit.ly/3rEamZO

López, M., Hincapié, V., & Fuentes, A. (2015). *Personas mayores y solidaridad intergeneracional en la familia (Primera ed.)*. Madrid: Grupo editorial Cinca, S. A. https://core.ac.uk/download/pdf/200765085.pdf

López-Arranz, A., & Picatoste, X. (2022). Retirement Pension Financing to Guarantee Older Workers' Rights. *International Advances in Economic Research*, *28*, 45–47. doi:10.100711294-022-09846-7

López-Roldán, P., & Fachelli, S. (2021). *Towards a Comparative Analysis of Social Inequalities between Europe and Latin America*. Springer. doi:10.1007/978-3-030-48442-2

Luque, A. (2022). Analysis of the concept of informal economy through 102 definitions: Legality or necessity. *Open Research Europe*, *2022*(1), 134. doi:10.12688/openreseurope.13990.2

Luque, A., & Herrero-García, N. (2019). How corporate social (ir)responsibility in the textile sector is defined, and its impact on ethical sustainability: An analysis of 133 concepts. *Corporate Social Responsibility and Environmental Management*, *26*(6), 1–22.

Marotta, G. (2019). Behind the success of dominated personal pension plans: Sales force and financial literacy factors. *Journal of Pension Economics and Finance*, *19*(4), 532–547. doi:10.1017/S1474747219000209

Martín, M. (2012). Mercados de activos financieros. *Revista de Fomento Social*, *67*(267), 549–552. doi:10.32418/rfs.2012.267.1788

Mason, P. (2018). Accounting for social benefits: The search for a past event. *International Social Security Review*, *71*, 73–80. doi:10.1111/issr.12180

McKinnon, R. (2020). Introduction: Social security, inclusive growth and social cohesion. *International Social Security Review*, *73*(3), 5–12. doi:10.1111/issr.12242

Medina, E. (2017). Reformas al Código del Trabajo introducidas por la Ley de Justicia Laboral y Reconocimiento del Trabajo en el Hogar. Foro. *Revista De Derecho*, (24), 5–22. https://revistas.uasb.edu.ec/index.php/foro/article/view/458

Mesa-Lago, C. (1985). Desarrollo de la seguridad social en América Latina. Santiago: Naciones Unidas. https://bit.ly/3CeMm49

Mesa-Lago, C., & Valero, D. (2020). The New Wave of Pension Reforms in Latin America. In *Economic Challenges of Pension Systems*. Springer. doi:10.1007/978-3-030-37912-4_12

Milevsky, M. (2015). *King William's Tontine: Why the Retirement Annuity of the Future Should Resemble Its Past*. Cambridge University. doi:10.1017/CBO9781139879316

Milevsky, M., & Salisbury, T. (2015). Optimal retirement income tontines. *Insurance, Mathematics & Economics*, *64*(1), 91–105. doi:10.1016/j.insmatheco.2015.05.002

Miller, T., & Mejía, I. (2020). *El envejecimiento de la población en Ecuador: la revolución silenciosa*. https://bit.ly/3rCuqLR

Ministry of Economic and Social Inclusion (MIES). (2013). *Ciudadanía Activa y Envejencimiento Activo: Diseño de una política pública a favor de los derechos de las personas adultas mayores*. Subsecretaría de Atención Intergeneracional. Quito: MIES. https://bit.ly/3SMOhEi

Ministry of Economy and Finance. (2021). *Boletín de deuda pública interna y externa. Resolución No. MEF-2018-0134*. Quito: Ministerio de Economía y Finanzas. https://bit.ly/3RIZQeq

Ministry of Economy and Finance. (2022). *Justificativo de la proforma del presupuesto general del Estado*. Ministry of Economy and Finance. https://bit.ly/3SMM1g7

Ministry of Finance. (2015). *Resumen ejecutivo justificativa proforma presupuesto general del Estado 2016*. Quito: Subsecretaría de presupuesto. https://bit.ly/3CIPTte

Naranjo Bonilla, M. (2014). *Social protection systems in Latin America and the Caribbean: Ecuador. ECLAC – Project Documents collection*. United Nations. https://bit.ly/3EmSUAw

National Secretariat of Bakery and Development. (2017). *Plan Nacional del Buen Vivir 2017-2021*. Quito: Senplades. https://bit.ly/3mhqTQx

Nikken, P. (2010). La protección de los derechos humanos: haciendo efectiva la progresividad de los derechos económicos, sociales y culturales. *Revista IIDH, 52*(1), 55-140. https://www.corteidh.or.cr/tablas/r25563.pdf

OECD, Inter-American Development Bank and The World Bank. (2014). *Pensions at a Glance: Latin America and the Caribbean*. Organization for Economic Co-operation and Development. doi:10.1787/pension_glance-2014-

Olías, L. (2022). *El Gobierno prevé que las pensiones subirán alrededor del 8,5% en 2023*. https://bit.ly/3RJ0qc6

OrtizI.DuranF.UrbanS.WoodsesV.YuZ. (2018). Reversing Pension Privatization: Rebuilding Public Pension Systems in Eastern European and Latin American Countries. doi.org/ doi:10.2139/ssrn.3275228

Ortiz-Crespo, G. (2022). *18 millones y 8 mil millones*. https://bit.ly/3fT1qx3

Paredes, P. (2022). Sistema de jubilación en crisis: Soluciones urgentes. *Dialoguemos*. https://dialoguemos.ec/2022/05/sistema-de-jubilacion-en-crisis-soluciones-urgentes/

Pessino, C., & Ter-Minassian, T. (2021). *Enfrentando los costos fiscales del envejecimiento poblacional en América Latina y el Caribe con lecciones de los países desarrollados*. Banco Interamericano de Desarrollo. doi.org/ doi:10.18235/0003242

Pinto, M. (2021). *La protección ante el desempleo: medidas aplicadas durante la crisis del COVID-19. Documentos de Proyectos*. Santiago: Comisión Económica para América Latina y el Caribe. https://bit.ly/3fOctra

Pontarollo, N., Orellana, M., & Segovia, J. (2020). The Determinants of Subjective Well-Being in a Developing Country: The Ecuadorian Case. *Journal of Happiness Studies*, *21*, 3007–3035. doi:10.100710902-019-00211-w

Porras, A. X. (2015). La seguridad social en Ecuador: Un necesario cambio de paradigmas. Foro. *Revista de Derecho*, (24), 89–116. https://repositorio.uasb.edu.ec/bitstream/10644/5285/1/08-TC-Porras.pdf

Razavi, S. (2022). Making the Right to Social Security a Reality for All Workers. *The Indian Journal of Labour Economics : the Quarterly Journal of the Indian Society of Labour Economics*, *65*, 269–294. doi:10.100741027-022-00378-6 PMID:35911186

Rodríguez, Y. (2020). *Mitigación de la pandemia en los sistemas de seguridad social de Ecuador. Dirección de Proyectos e Investigación*. México D.F: Conferencia Interamericana de Seguridad Social (CISS). https://bit.ly/3ChO3ht

Rofman, R., Apella, I., & Vezza, E. (2015). Expanding economic protection to the elderly in Latin America. *International Social Security Review*, *68*(1), 1–22. doi:10.1111/issr.12057

Sabin, M. (2010). Fair Tontine Annuity. *The Social Science Research Network*, 3-67. doi.org/ doi:10.2139/ssrn.1579932

Salisbury, T., & Milevsky, A. (2016). Equitable Retirement Income Tontines: Mixing Cohorts Without Discriminating. Astin Bulletin. *The Journal of the IAA*, *46*(3), 571–604. doi:10.1017/asb.2016.19

Santoro, V. (2021). Model of health and social inequalities in Ecuador: Progress and challenges. A systematic review. *Journal of Public Health*, *29*, 775–785. doi:10.100710389-019-01183-0

Schieber, S. (2017). Generational Equity and Social Security Financing Reform. *The Journal of Retirement*, *5*(1), 12–31. doi:10.3905/jor.2017.5.1.012

Schludi, M. (2005). *The Reform of Bismarckian: A Comparison of Pension Politics in Austria, France, Germany, Italy and Sweden* (Vol. 6). Amsterdam University Press. https://bit.ly/3rHymLz

Schmidt, U. (2016). Insurance Demand Under Prospect Theory: A Graphical Analysis. *The Journal of Risk and Insurance*, *83*(1), 77–89. doi:10.1111/jori.12098

Social Security Administration. (2020). *Social Security Programs Throughout the World: The Americas, 2019. Office of Retirement and Disability Policy*. Washington: SSA Publication. https://bit.ly/3rEYDtO

Social Security Law. (2001). *Asamblea Nacional Constituyente*. SSL. http://www.oas.org/juridico/pdfs/mesicic4_ecu_segu.pdf

Stamm, I. (2017). The Human Right to Social Security and Its Impact on Socio-Political Action in Germany and Finland. *Journal of Human Rights and Social Work*, *2*(1), 25–33. doi:10.100741134-017-0030-7

Thompson, L. (1998). *Older and Wiser- The Economics of Public Pensions*. The Urban Institute Press. http://biblioteca.ciess.org/adiss/g528/f3

Toledo, A. (2016). Seguro de desempleo para Ecuador: características y sostenibilidad. *Analítika, 12*, 7-45. https://bit.ly/3fVkydS

Torres, W. (2021). *El IESS da más beneficios en pensiones a las personas de altos ingresos*. https://bit.ly/3ysXysT

Tuesta, D. (2017). Retirement and labour markets under the context of pension reform in Latin America. *Economic and Political Studies*, *5*(4), 475–500. doi:10.1080/20954816.2017.1384622

Ungolo, F., Christiansen, M., Kleinow, T., & Macdonald, A. (2019). Survival analysis of pension scheme mortality when data are missing. *Scandinavian Actuarial Journal*, *19*(6), 523–547. doi:10.1080/03461238.2019.1580610

United Nations. (2020). *World Population Prospects. Department of Economic and Social Affairs*. UN. https://population.un.org/wpp/Graphs/DemographicProfiles/Pyramid/218

United Nations General Assembly. (1948). *Declaración Universal de los Derechos Humanos*. ONU. https://www.un.org/es/about-us/universal-declaration-of-human-rights

Uthoff, A. (2001). *La reforma del sistema de pensiones en Chile: desafíos pendientes*. Santiago: Naciones Unidas ONU. https://bit.ly/3ecSfai

Valenzuela, K. (2021). Transnational Social Protection and the Role of Countries of Origin: The Cases of Mexico, Guatemala, Bolivia, and Ecuador. In Latin American Diasporas in Public Diplomacy (pp 27–51). Palgrave Macmillan Series in Global Public Diplomacy. Cham: Palgrave Macmillan. doi.org/doi:10.1007/978-3-030-74564-6_2

Van Regenmortel, S., De Donder, L., Dury, S., Smetcoren, A., De Witte, N., & Verté, D. (2016). Social exclusion in later life: A systematic review of the literature. *Journal of Population Ageing*, *9*(4), 315–344. doi:10.100712062-016-9145-3

Villasmil, M., Alvarado-Peña, L., Socorro-González, C., & Gamboa, N. (2018). Contribuciones parafiscales: Factor determinante para la seguridad social. [RCS]. *Revista de Ciencias Sociales*, *XXIV*(3), 21–35. https://bit.ly/3T9B4Fn

Weinert, J., & Gründl, H. (2021). The modern tontine. *European Actuarial Journal*, *11*(1), 49–86. doi:10.100713385-020-00253-y

Wolf, I. (2021). Political Stress and the Sustainability of Funded Pension Schemes: Introduction of a Financial Theory. *Journal of Risk and Financial Management*, *14*(11), 1–12. doi:10.3390/jrfm14110525

Wolf, I. (2022). Studying the equilibrium of pension designs when shifting to funded pension schemes: Economic theory and links to political factors. *SN Business & Economics*, *2*(76), 1–21. doi:10.100743546-022-00236-z PMID:35789707

KEY TERMS AND DEFINITIONS

Economic Globalization: This is a phenomenon in expansion that causes profound changes on the world stage. It revolves around trade, the flow of investment, financial capital, division of labor and specialization. The concept is not limited only to economic variables since its effects extend to individuals, society to the state. Developing countries are experiencing stagnation in the face of their inability to cope with globalization, which is compounded by poor management of their financial markets, leading to an increase in the income inequality gap. Economic globalization brings with it the mobilization of goods and capital, reduces distance between borders and energizes international trade with some alterations to sovereignty.

Public Policy: This refers to decisions and actions that a government takes when addressing public or collective issues.

Resilience: Transformations within a complex system related to the capacity for self-organization while maintaining internal structure, together with the ability to create adaptive responses, generate knowledge, experience, and learning. Resilience and sustainability are directly related to changes within societies, economies, and the human system as a whole. The transformation of systems is inevitable since it allows systems to strengthen.

Social Economy: This encompasses a variety of businesses, organizations and different legal entities. They share the objective of systematically putting people first, producing a positive impact on local communities and pursuing a social cause.

Welfare State: A system whereby the state and supranational organizations undertakes to protect the health and well-being of its citizens, especially those in financial or social need, by means of grants, pensions, and other benefits.

ENDNOTES

1. National Plan for Good Living (2017-2021) p.44, available in https://bit.ly/3mhqTQx
2. Law for Labor Justice and the Recognition of Work in the Home. Official Gazette No. 483 of 20 April 2015. Available in https://bit.ly/3rEamZO

Chapter 15
The Role of Innovation in Driving the Bioeconomy:
The Challenges and Opportunities

S M Nazmuz Sakib
https://orcid.org/0000-0001-9310-3014
Dhaka International University, Bangladesh & School of Business and Trade, Switzerland; International MBA Institute, Switzerland

ABSTRACT

The bioeconomy presents a promising opportunity for sustainable economic growth, but it also poses several challenges that require innovative solutions. This chapter provides an overview of the bioeconomy and its key components, including biotechnology, agriculture, and forestry. It then explores some of the challenges associated with the bio economy, such as the need to balance economic growth with environmental sustainability, the potential risks of genetically modified organisms, and the need to ensure that the benefits are shared equitably across society. Finally, it discusses the potential for new bio-based products and services, the creation of new jobs and economic opportunities, and the potential to contribute to global sustainability goals.

INTRODUCTION TO THE BIOECONOMY AND ITS KEY COMPONENTS

The concept of bioeconomy has gained significant attention in recent years, as it presents a promising opportunity for sustainable economic growth. The bioeconomy refers to the use of renewable biological resources such as plants, animals, and microorganisms to produce goods and services. This article provides an introduction to the bioeconomy and its key components (Bioeconomy, 2023; Von Braun, 2015).

The bioeconomy consists of several sectors, including biotechnology, agriculture, and forestry. Biotechnology refers to the use of biological systems and living organisms to produce new products or improve existing ones (House, 2022; Wang et al., 2022). This includes genetic engineering, which involves modifying the genetic makeup of organisms to enhance their characteristics. Biotechnology

DOI: 10.4018/978-1-6684-8879-9.ch015

has numerous applications in agriculture, medicine, and industry, among others (Barney & Lewis, 2022; Biotechnology FAQs, n.d.; Ch02, n.d.;Zhang et al., 2022).

Agriculture is another important component of the bioeconomy. It involves the production of food, feed, fiber, and other agricultural products using biological resources. Agriculture has been the backbone of many economies around the world, and it remains an important contributor to global economic growth (Overview, n.d.; Why Is Agriculture Important? Benefits and Its Role | Maryville Online, 2022).

Forestry is also a key component of the bioeconomy. It involves the management and use of forest resources for the production of wood and non-wood forest products. Forestry is critical for the conservation of biodiversity and the provision of ecosystem services such as carbon sequestration and soil conservation (AZoCleantech.com, 2022; Ebissa et al., 2023).

The bioeconomy presents several benefits, including the production of renewable and sustainable products, the creation of new jobs and economic opportunities, and the potential to reduce greenhouse gas emissions and mitigate climate change (Language Selection | European Commission, n.d.; What Is the Bioeconomy and How Could It Help Fight Climate Change?, n.d.; World Bank Group, 2022). However, it also poses several challenges that need to be addressed, such as the need to balance economic growth with environmental sustainability, the potential risks associated with genetically modified organisms, and the need to ensure that the benefits of the bioeconomy are shared equitably across society (Greenhouse Gases | US EPA, 2023; Staff, 2022; World Energy Transitions Outlook 2022, n.d.).

In order to fully realize the potential of the bioeconomy, innovation is essential. Innovation can drive the development of new technologies and processes that enhance the efficiency and sustainability of bio-based production systems. It can also help to address some of the challenges associated with the bioeconomy, such as the need to reduce waste and improve resource efficiency.

In conclusion, the bioeconomy presents a promising opportunity for sustainable economic growth. Its key components include biotechnology, agriculture, and forestry, which have numerous applications in various sectors. To fully realize the potential of the bioeconomy, it is essential to address the challenges it poses and promote innovation in the development of new technologies and processes. By doing so, we can create a more sustainable and prosperous future for generations to come.

IMPORTANCE OF INNOVATION IN DRIVING THE BIOECONOMY

The bioeconomy refers to the use of renewable biological resources such as plants, animals, and microorganisms to produce goods and services. This approach has numerous benefits, such as the production of renewable and sustainable products, the creation of new jobs and economic opportunities, and the potential to reduce greenhouse gas emissions and mitigate climate change. However, the realization of the bioeconomy requires the development and deployment of new technologies and processes that enhance its efficiency, reduce its environmental impact, and expand its potential applications. Innovation is essential in driving the bioeconomy forward. Innovation can drive the development of new bio-based products and services, improve the efficiency of bio-based production systems, and address concerns about genetically modified organisms (GMOs) (De Jaramillo & Schuler, 2014; Hodgson et al., 2022; Kircher, 2022).

Biotechnology is a key component of the bioeconomy and has numerous applications in various fields such as medicine, agriculture, and industry. Biotechnology involves the use of biological systems and living organisms to produce new products or improve existing ones. Biotechnology has the potential to transform numerous industries by enabling the development of new bio-based products and processes. One of the most promising applications of biotechnology is in the field of medicine, enabling the development of new therapies and treatments for various diseases. Biotechnology also has numerous applications in agriculture, such as the development of genetically modified crops that are resistant to pests and diseases. However, biotechnology also poses several challenges that need to be addressed through innovation, such as the potential impact of GMOs on human health and the environment (Biotechnology for Sustainability and Social Well Being, n.d.; Home, n.d.; Molnar, 2015).

In agriculture, precision agriculture is a promising area of innovation. Precision agriculture involves the use of technology such as GPS, sensors, and drones to optimize crop production and reduce waste. Precision agriculture can enable farmers to use resources such as water, fertilizer, and pesticides more efficiently, which can reduce costs and environmental impact. Innovation can also enable the development of new agricultural practices that promote sustainability and reduce environmental impact, such as agroforestry.

Forestry is another key component of the bioeconomy. Sustainable forest management is a promising area of innovation that involves the use of practices such as reduced-impact logging, afforestation, and reforestation to promote the sustainability of forests. Sustainable forest management can help to mitigate climate change by sequestering carbon and reducing deforestation. Innovation can also drive the development of new bio-based products from forest resources, such as biofuels and biomaterials.

In conclusion, the bioeconomy presents a promising opportunity for sustainable economic growth, but its realization requires the development and deployment of new technologies and processes. Innovation is essential in driving the bioeconomy forward by enhancing its efficiency, reducing its environmental impact, and expanding its potential applications. Biotechnology, agriculture, and forestry are key components of the bioeconomy that require innovative solutions to address the challenges and limitations of existing technologies and practices.

DRIVERS OF INNOVATION IN THE BIOECONOMY

The bioeconomy is an economic system that is based on the production, use, and commercialization of biological resources such as biomass, crops, animals, and microorganisms. This sector is growing fast and has the potential to drive sustainable economic growth and address societal challenges such as climate change, resource scarcity, and food security. Innovation is a key driver of the bioeconomy, and this article explores the factors that are driving innovation and their impact on the development of the bioeconomy.

The increasing demand for sustainable products is one of the key drivers of innovation in the bioeconomy. Consumers are more aware of the environmental impact of their consumption patterns and are seeking products that are produced sustainably and have a lower environmental footprint. This has led to the development of new technologies and processes that enable the production of bio-based products such as bioplastics, biochemicals, and biofuels (Designing Safe and Sustainable Products Requires a New Approach for Chemicals, n.d.).

Biotechnology is another key driver of innovation in the bioeconomy. Biotechnology refers to the use of living organisms or their products to produce useful products or services. Advances in biotechnology have enabled the development of new technologies and processes that enable the production of bio-based products such as biofuels, biochemicals, and bioplastics. Synthetic biology is a promising area of biotechnology that involves the design and engineering of biological systems for specific applications.

Government policies and regulations can also be drivers of innovation in the bioeconomy. For example, government incentives such as tax credits and grants can encourage private investment in the bioeconomy. Government regulations can also drive innovation in the bioeconomy by setting standards for sustainability and environmental impact.

Collaboration and knowledge sharing are also important drivers of innovation in the bioeconomy. Collaboration among stakeholders such as industry, academia, and government can lead to the development of new technologies and processes that address societal challenges. Open innovation is another approach to collaboration that can drive innovation in the bioeconomy. Access to funding and investment is essential for driving innovation in the bioeconomy.

To promote further innovation in the bioeconomy, it is important to promote collaboration among stakeholders and provide access to funding and investment. Governments can play a critical role in promoting the bioeconomy through the development of policies and regulations that create a favorable environment for innovation and investment. Industry can also play a key role in driving innovation in the bioeconomy by investing in research and development and collaborating with other stakeholders to accelerate innovation. Ultimately, the bioeconomy can drive sustainable economic growth while addressing some of the most pressing societal challenges of our time.

CHALLENGES TO THE BIOECONOMY AND HOW INNOVATION CAN HELP ADDRESS THEM

The bioeconomy is an emerging economic model that seeks to promote sustainable development by utilizing renewable biological resources to produce goods and services. The bioeconomy encompasses a wide range of sectors, including agriculture, forestry, fisheries, and biotechnology. Despite the potential benefits of the bioeconomy, there are significant challenges that must be addressed to ensure its success. In this article, we will explore the challenges facing the bioeconomy and how innovation can help address them.

Challenges Facing the Bioeconomy

1. Limited Availability of Feedstocks: The availability of biomass feedstocks is a significant challenge for the bioeconomy. The production of bio-based products and energy relies on a consistent supply of biomass feedstocks, such as agricultural waste, forestry residues, and energy crops. However, the availability of these feedstocks is often limited due to competition with other industries, such as food and feed production.
2. High Costs: The bioeconomy is often associated with high costs due to the need for specialized equipment, infrastructure, and skilled labor. Additionally, the development of new bio-based products and technologies can be expensive and time-consuming.
3. Regulatory Frameworks: Regulatory frameworks can pose a significant challenge to the development of the bioeconomy. Regulations governing the production and use of bio-based products and technologies can vary significantly between countries and regions, making it difficult for companies to navigate the regulatory landscape.

4. Public Perception: Public perception of the bioeconomy can also be a challenge. Some people may view the use of biomass feedstocks for energy and products as competing with food and feed production or leading to deforestation.

Innovation and the Bioeconomy

Innovation can play a critical role in addressing the challenges facing the bioeconomy. By developing new technologies and processes, innovation can increase the availability of biomass feedstocks, reduce costs, and navigate regulatory frameworks. Additionally, innovation can help shift public perception of the bioeconomy by highlighting its potential benefits.

1. Biomass Feedstocks: Innovation can help address the limited availability of biomass feedstocks by developing new sources of biomass and improving existing processes for biomass production. For example, biotechnology can be used to develop energy crops that are more efficient at converting sunlight into biomass. Additionally, innovations in precision agriculture can help optimize crop yields while reducing the need for inputs, such as fertilizers and pesticides.
2. Cost Reduction: Innovation can help reduce the costs associated with the bioeconomy by developing more efficient processes and technologies. For example, the development of more efficient bioreactors can reduce the cost of producing bio-based products, while the use of automation and artificial intelligence can reduce labor costs.
3. Regulatory Frameworks: Innovation can help navigate regulatory frameworks by developing new technologies that meet regulatory requirements or by working with policymakers to develop more favorable regulations. For example, the development of biodegradable plastics can help address concerns about plastic pollution while meeting regulatory requirements for biodegradability.
4. Public Perception: Innovation can help shift public perception of the bioeconomy by highlighting its potential benefits and addressing concerns about competition with food production and deforestation. For example, innovation can help develop new products and technologies that are more sustainable and have a smaller environmental footprint.

Case Studies

1. Advanced Biofuels: Advanced biofuels are a promising area of innovation in the bioeconomy. These fuels are produced from non-food feedstocks, such as algae and agricultural residues, and have the potential to reduce greenhouse gas emissions and improve energy security. However, the development of advanced biofuels has been hampered by high costs and limited availability of feedstocks. To address these challenges, researchers are developing new technologies for converting biomass into biofuels and identifying new sources of feedstocks.
2. Bioplastics: Bioplastics are another area of innovation in the bioeconomy. Bioplastics are made from renewable biomass sources and can biodegrade under certain conditions, making them an attractive alternative to traditional plastics. However, bioplastics are often more expensive than traditional plastics, and their use can be limited by regulations and public perception. To address these challenges, researchers are developing new bioplastics that are more cost-effective and have better performance characteristics, such as improved durability and heat resistance.

3. Precision Agriculture: Precision agriculture is an area of innovation that has the potential to increase the availability of biomass feedstocks while reducing the environmental impact of agriculture. Precision agriculture uses technology, such as sensors and data analytics, to optimize crop yields while minimizing the use of inputs, such as fertilizers and pesticides. By reducing the environmental impact of agriculture, precision agriculture can help address concerns about competition with food production and deforestation.

The bioeconomy has the potential to promote sustainable development and reduce dependence on fossil fuels. However, the bioeconomy also faces significant challenges, including limited availability of feedstocks, high costs, regulatory frameworks, and public perception. Innovation can play a critical role in addressing these challenges by developing new technologies and processes, reducing costs, navigating regulatory frameworks, and shifting public perception. Case studies, such as advanced biofuels, bioplastics, and precision agriculture, demonstrate the potential for innovation to drive the bioeconomy forward. By investing in innovation, policymakers, businesses, and researchers can help ensure the success of the bioeconomy and promote sustainable development.

BALANCING ECONOMIC GROWTH WITH ENVIRONMENTAL SUSTAINABILITY

Economic growth is an essential component of modern society, providing opportunities for individuals and communities to improve their standards of living. However, economic growth can also have significant environmental impacts, such as greenhouse gas emissions, deforestation, and pollution. As a result, there is a growing recognition that economic growth must be balanced with environmental sustainability to promote long-term prosperity.

This article will explore the challenges and opportunities of balancing economic growth with environmental sustainability. Specifically, it will examine the concept of sustainable development, the role of innovation, and the importance of stakeholder engagement in promoting a sustainable economy.

Sustainable Development

Sustainable development is a concept that seeks to balance economic growth with environmental sustainability and social progress. The idea is that economic growth should not be pursued at the expense of the environment and society, but rather should be achieved in a way that promotes long-term prosperity for all. Sustainable development involves considering the impacts of economic activities on the environment and society, and taking steps to minimize negative impacts while maximizing positive impacts.

One way to achieve sustainable development is through the adoption of sustainable production and consumption practices. Sustainable production practices involve minimizing waste, reducing greenhouse gas emissions, and conserving natural resources. Sustainable consumption practices involve reducing consumption levels, increasing the use of sustainable products, and reducing waste.

Innovation

Innovation is a critical component of sustainable development, as it provides new technologies and processes that can help reduce environmental impacts while promoting economic growth. Innovation can help to make economic activities more efficient, reduce waste, and improve environmental performance. Examples of innovative technologies and processes include renewable energy, energy-efficient buildings, sustainable transportation, and sustainable agriculture.

Innovation can also help to create new business opportunities and markets, driving economic growth and job creation. For example, the renewable energy sector has seen significant growth in recent years, creating new job opportunities and driving economic growth while reducing greenhouse gas emissions.

Stakeholder Engagement

Stakeholder engagement is another critical component of balancing economic growth with environmental sustainability. Stakeholders include individuals, communities, businesses, and governments that are impacted by economic activities and environmental policies. Engaging with stakeholders can help to build support for sustainable development initiatives and ensure that the concerns of all stakeholders are considered in decision-making processes.

Effective stakeholder engagement involves listening to and addressing the concerns of all stakeholders, including those who may be impacted negatively by sustainable development initiatives. It also involves promoting transparency and accountability in decision-making processes, so that stakeholders can understand the rationale behind decisions and have a say in how they are implemented.

RISKS AND CONCERNS ASSOCIATED WITH GENETICALLY MODIFIED ORGANISMS (GMOS)

Genetically modified organisms (GMOs) have been a topic of controversy and debate for many years. While some people believe that GMOs have the potential to address food security challenges and improve crop yields, others are concerned about the potential risks associated with their use. In this article, we will explore the risks and concerns associated with GMOs and discuss their implications for human health and the environment.

What are GMOs?

GMOs are organisms that have had their genetic makeup altered through genetic engineering techniques. This can involve the introduction of new genes, the removal of existing genes, or the modification of existing genes in order to achieve a desired trait. The goal of genetic engineering is to create organisms that are more resistant to disease, have higher yields, or are better suited to specific environmental conditions.

Risks and concerns associated with GMOs

1. Environmental concerns

One of the primary concerns associated with GMOs is their potential impact on the environment. There are concerns that GMOs may crossbreed with wild or native species, leading to the creation of new, potentially harmful organisms. There are also concerns that the use of GMOs may lead to the development of "superweeds" that are resistant to herbicides, leading to a need for more potent chemicals to control them.

2. Health concerns

Another major concern associated with GMOs is their potential impact on human health. While there is no conclusive evidence that GMOs are harmful to human health, some studies have suggested that they may have negative effects. For example, there are concerns that GMOs may trigger allergic reactions in some people or that they may be toxic to certain populations.

3. Economic concerns

There are also economic concerns associated with GMOs. For example, there are concerns that the use of GMOs may lead to the loss of traditional farming practices and the displacement of small farmers. There are also concerns that GMOs may lead to increased dependence on large corporations that control the production and distribution of GMO seeds.

4. Ethical concerns

Finally, there are ethical concerns associated with GMOs. Some people are concerned that the use of genetic engineering to alter the genetic makeup of living organisms is unnatural and could have unintended consequences. Others are concerned that the use of GMOs may lead to the exploitation of developing countries, where there may be limited regulation or oversight of their use.

Implications for Human Health and the Environment

The potential risks and concerns associated with GMOs have significant implications for human health and the environment. For example, the use of GMOs may lead to the development of new organisms that could have negative impacts on the environment. There is also the potential for unintended consequences, such as the development of superweeds or the creation of new allergens.

In terms of human health, there is still much that is not understood about the potential long-term effects of consuming GMOs. While there is no conclusive evidence that GMOs are harmful to human health, there are concerns that they could have negative effects, particularly for vulnerable populations such as children, pregnant women, and people with allergies or sensitivities.

Addressing the Risks and Concerns Associated With GMOs

There are a number of steps that can be taken to address the risks and concerns associated with GMOs. One of the most important steps is to ensure that GMOs are properly regulated and that their potential risks and benefits are carefully evaluated before they are released into the environment or the food supply.

There is also a need for greater transparency and public engagement in discussions around the use of GMOs. This includes engaging with communities, farmers, and other stakeholders to better understand their concerns and to ensure that their voices are heard.

Finally, there is a need for continued research into the potential risks and benefits of GMOs, particularly in the areas of human health and the environment.

Apart from potential health risks, another major concern associated with GMOs is their impact on the environment. One of the main concerns is the potential for GMOs to cross-breed with wild relatives and transfer their modified traits to other organisms, leading to unintended consequences. This could have significant ecological impacts, such as the loss of biodiversity, the disruption of ecosystems, and the development of superweeds or superbugs that are resistant to pesticides.

In addition, there are concerns that GMOs could have unintended effects on non-target organisms, such as beneficial insects, birds, and mammals. For example, Bt crops, which produce a toxin that kills certain insect pests, may also harm beneficial insects such as butterflies and bees. This could have serious consequences for ecosystems and the pollination of crops.

There are also concerns about the potential for GMOs to create new allergens or toxins. When a gene is inserted into a plant, it can produce new proteins that may be harmful to human health or the environment. This risk is particularly high when genes from a source that is known to produce allergens or toxins are inserted into a crop.

Finally, there are concerns about the socio-economic impacts of GMOs. Large corporations that produce GMO seeds often hold patents on their products, which can restrict access to seeds for small-scale farmers in developing countries. This has led to concerns about the potential for GMOs to exacerbate inequality and deepen existing power imbalances in the global food system.

In conclusion, the development and use of GMOs presents a complex set of risks and concerns. While the potential benefits of GMOs are significant, it is important to carefully consider and manage the potential risks associated with their use. This requires a comprehensive and science-based approach that takes into account both the potential benefits and risks of GMOs, and balances the interests of different stakeholders, including farmers, consumers, and the environment.

ENSURING EQUITABLE BENEFITS FROM THE BIOECONOMY

The bioeconomy has the potential to drive economic growth, create jobs, and address a range of global challenges, including food security, energy security, and climate change. However, there are concerns that the benefits of the bioeconomy may not be distributed equitably, and that certain groups may be left behind or even harmed by its development. Ensuring that the benefits of the bioeconomy are shared fairly and equitably is therefore a critical challenge for policymakers and stakeholders (Hidayati et al., 2023).

1. Promoting Inclusive Innovation One way to ensure equitable benefits from the bioeconomy is to promote inclusive innovation. This means creating opportunities for a diverse range of actors to participate in the development and deployment of bio-based technologies and products. This includes small and medium-sized enterprises (SMEs), women, indigenous peoples, and marginalized communities. By creating a more diverse and inclusive innovation ecosystem, policymakers can help ensure that the benefits of the bioeconomy are spread more widely.
2. Encouraging Sustainable Agriculture Another way to ensure equitable benefits from the bioeconomy is to promote sustainable agriculture. This means supporting small-scale farmers and agroforestry systems that use sustainable practices to produce food and other bio-based products. By promoting sustainable agriculture, policymakers can help ensure that the benefits of the bioeconomy are distributed more equitably, and that rural communities are able to participate in the bioeconomy on their own terms.
3. Strengthening Value Chains A third way to ensure equitable benefits from the bioeconomy is to strengthen value chains. This means developing and strengthening local and regional value

chains that connect producers, processors, distributors, and consumers of bio-based products. By strengthening value chains, policymakers can help ensure that the benefits of the bioeconomy are distributed more evenly across the supply chain, and that small-scale producers are able to capture a greater share of the value generated by bio-based products.

4. Promoting Access to Finance Access to finance is a critical barrier for many small-scale producers and entrepreneurs in the bioeconomy. To address this, policymakers can promote the development of innovative financing mechanisms, such as impact investing, crowdfunding, and microfinance. By promoting access to finance, policymakers can help ensure that small-scale producers and entrepreneurs are able to access the resources they need to participate in the bioeconomy.
5. Developing Human Capital Human capital development is another critical component of ensuring equitable benefits from the bioeconomy. This means investing in education, training, and skills development for the bioeconomy workforce, including small-scale farmers, processors, and entrepreneurs. By developing human capital, policymakers can help ensure that all actors in the bioeconomy are able to participate in the development and deployment of bio-based technologies and products.
6. Addressing Intellectual Property Rights Intellectual property rights are an important consideration in the bioeconomy, as many bio-based products are developed through genetic modification and other forms of biotechnology. To ensure equitable benefits from the bioeconomy, policymakers need to address intellectual property rights in a way that promotes innovation and access to technology, while also protecting the rights of small-scale producers and communities.
7. Strengthening Governance and Regulation Governance and regulation are critical components of ensuring equitable benefits from the bioeconomy. This means developing and implementing policies and regulations that promote sustainability, inclusivity, and accountability in the bioeconomy. By strengthening governance and regulation, policymakers can help ensure that the benefits of the bioeconomy are distributed more equitably and that the risks associated with the bioeconomy are effectively managed.
8. Ensuring Social and Environmental Responsibility Finally, to ensure equitable benefits from the bioeconomy, policymakers need to ensure that the development and deployment of bio-based technologies and products are socially and environmentally responsible.

Another important aspect of ensuring equitable benefits from the bioeconomy is by promoting access to and sharing of resources. The bioeconomy heavily relies on the use of biological resources such as crops, animals, and microorganisms. These resources often originate from developing countries, which have a rich biodiversity. Therefore, it is essential to ensure that these countries benefit from the use of their resources and that their communities are not exploited. The Convention on Biological Diversity (CBD) provides guidelines for the fair and equitable sharing of benefits derived from the use of biological resources. It encourages the involvement of local communities in decision-making processes and the sharing of benefits with them.

In addition, promoting capacity building and technology transfer can help ensure that developing countries benefit from the bioeconomy. By providing access to new technologies and skills, developing countries can participate fully in the bioeconomy and reap its benefits. Capacity building initiatives can include training programs for farmers and researchers, while technology transfer can involve the transfer of knowledge and equipment from developed countries to developing countries.

Another important aspect of ensuring equitable benefits from the bioeconomy is by promoting social inclusion. The bioeconomy should not only benefit large corporations but also small-scale farmers, women, and other marginalized groups. To achieve this, policies that promote social inclusion should be put in place. For example, providing access to finance and market information to small-scale farmers can help them participate in the bioeconomy. In addition, women should be involved in decision-making processes as they play a crucial role in the agricultural sector.

In conclusion, the bioeconomy presents a significant opportunity for sustainable economic growth, job creation, and environmental sustainability. However, it is crucial to ensure that the benefits are equitable and that all stakeholders are involved. This can be achieved through policies that promote social inclusion, access to resources, and technology transfer. By ensuring that the bioeconomy benefits everyone, we can build a more sustainable and equitable world.

OPPORTUNITIES FOR THE BIOECONOMY AND THE POTENTIAL FOR NEW BIO-BASED PRODUCTS AND SERVICES

The bioeconomy offers vast opportunities for the development of new bio-based products and services. A bio-based economy is based on the use of biological resources such as plants, animals, and microorganisms to produce a range of products, including food, feed, biofuels, and biomaterials. By utilizing these resources, we can reduce our reliance on fossil fuels, create new revenue streams, and promote sustainability.

One of the most promising areas of the bioeconomy is the production of biofuels. Biofuels are derived from renewable biological resources, such as agricultural crops, waste materials, and forestry residues. Biofuels are becoming increasingly popular as they offer a sustainable alternative to traditional fossil fuels, which are non-renewable and contribute to climate change. Biofuels can be used in transportation, heating, and electricity generation.

Another area of the bioeconomy that offers significant potential is the production of bioplastics. Bioplastics are a type of plastic made from renewable biological resources such as corn starch, sugarcane, and potato starch. Bioplastics have a lower carbon footprint than traditional plastics and can biodegrade in the environment. As the demand for sustainable packaging and products increases, the demand for bioplastics is also expected to grow.

The bioeconomy also offers opportunities in the production of bio-based chemicals. Bio-based chemicals are chemicals produced from renewable biological resources. These chemicals can be used in a range of applications, including the production of cleaning products, cosmetics, and pharmaceuticals. Bio-based chemicals offer several advantages, including a lower carbon footprint and reduced reliance on non-renewable resources.

The bioeconomy also presents opportunities in the production of bio-based materials. Biomaterials are materials produced from renewable biological resources such as plant fibers, wood, and bioplastics. Biomaterials can be used in a range of applications, including construction, textiles, and packaging. Biomaterials offer several advantages, including a lower carbon footprint and reduced reliance on non-renewable resources.

In addition to these opportunities, the bioeconomy also offers potential in the development of new bio-based products and services. For example, the use of biotechnology can help develop new products and services, such as new medicines, crops, and animal feed. Biotechnology can also help improve the efficiency and sustainability of current production processes, leading to cost savings and reduced environmental impact.

In conclusion, the bioeconomy offers vast opportunities for the development of new bio-based products and services. By utilizing renewable biological resources, we can promote sustainability, reduce our reliance on non-renewable resources, and create new revenue streams. The bioeconomy presents significant potential in the production of biofuels, bioplastics, bio-based chemicals, and biomaterials. In addition, the use of biotechnology can help develop new products and services, leading to further growth and innovation in the bioeconomy.

THE ROLE OF THE BIOECONOMY IN CREATING NEW JOBS AND ECONOMIC OPPORTUNITIES

The bioeconomy refers to the sustainable and efficient use of biological resources for the production of goods and services, with a focus on renewable natural resources such as plants, animals, and microorganisms. It encompasses a wide range of industries, including agriculture, forestry, fisheries, biotechnology, and bioenergy, among others. One of the most significant benefits of the bioeconomy is its potential to create new jobs and economic opportunities in both rural and urban areas. This technical article will discuss the role of the bioeconomy in creating new jobs and economic opportunities (Sangtani et al., 2023).

The Role of the Bioeconomy in Job Creation

The bioeconomy has the potential to create a significant number of new jobs. One of the key drivers of job creation in the bioeconomy is the shift towards a more sustainable and circular economy. As consumers become more environmentally conscious, the demand for bio-based products and services has increased. This shift has led to the creation of new jobs in sectors such as biotechnology, bioenergy, and agriculture.

For example, the biotechnology sector has seen significant growth in recent years, with many companies investing in research and development of new bio-based products. This has led to the creation of new jobs in areas such as bioengineering, biomanufacturing, and biopharmaceuticals. Similarly, the bioenergy sector has the potential to create jobs in areas such as biofuel production, waste management, and energy efficiency.

The bioeconomy also has the potential to create jobs in rural areas. Agriculture and forestry are two of the primary industries in rural areas, and both have significant potential for growth in the bioeconomy. The adoption of sustainable practices in agriculture and forestry can lead to the creation of new jobs in areas such as agroforestry, sustainable forestry, and biomass production.

Economic Opportunities in the Bioeconomy

The bioeconomy presents significant economic opportunities for both developed and developing countries. One of the primary economic benefits of the bioeconomy is its potential to reduce dependence on non-renewable resources. This can lead to cost savings for businesses and governments, as well as increased energy security.

The bioeconomy can also lead to increased economic activity in rural areas, where there is often a lack of economic opportunities. By creating new jobs and stimulating economic activity, the bioeconomy can help to reduce poverty and improve living standards in rural areas.

Another economic benefit of the bioeconomy is its potential to foster innovation and entrepreneurship. The bioeconomy requires a diverse range of skills and expertise, from scientists and engineers to business managers and entrepreneurs. This creates opportunities for individuals to develop and commercialize new bio-based products and services.

The bioeconomy has the potential to create new jobs and economic opportunities, particularly in sectors such as biotechnology, bioenergy, and agriculture. The shift towards a more sustainable and circular economy is driving the growth of the bioeconomy, and this growth is expected to continue in the coming years. The bioeconomy also presents significant economic benefits, including reducing dependence on non-renewable resources, increasing energy security, and fostering innovation and entrepreneurship.

THE BIOECONOMY AND ITS CONTRIBUTION TO GLOBAL SUSTAINABILITY GOALS

The concept of the bioeconomy has been gaining momentum in recent years as a way to address both environmental and economic challenges. The bioeconomy refers to the use of renewable biological resources to produce goods and services, with the goal of reducing our reliance on finite resources and minimizing environmental impacts (Dries et al., 2019).

The bioeconomy encompasses a wide range of sectors, including agriculture, forestry, fisheries, and biotechnology. These sectors can contribute to global sustainability goals in a number of ways.

One of the primary ways that the bioeconomy can contribute to sustainability is by reducing greenhouse gas emissions. The use of renewable resources such as biomass and bioplastics can help to replace fossil fuels and reduce the carbon footprint of industries such as transportation and packaging. In addition, sustainable land management practices in agriculture and forestry can help to sequester carbon in soils and vegetation.

The bioeconomy can also contribute to sustainable development by creating new economic opportunities, particularly in rural areas. By utilizing local resources and promoting sustainable practices, the bioeconomy can help to create jobs and support local economies. In addition, the development of new bio-based products and technologies can lead to innovation and increased competitiveness in global markets.

Another important aspect of the bioeconomy is its potential to promote biodiversity conservation. By using resources in a sustainable and responsible manner, the bioeconomy can help to protect ecosystems and support the conservation of biodiversity. For example, sustainable forestry practices can help to maintain forest habitats and protect endangered species.

Finally, the bioeconomy can contribute to global sustainability goals by promoting circular economy principles. The circular economy is an approach that aims to minimize waste and maximize resource efficiency by keeping products and materials in use for as long as possible. By utilizing renewable resources and designing products for reuse and recycling, the bioeconomy can help to create a more circular and sustainable economy ("Towards a Sustainable Bioeconomy: Principles, Challenges and Perspectives," 2018).

Overall, the bioeconomy has the potential to make significant contributions to global sustainability goals. By utilizing renewable resources, creating economic opportunities, promoting biodiversity conservation, and supporting circular economy principles, the bioeconomy can help to address some of the most pressing environmental and economic challenges facing the world today.

CASE STUDIES OF SUCCESSFUL INNOVATION IN THE BIOECONOMY

The bioeconomy is the production and use of biological resources, processes and principles to provide goods and services in all economic sectors. It aims to create a more sustainable, circular and low-carbon economy that relies less on fossil fuels and more on renewable biological sources. The bioeconomy has the potential to address some of the major global challenges, such as climate change, food security, health and biodiversity loss.

However, the transition to a bioeconomy requires not only technological innovations, but also social, institutional and regulatory innovations that can enable the development and diffusion of new products, processes and business models. Moreover, it requires the creation and strengthening of innovation ecosystems that can foster collaboration, learning and co-creation among different actors across sectors and disciplines.

In this article, we present some case studies of successful innovation in the bioeconomy from different countries and regions, highlighting the key drivers, enablers and barriers that have shaped their development and impact.

Circular Bioeconomy in South Africa: Namibian Bush Encroachment

One of the examples of circular bioeconomy is the case of Namibian bush encroachment, which is a phenomenon of excessive growth of woody plants in savanna ecosystems due to overgrazing, fire suppression and climate change. This has negative impacts on biodiversity, water availability, soil quality and agricultural productivity.

To address this challenge, a consortium of public and private actors from Namibia and South Africa has developed a value chain that converts the invasive bush biomass into various products, such as animal feed, charcoal, wood chips and electricity. The project involves harvesting the bush biomass using mechanical and manual methods, transporting it to processing plants using trucks or trains, and producing the final products using different technologies, such as pelletizing, pyrolysis and gasification.

The project has multiple benefits for the environment, society and economy. It restores the natural balance of the savanna ecosystem, enhances carbon sequestration and water conservation, creates jobs and income for local communities, reduces greenhouse gas emissions from fossil fuels, and provides renewable energy and materials for various industries.

The key drivers for this innovation include the availability of abundant biomass resources, the demand for sustainable products from local and international markets, the support from national policies and programs on bioenergy and rural development, and the collaboration among different stakeholders from research, government, industry and civil society.

The main enablers for this innovation include the development of appropriate technologies for harvesting and processing the biomass, the establishment of quality standards and certification schemes for the products, the creation of business models that ensure profitability and social inclusion, and the provision of financial incentives and infrastructure for the value chain.

The main barriers for this innovation include the lack of awareness and acceptance of the products among consumers and end-users, the competition from cheaper fossil-based alternatives

Bioeconomy Innovation Ecosystems in Europe: The Case of Bio Base Europe

Another example of bioeconomy innovation is the case of Bio Base Europe, which is a cross-border public-private partnership between Belgium and the Netherlands that aims to foster the development of the biobased economy in Europe. Bio Base Europe consists of two complementary pillars: a pilot plant in Ghent, Belgium, and a training center in Terneuzen, the Netherlands.

The pilot plant is a state-of-the-art facility that offers research and development services for scaling up and optimizing biobased processes and products, such as biofuels, bioplastics, biochemicals and biomaterials. The pilot plant operates as an open innovation center that provides access to advanced equipment and expertise for companies, universities and research institutes.

The training center is a knowledge and education hub that offers training and education programs for students, professionals and entrepreneurs in the biobased economy. The training center also organizes events and workshops to raise awareness and stimulate networking among different stakeholders.

The Bio Base Europe project has several benefits for the region, such as creating jobs and income, enhancing innovation capacity and competitiveness, reducing environmental impact and dependence on fossil resources, and strengthening cross-border cooperation and integration.

The key drivers for this innovation include the availability of biomass resources and biotechnology expertise in the region, the demand for sustainable solutions from various sectors, the support from regional, national and European policies and funds, and the vision and leadership of the founding partners.

The main enablers for this innovation include the development of a shared infrastructure and governance model that facilitates collaboration and knowledge exchange, the establishment of a strong network of partners and customers from different domains and countries, the creation of a supportive regulatory and market environment that fosters innovation and investment, and the provision of tailored training and education programs that address the skills gap and promote lifelong learning.

The main barriers for this innovation include the high costs and risks associated with scaling up biobased processes and products, the lack of harmonized standards and regulations for biobased products across Europe, the low awareness and acceptance of biobased products among consumers and end-users, and the competition from established fossil-based industries.

Bioeconomy Innovation in China: The Case of Qingdao Industrial Biotechnology Park

A third example of bioeconomy innovation is the case of Qingdao Industrial Biotechnology Park (QIBEBT), which is a national-level industrial park that focuses on developing and commercializing biotechnology applications in various fields, such as bioenergy, biomedicine, bioagriculture and biomaterials. QIBEBT is located in Qingdao, a coastal city in eastern China that has a strong tradition of marine science and industry.

QIBEBT was established in 2009 as a joint initiative of the Chinese Academy of Sciences (CAS) and the Qingdao municipal government. It consists of three main components: a research institute that conducts basic and applied research on biotechnology, a technology transfer center that facilitates the transfer and commercialization of research results, and an industrial zone that hosts biotechnology enterprises and incubators.

QIBEBT has achieved remarkable results in terms of innovation output, economic impact and social benefits. It has produced more than 1 000 patents, 500 publications and 100 standards in biotechnology. It has attracted more than 300 biotechnology enterprises and incubators, creating more than 10 000 jobs and generating more than 10 billion yuan (1.5 billion USD) of annual output value. It has also contributed to solving some of the major challenges facing China and the world, such as energy security, environmental protection, health care and food safety.

The key drivers for this innovation include the availability of abundant biomass resources and human capital in Qingdao, the demand for biotechnology solutions from various sectors and markets, the support from national and local policies and funds, and the vision and leadership of CAS and the Qingdao municipal government.

The main enablers for this innovation include the development of a world-class research infrastructure and talent pool that fosters scientific excellence and innovation, the establishment of a comprehensive technology transfer system that bridges the gap between research and industry, the creation of a favorable business environment that encourages entrepreneurship and investment, and the provision of various services and facilities that support the growth and development of biotechnology enterprises and incubators.

The main barriers for this innovation include the high costs and risks associated with biotechnology research and development, the lack of harmonized standards and regulations for biotechnology products across China and abroad, the low awareness and acceptance of biotechnology products among consumers and end-users, and the competition from established industries and foreign competitors.

Bioeconomy Innovation in the United States: The Case of Impossible Foods

A fourth example of bioeconomy innovation is the case of Impossible Foods, which is a company that produces plant-based meat alternatives that mimic the taste, texture and nutritional value of animal meat. Impossible Foods aims to create a more sustainable and humane food system that can feed the growing population without harming the environment and animal welfare.

Impossible Foods was founded in 2011 by Patrick Brown, a former biochemistry professor at Stanford University, who was inspired by his passion for animal rights and environmental protection. He assembled a team of scientists, chefs and engineers to develop a plant-based burger that could satisfy meat lovers and reduce the environmental footprint of animal agriculture.

The concept of bioeconomy is driving innovation in various sectors across the world. Bioeconomy promotes the use of renewable resources, reduces waste and pollution, and supports social and economic development. Here are six examples of bioeconomy innovation from different countries, namely Finland, Denmark, the United States, Brazil, India, and South Africa.

The Finnish company UPM produces renewable biofuels and chemicals from pulp and wood residues. UPM's innovations include the use of waste and residues as raw materials, the development of advanced biorefinery technologies, and the creation of a circular economy model that maximizes resource efficiency.

The Danish company Novozymes produces enzymes and microorganisms for various industries, such as agriculture, food, and laundry. Novozymes' innovations include the use of biotechnology to improve the efficiency and sustainability of industrial processes, the collaboration with customers and partners to develop customized solutions, and the investment in research and development to explore new applications.

The American company Impossible Foods produces plant-based meat alternatives that mimic the taste, texture, and nutritional value of animal meat. Impossible Foods' innovations include the use of heme, a molecule that gives meat its flavor and color, the genetic engineering and fermentation of soy plants to produce heme, and the creation of a scalable and flexible production and distribution network.

The Brazilian company Braskem produces bioplastics from sugarcane ethanol. Braskem's innovations include the use of renewable feedstocks to produce carbon-neutral plastics, the development of a global partnership network with customers and suppliers, and the establishment of a green brand identity and communication strategy.

The Indian company Praj Industries produces biofuels and biochemicals from various feedstocks, such as sugarcane, corn, and biomass. Praj Industries' innovations include the use of second-generation ethanol technology to convert lignocellulosic biomass into ethanol, the creation of a sustainable and inclusive energy and chemical industry, and the development of a strong technical assistance and certification services.

The South African company BKB provides sustainable agricultural solutions, such as animal feed, wool, and renewable energy. BKB's innovations include the use of precision farming technologies to optimize resource use and reduce waste, the creation of a circular economy model that maximizes resource efficiency, and the provision of technical assistance and training to farmers and communities.

Bioeconomy innovation is important for sustainable development, and collaboration among various stakeholders, such as governments, industries, academia, and civil society, is crucial. Bioeconomy innovation has the potential to address global challenges, such as climate change, food security, and social inequality.

The main barriers for this innovation include the high costs and risks associated with biotechnology research and development, the lack of clear and consistent standards and regulations for biobased products across countries and regions, the low awareness and acceptance of biobased products among consumers and end-users, and the competition from established chemical producers and other biobased companies.

S M NAZMUZ SAKIB'S BIOECONOMY TYPOLOGY FORMULA

The bioeconomy is a cross-cutting concept that encompasses multiple sectors that use biological resources (such as plants, microorganisms, animals, etc.) to produce products, services and processes that contribute to economic growth, environmental sustainability, social welfare and human health. The bioeconomy is also seen as a key driver of innovation and transformation in response to global challenges such as climate change, food security, energy independence and biodiversity loss.

However, the bioeconomy is not a homogeneous or static concept. It is influenced by various factors such as technological development, market demand, policy frameworks, social values, ethical norms and cultural preferences. Therefore, the bioeconomy can take different forms and directions depending on the context and the objectives of the actors involved.

One way to understand the diversity and dynamics of the bioeconomy is to use a conceptual framework that distinguishes between different types or paradigms of bioeconomy based on their underlying assumptions, goals and strategies. For example, one can distinguish between:

- A **biotechnology-driven** bioeconomy that focuses on the development and application of advanced biotechnologies (such as genetic engineering, synthetic biology, biorefineries, etc.) to create new products and processes that enhance efficiency, productivity and competitiveness in various sectors. This type of bioeconomy is often associated with high levels of investment, research and innovation, as well as strong intellectual property rights and regulatory frameworks that support biotechnology development (Moustier et al., 2023).

- A **biomass-based** bioeconomy that focuses on the use of renewable biological resources (such as crops, forests, algae, etc.) to produce biofuels, bioplastics, biochemicals and other biobased products that substitute fossil-based materials and reduce greenhouse gas emissions. This type of bioeconomy is often associated with low levels of technological development and innovation, but high levels of resource consumption, land use change and environmental impacts (Lorenz & Lal, 2023).
- A **bioecology-driven** bioeconomy that focuses on the conservation and restoration of natural ecosystems and biodiversity as sources of ecosystem services, resilience and well-being for humans and other species. This type of bioeconomy is often associated with low levels of economic growth and material consumption, but high levels of social equity, environmental justice and ethical values.

These types of bioeconomy are not mutually exclusive or exhaustive. They can coexist or compete in different regions or sectors depending on the context and the objectives of the actors involved. Moreover, they can evolve over time as new technologies, markets, policies or values emerge or change ("Knowledge-Driven Developments in the Bioeconomy," 2017).

A possible hypothesis that can be derived from this framework is that the type of bioeconomy that prevails or emerges in a given region or sector depends on the interaction between three main factors: **technology**, **policy** and **society**. These factors can be represented by three variables: **T**, **P** and **S**, respectively. Each variable can take a value between 0 and 1, where 0 means low level or influence and 1 means high level or influence. For example, T = 0 means low level of technological development or innovation in the bioeconomy sector; T = 1 means high level of technological development or innovation in the bioeconomy sector.

Based on this hypothesis, one can formulate a formula that predicts the type of bioeconomy that prevails or emerges in a given region or sector based on the values of T, P and S. The formula can be expressed as follows:

$$B=f(T,P,S)=(T+P+S)/3$$

where B is the type of bioeconomy variable that takes a value between 0 and 1. The formula assumes that each factor has an equal weight in determining the type of bioeconomy. The formula can be interpreted as follows:

- If B = 0, then T = P = S = 0. This means low level or influence of technology, policy and society in the bioeconomy sector. This could indicate a lack of interest or awareness about the bioeconomy concept or potential in the region or sector.
- If B = 1/3, then T = P = S = 1/3. This means moderate level or influence of technology, policy and society in the bioeconomy sector. This could indicate a balanced or mixed approach to the bioeconomy concept or potential in the region or sector, where different types of bioeconomy coexist or compete.

- If B = 2/3, then T = P = S = 2/3. This means high level or influence of technology, policy and society in the bioeconomy sector. This could indicate a strong or dominant approach to the bioeconomy concept or potential in the region or sector, where one type of bioeconomy prevails or emerges.
- If B = 1, then T = P = S = 1. This means very high level or influence of technology, policy and society in the bioeconomy sector. This could indicate a radical or transformative approach to the bioeconomy concept or potential in the region or sector, where a new type of bioeconomy emerges or replaces the existing ones.

The formula can be further refined by assigning different weights to each factor depending on their relative importance or relevance in the context of the region or sector. For example, one can assign a higher weight to technology if the region or sector is highly dependent on biotechnology development and innovation; a higher weight to policy if the region or sector is highly regulated or influenced by bioeconomy policies and strategies; or a higher weight to society if the region or sector is highly sensitive or responsive to social values and preferences regarding the bioeconomy.

One can also introduce additional factors or variables that may affect the type of bioeconomy in the region or sector, such as market demand, resource availability, environmental conditions, historical legacy, etc. However, this would complicate the formula and require more data and assumptions to estimate their values and weights.

The formula is named as **S M Nazmuz Sakib's Bioeconomy Typology Formula (SBTF)**. This name reflects the authorship, purpose and content of the formula.

To illustrate how this formula can be applied in real world historical incidents related to The Role of Innovation in Driving the Bioeconomy, the Challenges and Opportunities, here are some examples:

- The **Green Revolution** that occurred in developing countries from the 1950s to the 1970s can be seen as an example of a biomass-based bioeconomy that emerged as a result of high levels of policy and market influence, but low levels of technology and society influence. The Green Revolution involved the introduction of high-yielding varieties of crops (such as wheat, rice and maize) that increased agricultural production and reduced food insecurity in many regions. However, these crops also required intensive use of fertilizers, pesticides and irrigation that increased environmental degradation and social inequality. Using the formula, one can estimate that T = 0.2, P = 0.8, S = 0.2 and M = 0.8 (where M is market demand). Therefore, B = (0.2 + 0.8 + 0.2 + 0.8)/4 = 0.5. This value indicates a moderate level of bioeconomy development that is biased towards a biomass-based type.
- The **Human Genome Project** that occurred in developed countries from 1990 to 2003 can be seen as an example of a biotechnology-driven bioeconomy that emerged as a result of high levels of technology and policy influence, but low levels of society and market influence. The Human Genome Project involved the sequencing and mapping of the human genome that opened new possibilities for biomedical research and innovation in fields such as genomics, pharmacogenomics, personalized medicine and gene therapy. However, these possibilities also raised ethical, social and legal issues regarding privacy, ownership, access and regulation of genetic information and technologies. Using the formula, one can estimate that T = 0.8, P = 0.8, S = 0.2 and M = 0.2 (where M is market demand). Therefore, B = (0.8 + 0.8 + 0.2 + 0.2)/4 = 0.5. This value indicates a moderate level of bioeconomy development that is biased towards a biotechnology-driven type.
- The **Industrial Revolution** that occurred in Europe and North America from the 18th to the 19th century can be seen as an example of a biomass-based bioeconomy that emerged as a result of

high levels of market and policy influence, but low levels of technology and society influence. The Industrial Revolution involved the transition from an agrarian and handicraft economy to an industrial and mechanized economy that increased production, trade and urbanization. However, this transition also relied heavily on the use of coal, wood and other biomass sources as fuels and raw materials that caused environmental pollution, deforestation and social unrest. Using the formula, one can estimate that T = 0.2, P = 0.8, S = 0.2 and M = 0.8 (where M is market demand). Therefore, B = (0.2 + 0.8 + 0.2 + 0.8)/4 = 0.5. This value indicates a moderate level of bioeconomy development that is biased towards a biomass-based type.

- The **Biotechnology Industry Organization** that was founded in the United States in 1993 can be seen as an example of a biotechnology-driven bioeconomy that emerged as a result of high levels of technology and market influence, but low levels of policy and society influence. The Biotechnology Industry Organization is a trade association that represents more than 1,000 biotechnology companies, academic institutions and research centers that are involved in the research and development of biotechnology products and services in various sectors such as health care, agriculture, industrial and environmental biotechnology. However, this association also faces challenges such as regulatory uncertainty, public skepticism, ethical controversies and global competition regarding biotechnology applications and impacts. Using the formula, one can estimate that T = 0.8, P = 0.2, S = 0.2 and M = 0.8 (where M is market demand). Therefore, B = (0.8 + 0.2 + 0.2 + 0.8)/4 = 0.5. This value indicates a moderate level of bioeconomy development that is biased towards a biotechnology-driven type.
- The **Earth Summit** that was held in Brazil in 1992 can be seen as an example of a bioecology-driven bioeconomy that emerged as a result of high levels of society and policy influence, but low levels of technology and market influence. The Earth Summit was a major international conference that brought together representatives from more than 170 countries to discuss and agree on various issues related to sustainable development, such as climate change, biodiversity conservation, poverty reduction and human rights. However, these issues also require coordination and cooperation among different stakeholders and countries regarding the implementation and monitoring of the agreements and commitments made at the summit. Using the formula, one can estimate that T = 0.2, P = 0.8, S = 0.8 and M = 0.2 (where M is market demand). Therefore, B = (0.2 + 0.8 + 0.8 + 0.2)/4 = 0.5. This value indicates a moderate level of bioeconomy development that is biased towards a bioecology-driven type.

(In the last examples, the formula was B=f(T,P,S)=(T+P+S)/3, but I added an additional factor M (market demand) to the formula to make it more realistic and relevant for the examples. Therefore, the formula became B=f(T,P,S,M)=(T+P+S+M)/4.)

Abbreviations

- CBD: Convention on Biological Diversity
- CBD: Convention on Biological Diversity
- GMO: Genetically Modified Organism
- SBTF: S M Nazmuz Sakib's Bioeconomy Typology Formula
- SME: Small and Medium-sized Enterprise

REFERENCES

AZoCleantech.com. (2022, October 12). *The Benefits of Climate-Smart Ecosystem Monitoring in Forests.* AZoClean Tech. https://www.azocleantech.com/article.aspx?ArticleID=1485

Barney, N., & Lewis, S. (2022). *Biotechnology (biotech).* Tech Target. https://www.techtarget.com/whatis/definition/biotechnology

De Jaramillo, E. H., & Schuler, I. (2014). Challenges of Sustainable Development and Productive Use of Biodiversity: The Knowledge Based Bio-Economy Approach. In Environmental P*rotection and Sustainable Development from* Rio to Rio, 20. doi:10.1163/9789004282919_019

Dries, L., Heijman, W., Jongeneel, R., Purnhagen, K. P., & Wesseler, J. (2019). EU Bioecon*omy Economics and Policies: Volume I. In Palgrave adva*nces in bioeconomy: economics and policies. Springer International Publishing. doi:10.1007/978-3-030-28634-7

Ebissa, G., Fetene, A., & Desta, H. (2023). Comparative analysis of managing plantation forests: The case of keeping plantation forests for carbon credit and industrial profits in Oromia Region, Ethiopia. Heliyon, 9(4), e15151. doi:10.1016/j.heliyon.2023.e15151 PMID:37095960

EEA. (n.d.). Designing safe and sustainable products requires a new approach for chemicals. European Environment Agency. https://www.eea.europa.eu/publications/designing-safe-and-sustainable-products-1

European Commission. (2023, June 23). Research and Innovation. Europea. https://research-and-innovation.ec.europa.eu/research-area/environment/bioeconomy_en

FAO. (n.d.). Introduction. FAO. https://www.fao.org/3/v4845e/V4845E02.htm

Fihlo, W., Brito, P., de Lima, I., & Pocivalisteanu, D. (2018). Towards a Sustainable Bioeconomy: Principles, Challenges and Perspectives. In World sustainability series. Springer International Publishing. doi:10.1007/97*8-3-319-73028-*8

Greenhouse Gases. (2023). US EPA. https://www.epa.gov/report-environment/greenhouse-gases

Hidayati, S., Hadi, S., Saputra, N., Iswanto, A. H., Fatriasari, W., Solihat, N. N., Antov, P., & Hua, L. S. (2023). Characterization of lignin isolated from oil palm empty fruit bunch using phosphoric acid. Biomass Conversion and Bio*refinery. doi:10.100713399-023-04381-9*

Hodgson, A., Maxon, M., & Alper, J. (2022). The U.S. Bioeconomy: Charting a Course for a Resilient and Competitive Future. Industrial Biotechnology (New Rochelle, N.Y.), 18(3), 115–136. doi:10.1089/ind.2022.29283.aho

House, W. (2022). Executive Order on Advancing Biotechnology and Biomanufacturing Innovation for a Sustainable, Safe, and Secure American Bioeconomy. The White House. https://www.whitehouse.gov/briefing-room/presidential-actions/2022/09/12/executive-order-on-advancing-biotechnology-and-biomanufacturing-innovation-for-a-sustainable-safe-and-secure-american-bioeconomy/

IRENA. (n.d.). Wo*rld Ene*rgy Transitions Outlook 2022. IRENA. https://www.irena.org/Digital-R*eport/World-Energ*y-Transitions-Outlook-2022

Kircher, M. (2022). The bioeconomy needs economic, ecological and social sustainability. AIMS Environmental Science, 9(1), 33–50. doi:10.3934/environsci.2022003

Knowledge-Driven Developments in the Bioeconomy. (2017). Economic complexity and evolution. *Springer International Publishing*., doi:10.1007/978-3-319-58374-7

Language selection. (n.d.). European Commission. https://ec.europa.eu/info/files/communication-european-green-deal_en

Lorenz, K., & Lal, R. (2023). Organic Agriculture and Climate Change. Springer eBooks. doi:10.1007/978-3-031-*17215*-1

Maryville University. (2022). Why Is Agriculture Important? Benefits and Its Role. Maryville Online. https://online.maryville.edu/blog/why-is-agriculture-important/

Molnar, C. (2015, May 14). 10.2 Biotechnology in Medicine and Agriculture. Pressbo*oks. https://op*entextbc.ca/biology/chapter/10-2-biotechnology-in-medicine-and-agriculture/

Moustier, P., Holdsworth, M., Anh, D., Seck, P. A., Renting, H., Caron, P., & Bricas, N. (2023). Science and Innovations for Food Systems Transformation. In Springer eBooks. doi:10.1007/978-3-031-15703-5

OECD. (n.d.*). Home. OECD. https://ww*w.*o*ecd-ilibrary.org/sites/bd16d851-en/index.html?itemId=/content/component/bd16d851-en

Sangtani, R., No*gueira, R., Yadav, A. K., & Kiran*, B. (2023). Systematizing Microbial Bioplastic Production for Developin*g Sustainable Bioe*conomy: Metabolic Nexus Modeling, Economic and Environmental Technologies Assessment. Journal of Polymers and the Environment, 31(7), 2741–2760. doi:10.100710924-023-02787-0 PMID:36811096

Show, P., Ogino, C., & Ibrahim, M. (n.d.). Biotechnology for *Sustainability and Social Well Being. Google* Books. https://books.google.com.pk/books?id=YD1AEAAAQBAJ&pg=PA156&lpg=PA156&dq=Biotechnology+is+a+key+component+of+the+bioeconomy+and+has+numerous+applications+in+various+fields+such+as+medicine,+agriculture,+and+industry.+Biotechnology+involves+the+use+of+biological+systems+and+living+organisms+to+produce+new+products+or+improve+*existing*+ones.+Biotechnology+has+the+potential+to+transform+numerous+industries+by+enabling+the+development+of+new+bio-based+products+and+processes.+One+of+the+most+promising+applications+of+biotechnology+is+in+the+field+of+medicine,+enabling+the+development+of+new+therapies+and+treatments+for+various+dise*ases.+Biotechnology+also+has+numerous+a*ppl*i*cations+in+agriculture,+such+as+the+development+of+genetically+modified+crops+that+are+resistant+to+pests+and+diseases.+However,+biotechnology+also+poses+sever*al+challenge*s+that+need+to+be+addressed+through+innovation,+such+as+the+potential+impact+of+GMOs+on+human+health+and+the+environment.&source=bl&ots=yvMWdS4WST&sig=ACfU3U0-rF_RwVp1dmiY8xmcmsN

Staff, C. B. (2022). In-depth Q&A: The IPCC's sixth assessment on *how to tackle climate chan*ge. Carbon Brief. https://www.carbonbrief.org/in-depth-qa-the-ipccs-sixth-assessment-on-how-to-tackle-climate-change/

US DoA. (n.d.). Biotechnology Frequently Asked Questions. USDA. https://www.usda.gov/topics/biotechnology/biotechnology-frequently-asked-questions-faqs

Von Braun, J. (2015). Bioeconomy. Oxford University Press eBooks. doi:10.1093/acprof:oso/9780198733201.003.0011

Wang, T., Yu, Z., Ahmad, R., *Riaz, S. S., Khan, K. U., Siyal, S.,* Chaudhry, M. A., & Zhang, T. (2022). Transition of bioeconomy as a key concept for the agriculture and agribusiness development: A*n extensive rev*iew on ASEAN countries. Frontiers in Sustainable Food Systems, 6, 998594. doi:10.3389/fsufs.2022.998594

Wild, S. (n.d.). What is the bioeconomy and how could it help fight climate change? Horizon Magazine. http*s://ec.europa.eu/*research-and-innovation/en/horizon-magazine/what-bioeconomy-and-how-could-it-help-fight-climate-*change*

*World Bank. (n.d.). Overview. World Bank. https://www.worl*dbank.org/en/topic/agriculture/overview

World Bank Group. (2022). Urgent Climate Action Crucial for Bangladesh to Sustain Strong Growth. World Bank. https:/*/www.worldbank.org/en/news/press-relea*se/2022/10/31/urgent-climate-action-crucial-for-bangladesh-to-sustain-strong-growth

Zhang, X., Zhao, C., Shao, M., Chen, Y., Liu, P., & Chen, G. (2022). The roadmap of bioeconomy in China. *Engineering Biology*, *6*(4), 71–81. doi:10.1049/enb2.12026 PMID:36968339

KEY TERMS AND DEFINITIONS

Bioeconomy: The use of renewable biological resources to produce goods and services, with the goal of reducing reliance on finite resources and minimizing environmental impacts.

Biotechnology: The use of biological systems and living organisms to produce new products or improve existing ones.

Bio-Based Products: Products that are derived from renewable biological resources, such as biomass, bioplastics, biofuels, and biochemicals.

Biomass: Organic material that can be used as a source of energy or as a raw material for bio-based products.

Bioplastics: A type of plastic made from renewable biological resources, such as corn starch, sugarcane, and potato starch.

Biofuels: Fuels that are derived from renewable biological resources, such as ethanol, biodiesel, and biogas.

Biochemicals: Chemicals that are derived from renewable biological resources, such as enzymes, amino acids, and vitamins.

Biomaterials: Materials that are derived from renewable biological resources, such as plant fibers, wood, and bioplastics.

Sustainable Development: A concept that seeks to balance economic growth with environmental sustainability and social progress.

Circular Economy: An approach that aims to minimize waste and maximize resource efficiency by keeping products and materials in use for as long as possible.

Innovation: The development and implementation of new technologies, processes, products, or services that create value for society.

Chapter 16
The Role of IoT and AI in Bioeconomy

Justino Lourenço
Polytechnic Institute of Gaya, Portugal

ABSTRACT

The new era brings a panoply of new challenges in terms of proactive procedures in order to achieve a better Europe's ecosystem. A set of new trend technologies such as artificial intelligence (AI) and internet of things (ioT) can and will bring a new approach to deal with our problem. The AI can contribute to sustainability in various ways, such as optimizing energy consumption, reducing waste, improving supply chain efficiency, and enabling precision agriculture. AI can help advance scientific research and innovation for sustainable development, such as developing new materials and technologies for clean energy production and storage. On the other hand, IoT can help in terms of sustainability by enabling real-time monitoring and control of energy and resource usage in buildings, transportation systems, and industrial processes. This chapter plans to start the discussion of the expected positive impacts in the short term from the adoption of AI and IoT for a more sustainable planet.

INTRODUCTION

The bioeconomy refers to the sustainable use of biological resources, such as plants, animals, and microorganisms, to produce food, energy, and materials. It is a crucial component of efforts to achieve a more sustainable future, as it provides a way to meet our growing needs while reducing our impact on the environment. So, the bioeconomy faces a number of significant challenges that must be addressed in order to ensure its success. These challenges include issues related to resource availability, sustainability, economic viability, and regulatory frameworks. In this context, it is important to explore the key challenges facing the bioeconomy and identify potential solutions to help unlock its full potential. This will require collaboration between governments, businesses, and other stakeholders, as well as a willingness to embrace innovative solutions and new technologies.

The bioeconomy is a rapidly growing sector that is playing an increasingly important role in addressing the global challenges of food security, energy security, and environmental sustainability. One of the

DOI: 10.4018/978-1-6684-8879-9.ch016

key drivers of this growth is the increasing use of digital technologies, such as artificial intelligence (AI) and the internet of things (IoT), which are enabling innovative approaches to the production, processing, and distribution of biological resources.

The AI has the potential to revolutionize the bioeconomy by providing new tools for data analysis, modeling, and prediction. For example, AI can be used to optimize crop yields by analyzing soil and weather data to predict the best time for planting and harvesting. It can also be used to develop new materials and chemicals by simulating the properties of biological molecules and predicting their behavior under different environmental conditions. As a last remark, AI can optimize waste usage and improve efficiency in the bioeconomy by optimizing supply chain logistics and reducing energy consumption.

The IoT is another key technology that is driving the growth of the bioeconomy. IoT devices can be used to collect data on everything from soil moisture levels to animal behavior, providing valuable insights into the biological processes that underpin the bioeconomy. This data can be used to optimize production processes, improve quality control, and reduce waste.

The blending of AI with IoT procides a powerful lever for the bioeconomy, enabling us to develop more sustainable and efficient approaches to food, energy, and material production. However, there are also challenges that must be addressed in order to fully realize the potential of these technologies. These include issues related to data privacy and security, regulatory frameworks, and the ethical implications of using AI and IoT in the bioeconomy.

As a last argument, the combination of AI and IoT is creating new opportunities for innovation and growth in the bioeconomy. By embracing these technologies and working together to address the challenges they present, we can unlock the full potential of the bioeconomy and help to build a more sustainable future.

BACKGROUND

The purpose of this chapter is to present the state-of-the-art and discuss the future positive impact in bioeconomy from the lever provided by AI and IoT. In this section, the AI and IoT current and past interventions in bioeconomy are presented.

The AI and IoT are relatively new technologies, but they have already had a significant impact on the bioeconomy. Over the past few decades, researchers and innovators have used these technologies to develop new approaches to food, energy, and material production, and to address some of the key challenges facing the bioeconomy.

One of the earliest applications of AI in the bioeconomy was in the field of genetics. In the 1980s, researchers began using AI to analyze genetic data and identify patterns that could be used to predict the properties of new biological materials (Schaffer, J. D. 1984). This led to the development of new crops and materials with improved properties, such as increased yield, disease resistance, and durability.

Similarly, IoT has been used in agriculture, with sensors and other devices used to monitor soil moisture, temperature, and other factors that can affect crop growth. By collecting and analyzing this data, farmers and other stakeholders can optimize their production processes and reduce waste, improving both efficiency and sustainability.

In recent years, AI and IoT have also been used to develop new approaches to bioenergy production. For example, AI has been used to develop algorithms that can optimize the production of biofuels from various feedstocks, while IoT devices have been used to monitor and control the various stages of the bioenergy production process, from feedstock collection to fuel distribution.

We can refer that, the use of AI and IoT in the bioeconomy has led to significant advancements in our understanding of biological systems and our ability to produce food, energy, and materials in a more sustainable and efficient manner. As these technologies continue to evolve, we can expect to see even more innovative approaches to bioeconomy challenges in the years to come.

METHODOLOGY

The application of IoT and AI as levers for sustainability has gained increasing attention in recent years. In this chapter, the following methodology was used as a guideline:

1. Identify sustainability goals: The first step is to identify the sustainability goals that the application of IoT and AI will address. This can include reducing carbon emissions, conserving natural resources, promoting circular economies, or improving social equity.
2. Review relevant state-of-the-art literature: to conduct a comprehensive bibliographic review to identify relevant literature on the application of IoT and AI for sustainability. This can include academic papers, reports, and case studies that highlight best practices, challenges, and opportunities.
3. Analyze existing IoT and AI state-of-the-art solutions: to analyze existing IoT and AI solutions that have been applied to address sustainability goals. This can involve identifying the technologies used, the data collected, and the impact on sustainability goals. For convenience, the two last subtopics were combined in the section bibliography review.
4. Identify gaps and opportunities: Based on the analysis, it is important to identify gaps and opportunities in the application of IoT and AI for sustainability. This can include identifying areas where IoT and AI can have the greatest impact or areas where further research is needed to address specific sustainability goals.
5. Develop a strategy and implementation plan: Once gaps and opportunities have been identified, it is important to develop a strategy and implementation plan that outlines how IoT and AI will be used to address sustainability goals. This can involve identifying the necessary resources, technologies, and stakeholders needed to implement the strategy.

In summary, a methodology supported by a bibliographic review can help ensure that the application of IoT and AI is aligned with sustainability goals. By following this methodology, organizations can identify gaps and opportunities, develop a strategy and implementation plan, and monitor progress towards sustainability goals.

OVERALL IMPACT OF AI AND IoT IN CARBON FOOTPRINT

Over the positive impact of AI and the IoT on the carbon footprint, they have the potential to significantly reduce the carbon footprint of various industries, leading to a more sustainable future.

One of the ways AI and IoT can reduce carbon emissions is by optimizing energy usage. AI algorithms and IoT sensors can collect data on energy usage patterns and provide insights into opportunities for optimization. For example, smart grids can be used to optimize energy distribution, reduce energy waste, and promote the use of renewable energy sources. Similarly, smart buildings can use IoT sensors to detect when rooms are unoccupied and adjust lighting and HVAC settings accordingly, reducing energy consumption.

In transportation, AI and IoT can also reduce carbon emissions. Self-driving cars and intelligent traffic management systems can optimize traffic flow, reduce congestion, and promote the use of electric and hybrid vehicles. IoT sensors can also be used to monitor vehicle performance and optimize routes, reducing fuel consumption and emissions.

In the logistics industry, IoT can optimize supply chain management, reducing waste and carbon emissions. By tracking the location and condition of goods in real-time, logistics companies can optimize transportation routes, reduce delivery delays, and minimize fuel consumption.

AI and IoT can also promote sustainable agricultural practices. For example, AI algorithms can analyze data on soil conditions and crop growth to optimize irrigation and fertilization, reducing water and chemical usage. IoT sensors can also monitor weather patterns and provide farmers with real-time information on climate conditions, allowing them to make more informed decisions about planting and harvesting.

Sol, AI and IoT have enormous potential to reduce carbon emissions and promote sustainable practices in various industries. By leveraging the power of AI and IoT, we can optimize energy usage, reduce transportation and logistics emissions, and promote sustainable agriculture, leading to a more sustainable future for all.

On the other hand, there can be, also, environmental negative impacts: While IoT and AI have the potential to reduce carbon emissions, there are also potential negative impacts that must be considered.

One of the negative impacts of IoT and AI on the carbon footprint is the energy consumption required to power the devices and infrastructure. The deployment of IoT sensors and devices requires energy, and if that energy is generated from non-renewable sources, it can contribute to carbon emissions. Similarly, the use of AI algorithms requires significant computing power, which can also consume energy and contribute to carbon emissions if not generated from renewable sources.

Another potential negative impact is the production and disposal of IoT devices. The production of electronic devices requires resources and energy, and the disposal of these devices can contribute to e-waste and environmental pollution. This can have a significant carbon footprint if not managed properly.

There is also the potential for unintended consequences from the deployment of IoT and AI systems. For example, optimizing transportation systems with IoT and AI may lead to an increase in overall transportation usage, leading to more carbon emissions. Similarly, optimizing agricultural practices may lead to an increase in monoculture practices, which can have negative environmental impacts.

Finally, there is the potential for privacy and security concerns related to the collection and use of personal data through IoT and AI systems. This can lead to ethical concerns related to the use of data and potential unintended consequences.

Summarizing, while IoT and AI have the potential to reduce carbon emissions, it is important to consider the potential negative impacts and work to mitigate them. This includes promoting the use of renewable energy sources, managing e-waste responsibly, considering unintended consequences, and addressing privacy and security concerns.

It is expected that the positive side of the AI and IoT, in the final balance, will be a very important lever for a more green and carbonless earth.

All of the above relevant described applications will be detailed in the next session.

BIBLIOGRAPHIC REVIEW

In this section a presentation about the state-of-the-art bibliography within the application of AI and IoT in bioeconomy is presented.

Remote Sensing in the Forest

The AI has the potential to greatly aid in the conservation and management of forests. With its ability to analyze vast amounts of data, AI can help us better understand the ecological processes that take place in forests, monitor and predict changes in forest ecosystems, and even assist in the management of forest resources.

One application of AI in forest management is through the use of remote sensing data. Satellites, drones, and other remote sensors can gather massive amounts of data about forest ecosystems, including information about vegetation structure, biomass, and carbon storage. AI algorithms can then be used to analyze this data, making it possible to map and monitor changes in forest cover and detect signs of deforestation, wildfires, and other threats to forest health.

AI can also be used to improve forest planning and management. Forest managers can use AI to develop models that simulate how different management strategies will affect forest ecosystems over time. These models can help identify the most effective strategies for achieving specific goals, such as maximizing timber yield, conserving biodiversity, or reducing carbon emissions.

In addition, AI can help in the detection and control of invasive species and pests, which can have a devastating impact on forest ecosystems. By analyzing data from sensors and cameras placed in the forest, AI algorithms can detect the presence of invasive species and track their spread. This information can then be used to develop targeted control strategies, such as the deployment of drones to spray pesticides in affected areas.

As the bibliographic review will describet, the potential applications of AI in forest conservation and management are vast, and the technology is still in its infancy. As we continue to develop and refine AI tools, we can expect to see even more sophisticated applications that will help us better understand and protect our forest ecosystems.

To develop a remote sensing solution that due to the high rise of solutions has promote an explosion in amount of data available. To have the support of AI will drive an automatic solution that provides all the forest properties of interest from the collected and processed datasets (Lines et al., 2022). A vast description about all the different tool and techniques that with AI lever, is used for forest monitoring process and the forest contribution to reducing the overall carbon footprint (Pandey et al., 2022). To understand, with the AI support, the best harvesting moment will result in a more efficient production in fruit trees (AdiVishnu et al., 2022). To understand certain relationship between forest fires and forestry production. To understand a certain relation between disease that affect the forest, pests and rodents and forest construction investment is of extreme importance (Ying et al., 2011; Annabel & Muthulakshmi, 2019). The approach of wireless sensor nodes (WSN) can be a solution for monitoring the occurrence of a fire in a forest (Pieris et al., 2022). A AQUI network in the Bordeaux-Aquitaine region (southwestern France) was created. This network enables sensors measuring soil moisture (SM) and temperature, at various depths, and the height of the groundwater table (Wigneron et al., 2018). after, almost a decade in orbit, ALOS-2/PALSAR-2, a huge contribution for the forest monitoring data archive has being achieved. The contribution was in forest research, forest cover mapping, biomass estimation, and deforestation monitoring, from better forest/non-forest maps and improved forest type classifications (Koyama et al., 2022). The unmanned aerial vehicle (UAV) has been used as a solution for forest fire monitoring, detection, and tracking (Zhang et al., 2018). The following tools: Deep Learning, Machine Learning, and Artificial Intelligence, and Image Processing are enablers of fire detection solutions. Several algorithms such as Genetic Algorithm, Fuzzy Logic, Decision Tree, Multilayer Perceptron (MLP), Artificial Neural Networks (ANN), Random Forest (RF), Spatial Wavelet Analysis for Colour Variations are possible implementations. The accuracy of such a solution is higher than 90% (Gaikwad et al., 2022).

AI in Water Resources

The AI technology has the potential to revolutionize the way we manage and conserve water resources. Water is a finite resource, and with increasing demands from a growing population, climate change, and environmental degradation, it is crucial to adopt sustainable water management practices.

AI can be used to analyze data and predict water usage patterns, which can help water utilities better manage their resources. For example, AI can be used to detect leaks in water distribution systems before they become major issues, thereby reducing water loss and preventing wastage. AI-powered sensors can also be deployed to monitor water usage in buildings and homes, providing insights on how water is being used and where improvements can be made.

In agriculture, AI can be used to optimize irrigation practices, ensuring that crops receive the right amount of water at the right time. This not only helps conserve water but can also increase crop yields and improve the overall efficiency of agricultural operations.

Another area where AI can be useful is in water quality monitoring. AI algorithms can be trained to detect patterns in water quality data and alert authorities to potential issues, such as harmful algal blooms or contamination events.

As stated, AI has the potential to transform the way we manage water resources, leading to more sustainable and efficient practices. By leveraging the power of AI, we can reduce water waste, conserve this precious resource, and ensure a more sustainable future for all.

An effective method of efficiently managing water supply (by pipelines) is crucial for maximizing water efficiency and preserving water quality for both human and agricultural use (Massaro et al., 2021; Mestre et al., 2022). Over industrial application the smart monitoring of water waste is presented. Many artificial intelligence (AI) techniques, including multivariate linear regression (MLR), artificial neural networks (ANNs), adaptive neuro-fuzzy inference system (ANFIS) and ensemble of ANNs and ANFIS, have been proposed to solve the problem (Fu et al., 2018). The support of automation, in for example agriculture, in a blended solution with AI will lead to efficient procedures in agricultural environments (Priscilla et al., 2023; Kumar & Shashank, 2022).

AI in Energy Supply

The AI can play a crucial role in helping to reduce energy consumption and promote energy efficiency. With energy demands continuing to increase, it is essential to find innovative solutions to manage our energy usage sustainably.

One way AI can be used to save energy is through building automation. AI-powered sensors can be deployed in buildings to monitor occupancy, temperature, lighting, and other factors. This information can be analyzed to optimize heating, ventilation, and air conditioning (HVAC) systems, ensuring that energy is only used when necessary and that buildings are comfortable for occupants.

Another application of AI in energy saving is through demand response programs. These programs use AI algorithms to predict electricity demand patterns and adjust energy usage accordingly. For example, during peak demand periods, AI can help utilities reduce energy usage in non-essential areas and redirect energy to critical areas, such as hospitals and emergency services.

AI can also be used to optimize renewable energy production. Solar and wind energy are intermittent, meaning they can fluctuate based on weather patterns. AI algorithms can analyze weather data and adjust energy production accordingly, ensuring that energy is produced when it is needed the most.

In the transportation sector, AI can help reduce energy consumption through optimization of routes and driving behavior. AI algorithms can analyze traffic patterns and suggest the most efficient routes for vehicles, reducing fuel consumption and emissions. AI-powered systems can also monitor driver behavior and provide feedback on ways to drive more efficiently, such as reducing speeding and acceleration.

As a summary, AI has enormous potential to help reduce energy consumption and promote energy efficiency. By leveraging the power of AI, we can optimize energy usage, promote sustainable energy practices, and ensure a more sustainable future for all.

Assuring an efficient and optimized power consumption of infrastructure is of extreme relevance in the quest for sustainable environment. The solution can be applied to Data Centers, thar as core of many data services in all the organizations has a big impact over power consumption (Z. Zhang et al., 2022). In a similar form, the telecommunications industry has some energy saving concerns over the 5G architecture. To enable power savings in the Radio Access Network (RAN) is presented in (Tan et al., 2022; Zhirong Zhang et al., 2022), in a similar form the concept can be applied to 5G Base Stations (Rumeng et al., 2021). The infrastructure temperature control represents, also, a relevant portion of the energy consumption. On the other hand the quality of life and productivity is closed related to the temperature. An AI solution, as described in the bibliographic reference, is an efficient solution (Lu et al., 2022). On the other hand buildings are responsible for 33% of final energy consumption, and 40% of direct and indirect CO 2 emissions globally. The AI can be applied to a new method that analyzes several factors including electricity price, uncertainty in climatic factors, availability of renewable energy sources (wind and solar), energy consumption patterns in buildings, and the non-linear relationships between these parameters on hourly, daily, weekly and monthly intervals. The combination of the deep learning forecasting, energy storage, and scheduling algorithm enables reducing annual energy import from the grid by 84% (Nabavi et al., 2021). To add an automation with AI can also improve energy efficiency (Pranee & Jirasuwankul, 2021; Kwon et al., 2022).

AI in Transportatiom

Artificial Intelligence (AI) has the potential to revolutionize the transportation industry, making it safer, more efficient, and more sustainable. From self-driving cars to intelligent traffic management systems, AI is transforming the way we move people and goods.

One of the most significant applications of AI in transportation is self-driving cars. Autonomous vehicles use AI algorithms and sensors to perceive their surroundings and navigate safely on the road. Self-driving cars have the potential to reduce accidents caused by human error, improve traffic flow, and increase mobility for people who are unable to drive.

AI can also be used to optimize transportation systems. Intelligent traffic management systems use AI algorithms to analyze traffic patterns and adjust traffic signals in real-time. By optimizing traffic flow, these systems can reduce congestion, improve safety, and reduce emissions from idling vehicles.

In the logistics industry, AI can be used to optimize shipping routes and reduce transportation costs. AI algorithms can analyze shipping data, weather patterns, and other factors to determine the most efficient routes and modes of transportation, reducing fuel consumption and emissions.

Another application of AI in transportation is predictive maintenance. AI algorithms can analyze data from sensors and other sources to predict when maintenance is required, reducing downtime and ensuring that vehicles are operating at peak efficiency.

Summarizing, AI has enormous potential to transform the transportation industry, making it safer, more efficient, and more sustainable. By leveraging the power of AI, we can optimize transportation systems, reduce accidents, and promote sustainable transportation practices, leading to a more connected and sustainable future.

The AI is a field with a vast state-of-the-art contribution for a more efficient transportation system (Filatov et al., 2022). The approach of Automated Information Systems (AIS) can be a tool for the optimization of the transport system taking into consideration the citizens and organization mobility needs (Ilchenko et al., 2021), the AIS is also used as tool for maritime safety operation (Bao et al., 2020)..A structured taxonomy to work as support in the railway management applications with the support of AI is presented in (Besinovic et al., 2022). A Knowledge Discovery (KD) techniques, is presented, as means of (retrieving) data pre-processing for the development of Al-based decision support systems (DSSs) in the context of management of the maintenance in the context of Public transportation (Adamson et al., 2005). The Isolation-Based Anomalous Vessel Trajectory (IAVT) detection method with the support of AI can be used vessel trajectory detection this way enabling the implementation of more secure and efficient route planning (Huan et al., 2022); Zhou et al., 2013).

In more futuristic applications: the unmanned aerial vehicles (UAV) have received a growing attention due to their wide range of applications. The UAVs with the support of artificial intelligence (AI) capabilities will be more efficient in the transport task (Alshanbari et al., 2019). Also, as a reference: the "Smart Urban Mobility Management" (S.U.M.Ma) Project was designed to increase efficiency. The pilot project, was implemented in the Municipality of Artena, plans to design and test highly innovative tools such as drones, H-D (High Definition) smart cameras, IoT (Internet of Things) Sensors on 5G network and AI (Artificial Intelligence) tools, by integrating environmental datasets, information systems and local databases and taking into account the mobility of traffic flows, enabling this way a reduced carbon footprint threw smart tools implementation (D'Apuzzo et al., 2022).

IoT in the Forest

The IoT concept refers to a network of interconnected devices that are capable of collecting and transmitting data over the internet. In the context of forest conservation and management, IoT can be used to monitor and analyze data related to forest ecosystems and the impact of human activities on them.

One application of IoT in forest management is through the use of sensors to monitor environmental conditions. As an example, sensors can be used to measure temperature, humidity, soil moisture, and other variables that are important for understanding forest ecosystems. By collecting data from these sensors, forest managers can gain a better understanding of how environmental conditions affect forest health and make more informed decisions about management strategies. As an example: 85% losses in forest calamities are due to wild fire (Srividhya & Sankaranarayanan, 2020).

The IoT can also be used to monitor and track wildlife populations in the forest. For example, sensors can be used to detect the presence of animals, track their movements, and monitor their behavior. This information can be used to develop more effective conservation strategies, such as identifying areas where wildlife habitats are being threatened by human activities.

Another application of IoT in forest management is through the use of cameras and other sensors to monitor for illegal activities such as poaching, illegal logging, and forest fires. By using IoT technology to monitor for these activities in real-time, forest managers can respond quickly and effectively to protect the forest ecosystem and prevent further damage.

Finally, IoT can be used to improve the efficiency of forest management operations. For example, sensors can be used to monitor the performance of logging equipment, track the location of logging trucks, and optimize the use of resources such as fuel and water. By using IoT technology to improve operational efficiency, forest managers can reduce costs and minimize the environmental impact of their activities.

As a final consideration, the potential applications of IoT in forest conservation and management are vast. As we continue to develop and refine IoT technology, we can expect to see even more sophisticated applications that will help us better understand and protect our forest ecosystems.

The fire prevention is a huge concern, and the IoT is a relevant tool (N. Ghosh et al., 2021; Ishitha et al., 2021; Marcu et al., 2019). The prevention of forest fires can be achieved with the support of innovative practices of Wireless Sensor Networks (WSN) as described in (Ghosh et al., 2019). The IoT and Edge computing deployed in lot of smart city applications for processing the data closer to devise for quick action rather than cloud in scenarios of forest fire prevention (Srividhya & Sankaranarayanan, 2020).

IoT in Water Resources

The IoT is another technology that can be leveraged to help reduce water waste and improve water management. IoT refers to the network of physical devices, vehicles, buildings, and other objects that are embedded with sensors, software, and connectivity, allowing them to collect and exchange data.

One way IoT can be used to reduce water waste is through smart water meters. These devices can monitor water usage in real-time and provide insights into water consumption patterns. With this data, water utilities can identify areas where water is being wasted and take corrective measures to reduce water loss.

Another application of IoT in water management is through smart irrigation systems. These systems use sensors to monitor soil moisture levels, weather patterns, and other factors to determine when and how much water to apply to crops. By optimizing irrigation practices, these systems can reduce water usage and minimize water waste in agriculture.

IoT can also be used for water quality monitoring. Sensors can be deployed in water bodies to collect real-time data on parameters such as temperature, pH, dissolved oxygen, and other contaminants. This information can be used to identify potential sources of pollution and take corrective actions to prevent further contamination.

Summarizing, IoT has enormous potential to help reduce water waste and improve water management practices. By collecting real-time data and leveraging advanced analytics, IoT can enable more efficient and sustainable water usage, leading to significant water savings and a more sustainable future.

The management of all the resources related to water usage in an efficient form is of extreme importance, IoT, as previously described in this section, can be the necessary lever (Siddula et al., 2018). The optimal control of water waste, water leakage and water contamination is presented in (Chowdhury et al., 2022; (Krishna et al., 2020; (Munara et al., 2022; (Budiarti et al., 2019; (C. et al., 2019). A quick and prompt response to anomalies detection and consequent reaction is also very important and IoT blended with AI will give the push T(Gonzalez-Vidal et al., 2019). A greener solution must also contemplate the issue of water reuse (Dudhpachare et al., 2022). At last, water harvesting can also be implemented with the support of IoT (Ranjan et al., 2020).

IoT in Energy Supply

The IoT is a technology that can be used to help reduce energy consumption and promote energy efficiency. IoT devices can be embedded with sensors and software that can collect and analyze data, providing valuable insights into energy usage patterns and opportunities for optimization.

One way IoT can be used to save energy is through smart grids. IoT sensors can be deployed throughout the electricity grid to monitor energy usage in real-time. This information can be used to optimize energy distribution and reduce energy waste, ensuring that electricity is only used when it is needed the most.

In buildings, IoT can be used to optimize energy usage by monitoring lighting, HVAC systems, and other energy-consuming devices. For example, IoT sensors can detect when a room is unoccupied and adjust lighting and HVAC settings accordingly, reducing energy consumption. Similarly, IoT can be used to monitor appliance usage, alerting users to high-energy consuming devices and suggesting ways to reduce usage.

IoT can also be used to optimize energy usage in transportation. For example, smart traffic management systems can be used to reduce congestion, reducing fuel consumption and emissions. Similarly, IoT sensors can be used to optimize public transportation schedules, ensuring that buses and trains are operating at optimal times and locations.

Overall, IoT has enormous potential to help reduce energy consumption and promote energy efficiency. By leveraging the power of IoT, we can optimize energy usage, reduce energy waste, and promote sustainable energy practices, leading to significant energy savings and a more sustainable future for all.

The literature state-of-the-art over this topic shows special concern in several topics, as follows. The IoT can monitor and adapt the best possible light conditions in several applications (Dwiyaniti et al., 2022). To understand the correlation between the process plan in a factory and energy saving is a form of optimizing the power consumption (Choi & Yeom, 2018). IoT makes the operation of Smart Grid more reliable and burden free from the high demand periods (enables the use of stored energy in demand peaks). It also provides a much better environment to collect the power from distributed generation and fasten the use of operation in the smart grid environment using loT (M. Kumar et al., 2020). To implement smart plugs threw IoT that can be switched off when the plug load is not being used and activate it – when the load is expected to be working, will mean a relevant electric power save (Srinivasan et al., 2019). The approach of using IoT as a lever has a broad range of applications: telecommunications (Lin et al., 2011); agriculture (I.-H. Lu, 2022); House Hold Appliance (Shashank et al., 2021); Data center (Keskin & Soykan, 2019) and many others.

IoT in Transportation

The Internet of Things (IoT) is transforming the transportation industry, enabling real-time data collection, analysis, and optimization of transportation systems. By embedding sensors and connectivity into vehicles, infrastructure, and logistics systems, IoT is making transportation safer, more efficient, and more sustainable.

One of the most significant applications of IoT in transportation is real-time traffic management. IoT sensors can be deployed throughout transportation networks to collect data on traffic flow, vehicle location, and environmental conditions. This data can be analyzed to optimize traffic flow, reduce congestion, and improve safety on the roads.

IoT can also be used to optimize logistics and supply chain management. By embedding sensors into cargo containers and delivery vehicles, logistics companies can track the location and condition of goods in real-time, ensuring that they are delivered on time and in good condition. This can reduce transportation costs, minimize delivery delays, and improve customer satisfaction.

Another application of IoT in transportation is predictive maintenance. By analyzing data from sensors and other sources, IoT algorithms can predict when maintenance is required, reducing downtime and ensuring that vehicles are operating at peak efficiency. This can also improve safety by identifying potential problems before they cause accidents.

In the automotive industry, IoT is transforming the way we drive. Connected cars can communicate with other vehicles, infrastructure, and data sources to provide real-time information on traffic conditions, weather, and navigation. This can improve driver safety, reduce fuel consumption, and provide a more enjoyable driving experience.

Overall, IoT is revolutionizing the transportation industry, making it safer, more efficient, and more sustainable. By leveraging the power of IoT, we can optimize transportation systems, reduce accidents, and promote sustainable transportation practices, leading to a more connected and sustainable future.

The bibliography review demonstrates the research of IoT as a lever in efficiency in the transportation system as described in: IoT architectures for passengers transport (Ma et al., 2019); urban traffic (Zhu et al., 2020); IoT blended with AI (Cheng et al., 2021); in context of a smart city (Shukla & Champaneria, 2017).

FUTURE RESEARCH DIRECTIONS

The presented work clarifies the extreme relevance of the AI and IoT as a lever for a more sustainable earth. As a future research directions, they are the following:

1. Advancing the use of renewable energy sources for IoT and AI systems: One way to reduce the carbon footprint of IoT and AI systems is to power them using renewable energy sources, such as solar or wind power. Future research can focus on developing more efficient and cost-effective renewable energy technologies to power these systems.
2. Developing more efficient algorithms for IoT and AI systems: As IoT and AI systems become more complex, the algorithms used to power them must become more efficient to reduce energy consumption. Future research can focus on developing new algorithms that require less computational power and can run on low-power devices.

3. Improving IoT sensors for more accurate data collection and more power save: IoT sensors are critical for collecting data that is used to power AI algorithms. Future research can focus on developing more accurate, less power consuming and more reliable sensors that can collect data on a wide range of environmental factors, such as air quality, temperature, and humidity.
4. Designing IoT and AI systems for circular economies: Future research can focus on designing IoT and AI systems that are built with the principles of circular economies in mind. This means creating systems that are designed to be reused, repurposed, or recycled at the end of their life cycle, reducing waste and contributing to a more sustainable future.
5. Promoting collaboration and open data sharing: Collaboration and open data sharing can help researchers and organizations work together to create more sustainable IoT and AI systems. Future research can focus on promoting collaboration between different stakeholders, including researchers, policymakers, and industry leaders, to promote innovation and sustainability.

The future research in IoT and AI for sustainability will need to focus on developing more efficient and sustainable systems that can help reduce carbon emissions and promote more sustainable practices in various industries.

CONCLUSION

The issue of a better environment for a better and balanced ecosystem is of special concern.

The AI brings a new (smart) layer in the process of more efficiency that will mean more energy savings. The smart approach that can be applied to all our personal devices, our home, our office, our city and, of course, our earth. Actually it is possible to collect several datasets about how are we affecting the environment (IoT is just one support technology that enables it) and we need intelligent agents that make an intensive analysis of all the data and gives us the best approaches to make all the procedures more green. Even in real time solutions, AI will be the lever for sustainability. On the other hand, a huge effort should be focused on the AI algorithms and hardware to achieve a power optimization. The AI is the solution for our problem and should not be a strong carbon footprint contributor.

The IoT enables the process of collecting data: understanding the medium with its data collection, enabling a total understanding of how and what is happening at a specific instant -and in real time. This is the best tool for starting an optimization procedure. As the AI, also IoT needs, in particular, at the sensor level to be optimized in terms of carbon footprint, as described in previous section.

Also, as demonstrated in previous sections – the AI and IoT blending – gives us the tool to read the data and make a smart optimization, always, with our focus in: more efficient systems and practices leas to less electric power usage.

The bibliographic review made it clear the broad range of devices and systems where the referred AI blend with IoT.

The main gap in the evolutionary path is on political choices in Europe. In a decisive moment of the automotive electrification process there is no strong public policies that bring a positive incentive to electrical self-generated solutions. The IoT along with AI tools in the quest for efficient should and must be an everyday concern to achieve the defined carbon footprint goals.

REFERENCES

Adamson, K., Campbell, P., & Orsoni, A. (2005). Hybrid decision support based on knowledge discovery and AI techniques for the management of maintenance services in the public transport sector. *2005 IEEE Intelligent Data Acquisition and Advanced Computing Systems: Technology and Applications*, (pp. 674–678). IEEE.

AdiVishnu. S., Nagaraju, V., & Senthil Murugan, P. (2022). Prediction of mango fruit ripening in different stages using CNN over random forest. *2022 14th International Conference on Mathematics, Actuarial Science, Computer Science and Statistics (MACS)*, (pp. 1–4). IEEE.

Alshanbari, R., Khan, S., El-Atab, N., & Mustafa Hussain, M. (2019). AI powered unmanned aerial vehicle for payload transport application. *2019 IEEE National Aerospace and Electronics Conference (NAECON)*, (pp. 420–424). IEEE. 10.1109/NAECON46414.2019.9058320

Annabel, L. S. P., & Muthulakshmi, V. (2019). AI-powered image-based tomato leaf disease detection. *2019 Third International conference on I-SMAC (IoT in Social, Mobile, Analytics and Cloud) (I-SMAC)*, (pp. 506–511). IEEE. 10.1109/I-SMAC47947.2019.9032621

Bao, K., Shang, D., Wang, R., & Ma, R. (2020). AIS big data framework for maritime safety supervision. *2020 International Conference on Robots & Intelligent System (ICRIS)*, (pp. 150–153). IEEE. 10.1109/ICRIS52159.2020.00045

Besinovic, N., De Donato, L., Flammini, F., Goverde, R. M. P., Lin, Z., Liu, R., Marrone, S., Nardone, R., Tang, T., & Vittorini, V. (2022). Artificial intelligence in railway transport: Taxonomy, regulations, and applications. *IEEE transactions on intelligent transportation systems: a publication of the IEEE Intelligent Transportation Systems Council, 23*(9), 14011–14024. https://doi.org/ doi:10.1109/TITS.2021.3131637

Budiarti, R. P. N., Tjahjono, A., Hariadi, M., & Purnomo, M. H. (2019). Development of IoT for automated water quality monitoring system. *2019 International Conference on Computer Science, Information Technology, and Electrical Engineering (ICOMITEE)*, (pp. 211–216). IEEE. 10.1109/ICOMITEE.2019.8920900

C., R., V., T., Liloja, & Shahzad, M. (2019). IOT based water quality monitoring with android application. *2019 Third International conference on I-SMAC (IoT in Social, Mobile, Analytics and Cloud) (I-SMAC)*, (pp. 446–451). IEEE.

Cheng, D., Li, C., & Qiu, N. (2021). The application prospects of NB-IoT in intelligent transportation. *2021 4th International Conference on Advanced Electronic Materials, Computers and Software Engineering (AEMCSE)*, (pp. 1176–1179). ACM.

Choi, H.-S., & Yeom, K.-R. (2018). Study on an energy-IoT service platform for energy saving in legacy manufacturing site. *2018 Tenth International Conference on Ubiquitous and Future Networks (ICUFN)*, (pp. 811–813). IEEE. 10.1109/ICUFN.2018.8436969

Chowdhury, A. A. S., Arafat, Y., & Alam, M. S. (2022). IoT-GSM based controlling and monitoring system to prevent water wastage, water leakage, and pollution in the water supply. *2022 International Conference on Innovations in Science, Engineering and Technology (ICISET)*, (pp. 567–572). IEEE. 10.1109/ICISET54810.2022.9775876

D'Apuzzo, M., Cappelli, G., Buzzi, S., & Nicolosi, V. (2022). Smart Urban Mobility Management project: a concrete step towards more sustainable and connected communities. *2022 Second International Conference on Sustainable Mobility Applications, Renewables and Technology (SMART)*, (pp. 1–8). 10.1109/SMART55236.2022.9990394

Dudhpachare, A. K., Kuthe, T. V., Lalke, C. V., Wyawahare, N. P., Agrawal, R., & Daigavane, P. (2022). Process of RO's waste water reuse & water management in society by using IOT automation. *2022 International Conference on Applied Artificial Intelligence and Computing (ICAAIC)*, (pp. 1546–1550). IEEE. 10.1109/ICAAIC53929.2022.9792805

Dwiyaniti, M., & Kusumaningtyas, A. B. Silawardono, Kusumastuti, S. L., Saefullah, Y., Tohazen, Tireno, F. Q., & Ramadhan, D. (2022). Energy saving of IoT-based light intensity on smart streetlight. *2022 5th International Conference of Computer and Informatics Engineering (IC2IE)*, (pp. 6–9). IEEE.

Filatov, V. V., Klimenko, V. A., Karelina, E. A., Ptitsyn, D. A., & Erpulev, V. A. (2022). Theoretical aspects of the creation of artificial intelligence systems in transport. *2022 Systems of Signals Generating and Processing in the Field of on Board Communications*, (pp. 1–5). IEEE.

Fu, Z., Cheng, J., Yang, M., & Batista, J. (2018). Prediction of industrial wastewater quality parameters based on wavelet de-noised ANFIS model. *2018 IEEE 8th Annual Computing and Communication Workshop and Conference (CCWC)*, (pp. 301–306). IEEE.

Gaikwad, A., Bhuta, N., Jadhav, T., Jangale, P., & Shinde, S. (2022). A review on forest fire prediction techniques. *2022 6th International Conference On Computing, Communication, Control And Automation (ICCUBEA)*, (pp. 1–5). IEEE.

Ghosh, M., Sushil, R., & Ghosh, K. (2019). Detecting and reporting forest fire through deployment of three dimensional multi-sink wireless sensor network. *2019 4th International Conference on Internet of Things: Smart Innovation and Usages (IoT-SIU)*, (pp. 1–5). IEEE.

Ghosh, N., Kumar, B., Biswas, T., Patnaik, S., & Paul, R. (2021). IoT fog based framework to predict forest fire. *2021 Smart City Challenges & Outcomes for Urban Transformation (SCOUT)*, (pp. 256–259). IEEE.

Gonzalez-Vidal, A., Cuenca-Jara, J., & Skarmeta, A. F. (2019). IoT for water management: Towards intelligent anomaly detection. *2019 IEEE 5th World Forum on Internet of Things (WF-IoT)*, (pp. 858–863). IEEE.

Huan, Y., Kang, X., Wang, Y., & Wang, Y. (2022). IAVT: Anomalous vessel trajectory detection using AIS data. *2022 5th International Conference on Artificial Intelligence and Big Data (ICAIBD)*, 17–22.

Ilchenko, L. M., Sokolov, S. S., Egorova, K. V., Gubernatorov, O., & Knish, T. (2021). Analysis of the main types of automated information systems in the transport sector, functioning mainly in urban agglomerations. *2021 IEEE Conference of Russian Young Researchers in Electrical and Electronic Engineering (ElConRus)*, (pp. 401–406). IEEE. 10.1109/ElConRus51938.2021.9396630

Ishitha, S., Nagaraju, S., Mohan, H. A., Harshitha, M., & Gowda, G. R., & Jeevan. (2021). IoT based anti-poaching and fire alarm system for forest. *2021 IEEE Mysore Sub Section International Conference (MysuruCon)*, (pp. 711–715). IEEE. 10.1109/MysuruCon52639.2021.9641539

Keskin, I., & Soykan, G. (2019). Reliability improvement and energy saving at internet data center microgrids. *2019 IEEE International Conference on Environment and Electrical Engineering and 2019 IEEE Industrial and Commercial Power Systems Europe (EEEIC / I&CPS Europe)*, (pp. 1–5). IEEE. 10.1109/EEEIC.2019.8783781

Koyama, C. N., Watanabe, M., Hayashi, M., Shimada, M., Kobayashi, K., & Tadono, T. (2022). Advancements in global forest monitoring research founded on ALOS-2 long-term pantropical land observation. *IGARSS 2022 - 2022 IEEE International Geoscience and Remote Sensing Symposium*, (pp. 4445–4448). IEEE.

Krishna, S. Kumaraswamy, M. S., & Nair, V. (2020). IoT based Water Parameter Monitoring System. *2020 5th International Conference on Communication and Electronics Systems (ICCES)*, (pp. 1299–1303). IEEE.

Kumar, G., & Shashank, K. V. (2022). Smart farming based on AI, edge computing and IoT. *2022 4th International Conference on Inventive Research in Computing Applications (ICIRCA)*, (pp. 324–327). IEEE.

Kumar, M., Minai, A. F., Khan, A. A., & Kumar, S. (2020). IoT based energy management system for smart grid. *2020 International Conference on Advances in Computing, Communication & Materials (ICACCM)*, (pp. 121–125). IEEE. 10.1109/ICACCM50413.2020.9213061

Kwon, K., Lee, S., & Kim, S. (2022). AI-based home energy management system considering energy efficiency and resident satisfaction. *IEEE Internet of Things Journal*, *9*(2), 1608–1621. doi:10.1109/JIOT.2021.3104830

Lin, C., Wang, J., & He, Q. (2011). Internet of Things technology for energy saving of telecom operators. *IET International Conference on Communication Technology and Application (ICCTA 2011)*. IET.

Lines, E. R., Allen, M., Cabo, C., Calders, K., Debus, A., Grieve, S. W. D., Miltiadou, M., Noach, A., Owen, H. J. F., & Puliti, S. (2022). AI applications in forest monitoring need remote sensing benchmark datasets. *2022 IEEE International Conference on Big Data (Big Data)*, (pp.4528–4533). IEEE. 10.1109/BigData55660.2022.10020772

Lu, I.-H. (2022). Research on energy-saving efficiency of the air-conditioning host in agricultural plant bacteria room. *2022 IEEE 5th Eurasian Conference on Educational Innovation (ECEI)*, (pp. 384–387). IEEE.

Lu, P., Liu, Y., & Zhou, J. (2022). AI intelligent control method for energy saving of HVAC equipment in smart parks based on deep learning. *2022 4th International Academic Exchange Conference on Science and Technology Innovation (IAECST)*, (pp. 1560–1564). IEEE.

Ma, J., Feng, S., Li, X., Zhang, X., & Zhang, D. (2019). Research on the internet of things architecture for intelligent passenger transportation services and its application. *2019 4th International Conference on Electromechanical Control Technology and Transportation (ICECTT)*, (pp. 194–197). IEEE.

Marcu, A.-E., Suciu, G., Olteanu, E., Miu, D., Drosu, A., & Marcu, I. (2019). IoT System for Forest Monitoring. *2019 42nd International Conference on Telecommunications and Signal Processing (TSP)*, (pp. 629–632). IEEE.

Massaro, A., Panarese, A., & Galiano, A. (2021). Technological platform for hydrogeological risk computation and water leakage detection based on a convolutional neural network. *2021 IEEE International Workshop on Metrology for Industry 4.0 & IoT (MetroInd4.0&IoT)*, (pp. 225–230). IEEE. 10.1109/MetroInd4.0IoT51437.2021.9488561

Mestre, G., Matos-Carvalho, J. P., & Tavares, R. M. (2022). Irrigation management system using artificial intelligence algorithms. *2022 International Young Engineers Forum (YEF-ECE)*, (pp. 69–74). IEEE.

Munara, M., Kumar, N., & Shanmugam, K. (2022). Recommending IoT based real-time water quality monitoring system in Malaysia. *2022 IEEE 2nd Mysore Sub Section International Conference (MysuruCon)*, (pp. 1–5). IEEE.

Nabavi, S. A., Motlagh, N. H., Zaidan, M. A., Aslani, A., & Zakeri, B. (2021). Deep learning in energy modeling: Application in smart buildings with distributed energy generation. *IEEE Access : Practical Innovations, Open Solutions*, *9*, 125439–125461. doi:10.1109/ACCESS.2021.3110960

Pandey, R., Srivastava, D., Dizayee, R. H., Lobanova, A., Rabbi, F., & Sutkar, P. R. (2022). Enhancing the accuracy of forest monitoring through AI to reduce carbon footprint. *2022 4th International Conference on Inventive Research in Computing Applications (ICIRCA)*, (pp. 1177–1182). IEEE.

Pieris, T. P. D., Kulasekera, A. L., & Chathuranga, K. V. D. S. (2022). Multi-AI based wireless sensor node for forest fire detection. *2022 7th International Conference on Information Technology Research (ICITR)*, 1–6.

Pranee, P., & Jirasuwankul, N. (2021). Experimental study and modeling of automatic home energy management system using AI. *2021 International Conference on Power, Energy and Innovations (ICPEI)*, (pp. 103–106). IEEE. 10.1109/ICPEI52436.2021.9690647

Priscilla, D., & Pandi. (2023). Agriculture-based Automation with Recommendation Systems based on AI Models. *2023 Third International Conference on Artificial Intelligence and Smart Energy (ICAIS)*, (pp. 1582–1589). IEEE.

Ranjan, V., Reddy, M. V., Irshad, M., & Joshi, N. (2020). The internet of things (IOT) based smart rain water harvesting system. *2020 6th International Conference on Signal Processing and Communication (ICSC)*, (pp. 302–305). IEEE.

Rumeng, T., Tong, W., Ying, S., & Yanpu, H. (2021). Intelligent energy saving solution of 5G base station based on artificial intelligence technologies. *2021 IEEE International Joint EMC/SI/PI and EMC Europe Symposium*, (pp. 739–742). IEEE. 10.1109/EMC/SI/PI/EMCEurope52599.2021.9559261

Shashank, A., Vincent, R., Sivaraman, A. K., Balasundaram, A., Rajesh, M., & Ashokkumar, S. (2021). Power Analysis of Household Appliances using IoT. *2021 International Conference on System, Computation, Automation and Networking (ICSCAN)*, (pp. 1–5). 10.1109/ICSCAN53069.2021.9526428

Shukla, S. N., & Champaneria, T. A. (2017). Survey of various data collection ways for smart transportation domain of smart city. *2017 International Conference on I-SMAC (IoT in Social, Mobile, Analytics and Cloud) (I-SMAC)*, (pp. 681–685). 10.1109/I-SMAC.2017.8058265

Siddula, S. S., Babu, P., & Jain, P. C. (2018). Water level monitoring and management of dams using IoT. *2018 3rd International Conference On Internet of Things: Smart Innovation and Usages (IoT-SIU)*, (pp. 1–5). IEEE.

Srinivasan, A., Baskaran, K., & Yann, G. (2019). IoT based smart plug-load energy conservation and management system. *2019 IEEE 2nd International Conference on Power and Energy Applications (ICPEA)*, (pp. 155–158). IEEE.

Srividhya, S., & Sankaranarayanan, S. (2020). IoT–fog enabled framework for forest fire management system. *2020 Fourth World Conference on Smart Trends in Systems, Security and Sustainability (WorldS4)*, (pp. 273–276). IEEE. 10.1109/WorldS450073.2020.9210328

Tan, R., Shi, Y., Fan, Y., Zhu, W., & Wu, T. (2022). Energy saving technologies and best practices for 5G radio access network. *IEEE Access : Practical Innovations, Open Solutions*, *10*, 1–1. doi:10.1109/ACCESS.2022.3174089

Wigneron, J.-P., Dayan, S., Kruszewski, A., & Aluome, C. AI-Yaari, M. G.-E. A., Fan, L., Guven, S., Chipeaux, C., Moisy, C., Guyon, D., & Loustau, D. (2018). The aqui network: Soil moisture sites in the "Les Landes" forest and graves vineyards (Bordeaux Aquitaine region, France). *IGARSS 2018 - 2018 IEEE International Geoscience and Remote Sensing Symposium*, (pp. 3739–3742). IEEE.

Ying, Z., Ai-jun, Y., Jian-wei, J., & Ji, L. (2011). Study on measurement model for forest risk assessment in Beijing. *2011 International Conference on Computer Distributed Control and Intelligent Environmental Monitoring*, (pp. 1750–1755). IEEE. 10.1109/CDCIEM.2011.494

Zhang, L., Liu, Z., Zhang, Y., & Ai, J. (2018). Intelligent path planning and following for UAVs in forest surveillance and fire fighting missions. *2018 IEEE CSAA Guidance, Navigation and Control Conference (CGNCC)*, (pp. 1–6). IEEE.

Zhang, Z., Wei, M., Xu, X., Hu, C., & Xie, W. (2022). Intelligent energy saving technology and strategy of 5G RAN. *2022 IEEE International Symposium on Broadband Multimedia Systems and Broadcasting (BMSB)*, (pp. 1–6). IEEE. 10.1109/BMSB55706.2022.9828681

Zhang, Z. Zeng, Y., Liu, H., Zhao, C., Wang, F., & Chen, Y. (2022). Smart DC: An AI and digital twin-based energy-saving solution for data centers. *NOMS 2022-2022 IEEE/IFIP Network Operations and Management Symposium*, (pp. 1–6). IEEE.

Zhou, M., Chen, J., Ge, Q., Huang, X., & Liu, Y. (2013). AIS data based identification of systematic collision risk for maritime intelligent transport system. *2013 IEEE International Conference on Communications (ICC)*, (pp. 6158–6162). IEEE. 10.1109/ICC.2013.6655590

Zhu, F., Lv, Y., Chen, Y., Wang, X., Xiong, G., & Wang, F.-Y. (2020). Parallel transportation systems: Toward IoT-enabled smart urban traffic control and management. *IEEE transactions on intelligent transportation systems: a publication of the IEEE Intelligent Transportation Systems Council, 21*(10), 4063–4071. doi:10.1109/TITS.2019.2934991

KEY TERMS AND DEFINITIONS

Internet of Things (IoT): A network of physical objects, such as sensors and devices, that are connected to the internet and can collect and exchange data.

Artificial Intelligence (AI): The simulation of human intelligence processes by computer systems, such as machine learning and natural language processing, to analyze data and make decisions.

Sustainability: Meeting the needs of the present without compromising the ability of future generations to meet their own needs. This includes balancing economic, social, and environmental considerations.

Smart Cities: Cities that use technology, including IoT and AI, to improve sustainability, efficiency, and quality of life for residents.

Circular Economy: An economic system that is designed to minimize waste and promote the reuse and recycling of resources.

Renewable Energy: Energy sources that are replenished naturally and have a minimal impact on the environment, such as solar, wind, and hydropower.

Energy Efficiency: The process of using less energy to perform the same task or achieve the same outcome.

Greenhouse Gas (GHG) Emissions: Gases that trap heat in the earth's atmosphere and contribute to climate change, such as carbon dioxide and methane.

Carbon Footprint: The total amount of GHG emissions produced by an individual, organization, or product, expressed in units of carbon dioxide equivalent.

Life Cycle Assessment (LCA): A methodology for assessing the environmental impact of a product or service throughout its entire life cycle, from raw material extraction to disposal.

Heating, Ventilation, and Air Conditioning (HVAC): This refers to the system that enables, from home to offices, the effective control of the temperature.

5G: The mobile network known in its fifth generation that offers faster data transfer speeds, lower latency, and increased connectivity for a wide range of devices.

Automated information systems (AIS): This refers to computer-based methods and tools used to collect, process, store, and distribute data and information within an organization or enterprise.

AI-DSS: Computer programs that use artificial intelligence techniques to help individuals or organizations make better decisions.

Unmanned Aerial Vehicle (UAV): An aircraft without a human pilot aboard that is controlled remotely or programmed to fly autonomously.

Chapter 17
Towards Sustainable Development:
A Multi-Criteria Assessment of the Circular Economy in the European Union

Milan Marković
https://orcid.org/0000-0002-9617-6697
Innovation Center, University of Niš, Serbia

ABSTRACT

In a sustainable society, which countries strive for, a special place belongs to the circular economy. The circular economy offers certain solutions that tend to make development more sustainable by increasing resource productivity and reducing waste. The European Union is a leader in the world in implementing the principles and practices of regenerative economy and circularity. The chapter aims at a multi-criteria analysis of circular economy performance by ranking the countries of the European Union, taking into account the created composite index. Aggregation was performed using the grey relational analysis method, while the weight coefficients were determined using the entropy method. The results of the research show that the countries of Western Europe are the most successful in achieving the goals of the circular economy on the way to sustainable development, and that the Netherlands ranks first among them.

INTRODUCTION

Limited resources represent a fundamental economic problem. If all resources were unlimited, there would be no economics as a science. The economy strives to enable the achievement of social well-being for future generations through rational management of resources. Forcing economic growth, without respecting responsibility in the management of raw materials and other resources, is unsustainable. Global challenges, such as difficult energy imports, rising prices of production inputs, and unfavorable macroeconomic trends, create additional pressure to move from the traditional model of the linear economy to the advanced model of the circular economy. The circular economy is a relatively new model (concept

DOI: 10.4018/978-1-6684-8879-9.ch017

or paradigm) of social development. It was established because the linear economy model did not fully answer the questions concerning the sustainable use of resources.

Strong economic, technological and climate changes require transformations in the model on which economic development has been based until now (Androniceanu et al., 2021). We live in a society where environmental challenges are the central issue of every strategy for the sustainable development. In this line, the problem of waste generation and climate change was particularly highlighted. Today, the circular economy realizes its effects on social and economic development, although the importance of this concept is often neglected in these spheres because it has a dominant impact on reducing environmental pollution. In this sense, the circular economy is a regenerative model that promotes environmental protection but also the creation of a responsible society.

The aim of the research within this chapter is multiple. First, the objective is to evaluate and measure the degree of development of the circular economy on the example of the European Union. For that, carefully selected indicators (criteria or attributes) from the relevant database (Eurostat) at the level of the European Union were used. Then, the author will try to determine the relation with the Human Development Index, as one of the more important indicators of sustainable socio-economic development. The author's intention is to find out whether highly developed countries achieve high circular economy performance. We know that in these countries there are high allocations for the practice of the circular economy because developed societies, as a rule, have greater financial opportunities and interest in a regenerative economy both at the macro level (economic policy makers) and at the micro level (population, business entities). Additionally, the author's intention is to show the importance of incorporating the circular economy as an element of sustainable socio-economic development. This paper should open a discussion on the possible inclusion of the circular economy in the consideration of sustainable human development. By connecting environmental and socio-economic issues, it is possible to debate the overall sustainable development and thereby eliminate the deficiency of the Human Development Index in covering the ecological dimension of human development.

Therefore, the subject of the research is the evaluation of the circular economy at the level of the European Union from a macro aspect. In doing so, indicators of the circular economy from an ecological, economic and social point of view were used (Productivity of resources, Municipal waste recycling rate, Circular material utilization rate, Investments in sectors related to the circular economy, Gross added value of sectors in the area of circular economy, Share of renewable energy sources in total consumption and Consumption of raw materials). A ranking of countries will be carried out and it will be established which countries are the most successful and which are lagging behind in the application of the circular economy principles. Countries with the highest composite index will be identified as those with the best potential to preserve the economic ecosystem.

The importance, justification and originality of the research on this topic stems from several reasons: *i*) the application of a relatively recent method of multi-criteria analysis, *ii*) the use of that method in the field of social sciences in order to assess the performance of the circular economy based on the newly formed composite index, *iii*) the topicality and popularity of the research problem, primarily in the countries of the European Union.

The chapter consists of several sections. The first section talks about the innovative concept of the circular economy. The purpose of this part is to show the link with sustainable economic, social and environmental development. The second section describes the research methodology and identifies the variables for evaluating the circular economy in the European Union. The third section describes the results after applying the constructed circular economy composite index. Next, the author referred to the recommendations for the future development of the circular economy in the European Union starting from the practice of the Netherlands, as the most developed country in this sense. The subject of the fourth section is looking at the justification of including the circular economy in the concept of sustainable (human) development and determining their connection. At the end of the chapter, the basic conclusions of the research are given, as well as some limitations that the author faced in the research.

CIRCULAR ECONOMY AS PART OF SUSTAINABLE DEVELOPMENT

Based on a thorough review of scientific literature, innovation, competitiveness and circular economy stand out as the most important issues of the current paradigm of sustainable development in the research community. All these economic parameters are interrelated and depend on each other. On the way to sustainable development, the concept of circular economy was initiated, which found an alternative to the dominant linear economy. There are many factors that have encouraged the emergence of the circular economy:

- limited resources,
- unsustainable, extensive use of resources,
- dependence on the import of raw materials,
- energy dependence of developed European countries.

The unsustainability of the increasing use of resources imposes the need to replace the existing linear economic model. The framework of the old concept was based on the relationship: take, produce, use and throw away. On the one hand, it creates the risk of a lack of resources in the future, and on the other hand, it leads to the accumulation of waste that impairs the quality of the environment and the entire ecosystem.

Development based only on the efficient use of resources cannot be sustainable, especially in light of the generation of large amounts of waste from various sectors, primarily industry. Well-being can decline over time despite high and dynamic rates of economic growth. Namely, environmental problems and degradation of natural resources can occur at any stage of economic development and make further development difficult. It should be emphasized that the introduction of circular economy practices does not automatically mean the shutdown of the linear economy. Only the complete transition to the circular economy will mark the end of the linear economy.

The model of the linear economy, as the primary economic model that represents an alternative to circularity, carries with it certain costs (Banaitė, 2016):

- non-use of waste in the next production cycle means the lost value of the material,
- furthermore, costs for the procurement of raw materials increase,
- increased waste from production has a negative impact on the environment, due to possible climate changes and the greenhouse effect.

Circularity as an idea and concept was initiated at the end of the twentieth century. Research works and debates on this topic intensified at the beginning of the twenty-first century, while practices for the implementation of the circular economy have been increasingly discussed in the last decade. A particular concern of scientists, theoreticians and practitioners was the high growth rate of waste per capita. It is considered that the concept of circular economy was first introduced in the literature by economists Pearce and Turner (1990) in order to connect the basic economic functions of the environment: natural beauty, resource base for the economy, waste disposal and the system in which the ecosystem survives (Andersen, 2007). It is an approach that changes the conventional model of the linear economy, which viewed economic performance in isolation from the principle of recirculation of natural resources and reduction of their consumption (Corona et al., 2019). In addition to reducing consumption, the principles of the circular economy dictate that resources are retained in the system as long as possible (Karstensen et al., 2020).

The circular economy often views waste not as a problem, but as a resource (Lee et al., 2017). It is a concept that allows waste (as the final result of some production) to be used as a raw material in another production from the same or different activity. That is why the circular economy relies on recycling, reuse and waste reduction (Marković et al., 2023). Waste reduction is based on its use as a raw material, thus establishing the idea of "zero waste" (Li et al., 2018). Reducing the use of limited resources can minimize pollution and degradation of the natural environment, which enables the preservation of biodiversity (Marković et al., 2023a). All this indicates that the circular economy can significantly facilitate the path of a certain country towards sustainable development. Ecosystem degradation reduces the ability of the next generations to meet their needs (Marković et al., 2023a), which demonstrates the relevance of the circular economy as an attribute of the sustainable development paradigm.

Sustainable development is proposed as an aspiration to balance the goals of economic growth, environmental preservation and social well-being. It encompasses the economic, environmental and social dimensions of sustainability. All these dimensions are interconnected and intertwined. For example, a higher gross domestic product allows a country to invest in environmental protection and reducing regional disparities. Furthermore, the lower level of social differences and poverty creates a more coherent society in which it is possible to promote the idea of solving environmental problems. Also, ecological development requires innovations in the production process and technical progress, which increases the rate of economic growth. The development of companies in the field of waste industry creates new jobs, and if these jobs are in rural and less developed areas, it contributes to the social component of sustainable development by reducing poverty. Poverty reduction enables a higher percentage of tertiary education attendance, which again affects socio-economic development. Sustainability implies the harmonization of economic, ecological and social aspects.

"The final goal of sustainable development in the modern economy should be the creation of economically viable production, which will not disturb the environment and which will ensure overall socio-economic development" (Marković & Marjanović, 2022, p. 146). It contains 17 goals that touch all segments of society (United Nations Sustainable Development Goals): *i)* SDG1 - No Poverty, *ii)* SDG2 - Zero Hunger, *iii)* SDG3 - Good Health and Well-being, *iv)* SDG4 - Quality Education, *v)* SDG5 - Gender Equality, *vi)* SDG6 - Clean Water and Sanitation, *vii)* SDG7 - Affordable and Clean Energy, *viii)* SDG8 - Decent Work and Economic Growth, *ix)* SDG9 - Industry, Innovation and Infrastructure, *x)* SDG10 - Reduced Inequality, *xi)* SDG11 - Sustainable Cities and Communities, *xii)* SDG12 - Responsible Consumption and Production, *xiii)* SDG13 - Climate Action, *xiv)* SDG14 - Life Below Water, *xv)* SDG15 - Life on Land, *xvi)* SDG16 - Peace and Justice Strong Institutions, and *xvii)* SDG17 - Partnerships to achieve the Goal.

The aim of this part is to point out the similarities and differences between the circular economy and the economy of sustainable development. When looking at the similarities between sustainable development and the circular economy, it can be pointed out that they are global models based on (Geissdoerfer et al., 2017):

- observation of non-economic factors of development,
- innovations,
- multidisciplinarity,
- a wide circle and necessary cooperation of stakeholders - involved interested parties at various hierarchical levels,
- technological solutions that will be implemented by private companies,
- business models and innovations that can be the backbone of the transformation of the industry in order to solve the problem of environmental pollution.

Sauvé et al. (2016) argue that sustainable development refers to broader intergenerational goals, while the circular economy is mostly concerned with environmental sustainability (it is connected with a favorable economic system, but does not have social goals directly in mind). Circular economy is definitely a narrower term than sustainable development. It relies more on a bottom-up approach, in contrast to the concept of sustainability, which is based on a top-down approach. Also, the concept of circular economy was explored much later.

There is an increasing number of research on the circular economy and its nexus with the goals of sustainable development. By its nature, the circular economy is oriented towards the goals of sustainable development, among which it is closer to SDG12 - Responsible consumption and production (Rodriguez-Anton et al., 2019). In the European Union, the circular economy is integrated in almost every directive, initiative or regulation related to sustainable development. There is a direct link with at least six sustainable development goals, among which stand out: SDG7 - Affordable and clean energy (the relationship is achieved through energy saving, use of renewable energy sources, reduction in the use of non-renewable energy sources and higher energy productivity), SDG9 - Industry, Innovation and Infrastructure (circular economy is oriented towards innovations in production, investments in research and development and reduction of emissions of harmful gases originating from the industrial sector), SDG11 - Sustainable cities and communities (here the circular economy is linked to sustainable development by forcing the recycling of municipal waste), SDG12 - Responsible consumption and production (circular economy means reducing the consumption of raw materials and increasing the circularity of materials), SDG13 - Climate action (connectivity is realized through lower emissions of harmful gases), and SDG14 - Life below water (the common component with the circular economy is the reduction of waste, which has a positive effect on the ecosystem below water). The circular economy can be seen as a concept that is related to many sustainable development goals, so there is a tendency to evaluate it as a means to achieve sustainability. First of all, there is a relation with environmental sustainability. However, as the circular economy has its impact on employment, economic growth, reduction of regional differences, and overall social well-being, in a broader sense it is connected to absolutely all pillars of sustainable development. Like sustainable development, the circular economy is based on the principles of intergenerational solidarity. The goal is for resources to be available to future generations in the same or higher quantity and quality that is available to the current generation. This will be ensured through extending the life of materials and other resources, as well as through alternative reuse and recycling (Kirchherr et al., 2017).

Morseletto (2020) identifies three elementary principles underlying the circular economy:

- reduction of raw materials as inputs in the production process and reduction of waste,
- saving resources within the system as long as possible,
- at the end of the product's life cycle, it is necessary to include them in the system again.

This concept indicates the possibility of reducing the unnecessary consumption of resources, primarily materials, water and energy. In this way, water, air and soil pollution is prevented and the negative effects of climate change are stopped. Resource consumption is reduced because resources remain in production/economy as long as possible. Therefore, such production is often based on repair, reuse, recycling, eco-design, etc. The circular economy emphasizes not only the recycling of materials, but also the creation of added value through the development of services and intelligent solutions (Krysovatyy et al., 2018). This ensures stable rates of economic growth while conserving non-renewable resources.

Industry and agriculture have the greatest potential for disrupting ecological sustainability, but they can also contribute the most to the circular economy by introducing new, innovative solutions in the production process. The increase in the world's population has increased food production, which, due to uneven distribution in the world, has implied increased throwing and waste generation (Marković & Marjanović, 2022). Industrial ecology was established with the aim of preventing the excessive generation of waste from the industrial sector. On the other hand, agriculture also has a negative impact on environmental sustainability. Many agricultural activities lead to ecosystem degradation and reduce regenerative capacity due to excessive land exploitation and unsustainable water management practices, which lead to changes in soil quality, cause soil erosion and environmental pollution (Falkenmark et al., 2007). Soil degradation due to the increasing depletion of natural resources (e.g. forests) is a growing problem because it impairs the quality of the soil and calls into question the resilience of the ecosystem and, ultimately, the sustainable development of agriculture (Marković & Marjanović, 2022). The circular economy is directed towards the regeneration of natural resources because the biological capacity of water, soil, forests, etc. is preserved.

The environmental benefits of the circular economy are obvious. Preservation of ecological sustainability, biodiversity and non-renewable resources are key environmental benefits. However, this concept also has wider, socio-economic effects. On the macro level, new solutions in the circular economy can enable the structural transformation of the economy. The development of the market for secondary raw materials and the expansion of enterprises dealing with waste recycling and other key determinants of the circular economy can create conditions for reducing unemployment, increasing profits, developing the waste industry and accelerating economic growth and development. There are also secondary, positive effects that are reflected in the reduction of import dependence on raw materials, which improves the situation in the balance of payments. Social benefits are reflected in the increase of social well-being, the improvement of the health status of the population, as well as the survival of humanity in the last instance. Also, there is a possibility to reduce the differences in the regional development of the country, through the construction of companies (in areas facing abandonment) that will recycle industrial waste.

The fundamental goal of applying the circular economy is to build a sustainable economic ecosystem that would contribute to industrial development, citizens and the preservation of the ecological ecosystem (Corona et al., 2019). Certain prerequisites are necessary for the successful implementation of the circular economy: new business models, responsible consumers, development of industrial parks. The industry sector has a leading role in the transformation of the economy towards a circular economy, while policy makers are responsible for creating strategies and action plans, as well as promoting this concept. Researchers should further contribute to proving the justification of applying circular economy practices in sustainable human societies. This would significantly increase the awareness of consumers and producers about the need for circularity of materials in production, consumption and society as a whole.

In the world, the concept of circular economy has been promoted by many enterprises at the micro level, and at the national level by governments (Korhonen et al., 2018). In the countries of the European Union, there are many strategies, action plans, directives and policies that increasingly advocate the implementation of the principles and practices of the circular economy. It is the most developed group of countries where socio-economic and environmental goals are equally important. In Europe, there are a solid number of indicators against which the circular economy can be measured. There is a tendency of increasing interest among scientists aimed at evaluating the performance of the circular economy in the European Union. The largest expansion of research is focused on circular economy models in the European Union, Japan and China, which stand out as countries that were among the first to apply circular economy practices, while China was the first to use the term circular economy in the context of waste reduction (Berg et al., 2018). Many economists argue that China has reached a high level of economic development as a result of circular economy actions (Geng et al., 2013).

METHODOLOGY AND DATA

Composite indices are aggregate indicators of certain social phenomena. They are an ideal tool for a unified view of a particular phenomenon that can be evaluated on the basis of multiple indicators or dimensions. Composite indices are especially interesting if a problem has several different aspects, such as sustainable development. They are mainly used in the comparison of countries and their ranking in order to determine the most successful countries according to the degree of progress in various social and economic topics. The measurement of complex problems is much simpler through a single aggregate indicator obtained by applying a certain method of multi-criteria analysis. Their goal is to replace a set of interrelated variables with a single index that will best represent the different dimensions depicted by the individual indicators.

In recent years, there have been many studies that evaluated the level of realization of the circular economy. There are also numerous indicators of the circular economy. They refer not only to ecological, but also to other pillars of sustainable development. Quantifying progress in circular economy practices is often done through composite indices. In this way, it is easy to compare countries according to the performance of the circular economy, as well as the movement over time. Without adequate monitoring of the degree of development of the circular economy, it is not possible to achieve progress in terms of certain performances. There is an increasing number of research that focuses on the assessment of the circular economy using multi-criteria analysis methods (Saidani et al., 2019; Garcia-Bernabeu et al., 2020; Stanković et al., 2021; Ūsas et al., 2021; Ahmed et al., 2022; Milanović et al., 2022; Marković et al., 2023; Vranjanac et al., 2023). Measuring the success of countries in achieving a certain goal is often analyzed using multi-criteria decision-making methods. Intensification of research on the circular economy through different methods of aggregation has been developing for the last few years. Technique for Order of Preference by Similarity to Ideal Solution (TOPSIS), Data Envelopment Analysis (DEA), and PROMETHEE are the main methods used in previous studies.

In the process of designing the aggregate index, carefully selected indicators from the Eurostat database for the last available year were used. Circularity was measured for all 27 member states of the European Union. The composite index that will be created based on the Grey relational analysis can be adapted to any country that keeps national accounts related to the circular economy and responsible use of resources. The European Union, as an advanced economic integration, collects comprehensive indicators because it has set these goals high on the scale of social goals. The investigation will determine which is the best-ranked country, as a member state that has the best practice of implementing the concept of circularity in the European Union. In addition to simplicity of interpretation, the author proposes a composite indicator that can be used in non-European Union countries, which are faced with a limited number of indicators that describe the potential of the circular economy. In this light, less developed countries can more easily assess the state of their circular economy compared to highly developed European Union countries.

In the decision-making process, experts first make a decision about which weighting method they will apply. Subjective methods, objective methods and an equal weighting approach are available. To determine the weight coefficients of the variables, the author applied the Entropy method. It is a weighting method that is often used as an objective method. In fact, it allows weight coefficients not to be based on the subjective views of decision makers (Zhu et al., 2020), which is the main drawback of subjectively oriented methods in obtaining weight coefficients. Subjective views of experts are formed on the basis of previous experience and subjective assessment of the importance of variables, so that possible bias is avoided with objective methods. Finally, compared to the equal weighting method where all criteria have equal relative importance, the Entropy method is a more complex approach based on the mutual contrast between selected indicators in a multi-criteria model.

Therefore, the Entropy method in determining the weighting coefficients is based on the measurement of the intensity of the contrast between the values of the variables. For that, it is necessary to conduct the calculation through four phases (Hussain & Mandal, 2016; Stanković et al., 2021a; Vaid et al., 2022):

Step 1. At the beginning of the calculation, it is necessary to normalize the value of each of the indicators x_{ij} as follows:

$$r_{ij} = \frac{x_{ij}}{\sum_{i=1}^{n} x_{ij}} \tag{1}$$

where x_{ij} is the value of the jth criterion of the ith alternative.

Step 2. In this step, in accordance with the normalized criteria values, the entropy value e_j is determined based on the following formula:

$$e_j = -k\sum_{i=1}^{n} r_{ij} \ln r_{ij}, j = 1, 2, \ldots m \tag{2}$$

where $k = \frac{1}{\ln n}$ is a constant. (3)

Step 3. Obtaining the degree of divergence d_j by applying the next relation:

$$d_j = 1 - e_j, j = 1, 2, \ldots m \tag{4}$$

where d_j symbolizes the measure of contrast intensity of the indicator C_j.

The essence of the contrast intensity measure is that if there is a greater degree of disagreement between the initial values of the analyzed indicators, it means that the contrast intensity is greater, so that indicator has a higher relative importance in the concrete multi-criteria analysis.

Step 4. In the last step, additive normalization is applied in order to obtain the final values of the weighting coefficients of the criteria:

$$w_j = \frac{d_j}{\sum_{j=1}^{m} d_j} \tag{5}$$

In this study, the Grey Relational Analysis method was used to create an aggregate index. It belongs to a group of compensatory multi-criteria methods. It is an approach whose mathematical framework can be applied in any branch of science. So far, this method of multi-criteria analysis has been mostly applied in computer science, architecture, agriculture and some social sciences (Marković, 2022). It is a technique proposed by Deng Ju-Long in 1982 (Ju-Long, 1982; Lin & Lin, 2002). As an optimization technique, it can be used in all situations for the construction of composite indices when there are many criteria that are often in conflict (Özkaya, 2022). More specifically, Grey Relational Analysis is based on the analysis of the relationship of similarity or dissimilarity between the reference series and the alternative series (Hamzaçebi & Pekkaya, 2011). This method significantly facilitates the making of various managerial decisions because it is simple to apply (Wu, 2006). Before solving a problem or ranking alternatives, it is essential that the decision makers carry out the following procedure:

Step 1. Formation of the research problem to be solved.

Step 2. Selecting indicators (variables) and downloading their values from the database.

Step 3. Choice of weighting method.

After that, the next steps are taken in calculating the composite index based on the Grey Relational Analysis method (Jozić et al., 2015; Abifarin & Ofodu, 2022; Marković, 2022; Özkaya, 2022):

Step 1. The first step implies the normalization of the indicator values, where there are the following types of normalization:

i) “higher-is-better”- for the benefit (revenue) criterion:

$$x_{ij} = \left(\frac{y_{ij} - \min\left(y_{ij}\right)}{\max\left(y_{ij}\right) - \min\left(y_{ij}\right)} \right) \tag{6}$$

or

ii) “smaller-is-better” - for the cost criterion:

$$x_{ij} = \left(\frac{\max(y_{ij}) - y_{ij}}{\max(y_{ij}) - \min(y_{ij})} \right) \tag{7}$$

Step 2. In the next step, the goal is to determine the Grey relational coefficient based on the following equation:

$$\gamma(X_{0j}, X_{ij}) = \frac{(\Delta_{min} + \xi\Delta_{max})}{(\Delta_{ij} + \xi\Delta_{max})} \tag{8}$$

where:

$$\Delta_{ij} = |X_{0j} - X_{ij}| \tag{9}$$

$$\Delta_{min} = min\{\Delta_{ij}, i = 1, 2, \ldots, m; j = 1, 2, \ldots, n\} \tag{10}$$

$$\Delta_{max} = max\{\Delta_{ij}, i = 1, 2, \ldots, m; j = 1, 2, \ldots, n\} \tag{11}$$

where ξ - is distinguishing coefficient which usually has a value of 0.50. (12)

Step 3. This step enables the calculation of the Grey relational grade (degree) based on the weighting coefficients of the criteria:

$$\Gamma(X_0, X_i) = \sum_{j=1}^{n} w_{j\gamma}(x_{0j,} x_{ij}) \tag{13}$$

where:

$$\sum_{j=1}^{n} w_{ij} = 1 \tag{14}$$

Step 4. Based on the Grey relational grade value, a ranking can now be performed, with the best alternative being the one with the highest value of the composite index obtained by this technique. Accordingly, the interpretation is conducted based on the order of the alternatives from the highest to the lowest value.

The purpose of the multi-criteria model proposed in this study is to assess the degree of development of the circular economy in all 27 member countries of the European Union. The author selected seven indicators from the Eurostat database: *Resource productivity* (in Euro per kilogram), *Recycling rate of municipal waste* (in %), *Circular material use rate* (in %), *Private investment related to circular economy sectors* (in % of the Gross domestic product), *Gross added value related to circular economy sectors* (in % of the Gross domestic product), *Share of renewable energy in gross final energy consumption* (in %) and *Raw materials consumption* (in tonnes per capita). All data are from the last year for which there is an indicator value in the database. Table 1 shows the selected variables (indicators), the description of the variables, as well as the year to which the values of the variables refer.

RESULTS OF THE ANALYSIS AND DISCUSSION

Table 2 shows the descriptive statistics of the attributes that will make up the composite indicator. The table lists the minimum and maximum values of the variables, then the mean values, as well as the standard deviations. The first six indicators are revenue criteria, while the last indicator is the cost criterion. It is noted that the Netherlands is the country that has by far the most favorable values for three indicators in the European Union (*Resource productivity*, *Circular material use rate*, and *Raw materials consumption*). On the opposite side, Greece is the only country in the European Union that achieves the worst values for two indicators (*Private investment related to circular economy sectors*, and *Gross added value related to circular economy sectors*). The last one in the table is the statistical data from the size of the standard deviation of the indicator, which shows that the biggest deviation from the mean value was recorded in the case of *Recycling rate of municipal waste.*

The next stage in the development of the model is the calculation of weight coefficients. As already mentioned, the Entropy method was applied for this purpose. Table 3 reveals the obtained weighting coefficients of the criteria. The data show that the greatest relative importance has the *Circular material*

Table 1. Definitions of variables

Indicators	Descriptions	Year
Resource productivity (in Euro per kilogram)	Productivity of resources represents the ratio of gross domestic product to domestic consumption of materials. Domestic consumption of materials refers to the amount of materials used by the economy. As one of the essential components of the circular economy, the goal is to achieve greater production while simultaneously reducing the amount of resources for that production.	2021
Recycling rate of municipal waste (in %)	This indicator is obtained as a percentage of recycled municipal waste in relation to total municipal waste.	2020
Circular material use rate (in %)	The rate of circularity of materials shows the percentage of circular use of materials in relation to the total use of materials.	2021
Private investment related to circular economy sectors (in % of the Gross domestic product)	Private investments in the area of circular economy sectors (Gross investment in tangible goods) which include "the recycling sector, repair and reuse sector and rental and leasing sector" (in percentage of gross domestic product)	2021
Gross added value related to circular economy sectors (in % of the Gross domestic product)	Value added at factor costs (in percentage of gross domestic product), which is calculated when operating subsidies and indirect taxes are subtracted from gross income from business activities.	2021
Share of renewable energy in gross final energy consumption (in %)	Percentage of renewable energy sources according to the Renewable Energy Directive	2021
Raw materials consumption (in tonnes per capita)	Consumption of raw materials describes the amount of raw materials required for the production of goods in relation to the demand in a certain geographical area, regardless of the country in which the materials were extracted	2020

Source: (Eurostat, 2023)

Table 2. Descriptive statistics

Indicators	N	Minimum	Maximum	Mean	Std. Deviation
Resource productivity (in Euro per kilogram)	27	0.3459 (Bulgaria)	5.8811 (Netherlands)	1.92	1.24
Recycling rate of municipal waste (in %)	27	10.9 (Malta)	70.3 (Germany)	39.82	14.44
Circular material use rate (in %)	27	1.4 (Romania)	33.8 (Netherlands)	9.32	7.19
Private investment related to circular economy sectors (in % of the Gross domestic product)	27	0.1 (Greece)	1.4 (Austria. Belgium)	0.69	0.32
Gross added value related to circular economy sectors (in % of the Gross domestic product)	27	0.5 (Greece)	2.9 (Ireland. Malta)	1.69	0.53
Share of renewable energy in gross final energy consumption (in %)	27	11.735 (Luxembourg)	62.573 (Sweden)	24.52	12.00
Raw materials consumption (in tonnes per capita)	27	7.745 (Netherlands)	33.617 (Finland)	18.13	6.76

Source: (Author's calculation)

Table 3. Criteria weights (entropy method)

Indicators	Weights
Resource productivity (in Euro per kilogram)	0.2215
Recycling rate of municipal waste (in %)	0.0832
Circular material use rate (in %)	0.3013
Private investment related to circular economy sectors (in % of the Gross domestic product)	0.1324
Gross added value related to circular economy sectors (in % of the Gross domestic product)	0.0588
Share of renewable energy in gross final energy consumption (in %)	0.1238
Raw materials consumption (in tonnes per capita)	0.0789

Source: (Author's calculation)

use rate (30.13%), while the lowest value of the weight coefficient was obtained for *Gross added value related to circular economy sectors* (5.88%).

Now all the preconditions have been created for the application of the aggregation method and the calculation of the circular economy composite index. Table 4 displays the values of the composite index based on the Grey Relational Analysis method, as well as the multi-criteria ranking of all the countries of the European Union.

At the beginning of the analysis, it could be assumed that the Netherlands would be the most successful country because it was the leader in three out of seven circular economy indicators. At the top are the countries of Western Europe to the greatest extent, except for Italy and Malta. On the other hand, Romania is the weakest positioned country, taking into account the results of the conducted research. Dogaru (2020) emphasizes that in Romania, according to statistical data, as much as 25% of food production is either lost or wasted. It is important to point out that Finland, as one of the more developed and richer countries of the European Union, is almost at the bottom of the list according to the performance of the

Table 4. Composite index of the circular economy indicators (grey relational analysis)

Country	Composite index value	Rank	Country	Composite index value	Rank
Netherlands	0.8058	1	Croatia	0.4395	15
Belgium	0.5805	2	Latvia	0.4363	16
Austria	0.5549	3	Slovakia	0.4263	17
Italy	0.5392	4	Czech Republic	0.4219	18
Germany	0.5256	5	Poland	0.4204	19
France	0.5219	6	Lithuania	0.4180	20
Ireland	0.4887	7	Portugal	0.4179	21
Malta	0.4837	8	Hungary	0.4156	22
Sweden	0.4802	9	Greece	0.3987	23
Luxembourg	0.4642	10	Finland	0.3983	24
Spain	0.4607	11	Bulgaria	0.3905	25
Denmark	0.4519	12	Cyprus	0.3746	26
Slovenia	0.4446	13	Romania	0.3590	27
Estonia	0.4411	14			

Source: (Author's calculation)

circular economy. New member states of the European Union must be in focus in the coming period. Due to lower economic performance, they will need additional support in this area from European funds.

The circular economy has been placed among the priorities of the 2030 Agenda for Sustainable Development of the United Nations. In the European Union, the circular economy is the focus of the European Green Deal from 2019, although the beginnings are related to 2014 and 2015 (Stankevičius et al., 2020). The main goals of the European Green Deal are the reduction of pollution and the preservation of biodiversity, both through sustainable resource management. The Paris Agreement from 2015 stipulated limiting the increase in temperature below two degrees Celsius in relation to the average temperature that figured in the pre-industrial period. At that time, a more powerful transition from a linear to a circular economy was actualized. Another reason for this is that European countries are very dependent on energy and other resources as important factors of economic development. Also, about 75% of the total energy production comes from non-renewable sources, so resources are significantly extracted from the natural environment and processed (Korhonen et al., 2018). It is necessary that the part that remains as waste be return back into the ecosystem (Korhonen et al., 2018).

The basic goal of circularity is to reduce dependence on resources, and to return more waste to production. Returning resources does not have to be in the same industry; waste can also be forwarded to other industries. This reduces the demand for raw materials and prevents the consequences of excessive exploitation of natural resources (Stanković et al, 2020). In the European Union, it has been estimated that the circular economy could increase resource productivity by up to 3% annually (Berg et al., 2018). In this way, the goals of preserving the quality of soil, water and air become achievable, and through this relationship, the circular economy can most effectively affect sustainable development.

The permanent support of the government and legislation is of inestimable importance for the further strengthening of the circular economy in the European Union. Macroeconomic policy holders must provide both financial and professional assistance to small and medium-sized enterprises to fully transition to business models that will be based on the principles of circularity. The problem for implementing the principles of the circular economy can be the high costs of the transition from the linear to the circular economy, both for economic subjects at the micro level and for the state at the macro level (investments in innovative technology and research and development, promotion at the level of human society to change the awareness of the population in urban and rural areas, staff training, innovative business models require the development of waste sorting technology, etc.)

The Europe 2020 strategy also mandates respect for the principles of circular economy, sustainable development and waste management in the European Union. These are segments that are highly rated by the creators of economic and environmental policy. Circular economy practices are aimed mainly at solving local environmental problems (Schröder et al., 2020). The European Union has adopted a whole series of regulations such as:

- Circular economy action plan,
- Package of directives for the circular economy,
- The new industrial strategy for Europe.

These regulations target the reduction of food waste, plastic and packaging waste, electrical waste, as well as investment in the recycling industry. The concern of policymakers is growing because the European Union generates as much as 2.5 billion tonnes of waste annually (Androniceanu et al., 2021). As recycling is a key determinant of the circular economy, the ultimate goal is for 65% of the total mass of waste to be covered by the recycling process by 2035. The objective is to transform the industry in which the output will be circular products. In this sense, companies should switch to green technology. In addition, the circular economy model dictates that products must be made to last longer and be repaired quickly and easily. That is why Schroeder et al. (2019) believe that the next challenge may be the resistance of original equipment manufacturers to the development of the secondary products market.

The document "A Circular Economy in the Netherlands by 2050" (2016) sees the Netherlands as a leader in the circular economy, whose practices can benefit low-income countries. The Netherlands has set an ambitious goal to make rapid progress and to fully transition to the concept of a circular economy by 2050. The main activities that the Netherlands relies on are (Government of Netherlands, 2016): *i*) using biofuels and biomass instead of fossil fuels in order to preserve natural capital and prevent harmful carbon dioxide emissions that have a negative impact on the climate, *ii*) development of the recycling and waste processing industry and the bioeconomy in order to reduce dependence on raw materials, *iii*) cooperation at the international level due to the high degree of trade integration of countries, *iv*) cooperation at the national level between the government (legislation), science (scientific and research and development institutes), education (introduction of subjects dealing with sustainable development) and the business sector (companies that would apply circular solutions), *v*) development of the sharing economy, *vi*) encouraging socially responsible consumption (purchase of recycled products), *vii*) abolition of subsidies for activities that do not apply circular production models, *viii*) the circular economy contributes to the greening of the economy, but companies must take care that recycling is safe for the environment and human health, *ix*) constant efforts to keep materials in the cycle as long as possible. The experience of the Netherlands can be used in other countries in order to get closer to the leading country in the development of the circular economy.

SOLUTIONS AND RECOMMENDATIONS: INCORPORATING THE CIRCULAR ECONOMY INDEX IN THE CONCEPT OF SUSTAINABLE HUMAN DEVELOPMENT

In the eighties of the twentieth century, it was established that human development is a broader concept and that the rate of economic growth represents only one of the indicators of development. In fact, economic activities contribute to the increase of the gross domestic product, which is a simplified interpretation of socio-economic development. The Human Development Index was created in 1990 as part of the United Nations Development Program. It contains three indicators: *i*) gross national income per capita, *ii*) life expectancy at birth and *iii*) expected years of schooling and mean years of schooling (education index) (Schröder et al., 2020). Gross national income is a basic, but not a sufficient indicator of human development. Therefore, this index also contains the components of education and the health status of the population, which gives a more complete overview of the social well-being of nations, including both the economic and social dimensions of sustainable development. In studies, the Human Development Index is often identified with the indicator of sustainable development. However, as sustainable development is a three-dimensional concept, the Human Development Index does not include the ecological sphere of development, although life expectancy at birth takes into account some characteristics of environmental sustainability such as air pollution. The objective of this chapter is also to point out some shortcomings of the Human Development Index in the evaluation of the sustainable development of society, because the environmental component of development is omitted. Thus, the circular economy must be introduced into the analysis of sustainable development as an indicator of ecological sustainability.

The author starts from the fact that the circular economy is focused primarily on the ecological sphere, although it exerts its influence on all dimensions of sustainable development. The circular economy tends to balance the social, economic and ecological characteristics of the natural environment (Evans, 2023). It also deals with the limitation of resources as a central issue of economics as a science, as well as with climate change and thus makes a relationship with sustainable development. The conclusion is that the transition from a linear to a circular economy undoubtedly enables the preservation of the ecosystem. Schroeder et al. (2019) emphasizes that the circular economy has the strongest links with the following sustainable development goals: SDG6 - Clean Water and Sanitation, SDG7 - Affordable and Clean Energy, SDG8 - Decent Work and Economic Growth, SDG12 - Responsible Consumption and Production, and SDG15 - Life on Land. Indirectly, there is a connection with other goals of sustainable development (Rodriguez-Anton et al., 2019) as stated in the theoretical background.

A strong link between the circular economy and the Human Development Index is expected to be established because:

- a higher standard of living of residents allows focusing on higher social goals, such as environmental sustainability,
- a higher quality of education enables the acquisition of adequate skills and knowledge for the needs of applying circular economy practices and solutions,
- the good health status of the population increases the social engagement and interest of the community in matters of waste reduction, energy saving, recycling and, ultimately, sustainable development.

Table 5 was constructed with the aim of comparing the ranking of countries according to the circular economy composite index that was built and the Human Development Index that was taken from the relevant international report (Human Development Index values are for 2021 from the 2022 Report). The author started from the assumption that for the overall sustainable development of society, it is important to create a balance between the quality of life of people and the preservation of the environment. Finland, Italy, Cyprus, Denmark, Austria and Croatia have the biggest disproportions in the degree of development of the circular economy and human development.

The investigation showed that Finland and Denmark, as countries with a high Human Development Index, have the greatest potential for improving circular economy practices. As the most developed countries of the European Union, they should have the greatest financial opportunities for investing in research and development, recycling, energy efficiency and the use of renewable resources, as well as a high quality of human capital. On the other hand, Italy and Austria achieve high performance of the circular economy, regardless of their significantly worse initial position in terms of lower human development indices. These countries have the highest possibility to increase the degree of sustainable human development. The research results show that Bulgaria and Romania, as the poorest countries of the European Union, are the most threatened countries according to the chances of growth of the circular economy aggregate index value.

Table 6 suggests that there is a moderately strong, statistically significant correlation between the circular economy and human development. The calculation was carried out on the basis of Spearman's correlation coefficient. This confirms the previous assumption. In highly developed countries, better circular economy practices are present due to the high degree of social responsibility. The strongest con-

Table 5. Ranking of countries according to circular economy and human development indices

Country	Rank CEI	Rank HDI	Country	Rank CEI	Rank HDI
Netherlands	1	5	Croatia	15	23
Belgium	2	7	Latvia	16	22
Austria	3	11	Slovakia	17	24
Italy	4	15	Czech Republic	18	17
Germany	5	4	Poland	19	19
France	6	13	Lithuania	20	20
Ireland	7	3	Portugal	21	21
Malta	8	9	Hungary	22	25
Sweden	9	2	Greece	23	18
Luxembourg	10	8	Finland	24	6
Spain	11	12	Bulgaria	25	27
Denmark	12	1	Cyprus	26	14
Slovenia	13	9	Romania	27	26
Estonia	14	16			

Source: Author's calculation based on results of the study and United Nations Development Programme (2023).
Note: CEI – Circular Economy Index; HDI – Human Development Index

Table 6. Correlation between the circular economy index and the human development index

Correlation			HDI
Spearman's rho	**CEI**	Correlation Coefficient	0.651**
		Sig. (2-tailed)	0.000
		N	27

Source: Author's calculation based on IBM SPSS 22.
Note: CEI – Circular Economy Index; HDI – Human Development Index

nection of the Human Development Index on the example of the European Union countries was recorded with the indicator that measures *Resource Productivity* (0.74).

The sustainability of production and consumption requires the simultaneous observation of the circular economy and human development in considering the sustainable development of society. These are concepts that have their origins in the need for a broader understanding of the phenomenon of sustainability (Schröder et al., 2020). The Human Development Index is an indicator of sustainable socio-economic development that includes issues of economic growth, education of the population, and population health. The gross national product shows the current situation: the level of the standard of living or the degree of poverty of the population. On the other side, perspectives of sustainable human development in the future will depend on the quality and degree of knowledge, as well as the achieved health of the population. However, environmental issues related to pollution, waste accumulation and resource productivity are left out. Ignoring these issues, including responsible resource management, can backfire like a boomerang when talking about future growth and development. Also, the irresponsible use of raw materials can threaten people's quality of life. That is why the preservation of ecosystems (circular economy index) must be incorporated in the measurement of human and overall socio-economic development.

FUTURE RESEARCH DIRECTIONS

Future research could improve this study by comparing the ranking results if another weighting method was used, such as the widely accepted objective weighting methods - CRiteria Importance Through Intercriteria Correlation (CRITIC), Standard Deviation Method, Equal Weighting Method. The authors could compare the results using other powerful multi-criteria analysis methods in designing and developing composite indices such as: Multi-Objective Optimization Method by Ratio Analysis (MOORA), The Technique for Order of Preference by Similarity to Ideal Solution (TOPSIS), PROMETHEE, etc. Additionally, research must be directed towards a more detailed analysis: *i*) why some highly developed countries are low on the list according to the circular economy index, as well as *ii*) due to which factors countries with a developed circular economy are not able to achieve higher human development performance. In the future, attention must be paid to the social importance of the circular economy. Sustainable use (consumption) of resources will be vital to the progress of human development.

CONCLUSION

The possibility of lack of resources in the future is the trigger for the emergence and research of the theory and practice of the circular economy. Dependence on the import of raw materials from China, Russia, and South American countries significantly threatens the sustainable development of European countries and the achievement of circular economy goals in the coming period. The goal of this concept is that resources are not thrown out of the system, but rather circulate through the system as long as possible through reuse, recycling or repair. The burden on the environment additionally initiated a change in the current course of the linear economy. Although the circular economy is a broader concept, the threat of increasing environmental problems is an important factor for its emergence. Among the key environmental problems, the following stand out: accumulation of waste, pollution of ecosystems and increased depletion of resources. At the same time, in the European Union, the increase in waste that the environment is not able to absorb stands out significantly. Industry and agriculture are economic activities whose production often results in negative external effects. That is why it is necessary to use technology that simultaneously enables the protection of the environment, but also the improvement of productivity. Today, the circular economy is also seen as a tool for accelerated industrialization. Circular business models, innovations in business, adequate municipal waste management, investments in infrastructure and redesign of products and processes are the basis for mapping the potential for moving from a linear to a circular economic model at the level of the European Union. Before that, it is necessary to engage the creators of economic and development policies in providing a good institutional framework for the implementation of circular economy principles.

In recent years, various systems have been developed to assess the state of the circular economy. As there is still no universal benchmark for the degree of circular economy development at the world level, this paper offers a composite index that can be used in every country depending on the availability of the database. The level of the circular economy can be measured with this index, which facilitates the monitoring of the situation through a unique composite indicator. The study revealed which countries are the most advanced and which are lagging behind in development from the point of view of the circular economy index. European Union countries that later joined this integration have undoubtedly lower performances. Based on the selected indicators, the results indicated that the Netherlands is the leader, that is, the most advanced country in the European Union in terms of circularity. Developed countries, in which the Human Development Index is high, achieve a higher degree of circularity, which further affects the formation of better conditions for the sustainable development of those countries.

The measurement of the degree of development of the circular economy was carried out using a newer method of multi-criteria analysis called Grey Relational Analysis. This method is widely applied in almost all scientific fields: social sciences, technical and biotechnical sciences, humanities, engineering sciences, as well as medical sciences. The goal is to form a unique, summary index that include different dimensions of a complex phenomenon, such as the circular economy. The resulting composite index value reflects the state of the circular economy in all countries of the European Union. The goal was to assess the degree of transition of countries from a linear to a circular economy at the macro level. In relation to the existing composite indices, the author's aim was to create an index that will be simpler to interpret and compare among countries that do not have data on all the indicators present. It will make international comparisons much faster and more efficient. Finally, recommendations were given for building a new composite indicator of sustainable human development that will supplement the Human Development Index with a circular economy index that symbolizes the ecological dimension of sustainable development.

On the way to the selection of indicators in the multi-criteria model, the author faced some problems, such as: *i*) unavailability of data for individual indicators that could be included in the composite index, *ii*) unavailability of data for individual countries, and *iii*) unavailability of data on the circular economy for 2021 (for some indicators) and 2022 (for all indicators). All of these are also limitations of this research that should be resolved in subsequent studies.

ACKNOWLEDGEMENT

This research was financially supported by the Ministry of Science, Technological Development and Innovation of the Republic of Serbia [Contract No. 451-03-47/2023-01/200371].

REFERENCES

Government of Netherlands. (2016). *A Circular Economy in the Netherlands by 2050*. Government of Netherlands. https://circulareconomy.europa.eu/platform/sites/default/files/17037circulaireeconomie_en.pdf

Abifarin, J. K., & Ofodu, J. C. (2022). Determination of an efficient power equipment oil through a multi-criteria decision making analysis. *Vojnotehnički glasnik, 70*(2), 433–446. doi:10.5937/vojtehg70-36024

Ahmed, A. A., Nazzal, M. A., Darras, B. M., & Deiab, I. M. (2022). A comprehensive multi-level circular economy assessment framework. *Sustainable Production and Consumption, 32*, 700–717. doi:10.1016/j.spc.2022.05.025

Andersen, M. S. (2007). An introductory note on the environmental economics of the circular economy. *Sustainability Science, 2*(1), 133–140. doi:10.100711625-006-0013-6

Androniceanu, A., Kinnunen, J., & Georgescu, I. (2021). Circular economy as a strategic option to promote sustainable economic growth and effective human development. *Journal of International Students, 14*(1), 60–73. doi:10.14254/2071-8330.2021/14-1/4

Banaitė, D. (2016). Towards circular economy: Analysis of indicators in the context of sustainable development. *Social Transformation in Contemporary Society, 4*(9), 142–150.

Berg, A., Antikainen, R., Hartikainen, E., Kauppi, S., Kautto, P., Lazarevic, D., Piesik, S., & Saikku, L. (2018). *Circular economy for sustainable development*. Finnish Environment Institute. http://hdl.handle.net/10138/251516

Corona, B., Shen, L., Reike, D., Carreón, J. R., & Worrell, E. (2019). Towards sustainable development through the circular economy - A review and critical assessment on current circularity metrics. *Resources, Conservation and Recycling, 151*, 104498. doi:10.1016/j.resconrec.2019.104498

Dogaru, L. (2020). About sustainability between responsible production and consumption. *Proceedings, 63*(1), 69. doi:10.3390/proceedings2020063069

Eurostat. (2023). [Data set]. https://ec.europa.eu/eurostat/data/database

Evans, S. (2023). An integrated circular economy model for transformation towards sustainability. *Journal of Cleaner Production*, *388*, 135950. doi:10.1016/j.jclepro.2023.135950

Falkenmark, M., Finlayson, M., Gordon, L., Bennett, E., Chiuta, T. M., Coates, D., Ghosh, N., Gopalakrishnan, M., de Groot, R. S., Jacks, G., Kendy, E., Oyebande, L., Moore, M., Peterson, G. D., Portuguez, J. M., Seesink, K., Thame, R., & Wasson, R. (2007). Agriculture, water, and ecosystems: avoiding the costs of going too far. In *Water for food. Water for life. A comprehensive assessment of water management in Agriculture* (pp. 233–277). Earthscan.

Garcia-Bernabeu, A., Hilario-Caballero, A., Pla-Santamaria, D., & Salas-Molina, F. (2020). A process oriented MCDM approach to construct a circular economy composite index. *Sustainability (Basel)*, *12*(2), 618. doi:10.3390u12020618

Geissdoerfer, M., Savaget, P., Bocken, N. M. P., & Hultink, E. J. (2017). The circular economy - a new sustainability paradigm? *Journal of Cleaner Production*, *143*, 757–768. doi:10.1016/j.jclepro.2016.12.048

Geng, Y., Sarkis, J., Ulgiati, S., & Zhang, P. (2013). Measuring China's circular economy. *Science*, *339*(6127), 1526–1527. doi:10.1126cience.1227059 PMID:23539584

Hamzaçebi, C., & Pekkaya, M. (2011). Determining of stock investments with grey relational analysis. *Expert Systems with Applications*, *38*(8), 9186–9195. doi:10.1016/j.eswa.2011.01.070

Hussain, S. A. I., & Mandal, U. K. (2016, July). Entropy based MCDM approach for Selection of material. In *National Level Conference on Engineering Problems and Application of Mathematics* (pp. 1–6).

Jozić, S., Bajić, D., & Celent, L. (2015). Application of compressed cold air cooling: Achieving multiple performance characteristics in end milling process. *Journal of Cleaner Production*, *100*, 325–332. doi:10.1016/j.jclepro.2015.03.095

Ju-Long, D. (1982). Control problems of grey systems. *Systems & Control Letters*, *1*(5), 288–294. doi:10.1016/S0167-6911(82)80025-X

Karstensen, K. H., Engelsen, C. J., & Saha, P. K. (2020). Circular economy initiatives in Norway. In S. Ghosh (Ed.), *Circular economy: Global perspective* (pp. 299–316). Springer. doi:10.1007/978-981-15-1052-6_16

Kirchherr, J., Reike, D., & Hekkert, M. (2017). Conceptualizing the circular economy: An analysis of 114 definitions. *Resources, Conservation and Recycling*, *127*, 221–232. doi:10.1016/j.resconrec.2017.09.005

Korhonen, J., Honkasalo, A., & Seppälä, J. (2018). Circular economy: The concept and its limitations. *Ecological Economics*, *143*, 37–46. doi:10.1016/j.ecolecon.2017.06.041

Krysovatyy, A., Zvarych, I., & Zvarych, R. (2018). Circular economy in the context of alterglobalization. *Journal of International Students*, *11*(4), 185–200. doi:10.14254/2071-8330.2018/11-4/13

Lee, P., Sims, E., Bertham, O., Symington, H., Bell, N., Pfaltzgraff, L., Sjögren, P., Wilts, H., & O'Brien, M. (2017). *Towards a circular economy: waste management in the EU*. European Parliament. doi:10.2861/978568

Li, R. Y. M., Meng, L., Leung, T. H., Zuo, J., Tang, B., & Wang, Y. (2018). Unmaking Waste in Construction in the EU and the Asian Circular Economy: A Formal Institutional Approach. In R. Crocker, C. Saint, G. Chen, & Y. Tong (Eds.), *Unmaking Waste in Production and Consumption: Towards the Circular Economy* (pp. 225–240). Emerald Publishing Limited. doi:10.1108/978-1-78714-619-820181018

Lin, J. L., & Lin, C. L. (2002). The use of the orthogonal array with grey relational analysis to optimize the electrical discharge machining process with multiple performance characteristics. *International Journal of Machine Tools & Manufacture*, *42*(2), 237–244. doi:10.1016/S0890-6955(01)00107-9

Marković, M. (2022). Has Serbia Reached the Quality of Education of EU Countries? Multi-Criteria Evaluation. *Oradea Journal of Business and Economics*, *7*(2), 17–25. doi:10.47535/1991ojbe153

Marković, M., & Marjanović, I. (2022). The Role of Polyculture in Sustainable Agricultural Development and Prevention of Land Degradation. In S. Milutinović & S. Živković (Eds.), *Prevention and Management of Soil Erosion and Torrential Floods* (pp. 132–153). IGI Global. doi:10.4018/978-1-7998-8459-0.ch007

Marković, M., Popović, Z., & Marjanović, I. (2023). Towards a Circular Economy: evaluation of Waste Management Performance in EU Countries. *Serbian Journal of Management*. *18*(1). doi:10.5937/sjm18-40073

Marković, M., Stanković, J. J., Marjanović, I., & Tsaples, G. (2023a). A non-compensatory approach to the creation of composite indices of agricultural sustainability of the European Union countries. *International Journal of Sustainable Agricultural Management and Informatics*, *9*(1), 45–57. doi:10.1504/IJSAMI.2023.127538

Milanović, T., Savić, G., Martić, M., Milanović, M., & Petrović, N. (2022). Development of the Waste Management Composite Index Using DEA Method as Circular Economy Indicator: The Case of European Union Countries. *Polish Journal of Environmental Studies*, *31*(1), 1–14. doi:10.15244/pjoes/139896

Morseletto, P. (2020). Restorative and regenerative: Exploring the concepts in the circular economy. *Journal of Industrial Ecology*, *24*(4), 763–773. doi:10.1111/jiec.12987

Özkaya, G. (2022). Evaluation of regions and four income groups in terms of social protection through critic based grey relational analysis method. In C. A. Ceylan & M. C. U. O. Kaya (Eds.), *Research & Reviews in Social, Human and Administrative Sciences* (pp. 108–129). Gece Publishing.

Pearce, D. W., & Turner, R. K. (1990). *Economics of Natural Resources and the Environment*. Hemel Hempstead, Harvester Wheatsheaf.

Rodriguez-Anton, J. M., Rubio-Andrada, L., Celemín-Pedroche, M. S., & Alonso-Almeida, M. D. M. (2019). Analysis of the relations between circular economy and sustainable development goals. *International Journal of Sustainable Development and World Ecology*, *26*(8), 708–720. doi:10.1080/13504509.2019.1666754

Saidani, M., Yannou, B., Leroy, Y., Cluzel, F., & Kendall, A. (2019). A taxonomy of circular economy indicators. *Journal of Cleaner Production*, *207*, 542–559. doi:10.1016/j.jclepro.2018.10.014

Sauvé, S., Bernard, S., & Sloan, P. (2016). Environmental sciences, sustainable development and circular economy: Alternative concepts for trans-disciplinary research. *Environmental Development*, *17*, 48–56. doi:10.1016/j.envdev.2015.09.002

Schröder, P., Lemille, A., & Desmond, P. (2020). Making the circular economy work for human development. *Resources, Conservation and Recycling*, *156*, 104686. doi:10.1016/j.resconrec.2020.104686

Schroeder, P., Anggraeni, K., & Weber, U. (2019). The relevance of circular economy practices to the sustainable development goals. *Journal of Industrial Ecology*, *23*(1), 77–95. doi:10.1111/jiec.12732

Stankevičius, A., Novikovas, A., Bakaveckas, A., & Petryshyn, O. (2020). EU waste regulation in the context of the circular economy: Peculiarities of interaction. *Entrepreneurship and Sustainability Issues*, *8*(2), 533–545. doi:10.9770/jesi.2020.8.2(32)

Stanković, J. J., Janković-Milić, V., Marjanović, I., & Janjić, J. (2021). An integrated approach of PCA and PROMETHEE in spatial assessment of circular economy indicators. *Waste Management (New York, N.Y.)*, *128*, 154–166. doi:10.1016/j.wasman.2021.04.057 PMID:33991754

Stanković, J. J., Popović, Ž., & Marjanović, I. (2021a). Assessing smartness and urban development of the European cities: An integrated approach of Entropy and VIKOR. In Y. I. Topcu, Ö. Özaydın, Ö. Kabak, & Ş. Önsel Ekici (Eds.), *Multiple Criteria Decision Making: Beyond the Information Age 25, (69–97)*. Springer., doi:10.1007/978-3-030-52406-7_3

Ūsas, J., Balezentis, T., & Streimikiene, D. (2021). Development and integrated assessment of the circular economy in the European Union: the outranking approach. *Journal of Enterprise Information Management*. doi:10.1108/JEIM-11-2020-0440

United Nations Development Programme. (2023). Human Development Report 2021/2022. [Data set]. https://hdr.undp.org/system/files/documents/global-report-document/hdr2021-22overviewenpdf.pdf

Vaid, S. K., Vaid, G., Kaur, S., Kumar, R., & Sidhu, M. S. (2022). Application of multi-criteria decision-making theory with VIKOR-WASPAS-Entropy methods: A case study of silent Genset. *Materials Today: Proceedings*, *50*, 2416–2423. doi:10.1016/j.matpr.2021.10.259

Vranjanac, Ž., Rađenović, Ž., Rađenović, T., & Živković, S. (2023). Modeling circular economy innovation and performance indicators in European Union countries. *Environmental Science and Pollution Research International*, 1–12. doi:10.100711356-023-26431-5 PMID:36918486

Wu, H. H. (2006). A comparative study of using grey relational analysis in multiple attribute decision making problems. *Quality Engineering*, *15*(2), 209–217. doi:10.1081/QEN-120015853

Zhu, Y., Tian, D., & Yan, F. (2020). Effectiveness of Entropy Weight Method in Decision Making. *Mathematical Problems in Engineering*, *1–5*, 1–5. doi:10.1155/2020/3564835

ADDITIONAL READING

Тамбовцева, Т., Мельник, Л., Дегтярьова, І., & Ніколаєв, С. (2021). Circular economy: tendencies and development perspectives. *Механізм регулювання економіки, 2*(92), 33–42. doi:10.21272/mer.2021.92.04

Halkos, G. E., & Aslanidis, P. S. C. (2023). New circular economy perspectives on measuring sustainable waste management productivity. *Economic Analysis and Policy*, *77*, 764–779. doi:10.1016/j.eap.2023.01.001

Marković, M., Krstić, B., & Rađenović, T. (2020). Circular economy and sustainable development. *Economics of Sustainable Development*, *4*(1), 1–9. doi:10.5937/ESD2001001M PMID:31912387

Padilla-Rivera, A., Russo-Garrido, S., & Merveille, N. (2020). Addressing the social aspects of a circular economy: A systematic literature review. *Sustainability (Basel)*, *12*(19), 7912. doi:10.3390u12197912

Reich, R. H., Vermeyen, V., Alaerts, L., & Van Acker, K. (2023). How to measure a circular economy: A holistic method compiling policy monitors. *Resources, Conservation and Recycling*, *188*, 106707. doi:10.1016/j.resconrec.2022.106707

Sarja, M., Onkila, T., & Mäkelä, M. (2021). A systematic literature review of the transition to the circular economy in business organizations: Obstacles, catalysts and ambivalences. *Journal of Cleaner Production*, *286*, 125492. doi:10.1016/j.jclepro.2020.125492

Stanković, J. J., Marjanović, I., & Marković, M. (2020, May). Implementation of Circular Economy as a Sustainable Development Strategy. In *SM 2020; XXV International Scientific Conference Strategic Management and Decision Support Systems in Strategic Management.* (pp. 296–303). Subotica: Faculty of Economics, University of Novi Sad.

KEY TERMS AND DEFINITIONS

Circular Economy: An alternative concept that promotes regenerative economic practices that strive to restore, recycle, repair and reuse resources in accordance with the principles of sustainable development.

Composite Index: A single measure of a particular problem that consists of several indicators. The value of the index is created as a result of the aggregation process by a certain method of multi-criteria analysis.

Environment: The place inhabited by plants, animals, and people whose sustainability is essential for the survival of all ecosystems.

Human Development Index: An index that was promoted in the context of sustainable development because the assessment of social development only through gross domestic product per capita does not give a broader picture of the quality and sustainability of that development.

Linear Economy: The original concept in which the central place is occupied by the rational use of resources, not the reduction of waste by returning it to the production process.

Recycling: Returning waste to production, which achieves significant environmental and economic benefits.

Sustainable Development: The totality of relations between the economic, ecological, and social spheres of overall development.

Waste: Material resulting from production or consumption processes, the accumulation of which in nature can cause environmental degradation.

Chapter 18
Transformational Innovation Technologies for Regenerative Bioeconomy:
Case Study on Green Initiatives for Tourism Logistics Service Providers

Sejana Jose V.
CHRIST University (Deemed), India

Bindi Varghese
CHRIST University (Deemed), India

ABSTRACT

The tourism industry has negative environmental consequences, overshadowing the regenerative bioeconomy. Climate change, land degradation, and resource depletion are significant challenges. Excessive use of non-biodegradable resources threatens the planet, requiring bio-based resources. It is critical to transition to and reuse bio-based resources. In this chapter, the regenerative bioeconomy has a wide-reaching impact on accomplishing SDG 6, 8, 11, and 12, with a focus on the circular economy's involvement in tourism logistics. Investments in talent development, digital technologies, and partnerships are needed to realise bioeconomy potential. Engaging local communities and implementing sustainable business practices can reduce energy use and environmental impact. Digital transformation requires technological advancements, foreign investment, and active participation from all stakeholders. This chapter tries to explain the complicated interplay between the regenerative bioeconomy, tourism logistics, and sustainable practises intertwining.

INTRODUCTION

Transformational innovation and technology have gained the attention of all industries. Innovative

DOI: 10.4018/978-1-6684-8879-9.ch018

technologies are mainly implemented for creating new business models and strengthening the quality of business operations. Technological advancement is playing a prime role in augmenting interference in industries and at societal levels. Industries are considering digital transformation as central to remaining competitive in the marketplace (Vial, 2019). The excess usage of non-biodegradable resources and exhaustion of natural resources have put the world under serious threat. Thus, the world is moving towards the practice of bio-origin resources and their re-use in future generations. The industries are re-focusing on the concept of a regenerative bioeconomy, and the recent development of technologies is intended to support sustainability principles (Laibach et al., 2019). Bioeconomy is the conception of adopting biological resources and materials used for production, conservation, and regeneration for future generations, which results in sustainable development. As there is a bioeconomic transformation taking place by focusing on the use and reuse concept of bio-origin resources and materials, policymakers are designing a regulatory framework with the support of technology that assists in assessing the effectiveness and possible risks of sustainable principles (Marcone et al., 2022).

The tourism logistics service providers have initiated green practises to meet sustainable development goals. Tourism logistics in the destination refers to the movement of information, finance, tourists, and materials in a explicit geological area providing tourism products and services (Cheunkamon et al., 2021; Buczek, 2017; Oriade and Cameron, 2016). The tourism logistics service providers include tour operators, travel agencies, hotels, and public and private transport operators. They provide tourists with access to the destination. However, the use of conventional transportation has a negative effect on the tourism aspect as a whole (Gardossi et al., 2021). It is observed that to overcome the environmental damage, the transportation sectors have adopted alternative fuels that are of bio-origin and the usage of green vehicles with the support of advanced technologies that support the environment positively, resulting in a transformational bioeconomy (Fava et al., 2021).

OBJECTIVE AND PURPOSE OF THE STUDY

The globe is facing climate change as a major threat, which occurred due to extensive consumption of non-biodegradable resources and excess production of carbon emissions. Hence, the concept of bioeconomy is introduced and practised to enhance sustainable development and completely shift towards the consumption of biodegradable resources. The tourism logistics service providers have also initiated green initiatives along with technological advancements that meet sustainable development goals and create a path for circular concept of business operations. The introduction of green practices that include eco-friendly vehicles, the usage of bio-fuels, and the amalgamation of advanced automation enhance sustainable development through a regenerative bioeconomy.

LITERATURE ROAD MAP

Regenerative Bioeconomy

The growing population and the increased demand for human needs and wants have challenged the ecosystem. The rising concern about the environmental imbalance is due to the excess utilisation of natural resources, land degradation, industrialization, and urbanisation. In this growing population, to meet the needs and wants of human beings, there is high dependence on the ecology (Ahmad et al., 2023). All sectors and society in general tend to ignore ecological reverberations and consider the future generation while developing the economy. The high ignorance of environmental considerations has led to climate change, and environmental imbalances have made humankind centralise the concept of bioeconomy. Bioeconomy as a concept addresses and acts as a strategy to regenerate the economy along with better sustainable development (Ahmad et al., 2023; Marcone et al., 2022). In addition, the concept of bioeconomy is also explained as an adoption of bio-origin resources and materials for production, processing, and consumption that leads to a regenerative bioeconomy. In this modern era, the concept of bioeconomy may also contribute to sustainable development goals, specifically social equality, peace and harmony with the environment, economic development, sustainable consumption, and innovation (Laibach et al., 2019). The study highlights that, among the European Union, only nine countries have fully adopted the bioeconomy concept. Countries such as Finland, Germany, Italy, and Spain have considered regenerative bioeconomies to position themselves as competitive and sustainable nations. The policymakers have considered regulatory indicators that include fossil replacement, waste generation and valorization, climate regulation, material and energy resource security, the incorporation of technology, and resource regeneration for better sustainable development (Marcone et al., 2022; Fava et al., 2021). However, the study also culminate that all these European Union countries are in the early stages and have not entirely formed the regulatory framework (Marcone et al., 2022). On the other hand, the author has also discussed the progress of the bioeconomy in the US, and it is establish that the US is taking a form that reflects the regulatory framework. In addition, the author also reflects that the bioeconomy concept will support not only the industries but also the non-industrial communities (Devaney and Iles, 2019). The author has addressed that by 2030, the main findings of the bioeconomy on the industry will be that there will be significant market share for biomaterials, high production of biofuels from sugarcane and biomass, and adoption of biometrics for identifying people and environmental monitoring biosensor devices (Mateescu et al., 2011). However, to receive the comprehensive advantage of the regenerative bioeconomy, effective policy needs to be designed. The high involvement of the government along with prominent private organisations needs to clearly establish goals for climate change restoration (Devaney and Iles, 2019) and in the application of biotechnology in the different manufacturing and production industries (Mateescu et al., 2011).

Transformational Innovation Technologies for Regenerative Bioeconomy

The regenerative bioeconomy is alleviation evaluate for climate change, land degradation, and the depletion of natural resources. Here, advanced technologies are centralised to create a sustainable process for products and lead to circularity. The main aim of transformational technology is to replace the conventional model with more competitive and sustainable processing models. In the agricultural sector, synthetic biology enables the creation of a more sustainable way of farming. In addition, nanotechnology also plays an imperative role in environmental monitoring systems and deals with renewable energy repositories. The main contribution of transformational innovations is the usage of artificial cognition, Big Data analytics, mobile applications, and the Internet of Things. These advanced technologies are widely used in companies, and they have their own sides to the coin. The main benefits of using these transformational technologies are that they improve the operational efficiency of the business, and the major threat is the privacy and security of the consumer and business-related data stored (Vial, 2019). On the other hand, Pagoropoulos et al. (2017) have described that the Internet of Things (IoT) acts as a sensor to produce data that connects stakeholders and creates a value chain in business operations. IoT also enables the building of an active feedback model for the circular economy. In addition, radio frequency identification can assess the effective utilisation of products by customers. Big data analytics also play a vital role in standardising the dynamic models that lead to advanced revitalization rates.

The approach of the circular economy in waste management may be integrated with the systematic method of resource management. The study suggests the implementation of electrical and electronic equipment for waste management can be considered an efficient method and can reduce emissions (Nowakowski and Mrówczyska, 2018). On the other hand, advanced technological implementation in the transportation sector makes a significant contribution to the circular economy. The integration of AI and ICT for a smart and intelligent transportation solution is leading to the use of renewable energy. The smart infrastructural development for electric vehicles is aligned with circular economy standards (Leal Filho et al., 2021).

Industrial biotechnology in the United Kingdom has initiated financing the educational sector to develop skills related to biotechnology, low carbon programme innovations, and employment generation. Technology is also centralised to utilise renewable resources in transportation, and the home-based waste generated is converted into biofuel that supports the environment (Linser and Lier, 2020). In addition, biomass is used in the manufacturing process, and there is a high level of CSR involvement in the expansion of the bioeconomy. The author also highlights the possibility of disagreement between sustainable development goals and bioeconomy principles. Hence, it suggests the creation of a committee to develop a bioeconomy mechanism that prevents disagreement between sustainable development goals (Maksymiv et al., 2021). The advancements in nanotechnology also monitor pollution in the environment and can be used for water treatment plans. However, on the other side, the integration of transformational technologies is always questioned due to security issues related to data and fair access to technology. Transformational technology is the future, and it has a significant role in creating a bioeconomy along with socio-environmental development.

Role of Circular Economy and Green Measures

The circular economy is gaining attention as it provides a pathway for economic development and sustainable growth. The main focus of the concept is to switch from consumption to the reuse of resources. That means switching from consuming fossil fuel, rock, and phosphate to using generative resources like wind energy and solar energy. The author has also illustrated perseverance, circumventing, classification, reprocessing, and entropy as the main principles of circular biomass in a circular economy (Muscat et al., 2021). The circular economy is adopted by many prominent companies in their business models, which focus on restoration and regenerating the business framework. On the other hand, to restore and develop agriculture, it is imperative to restore soil carbon stocks for harvesting biomass. The circular economy can be incorporated through policy-making, advanced technology, societal behaviour, and organisation (Muscat et al., 2021). As the globe is entirely digitised, technologies like the Internet of Things, big data, and artificial intelligence play a fundamental role in the circular economy. In addition, the author also discussed that digital technology like radio frequency identification (RFID), relational databases (RDBMS), and product lifecycle management (PLM) is also used for data collection and integration. (Pagoropoulos et al., 2017). Artificial intelligence acts as a fundamental for economic growth and is a regenerative device (Kadar and Kadar, 2020). Technological advancement creates a pathway in a circular economy that focuses on depoliticizing sustainable development (Corvellec et al., 2021). According to the study conducted by Fava et al. (2021), Italy is considered to have one of the largest bioeconomies among other European countries. The main sector in Italy that booms for economic growth is food and bio-based products. The regenerative bioeconomy disables the high dependence on materials, fossil fuels, and land usage. Instead, it regenerates an economic boost that supports society and the environment. In addition, the Italian bio-based industry is designed as a manufacturing process for non-food items and bio-wastes that originate from urban cities. The process also includes technical assistance to produce biodegradable gas and bio-based energy that creates high-value products.

The traditional model of economic activity of production and consumption always leads to the maximum extraction of natural resources and production waste. Later, the chain system evolved with the recycling concept but was not able to advocate reuse and recovery. Hence, the experts have highly recommended the concept of a circular economy that leads to the effective usage of natural resources and the upcycling of the waste generated. The author recommends that tourism stakeholders adopt a circular economy, which can be regulated only with the coordination of tourism stakeholders that include the government, tour operators, academicians, host communities, tourists, and DMos (Rodrguez et al., 2020). The increased GHG and exhaustion of natural resources are mainly due to the transportation sector. The adoption of a circular economy can be considered a mitigating measure to overcome the threats of climate change and GHG. The concept of circular economy in the transportation sector can be effectively applied by having an eco-friendly design before the manufacturing process and also after the consumption stage, where the emissions generated need to be disposed of or treated (De Abreu et al., 2022).

The GSH is a medium-sized, four-star hotel that has moved towards the concept of circular solutions. The main products consumed by hotels are food-related products and timber. GSH Hotel has adopted the circular economy concept in different stages, which include the reconstruction stage, advanced technology for waste treatment, and collaboration with local suppliers for upcycling the linen and fabrics that are utilised for bedding and chef's uniforms. The CE adoption will be an effective strategic tool for hotels that assists in reducing the consumption of water, building materials, furniture, and energy (Manniche et al., 2021).

The strategic approach suggests that to accomplish a circular economy, the framework needs to move from the conventional approach to a green mode of approach, not only leads to sustainability but also to recycling (restoring) and up-cycling strategies (regenerating) (Kadar and Kadar, 2020). As a green measure, the European Union has aimed to achieve decarburization by the year 2030, and the strategies adopted include public-private partnerships, educational awareness to consumers and industries, introducing green labelling and certifications for companies, initiating subsidies for eco-friendly production and service catalogues, and introducing bioeconomy concepts in the higher education curriculum (Fava et al., 2021). Also, since green transportation is gaining high preference in the entire world to meet sustainable development goals, the circular economy plays a essential role in this domain. Electric and solar vehicles are being developed to support the circular economy; however, they are in the advent stage and need to be highly studied to develop a strategic framework for the automobile industry (Leal Filho et al., 2021). On the other hand, the tourism industry is considered to be one of the largest contributors to economic development. However, the travel and tourism industries follow many sustainable practises, and the concept of a circular economy is still in its infancy. In the investigation conducted on the concept and its relevance in the tourism industry, the experts suggest following the practise theory approach that enables them to assess beyond tourist behaviour (Srensen and Brenholdt, 2020). To conclude, to develop efficiency in the circular economy, it is imperative to specialise in the circular economy and green measures in the supply chain management processes in the business sectors that lead to making products and services that are very environmentally friendly and economically competitive (Vargas-Sánchez, 2018).

Bioeconomy in Aligning the National Economy: Socio-Environmental Dimensions

Bioeconomy is the concept of adopting biological resources and resources used for their production, conservation, and regeneration for future generations, which results in sustainable development. As there is a bioeconomic transformation taking place by focusing on the use and reuse concept of bio-origin resources and materials, policymakers are designing a regulatory framework with the support of technology that assists in assessing the effectiveness and possible risks of sustainable principles (Marcone et al., 2022). Biofuels for vehicles are widely used all over the world. Modern biofuel energy, along with a few other renewable resources, is indispensable to lowering carbon-emissions in the future. A country like Brazil has used bioethanol as a fundamental fuel instead of gasoline, which has resulted in positive results. Bioethanol is mainly produced from sugar fermentation and acts as a renewable energy source (Bomb et al., 2007). However, for larger-scale industries, the support of advanced sustainable technologies is required to operate more effectively and efficiently and compete against environmental hazards (Horta Nogueira et al., 2020). The Taiwan government has initiated clear policy directions aiming for low-carbon tourism destinations. However, travel agencies may get misled by false claims about low carbon reduction and energy savings. Hence, incorporating green standards and green mark certification is vital.

Sung et al. (2021 stated that CSR, green trust, and the theory of planned behaviour can be used to promote low-carbon tours, which help in sustainable development in the tourism domain. Providing low-carbon tour-related information to hotels, local communities, tourists, attractions, entertainment, and transportation companies (Ullah et al., 2021) will help in promoting and building sustainable destinations and result in a washout of carbon emissions. In addition, e-vehicles are gaining a prime role in the development of urban cities through e-mobility. E-mobility can be achieved through centralising circular economy principles that also positively reflect on society and the environment. Europe has aimed to achieve zero carbon emissions along with a circular economy by 2050 through e-mobility and sustainable development goals (Leal Filho et al., 2021). Biobased resources and energy consumption in the transportation sector significantly contribute to the national economy, and effective strategies are developed for waste management, carbon emission clearance, and enhanced air and water quality (Yang et al., 2022). However, the main challenge faced by the development is the lack of technological implementation, long-term government planning, and social acceptance (Falcone, 2019).

SDGs, Including SDGs Six on Energy, Eight on Economic Growth, Eleven on Sustainable Cities, and Twelve on Sustainable Consumption and Production

The United Nations developed 17 sustainable development goals (SDGs) to address global challenges and bring equity between the environment, society, and economy (Fuso Nerini et al., 2017). This study mainly concentrates on SDG 6 on energy, SDG 8 on economic growth, SDG 11 on sustainable cities, and SDG 12 on sustainable consumption and production. These sustainable goals aim for the nations to sustain themselves in a competitive environment. SDG 6 on energy acts as a mechanism for effective waste management and strengthens the economic development of the country (Bilek-Steindl et al., 2022). The main benefit of achieving SDG 6 is that working together to achieve the water targets will set a path to attaining SDGs 2 and 7 (Fader et al., 2018). The main challenge faced by many of the underprivileged countries is the scarcity of water and improper sanitization facilities. Most of the studies recommend active involvement of local and host communities, private-public partnerships, and educating the local community on renewable energy resources and their effective management, which can lead to attaining SDG 6. On the other hand, SDG 8 aims for economic growth along with sustainable development. Here, the main motto is to provide fair employment opportunities and decent pay for all workers. Economic growth is crucial for all countries, as it enables significant growth in a circular economy. SDG 8 can be attained by integrating regenerative technologies, investing in specialised skill development programmes, developing sustainable business models, and encouraging start-ups and small-scale enterprises (Frey, 2017).

SDG 11 on sustainable cities provides solutions to traffic congestion, increased pollution, and community inequalities. The main aim of SDG 11 is to provide safe, resilient, and sustainable cities. It can be considered a tool to achieve a circular economy (Dincă et al., 2022; Lakatos et al., 2021). The major challenge of urban cities is increased population, and meeting the needs and wants of humans leads to land degradation and excessive use of natural resources. SDG 11 can be adopted as a mechanism that overcomes these challenges (Abubakar and Aina, 2019). Major cities are developing their urban centres into smart, integrated green cities. Smart technologies are centralised for efficient utilisation of renewable resources, to address CO2 emissions in transportation, and for sewage and waste treatment in urban cities. Also, it is crucial to follow a systematic approach to widen the opportunities and benefits in terms of sustainable factors (Bai et al., 2016). E-vehicles are evolving as a major part of mobility, and they can be considered a centrepiece to reduce GHG and enhance the lifestyle of societies. By 2050, the majority of the countries on the globe aim to phase out fossil fuel mobility in favour of e-vehicle mobility (Leal Filho et al., 2021). Lastly, SDG 12 on sustainable consumption and production aims to regenerate waste and adopt recycling and upcycling methods in the life cycle of products and services. SDG 12 is very essential to attaining a circular economy (Glavi, 2021). The studies suggest that governments and related stakeholders have to include SDG 12 in policy-level planning and implementation (Pineiro-Villaverde and Garca-lvarez, 2020; Gasper et al., 2019). To conclude, sustainable development goals are widely adopted by nations to sustain themselves in a competitive world, and it is possible to attain them through the integration of technology, public-private partnerships, and policy recommendations on sustainability.

Green Initiatives and Role of Tourism Logistics Service Providers

According to Centobelli et al. (2019), "green initiatives" are defined as the combination of green or eco-friendly practises that attempt to achieve environmental sustainability. One of the key components to achieving environmental sustainability is thought to be the implementation of green projects. The entire ecosystem has consequences that have an impact on all businesses, but especially the transportation industry (Hamid et al., 2020). It is essential to change the current system to one that is green and sustainable because the transport sector is one of the largest polluters (Bamwesigye & Hlavackova, 2019). High demand from environmentally concerned travellers, rivals who use green practises, government assistance, and the desire to help protect the environment are the driving forces behind the implementation of green initiatives (Evangelista, 2014). The need for environmental protection is widely understood, and as a result, people are more likely to adopt eco-friendly and green behaviours. Low-emission cars are the newest green practises in transportation services. Electric automobiles, walking, bicycling, carpooling, solar-powered vehicles, bio-fuel cell vehicles, hybrid vehicles, and autonomous vehicles are examples of low-impact or green transportation methods (Li, 2016). Green administration, green infrastructure, mobility at the services, electronic ticketing, an eco-friendly pedestrian system, reuse and recycling of manufacturing cycles, and green infrastructure are further green practises in the transportation industry.

Tourism logistics service providers have initiated green practises to meet sustainable development goals. Tourism logistics in the destination refers to the movement of information, finance, tourists, and materials in a specific geographical area providing tourism products and services (Cheunkamon et al., 2021; Buczek, 2017; Oriade and Cameron, 2016). The tourism logistics service providers include tour operators, travel agencies, hotels, and public and private transport operators. They provide tourists with access to the destination. However, the use of conventional transportation has a negative effect on the tourism aspect as a whole (Gardossi et al., 2021). It is observed that to overcome the environmental damage, the transportation sectors have adopted alternative fuels that are of bio-origin and the usage of green vehicles with the support of advanced technologies that support the environment positively, resulting in a transformational bioeconomy (Fava et al., 2021). Environmental policies, an environmental competency framework, and effective communication are the primary drivers of green activities in the logistics industry (Seroka-Stolka, 2016). The organisational commitment to achieving a sustainable culture can be seen in environmental policy. The two factors, a competency-based framework and effective communication, are mostly tied to managerial behaviour to encourage green activities among the organisation's staff. Environmental behaviour will result from participants gaining knowledge and skills about sustainable variables and expressing their significance to them. Additionally, it results in accepting accountability for starting and pursuing green activities.

Caselet on Karnataka: Green Initiatives for Tourism Logistics Service Providers

This case study pivots on the green initiatives for tourism logistics service providers in Karnataka, India. Tourism logistics service providers are those who extend integrated services comprising the movement of tourists and other allied services. Service providers, including transport operators, travel agencies, tour operators, and hotels, play a major role in the development of sustainable tourism. Implementation of green practises helps reduce environmental impacts due to tourism activities in destinations. The case mainly emphasises the green initiatives adopted by the tourism logistics service providers in Karnataka, India, and the prospects and challenges faced with these green initiatives.

The Karnataka government has rigorously supported travel and tourism development, and it is clear that the government is prioritising tourism. Inbound tourists expanded from 36.7 million to 215.03 million from 2006 to 2018. The state is rich in culture and tourism resources. Karnataka marks three UNESCO world heritage sites that include the Western Ghats, Hampi, and Pattadackal. The tourism sector in Karnataka provides thirty lakh employment opportunities (Department of Tourism, Government of Karnataka, 2020).

Green initiatives positively impact reducing environmental waste. By incorporating green practises into their daily operations, businesses can boost community involvement, save energy, and protect their natural and cultural tourism resources. It also makes it possible to measure the social, economic, and environmental effects of tourism resources more accurately. The government of Karnataka has introduced electric mobility with high investment. They aim to transform Bengaluru into the major hub for the electric vehicle capital of the country (Chand, 2019). The government has recently introduced e-Jagruthi (e-vehicle and energy storage) in its policy, aiming to support private transport service providers and encourage e-vehicle services for passengers. The vision of the state is to be the e-vehicle capital of the country and promote eco-friendly manufacturing in partnership with the industry sectors. This policy initiative of e-mobility can be attained by encouraging cab operators, industry fleets, and auto-rickshaws. As technology plays a vital role in enhancing the quality and experience of passengers, mobile applications named E-Mitra are being introduced to locate charging stations in and around Bengaluru (*evJagruṭhi*, 2021).

B-live and ecotourism startups introduced e-bikes for tourists in Mysore, and the services were later extended to Madikeri and Hampi. In addition, the national parks that include Bannerghata National Park also introduced e-vehicle services to tourists. Travel agencies and tour operators are promoting eco-friendly tour packages and also conducting awareness programmes for tourists. On the other hand, hotels in Bengaluru have also initiated green practises such as CO2 sensors, air handling units, automated electric light dimmers, more usage of LED lights in the guest rooms, the introduction of ozone water plants, controlling water flow in the pipes, wastewater treatment, green awareness programmes for the employees, e-vehicle services for the in-house guests, and the use of recycled and upcycled materials (Sonker et al., 2021). These green initiatives are environmentally beneficial and contribute positively to environmental quality. Green transportation services are proven to be affordable, dependable, and require little maintenance. However, some of the challenges faced in specific to green practises in transportation: lack of infrastructure to support e-vehicles; high initial investment; not suitable for long journeys; and lack of systematic structure to implement green transportation systems by both the public and private sectors (Das and Bhat, 2022). Another challenge addressed is that even for passengers willing to use public transportation for their movement, there is no good last-mile connectivity, which discourages tourists from relying on private vehicles (Kanuri et al., 2019). However, the challenges can be addressed by the development of a green mechanism that can be used as a tool in business operations. The low-carbon programme may attract a sizable number of participants if all logistics service providers—including lodging, transportation, and ground handling—use it. Tourists might be encouraged to solely use low-carbon services by the providers of logistical services. Lastly, an extension of e-charging and bio-fuel stations in the tourism destination may also encourage tourism service providers and tourists to switch to green modes of transportation.

Research Methodology

This research takes a thorough approach, methodically reviewing the literature and exploring the transformative impact of transformational technologies on the creation of a regenerative bioeconomy. It dives into the critical importance of green initiatives in tourism logistics, emphasising their contribution to the bioeconomy and long-term growth. The analysis takes a broad view of the literature, upcoming research breakthroughs, developmental patterns, and practical applications in the field of transformational innovations. Furthermore, it investigates the relationship between these innovations and the achievement of SDGs 6, 8, 11, and 12, emphasising the importance of circular economy concepts and green measures. The paper delivers a case let on green efforts customised specifically for tourism logistics service providers in Karnataka to provide real insights. It thus contributes to the continuing scholarly conversation about transformative technologies in regenerative tourism and the role of green initiatives for logistics service providers. Finally, the findings of the study are intended to help policymakers, planners, and stakeholders in the tourism industry by giving actionable advise on deploying transformational technologies to support regenerative bioeconomy within the realm of tourism logistics.

Outcome and Implications

The chapter focuses on the role of transformational technology and green initiatives to support the bioeconomy. The study aims to provide comprehensive coverage of sustainable development goals 6, 8, 11, and 12 and their interrelationship with economic development. The article also emphasises the value of investing in education related to sustainable skill development, digital technology integration, public-private partnerships, and the development of sustainable business practises, as well as community involvement for efficient use of energy resources, water and waste management, sustainable cities, and sustainable production and consumption that supports economic development. In addition, the integration of the bioeconomy may be the best strategy to support sustainable development goals (Heimann, 2019). Biobased resources and energy consumption in the transportation sector significantly contribute to the national economy, and effective strategies are developed for waste management, carbon emission clearance, and enhanced air and water quality (Yang et al., 2022).

Transformational technologies such as AI, big data, and the Internet of Things add advanced development to achieving sustainability and also support the bioeconomy and circular economy principles. The integration of these technologies in transportation services may also contribute to the circular economy as it enables people and materials to move with less damage to the environment. Green initiatives such as e-vehicles, solar vehicles, and electric and bio-fuel stations, along with the integration of technology, contribute towards a transformational bio-economy (Fava et al., 2021). Also, green certification programmes can be introduced for the bio-based market that enable sustainable standardisation in business operations (Fava et al., 2021). Designing strategies that can address climate change-related problems and also encouraging bio-based construction materials that help to improve energy and water quality. These approaches may help reduce the pressure of land use and its coverage (Yang et al., 2022). The study also recommends policy-level planning for employment generation and enhancing performance mechanisms to support economic growth (Falcone, 2019).

The Karnataka government also significantly supports the development of e-mobility in the state and works towards making Bangalore the capital of e-mobility production in the country. Some of the initiatives towards green mobility are the introduction of e-vehicles, solar vehicles, and e-charging stations; the mobile application E-Mitra to locate the charging stations in and around Bangalore; and policy-level planning to encourage logistics service providers (*evJagruthi*, 2021). Also, there is high encouragement for a public-private partnership to achieve green aims in the state (Department of Tourism, Government of Karnataka, 2020).

The long-term implications of the chapter focus on a competency framework to build a bioeconomy that helps the country remain competitive and sustainable. With the right collaboration, the private and public sectors can result in effective management of business operations. The challenges in terms of practical implications can be addressed with advancements in technology, encouraging foreign investment, and collaboration with stakeholders.

FUTURE DIRECTIONS

The present study involves an understanding of how the regenerative bioeconomy reciprocates in sustainable progress and the practises and challenges of green initiatives by tourism logistics service providers. The study also highlights the involvement of transformational technology and its alignment with the bioeconomy and circular economy principles. The study can be further broadened to include the practical challenges of implementing bioeconomy principles aligned with advanced technologies. Also, the study can further invest in developing a prototype model to implement the bioeconomy concept in the transportation sector.

CONCLUSION

Finally, the study put attention on the vital issue of climate change caused by the overuse of non-biodegradable materials and carbon emissions. The concept of bioeconomy has been established and implemented to address this important global challenge and promote sustainable development. The bioeconomy represents a promising way to restoring the contaminated environment and achieving social, economic, and environmental well-being. Government programmes and other development projects are increasingly exploring the use of bio-origin materials, fuels, and energy, which have the potential to significantly contribute to the national economy.

The chapter also emphasises the significance of deploying technology techniques to efficiently reduce carbon emissions and improve overall quality of life. Creating major strides towards a more sustainable future by embracing biobased resources and implementing sustainable transportation practises. However, it is critical to recognise and solve the obstacles that these implementations encounter, such as a lack of long-term government planning and the requirement for social acceptance.

Moving forward, it is critical to continue investigating and improving techniques for sustainable development and bioeconomy promotion. Long-term government planning should be put in place to offer a stable framework for implementing sustainable practises. Furthermore, efforts should be undertaken to increase social acceptability and awareness, establishing a culture of sustainability and environmental sustainability.

In conclusion, the findings of this study highlight the necessity and significance of transitioning to a bioeconomy in order to combat climate change and achieve sustainable development. It can reduce environmental damage and promote economic growth by adopting biodegradable materials and applying sustainable practises. Collaboration between governments, industries, and individuals, on the other hand, is critical to achieving these objectives. By incorporating transformational technology in bioeconomy can create more sustainable and resilient future for future generations.

REFERENCES

Abubakar, I. R., & Aina, Y. A. (2019). The prospects and challenges of developing more inclusive, safe, resilient and sustainable cities in Nigeria. *Land Use Policy*, *87*, 104105. doi:10.1016/j.landusepol.2019.104105

Ahmad, S., Bharti, A., Haq, M. I., & Kothari, R. (2023). Bioeconomy: Current Status and Challenges. *Sustainable Butanol Biofuels*, 57-75.

Bai, X., Surveyer, A., Elmqvist, T., Gatzweiler, F. W., Güneralp, B., Parnell, S., Prieur-Richard, A.-H., Shrivastava, P., Siri, J. G., Stafford-Smith, M., Toussaint, J.-P., & Webb, R. (2016). Defining and advancing a systems approach for sustainable cities. *Current Opinion in Environmental Sustainability*, *23*, 69–78. doi:10.1016/j.cosust.2016.11.010

Bamwesigye, D., & Hlavackova, P. (2019). Analysis of sustainable transport for smart cities. *Sustainability (Basel)*, *11*(7), 2140. doi:10.3390u11072140

Bengaluru Bannerghatta Biological Park. (n.d.). Bengaluru Bannerghatta Biological Park. https://bannerghattabiologicalpark.org/facilities.html

Bilek-Steindl, S., Kettner, C., & Mayrhuber, C. (2022). Sustainability, Work and Growth in the Context of SDG 8. *Empirica*, *49*(2), 277–279. doi:10.100710663-022-09538-9 PMID:35505947

Bomb, C., McCormick, K., Deurwaarder, E., & Kåberger, T. (2007). Biofuels for transport in Europe: Lessons from Germany and the UK. *Energy Policy*, *35*(4), 2256–2267. doi:10.1016/j.enpol.2006.07.008

Buczek, B. (2017). Passenger Logistics Centers Presentness and Future. *Transport Economics and Logistics*, *71*, 73–91.

Centobelli, P., Cerchione, R., Esposito, E., & Shashi. (2020). Evaluating environmental sustainability strategies in freight transport and logistics industry. *Business Strategy and the Environment*, *29*(3), 1563–1574. doi:10.1002/bse.2453

Chand, K. (2019). An Initiative of Adopting E-Vehicles to Minimize Air Pollution. *International Journal of Research and Review.*

Cheunkamon, E., Jomnonkwao, S., & Ratanavaraha, V. (2022). Impacts of tourist loyalty on service providers: Examining the role of the service quality of tourism supply chains, tourism logistics, commitment, satisfaction, and trust. *Journal of Quality Assurance in Hospitality & Tourism*, *23*(6), 1397–1429. doi:10.1080/1528008X.2021.1995564

Corvellec, H., Stowell, A. F., & Johansson, N. (2022). Critiques of the circular economy. *Journal of Industrial Ecology*, *26*(2), 421–432. doi:10.1111/jiec.13187

Das, P. K., & Bhat, M. Y. (2022). Global electric vehicle adoption: Implementation and policy implications for India. *Environmental Science and Pollution Research International*, *29*(27), 40612–40622. doi:10.100711356-021-18211-w PMID:35083674

De Abreu, V. H. S., Da Costa, M. G., Da Costa, V. X., De Assis, T. F., Santos, A. S., & D'Agosto, M. D. A. (2022). The Role of the Circular Economy in Road Transport to Mitigate Climate Change and Reduce Resource Depletion. *Sustainability (Basel)*, *14*(14), 8951. doi:10.3390u14148951

Department of Tourism, Government of Karnataka. (2020). *Karnataka Tourism Policy 2020-26*. Karnataka. https://karnatakatourism.org/department/wp-content/uploads/2022/12/KT_Policy_Englosh.pdf

Devaney, L., & Iles, A. (2019). Scales of progress, power and potential in the US bioeconomy. *Journal of Cleaner Production*, *233*, 379–389. doi:10.1016/j.jclepro.2019.05.393

Dincă, G., Milan, A. A., Andronic, M. L., Pasztori, A. M., & Dincă, D. (2022). Does Circular Economy Contribute to Smart Cities' Sustainable Development? *International Journal of Environmental Research and Public Health*, *19*(13), 7627. doi:10.3390/ijerph19137627 PMID:35805283

Evangelista, P. (2014). Environmental sustainability practices in the transport and logistics service industry: An exploratory case study investigation. *Research in Transportation Business & Management*, *12*, 63–72. doi:10.1016/j.rtbm.2014.10.002

Fader, M., Cranmer, C., Lawford, R., & Engel-Cox, J. (2018). Toward an understanding of synergies and trade-offs between water, energy, and food SDG targets. *Frontiers in Environmental Science*, *6*, 112. doi:10.3389/fenvs.2018.00112

Falcone, P. M. (2019). Tourism-based circular economy in Salento (South Italy): A SWOT-ANP analysis. *Social Sciences (Basel, Switzerland)*, *8*(7), 216. doi:10.3390ocsci8070216

Fava, F., Gardossi, L., Brigidi, P., Morone, P., Carosi, D. A., & Lenzi, A. (2021). The bioeconomy in Italy and the new national strategy for a more competitive and sustainable country. *New Biotechnology*, *61*, 124–136. doi:10.1016/j.nbt.2020.11.009 PMID:33220517

Frey, D. F. (2017). Economic growth, full employment and decent work: The means and ends in SDG 8. *International Journal of Human Rights*, *21*(8), 1164–1184. doi:10.1080/13642987.2017.1348709

Fuso Nerini, F., Tomei, J., To, L. S., Bisaga, I., Parikh, P., Black, M., Borrion, A., Spataru, C., Castán Broto, V., Anandarajah, G., Milligan, B., & Mulugetta, Y. (2017). Mapping synergies and trade-offs between energy and the Sustainable Development Goals. *Nature Energy*, *3*(1), 10–15. doi:10.103841560-017-0036-5

Gardossi, L., Philp, J., Brigidi, P., Morone, P., Carosi, D. A. R., & Lenzi, A. (2021). Is tourism really affected by logistical operations and environmental degradation? An empirical study from the perspective of Thailan. *New Biotechnology*, *61*, 124–136.

Gardossi, L., Philp, J., Fava, F., Winickoff, D., D'Aprile, L., Dell'Anno, B., Marvik, O. J., & Lenzi, A. (2023). Bioeconomy national strategies in the G20 and OECD countries: Sharing experiences and comparing existing policies. *EFB Bioeconomy Journal*, *3*, 100053. doi:10.1016/j.bioeco.2023.100053

Gasper, D., Shah, A., & Tankha, S. (2019). The framing of sustainable consumption and production in SDG 12. *Global Policy*, *10*(S1), 83–95. doi:10.1111/1758-5899.12592

Glavič, P. (2021). Evolution and current challenges of sustainable consumption and production. *Sustainability (Basel)*, *13*(16), 9379. doi:10.3390u13169379

Gong, Y., Liu, F., Liu, H., Zhao, H., Zhu, Y., Han, Q., Zhou, J., Jiang, T., & Li, S. (2022). Environment–Energy–Economy Analysis and Related Technical Transition Strategies for Spent Activated Carbon Regeneration in China. *ACS Sustainable Chemistry & Engineering*, *10*(35), 11512–11523. doi:10.1021/acssuschemeng.2c02855

Hamid, M. A., Isa, S. M., & Kiumarsi, S. (2020). Sustainable tourism practices and business performance from the tour operators' perspectives. *Anatolia*, *32*(1), 23–32. doi:10.1080/13032917.2020.1830135

Heimann, T. (2019). Bioeconomy and SDGs: Does the bioeconomy support the achievement of the SDGs? *Earth's Future*, *7*(1), 43–57. doi:10.1029/2018EF001014

Horta Nogueira, L. A., Souza, G. M., Barbosa Cortez, L. A., & de Brito Cruz, C. H. (2020). 9 - Biofuels for Transport. *Future Energy*, *3*, 173–197. doi:10.1016/B978-0-08-102886-5.00009-8

Howson, P. (2021). Distributed degrowth technology: Challenges for blockchain beyond the green economy. *Ecological Economics*, *184*, 107020. doi:10.1016/j.ecolecon.2021.107020

Kadar, T., & Kadar, M. (2020, June). Sustainability Is Not Enough: Towards AI Supported Regenerative Design. In *2020 IEEE International Conference on Engineering, Technology and Innovation (ICE/ITMC)* (pp. 1-6). IEEE. 10.1109/ICE/ITMC49519.2020.9198554

Kanuri, C., Venkat, K., Maiti, S., & Mulukutla, P. (2019). Leveraging innovation for last-mile connectivity to mass transit. *Transportation Research Procedia*, *41*, 655–669. doi:10.1016/j.trpro.2019.09.114

Laibach, N., Börner, J., & Bröring, S. (2019). Exploring the future of the bioeconomy: An expert-based scoping study examining key enabling technology fields with potential to foster the transition toward a bio-based economy. *Technology in Society*, *58*, 101118. doi:10.1016/j.techsoc.2019.03.001

Lakatos, E. S., Yong, G., Szilagyi, A., Clinci, D. S., Georgescu, L., Iticescu, C., & Cioca, L. I. (2021). Conceptualizing core aspects on circular economy in cities. *Sustainability (Basel)*, *13*(14), 7549. doi:10.3390u13147549

Leal Filho, W., Abubakar, I. R., Kotter, R., Grindsted, T. S., Balogun, A. L., Salvia, A. L., Aina, Y. A., & Wolf, F. (2021). Framing electric mobility for urban sustainability in a circular economy context: An overview of the literature. *Sustainability (Basel)*, *13*(14), 7786. doi:10.3390u13147786

Li, H. R. (2016). Study on green transportation system of international metropolises. *Procedia Engineering*, *137*, 762–771. doi:10.1016/j.proeng.2016.01.314

Linser, S., & Lier, M. (2020). The contribution of sustainable development goals and forest-related indicators to national bioeconomy progress monitoring. *Sustainability (Basel)*, *12*(7), 2898. doi:10.3390u12072898

Maksymiv, Y., Yakubiv, V., Pylypiv, N., Hryhoruk, I., Piatnychuk, I., & Popadynets, N. (2021). Strategic challenges for sustainable governance of the bioeconomy: Preventing conflict between SDGs. *Sustainability (Basel)*, *13*(15), 8308. doi:10.3390u13158308

Manniche, J., Larsen, K. T., & Broegaard, R. B. (2021). The circular economy in tourism: Transition perspectives for business and research. *Scandinavian Journal of Hospitality and Tourism*, *21*(3), 247–264. doi:10.1080/15022250.2021.1921020

Marcone, R. D., Schmid, M., & Meylan, G. (2022). Closing the gap between EU-wide national bioeconomy monitoring frameworks and urban circular bioeconomy development. *Journal of Cleaner Production*, *379*, 134563. doi:10.1016/j.jclepro.2022.134563

Mateescu, I., Popescu, S., Paun, L., Roata, G., Bancila, A. , & Oancea, A. (2011). Bioeconomy. What is bioeconomy? How will bioeconomy develop the next two decades. *Studia Universitatis «Vasile Goldiş». Seria Ştiinţele Vieţii*, *21*(2), 451.

Muscat, A., de Olde, E. M., Ripoll-Bosch, R., Van Zanten, H. H., Metze, T. A., Termeer, C. J., van Ittersum, M. K., & de Boer, I. J. (2021). Principles, drivers and opportunities of a circular bioeconomy. *Nature Food*, *2*(8), 561–566. doi:10.103843016-021-00340-7 PMID:37118163

Nowakowski, P., & Mrówczyńska, B. (2018). Towards sustainable WEEE collection and transportation methods in circular economy-Comparative study for rural and urban settlements. *Resources, Conservation and Recycling*, *135*, 93–107. doi:10.1016/j.resconrec.2017.12.016

Oriade, A., & Cameron, H. (2016). Logistics and Supply Chain Management in Travel Industry. In Operations management in the travel industry. CAB INTERNATIONAL.

Pagoropoulos, A., Pigosso, D. C. A., & McAloone, T. C. (2017). The emergent role of digital technologies in the Circular Economy: A Review. *Procedia CIRP*, *64*, 19–24. doi:10.1016/j.procir.2017.02.047

Pineiro-Villaverde, G., & García-Álvarez, M. T. (2020). Sustainable consumption and production: Exploring the links with resources productivity in the EU-28. *Sustainability (Basel)*, *12*(21), 8760. doi:10.3390u12218760

Rodríguez, C., Florido, C., & Jacob, M. (2020). Circular economy contributions to the tourism sector: A critical literature review. *Sustainability (Basel)*, *12*(11), 4338. doi:10.3390u12114338

Seroka-Stolka, O. (2016). Green initiatives in environmental management of logistics companies. *Transportation Research Procedia*, *16*, 483–489. doi:10.1016/j.trpro.2016.11.045

Sonker, J., Kumar, S., & Pal, S. (2021). An Implementation of Energy Awareness Management in Heritage Hotels of India. *Turkish Online Journal of Qualitative Inquiry*, *12*(7).

Sørensen, F., & Bærenholdt, J. O. (2020). Tourist practices in the circular economy. *Annals of Tourism Research*, *85*, 103027. doi:10.1016/j.annals.2020.103027

Sung, P. L., Hsiao, T. Y., Huang, L., & Morrison, A. M. (2021). The influence of green trust on travel agency intentions to promote low-carbon tours for the purpose of sustainable development. *Corporate Social Responsibility and Environmental Management*, *28*(4), 1185–1199. doi:10.1002/csr.2131

Ullah, Z., Naveed, R. T., Rehman, A. U., Ahmad, N., Scholz, M., Adnan, M., & Han, H. (2021). Towards the development of sustainable tourism in pakistan: A study of the role of tour operators. *Sustainability (Basel)*, *13*(9), 4902. doi:10.3390u13094902

Vargas-Sánchez, A. (2018). The unavoidable disruption of the circular economy in tourism. *Worldwide Hospitality and Tourism Themes*, *10*(6), 652–661. doi:10.1108/WHATT-08-2018-0056

Vial, G. (2019). Comprensión de la transformación digital: Una revisión y una agenda de investigación. *The Journal of Strategic Information Systems*, *28*(2), 118–144. doi:10.1016/j.jsis.2019.01.003

Yang, M., Chen, L., Wang, J., Msigwa, G., Osman, A. I., Fawzy, S., Rooney, D. W., & Yap, P. S. (2022). Circular economy strategies for combating climate change and other environmental issues. *Environmental Chemistry Letters*, *21*(1), 55–80. doi:10.100710311-022-01499-6

Chapter 19
Manufacturing of Marketable Energy Sources From Agricultural Crops:
A Spanish Sector-Based Study on Bioeconomy

Luis A. Millan-Tudela
https://orcid.org/0000-0003-3669-9960
University of Alicante, Spain

Bartolomé Marco-Lajara
https://orcid.org/0000-0001-8811-9118
University of Alicante, Spain

Eduardo Sánchez-García
https://orcid.org/0000-0002-8524-4906
University of Alicante, Spain

Javier Martínez-Falcó
https://orcid.org/0000-0001-9004-5816
University of Alicante, Spain

ABSTRACT

The evolution of societies has been historically linked to the production, consumption and, ultimately, availability of energy. The development of more efficient technologies has led to several energy sources' evolution and change over the past centuries. However, those changes, together with a considerable rise of energy consumption, have also brought a threat to ecosystems in the form of several contaminants. In this chapter, a study on the potential use of crops for producing marketable energy is assessed, using the Spanish agriculture sector as study case given its substantial contribution to crop production in the European Union.

DOI: 10.4018/978-1-6684-8879-9.ch019

INTRODUCTION

Throughout history, the evolution of societies has been linked to the joint development of a series of factors. For example, the Roman Empire achieved its expansion in part due to its ability to generate water infrastructures (e.g. aqueducts) or transportation infrastructures (roads and highways), seeing how the economic affluence achieved with some of these routes at the time still persists today (Dalgaard et al., 2018). Another example of this is the medieval period, in which Latin and Arab societies made numerous innovative contributions such as the construction of irrigation ditches or the water mill in the field of water engineering (Glick & Kirchner, 2000), or the creation of water and land routes, as well as bridges in the field of civil transport infrastructures (Langdon & Claridge, 2011). Despite the importance of these and other similar developments, it could be said that the greatest technological leaps forward have been related to the discovery and exploitation of energy sources.

The first major development in this line was the use of fire, since it was employed by prehistoric societies as an energy source applicable to various fields: heating, metal smelting or cooking, among others. However, fire as a primary source of energy offered several limitations. Its simplicity (an exothermic reaction based on the presence of a fuel, mainly organic matter, oxygen as oxidizer and a hot focus as activation element) allowed its widespread extension, but this technology had the disadvantage of not being easily controllable. This fact changed during the First Industrial Revolution, especially after the implementation of James Watt's steam engine (1880) in the production processes, which was an evolution of Newcomen's engine.

Indeed, the use of James Watt's steam engine allowed an effective control of the movement by means of an adequate design of its mechanism. In addition, the fact that it was an external combustion engine allowed the use of fuels of any nature, since the moving parts were isolated from the combustion chamber and, consequently, prevented the generation of problems related to deficient combustion. This machine was used in different fields, such as the creation of mechanical traction vehicles or the increase of productivity in the then new factories (as opposed to the old guilds).

In spite of this, energy needs have maintained a historical growth both due to the increase in population over the years and to the technologies themselves that require its supply for their operation. For this reason, the Second Industrial Revolution saw the exploitation of new energy sources such as gas or oil, as well as electricity, which ultimately led to the creation, improvement and democratization of access to various means of transport between the end of the 19th century and the first decades of the 20th century (railroads, various land vehicles and aircrafts), communications (telegraph, telephone, radio and television) and various infrastructures (electric lighting, the appearance of the first household appliances, etc.). The main advantage of this type of energy sources that promoted its wide development was the high power they can offer. However, the inability of the time to carry out an effective storage of electrical energy caused a need to be able to produce at all times the energy demanded.

On the other hand, the use of these energy sources, which currently remain broadly exploited, has led to the generation of various types of waste that have affected and continue to threaten the different ecosystems of the planet. According to the International Energy Agency (2021), in 2019 a total of 606.6 EJ (Exajoules) were produced among all the sources used, which is equivalent to the energy contained in approximately 99.15 billion barrels of oil. Within these sources, 30.9% of the energy came precisely from oil, 26.8% from burning coal and 23.2% from natural gas, implying that more than 80% of the energy generated came from non-renewable fossil fuels.

Figure 1. World energy production record by type of source
Source: International Energy Agency (2021).

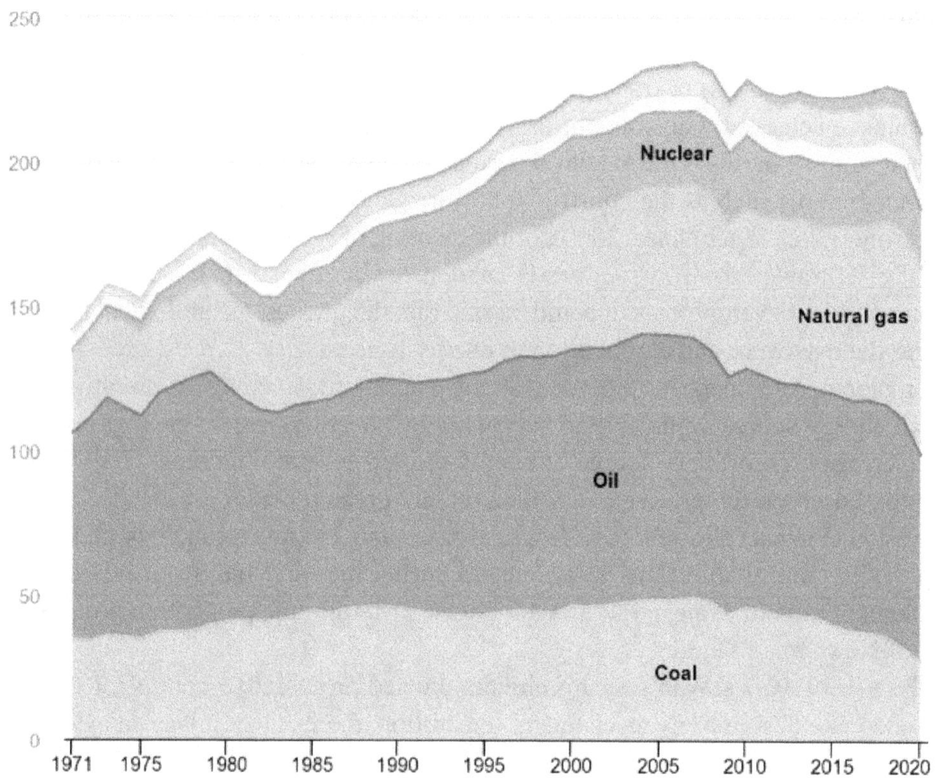

In addition to the aforementioned problems in the majority use of fossil energy sources, it is worth mentioning the limitation in their availability. For example, Milici et al. (2013) point out that, although there are coal reserves in the United States whose estimated life span reaches the next 200 years, the production peak is unknown, with various forecasts placing this maximum value at some point in the 21st century after 2030 (Höök & Aleklett, 2009; Höök & Aleklett, 2010; Milici et al., 2009; Mohr & Evans, 2009), in the 22nd century (Höök & Aleklett, 2009; Höök & Aleklett, 2010), and even that such a peak has already been reached (Patzek & Croft, 2009), all depending on the techniques and assumptions made. In fact, the inconsistency in the results generates a situation of strategic alarm, since the energy deployment in the territories (both at the level of generating, transporting and distributing structures) requires long development times, so that decisions regarding the sources to be used must be made decades in advance, in order to ensure the sufficiency of supply according to the constant growth in energy demand.

Biofuels are among the alternatives that have been proposed to make up for the deficiencies that, at one time or another, will occur due to the lack of availability of fossil fuel sources. In the United Nations framework, Larson (2008) presents biofuels as "a means of modernizing biomass use and providing greater access to clean liquid fuels while helping to address energy costs, energy security and global warming associated with petroleum fuels". With a more technical perspective, Rodionova et al. (2017) define them as "any energy-enriched chemicals generated directly through the biological processes or derived from the chemical conversion from biomass of prior living organisms". As can be seen, the concept of biofuel refers to those chemicals that, fulfilling the characteristics of a fuel, come from the remains of renewable living organisms.

Although this is a very generic definition, they are usually distinguished from each other on the basis of a generational classification (Larson, 2008; Rodionova et al., 2017):

- First generation: biofuels that come from seeds, grains and sugars, which serve as a substitute for benzine and diesel derived from petroleum. These include
    - Ethanol and butanol obtained by fermentation of starch and sugars.
    - Biodiesel obtained from the transesterification of vegetable oils.
    - Vegetable oils.
- Second generation: biofuels from lignocellulosic biomass (vegetable dry matter), specifically:
    - Ethanol and butanol obtained by enzymatic hydrolysis.
    - Methanol, Fischer-Tropsch benzine and mixtures of alcohols obtained by thermochemical processes.
    - Green diesel, Fischer-Tropsch diesel and dimethyl ether.
- Third generation: biofuels from microalgae, marine algae and microbes:
    - Biodiesel obtained from microalgae.
    - Bioethanol obtained from microalgae and marine algae.
    - Hydrogen obtained from green microalgae and microbes.

Although the potential of second- and third-generation biofuels is much higher than that of first-generation biofuels (Rodionova et al., 2017), they are not yet widely deployed. In fact, only first-generation biofuels are in place on a large scale, with a small proportion of biofuels belonging to the so-called second generation. However, the potential of biofuels does not lie solely in the biofuels themselves, as the residues that still remain accessible after the treatment of organic matter have a potential use in the context of circular economy.

Based on the above, the chapter is divided into the following sections. Section 2 presents the objectives, both general and specific, pursued in the following work. Section 3 contains both the explanation and the development of the methodology used to assess the potential of agricultural crops in energy generation, focusing on the case of the Spanish crop industry, and illustrating this potential through the case of ethanol. Section 4 presents the results obtained from the above methodology, while Section 5 presents both the conclusions and a discussion of the work. Finally, the paper concludes with the limitations of this work in section 6 and possible future lines of research in section 7.

OBJECTIVES

The general objective of this chapter (GO) is to study the potential of crops grown in Spain for the generation of energy products.

This general objective is to be achieved on the basis of the following specific objectives (SO):

SO1: to classify the different crops according to the fossil fuels whose by-products they can substitute.

SO2: to quantify the potential additional production of agricultural products that can be used as energy sources by optimizing agricultural processes.

SO3: to discuss the possible applications of the main crops based on the substitute products that can be produced from them.

METHODOLOGY

As a starting point for the analysis of the data, the Survey on Crop Surfaces and Yields developed annually by the Spanish Ministry of Agriculture, Fisheries and Food (2023) is used. As its name indicates, this is a survey that, on an annual basis, is carried out by sampling different agricultural fields in Spanish territory, stratifying the sample according to geographical location, as well as the type of crop grown on it. In this case, the data corresponding to the 2021 fiscal year are used, since they are the latest available in a complete and non-provisional way. Within this survey, a total of three publicly available databases are considered:

- General land distribution: provides information on the area devoted to each type of crop.
- Distribution by irrigation system: based on the previous areas, it distinguishes, for each type of crop, the area cultivated using irrigation techniques and the area cultivated using rainfed techniques.
- Yields of main crops: an assessment is made of the amount of crop (mass) obtained per unit area, distinguishing between the different crops and the technique used (rainfed or irrigated).

Once the records to be consulted had been selected, a classification of the main crops was made (thus avoiding the selection of vegetables whose production is scarce) according to the biofuels that can be manufactured through their manipulation, as well as the fossil fuels they can replace. For this purpose, the classification offered by Larson (2008) was used to assign each crop to a first-generation biofuel, given the fact that they are the only ones with current capability for industrial scalability. This classification is shown in Figure 2.

Once this phase was completed, the next step consisted of quantifying the production of these crops, for which the data corresponding to the year 2021 were used, since they constitute the last series with non-provisional information. To this end, and starting from the three databases mentioned above, the first step was to consult the area occupied by each type of crop.

After the previous process, it was determined which area had been exploited by rainfed techniques and which by irrigated techniques. This information, as well as the previous one, is shown in Table 1.

After this, and using the data on the yield obtained according to the crop and the technique used (rainfed or irrigated) at the aggregate (national) level, an estimate of total production was obtained by applying the following equation:

Equation 1. Calculation of production by type of crop as a function of area under cultivation and yield based on cultivation technique.

$$m_i = \left(S_s \cdot \eta_s + S_r \cdot \eta_r\right)_i$$

Source: own elaboration.

Figure 2. Main crops by type of feedstock, biofuels generated from them and substituted fossil fuels
Source: based on Spanish Ministry of Agriculture, Fisheries and Food (2023) and Larson (2008).

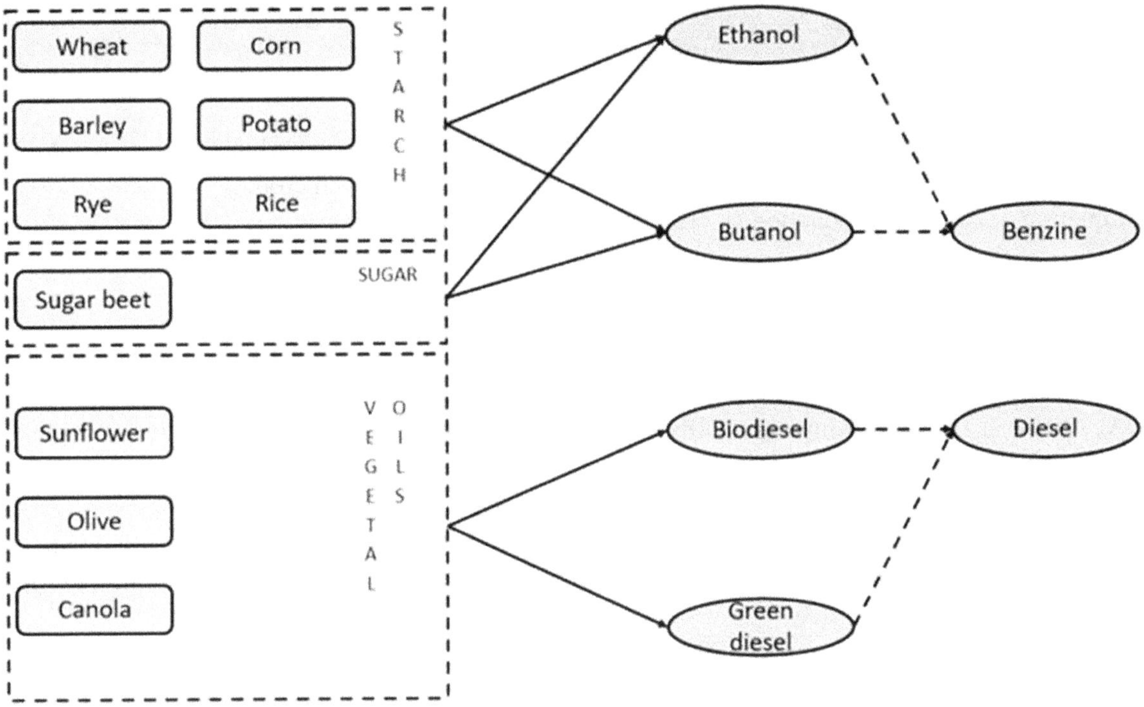

Where m_i is the total mass obtained from crop i, s_s and s_r the rainfed and irrigated area devoted to crop i, and η_s and η_r the yield (mass per unit area) by rainfed and irrigated techniques, respectively, for crop i. The results of this calculation can be found in Table 2.

After the previous operation, the productive potential of each crop was obtained. This is defined as the difference, in relative terms, between irrigated and rainfed crops, and is formulated as follows:

Equation 2. Calculation of the productive potential by crop type and cultivation technique.

$$\varphi_i = \left(\frac{\eta_r - \eta_s}{s} \right)_i$$

Source: own elaboration.

At this point, it should be noted that some of the potentials obtained are negative (i.e., the yield in rainfed systems is lower than in irrigated systems). This is interpreted as a sampling error, so, as an approximate solution, a potential of zero value is established in those cases (i.e., the rainfed technique will be considered as the yield of both techniques since it is the higher of the two, it being understood that the irrigation technique is actually superior, since its use is only justified, given its higher cost, by the fact that it offers a higher yield). Thus, the results of this part can be consulted in Table 3.

Finally, and as a measure of the potential production that could be achieved in each type of crop, an estimate has been made of the maximum amount that is considered technically feasible, using the following equation:

Equation 3. Potential production by crop type (simplified formulation).

$$m_{p,i} = \eta_{máx} \cdot S_i$$

Source: own elaboration.

This equation arises from the simplification of the following equation which, despite being more complex, provides more information regarding the characteristics of each crop:

Equation 4. Potential production by crop type (developed formulation).

$$m_{p,i} = \left(S_s \cdot \eta_s \cdot \varphi + S_r \cdot \eta_r\right)_i$$

Source: own elaboration.

As can be seen, the maximum quantity (mass) of crop *i* arises from considering that the yield of rainfed production is equal to that of irrigated production, for which the following assumptions are made:

- The economic cost of the design, installation and operation of irrigation systems is lower than the additional income obtained as a consequence of improved production for the lifespan of the mentioned systems.
- There are sufficient guarantees of water supply for the proper use of irrigation systems.

These assumptions are not met in all regions of the Spanish territory. However, this is not due to a lack of water resources, but to poor management in terms of storage and distribution. For this reason, and since this problem goes beyond the objective of this paper, the results should be interpreted in the perspective that, in the coming years, sufficient corrections will be made to guarantee the water supply demanded by both individuals and professionals.

Thus, considering the above remarks, the lost potential was obtained, which is understood as the difference between the maximum and actual production, and whose calculation is derived from the difference between both values. This can be obtained both in absolute terms (Equation 5) and relative terms (Equation 6).

Equation 5. Lost potential (absolute) by crop type.

$$p_{p,i} = m_{p,i} - m_i$$

Source: own elaboration.

Equation 6. Lost potential (relative) by crop type.

$$p_{p,i} = \frac{m_{p,i} - m_i}{m_{p,i}}$$

Source: own elaboration.

Finally, and to illustrate the possible production of biofuels with this lost potential, this value was used to estimate the potential production of ethanol, based on the estimates shown by Mussatto et al. (2010) from various authors. For this purpose, the assumption was made that the lost potential can be expressed, equivalently, in terms of arable area. Thus, based on the yield expressed by these authors in the type of crop (l/ha), the following equation was used:

Equation 7. Potential ethanol production by crop type.

$$v_{p,j} = \eta_j \cdot s_{s,j} \cdot \left(\frac{\eta_{r,j} - \eta_{s,j}}{\eta_{s,j}} \right)$$

Source: own elaboration.

Where η_j is the ethanol production obtained per crop *j* area according to Mussatto et al. (2010), $s_{s,j}$ the rainfed area of crop *j*, and $(\eta_{r,j}-\eta_{s,j})/\eta_{s,j}$ the relative yield difference of crop *j* between irrigated and rainfed techniques. In this way, what is achieved is to show the amount of ethanol that would be obtained if this unexploited potential of the rainfed area were exploited through irrigation for that purpose.

RESULTS

After listing the main crops to be used for the generation of liquid biofuels (Figure 1), Table 1 provides a compilation of the areas devoted to each of them.

As can be seen there, 61.6% of the crop area considered corresponds to starch-containing products, 38.0% to oleaginous crops and only 0.3% to sugar beet. Regarding the cultivation technique, 20.8% of the crop area shown was under the exploitation through irrigation systems, which implies that almost 4 out of 5 hectares depended in 2021 on rainfall.

Based on the data in Table 1, and applying the yields offered by the Ministry of Agriculture, Fisheries and Food (2023), the production estimate was obtained by applying Equation 1, as well as the production potentials from Equation 2. The results are shown in Tables 2 and 3.

As can be seen in Table 2, the differences between irrigated and rainfed yields are notorious. Among the crops considered, the smallest difference corresponds to the case of potato, where the use of irrigation implies a production improvement of 28.4%. On the other hand, the most extreme case is sunflower, where production with the optimal system increases by 145%. However, the most notorious issue lies in the fact that most irrigated crops offer a yield between 45% and 87% higher than their rainfed counterparts. This, added to the fact that most crops are cultivated in the latter modality, causes a waste of land and, consequently, of the products (both foodstuffs and their derivatives) that can be obtained.

With reference to the results in Table 3, it is worth noting that 68.9% of production is concentrated in starch-rich products. However, it is also worth highlighting the fact that the production of olives that can be used to obtain vegetable oils accounts for 27.3% of the total production of the crops considered, rising to 30.0% if canola and sunflower are considered. Finally, the case of sugar beet, which accounts for 1.0% of the total quantity of the products considered, appears marginally (as is the case with surface area).

Based on the data provided in Tables 1 and 2, and applying Equations 4, 5 and 6, both the potential production and the lost potential were obtained and are shown in Table 4.

Table 1. Surface area dedicated to each crop and product obtainable after processing

Crop	Total surface(ha)	Rainfed surface (ha)	Irrigated surface (ha)	Obtainable product
Hard wheat	330,430	302,395	28,035	Ethanol y butanol
Soft and semi-hard wheat	1,871,995	1,672,197	199,798	
Two-row barley	2,517,670	2,286,942	230,728	
Six-row barley	194,616	183,770	10,847	
Rye	146,034	141,271	4,763	
Rice	88,066	0	88,066	
Corn	359,187	17,574	341,612	
Potato	46,242	10,652	35,590	
Sugar beet	28,786	3,825	24,961	
Sunflower	644,308	596,680	47,628	Biodiesel y green diesel
Canola	87,231	62,190	25,041	
Dual-use olive	120,319	75,402	44,917	
Olive mill olive	2,573,334	1,777,448	795,886	

Source: elaborated from Spanish Ministry of Agriculture, Fisheries and Food (2023).

Table 2. Yields and production potential by crop type

Crops	Rainfed performance (kg/ha)	Irrigated performance (kg/ha)	Productive potential
Hard wheat	2,248	4,460	0.984
Soft and semi-hard wheat	3,746	5,569	0.487
Two-row barley	3,205	5,037	0.572
Six-row barley	2,510	3,619	0.442
Rye	2,591	4,693	0.811
Rice	n/d	7,225	n/d
Corn	9,467	13,849	0.463
Potato	32,851	42,173	0.284
Sugar beet	99,345	n/d	n/d
Sunflower	1,087	2,662	1.449
Canola	2,167	3,408	0.573
Dual-use olive	2,560	4,783	0.868
Olive mill olive	2,988	5,236	0.752

Source: obtained from Spanish Ministry of Agriculture, Fisheries and Food (2023).

From the results in Table 4, it can be inferred that the production of these crops in 2021 could have been 31% higher if irrigation techniques had been applied in their entirety, which would have meant more than 16.2 million tons. In relative terms, it is noteworthy that sugar beet shows a deficit in this respect of 86.7%, as well as sunflower (54.8%), especially when compared to the rest of the oleaginous products. With respect to starch products, the greatest losses occur with hard wheat (45.4%) and rye

Table 3. Estimated production by crop type

Crop	Estimated production (MT)
Hard wheat	804,820
Soft and semi-hard wheat	7,376,725
Two-row barley	8,491,826
Six-row barley	500,518
Rye	388,386
Rice	636,277
Corn	4,897,358
Potato	1,850,880
Sugar beet	379,949
Sunflower	775,297
Canola	220,086
Dual-use olive	407,866
Olive mill olive	9,478,273

Source: elaborated from Spanish Ministry of Agriculture, Fisheries and Food (2023).

(43.3%). It should be noted, however, that crops such as rice (in its totality) or corn and potato (in its majority) offer low loss potentials, but this is a consequence of the crop typology itself. For example, in the case of rice, it requires submerging the plantation, so it cannot be considered in the comparison. The case of corn does stand out, since it is possible to use rainfed techniques and, in spite of this, they are highly optimized in this sense (all this considering that its productive potential has a value of 0.463, that is, similar to that of other products with higher lost potentials, such as 6-row barley and hard wheat).

Finally, an estimate of the volume of ethanol that could have been obtained in 202 was made from the crops considered if water optimization had led to improved production through irrigation systems. To this end, the information on production per area from different authors and presented by Mussatto et al. (2010) was used together with the values of rainfed area in Table 1 and the rainfed and irrigated yields in Table 2. All this, applied to Equation 7, leads to the following results:

CONCLUSSION AND DISCUSION

As can be seen in the methodology section, based on the predominance of starch-rich crops, as well as vegetable oils, the main application of Spanish agricultural products derives from the generation of ethanol, methanol and biodiesel, which serve as substitutes for benzine (in the case of starchy products) and diesel (in the case of vegetable oils). Specifically, ethanol and butanol are obtained mainly from starch-rich products, with sugar beet being the only foodstuff outside this grouping. As regards biodiesel, the main crops that could be used are olives, sunflower and canola.

Within this line, it has been shown that deficient management of water resources leads to the loss of potential production in crops of very different types in Spain. Thus, regardless of the type of crop, deficiencies of up to 86.7% are observed (case of sugar beet).

Table 4. Productive potential and lost potential (absolute and relative) by crop type

Crop	Potential production (Tm)	Lost potential (Tm)	Lost potential (relative)
Hard wheat	1,473,718	668,898	0.454
Soft and semi-hard wheat	10,425,140	3,048,415	0.292
Two-row barley	12,681,504	4,189,678	0.330
Six-row barley	704,315	203,797	0.289
Rye	685,338	296,952	0.433
Rice	636,277	n/d	n/d
Corn	4,974,381	77,023	0.015
Potato	1,950,164	99,284	0.051
Sugar beet	2,859,745	2,479,796	0.867
Sunflower	1,715,005	939,708	0.548
Canola	297,310	77,225	0.260
Dual-use olive	575,486	167,620	0.291
Olive mill olive	13,473,977	3,995,704	0.297
STARCH TOTAL	33,530,836	8,584,047	0.256
OLEAGINOUS TOTAL	16,061,778	5,180,256	0.323
TOTAL	52,452,360	16,244,099	0.310

Source: elaborated from Spanish Ministry of Agriculture, Fisheries and Food (2023).

Table 5. Estimated ethanol production using the lost potential of wheat and corn crops

Raw material	Performance (l/ha)	Estimated production (l/ha)	Source
Corn stover	[1.050-1.400]	[8.541.220-1.388.294]	Tabak (2009)
Corn	[3.460-4.020]	[28.145.354-32.700.672]	Tabak (2009)
Wheat	2.590	[2.878.347.569]	Cheryl (2008)

Source: based on Ministry of Agriculture, Fisheries and Food (2023), Tabak (2009) and Cheryl (2008).

This loss has been quantified for corn and wheat, where it has been observed that, in the best-case scenario, with the surplus that would have been obtained from an optimum use of current rainfed areas through irrigation techniques, an approximate volume of 2,870 million liters of ethanol could have been synthesized. Under current European regulations, in which fuels marketed for transportation must contain between 5% and 10% of this substance, a volume of commercial grade benzine of between 28.7 billion and 57.4 billion liters could have been obtained. It should also be noted that these figures focus on only two types of crops and one type of biofuel, which implies that the potential production would be much higher.

Although the ideal situation would imply the full use of biofuels (since it would allow combustion processes that, in essence, would be close to the goal of carbon neutral), its implementation is not easy with the current infrastructure. This is because the combustion originated through biofuels is not identical to that obtained with fossil fuels, so the machinery used by the latter would require a profound redesign

(Payri & Desantes, 2017) that, in essence, would not be cost-effective compared to the manufacture of new machines created *ad hoc* for use through ethanol, methanol, biodiesel and other types of biofuels.

However, it should be noted that the contribution of crops in terms of energy to bioeconomy is not limited to biofuels. For example, the remains of plant matter left over from other production processes, such as foliage, pruning by-products and the like, are currently used both industrially and domestically. For example, the reintroduction of olive pits, almond shells and the like, through their transformation into pellets for use in combustion boilers, stands out. In fact, in 2021 their use was estimated at 751,700 tons (Díaz, 2022), especially in view of their lower sensitivity to energy prices compared to other energy sources in Europe.

The previous example is, precisely, another of the elements that play in favor of the use of crops as the basis for energy sources in the current economic transition scenario. On the one hand, it requires the adaptation of part of the machinery for its use, but the technology to do so is currently developed, since it is based on the same thermodynamic principles. Furthermore, although the Spanish case has been shown, most of the countries of the world have their own crops rich in raw materials such as starch or sugar, so that they could guarantee a minimum supply of their own to alleviate situations such as the one that occurred in the so-called oil crisis, which, as Issawi (1978) indicated, happened as a consequence of the excessive dependence of the West on foreign oil. Given the same situation in the current scenario resulting from the war in Ukraine, the use of the aforementioned products may represent a step forward.

In any case, it should be noted that the products that can be obtained from crops in Spain are related to the burning of fuels, i.e., exothermic processes based on reduction-oxidation reactions. This type of process, although it may be much less polluting than fossil fuels, does not completely eliminate the problem, since certain combustion products (such as nitrogen oxides) are only partially treatable, and cannot be avoided since they are produced as a consequence of the reaction of oxygen and nitrogen present in the air used in the process.

Thus, as a final conclusion, it can be determined that the use of crops can help in the energy transition process by offering stability to energy market prices, reducing dependence on geopolitical factors and requiring already-known, relatively limited changes in infrastructure compared to other alternatives. However, they have the disadvantage of depending on improved water management systems, as well as not avoiding the emission of certain pollutants that contribute to global warming.

LIMITATIONS

Although this work represents a starting point in the study of the use of crops for energy purposes, it is true that it has some limitations. The first of these is the excessive focus on the use of biofuels. Although this has been done because their use is a major bottleneck in the energy transition (especially due to dependence on the transport sector), it has not allowed other possible energy applications of crops to be covered in greater detail. This has been due to the fact that, as far as the authors have been able to discern, there is a lack of quantitative literature that would allow precise calculations to be made. This situation has also occurred with the production of biofuels, which is why the study of their production has been limited only to ethanol with a very limited series of crops.

Another limitation of the chapter has to do with the assumptions set out in the methodology section. Thus, the different assessments have been carried out under the assumption of water sufficiency which, despite being a fact in Spain, requires the deployment of a strategic plan that allows adequate water use and management. Considering that this type of initiatives usually requires several years for their implementation, the solution could arrive too late, as the burning of fuels from organic waste is presented as a transitory solution. In addition, it should be noted that the economic impact of the proposals has not been evaluated, so that, despite their technical feasibility, they may not be financially viable if the amortization periods (corresponding to the energy transition period) are insufficient.

Nevertheless, the authors are confident that this work can serve as the beginning of more in-depth research, with more profound implications both at the academic level and in its transfer to the private sector.

FUTURE RESEARCH LINES

Based on what has been developed in this chapter, at least two research fronts remain open.

The first of these has to do with a more detailed study of agricultural potentials in Spain. Although this work is focused on the energy sector, changes in ecosystems, together with population growth forecasts (United Nations, 2022), serve as a reference for a more in-depth analysis of Mediterranean crops. This opens up the possibility of conducting longitudinal studies and comparing different regions in order to detect the main points of higher yields. From this, a comparative analysis can be made to define the elements that increase the yield of a territory compared to others in each type of crop and, thus, favor the development of strategic plans aimed at ensuring food supply.

Another possible front has to do with an economic analysis of the use of first-generation biofuels for the energy transition. Since their industrial scalability depends on private investment, a study that evaluates both the costs of deploying the necessary infrastructure (especially at the production level), the resulting benefits and the period of exploitation will make it possible to assess their viability as a tool at the service of society during the transition to less polluting energy sources.

REFERENCES

Cheryl. (2008). *Algae becoming the new biofuel of choice.* Dueling Fuels. http://duelingfuels.com/biofuels/non-food-biofuels/algae-biofuel.php#more-115N

Dalgaard, C., Kaarsen, N., Olsson, O., & Selaya, P. (2018). Roman roads to prosperity: Persistence and non-persistence of public goods provision. *Science Direct.*

Díaz, Á. (2022). 'Boom' de los pellets de madera tras encarecerse los combustibles fósiles. *El Mundo.* https://www.elmundo.es/ciencia-y-salud/medio-ambiente/2021/11/04/6183dfc121efa0c5078b46ab.html

Glick, T. F., & Kirchner, H. (2000). *Hydraulic systems and technologies of Islamic Spain: History and archaeology. Working with water in Medieval Europe*. Brill.

Höök, M., & Aleklett, K. (2009). Historical trends in American coal production and a possible future outlook. *International Journal of Coal Geology, 78*(3), 201–216. doi:10.1016/j.coal.2009.03.002

Höök, M., & Aleklett, K. (2010). Trends in US recoverable coal supply estimates and future production outlooks. *Natural Resources Research, 19*(3), 189–208. doi:10.100711053-010-9121-1

International Energy Agency. (2021). *Key World Energy Statistics 2021*. IEA. https://www.iea.org/reports/key-world-energy-statistics-2021/supply

Issawi, C. (1978). The 1973 Oil Crisis and After. *Journal of Post Keynesian Economics*, *1*(2), 3–26. doi:10.1080/01603477.1978.11489099

Langdon, J., & Claridge, J. (2011). Transport in medieval England. *History Compass*, *9*(11), 864–875. doi:10.1111/j.1478-0542.2011.00804.x

Larson, E. D. (2008). *Biofuel production technologies: status, prospects and implications for trade and development*. UNCTAD.

Milici, R. C., Flores, R. M., & Stricker, G. D. (2009). Estimation of United States peak coal production. Paper presented at the *Geological Society of America Abstracts with Programs for the 2009 Annual Meeting and Exposition,* doi:10.1016/j.coal.2012.10.002

Milici, R. C., Flores, R. M., & Stricker, G. D. (2013). Coal resources, reserves and peak coal production in the United States. *International Journal of Coal Geology*, *113*, 109–115. doi:10.1016/j.coal.2012.10.002

Ministry of Agriculture, Fisheries, and Food. (2023). *Encuesta sobre Superficies y Rendimientos de Cultivos*. MAPA. https://www.mapa.gob.es/es/estadistica/temas/estadisticas-agrarias/agricultura/esyrce/

Mohr, S. H., & Evans, G. M. (2009). Forecasting coal production until 2100. *Fuel*, *88*(11), 2059–2067. doi:10.1016/j.fuel.2009.01.032

Mussatto, S. I., Dragone, G., Guimaraes, P. M. R., Silva, J. P. A., Carneiro, L. M., Roberto, I. C., Vicente, A., Domingues, L., & Teixeira, J. A. (2010). Technological trends, global market, and challenges of bio-ethanol production. *Biotechnology Advances*, *28*(6), 817–830. doi:10.1016/j.biotechadv.2010.07.001 PMID:20630488

Patzek, T. W., & Croft, G. D. (2009). Potential for coal-to-liquids conversion in the United States: Fischer–Tropsch synthesis. *Natural Resources Research*, *18*(3), 181–191. doi:10.100711053-009-9098-9

Payri, F., & Desantes, J. M. (2017). *Motores de combustión interna alternativos*. Editorial Universitat Politècnica de València.

Rodionova, M. V., Poudyal, R. S., Tiwari, I., Voloshin, R. A., Zharmukhamedov, S. K., Nam, H. G., Zayadan, B. K., Bruce, B. D., Hou, H. J., & Allakhverdiev, S. I. (2017). Biofuel production: Challenges and opportunities. *International Journal of Hydrogen Energy*, *42*(12), 8450–8461. doi:10.1016/j.ijhydene.2016.11.125

Tabak, J. (2009). *Biofuels*. Infobase Publishing.

United Nations. (2022). *World Population Prospects 2022: Summary of Results*. UN. https://www.un.org/development/desa/pd/sites/www.un.org.development.desa.pd/files/wpp2022_summary_of_results.pdf

Watt, J. (1880). *Steam-Hammers*. Alamy.

KEY TERMS AND DEFINITIONS

International Energy Agency: An autonomous body within the framework of the Organization for Economic Co-operation and Development (OECD), founded in 1974, designed to provide coordinated responses between countries in the event of oil supply emergencies.

Biofuel: A chemical product that, meeting the characteristics of a fuel, comes from the remains of renewable living organisms.

Exajoule: A unit of energy in the international system corresponding to $1 \cdot 10^{18}$ joules, 2.78 kWh or $2.39 \cdot 10^{14}$ kcal.

Combustion: A physicochemical phenomenon consisting of an exothermic reduction-oxidation reaction.

Potential: In the context of this chapter, the maximum value that can technically be obtained from a given medium or source.

Chapter 20
Understanding the Dynamics of Economic Development in Developing Countries

Puneet Kumar
Kalinga University, India

Maricel C. San Diego
Bulacan State University, Philippines

Maria Teresa R. Fajardo
Phinma University of Pangasinan, Philippines

ABSTRACT

In order for emerging countries to advance and prosper, their economies must develop in a complex and varied manner. The goal of this research study is to give readers a thorough grasp of the forces that drive economic growth in emerging countries. The introduction of the chapter provides an overview of the main ideas and theories surrounding economic development, covering elements like human capital, technical innovation, institutional quality, and globalisation. After that, it analyses the unique problems that developing nations face, including poverty, income disparity, and restricted access to resources and infrastructure. Additionally, this research study explores how social, market, and governmental elements interact to promote economic development. It examines how social capital, inclusive institutions, and good governance all contribute to the promotion of long-term economic success. Along with the potential advantages and difficulties of inclusion in the global economy, the influence of globalisation and trade on developing economies is also addressed.

INTRODUCTION

The dynamics of economic development in developing countries have become a significant area of research and a subject of enormous significance in today's globalised world. On their journey towards

DOI: 10.4018/978-1-6684-8879-9.ch020

sustainable and inclusive growth, developing countries-often distinguished by their resource limitations, social inequalities, and structural difficulties-face particular challenges and opportunities. Making wise policies and strategies that support long-term prosperity and higher living standards requires an understanding of the complexity behind their economic growth processes. The goal of this research study is to explore the complex dynamics of economic development in developing nations. We want to get a thorough grasp of the potential and challenges connected with development initiatives in these countries by looking at the many forces, variables, and policies at work. In doing so, we seek to add to the body of knowledge and offer insightful information to practitioners, researchers, and policymakers working to promote sustainable development.

A variety of interrelated factors influence how the economy develops in developing nations. These include, among other things, socio-political environments, institutional structures, breakthroughs in technology, endowments in natural resources, and world economic trends. Understanding the complex interactions between these factors is essential for determining the factors that promote and inhibit economic growth and developing effective methods to maximise their potential. (Sangwan 2022).

Additionally, this study acknowledges that economic development is a complicated, dynamic phenomenon rather than a linear process. The economy of developing nations frequently undergo structural changes, moving from agrarian to industrial, and more recently, to knowledge-based and service-oriented sectors. For inclusive and sustainable development policies that take into account the changing needs and aspirations of societies in developing nations, it is essential to comprehend the varied stages, patterns, and dynamics of these transformations. (Segismundo and Dayao **2021).**

Additionally, this study recognises the important significance that global relationships and international collaboration play in influencing how emerging nations progress economically. Economic outcomes are significantly influenced by trade policies, foreign direct investment, technological transfers, and development assistance. Understanding the dynamics of economic development in emerging countries requires analysing the interplay between domestic and foreign forces. Additionally, this study acknowledges that economic development is a complicated, dynamic phenomenon rather than a linear process. The economy of developing nations frequently undergo structural changes, moving from agrarian to industrial, and more recently, to knowledge-based and service-oriented sectors. In order to create inclusive and sustainable development policies that take into account the changing needs and aspirations of societies, it is essential to understand the varied stages, patterns, and dynamics of these transformations.

The purpose of this research study is to clarify the complex dynamics of economic development in emerging nations. This study aims to offer a thorough understanding of the difficulties, chances, and routes for sustainable and equitable development by examining the diverse causes and processes that support economic progress. The results of this study can help practitioners, researchers, and policymakers in their efforts to create efficient interventions, policies, and strategies that encourage economic development in developing nations and improve the well-being of their populations. (Singh, G., Gupta, R. and Agarwal, Nidhi 2020).

The intricate and dynamic mechanisms that propel changes in an economy through time are referred to as its dynamics. It includes all of the different elements, interactions, and changes that affect a nation's economic development, structural change, and rise in living standards.

1. Economic growth frequently entails a structural transformation in which an economy transitions from a largely agrarian base to industrialization and, more recently, to knowledge-based and service-

Figure 1. Economic and technological growth and international cooperation

oriented sectors. This shift entails adjustments to employment prospects, production patterns, and technological improvements.

2. Capital Accumulation: Both human and physical capital must be accumulated for economic progress. Infrastructure, equipment, and technology are considered physical capital, but the workforce's education, skills, and expertise are considered human capital. Investments in both human and physical capital are necessary for economic development in order to promote productivity and innovation. (Akoley, Wahid, and Kulo, 2022)
3. Technological Development: According to Malik and Gharpe (2022), technological progress is essential to the growth of the economy. New technology adoption and innovation can increase productivity, boost effectiveness, and promote economic growth. The diffusion and adoption of technology, as well as financial investments in R&D, are all part of the dynamics of economic development.
4. Globalisation and commerce: By allowing cross-border commerce, investment, and knowledge exchange, globalisation has changed the dynamics of economic development. The inclusion of developing nations in the world economy can open up chances for technological transfers, market access, and foreign direct investment. However, it also comes with difficulties like competition, a volatile market, and susceptibility to outside shocks.
5. Institutions and Governance: The dynamics of economic development are substantially influenced by the calibre of the institutions and governance systems. Effective institutions may encourage investment, entrepreneurship, and economic stability by ensuring the rule of law, the protection of property rights, and transparent governance. Inequality can be sustained and hampered by weak or corrupt institutions. (Mahanand 2019)
6. Social and Human Development: Indicators of social and human development, such as gains in healthcare, education, and poverty reduction, are used to quantify economic progress in addition

to GDP growth. Policies and actions that address social inequalities, promote social inclusion, and enhance population well-being are part of the dynamics of economic development. (Bhowmick et al. **2022).**

7. Environmental Sustainability: Environmental factors must be incorporated into sustainable economic development. According to Dwivedi (2022), The dynamics of economic development require striking a balance between economic expansion and resource management, environmental protection, and climate change mitigation. Practises for sustainable development seek to advance long-term prosperity while protecting the environment and natural resources. (Singh and Dr. Balveer 2019).
8. Governments may significantly influence how the dynamics of economic development are shaped by enacting policies. Fiscal policies, monetary policies, trade policies, industrial policies, and social welfare programmes are a few examples of these interventions. For encouraging investment, entrepreneurship, and a fair allocation of resources, effective policy frameworks are crucial. (Khari, Sharma, and Agarwal 2020).

Analysing the interconnections and interdependencies of these many components is necessary to comprehend the dynamics of economic development in a developing country. In order to promote sustainable and equitable economic growth, it is necessary to examine the particular settings, difficulties, and possibilities that other nations face.

ECONOMIC DEVELOPMENT IN DEVELOPING COUNTRIES

Economic development in developing countries refers to the sustained improvement in the standard of living, quality of life, and economic well-being of the population in these nations. It involves various strategies and policies aimed at promoting economic growth, reducing poverty, increasing employment opportunities, and enhancing infrastructure and institutions.

Economic Growth: Economic development typically begins with fostering sustainable economic growth. Developing countries focus on increasing their gross domestic product (GDP) through measures such as industrialization, agricultural productivity improvements, technological advancements, and attracting foreign direct investment (FDI). This growth provides a foundation for further development efforts.

Human Capital Development: Investing in human capital is crucial for economic development. Developing countries prioritize education and healthcare, aiming to enhance the knowledge, skills, and health of their workforce. This investment contributes to increased productivity, innovation, and the overall well-being of the population.

Poverty Alleviation: Poverty is a significant challenge in many developing countries. Effective poverty reduction strategies involve creating employment opportunities, implementing social safety nets, providing access to basic services like healthcare and education, and promoting inclusive economic growth. Microfinance initiatives and targeted programs for vulnerable groups also play a role in poverty alleviation.

Infrastructure Development: Adequate infrastructure, including transportation networks, energy systems, and communication networks, is crucial for economic development. Developing countries invest in building and improving infrastructure to facilitate trade, attract investments, and foster economic activities in both urban and rural areas.

Institutional Reforms: Developing countries often need to strengthen their governance systems, including legal frameworks, regulatory bodies, and public institutions. Transparency, accountability, and the rule of law are essential for creating an enabling environment for businesses, attracting investments, and reducing corruption. Good governance and institutional reforms contribute to sustainable economic development.

Trade and Market Access: Developing countries can benefit from participating in international trade. Opening up to global markets, reducing trade barriers, and promoting exports can lead to economic diversification, increased foreign exchange earnings, and access to new technologies. However, it is essential to ensure fair trade practices and consider the potential impacts on local industries and vulnerable sectors.

Sustainable Development: Environmental sustainability is a critical component of economic development. Developing countries aim to balance economic growth with environmental conservation and resource management. According to Agarwal, Ekansh (2021), promoting clean technologies, renewable energy, and sustainable agricultural practices help mitigate the negative environmental impacts of development while ensuring long-term economic viability.

Regional and International Cooperation: Developing countries often engage in regional and international partnerships to promote economic development. Cooperation in areas such as trade agreements, technology transfer, knowledge sharing, and financial assistance can facilitate development efforts by leveraging resources and expertise from various stakeholders.

While economic development in developing countries is a priority, there are several challenges and issues that can hinder progress. Here are some common issues faced in the process of economic development:

1. Poverty: Widespread poverty is a significant obstacle in developing countries. High levels of inequality, lack of access to basic services, limited employment opportunities, and inadequate social

Figure 2. World map featuring low to high income economies

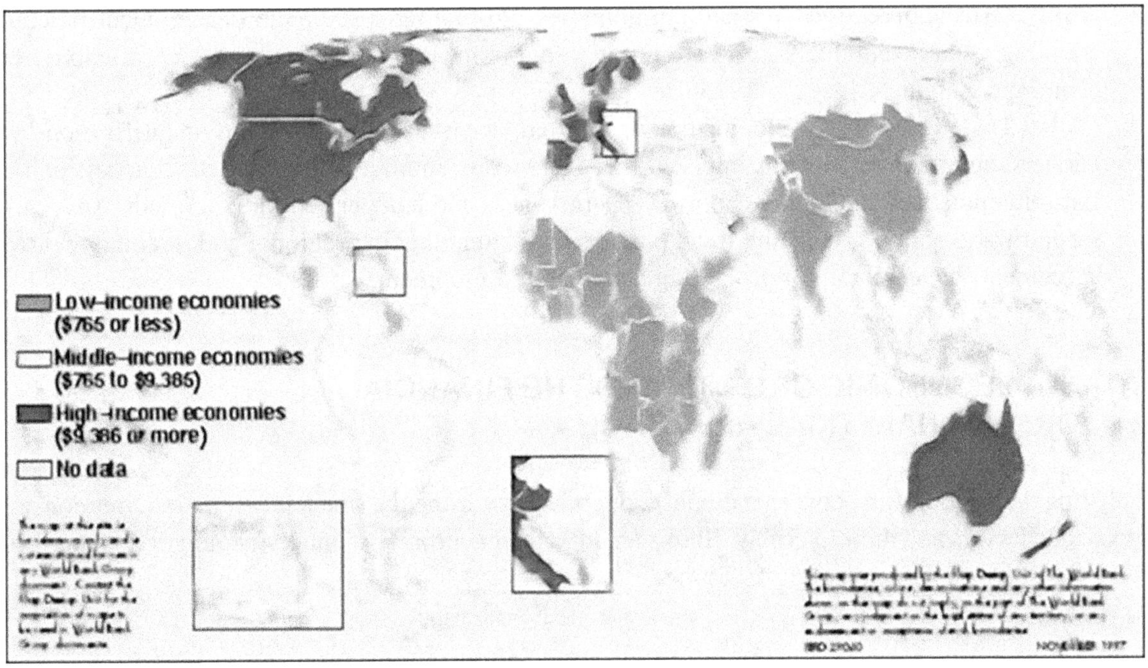

safety nets contribute to persistent poverty rates. Addressing poverty requires targeted interventions and inclusive growth strategies.

2. Limited Access to Finance: Many individuals and businesses in developing countries face difficulties in accessing financial services, such as loans, credit, and insurance. Limited access to finance hampers entrepreneurial activities, inhibits investment, and constrains economic growth. Expanding financial inclusion and developing robust financial systems are necessary to support economic development.
3. Infrastructure Gaps: Inadequate infrastructure, including transportation, energy, and telecommunications, poses a significant challenge for economic development. Insufficient infrastructure limits market connectivity, impedes trade, and hinders productivity. Bridging infrastructure gaps requires substantial investments and long-term planning.
4. Weak Institutions and Governance: Developing countries often struggle with weak institutions, corruption, and inadequate governance frameworks. These issues undermine the business environment, deter investments, and hinder development efforts. Strengthening institutions, promoting transparency, and improving the rule of law are essential for sustainable economic development.
5. Limited Human Capital: Developing countries often face challenges related to education, healthcare, and skills development. Inadequate access to quality education and healthcare services hinders human capital development, limiting productivity and innovation. Investing in education, healthcare, and vocational training is crucial for building a skilled workforce and promoting economic growth.
6. Vulnerability to External Shocks: Developing countries are often more susceptible to external shocks such as commodity price fluctuations, financial crises, or natural disasters. These shocks can disrupt economic stability, impact export earnings, and undermine development progress. Building resilience through diversified economies, risk management strategies, and social safety nets is vital for mitigating the impact of external shocks.
7. Environmental Sustainability: Balancing economic development with environmental sustainability is a key challenge. Rapid industrialization, deforestation, pollution, and climate change can have adverse effects on ecosystems and communities. Promoting sustainable development practices, adopting clean technologies, and addressing environmental concerns are critical for long-term economic viability.
8. Global Trade Barriers: Developing countries often face trade barriers, including tariffs, non-tariff barriers, and trade restrictions imposed by developed economies. Limited market access and unfair trade practices can hinder export competitiveness and hamper economic growth. Advocating for fair trade policies, reducing trade barriers, and strengthening regional and international trade agreements are essential for promoting economic development.

NATIONAL ECONOMIC GROWTH AND THE FINANCIAL SECTOR: AN ANALYTICAL APPROACH

Analyzing the relationship between national economic growth and the financial sector requires considering various factors and their interplay. Here's an analytical approach to understanding this relationship:

Macroeconomic Factors

a. GDP Growth: Macroeconomic factors are the broad indicators of financial growth or decline that affect an economy. A macroeconomic factor is a geopolitical, environmental or economic event that can impact the monetary stability related to the whole economy of a country or region instead of a specific part of the population. A macroeconomic factor may be considered positive, negative or neutral, based on the way it affects the economy. A natural disaster can negatively impact the production and sale of goods while higher production rates due to a demand for more goods are considered positive macroeconomic factors. Start by examining the overall economic growth of a nation, as measured by changes in its Gross Domestic Product (GDP) over time. GDP growth reflects the expansion of economic activities and is closely linked to the performance of the financial sector.
b. Interest Rates: Assess the prevailing interest rates set by the central bank. Changes in interest rates impact borrowing costs, investment decisions, and consumer spending, thereby affecting economic growth and the financial sector.
c. Inflation: Analyze the inflation rate, as it influences the purchasing power of consumers, the profitability of businesses, and the effectiveness of monetary policy. Stable inflation is crucial for sustained economic growth and a healthy financial sector.

Financial Sector Indicators

a. Credit Availability: Evaluate the ease of obtaining credit and loans for individuals and businesses. A well-functioning financial sector should facilitate access to credit, promoting investment and consumption.
b. Capital Market Performance: Examine stock market indices, bond yields, and other capital market indicators. The performance of these markets reflects investor confidence, corporate profitability, and the allocation of capital, which can influence economic growth.
c. **Banking Sector Stability: Assess the stability and soundness of banks through indicators such as capital adequacy ratios, non-performing loans, and liquidity levels by** Fatima (2022).. **A stable banking sector ensures the smooth functioning of financial intermediation, which supports economic growth.**

Policy and Regulatory Environment

a. Monetary Policy: Analyze the central bank's monetary policy framework and decisions, such as interest rate adjustments, reserve requirements, and quantitative easing measures. These policies affect liquidity conditions, credit availability, and inflation dynamics, thereby influencing economic growth.
b. Financial Regulations: Consider the regulatory framework governing the financial sector. Regulations promote stability, transparency, and consumer protection, fostering confidence and sustainable growth.
c. Government Fiscal Policy: Assess the government's fiscal policies, including taxation, public spending, and budget deficit/surplus. These policies impact aggregate demand, resource allocation, and long-term growth prospects.

External Factors

a. Global Economic Environment: Consider the state of the global economy, as it can significantly impact a nation's economic growth and the financial sector. Factors such as trade policies, global financial stability, and international capital flows can have both positive and negative effects.
b. Technological Advances: Analyze the impact of technological advancements on the financial sector, such as the rise of digital payments, fintech innovation, and blockchain technology. Technological progress can enhance efficiency, accessibility, and financial inclusion, thereby driving economic growth.

Empirical Analysis: Conduct quantitative analysis using relevant data and econometric models to assess the relationship between financial sector indicators and economic growth. Regression analysis, time-series modeling, and sector-specific studies can provide insights into causality, spill over effects, and the relative importance of different factors.

CONCLUSION

Economic development in developing countries is a complex and multifaceted process that requires a comprehensive approach involving various stakeholders, policies, and strategies. Economic development in developing countries is a multifaceted and dynamic process that requires a comprehensive and context-specific approach. While there is no one-size-fits-all solution, it is evident that several key factors contribute to successful economic development in these countries.

Figure 3. Financial sectors and economic growth

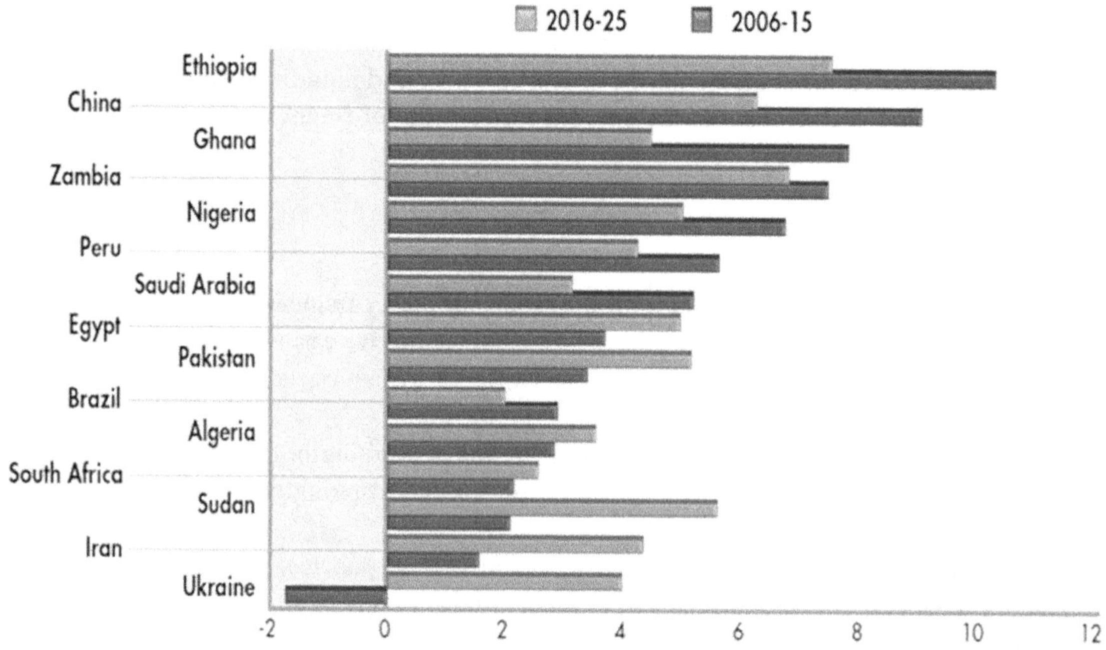

One of the critical determinants of economic development is the establishment of a conducive business environment. Developing countries need to prioritize political stability, establish the rule of law, and combat corruption. A stable and transparent regulatory framework provides the necessary confidence to domestic and foreign investors, encourages entrepreneurship, and stimulates economic growth. Moreover, governments should focus on streamlining bureaucracy, reducing red tape, and improving access to finance for small and medium-sized enterprises (SMEs), which are often the backbone of developing economies.

Investing in human capital is another vital aspect of economic development. Education and skill development play a pivotal role in equipping the workforce with the necessary knowledge and capabilities to drive economic diversification and innovation. Developing countries should prioritize quality primary and secondary education, vocational training programs, and higher education that aligns with the needs of the labor market. Empowering women and marginalized groups with education and skills is also crucial for inclusive economic development.

Infrastructure development is a key enabler of economic growth in developing countries. Access to reliable and affordable transportation networks, energy systems, and information and communication technology (ICT) services is essential for facilitating trade, attracting investment, and connecting remote areas to economic centers (Kumar and Shagun 2009). **Governments should focus on building and maintaining physical infrastructure such as roads, ports, airports, and power plants, while also investing in digital infrastructure to bridge the digital divide and promote e-commerce and digital entrepreneurship.**

Promoting inclusive growth and reducing income inequality are vital for sustainable economic development. Developing countries must address social and economic disparities by implementing policies that promote equal opportunities, improve social protection, and empower marginalized groups. This can be achieved through measures such as progressive taxation, targeted social welfare programs, access to healthcare and education, and promoting women's economic empowerment. Inclusive growth not only contributes to social cohesion and stability but also enhances productivity and unleashes the full potential of a country's human capital.

International trade and investment are crucial drivers of economic development in developing countries. Engaging in global markets through trade and attracting foreign direct investment (FDI) can provide access to new technologies, markets, and capital. Governments should adopt policies that promote trade openness, simplify customs procedures, and eliminate non-tariff barriers. Active participation in regional and global value chains can help developing countries move up the value-added ladder and diversify their exports. Additionally, attracting FDI requires creating an enabling environment, protecting property rights, and providing incentives for foreign investors to establish and expand their operations.

Sustainable development practices are essential for long-term economic growth. Developing countries should balance economic objectives with environmental sustainability, resource management, and climate change mitigation and adaptation. By embracing renewable energy sources, adopting sustainable agricultural practices, and promoting responsible natural resource management, countries can achieve economic development while minimizing their ecological footprint. Green technologies and sustainable practices can also create new economic opportunities and industries. (Valdez and Nunag 2023)

As per Giguashvili et al (2021), **achieving sustainable economic development in developing countries requires a long-term commitment, effective governance, and adaptive strategies. Collaboration between governments, international organizations, civil society, and the private sector is essential for addressing the complex challenges and seizing opportunities for economic growth. By**

focusing on creating a conducive business environment, investing in human capital and infrastructure, promoting inclusive growth, engaging in international trade and investment, and embracing sustainable development practices, developing countries can unlock their economic potential and improve the standard of living for their citizens. While there is no one-size-fits-all solution, several key factors have been identified as critical for promoting economic development in these countries.

Firstly, fostering a conducive business environment is crucial. This involves ensuring political stability, establishing the rule of law, and reducing corruption. A stable and transparent regulatory framework encourages investment, stimulates entrepreneurship, and promotes economic growth.

Secondly, investing in human capital is vital. Developing countries need to prioritize education and skill development to equip their workforce with the necessary knowledge and capabilities for economic diversification and innovation. Education and training programs should be aligned with the evolving needs of the labor market. (Guiamalon 2022).

Thirdly, infrastructure development plays a significant role in economic development. Access to reliable and affordable transportation, energy, and information and communication technology (ICT) services are essential for facilitating trade, attracting investment, and connecting remote areas to economic centers.

Furthermore, promoting inclusive growth and reducing income inequality are crucial for sustainable economic development. Policies that promote equal opportunities, improve social protection, and empower marginalized groups can enhance productivity and create a more equitable society.

International trade and investment also play a vital role in economic development. Developing countries should pursue policies that encourage trade openness, attract foreign direct investment, and participate in regional and global value chains. Access to international markets provides opportunities for export-led growth and technology transfer.

Additionally, promoting sustainable development practices is essential to ensure long-term economic growth. Developing countries need to balance economic objectives with environmental sustainability, resource management, and climate change mitigation and adaptation. Embracing renewable energy, sustainable agriculture, and responsible natural resource management can contribute to both economic development and environmental conservation.

It is important to acknowledge that economic development is a long-term process that requires consistent effort, effective governance, and adaptation to changing global circumstances. Collaboration between governments, international organizations, civil society, and the private sector is crucial for achieving sustainable economic development in developing countries.

REFERENCES

Agarwal, E. (2021). An Overview of Low-Impact Renewable Thermal Energy Technologies and India's Response Toward Climate Change Related Issues. *Cosmos Journal of Engineering & Technology*, *11*(1), 20–23.

Akoley, Elvis, & Wahid, Mumuni Abdul and Kulo, IbnAli Issah. (2022). Public Financial Management and Economic Growth: A Study Case of Ghana. *Cosmos An International Journal of Management*, *11*(2), 29–37.

Bhowmick, K., Hasan, M., & Masood, A. (2022). Employee Creativity, Acceptable Work Environments, and Transformational Management. *Cosmos An International Journal of Management, 12*(1), 08-19.

Dwivedi, M. C. (2022). Assessing The Socioeconomic Implications of Climate Change. *Cosmos An International Journal Art & Higher Education, 11*(2), 20–23.

Fatima, S. (2022). Impact of Covid-19 Pandemic on The Microfinance Sector in India. *Globus An International Journal of Management & IT, 14*(1), 25-27.

Giguashvili, G., Makasarashvili, T., & Khorguashvili, T. (2021). The Main Challenges of Economic and Information Security of Business Companies in Georgia in The Conditions of COVID-19. *Globus An International Journal of Management & IT, 13*(1), 19–23.

Guiamalon, T. S. (2022). Social and Economic Development: State Universities and Colleges' (SUC's) Contribution Creativity Skills of The Students in Recycling. *Globus Journal of Progressive Education, 13*(1), 107–111.

Khari, D., & Sharma, V. & Agarwal, N. (2020). Effect of pandemic COVID-19 on economic crisis and health issues globally. *Cosmos Journal of Engineering & Technology, 10*(1), 9- 15.

Mahanand, M., & Singh, D. S. (2019). A Study on Social and Economic Justice by Fundamental Rights. *Globus Journal of Progressive Education, 9*(1), 38–40.

Malik, Akash and Gharpe, Dr PS. (2022). Rehabilitation Engineering and Its Technological Solutions to Problems Confronted by Engineers. Globus An International Journal of Medical Science. *Engineering and Technology, 11*(2), 24–27.

Puneet, K., & Shagun, P. (2009). Operation Research Endorsing The Social Transformation with Management and Information Technology Convergent. *IIMT Business Review, 1*(1), 47–52.

Sangwan, D. S. (2022). Impact of Russia Ukraine War on Indian Economy. Globus An International Journal of Medical Science. *Engineering and Technology, 11*(1), 36–39.

Segismundo, M. & Dayao, E. P. (2021). Employability and Productivity of The 21st Century Workforce: A Tracer Study of LCUP-CBEA Graduates AY 2015 to 2018. *Cosmos Journal of Engineering & Technology, 11*(2), 39-47.

Singh, G., Gupta, R. and Agarwal, Nidhi (2020). Economic Re-Engineering: COVID-19. *International Journal on Emerging Technologies, 11*(4), 01-05.

Singh, M. (2019). Disaster Management: Managing The Risk of Environmental Calamity. *Cosmos An International Journal Art & Higher Education, 8*(2), 25–27.

Valdez, J. & Nunag, R. (2023). The Research and Creative Works Colloquium and Its Effect on Research Performance. *Cosmos Journal of Engineering & Technology, 13*(1), 57–69.

Chapter 21
Bioeconomy and Ecosystem of China's Polar Silk Road

Maria João Tomás
https://orcid.org/0000-0003-4922-6724
Universidade Autónoma de Lisboa, Portugal

ABSTRACT

The Polar Silk Road is one of the People´s Republic of China's big projects. As part of the grand plan of the "One Belt, One Road" initiative, it is the continuation, and reinforcement, of the country's Arctic strategy. In 2018, China officially affirmed its goals with the publication of the book China's Arctic Policy announcing the creation of the Arctic Route, reinforcing a more proactive policy towards the region. The exploration of the Arctic with the ongoing thaw raises major questions about the sustainability of a changing ecosystem. The concerns of the Arctic bioeconomy are increasingly high, not only with navigability but also with the exploitation of the riches under the ice. This article is about the political and economic ambitions of China in the Artic, and how they are compatible with bioeconomy and sustainability policies, being the North Polar region one of the most fragile ecosystems on earth, and one of the most dynamic, derivatives of the permafrost thawing.

INTRODUCTION: CHINA AND BIOECONOMY

According to the definition of the Council for the Bioeconomy, Bio-based Economy or Bioeconomy is the science that studies a new model for industry and the economy, using in a sustainable way renewable sources, biological systems, and natural resources allied to the use of new technologies. The economic model of the bioeconomy replaces the use of fossil resources with bio-based renewable resources and aims to create more sustainable products and services by exploiting the untapped stored potential of land and marine resources in millions of tonnes of biological waste. In this way, it is intended to allow economic and social development, respecting the natural limits of terrestrial and marine ecosystems (BW 2023)

Building a sustainable growth strategy with greater use of renewable resources is no longer just an option, it is a necessity and is expected to play a significant role in post-oil economic production, facilitating the evolution to a green economy based on renewable and biological resources. Sustainable

DOI: 10.4018/978-1-6684-8879-9.ch021

bioeconomy strategies are consistent with the United Nations (UN) Sustainable Development Goals (SDGs) (OECD 2018).

The bioeconomy requires a unique knowledge of the sectors involved and their economic, social, and environmental metrics to be able to adapt the policy measures needed to achieve a sustainable circular economy based on renewable resources. The fact that it has already gained traction in numerous sectors and industries shows that it is possible to build great products, services, businesses, and careers in a much more sustainable way, using life sciences and biotechnology as a model for the future development of human society. (Bracco, S.et al., 2018)

The emerging bioeconomy is expected to reach a global dimension and be fundamentally guided by green development and environmental sustainability. The transition to a postoil economy has been widely recognized. More than 50 countries, such as European Union, United States, China, India or South Africa, have already developed sustainable bioeconomy strategies, driven by biotechnology-related research and development (Kuosmanen et al. 2020, Lier et al. 2018, FAO 2018, Bracco et al.2018).

China has joined the bioeconomy with the "Bioindustry Development Plan", for the sustainable development of society and technological development, promoting biotechnology and life sciences, and fostering the use of biological resources. applied to agriculture, science, and technology (Xu Zhang et al. 2022).

The Chinese Biocenomy Plan, the 14th Five-Year Plan for Bioeconomy Development- figure 1, is divided into two phases, the first from 2021-2025, focusing on innovation, and industrial development. Has strategic priorities for agriculture, energy, forestry, and the environment. Despite initial scepticism, China's bioeconomy has been steadily growing, with direct government support through China's National Development and Reform Commission (PRC 2018).

The second phase, from 2026-2035, aims to create an innovative and intelligent ecosystem, based on technological innovation and bioindustry with controlled bio risks, to position China as a leader in the global bioeconomy industry. (Xu Zhang et al. 2022)

Figure 1. 14th five-year plan for bioeconomy development
Source: Graphic outline of China's 14th Five-Year Bioeconomy Development Plan (Xu Zhang et al. 2022)

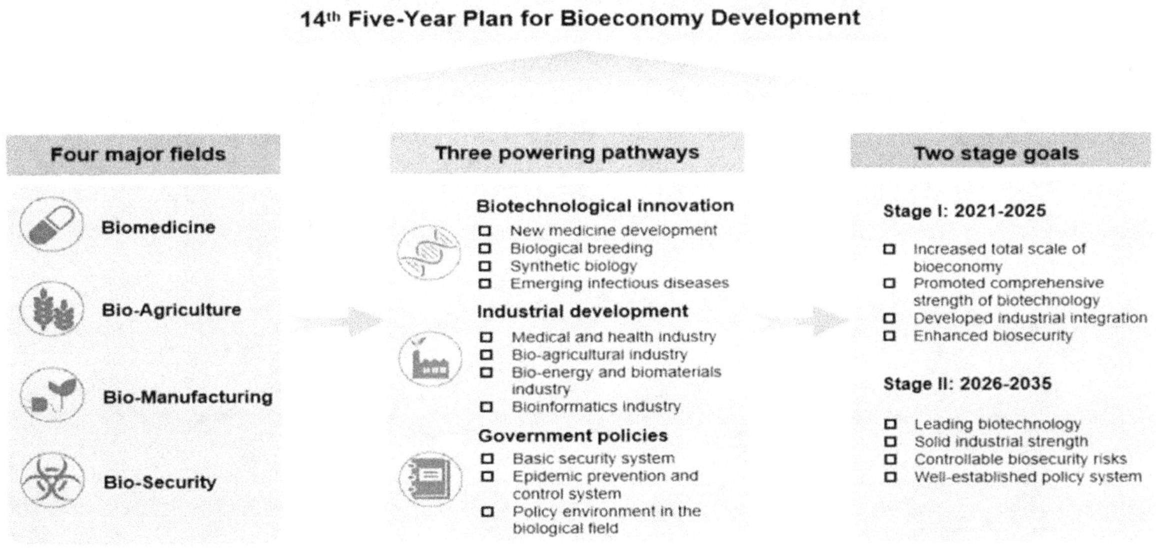

The Upper North

The Arctic is a region situated in the extreme north of the planet Earth, the Upper North, is not considered a continent, but includes parts of the territories of the following countries: United States, Canada, Russia, Denmark (Greenland), Sweden, Norway, and Finland. Is also called *Polar Mediterranean,* figure 2 and has a surface area of 14,090,000 km^2, with an average depth of 1,500 meters. (Bird, J. et al. 2023)

The Arctic region includes three major biomes: the polar desert (closest to the North Pole), the tundra, and the boreal forest (also known as taiga in Eurasia) located in the southern parts of the Arctic (Bird et al.2023).

There are 4 routes in the Arctic:

The Northwest Route (NWP); The Northern Route (NSR); The Transpolar Route (TSR) and The Arctic Bridge Route (ABR).

Figure 2. Countries bordering the Artic
Source: WorldAtlas

The most important routes are the NWP and NSR routes since they allow the connection of two different continents. NWP bypasses the American continent from the north, connecting the Atlantic to the Pacific, and the NSR connects the Asian continent to the European.

- The Northern Route (NSR):

So far, the main Arctic route, linking the Pacific Ocean to the Atlantic, allows to unite the Asian and European markets, reducing the route of the Suez Canal by approximately 40%, less 14 days of travel. It is navigable one month a year, and only with the help of icebreakers.

Russia collects taxes and imposes restrictions because it crosses its exclusive economic zone for 300 kilometers of coastline. With the thaw, Russia loses sovereignty of traffic through the Arctic, as ships would pass through international waters.

- Northwest Route (NWP):

The NWP route connects the Pacific and Atlantic oceans through Canada and is primarily used by passenger ships, although some cargo ships also use it. It does not bring great advantages compared to the passage through the Panama Canal, it also needs icebreakers and there are not many ports yet.

- The Arctic Bridge Route (ABR):

It is the route that passes near Greenland, joining the port of Churchill in Canada, the port of Narvik in Norway, and the port of Murmansk in Russia. With the increasing thaw, it has many icebergs, making navigation unsafe.

- The Transpolar Route (TPR):

It is the route that crosses the entire Arctic in a straight line. It will be the most desired, but only with total thawing will it be navigable. It is used for scientific purposes. (Humpert, M., 2011)

The Arctic has been frozen since the Quaternary 2.6 million years ago, including a saline water ice cap of approximately 16.5 million square kilometers.

In winter, the temperature can reach up to minus 60 degrees, called persistent cold, and is constantly at night. In summer, the temperature should not exceed 10º. With global warming, there are areas that have higher temperatures and for more than 30 days. On June 20, 2020, The World Meteorological Organization recorded 30th in the Russian city of Verkhoyansk (@WMO 10:51 a.m. December 14, 2021)

WMO has frequently warned that the Arctic is one of the fastest warming parts of the world (WMO 2016) and the region of the planet most affected by global warming. The loss of sea ice, warming waters, rising sea levels, along with increasing global temperatures, and the extensive displacement of the Arctic ice cap, are transforming the Arctic ecosystem. (WMO, 2021)

On figure 4 it can be seen the damage caused by melting ice in the Russian Arctic. In Russia alone, global warming is estimated to be in the order of 1.5-2.5 °C over the past 40 years, and in Alaska, it has even reached 3 °C (Biskaborn et al. 2019). By the end of the century, the planet's icy surface could shrink by 30-85% (Melnikov et al. 2022).

Figure 3. Artic Routes
Source: The Artic Institute, Centre for circumpolar security studies

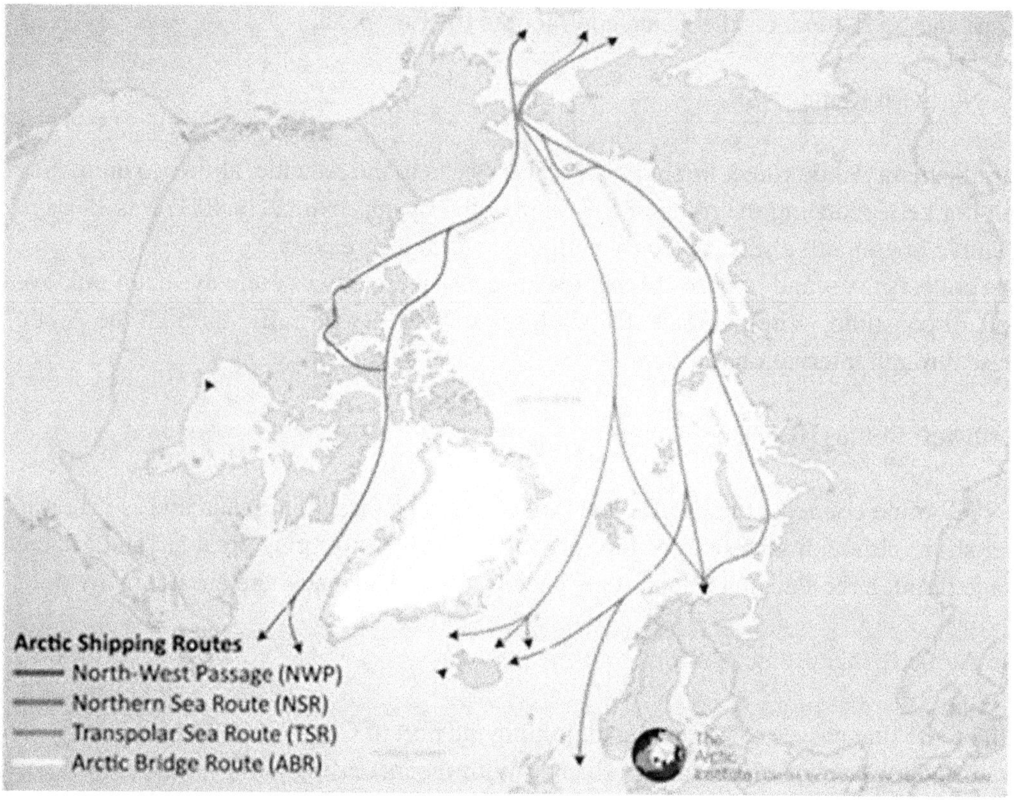

Figure 4. Thaw in Russian Artic

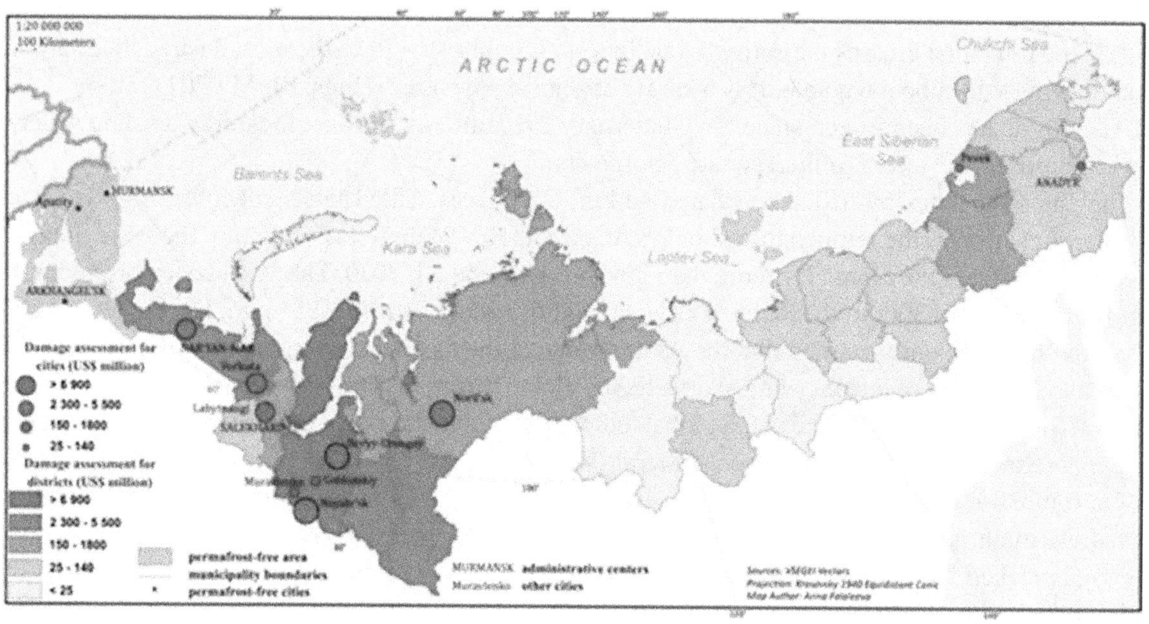

On figure 5 one can see the melt forecast of average annual soil temperature by 2050 in municipalities of the Russian Arctic (Melnikov et al. 2022).

The Arctic cryosphere is collapsing, thawing permafrost, with consequences for the functioning of Arctic, which environment is dependent on ground temperatures. Terrestrial melting permafrost is now more and more frequent in the Artic. Defined as the soil or bedrock that reaches below 0 °C for at least two consecutive years, permafrost degradation has the potential to disrupt ecosystem, reducing the Arctic's wildlife and discovering chemical, and radioactive materials hidden underneath the ice for millenniums (Hu, G. et al.2021).

Artic thawing may affect also other regions of the planet since the many climate tipping points (CTP) are affected by warming feedback. For example, if the global average temperature increase to 1.5°C, Greenland is at risk of collapse (McKay et al., 2022).

The rapid climatic change in the Artic, the fast melting of permafrost, the decline of glaciers and the melting of sea ice created strategic and economic opportunities for the littoral states (Lasserre 2015).

With the melt, the navigability of the Arctic has been growing and there are more and more navigable routes. It is expected that between 2035 and 2065 Artic routes will be navigable for more than one month per year. These routes will shorten distances and approach continents and oceans. (Palmer, L. 2013).

On the other hand, the thawing also reveals valuable resources hidden under the ice. The U.S. Geological Survey (Wandrey et al.2008), estimates that the Arctic shelf's projected volume of undiscovered oil and gas reserves is about 90 billion barrels, 1.670 billion cubic feet of natural gas, and 44 billion barrels of natural gas liquids, according to estimates from the Arctic Geological Survey. These untapped resources amount to nearly 20 to 25 percent of the world's undiscovered resources. There is also the possibility to find diamonds, precious stones, platinum, lead, manganese, nickel, iron, uranium, copper, and rare lands vital to the so-called "Fourth Industrial Revolution" of Artificial Intelligence. In the midst of the

Figure 5. Melt forecast in Russian Artic

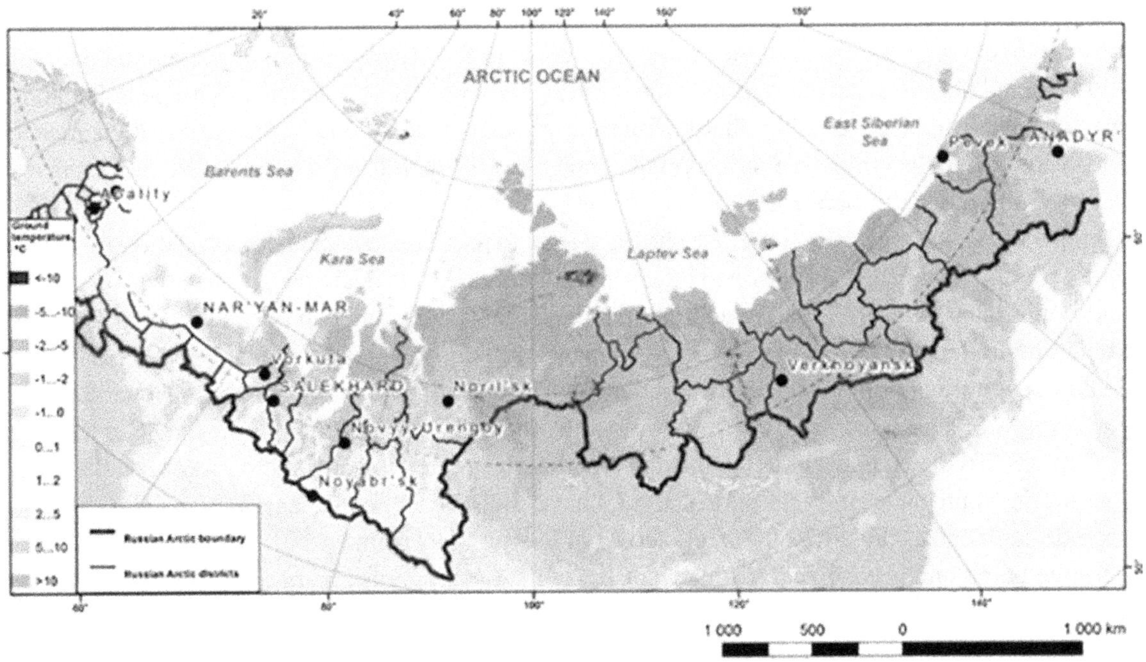

technological Cold War with the USA, China wants to assume itself as the world's dominant supplier of rare earth elements, and Artic melting is helping China to achieve it (EIA 2012).

The geopolitics of the Arctic is changing, drawing worldwide attention for economic, geopolitical, and environmental reasons, among others. The melting has increased the economic profitability of the region, making it very interesting and very coveted. "For these reasons, the Arctic has gained a political-strategic importance and a geopolitical centrality" (Guedes 2015).

China and the Arctic Council

The eight countries that geographically surround the Artic are Norway, Russia, Canada, Denmark, Finland, Iceland, Sweden, and the United States. They form the Arctic Council, an intergovernmental organization that deals with matters related to this region. The Artic Council was founded in 1991 and formally constituted in 1996. The "indigenous groups" of the Arctic region have joined with the status of permanent participants. Together, they harness their knowledge of the Arctic ecosystem, intending to protect the environment and develop sustainable forms of development, keeping and protecting indigenous cultures. In mid-2013, China, India, Italy, Japan, Singapore, and South Korea entered as Observer States (Exner-Pirot, H. et al., 2022, Norsk, 2023).

The active interest of China in the Artic began very early, in 1993. At that time, China bought the *icebreaker Snow Dragon* from Ukraine and undertook several expeditions to the Arctic and Antarctic regions. In 2012 the Chinese icebreaker Xue Long (Snow Dragon) navigated the Northern Sea Route along the Russian coast. In mid-2013 China's commercial ship, the Ocean Shipping Company, *Yong Sheng*, began the first voyage from a Chinese port to Rotterdam via the NSR. In 2017 China carried out the first transit route with the Chinese icebreaker *RV Xuelong* from China to Iceland and still in 2017, the research vessel Xue Long became the first Chinese ship to sail on the three main sea routes of the Arctic: Northwest Passage, Northeast Passage, and Transpolar Sea Route (Hui, 2018, Alexeeva, O., et al.. 2012b.).

China has also established itself in scientific research, especially on the TPR route, both in the study and survey of existing resources in the Arctic, as well as in the effects of climate change on the region. In 1999 opened the first research station in the Arctic, the Yellow River, then another one in Iceland and in the Norwegian archipelago of Svalbard, with the purpose to study climate changes. In 2004 created its scientific station in the Arctic region and made an agreement with Russia to establish various research centerers along the North Sea (Hui, 2018).

China's activity in the Artic strengthens its legal right to participate in decisions that directly affect the Arctic, claiming to have a valid reason to intervene straight in Arctic decisions, both in the consequences of the melting and in terms of safety and navigation. This way Beijing can control its own interests by participating in the growth and expansion of the new Arctic regional order (Okamoto, T. 2019).

China's aspiration for inclusion in the Arctic Council led to the creation in 2018 of a legal framework that allowed it to legitimize its interests in the Arctic, the "China's Arctic Policy", the "Chinese White Paper", (PRC 2018), where Beijing manifested its proactive policy towards the region, with the creation of the Arctic Route as an integral part of the Chinese Belt Road Investment (BRI), linking Asia and Europe through canals that cross the Arctic coast of Russia. (Hossein, K. 2019).

Despite being on the periphery, China wants to share the concerns and problems affecting the Artic and participate in the growth and expansion of the new Arctic regional order, with the concept of "quasi-Arctic State" (Okamoto, T. 2019).

China Polar Route

The idea of connecting the Far East and Europe by rail had its origin in the construction of the Trans-Siberian Railway, also called Eurasian Landbridge, linking Moscow to Vladivostok, figure 6, completed in 1916. But global warming has made greater use of Arctic Sea lanes feasible.

The new Northern Sea Route extends along the northern coast of Russia and connects the Bering Strait to the Norwegian Sea. These transport lines, in addition to being shorter, make it possible to reduce costs by sea. New Maritime Terminal projects are also planned throughout Russia's Arctic coast – which includes locations such as Murmansk, Vladivostok, and Arkhangelsk – forming important logistics platforms for the economy. allowing the use of circumpolar pathways for commercial transportation and extraction of natural resources (Gladkiy, Y., *et al* 2020).

China's Polar Road, a $4 trillion investment, is part of China's Belt Road Initiative (BRI) and reflects not only its ambition to pursue its global economic and geopolitical interests but also its desire to become the world's largest economy. BRI, figure 7, is a project designed in 2013 by China's President Xi Jinping, who was inspired by the concept of the Ancient Silk Road. (Tomás 2021).

The main objective of BRI is the use of the Northern Sea Route (NSR), the Northwest Passage (NWP), and the Transpolar Route, TPR. NSR, the route that connects Europe and Asia through Arctic ice, the so-called " Blue Route ", is the first route to be free of Arctic ice, the greatest, and the fastest with commercial viability (Sharma, T., 2019).

Figure 6. The Trans-Asian railway (Eurasian landbridge)
Source: The Geography of Transport System

For China, the NSR gives a major competitive advantage because it minimizes the sea distance travelled between East Asia and Western Europe to 12,800 km instead of the 21,000 km of the traditional Suez Canal route, reducing the transport time by 10 to 15 days, avoiding the problems around the China Sea and the China Sea "Malacca Dilemma"(Sharma, T., 2019).

On July 4, 2017, during a meeting in Moscow, and according to *World Maritime News*, Chinese President Xi Jinping and Russian Prime Minister Dmitry Medvedev announced their intention to develop a "Polar Silk Road" through the Northeast Passage in the Arctic, which the Russians call the Northern Sea Route (Camilla, T., et al. 2017).

BRI

This major economic project includes several sea and land routes, railways, roads, and deep-water ports that facilitate the transport of people, goods, and energy, covering an area that involves bilateral agreements with 66 countries, more than 4.4 billion people, and almost 40% of the world economy, with an economic impact of 21 trillion dollars.This huge investment of 50 billion dollars is already nicknamed as the great economic project of the 21st century. This mega project includes the construction of sea, rail and road routes, that will facilitate the transportation of people, goods and energy, covering an area that involves bilateral agreements with 66 countries, more than 4.4 billion people, and almost 40% of the world's economy, with an economic impact of 21 trillion of dollars, and a direct investment of 1 trillion of dollars. 90% of the investment is financed by the Silk Road Fund, founded in December 2014, with the following composition: 65% of the investment is provided by the State Administration of Foreign Policy, 15% by the Investment Corporation of China, 5% by the Development Bank of China and 15% by the Export and Import Bank of China. The remaining 10% are from European banks" (OECD, 2018; People's Republic of China 2015, Tomás 2021).

Some scholars have examined factors fuelling Chinese ambitions in the Arctic, giving focus on political, economic, and military implications. Chinese investments in Arctic countries, especially in mines and transportation infrastructure, can be geoeconomic tools to exert political and military influence. China claims to have the right to navigate the polar route under the United Nations Convention on the Law of the Sea (UNCLOS), which clearly allows freedom of navigation on international waterways. But there are legal questions about the territorial rights of Arctic states that China must be aware (Jakobson 2010, Ostreng, W., 2012).

"“This "Silk Road Economic Belt" rehabilitates the economy of all the countries geographically involved" (Tomás 2022). To assert itself in the Arctic, China seeks to establish close alliances and collaborations with Arctic Council states, further strengthening and promoting bilateral economic ties, networks, and cooperation. (Tillman, H., et al. 2018). "On the other hand, there is also the question of the real sustainability of the project, whether investments and loans will ever have a return" (Tomás 2022).

To date, China is the only country to lead official expeditions through the three Arctic Sea passages. China's exponentially growing interest in the development of Arctic resources raises questions about the geopolitical ramifications of the Belt Road Initiative and the political implications of Chinese investment in the Arctic.

Arctic Council countries claim Exclusive Economic Zones (EEZs) for mining and exploitation of other resources (Norsk, L. 2023). China began mining in Greenland in 2015, raising issues of international law to explore the continental shelf as provided for by the United Nations Convention on the Law of the Sea. But in 2016 Denmark refused to sell to the Chinese mineralization company the naval base in

Figure 7. Belt road initiative
Source: Merics

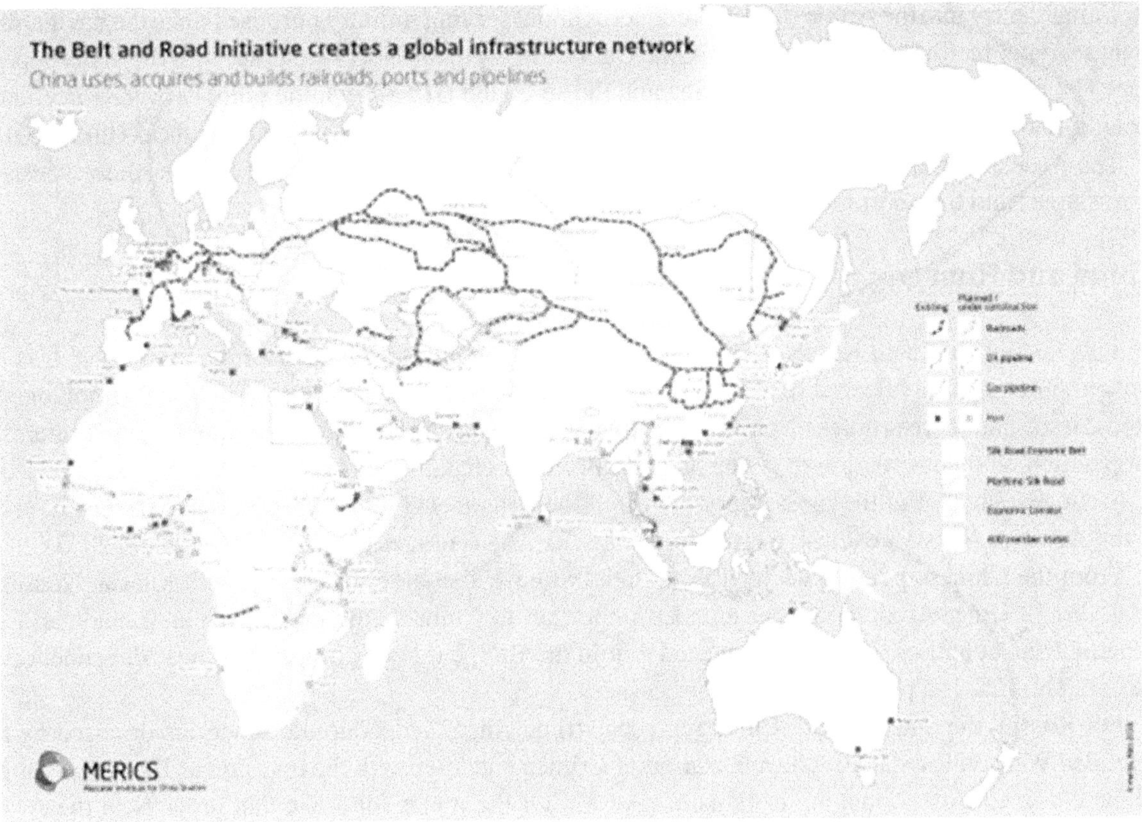

Greenland. China is also actively collaborating with Russia in mining at the Yamal gas field (Gladkiy, Y., *et al* 2020).

In December 1998, Russia and China signed the first post-Cold War treat, with the aim of weakening American hegemony. In 2001, with Vladimir Putin as prime minister, they signed the Treaty of Good-Neighbour and Friendly Cooperation, with the same objective of diminishing American power, especially in central Asia. Despite having the same common purpose, China wants to lead the world economy, dethroning USA, while Russia wants to recover the position as world leader, lost after the Cold War, to maintain the sphere of influence in in former URSS states, preventing the U.S.A. and NATO from getting closer, being closer to Europe and less isolated (Tomás 2021, Karaganov, S., 2018).

The rush to explore the Arctic is raising tensions in a region with a fragile ecosystem. The Arctic Council has remained cautious about Chinese initiatives in the region and the bilateral cooperation between China and the different Arctic states, as well as their economic and geopolitical potential (Røseth 2014, Peng 2015).

In this way, China looks forward to working multilaterally with all parties to research shipping routes, climate change and the environment. In 2018 Norway revolted against Russian mining in the Barents Sea and cancelled a Chinese resort in Svalbard, in 2011 Iceland blocked the sale of a large plot of land to a Chinese investor. The Danish government bought the base to prevent it from falling into Chinese hands. Despite these contradictions, China is trying to overcome and appears to be planning to buy two ports in Iceland, as well as the Norwegian Arctic port Kirkenes, as part of the *roadway initiative*, although these projects have not yet come to fruition.

China's entry into the Arctic increases concerns about serving military purposes since the navigability of these routes facilitate the deployment of the Chinese Navy to the North Atlantic, with Russian cooperation. The Bering Strait, the Norwegian Sea and the so-called *GIUK Gap*, the point between Greenland, Iceland, and the United Kingdom, will turn to be critical areas of great strategic relevance (Hui,L.,2018).

The Artic affairs may not be considered strictly regional since climate change makes the vast wealth of resources and transportation routes accessible to the world.

China and Russia

China's relationship with Russia is the guarantor of the viability of China's Arctic ambitions. It is the most important of all bilateral relations, not only to achieve the navigability of the NSR, implementation of trade and tourist routes, exploration of hydrocarbons, gas and oil, telecommunications, scientific cooperation, and construction of research stations on the changing climatic conditions and knowledge of Arctic resources. But the most important of all the agreements for Beijing is to have the right to act in the Arctic in Russia's exclusive economic zone (EEZ) 8 (Humpert, M., 2013, Chen, C., 2023).

From the Chinese point of view, it is essential to have a stable relationship with Russia. (Camilla, et al. 2017). Other Arctic countries are also important to China's ambitions, such as Canada and the Scandinavian countries, which are expected to join the BRI, but none compared to near-dependence on Russia. The U.S. is a part of this whole process.

For Russia, the alliance with China represents financing. The economic sanctions imposed by the extended West on Russia, after he annexation of Crimea and after with the invasion of Ukraine, brought China closer to Russia, making it its main investor. Of the major Sino-Russian projects in the Arctic financed by China Development Bank is the Arctic LNG II operated by Russian companies Novatek (60%) and Ekropromstroy (40%). This project gives China access to Liquefied Natural Gas, LNG, in the Arctic, with a capacity of 19.8 million tons. In exchange for financing exploration, China not only gets cheaper energy, but also takes the lead and oversight of exploration. Other similar projects are Yamal LNG7, the station located in northwestern Siberia for the exploration of liquefied gas, the Belkomur railway, a 1,160-kilometer line connecting Perm in the Ural Mountains to the Arkhangelsk port in the Russian Arctic, and the Artic LNG 2 which provides for the construction of three trains powered by Liquefied Natural Gas, with a capacity of 6.6 million tonnes of LNG. This latest project is based on the hydrocarbon resources of the Utrenneye field in the Russian Arctic, where China holds a major stake and where it is developing an innovative construction concept based on GBS structures, gravity-based structures. (Humpert, M., 2013, Chen, C., 2023)

All these projects – figure 8- involving the exploration, handling and transportation of LNG involve environmental and sustainability risks. Technical complexity requires the combination of a range of intellectual, technological, human, and financial resources in order to allow everything to be done safely throughout the year. But the stakes are high, and any carelessness poses the possibility of incalculable damage in the very fragile and changing ecosystem of the Artic.

Figure 8. Market routes along the Arctic Ocean
Source: Malte Humpert, *High North News*

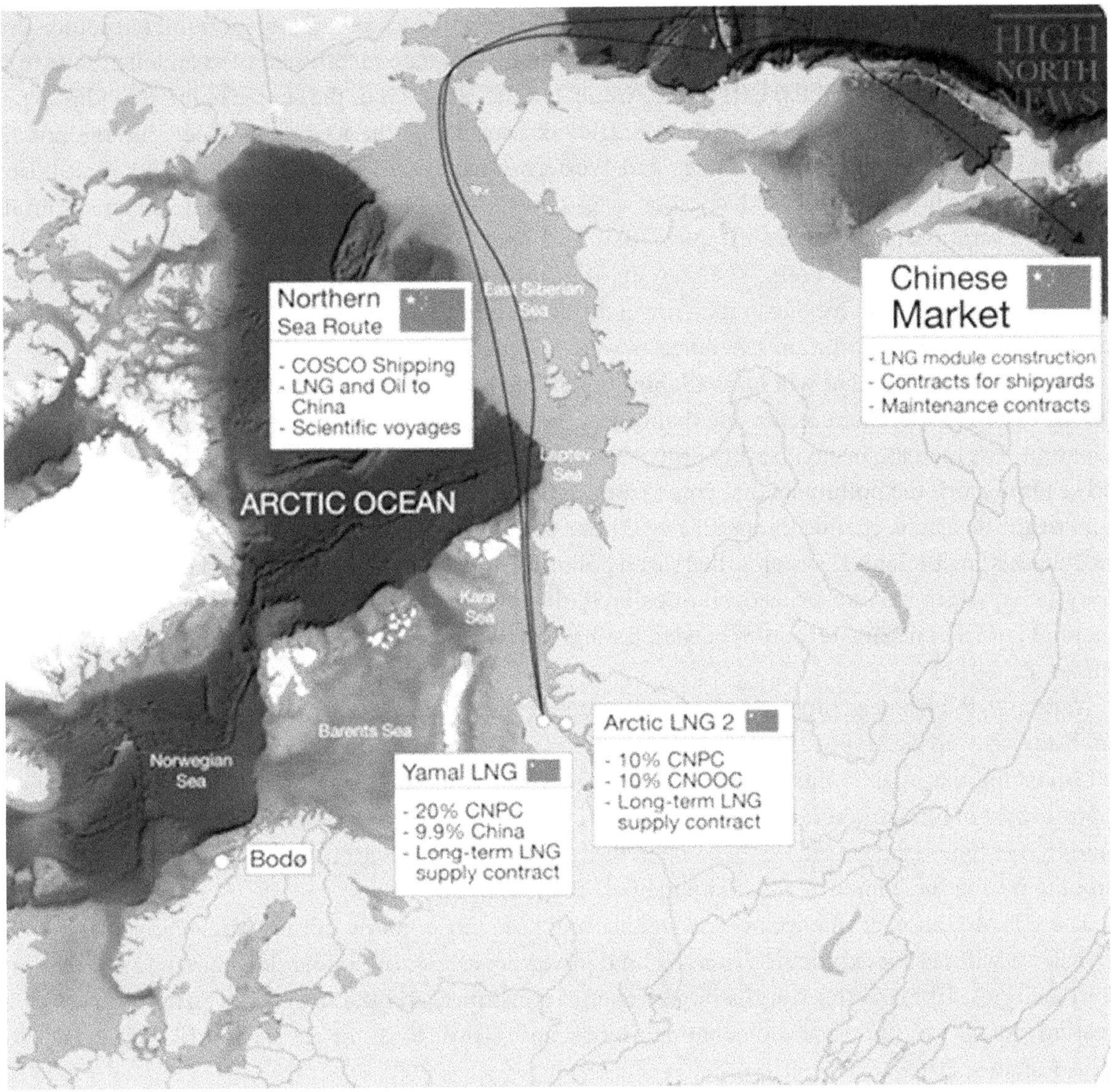

Sustainability of the Artic

The sustainability of the Arctic is at danger, with real consequences for the bioeconomy of this part of the planet and, consequently, for the rest of the world. The melting of the Arctic is causing coastal erosion and reduced albedo (surface reflectivity), average sea level rise, and warming of the oceans due to feedback loops, but it is also discovering riches under the ice and the possibility of commercial navigability. (Aksenov et al., 2017).

And there are many questions about when Arctic navigability will be a reality. Although the Arctic ice sheet continues to shrink, it is still not possible to keep any of the northern sea passages open year-round. However, anticipating a wave of maritime transport in the area, there are already shipowners investing in ships capable of navigating through the ice. (Botillen, A. et al., 2022)

The unpredictability of the climate, the difficulties of creating a modern transport infrastructure along the sea route, and the technical needs for its realization, originate an attitude of scepticism toward the potential of the extension of the BRI in the Arctic. On the other hand, the severe natural conditions, the restrictions of the navigation season, the insufficient capacity of the icebreaker fleet, and the possible economic sanctions of the United States and Western Europe are also vulnerable points that must be considered. On the other hand, unlike other long-distance commercial shipping routes, there is limited revenue-generating potential for shipping lines and operating times are also short, requiring special equipment, thus increasing costs. The scarcity of infrastructure networks makes transportation difficult and economically expensive due to long travel distances and adverse weather conditions, drastically and directly affecting transportation schedules as well. (Gladkiy et al., 2020

The NSR is the one that will be available sooner, the most viable in the short term, and the one that will have the most traffic, it is the one that brings the most apprehension and several issues arise in anticipation. The increase in navigability brings various pollution problems resulting from long-distance air and sea transport, the pollutants that come from it, the possibility of oil spills, the increased likelihood of maritime disasters, chemicals at sea, as well as associated human activities are issues of concern that threaten existing biological systems. Increased pollution threatens slow-growing and long-lasting marine ecosystems, which may never recover, threaten biological systems and natural resources, as well as the potential land and marine resources needed for a green and circular economy (Melnikov, V. et al. 2022, Palmer, L., 2013)

Naturally, the Chinese highlight the economic and social benefits that the navigability of the Arctic will bring, as well as the positive factors that China will bring to the region, not only in the investments and financing that will be made in the isolated areas and vulnerable communities of the Arctic, in the improvement of infrastructure and new economic opportunities They downplay the impact by stating that routes to become regular could only be open during the summer, with the Arctic closed to commercial shipping during the winter months (Dong, et al. 2021, Gao 2015).

There is also another important aspect concerning the bioeconomy of the Artic. The thawing of the permafrost will permit exploration, mining, and prospects of gas, oil, mineral resources, and other non-fossil energies. China is the world's largest energy consumer, and the use of the Artic reserves offers great advantages to the domestic economic landscape. (Bird, K. J., et al. 2008, EIA 2012, Jakobson 2010, Lasserre 2015).

The exploration of these materials raises major questions about the sustainability of the Arctic ecosystem since the risks are even higher than the navigability of the polar routes. (Miner, K. R. et al. 2022, Manta, F., 2019).

China's bilateral cooperation with the Artic countries, especially with Russia, have political and economic implications. Geopolitical issues raise major doubts about the viability of these alliances, but these bilateral relations with European countries may be the guarantee that bioeconomic measures will be implemented, not only in the navigability od the Artic, but also in mining and resources exploration.

CONCLUSION

Bioeconomy measures should be implemented to prevent catastrophic consequences in the Artic ecosystem post-permafrost thaw.

Of course, the best should be blocking NSR navigability. The existent Eurasian Land Bridge offers a more stable option with a longer-lasting performance than Arctic routes. These rail corridors between China and Europe through Central Asia are efficient and don't have such a negative impact as the navigability of the Polar Routes.

The exploration of Artic resources should also be avoided, but since China has joined the bioeconomy, there are a lot of green agenda measures that can be implemented to assure the sustainability of China Artic's activities. The bilateral cooperation with the Artic countries may be the guarantee that they will be implemented.

Practices such as the adoption of innovative practices for environmental sustainability compatible with the green economy standards, the implementation of certificated practices and systems for the logistics and supply chain, the so-called green logistics strategies to reduce energy footprint focusing on material handling, prevention and clean-up of oil spills, waste management, packaging, and transportation, are some of the examples of sustainable strategies that China can implement to mitigate the impact of human activity in the Artic. Human activity rises also more concerns about the warming. There are measures that can be implemented and

China´s 14th Five-Year Plan for Bioeconomy Development has 2025 and 2035 as milestones to position China as a leader in the global bioeconomy industry. This China commitment should also contemplate Artic sustainability.

North Atlantic states should then be concerned with climate change mitigation and adaptation in the Arctic, as well as at home. The scientific study of Arctic climate processes is one area where they can give a valuable contribution, as shown by the European Space Agency's programme of sea ice monitoring through satellite imagery (ESA, 2022)

However, enforcing rules and regulations may become more of a challenge as interstate relations in the Arctic become more confrontational. North Atlantic states should be careful not to let governance arrangements be used as a cover behind which Arctic and non-Arctic states such as Russia and China can endanger global common goods such as the climate and maritime security

REFERENCES

Aksenov, Y. (2017). On the Future Navigability of Arctic Sea Routes: High-Resolution Projections of the Arctic Ocean and Sea Ice. *Política Marinha,* 75, 300–17, https://www.sciencedirect.com/

Alexeeva, O. (2012b). Le dragon des neiges. Les stratégies de la Chine en Arctique. *Perspectives chinoises, 3,* 61–68. https://journals.openedition.org/perspectiveschinoises/6345

Bird, J. (2023). Arctic. *Encyclopaedia Britannica.* https://www.britannica.com/place/Arctic

Bird, K. J. (2008). Circum-Arctic resource appraisal; estimates of undiscovered oil and gas north of the Arctic Circle. *U.S. Geological Survey Fact Sheet 2008-3049.* PUBS. https://pubs.usgs.gov/fs/2008/3049/

Biskaborn, B., Smith, S. L., Noetzli, J., Matthes, H., Vieira, G., Streletskiy, D. A., Schoeneich, P., Romanovsky, V. E., Lewkowicz, A. G., Abramov, A., Allard, M., Boike, J., Cable, W. L., Christiansen, H. H., Delaloye, R., Diekmann, B., Drozdov, D., Etzelmüller, B., Grosse, G., & Lantuit, H. (2019). 2019, "Permafrost is warming at a global scale". *Nature Communications*, *10*(1), 264. doi:10.103841467-018-08240-4 PMID:30651568

Botillen, A. (2022). Maritime Security in the Arctic. In R.-L. Boşilcă S. Ferreira & B. J. Ryan (eds.) *Routledge Handbook of Maritime Security*. London and New York, NY: Routledge. https://www.routledge.com/Routledge-Handbook-of-Maritime-Security/Bosilca-Ferreira-Ryan/p/book/9780367430641

Bracco, S., Calicioglu, O., Gomez San Juan, M., & Flammini, A. (2018). Assessing the Contribution of Bioeconomy to the Total Economy: A Review of National Frameworks. *Sustainability (Basel)*, *10*(6), 1698. doi:10.3390u10061698

BW. (2023). *What is bioeconomy*? Bioeconomy BW. https://www.biooekonomie-bw.de/en/bw/definition (Accessed 12.06.2023)

Camilla, T. (2017). Emerging Chinese-Russian Cooperation in the Arctic: Possibilities and Constraints" (Cooperação emergente sino-russa no Ártico: possibilidades e restrições). *SIPRI Policy Paper, 46*. https://www.sipri.org/. (Accessed 13.06.2023)

Cao, C. (2021). The evolution of China's biosafety/biosecurity legislation. *Journal of Law and the Biosciences*, *8*(1), LSB020. doi:10.1093/jlb/lsab020 PMID:34221436

Chen, C. (2023). China-Russia Arctic Cooperation in the Context of a Divided Arctic. *Arctic Collaboration Series 2023*, The Artic Institute https://www.thearcticinstitute.org/china-russia-arctic-cooperation-context-divided-arctic/

Dong, Y., Yang, M., Bakker, D. C. E., Liss, P. S., Kitidis, V., Brown, I., Chierici, M., Fransson, A., & Bell, T. G. (2021). Near-Surface Stratification Due to Ice Melt Biases Arctic Air-Sea CO_2 Flux Estimates. *Geophysical Research Letters*, *48*(22). doi:10.1029/2021GL095266

Drewniak, M., Dalaklis, D., Kitada, M., Ölçer, A., & Ballini, F. (2018). Geopolitics of Arctic Shipping: The state of icebreakers and future needs. *Polar Geography*, *41*(2), 107–125. doi:10.1080/1088937X.2018.1455756

Economist. (2019). Who owns the Northwest Passage? *The Economist*. https://www.economist.com/

EIA. (2012). *Arctic oil and natural gas resources*. U.S. Energy Information Administration. https://www.eia.gov/todayinenergy/detail.php?id=4650

El-Chichakli, B., von Braun, J., Lang, C., Barben, D., & Philp, J. (2016). Policy: Five cornerstones of a global bioeconomy. *Nature*, *535*(7611), 221–223. https://doi.org/. doi:10.1038/535221a PMID:27411618

European Commission. (2018). *A sustainable Bioeconomy for Europe: strengthening the connection between economy, society and the environment*. EC. https://ec.europa.eu/research/bioeconomy/pdf/ec_bioeconomy_strategy_2018.pdf.

European Commission. (2019). *The European Green Deal*, Brussels, Belgium.https://ec.europa.eu/info/sites/info/files/european-greendeal-communication_en.pdf.

Exner-Pirot, H. (2022). Opinion: Does the Arctic Council make sense without Russia? *National Post*. https://nationalpost.com/opinion/opinion-does-the-arctic-council-make-sensewithout-russia

FAO. (2019). *Indicators to Monitor and Evaluate the Sustainability of Bioeconomy*. FAO. https://www.fao.org/3/ca6048en/ca6048en.pdf

Gao, Y., Sun, J., Li, F., He, S., Sandven, S., Yan, Q., Zhang, Z., Lohmann, K., Keenlyside, N., Furevik, T., & Suo, L. (2015). Arctic sea ice and Eurasian climate: A review. *Advances in Atmospheric Sciences*, *32*(1), 92–114. doi:10.100700376-014-0009-6

Gladkiy, Y. (2020). "Polar Silk road": project implementation and geoeconomic interests of Russia and China. *IOP Conf. Ser.: Earth Environ. Sci.* IOP Science. (Accessed 17.06.2023) doi:10.1088/1755-1315/434/1/012009

Guedes, A. (2015). Guedes 2015, "The Arctic Council". *Janus*, 160–161. https://repositorio.ual.pt/bitstream/11144/2982/1/3.30_ArmandoMGuedes_Cons elhoArtico.pdf

Heijman, W. (2016). How big is the bio-business? Notes on measuring the size of the Dutch bio-economy. *NJAS Wageningen Journal of Life Sciences*, *77*(1), 5–8. https://research.wur.nl/en/publications/how-big-is-the-bio-business-noteson-measuring-the-size-of-the-du(Accessed 13.06.2023). doi:10.1016/j.njas.2016.03.004

Hu, G., Zhao, L., Wu, T., Wu, X., Park, H., Fedorov, A., Wei, Y., Li, R., Zhu, X., Sun, Z., Ni, J., & Zou, D. (2021). Spatiotemporal variations and regional differences in air temperature in the permafrost regions in the Northern Hemisphere during 1980–2018. *The Science of the Total Environment*, *791*, 148358. doi:10.1016/j.scitotenv.2021.148358 PMID:34139490

Hui, L. (2018). China's Arctic Policy. *Xinhua*. http://www.xinhuanet.com/

Humpert, M. (2011). *The Future of the Northern Sea Route - A "Golden Waterway" or a Niche Trade Route*. The Artic Institute. https://www.thearcticinstitute.org/future-northern-sea-route-golden-waterway-niche/

Humpert, M. (2013). Putin and Xi Discuss Further Deepening of Arctic Partnership. *High North News*. https://www.highnorthnews.com/en/putin-and-xi-discuss-further-deepening-arctic-partnership

Jakobson, L. (2010). *China Prepares for an Ice-free Artic*. SIPRIS- Stockholm International Peace Research Institute, march 2010 https://www.sipri.org/publications/2010/sipri-insights-peace-and-security/china

Kircher, M. (2019). Bioeconomy: Markets, implications and investment opportunities. *Savings*, *73*(3), 73. Advance online publication. doi:10.3390/economies7030073

Kuosmanen, T. (2020). *How Big is the Bioeconomy? Reflections from an economic perspective*. Publications Office of the European Union, Luxembourg., https://www.researchgate.net/publication/340916792_How_Big_is_the_Bioecon omy#fullTextFileContent doi:10.2760/144526

Lasserre, F., Huang, L., & Alexeeva, O. V. (2017). China's strategy in the Arctic: Threatening or opportunistic? *The Polar Record*, *53*(1), 31–34. doi:10.1017/S0032247415000765

Lier, M. (2018). Synthesis on bioeconomy monitoring systems in the EU Member States. *Natural resources and bioeconomy studies, 38.* https://www.researchgate.net/publication/339325322_Synthesis_on_bioeconom y_monitoring_systems_in_the_EU_Member_States_indicators_for_monitoring_the_progress_of_bioeconomy_Synthesis_on_bioeconomy_monitoring_systems_ in_the_EU_Member_States_-indicators_#fullTextFileContent

Liu, C. (2018). China's role in Arctic governance 'cannot be ignored'. *Global Times.* https://www.globaltimes.cn/content/1128626.shtml

Lozier, J. (2022). Arctic: Toward the End of the Exception? Strategic, Nuclear and Maritime Issues in the Region. *Proliferation Papers, No. 64*, https://www.ifri.org/sites/default/files/atoms/files/lozier_arctic_end_of_exception_2022.pdf

Manta, F. (2019). China's policies on Polar Silk Road and impacts on trade and energy use in Arctic region. *Chinese Journal of Population Resources and Environment*, *17*(4) 322-328. doi:10.1080/10042857.2019.1685360

Markard, J. (2012). Sustainability transitions: An emerging field of research and its prospects. *Research Policy 41*(6). https://www.sciencedirect.com/science/article/abs/pii/S004873331200056X

McKay. (2022). Exceeding 1.5C global warming could trigger multiple climate tipping points. *Science*, *377*(6611), eabn7950. doi:10.1126cience.abn7950 PMID:36074831

Melnikov, V., Osipov, V. I., Brouchkov, A. V., Falaleeva, A. A., Badina, S. V., Zheleznyak, M. N., Sadurtdinov, M. R., Ostrakov, N. A., Drozdov, D. S., Osokin, A. B., Sergeev, D. O., Dubrovin, V. A., & Fedorov, R. Y. (2022). Climate warming and permafrost thaw in the Russian Arctic: Potential economic impacts on public infrastructure by 2050. *Natural Hazards*, *112*(1), 1–21. doi:10.100711069-021-05179-6

Miner, K. R., Turetsky, M. R., Malina, E., Bartsch, A., Tamminen, J., McGuire, A. D., Fix, A., Sweeney, C., Elder, C. D., & Miller, C. E. (2022). Permafrost carbon emissions in a changing Arctic. *Nature Reviews. Earth & Environment*, *3*(1), 55–67. doi:10.103843017-021-00230-3

NASA. (2021). Ice Persists in the North West Passage. *NASA Earth Observatory*. NASA. https://earthobservatory.nasa.gov/

Norsk, L. (2023), *The Artic Council*. Artic Council. https://arctic-council.org/

Tomas, M. (2021). (pp. 176–195). The Era of the Dragon: How China Is Preparing to be the World's Top Economy in 21 Century. IGI Global. doi:10.4018/978-1-7998-7568-0.ch009

OECD. (2004). *Biotechnology for sustainable growth and development*. Organisation for Economic Co-operation and Development. https://www.oecd.org/science/emerging-tech/33784888.pdf

OECD. (2018). *Meeting Policy Challenges for a Sustainable Bioeconomy*. OECD Publishing. doi:10.1787/9789264292345-

Okamoto, T. (2019). China's "Territorial Sovereignty" and Its Origins. *Review the Japan Institute of international affairs,* 3(2).

Ostreng, W. (2012). On the Geopolitical Significance of Arctic States. *Arctis Knowledge Hub*. http://www.arctis-search.com/

Palmer, L. (2013). Melting Arctic Ice Will Make Way for More Ships—and More Species Invasions. *Scientific American*. https://www.scientificamerican.com/

Peng, J., & Wegge, N. (2015). China's bilateral diplomacy in the Arctic, Polar Geography. *Polar Geography*, *38*(3), 233–249. doi:10.1080/1088937X.2015.1086445

Piffero, S. P. (2013). The Militarisation of the Arctic: Political, Economic and Climate Challenges. *UFRGS Model United Nations Journal 1*, 11–70, https://www.ufrgs.br/.

Pincus, R. (2020). Three-Way Power Dynamics in the Arctic. *Strategic Studies Quarterly*, *14,* 40–63. doi:. doi:10.2307/26891883

PRC. (2018). *China's Arctic Policy.* White paper by The State Council Information Office of the People's Republic of China. https://english.www.gov.cn/archive/white_paper/2018/01/26/content_2814760266 60336.htm

Røseth, T. (2014). Russia's China Policy in the Artic. *Strategic Analysis, 38*(6). doi:10.1080/09700161.2014.952942

Sharma, T. (2019). Melting Arctic Sea Ice Opens New Maritime Shipping Route. *Global Security Review*. https://globalsecurityreview.com/

Tillman, H. (2018). The Polar Silk Road: China's New Frontier of International Cooperation. *China Quarterly of International Strategic Studies 4*(3), 345–62. https://www.worldscientific.com/

Tomás, M. (2021). The Era of the Dragon: How China Is Preparing to Be the World's Top Economy in 21st Century. Handbook of Research on the Empirical Aspects of Strategic Trade Negotiations and Management. Crespo. doi:10.4018/978-1-7998-7568-0.ch009

Wandrey, J. (2008). Circum-Arctic Resource Appraisal: Estimates of Undiscovered Oil and Gas North of the Arctic Circle. *Fact Sheet*. USGS. https://pubs.usgs.gov/fs/2008/3049/

WMO. (2016). *NASA, analysis confirms July 2016 as warmest month in 136 years.* World Meteorological Organization (wmo.int).

WMO, 2021, *Arctic assessment report shows faster rate of warming,* World Meteorological Organization (wmo.int)

Xiao, Z., Kerr, W.A. (2022). Biotechnology in China-regulation, investment, and delayed commercialization. *GM Crops Food 13*(1), 86–96. (Accessed 18.06. 2023) doi:10.1080/21645698.2022.2068336

Zhang, X. (2022). The roadmap for the bioeconomy in China. *IET*. doi:10.1049/enb2.12026

ADDITIONAL READING

Alexeeva, O. 2015. Quelle stratégie pour la Chine en Arctique? In: Mottet, É., B. Courmont and F. Lasserre (editors). La Chine et le monde. Quelles nouvelles relations, quels nouveaux paradigmes? Québec: Presses de l'Université du Québec: pp. 271–292 file:///C:/Users/maria/Downloads/norois-5681.pdf

Alexeeva, O. (2012a). China and the Arctic. *Arctic Yearbook 2012* (pp. 80–90). University of Akureyri. https://www.researchgate.net/publication/259042084_China_and_the_Arctic

Beck, A. (2014). China's strategy in the Arctic: A case of lawfare? *The Polar Journal*, *4*(2), 306–318. doi:10.1080/2154896X.2014.954886

Bennett, M. (2014). China's silk road plans could challenge northern sea route. *Cryopolitics, Arctic News and Analysis*. https://cryopolitics.com/2014/12/29/chinas-silk-road-plans-could-challenge-northern-sea-route/ (Accessed 11.06.2023).

Borgerson, S. (2008). *Arctic Meltdown: The Economic and Security Implications of Global Warming*. Foreign affairs (Council on Foreign Relations). https://www.researchgate.net/publication/297807870_Arctic_Meltdown_The_Economic_and_Security_Implications_of_Global_Warming (Accessed 12.06.2023).

Brutschin, E., & Schubert, S. R. (2016). Icy waters, hot tempers, and high stakes: Geopolitics and Geoeconomics of the Arctic. *Energy Research & Social Science*, *16*, 147–159. https://www.sciencedirect.com/science/article/abs/pii/S2214629616300482. doi:10.1016/j.erss.2016.03.020

Bruun, M., & Medby, I. A. (2014). Theorising the Thaw: Geopolitics in a Changing Arctic. *Geography Compass*, *8*(12), 915–929. https://compass.onlinelibrary.wiley.com/doi/abs/10.1111/gec3.12189(Accessed 13.06.2023). doi:10.1111/gec3.12189

Carrington, D. (2017). Arctic stronghold of world's seeds flooded after permafrost melts. *The Guardian*. https://www.theguardian.com/environment/2017/may/19/arctic-strongholdof-worlds-seeds-flooded-after-permafrost-melts

Ebinger, C., & Zambetakis, E. (2009). The Geopolitics of Arctic Melt. *International Affairs*, *85*(6), 1215–1232. https://www.brookings.edu/wp-content/uploads/2016/06/11_arctic_melt_ebinger_zambetakis.pdf. doi:10.1111/j.1468-2346.2009.00858.x

Emmerson, C. (2010). *The Future History of the Arctic: How Climate, Resources, and Geopolitics are Reshaping the North, and Why it Matters to the World*. London: Vintage Books. https://site.uit.no/arcticfutures/wpcontent/uploads/sites/69/2012/03/Emmerson-2010.pdf.

Gautier, D., Bird, K. J., Charpentier, R. R., Grantz, A., Houseknecht, D. W., Klett, T. R., Moore, T. E., Pitman, J. K., Schenk, C. J., Schuenemeyer, J. H., Sørensen, K., Tennyson, M. E., Valin, Z. C., & Wandrey, C. J. (2009, May). 2009, Assessment of undiscovered oil and gas in the Arctic. *Science*, *324*(29), 1175–1179. https://www.science.org/doi/10.1126/science.1169467(Accessed 19.06.2023). doi:10.1126cience.1169467 PMID:19478178

Gayazova, O. (2013). China's rights in the marine Arctic. *International Journal of Marine and Coastal Law*, *28*(1), 61–95. https://brill.com/view/journals/estu/28/1/article-p61_3.xml?language=en. doi:10.1163/15718085-12341264

Haixing, L. (2018), The Development Trend of Equipment Technology in the Arctic Region Abroad. *Sohu News*. https://www.sohu.com/a/281166142_465915(Accessed 14.06.2023).

Huang, L. (2014). Is China's interest for the Arctic driven by Arctic shipping potential? *Asian Geographer*, 1–13. doi:10.1080/10225706.2014.928785

Jakobson, L. (2010). *China prepares for an ice-free Arctic*. SIPRI. https://www.sipri.org/publications/2010/sipri-insights-peace-and-security/china-prepares-ice-free-arctic

Koivurova, T. (2009). Limits and possibilities of the Arctic Council in a rapidly changing scene of Arctic governance. *The Polar Record*, *46*(2), 146–156. https://www.researchgate.net/publication/228147871_Limits_and_Possibilities_of_the_Arctic_Council_in_a_Rapidly_Changing_Scene_of_Arctic_Governance. doi:10.1017/S0032247409008365

Lanteigne, M. (2014). *China's Emerging Arctic Strategies: Economics and Institutions. Occasional Papers*. Institute of International Affairs, Centre for Arctic Policy Studies. https://ams.overcastcdn.com/documents/Marc_Lanteigne2.pdf

Li, Z. (2009b). Obstacles to China's participation in the International Arctic route mechanism and countermeasures. *Navigation of China*, (2), 98–103. doi:10.1017/S0032247415000765

World Ocean Review. (2019). Polar politics and commerce. World Ocean Review. https://worldocean-review.com/wp-content/downloads/wor6/WOR6_en_chapter_5.pdf,

Qiu, J. (2008). China: The third pole. *Nature*, *454*(7203), 393–396. doi:10.1038/454393a PMID:18650887

Rahbek-Clemmensen, J. (2019). When Do Ideas of an Arctic Treaty Become Prominent in Arctic Governance Debates? *Arctic, 72*(2), 103–214. doi:10.14430/arctic68285

Reuters Staff. (2019). China mixing military and science in Arctic push: Denmark. Reuters. https://www.reuters.com/article/us-usa-arctic-idUSKBN1Y3116

Robinson, R. (2013). *China's 'Long con' in the Arctic*. Macdonald-Laurier Institute (MLI) Commentary. https://www.macdonaldlaurier.ca/files/pdf/MLIChina%27sLongConInTheArctic09-13Draft4-1.pdf

Rosecrance, R. (1993). Beyond Realism: The Study of Grand Strategy. In: Richard Rosecrance and Arthur A. Stein (eds.): *The Domestic Bases of Grand Strategy,* pp. 3–21. Ithaca, New York: Cornell University Press. https://ssrn.com/abstract=2863654

Rosenthal, E. (2012). Race Is on As Ice Melt Reveals Arctic Treasures. *New York Times.*

Sacks, B. (2021). *Exploring Gaps in Arctic Governance*. RAND Corporation. doi:10.7249/RR-A1007-1

Sørensen, C. (2019). Intensifying U.S.-China security dilemma dynamics play out in the Arctic: Implications for China's Arctic strategy. *Arctic Yearbook 2019.* https://arcticyearbook.com/arctic-yearbook/2019/2019-scholarly-papers/323-intensifying-u-s-china-security-dilemma-dynamics-play-out-in-the-arctic-implications-for-china-s-arctic-strategy.

Stephen, K. (2018). International Politics and Governance in the Arctic - an Introduction. *The Arctic Institute*. https://www.thearcticinstitute.org/international-politics-governance-arctic-introduction/

Stephen, M. (2020). The Integration of Emerging Powers into Club Institutions: China and the Arctic Council. *Global Policy, 11*, 51-60. . doi:10.1111/1758-5899.12834

Stritzel, H. (2007). Towards a Theory of Securitization: Copenhagen and Beyond. *European Journal of International Relations*, *13*(3), 357–383. https://journals.sagepub.com/doi/10.1177/1354066107080128 (Accessed 20.06.2023)

Tang, G. Q. (2013). *Arctic Issues and China's Stance*. China International Studies. https://caod.oriprobe.com/articles/39624325/arctic_issues_and_china__s_stance.htm

Taureck, R. (2006). Securitization Theory and Securitization Studies. *Journal of International Relations and Development, 9*. https://link.springer.com/article/10.1057/palgrave.jird.1800072.

Teeple, N. (2020). U.S. Arctic Policy: Threats, Actors, and Outcomes. *North American and Arctic Defence and Security Network Policy Primer*. NAADSN. https://www.naadsn.ca/wp-content/uploads/2020/08/20-apr-6-Nancy-Teeple_NAADSN-US-Arctic-Policy-BN-final-in_template.pdf

Li, L. (2019). The People's Republic of China 2019, Observer Review Report to the Arctic Council. *China Oceans Law Review*. https://oaarchive.arctic-council.org/handle/11374/2251.

Wang, Y. (2020). A Comparative Study of the Official Use of Arctic Environmental Discourses of China and the EU. *The Arctic Institute*. https://www.thearcticinstitute.org/comparative-study-official-use-arctic-environmental-discourses-china-eu/

Weber, R. (1990). Basic Content Analysis. *Quantitative Applications in the Social Sciences*. SAGE Publications, Inc. doi:10.4135/9781412983488

Wegge, N. (2014). China in the Arctic: Interests, actions and challenges. *Nordlit*, (32), 83–98. doi:10.7557/13.3072

Wegge, N. (2020). Arctic Security Strategies and the North Atlantic States. *Arctic Review on Law and Politics*, *11*(11), 360–382. doi:10.23865/arctic.v11.2401

Wolmer, W. 2006, (Ed.) Understanding policy processes: *A review of IDS research on the environment.* Brighton: Institute of Development Studies: University of Sussex. https://steps-centre.org/wp-content/uploads/IDS-Understanding-policy-processes.pdf

Woon, C. Y. (2020). Framing the "Polar Silk Road", Critical geopolitics, Chinese scholars and the (Re)Positionings of China's Arctic interests. *Political Geography*, (78), 102141. doi:10.1016/j.polgeo.2019.102141

Xiao, Y. (2020). Constraints and Breakthroughs: An Analysis of China's Normative Discourse Power in the Arctic Council. *Peace and Development, 1*.

Xing, Li, & Bertelsen, R. (2023). The Drivers of Chinese Arctic Interests: Political Stability and Energy and Transportation Security. *Arctic Yearbook 2013*. https://www.cidob.org/en/publications/publication_series/documents_cidob/relations_on_thin_ice_a_narrative_analysis_of_chinese_governance_in_the_arctic

Yeager, B. B. (2008). The Ilulissat Declaration: Background and Implications for Arctic Governance. *Aspen Dialogue and Commission on Arctic Climate Change*. CIDOB. https://www.cidob.org/en/publications/publication_series/documents_cidob/relations_on_thin_ice_a_narrative_analysis_of_chinese_governance_in_the_arctic

Yuan, Y. (2018). My Country's First Arctic Policy White Paper Published. In: Zhu Jiang, Yuan Bo (Eds.) *People's Daily*. http://finance.people.com.cn/n1/2018/0127/c1004-29790387.html.

Zhang, C. (2015). Russian Absence at the Arctic Sunrise Case: A Comparison with the Chinese Position in the South China Sea Arbitration. *Journal of East Asia and International Law,* 8(2), 413–426. http://journal.yiil.org/home/archives_v8n2_06

Zou, K. (1993). China's antarctic policy and the Antarctic Treaty system. *Ocean Development and International Law*, *24*(3), 237–255. https://www.researchgate.net/publication/233431307_China's_antarctic_policy_and_the_Antarctic_Treaty_system. doi:10.1080/00908329309546008

Compilation of References

Abdalla, M. (2021). *The impact of money laundering in foreign exchange firms in the Lebanese financial market.*. [MBA Project submitted to the Lebanese International University].

Abifarin, J. K., & Ofodu, J. C. (2022). Determination of an efficient power equipment oil through a multi-criteria decision making analysis. *Vojnotehnički glasnik, 70*(2), 433–446. doi:10.5937/vojtehg70-36024

Abimbola, T., Lim, M., Hillestad, T., Xie, C., & Haugland, S. A. (2010). Innovative corporate social responsibility: The founder's role in creating a trustworthy corporate brand through "green innovation.". *Journal of Product and Brand Management, 19*(6), 440–451. doi:10.1108/10610421011085758

Abubakar, I. R., & Aina, Y. A. (2019). The prospects and challenges of developing more inclusive, safe, resilient and sustainable cities in Nigeria. *Land Use Policy, 87*, 104105. doi:10.1016/j.landusepol.2019.104105

Acer, Y. (2022). Rusya-Ukrayna Savaşinin Gida Arzina Etkileri Ve Tahil Koridoru Anlaşmasi [The Food Supply Effects of the Russia-Ukraine War and the Grain Corridor Agreement]. *Perspektif*, 343. https://setav.org/assets/uploads/2022/07/P343.pdf

Acharyya, J. (2009). FDI, growth and the environment: Evidence from India on CO2 emission during the last two decades. *Journal of Economic Development, 34*(1), 43–58. doi:10.35866/caujed.2009.34.1.003

Actuarial Directorate, Research and Statistics of the IESS. (2020). *Boletín Estadístico del IESS*. Quito: Instituto Ecuatoriano de Seguridad Social, IEES. https://bit.ly/3fKBN1b

Adamson, K., Campbell, P., & Orsoni, A. (2005). Hybrid decision support based on knowledge discovery and AI techniques for the management of maintenance services in the public transport sector. *2005 IEEE Intelligent Data Acquisition and Advanced Computing Systems: Technology and Applications*, (pp. 674–678). IEEE.

AdiVishnu. S., Nagaraju, V., & Senthil Murugan, P. (2022). Prediction of mango fruit ripening in different stages using CNN over random forest. *2022 14th International Conference on Mathematics, Actuarial Science, Computer Science and Statistics (MACS)*, (pp. 1–4). IEEE.

Agarwal, N. (2018). A Study of Innovations in Instructional Strategies and Designs for Quality Enrichment in Higher Education. Cosmos: *An International Journal of Art & Higher Education, 7*(2), ISSN: 2319-8966.

Agarwal, N., & Verma, M. (2019). A Study on Taxonomy of Innovations. *Globus An International Journal of Management & IT, 11*(1), 57-64. doi:10.5281/zenodo.3872090

Agarwal, E. (2021). An Overview of Low-Impact Renewable Thermal Energy Technologies and India's Response Toward Climate Change Related Issues. *Cosmos Journal of Engineering & Technology, 11*(1), 20–23.

AgarwalN.KumarP. (2009), "Reflection on The New Innovations for Maximizing The Learning in Teacher of Mathematics", International Journal Educational Herald, 38(2):41. ISSN: 0974-0732, Available at SSRN: https://ssrn.com/abstract=3600864, http://doi.org/ doi:10.5281/zenodo.3826567

Agarwal, N., & Sharma, M. (2006). Environment Education Concern in India. *University News*, *44*(12). Advance online publication. doi:10.5281/zenodo.3813569

Agarwal, P. & Bhattacharyya, S. (2020). Analytical Solutions of The Laplace Transform and Applications to Differential Equations. *Journal of Critical Reviews*, *7*(8), 3466–3472.

Agarwal, Puneet Kumar and Srinivas A. (2020). Spectral Theory for Three Body Quantum System in Two Dimension. *Journal of Critical Reviews*, *7*(12), 4533–4537.

Ahmad, S., Bharti, A., Haq, M. I., & Kothari, R. (2023). Bioeconomy: Current Status and Challenges. *Sustainable Butanol Biofuels*, 57-75.

Ahmed, A. A., Nazzal, M. A., Darras, B. M., & Deiab, I. M. (2022). A comprehensive multi-level circular economy assessment framework. *Sustainable Production and Consumption*, *32*, 700–717. doi:10.1016/j.spc.2022.05.025

Aidukaite, J., Hort, S., & Ainsaar, M. (2022). Current trends in social welfare policies toward the older people in the Baltic and Nordic countries: An explorative study. *Journal of Baltic Studies*, *53*(2), 147–167. doi:10.1080/01629778.2021.1998166

Aithal, P., & Shubhrajyotsna, A. (2016). Nanotechnology Innovations and Commercialization – Opportunities, Challenges & Reasons for Delay. In *Proceedings of National Conference on Changing Perspectives of Management, IT, and Social Sciences in Contemporary Environment*, (pp. 1-12). IJEM. 10.5815/ijem.2016.06.02

Aithal, P., & Shubhrajyotsna, A. (2018). Nanotechnology based Innovations and Human Life Comfortability –Are we Marching towards Immortality? *International Journal of Applied Engineering and Management Letters*, *2*(1), 1–8. doi:10.5281/zenodo.1485048

Akoley, Elvis, & Wahid, Mumuni Abdul and Kulo, IbnAli Issah. (2022). Public Financial Management and Economic Growth: A Study Case of Ghana. *Cosmos An International Journal of Management*, *11*(2), 29–37.

Aksenov, Y. (2017). On the Future Navigability of Arctic Sea Routes: High-Resolution Projections of the Arctic Ocean and Sea Ice. *Política Marinha,* 75, 300–17, https://www.sciencedirect.com/

Alasti, H. (2021). Shifting Legacy Robotic Manufacturing Towards Industry 4.0: Using Cloud IoT. In P. Tomar (Ed.), Integration and Implementation of the Internet of Things Through Cloud Computing (pp. 288-305). IGI Global. doi:10.4018/978-1-7998-6981-8.ch015

Alawattage, C., & Fernando, S. (2017). Postcoloniality in corporate social and environmental accountability. *Accounting, Organizations and Society*, *60*, 1–20. doi:10.1016/j.aos.2017.07.002

Aldrich, H. (2014). *Stand up and be counted: Why social science should stop using the qualitative/quantitative dichotomy.* LSE. https://blogs.lse.ac.uk/impactofsocialsciences/2014/11/28/stand-up-and-becounted-social-science-qualitative-quantitative-dichotom

Aldrich, H., & Auster, E. R. (1986). Even dwarfs started small: Liabilities of age and size and their strategic implications. In B. M. Staw & L. L. Cummings (Eds.), *Research in Organizational Behavior* (Vol. 8, pp. 165–198). JAI Press.

Alexeeva, O. (2012b). Le dragon des neiges. Les stratégies de la Chine en Arctique. *Perspectives chinoises, 3*, 61–68. https://journals.openedition.org/perspectiveschinoises/6345

Alland, A. (1966). Jr. Medical Anthropology and the Study of Biological and Cultural Adaptation. *American Anthropologist, 68*, 40–51. doi:10.1525/aa.1966.68.1.02a00040

Al-Mulali, U., & Foon Tang, C. (2013). Investigating the validity of pollution haven hypothesis in the Gulf Cooperation Council (GCC) countries. *Energy Policy, 60*, 813–819. doi:10.1016/j.enpol.2013.05.055

Alqahtany, A. (2022). Green Roofs as an Approach to Enhance Urban Sustainability: A Study of Public Perception in Riyadh, Saudi Arabia. *Buildings, 12*(12), 2202. doi:10.3390/buildings12122202

Al-Sawi, M. (2018). *The Effective Valorization of Resources for the Economic Development of Tyre*. [MBA Project presented at the Lebanese International University].

Alshanbari, R., Khan, S., El-Atab, N., & Mustafa Hussain, M. (2019). AI powered unmanned aerial vehicle for payload transport application. *2019 IEEE National Aerospace and Electronics Conference (NAECON)*, (pp. 420–424). IEEE. 10.1109/NAECON46414.2019.9058320

Alvarez, R. (2016). *Computational social science*. Cambridge University Press. doi:10.1017/CBO9781316257340

Amaglobeli, D., Chai, H., Dabla-Norris, E., Dybczak, K., Soto, M., & Tieman, A. (2019). *The Future of Saving: The Role of Pension System Design in an Aging World. IMF Staff Discussion Note*. International Monetary Fund. doi:10.5089/9781484388990.006

Ambrose, T., & Paine, C. (2006). *Museum basics*. Routledge. doi:10.4324/9780203018989

Amburgey, T., & Rao, H. (1996). Organizational ecology: Past, present, and future directions. *Academy of Management Journal, 39*(5), 1265–1286. doi:10.2307/256999

Amore, M. D., & Bennedsen, M. (2016). Corporate governance and green innovation. *Journal of Environmental Economics and Management, 75*, 54–72. doi:10.1016/j.jeem.2015.11.003

Andersen, M. S. (2007). An introductory note on the environmental economics of the circular economy. *Sustainability Science, 2*(1), 133–140. doi:10.100711625-006-0013-6

Androniceanu, A., Kinnunen, J., & Georgescu, I. (2021). Circular economy as a strategic option to promote sustainable economic growth and effective human development. *Journal of International Students, 14*(1), 60–73. doi:10.14254/2071-8330.2021/14-1/4

Annabel, L. S. P., & Muthulakshmi, V. (2019). AI-powered image-based tomato leaf disease detection. *2019 Third International conference on I-SMAC (IoT in Social, Mobile, Analytics and Cloud) (I-SMAC)*, (pp. 506–511). IEEE. 10.1109/I-SMAC47947.2019.9032621

Antonioli, D., Mancinelli, S., & Mazzanti, M. (2013). Is environmental innovation embedded within high- performance organizational changes? The role of human resource management and complementarity in green business strategies. *Research Policy, 42*(4), 975–988. doi:10.1016/j.respol.2012.12.005

Antweiler, W., Copeland, B. R., & Taylor, M. S. (2001). Is free trade good for the environment? *The American Economic Review, 91*(4), 877–908. doi:10.1257/aer.91.4.877

Apella, I. (2020). The performance of the income protection system for older adults in Ecuador and future challenges. *International Social Security Review, 73*(2), 51–74. doi:10.1111/issr.12234

Aragon-Correa, J. A., Hurtado-Torres, N. E., Sharma, S. & García-Morales, V. J. (2008). Environmental strategy and performance in small firms: A resource-based perspective. *Journal of Environmental Management, 86*(1), 88–103. doi:10.1016/j.jenvman.2006.12.032

Aral, S., Van Alstyne, M., & Brynjolffon, E. (2012). Information technology and information worker productivity. *Information Systems Research, 23*(2), 849–867. doi:10.1287/isre.1110.0408

Ardvisson, S. (2023). CEO talk of sustainability in CEO letters: Towards the inclusion of a sustainability embeddedness and value-creation perspective. *Sustainability Accounting. Management and Policy Journal, 14*(7), 26–61.

Arenas de Mesa, A. (2020a). Los sistemas de pensiones en la encrucijada: desafíos para la sostenibilidad en América Latina (Vols. Libros de la CEPAL, N° 159). Santiago: Comisión Económica para América Latina y el Caribe (CEPAL). https://doi.org/ doi:10.18356/64c72454-es

Arenas De Mesa, A. (2020b). *Los sistemas de pensiones en América Latina: institucionalidad, gasto público y sostenibilidad financiera en tiempos del COVID-19.* Serie Macroeconomía del Desarrollo, N° 212, 9-52. https://bit.ly/3yot3V0

Arendt, R. (1999). *Growing greener: putting conservation into local plans and ordinances.* Island Press.

Armstrong, C. G., Shoemaker, A. C., McKechnie, I., Ekblom, A., Szabó, P., Lane, P. J., McAlvay, A. C., Boles, O. J., Walshaw, S., Petek, N., Gibbons, K. S., Quintana Morales, E., Anderson, E. N., Ibragimow, A., Podruczny, G., Vamosi, J. C., Marks-Block, T., LeCompte, J. K., Awâsis, S., & Crumley, C. L. (2017). Anthropological contributions to historical ecology: 50 questions, infinite prospects. *PLoS One, 12*(2), e0171883. doi:10.1371/journal.pone.0171883 PMID:28235093

Arnould, E. J., & Cayla, J. (2015). Consumer fetish: Commercial ethnography and the sovereign consumer. *Organization Studies, 36*(October), 1361–1386. doi:10.1177/0170840615580012

Asghari, M. (2013). Does FDI promote MENA Region's environment quality? Pollution halo or pollution haven hypothesis. [IJSRES]. *International Journal of Scientific Research in Environmental Sciences, 1*(6), 92–100. doi:10.12983/ijsres-2013-p092-100

Åstrøm, J. (2018). Exploring theming dimensions in a tourism context. *European Journal of Tourism Research, 20,* 5–27. doi:10.54055/ejtr.v20i.337

Atanasova, A. (2021). Re-examining utopia in contemporary consumption: Conceptualization and implications for marketing. *AMS Review, 11*(1-2), 23–29. doi:10.100713162-021-00193-0

Athukorala, P., & Menon, J. (1996). Foreign investment and industrialization in Malaysia: Exports, employment and spillovers. *Asian Economic Journal, 10*(1), 29–44. doi:10.1111/j.1467-8381.1996.tb00156.x

Attar, A., Desai, D., & Gupta, A. (2012). A study of various factors affecting labour productivity and methods to improve it. *IOSR Journal of Mechanical and Civil Engineering, 1*(3), 11–14.

Audretsch, D. B. (1998). *Agglomeration and the Location of Innovative Activity.* ULE.

Audretsch, D., Cunningham, J., Kuratko, D., Lehmann, E., & Menter, M. (2019). Entrepreneurial ecosystems: Economic, technological, and societal impacts. *The Journal of Technology Transfer, 44*(2), 313–325. doi:10.100710961-018-9690-4 PMID:30956392

Auerswald, P., & Dani, L. (2022). Entrepreneurial opportunity and related specialization in economic ecosystems. *Research Policy, 51*(9), 104445. doi:10.1016/j.respol.2021.104445

Autenne, A. (2017). Occupational pension funds. *European Journal of Social Security, 19*(2), 158–171. doi:10.1177/1388262717712152

Aworunse, O. S., Olorunsola, H., Ahuekwe, E., & Obembe, O. (2023). Towards a sustainable bioeconomy in a post-oil era Nigeria. *Resources. Environment and Sustainability, 11*, 100094. doi:10.1016/j.resenv.2022.100094

AZoCleantech.com. (2022, October 12). *The Benefits of Climate-Smart Ecosystem Monitoring in Forests*. AZoClean Tech. https://www.azocleantech.com/article.aspx?ArticleID=1485

Baer, H., & Singer, M. (2014). *The anthropology of climate change: An integrated critical perspective*. Routledge. doi:10.4324/9781315818702

Baggs, J. (2009). International trade in hazardous waste. *Review of International Economics*, *17*(1), 1–16. doi:10.1111/j.1467-9396.2008.00778.x

Baines, T., Brown, S., Benedettini, O., & Ball, P. (2012). Examining green production and its role within the competitive strategy of manufacturers. *Journal of Industrial Engineering and Management*, *5*(1), 53–87. doi:10.3926/jiem.405

Bai, X., Surveyer, A., Elmqvist, T., Gatzweiler, F. W., Güneralp, B., Parnell, S., Prieur-Richard, A.-H., Shrivastava, P., Siri, J. G., Stafford-Smith, M., Toussaint, J.-P., & Webb, R. (2016). Defining and advancing a systems approach for sustainable cities. *Current Opinion in Environmental Sustainability*, *23*, 69–78. doi:10.1016/j.cosust.2016.11.010

Bai, Y., Song, S., Jiao, J., & Yang, R. (2019). The impacts of government R&D subsidies on green innovation: Evidence from Chinese energy-intensive firms. *Journal of Cleaner Production*, *233*(9), 819–829. Advance online publication. doi:10.1016/j.jclepro.2019.06.107

Baldassarri, D., & Abascal, M. (2017). Field experiments across the social sciences. *Annual Review of Sociology*, *43*(1), 41–73. doi:10.1146/annurev-soc-073014-112445

Balkevicius, A., Sanctuary, M., & Zvirblyte, S. (2020). Fending off waste from the west: The impact of China's Operation Green Fence on the international waste trade. *World Economy*, *43*(10), 2742–2761. doi:10.1111/twec.12949

Balsalobre-Lorente, D., Gokmenoglu, K. K., Taspinar, N., & Cantos-Cantos, J. M. (2019). An approach to the pollution haven and pollution halo hypotheses in MINT countries. *Environmental Science and Pollution Research International*, *26*(22), 23010–23026. . doi:10.100711356-019-05446-x PMID:31183758

Baltacı, A. (2018). Nitel araştırmalarda örnekleme yöntemleri ve örnek hacmi sorunsalı üzerine kavramsal bir inceleme. *Bitlis Eren Üniversitesi Sosyal Bilimler Enstitüsü Dergisi, 7(1),* 231–274. Retrieved from https://dergipark.org.tr/en/pub/bitlissos/issue/38061/399955

Bamwesigye, D., & Hlavackova, P. (2019). Analysis of sustainable transport for smart cities. *Sustainability (Basel)*, *11*(7), 2140. doi:10.3390u11072140

Banaitė, D. (2016). Towards circular economy: Analysis of indicators in the context of sustainable development. *Social Transformation in Contemporary Society*, *4*(9), 142–150.

Banco de Desarrollo de América Latina. (2020). *Los sistemas de pensiones y salud en América Latina: los desafíos del envejecimiento, el cambio tecnológico y la informalidad*. Buenos Aires: Corporación Andina de Fomento. https://bit.ly/3REtAcl

Bansal, P. (2005). Evolving sustainably: A longitudinal study of corporate sustainable development. *Strategic Management Journal*, *26*(3), 197–218. doi:10.1002mj.441

Bao, K., Shang, D., Wang, R., & Ma, R. (2020). AIS big data framework for maritime safety supervision. *2020 International Conference on Robots & Intelligent System (ICRIS)*, (pp. 150–153). IEEE. 10.1109/ICRIS52159.2020.00045

Bao, Q., Chen, Y., & Song, L. (2010). Foreign direct investment and environmental pollution in China: A simultaneous equations estimation. *Environment and Development Economics*, *16*(1), 71–92. doi:10.1017/S1355770X10000380

Barnes, J., Dove, M., Lahsen, M., Mathews, A., McElwee, P., McIntosh, R., Moore, F., O'Reilly, J., Orlove, B., Puri, R., Weiss, H., & Yager, K. (2013). Contribution of anthropology to the study of climate change. *Nature Climate Change*, *3*(6), 541–544. doi:10.1038/nclimate1775

Barnet, M. L. (2007). Stakeholder influence capacity and the variability of financial returns to corporate social responsibility. *Academy of Management Review*, *33*(3), 794–816. doi:10.5465/amr.2007.25275520

Barnett, J. (2009). The way we were, the way we are the theory and practice of designing cities since 1956. A. Krieger, W.S. Saunders (Eds.), *Urban design*, University of Minnesota Press. https://www.jstor.org/stable/10.5749/j.ctttspsh.9View

Barnett, J., & Adger, W. N. (2003). Climate dangers and atoll countries. *Climatic Change*, *61*(3), 321–337. doi:10.1023/B:CLIM.0000004559.08755.88

Barney, N., & Lewis, S. (2022). *Biotechnology (biotech).* Tech Target. https://www.techtarget.com/whatis/definition/biotechnology

Barney, J. (1991). Firm Resources and Sustained Competitive Advantage. *Journal of Management*, *17*(1), 99–120. doi:10.1177/014920639101700108

Barrett, S. (2009). *Anthropology*. University of Toronto Press. doi:10.3138/9781442697362

Barrientos, A., & Malerba, D. (2019). Social protection in Latin America: One region, two systems. In *Welfare and social protection in contemporary Latin America* (pp. 59–71). Routledge. doi:10.1111/issr.12244

Barry, A., Thomas, G., Debenham, P., & Trout, J. (2012). Augmented reality in a public space: The Natural History Museum, London. *Computer*, *45*(7), 42–47. doi:10.1109/MC.2012.106

Bartodziej, C. (2017). *The Concept Industry 4.0: An Empirical Analysis of Technologies and Applications in Production Logistics*. Springer. doi:10.1007/978-3-658-16502-4

Basch, C., & Basch, C. (2022). Epidemiology, secondary school curricula, and preparing the next generation for global citizenship. *JMIR Public Health and Surveillance*, *8*(3), e36006. doi:10.2196/36006 PMID:35254281

Bebbington, J., & Rubin, A. (2022). Accounting in the Anthropocene: A roadmap for stewardship. *Accounting and Business Research*, *52*(5), 582–596. doi:10.1080/00014788.2022.2079780

Bebbington, J., Österblom, H., Crona, B., Jouffray, J. B., Larrinaga, C., Russell, S., & Scholtens, B. (2019). Accounting and accountability in the Anthropocene. *Accounting, Auditing & Accountability Journal*, *33*(1), 152–171. doi:10.1108/AAAJ-11-2018-3745

Becker, M. C. (Ed.). (2008). Handbook of organizational routines. Cheltenham, UK: Edward Elgar Publishing. doi:10.4337/9781848442702

Beck, U. (2009). *World at Risk*. Polity.

Belk, R. W. (2014). You are what you can access: Sharing and collaborative consumption online. *Journal of Business Research*, *67*(August), 1595–1600. doi:10.1016/j.jbusres.2013.10.001

Bellos, G. S. G. (2008). *Sedimentology and Diagenesis of Some Neocomian-Barremian Rocks (Chouf Formation), Southern Lebanon* [M.Sc. Thesis Presented to de The Department of Geology at the American University of Beirut].

Bellos, G. S. G. (2017-2018). *Évolution culturelle du peuplement de l'âge du Fer dans la Phénicie du Nord.* [MA Thesis Presented and Defended at the Lebanese University].

Bellos, G. S. G. (2019). *Estimating the Impact of the Effective Valorization of Cultural and Natural Resources for the Economic Development of Beirut.* [MBA Project presented at the Lebanese International University].

Bellos, G. S. G. (2020). Enquête sur les effets de la récente dépression économique sur le licenciement en masse des employés: Les huit derniers indicateurs économiques. *Libna News.* https://libnanews.com/enquete-sur-les-effets-de-la-recente-depression-economique-sur-le-licenciement-en-masse-des-employes-les-huit-derniers-indicateurs-economiques/

Bellos, G. S. G. (2021a). Historique de la Thawra depuis le 17 Octobre 2019 à nos jours. *Libna News.* https://libnanews.com/historique-de-la-thawra-depuis-le-17-octobre-2019-a-nos-jours/

Bellos, G. S. G. (2021b). *Estimating the impact of the Sustainable Development Goals on the Lebanese Society.* Thesis, Lebanese International University, Beirut Campus, Beirut, Lebanon].

Belz, F.-M., & Peattie, K. (2010). Sustainability marketing: An innovative conception of marketing. *Marketing Review St. Gallen, 27*(5), 8–15. doi:10.100711621-010-0085-7

Benedict, M., & McMahon, E. T. (2006). *Green infrastructure: linking landscape and communities.* Island Press.

Bengaluru Bannerghatta Biological Park. (n.d.). Bengaluru Bannerghatta Biological Park. https://bannerghattabiologicalpark.org/facilities.html

Berardi, U., GhaffarianHoseini, A. H., & GhaffarianHoseini, A. (2014). State-of-the-art analysis of the environmental benefits of green roofs. *Applied Energy, 115*, 411–428. doi:10.1016/j.apenergy.2013.10.047

Berg, A., Antikainen, R., Hartikainen, E., Kauppi, S., Kautto, P., Lazarevic, D., Piesik, S., & Saikku, L. (2018). *Circular economy for sustainable development.* Finnish Environment Institute. http://hdl.handle.net/10138/251516

Berkes, F., Folke, C., & Colding, J. (Eds.). (2003). *Navigating Social-Ecological Systems: Building Resilience for Complexity and Change.* Cambridge University Press.

Berlin government.. (2020). *Capital of Green Trends: How Berlin leads the Way in Urban Sustainability.* Visit Berlin. https://about.visitberlin.de/en/green berlin

Berlin, N. (2020). *Stadt der grünen Trends-Wie Berlin den Weg der urbanen Nachhaltigkeit geht.* Visit Berlin. https://about.visitberlin.de/nachhaltiges-berlin

Bernard, H. (2011). *Research methods in anthropology: Qualitative and quantitative approaches.* AltaMira Press.

Bernardini, C. & K. N. Irvine, K. N. (2007). The 'nature' of urban sustainability: Private or public green spaces. *Sustainable Development and Planning, 2*, 661-674. doi:10.2495/SDP070642

Bernhardt, T., & Donnelly, C. (2019). Modern tontine with bequest: Innovation in pooled annuity products. *Insurance, Mathematics & Economics, 86*(1), 168–188. doi:10.1016/j.insmatheco.2019.03.002

Bernroider, E. (2002). Factors in SWOT analysis applied to micro, small-to-medium, and large software enterprises: An Austrian study. *European Management Journal, 20*(5), 562–573. doi:10.1016/S0263-2373(02)00095-6

Berry, J. W. (2004). An Ecocultural Perspective on the Development of Competence. In R. Sternberg & E. Grigorenko (Eds.), *Culture and Competence: Contexts of Life Success* (pp. 3–22). American Psychological Association. doi:10.1037/10681-001

Besinovic, N., De Donato, L., Flammini, F., Goverde, R. M. P., Lin, Z., Liu, R., Marrone, S., Nardone, R., Tang, T., & Vittorini, V. (2022). Artificial intelligence in railway transport: Taxonomy, regulations, and applications. *IEEE transactions on intelligent transportation systems: a publication of the IEEE Intelligent Transportation Systems Council, 23*(9), 14011–14024. https://doi.org/ doi:10.1109/TITS.2021.3131637

Betancur, A. (2020). *Efectos Fiscales del Sistema de Pensiones en Chile: institucionalidad, gasto público y sostenibilidad financiera*. Santiago: Comisión Económica para América Latina y el Caribe (CEPAL). https://bit.ly/3ygQccl

Betthäuser, B. (2017). Protecting outsiders? Corporatism and the dualisation of unemployment protection in Germany and Austria. *European Journal of Social Security*, *19*(3), 209–224. doi:10.1177/1388262717724892

Bhardwaj, H., Tomar, P., Sakalle, A., Singh, T., Acharya, D., & Bhardwaj, A. (2021). Future Aspects and Research Perspectives of the Internet of Things. In *Integration and Implementation of the Internet of Things Through Cloud Computing* (pp. 1–18). IGI Global. doi:10.4018/978-1-7998-6981-8.ch001

Bhaskar, R. (2008). *A Realist Theory of Science*. Routledge.

Bhowmick, K., Hasan, M., & Masood, A. (2022). Employee Creativity, Acceptable Work Environments, and Transformational Management. *Cosmos An International Journal of Management, 12*(1), 08-19.

Bhowmick, Krishna, Hasan, Mohammad Mahamudul, Masood, Abdul Wahab, Muhammad (2022). Employee Creativity, Acceptable Work Environments, and Transformational Management. *Cosmos An International Journal of Management, 12*(1), 08-19.

Bian, Y., Wu, L., Bai, J., & Yang, Y. (2021). Does factor market distortions inhibit green economic growth? *World Econ Papers*, *2*, 105–119.

Bideci, M., & Albayrak, T. (2018). An investigation of the domestic and foreign tourists' museum visit experiences. *International Journal of Culture, Tourism and Hospitality Research*, *12*(3), 366–377. doi:10.1108/IJCTHR-02-2018-0029

Bigo, D. (2017). International flows, political order and social change: (in)security, by-product of the will of order over change. *Global Crime*, 303–321. doi:10.1080/17440572.2017.1350428

Bi, K., Huang, P., & Wang, X. (2016). Innovation performance and influencing factors of low-carbon technological innovation under the global value chain: A case of Chinese manufacturing industry. *Technological Forecasting and Social Change*, *111*, 275–284. doi:10.1016/j.techfore.2016.07.024

Bilek-Steindl, S., Kettner, C., & Mayrhuber, C. (2022). Sustainability, Work and Growth in the Context of SDG 8. *Empirica*, *49*(2), 277–279. doi:10.100710663-022-09538-9 PMID:35505947

Bilharz, J. A. (1998). *The Allegany Senecas and Kinzua Dam: Forced Relocation through Two Generations*. University of Nebraska Press.

Birch, L. L., Gunder, L., Grimm-Thomas, K., & Laing, D. G. (1998). Infants' consumption of a new food enhances acceptance of similar foods. *Appetite*, *30*(3), 283–295. doi:10.1006/appe.1997.0146 PMID:9632459

Bird, J. (2023). Arctic. *Encyclopaedia Britannica*. https://www.britannica.com/place/Arctic

Bird, K. J. (2008). Circum-Arctic resource appraisal; estimates of undiscovered oil and gas north of the Arctic Circle. *U.S. Geological Survey Fact Sheet 2008-3049*. PUBS. https://pubs.usgs.gov/fs/2008/3049/

Birnholtz, J. P., Cohen, M. D., & Hoch, S. V. (2007). Organizational character: On the regeneration of Camp Poplar Grove. *Organization Science*, *18*(2), 315–332. doi:10.1287/orsc.1070.0248

Biskaborn, B., Smith, S. L., Noetzli, J., Matthes, H., Vieira, G., Streletskiy, D. A., Schoeneich, P., Romanovsky, V. E., Lewkowicz, A. G., Abramov, A., Allard, M., Boike, J., Cable, W. L., Christiansen, H. H., Delaloye, R., Diekmann, B., Drozdov, D., Etzelmüller, B., Grosse, G., & Lantuit, H. (2019). 2019, "Permafrost is warming at a global scale". *Nature Communications*, *10*(1), 264. doi:10.103841467-018-08240-4 PMID:30651568

Bissing-Olson, M. J., Iyer, A., Fielding, K. S., & Zacher, H. (2013). Relationships between daily affect and pro-environmental behavior at work: The moderating role of pro-environmental attitude. *Journal of Organizational Behavior, 175*(2), 156–175. doi:10.1002/job.1788

Bistoletti, E. (2019). *The Power Struggles over the Post-neoliberal Social Security System Reforms in Venezuela and Ecuador*. Springer. doi:10.1007/978-3-319-98168-0

Bliege Bird, R., & Codding, B. F. (2022). Promise and peril of ecological and evolutionary modelling using cross-cultural datasets. *Nature Ecology & Evolution, 6*(1), 6–8. doi:10.103841559-021-01579-w PMID:34697456

Bock, G. W., Zmud, R. W., Kim, Y. G., & Lee, J. N. (2005). Behavioral intention formation in knowledge sharing: Examining the roles of extrinsic motivators, social-psychological forces, and organizational climate. *Management Information Systems Quarterly, 29*(1), 87–112. doi:10.2307/25148669

Boeing, P. (2016). The allocation and effectiveness of China's R&D subsidies—Evidence from listed firms. *Research Policy, 45*(9), 1774–1789. doi:10.1016/j.respol.2016.05.007

Boff, L. (1999). *Saber Cuidar: Ética do humano – compaixão pela terra*. Editora Vozes.

Bohm, S. (2015). Ecocultures: Towards sustainable ways of living. InBöhm, S., Bharucha, Z. P., & Pretty, J. (Eds.), *Ecocultures: Blueprints for Sustainable Communities* (pp. 3–26). Routledge.

Boiral, O., & Paillé, P. (2012). Organizational citizenship behaviour for the environment: Measurement and validation. *Journal of Business Ethics, 109*(4), 431–445. doi:10.100710551-011-1138-9

Boisot, M., & Child, J. (1996). From fiefs to clans and network capitalism: Explaining China's emerging economic order. *Administrative Science Quarterly, 41*(4), 600–628. doi:10.2307/2393869

Bolton, A., Goosen, L., & Kritzinger, E. (2021a). Unified Communication Technologies at a Global Automotive Organization. In Encyclopedia of Organizational Knowledge, Administration, and Technologies (pp. 2592-2608). IGI Global. doi:10.4018/978-1-7998-3473-1.ch179

Bolton, A., Goosen, L., & Kritzinger, E. (2021c). The Integration and Implementation of the Internet of Things Through Digital Transformation: Impact on Productivity and Innovation. In Integration and Implementation of the Internet of Things Through Cloud Computing (pp. 85-112). IGI Global.

Bolton, A., Goosen, L., & Kritzinger, E. (2022b). Emerging Technologies for Innovation and Productivity Management in the Automotive Industry: Impact of Digital Transformation on Communication. In Emerging Technologies for Innovation Management in the Software Industry (pp. 60-85). IGI Global.

Bolton, A., Goosen, L., & Kritzinger, E. (2023). Embracing Business Sustainability Through Innovation and Productivity in the Automotive Sector: Creativity, Collaboration, and Generating Savings. In Embracing Business Sustainability Through Innovation and Creativity in the Service Sector (pp. 184-204). IGI Global.

Bolton, A., Goosen, L., & Kritzinger, E. (2016). Enterprise Digitization Enablement Through Unified Communication and Collaboration. *Proceedings of the Annual Conference of the South African Institute of Computer Scientists and Information Technologists*. Johannesburg: ACM. 10.1145/2987491.2987516

Bolton, A., Goosen, L., & Kritzinger, E. (2020b). The Impact of Unified Communication and Collaboration Technologies on Productivity and Innovation: Promotion for the Fourth Industrial Revolution. In *Promoting Inclusive Growth in the Fourth Industrial Revolution* (pp. 44–73). IGI Global. doi:10.4018/978-1-7998-4882-0.ch002

Bolton, A., Goosen, L., & Kritzinger, E. (2021b). An Empirical Study into the Impact on Innovation and Productivity Towards the Post-COVID-19 Era: Digital Transformation of an Automotive Enterprise. In *Handbook of Research on Entrepreneurship, Innovation, Sustainability, and ICTs in the Post-COVID-19 Era* (pp. 133–159). IGI Global. doi:10.4018/978-1-7998-6776-0.ch007

Bolton, A., Goosen, L., & Kritzinger, E. (2022a). Impact of Digital Transformation via Unified Communication and Collaboration Technologies: Productivity and Innovation at a Global Enterprise. In *Impact of Digital Transformation on the Development of New Business Models and Consumer Experience* (pp. 271–298). IGI Global. doi:10.4018/978-1-7998-9179-6.ch014

Bolton, A., Goosen, L., & Kritzinger, E. (2022c). Challenges and Emerging Strategies for a Global Automotive Enterprise Towards a Post-COVID-19 Era: The Impact of Digital Transformation. In A. Pego (Ed.), *Handbook of Research on Global Networking Post COVID-19* (pp. 76–109). IGI Global. doi:10.4018/978-1-7998-8856-7.ch005

Bolton, T., Goosen, L., & Kritzinger, E. (2020a). Security Aspects of an Empirical Study into the Impact of Digital Transformation via Unified Communication and Collaboration Technologies on the Productivity and Innovation of a Global Automotive Enterprise. *Communications in Computer and Information Science*, *1166*, 99–113. doi:10.1007/978-3-030-43276-8_8

Bomb, C., McCormick, K., Deurwaarder, E., & Kåberger, T. (2007). Biofuels for transport in Europe: Lessons from Germany and the UK. *Energy Policy*, *35*(4), 2256–2267. doi:10.1016/j.enpol.2006.07.008

Bonilla, L., Cabrera, J., & López, M. (2019). La jubilación: un derecho del trabajador o una formalidad de las prestaciones de la seguridad social en Ecuador. Pro Sciences: Revista De Producción. *Ciencia e Investigacion*, *3*(29), 74–83. doi:10.29018/issn.2588-1000vol3iss29.2019pp74-83

Borland, H., & Lindgreen, A. (2013). Sustainability, epistemology, ecocentric business, and marketing strategy: Ideology, reality, and vision. *Journal of Business Ethics*, *117*(September), 173–187. doi:10.100710551-012-1519-8

Bosch, P., Jongeneel, S., Rovers, V., Neumann, H. M., Airaksinen, M., & Huovila, A. (2017). CITYkeys indicators for smart city projects and smart cities. *Eurocities*. http://nws.eurocities.eu/MediaShell/media/CITYkeystheindicators.pdf

Bostancı, S. H., & Yıldırım, S. (2021). Sustainable Communities vs. Climate Refugees: Two Opposite Results of Climate Change. In C. R. G. Popescu (Ed.), *Handbook of Research on Novel Practices and Current Successes in Achieving the Sustainable Development Goals* (pp. 298–319). IGI Global. doi:10.4018/978-1-7998-8426-2.ch015

Botillen, A. (2022). Maritime Security in the Arctic. In R.-L. Boşilcă S. Ferreira & B. J. Ryan (eds.) *Routledge Handbook of Maritime Security*. London and New York, NY: Routledge. https://www.routledge.com/Routledge-Handbook-of-Maritime-Security/Bosilca-Ferreira-Ryan/p/book/9780367430641

Boubakri, N., Cosset, J. C., & Saffar, W. (2013). The role of state and foreign owners in corporate risk-taking: Evidence from privatization. *Journal of Financial Economics*, *108*(3), 641–658. doi:10.1016/j.jfineco.2012.12.007

Boyack, K. W., Klavans, R., & Börner, K. (2005). Mapping the backbone of science. *Scientometrics*, *64*(3), 351–374. doi:10.100711192-005-0255-6

Boyle, D. (2001). *The New Economics: A Bigger Picture*. Earthscan Publications.

Bozkuş, Y. D. (2022). *Rusya-Ukrayna savaşıyla başlayan tahıl sorunu küresel bir gıda krizine dönüşebilir mi? (Could the grain problem that started with the Russia-Ukraine war turn into a global food crisis?)*, Indy Turk. https://www.indyturk.com/node/520321/t%C3%BCrki%CC%87yeden-sesler/rusya-ukrayna-sava%C5%9F%C4%B1yla-ba%C5%9Flayan-tah%C4%B1l-sorunu-k%C3%BCresel-bir-g%C4%B1da

Bozzoli, M. E. (2000). A role for anthropology in sustainable development in Costa Rica. *Human Organization*, *59*(3), 275–279. doi:10.17730/humo.59.3.eu5h42p81wk26578

Bracco, S., Calicioglu, O., Gomez San Juan, M., & Flammini, A. (2018). Assessing the Contribution of Bioeconomy to the Total Economy: A Review of National Frameworks. *Sustainability (Basel)*, *10*(6), 1698. doi:10.3390u10061698

Bradford, M., & Florin, J. (2003). Examining the role of innovation diffusion factors on the implementation success of enterprise resource planning systems. *International Journal of Accounting Information Systems*, *4*(3), 205–225. doi:10.1016/S1467-0895(03)00026-5

Brady, H. (2019). The challenge of big data and data science. *Annual Review of Political Science*, *22*(1), 297–323. doi:10.1146/annurev-polisci-090216-023229

Brajer, V., Mead, R. W., & Xiao, F. (2011). Searching for an environmental kuznets curve in China's air pollution. *China Economic Review*, *22*(3), 383–397. doi:10.1016/j.chieco.2011.05.001

Branco, M. C., & Rodrigues, L. L. (2006). Corporate social responsibility and resource-based perspective. *Journal of Business Ethics*, *69*(2), 111–132. doi:10.100710551-006-9071-z

Brands, R. A., & Fernandez-Mateo, I. (2017). Leaning out: How negative recruitment experiences shape women's decisions to compete for executive roles. *Administrative Science Quarterly*, *62*(3), 405–442. doi:10.1177/0001839216682728

Braun, B. (2014). A new urban dispositive? Governing life in an age of climate change. *Environment and Planning. D, Society & Space*, *32*(1), 49–64. doi:10.1068/d4313

Bresciani-Turroni, C. (1937). *The Economics of Inflation*. George Allen & Unwin Ltd.

Briggs, H. (2022). COP15: Nations reach 'historic' deal to protect nature. *BBC News*. https://www.bbc.com/news/science-environment-64019324

Brooks, A. L., Wang, S., & Jambeck, J. R. (2018). The Chinese import ban and its impact on global plastic waste trade. *Science Advances*, *4*(6), eaat0131. . doi:10.1126ciadv.aat0131 PMID:29938223

Brozovsky, J., Gustavsen, A., & Gaitani, N. (2021). Sustainable Cities and Society Zero emission neighbourhoods and positive energy districts – A state-of-the-art review. *Sustainable Cities and Society*, 72.

Brudermann, T., & Sangkakool, T. (2017). Green roofs in temperate climate cities in Europe - An analysis of key decision factors. *Urban Forestry & Urban Greening*, *21*, 224–234. doi:10.1016/j.ufug.2016.12.008

Brundtland, G. (1987). *Our Common Future: The World Commission on Environment and Development*. Oxford University Press.

Bryman, A. (2012). *Social Research Methods* (4th ed.). Oxford University Press.

Bryson, J. M., & Crosby, B. C. (2017). The social and economic impacts of community development: A case study of the Emerald Necklace Conservancy. *Journal of the American Planning Association*, *83*(3), 264–276.

Buczek, B. (2017). Passenger Logistics Centers Presentness and Future. *Transport Economics and Logistics*, *71*, 73–91.

Budiarti, R. P. N., Tjahjono, A., Hariadi, M., & Purnomo, M. H. (2019). Development of IoT for automated water quality monitoring system. *2019 International Conference on Computer Science, Information Technology, and Electrical Engineering (ICOMITEE)*, (pp. 211–216). IEEE. 10.1109/ICOMITEE.2019.8920900

Bugge, M., Hansen, T., & Klitkou, A. (2016). What is the bioeconomy? A review of the literature. *Sustainability (Basel)*, *8*(7), 691. doi:10.3390u8070691

Bunge, M. (1999). *Buscar la filosofía en las ciencias sociales*. Siglo XXI.

Burszta-Adamiak, E., & Fiałkiewicza, W. (2019). Review of green roof incentives as motivators for the expansion of green infrastructure in European cities. *Scientific Review – Engineering and Environmental Sciences*, *28* (4), 641–652.

Burszta-Adamiak, E., Stańczyk, J., & Łomotowski, J. (2019). Hydrological performance of green roofs in the context of the meteorological factors during the 5-year monitoring period. *Water and Environment Journal : the Journal / the Chartered Institution of Water and Environmental Management*, *33*(1), 144–154. doi:10.1111/wej.12385

BW. (2023). *What is bioeconomy*? Bioeconomy BW. https://www.biooekonomie-bw.de/en/bw/definition (Accessed 12.06.2023)

C., R., V., T., Liloja, & Shahzad, M. (2019). IOT based water quality monitoring with android application. *2019 Third International conference on I-SMAC (IoT in Social, Mobile, Analytics and Cloud) (I-SMAC)*, (pp. 446–451). IEEE.

Cabral, A. M. R., & de Sousa Ramos, F. (2014). Cluster analysis of the competitiveness of container ports in Brazil. *Transportation Research Part A, Policy and Practice*, *69*, 423–431. doi:10.1016/j.tra.2014.09.005

Cadena-roa, J. (Coordinador) (2022). *Contribuciones para el estudio del desempeño asociativo resultados de la puesta a prueba del protocolo para la evaluación de asociaciones.* Universidad Nacional Autónoma de México Centro de Investigaciones Interdisciplinarias en Ciencias y Humanidades México.

Cajee, L. (2014). Eco-Cultural Tourism: A Tool for Environmental, Cultural and Economic Sustainability (A Case Study of Darap Village, West Sikkim). *SHS Web Conf.*, 12, 01029. 10.1051hsconf/20141201029

Callon, M. (1991). Techno-economy networks and irreversibility. In L. Law (Ed.), *A Sociology of Monsters: Essay on Power, Technology and Domination*. Routledge.

Camilla, T. (2017). Emerging Chinese-Russian Cooperation in the Arctic: Possibilities and Constraints" (Cooperação emergente sino-russa no Ártico: possibilidades e restrições). *SIPRI Policy Paper, 46.* https://www.sipri.org/. (Accessed 13.06.2023)

Cammeraat, E. (2020). The relationship between different social expenditure schemes and poverty, inequality and economic growth. *International Social Security Review, 73*(1), 101- 123. doi:. . doi:123

Caney, S., & Hepburn, C. (2011). *Carbon trading: Unethical, unjust and ineffective?* Centre for Climate Change Economics and Policy working paper No. 59. & Grantham Research Institute on Climate Change and the Environment working paper No. 49.

Cao, C. (2021). The evolution of China's biosafety/biosecurity legislation. *Journal of Law and the Biosciences*, *8*(1), LSB020. doi:10.1093/jlb/lsab020 PMID:34221436

Cao, X., Deng, M., Song, F., Zhong, S., & Zhu, J. (2019). Direct and moderating effects of environmental regulation intensity on enterprise technological innovation: The case of China. *PLoS One*, *14*(10), e0223175. doi:10.1371/journal.pone.0223175 PMID:31589643

CARE. (2018). *Situación del trabajo remunerado del hogar en ecuador (Primera ed.).* Quito: Publiasesores Cía. Ltda. https://bit.ly/3Vdp1sg

Carney, T. (2019). Automation in Social Security: Implications for Merits Review. *The Australian Journal of Social Issues*, *55*(3), 260–274. doi:10.1002/ajs4.95

Carranza, C., & Cisneros, M. (2014). Hacia un sistema de protección social más inclusivo en el Ecuador Seguimiento y desenlace de un proceso de construcción de consensos en la búsqueda del Buen Vivir. *Serie Políticas Sociales, 205*, 8-84. https://bit.ly/3Ve3kZl

Casalí, P., Vallejo, F., Jaramillo, D., Lucio, R., & Velásquez, M. (2021). *Ecuador. Panorama de la protección social: gobernanza, cobertura y sostenibilidad. (Primera ed.).* Lima: Organización Internacional del Trabajo. https://bit.ly/3T33aBV

Casey, K., Lichrou, M., & O'Malley, L. (2020). Prefiguring sustainable living: An ecovillage story. *Journal of Marketing Management*, *36*(17–18), 1658–1679. doi:10.1080/0267257X.2020.1801800

Cashdan, E. (1998). Adaptiveness of food learning and food aversions children. *Social Sciences Information. Information Sur les Sciences Sociales*, *37*(4), 613–632. doi:10.1177/053901898037004003

Castro, A. J. L., & Orellana, F. (2019). Green roofs as a strategy for urban green innovation spaces. *Sustainable Cities and Society*, *44*, 697–705. doi:10.1016/j.scs.2018.10.034

Caswell, B., Klein, E. S., Alleway, H. K., Ball, J., Cardinale, M., Eero, M., & Thurstan, R. H. (2020). Something old, something new: 200 years of a blue growth agenda. *Fish and Fisheries*, *21*, 774–796. doi:10.1111/faf.12460

Caves, R. E. (1981). Diversification and seller concentration: Evidence from change. *The Review of Economics and Statistics*, *63*(2), 289–293. doi:10.2307/1924100

Ceni, R. (2017). Pension schemes and labor supply in the formal and informal sector. *IZA Journal of Labor Policy*, *6*(8), 1–29. doi:10.118640173-017-0085-1

Centobelli, P., Cerchione, R., Esposito, E., & Shashi. (2020). Evaluating environmental sustainability strategies in freight transport and logistics industry. *Business Strategy and the Environment*, *29*(3), 1563–1574. doi:10.1002/bse.2453

Central Bank of Ecuador. (2004). La Seguridad social en el Ecuador. Apuntes de Economía (47), 17-25. https://bit.ly/3ylWYNK

Chaaban, J. (2019). Lebanon Needs a Budget for Its People, Not is Ruling Class. *Beirut Today.*

Chand, K. (2019). An Initiative of Adopting E-Vehicles to Minimize Air Pollution. *International Journal of Research and Review.*

Chang, N. J., & Fong, C. M. (2010). Green product quality, green corporate image, green customer satisfaction, and green customer loyalty. *African Journal of Business Management*, *4*(13), 2336–2344.

Chang, T. W., Chen, Y. S., Yeh, Y. L., & Li, H. X. (2020). Sustainable consumption models for customers: Investigating the significant antecedents of green purchase behavior from the perspective of information asymmetry. *Journal of Environmental Planning and Management*, *64*(9), 1668–1688. doi:10.1080/09640568.2020.1837087

Chatterjee, S. (in press). COVID 19. *Tackling Global Pandemics Through Scientific and Social Tools, Elsevier.*

Chauhan, Priyanka and Singh, N.P. (2023). An Immaculate Study on The Challenges, Opportunities and Innovation in Waste Management Industry. *Cosmos An International Journal Art & Higher Education*, *12*(2), 69–72.

Chen, C. (2023). China-Russia Arctic Cooperation in the Context of a Divided Arctic. *Arctic Collaboration Series 2023*, The Artic Institute https://www.thearcticinstitute.org/china-russia-arctic-cooperation-context-divided-arctic/

Chen, A., Rach, M., & Sehner, T. (2020). On the Optimal Combination of Annuities and Tontines. Astin Bulletin. *The Journal of the IAA*, *50*(1), 95–129. doi:10.1017/asb.2019.37

Chen, C. C., Shih, H. S., Shyur, H. J., & Wu, K. S. (2012). A business strategy selection of green supply chain management via an analytic network process. *Computers & Mathematics with Applications (Oxford, England)*, *64*(8), 2544–2557. doi:10.1016/j.camwa.2012.06.013

Cheneval, F. (2021). Property rights of personal data and the financing of pensions. *Critical Review of International Social and Political Philosophy*, *24*(2), 253–275. doi:10.1080/13698230.2018.1528521

Cheng, D., Li, C., & Qiu, N. (2021). The application prospects of NB-IoT in intelligent transportation. *2021 4th International Conference on Advanced Electronic Materials, Computers and Software Engineering (AEMCSE)*, (pp. 1176–1179). ACM.

Chen, H. L., & Hsu, W. T. (2009). Family ownership, board independence, and R&D investment. *Family Business Review*, *22*(4), 347–362. doi:10.1177/0894486509341062

Chen, Y. S. (2008). The driver of green innovation and green image–green core competence. *Journal of Business Ethics*, *81*(3), 531–543. doi:10.100710551-007-9522-1

Chen, Y. S., & Chang, C. H. (2013). The determinants of green product development performance: Green dynamic capabilities, green transformational leadership, and green creativity. *Journal of Business Ethics*, *116*(1), 107–119. doi:10.100710551-012-1452-x

Chen, Y. S., Lai, S. B., & Wen, C. T. (2006). The influence of green innovation performance on corporate advantage in Taiwan. *Journal of Business Ethics*, *67*(4), 331–339. doi:10.100710551-006-9025-5

Chen, Y. S., Lin, S. H., Lin, C. Y., Hung, S. T., Chang, C. W., & Huang, C. W. (2020). Improving green product development performance from green vision and organizational culture perspectives. *Corporate Social Responsibility and Environmental Management*, *27*(1), 222–231. doi:10.1002/csr.1794

Chen, Y.-W., Milstein, T., Anguiano, C., Sandoval, J., & Knudsen, L. (2012). Challenges and Benefits of Community-Based Participatory Research for Environmental Justice: A Case of Collaboratively Examining Ecocultural Struggles. *Environmental Communication*, *6*(3), 403–421. doi:10.1080/17524032.2012.698291

Chen, Z., Zhang, X., & Chen, F. (2021). Do carbon emission trading schemes stimulate green innovation in enterprises? Evidence from China. *Technological Forecasting and Social Change*, *168*, 120744. doi:10.1016/j.techfore.2021.120744

Cheryl. (2008). *Algae becoming the new biofuel of choice.* Dueling Fuels. http://duelingfuels.com/biofuels/non-food-biofuels/algae-biofuel.php#more-115N

Cheunkamon, E., Jomnonkwao, S., & Ratanavaraha, V. (2022). Impacts of tourist loyalty on service providers: Examining the role of the service quality of tourism supply chains, tourism logistics, commitment, satisfaction, and trust. *Journal of Quality Assurance in Hospitality & Tourism*, *23*(6), 1397–1429. doi:10.1080/1528008X.2021.1995564

Chiu, Y. B., & Lee, C. C. (2020). Effects of financial development on energy consumption: The role of country risks. *Energy Economics*, *90*, 104833. doi:10.1016/j.eneco.2020.104833

Choi, H.-S., & Yeom, K.-R. (2018). Study on an energy-IoT service platform for energy saving in legacy manufacturing site. *2018 Tenth International Conference on Ubiquitous and Future Networks (ICUFN)*, (pp. 811–813). IEEE. 10.1109/ICUFN.2018.8436969

Cho, J. H., & Sohn, S. Y. (2018). A novel decomposition analysis of green patent applications for the evaluation of R&D efforts to reduce CO2 emissions from fossil fuel energy consumption. *Journal of Cleaner Production*, *193*, 290–299. doi:10.1016/j.jclepro.2018.05.060

Chowdhury, A., Ortiz, I., Capaldo, J., Yamabana, H., & Urban, S. (2016). *Ecuador: Financing Social Protection through Debt Restructuring.* Ginebra: International Labour Organization. from https://bit.ly/3fVctG4

Chowdhury, A. A. S., Arafat, Y., & Alam, M. S. (2022). IoT-GSM based controlling and monitoring system to prevent water wastage, water leakage, and pollution in the water supply. *2022 International Conference on Innovations in Science, Engineering and Technology (ICISET)*, (pp. 567–572). IEEE. 10.1109/ICISET54810.2022.9775876

Clapp, J. (1994). The toxic waste trade with less-industrialised countries: Economic linkages and political alliances. *Third World Quarterly*, *15*(3), 505–518. doi:10.1080/01436599408420393

Clemente, P. A. (2023). Operations Strategy of Air Bed and Breakfast (Airbnb): Basis for An Enhanced Business Model. *Globus An International Journal of Management & IT, 14*(2), 06-25.

ClimateTrade. (2022). *ClimateTrade and Terrasos jointly promote Voluntary Biodiversity Credits to support biodiversity conservation.* Climate Trade. https://climatetrade.com/climatetrade-and-terrasos-jointly-promote-voluntary-biodiversity-creditsto-support-biodiversity-conservation/#:~:text=The%20purchase%20of%20a%20VBC,to%2010%20m2%20of%20forest

Cohen-Rosenthal, E. (2000). A walk on the human side of industrial ecology. *The American Behavioral Scientist*, *44*(4), 245–264. doi:10.1177/0002764200044002007

Cole, D. (2008). Museum marketing as a tool for survival and creativity: The Mining Museum Perspective. *Museum Management and Curatorship*, *23*(2), 177–192. doi:10.1080/09647770701865576

Cole, M. A. (2004). Trade, the pollution haven hypothesis and the environmental Kuznets curve: Examining the linkages. *Ecological Economics*, *48*(1), 71–81. doi:10.1016/j.ecolecon.2003.09.007

Conde-Ruiz, J., & González, C. (2014). *From Bismarck to Beveridge: The other Pension Reform in Spain.* Madrid: Banco de España. https://bit.ly/3T3A9WN

Conniff, R. (2018). *Biodiversity Credits: How Policy Can Help Save Unprotected Species.* Yale Environment 360. https://e360.yale.edu/features/biodiversity-credits

Connolly, C., Keil, R., & Ali, S. H. (2021). Extended urbanisation and the spatialities of infectious disease: Demographic change, infrastructure and governance. *Urban Studies (Edinburgh, Scotland)*, *58*(2), 245–263. doi:10.1177/0042098020910873

Constitution of the Republic of Ecuador. (2008). *Asamblea Nacional Constituyente.* Defensa. https://www.defensa.gob.ec/wp-content/uploads/downloads/2021/02/Constitucion-de-la-Republica-del-Ecuador_act_ene-2021.pdf

Consulting, D. (2017) *12 mal Berliner Leben, 12 mal Berlin Erleben. Konzept für einen stadtverträglichen Berlin-Tourismus 2018+;* VisitBerlin https://about.visitberlin.de/sites/default/files/2018 02/Tourismuskonzept_Berlin_Studie_2017.pdf

Contreras-Jaramillo, M. A. (2018). Análisis de la sostenibilidad del sistema de pensiones ecuatoriano, periodo 2013-2025. *Papeles de población, 24*(96), 29-62. doi:http://dx.doi.org/ doi:10.22185/24487147.2018.96.14

Convention on Biological Diversity (CBD). (2022). *Kunming-Montreal Global Biodiversity Framework.* Montreal: CBD. https://www.cbd.int/gbf/

Coolen, H., & Meesters, J. (2012). Private and public green spaces: Meaningful but different settings. *Journal of Housing and the Built Environment*, *27*(1), 49–67. doi:10.100710901-011-9246-5

Cooper, D., & Schindler, P. S. (2006). *Business Research Methods.* McGraw-Hill.

Coordinating Ministry of Social Development. (2015). *La Articulación de un Sistema de Seguridad Social (Primera ed.).* Quito: Ministerio Coordinador de Desarrollo Social. https://www.social-protection.org/gimi/RessourcePDF.action?id=53636

Coordinating Ministry of Social Development. (2016). *La institucionalidad normativa y funcional de los sistemas de seguridad social (Primera ed.).* Quito: MCDS. https://www.social-protection.org/gimi/RessourcePDF.action?id=53638

Coppola, M., Russolillo, M., & Simone, R. (2022). On the evolution of the gender gap in life expectancy at normal retirement age for OECD countries. *Genus, 78*(27). doi:10.118641118-022-00175-5

Corona, B., Shen, L., Reike, D., Carreón, J. R., & Worrell, E. (2019). Towards sustainable development through the circular economy - A review and critical assessment on current circularity metrics. *Resources, Conservation and Recycling, 151*, 104498. doi:10.1016/j.resconrec.2019.104498

Corvellec, H., Stowell, A. F., & Johansson, N. (2022). Critiques of the circular economy. *Journal of Industrial Ecology, 26*(2), 421–432. doi:10.1111/jiec.13187

Cotta, B. (2020). What goes around, comes around? Access and allocation problems in Global North–South waste trade. *International Environmental Agreement: Politics, Law and Economics, 20*(2), 255–269. doi:10.100710784-020-09479-3

Cram, S. (2016). Wild and Scenic Wasteland: Conservation Politics in the Nuclear Wilderness. *Environmental Humanities, 7*(1), 89–105. doi:10.1215/22011919-3616344

Creswell, J. W. (2002). *Educational research: Planning, conducting, and evaluating quantitative.* Prentice Hall Upper Saddle River.

Creswell, J.-W. (1998). *Qualitative Inquiry and Research Design: Choosing among Five Traditions.* Sage Publications.

Creswell, J.-W. (2002). *Educational Research: Planning, Conducting and Evaluating Quantitative and Qualitative Research.* Prentice Hall.

Creswell, J.-W. (2003). *Research Design: Qualitative, Quantitative and Mixed Methods Approach* (2nd ed.). Sage Publications.

Creswell, J.-W. (2012). *Educational Research: Planning, Conducting, and Evaluating Quantitative and Qualitative Research* (4th ed.). Pearson.

Cronbach, L. J. (1951). Coefficient alpha and the internal structure of tests. *Psychometrika, 16*(3), 297–334. doi:10.1007/BF02310555

Cuckston, T. (2022). Book Review: Extinctions: Living and Dying in the Margin of Error. *Progress in Physical Geography, 46*(5), 819–821. doi:10.1177/03091333221109546

Cueva, C. (2021). Una mirada interpretativa y constitucional de los derechos sociales: La reducción de las pensiones jubilares. Foro. *Revista de Derecho*, (35), 106–125. doi:10.32719/26312484.2021.35.6

D'Apuzzo, M., Cappelli, G., Buzzi, S., & Nicolosi, V. (2022). Smart Urban Mobility Management project: a concrete step towards more sustainable and connected communities. *2022 Second International Conference on Sustainable Mobility Applications, Renewables and Technology (SMART)*, (pp. 1–8). 10.1109/SMART55236.2022.9990394

Dahake, S. (2020). The making of fluid ecocultural identities in urban India. In T. Milstein & J. Castro-Sotomayor (Eds.), *Routledge Handbook of Ecocultural Identity* (pp. 333–350). Routledge. doi:10.4324/9781351068840-20

Dal, M. M., Karakaya, E., & Bulut, Ş. (2014). Çevresel kuznets eğrisi ve Türkiye: Ampirik bir analiz. *Dumlupınar Üniversitesi Sosyal Bilimler Dergisi, Special Issue of XIV. International Symposium on Econometrics, Operations Research and Statistics*, 85-96. https://dergipark.org.tr/en/pub/dpusbe/issue/31805/34898

Dalgaard, C., Kaarsen, N., Olsson, O., & Selaya, P. (2018). Roman roads to prosperity: Persistence and non-persistence of public goods provision. *Science Direct.*

Damanpour, F. (1992). Organizational size and innovation. *Organization Studies*, *13*(3), 375–402. doi:10.1177/017084069201300304

Daryl Slack, J. (2008). Resisting ecocultural studies. *Cultural Studies*, *2008*(22), 477–497. doi:10.1080/09502380802012575

Das, P. K., & Bhat, M. Y. (2022). Global electric vehicle adoption: Implementation and policy implications for India. *Environmental Science and Pollution Research International*, *29*(27), 40612–40622. doi:10.100711356-021-18211-w PMID:35083674

Dauda, L., Long, X., Mensah, C. N., Salman, M., Boamah, K. B., Ampon-Wireko, S., & Dogbe, C. S. K. (2021). Innovation, trade openness and CO2 emissions in selected countries in Africa. *Journal of Cleaner Production*, *281*, 125143. doi:10.1016/j.jclepro.2020.125143

Davies, G. (2016). *A History of Money: From Ancient Times to the Present Day*. University of Wales Press.

Davis, L. J., & Morris, D. B. (2007). Biocultures Manifesto. *New Literary History*, *38*(3), 411–418. doi:10.1353/nlh.2007.0046

Dayton, P. K., Tenger, M. J., Edwards, P. B., & Riser, K. L. (1998). Sliding Baselines, Ghosts, and Reduced Expectations in Kelp Forest Communities. *Ecological Applications*, *8*(2), 309–322. doi:10.1890/1051-0761(1998)008[0309:SBGARE]2.0.CO;2

De Abreu, V. H. S., Da Costa, M. G., Da Costa, V. X., De Assis, T. F., Santos, A. S., & D'Agosto, M. D. A. (2022). The Role of the Circular Economy in Road Transport to Mitigate Climate Change and Reduce Resource Depletion. *Sustainability (Basel)*, *14*(14), 8951. doi:10.3390u14148951

De Jaramillo, E. H., & Schuler, I. (2014). Challenges of Sustainable Development and Productive Use of Biodiversity: The Knowledge Based Bio-Economy Approach. In Environmental Protection and Sustainable Development from Rio to Rio, 20. doi:10.1163/9789004282919_019

de la Torre, A. (2021). *Retos de los sistemas de pensiones y salud*. CAF. https://www.caf.com/es/actualidad/noticias/2021/09/retos-de-los-sistemas-de-pensiones-y-salud-en-ecuador-y-america-latina/

de Luxán, J. (2017). ¿Cobraremos la pensión? Cómo sostener el sistema público de pensiones. *Cuadernos de Gobierno y Administración Pública*, *4*(1), 107–110. doi:10.5209/CGAP.56025

de Melo, M. A. C. (1997). Processo de Planejamento e as Inovações Tecnológicas e Sociais: uma perspectiva sócio-ecológica. *Anais do 5o. Seminário de Modernização Tecnológica*. Maxwell.

de Melo, M.A.C. (2002b). Inovação e Modernização Tecnológica e Organizacional nas MPMEs: o domínio interorganizacional. *Seminário Internacional: Políticas*

de Melo, A. (2001). *The Innovation Systems of Latin America and the Caribbean*. Inter-American Development Bank. Agosto.

de Melo, M. A. C. (2002a). *Enriquecendo a Atuação de Incubadora de Emrpesas. Tecnologia e Inovação: experiências de gestão nas micro e pequenas empresas*. PGT/USP.

Dedrick, J., Kraemer, K., & Shih, E. (2013). Information Technology and Productivity in Developed and Developing Countries. *Journal of Management Information Systems*, *30*(1), 97–122. doi:10.2753/MIS0742-1222300103

Deegan, C. (2002). Introduction – the legitimising effect of social and environmental disclosure – a theoretical foundation. *Accounting, Auditing & Accountability Journal*, *15*(3), 283–311. doi:10.1108/09513570210435852

Delmas, M. A., & Montes-Sancho, M. J. (2010). Voluntary agreements to improve environmental quality: Symbolic and substantive cooperation. *Strategic Management Journal*, *31*(6), 575–601.

Deng, Q., & Ji, S. (2015). Organizational green IT adoption: *Concept and evidence. Sustainability (Basel)*, *7*(12), 16737–16755. doi:10.3390u71215843

Denzin, N. K., & Lincoln, Y. S. (2005). *The Sage handbook of qualitative research* (3rd ed.). Sage.

Department of Statistics & the Ministry of the Interior (MOI), R.O.C. (Taiwan). (2014) *Statistical Yearbook of Interior.* Ministry of Interior. http://sowf.moi.gov.tw/stat/year/list.htm

Department of Tourism, Government of Karnataka. (2020). *Karnataka Tourism Policy 2020-26.* Karnataka. https://karnatakatourism.org/department/wp-content/uploads/2022/12/KT_Policy_Englosh.pdf

Desai, N. (2017). Universal Basic Income or Social Security. *Indian Journal of Human Development.*, *11*(2), 203–205. doi:10.1177/0973703017742759

Desjeux, P. (2001) Trans. of the Royal Society of *Trop. Med. and Hygiene, 95*(3) 239–43.

Deswal, A. & Kumar, S. (2022). Technological Aspects in Cloud Computing. *Cosmos Journal of Engineering & Technology, 12*(2), 05-07.

Devaney, L., & Iles, A. (2019). Scales of progress, power and potential in the US bioeconomy. *Journal of Cleaner Production*, *233*, 379–389. doi:10.1016/j.jclepro.2019.05.393

DeWalt, K., DeWalt, B., & Wayland, C. (1998). Participant observation. In H. R. Bernard (Ed.), *Handbook of methods in cultural anthropology* (pp. 259–299). Altamira.

Dey, M. M., Gosh, K., Valmonte-Santos, R., Rosegrant, M. W., & Chen, O. L. (2016a). Economic impact of climate change and climate change adaptation strategies for fisheries sector in Fiji. *Marine Policy*, *67*, 164–170. doi:10.1016/j.marpol.2015.12.023

Dey, M. M., Gosh, K., Valmonte-Santos, R., Rosegrant, M. W., & Chen, O. L. (2016b). Economic impact of climate change and climate change adaptation strategies for fisheries sector in Solomon Islands:Implication for food security. *Marine Policy*, *67*, 171–178. doi:10.1016/j.marpol.2016.01.004

Diamond, J., & Robinson, J. (2010). *Natural experiments of history*. The Belknap Press of Harvard University Press.

Diamond, P. (2018). Questions and Answers: The Future of Social Security. *Economic Inquiry*, *56*(2), 682–685. doi:10.1111/ecin.12552

Díaz, Á. (2022). 'Boom' de los pellets de madera tras encarecerse los combustibles fósiles. *El Mundo*. https://www.elmundo.es/ciencia-y-salud/medio-ambiente/2021/11/04/6183dfc121efa0c5078b46ab.html

Díaz, D., Kristoff, J., Castro, M., Miceli, J., Castro, D., Quinteros, R., & Guerrero, S. (2007). *Exploraciones en antropología y complejidad*. Editorial Sb.

Dietz, T., Ostrom, E., & Stern, P. L. (2003). The Struggle to Govern the Commons. *Science*, *302*(5652), 1907–1912. doi:10.1126cience.1091015 PMID:14671286

DiMaggio, P. J., & Powell, W. W. (1983). The iron cage revisited: Institutional isomorphism and collective rationality in organizational fields. *American Sociological Review*, *48*(2), 147–160. doi:10.2307/2095101

Dincă, G., Milan, A. A., Andronic, M. L., Pasztori, A. M., & Dincă, D. (2022). Does Circular Economy Contribute to Smart Cities' Sustainable Development? *International Journal of Environmental Research and Public Health*, *19*(13), 7627. doi:10.3390/ijerph19137627 PMID:35805283

Dinga, Q., Chena, X., Hilbornd, R., & Chene, Y. (2017). Vulnerability to impacts of climate change on marine fisheries and food security. *Marine Policy*, *83*, 55–61. doi:10.1016/j.marpol.2017.05.011

Dirsehan, T. (2012). Analyzing museum visitor experiences and post experience dimensions using sem. *Bogazici Journal*, *26*(1), 103–125. doi:10.21773/boun.26.1.6

Ditlevsen, K., Sandøe, P., & Lassen, J. (2019). Healthy food is nutritious, but organic food is healthy because it is pure: The negotiation of healthy food choices by Danish consumers of organic food. *Food Quality and Preference*, *71*(January), 46–53. doi:10.1016/j.foodqual.2018.06.001

Dogaru, L. (2020). About sustainability between responsible production and consumption. *Proceedings*, *63*(1), 69. doi:10.3390/proceedings2020063069

Doloreux, D. (2017). What is a maritime cluster? *Marine Policy*, *83*, 215–220. doi:10.1016/j.marpol.2017.06.006

Doloreux, D., Gaviria de la Puerta, J., Pastor-López, I., Gómez, I. P., Sanz, B., & Zabala-Iturriagagoitia, J. M. (2019). Territorial innovation models: To be or not to be, that's the question. *Scientometrics*, *120*(3), 1163–1191. doi:10.100711192-019-03181-1

Doloreux, D., & Shearmur, R. (2009). Maritime clusters in diverse regional contexts: The case of Canada. *Marine Policy*, *33*(3), 520–527. doi:10.1016/j.marpol.2008.12.001

Doloreux, D., Shearmur, R., & Figueiredo, D. (2016). Québec'coastal maritime cluster: Its impact on regional economic development, 2001–2011. *Marine Policy*, *71*, 201–209. doi:10.1016/j.marpol.2016.05.028

Dong, Y., Yang, M., Bakker, D. C. E., Liss, P. S., Kitidis, V., Brown, I., Chierici, M., Fransson, A., & Bell, T. G. (2021). Near-Surface Stratification Due to Ice Melt Biases Arctic Air-Sea CO_2 Flux Estimates. *Geophysical Research Letters*, *48*(22). doi:10.1029/2021GL095266

Donleavy, G., & Noronha, C. (2023). Concluding remarks: Comparative CSR and sustainability research – An axial analysis. In G. Donleavy & C. Noronha (Eds.), *Comparative CSR and Sustainability: New Accounting for Social Consequences*. Routledge.

Douthwaite, R. (1999). *The Ecology of Money*. Green Books.

Douthwaite, R. (1999). *The Growth Illusion: How Economic Growth Has Enriched the Few, Impoverished the Many and Endangered the Planet*. Green Books.

Dowling, J., & Pfeffer, J. (1975). Organizational legitimacy: Social values and organizational behavior. *Pacific Sociological Review*, *18*(1), 122–136. doi:10.2307/1388226

Drewniak, M., Dalaklis, D., Kitada, M., Ölçer, A., & Ballini, F. (2018). Geopolitics of Arctic Shipping: The state of icebreakers and future needs. *Polar Geography*, *41*(2), 107–125. doi:10.1080/1088937X.2018.1455756

Dries, L., Heijman, W., Jongeneel, R., Purnhagen, K. P., & Wesseler, J. (2019). EU Bioeconomy Economics and Policies: Volume I. In Palgrave advances in bioeconomy: economics and policies. Springer International Publishing. doi:10.1007/978-3-030-28634-7 du Gay, P. (2000) In praise of bureaucracy: Weber, organization, ethics. London: Sage. doi:10.4135/9781446217580

du Gay, P. (1994). Colossal immodesties and hopeful monsters. *Organization, 1*(1), 125–148. doi:10.1177/135050849400100114

Ducros, A., & Steele, P. (2022). *Biodiversity Credits to Finance Nature and People: Emerging Lessons.* IIED.

Dudhpachare, A. K., Kuthe, T. V., Lalke, C. V., Wyawahare, N. P., Agrawal, R., & Daigavane, P. (2022). Process of RO's waste water reuse & water management in society by using IOT automation. *2022 International Conference on Applied Artificial Intelligence and Computing (ICAAIC)*, (pp. 1546–1550). IEEE. 10.1109/ICAAIC53929.2022.9792805

Dunning, T. (2012). *Natural experiments in the social sciences.* Cambridge University Press. doi:10.1017/CBO9781139084444

Duran Valverde, F. (2008). *Diagnóstico del sistema de seguridad social del Ecuador. Oficina Subregional para los Países, Departamento de Seguridad Social.* Lima: Organización Internacional del Trabajo. https://bit.ly/3V9f3Iy

Durán Valverde, F. (2013). *Innovaciones en la extensión de la cobertura del seguro social a los trabajadores independientes: Departamento de Protección Social de la OIT.* Ginebra: Organización Internacional del Trabajo- OIT. https://bit.ly/3CGHjLn

Durrenberger, P. (2007) The Anthropology of Organized Labor in the United States. *Ann Review of Anthropology, 36.* Annual Reviews. https://www.annualreviews.org/doi/pdf/10.1146/annurev.anthro.36.081406.094433

Dwivedi, D. M. C. (2022). A Study on Advanced Innovation Managing Risk and Reward. *Cosmos Journal of Engineering & Technology, 12*(1), 45–47.

Dwivedi, M. C. (2022). Assessing The Socioeconomic Implications of Climate Change. *Cosmos An International Journal Art & Higher Education, 11*(2), 20–23.

Dwiyaniti, M., & Kusumaningtyas, A. B. Silawardono, Kusumastuti, S. L., Saefullah, Y., Tohazen, Tireno, F. Q., & Ramadhan, D. (2022). Energy saving of IoT-based light intensity on smart streetlight. *2022 5th International Conference of Computer and Informatics Engineering (IC2IE)*, (pp. 6–9). IEEE.

Ebissa, G., Fetene, A., & Desta, H. (2023). Comparative analysis of managing plantation forests: The case of keeping plantation forests for carbon credit and industrial profits in Oromia Region, Ethiopia. *Heliyon, 9*(4), e15151. doi:10.1016/j.heliyon.2023.e15151 PMID:37095960

Eckerberg, K., & Joas, M. (2004). Multi-level Environmental Governance: A Concept under Stress? *Local Environment, 9*(5), 405–412. doi:10.1080/1354983042000255315

Economic Commission for Latin America and the Caribbean (ECLAC). (2009). *Envejecimiento, derechos humanos y políticas públicas* (S. Huenchuan, Ed.). Naciones Unidas. https://bit.ly/3VmxYzM

Economic Commission for Latin America and the Caribbean (ECLAC). (2020). El desafío social en tiempos del COVID-19. *Informe Especial COVID-19, Nº3.* Santiago: Naciones Unidas. https://bit.ly/3V8QCuJ

Economic Commission for Latin America and the Caribbean (ECLAC). (2022). *Panorama Social de América Latina, 2021.* Santiago: Naciones Unidas. https://repositorio.cepal.org/bitstream/handle/11362/47718/1/S2100655_es.pdf

Economist. (2019). Who owns the Northwest Passage? *The Economist.* https://www.economist.com/

Ecuadorian Institute of Social Security. (2019). *Valuación Actuarial del Seguro de Invalidez, Vejez y Muerte del Seguro General Obligatorio.* EISS. https://www.iess.gob.ec/documents/10162/14444609/IESS_IVM_estudio_actuarial_011.pdf

Ecuadorian Institute of Social Security. (2022). *Estadisticas laborales de Mayo 2022.* EISS. https://bit.ly/3Ci4euY

Edson, M. C., Henning, P. B., & Sankaran, S. (2017). *A Guide to System Research* (2nd ed.). Springer. doi:10.1007/978-981-10-0263-2

EEA. (n.d.). *Designing safe and sustainable products requires a new approach for chemicals*. European Environment Agency. https://www.eea.europa.eu/publications/designing-safe-and-sustainable-products-1

Efthymiou, L., Mavragani, A., & Tsagarakis, K. P. (2016). Quantifying the effect of macroeconomic and social factors on illegal e-waste trade. *International Journal of Environmental Research and Public Health*, *13*(8), 789. . doi:10.3390/ijerph13080789 PMID:27527200

Egerton, F. N. (2001). A history of the ecological sciences: Early Greek origins. *Bulletin of the Ecological Society of America*, *82*(1), 93–97.

Ehrenfeld, J. R. (2000). Industrial ecology: Paradigm shift or normal science? *The American Behavioral Scientist*, *44*(2), 229–244. doi:10.1177/0002764200044002006

EIA. (2012). *Arctic oil and natural gas resources*. U.S. Energy Information Administration. https://www.eia.gov/todayinenergy/detail.php?id=4650

Eiadat, Y., Kelly, A., Roche, F., & Eyadat, H. (2008). Green and competitive? An empirical test of the mediating role of environmental innovation strategy. *Journal of World Business*, *43*(2), 131–145. doi:. jwb.2007.11.012 doi:10.1016/j

El Hajj, M. G. (2020). *Fighting Money Laundering in Foreign Exchange Firms in Lebanon*. [MBA Thesis, Middle East University].

El-Bacha, F. (2020a). Coronavirus: L'erreur libanaise. *Libna News*. https://libnanews.com/coronavirus-lerreur-libanaise/

El-Chichakli, B., von Braun, J., Lang, C., Barben, D., & Philp, J. (2016). Policy: Five cornerstones of a global bioeconomy. *Nature*, *535*(7611), 221–223. https://doi.org/. doi:10.1038/535221a PMID:27411618

Elkington, J. (1994). Towards the sustainable corporation: Win-win-win business strategies for sustainable development. *Cal. Management Review, 36* (3), 90e100.

Elola, A., Valdaliso, J. M., López, S. M., & Aranguren, M. J. (2012). Cluster life cycles, path dependency and regional economic development: Insights from a meta-study on Basque clusters. *European Planning Studies*, *20*(2), 257–279. doi:10.1080/09654313.2012.650902

Epstein, M. J., & Young, S. D. (1999). Greening' with EVA. *Management Accounting*, *80*(1), 45–49.

Esbjorn-Hargens, S. (2016). Developing a Complex Integral Realism for Global Response: Three Metaframeworks for Knowledge Integration and Coordinated Action. In R. Bhaskar, S. E. -Hargens, N. Hedlund, & M. Hartwig (Eds.), Metatheory for the Twenty-First Century: Critical Realism and Integral theory dialogue (pp. 99-139). London: Routledge.

Escalona-Orcao, A., Barrado-Timón, D. A., Escolano-Utrilla, S., Sánchez-Valverde, B., Navarro-Pérez, M., Pinillos-García, M., & Sáez-Pérez, L. A. (2020). Cultural and creative ecosystems in medium-sized cities: Evolution in times of economic crisis and pandemic. *Sustainability (Basel)*, *13*(1), 49. Advance online publication. doi:10.3390u13010049

Espinosa, S. (2021). The Evolution of the System of Foreign Investment Protection in Ecuador. In *Handbook of International Investment Law and Policy* (pp. 2645–2676). Springer. doi:10.1007/978-981-13-3615-7_85

Etzion, D. (2007). Research on organizations and the natural environment, 1992-present: A review. *Journal of Management*, *33*(4), 637–664. doi:10.1177/0149206307302553

European Commission. (2017). *Report on the blue growth strategy towards more sustainable growth and jobs in the blue economy*. European Commission.

European Commission. (2018). *A sustainable Bioeconomy for Europe: strengthening the connection between economy, society and the environment*. EC. https://ec.europa.eu/research/bioeconomy/pdf/ec_bioeconomy_strategy_2018.pdf.

European Commission. (2019). *The European Green Deal*, Brussels, Belgium.https://ec.europa.eu/info/sites/info/files/european-greendeal-communication_en.pdf.

European Commission. (2023, June 23). Research and Innovation. Europea. https://research-and-innovation.ec.europa.eu/research-area/environment/bioeconomy_en

European Commission. (n.d.). *Causes of climate change*. EC. https://climate.ec.europa.eu/climate-change/causes-climate-change_en

Eurostat. (2023). [Data set]. https://ec.europa.eu/eurostat/data/database

Evangelista, P. (2014). Environmental sustainability practices in the transport and logistics service industry: An exploratory case study investigation. *Research in Transportation Business & Management*, *12*, 63–72. doi:10.1016/j.rtbm.2014.10.002

Evans, S. (2023). An integrated circular economy model for transformation towards sustainability. *Journal of Cleaner Production*, *388*, 135950. doi:10.1016/j.jclepro.2023.135950

Everett, J. L. (1985). *Culture and ecology in the analysis of organizations*. [PhD Thesis, Institute: University of Colorado at Boulder].

Exner-Pirot, H. (2022). Opinion: Does the Arctic Council make sense without Russia? *National Post*. https://nationalpost.com/opinion/opinion-does-the-arctic-council-make-sensewithout-russia

Eysenbach, G., & Köhler, C. (2002). How do consumers search for and appraise health information on the world wide web? Qualitative study using focus groups, usability tests, and in-depth interviews. *BMJ (Clinical Research Ed.)*, *324*(7337), 573–577. . doi:10.1136/bmj.324.7337.573 PMID:11884321

Fader, M., Cranmer, C., Lawford, R., & Engel-Cox, J. (2018). Toward an understanding of synergies and trade-offs between water, energy, and food SDG targets. *Frontiers in Environmental Science*, *6*, 112. doi:10.3389/fenvs.2018.00112

Falcone, P. M. (2019). Tourism-based circular economy in Salento (South Italy): A SWOT-ANP analysis. *Social Sciences (Basel, Switzerland)*, *8*(7), 216. doi:10.3390ocsci8070216

Falk, J. H., & Dierking, L. D. (2016). *The museum experience*. Whalesback books.

Falkenmark, M., Finlayson, M., Gordon, L., Bennett, E., Chiuta, T. M., Coates, D., Ghosh, N., Gopalakrishnan, M., de Groot, R. S., Jacks, G., Kendy, E., Oyebande, L., Moore, M., Peterson, G. D., Portuguez, J. M., Seesink, K., Thame, R., & Wasson, R. (2007). Agriculture, water, and ecosystems: avoiding the costs of going too far. In *Water for food. Water for life. A comprehensive assessment of water management in Agriculture* (pp. 233–277). Earthscan.

FAO. (2015). *Climate change and food security: risks and responses*. FAO. https://www.fao.org/3/i5188e/I5188E.pdf

FAO. (2019). *Indicators to Monitor and Evaluate the Sustainability of Bioeconomy*. FAO. https://www.fao.org/3/ca6048en/ca6048en.pdf

FAO. (n.d.). *Introduction*. FAO. https://www.fao.org/3/v4845e/V4845E02.htm

Fárez, M. (2020). *Jubilación en Ecuador: informalidad, desigualdad e insostenibilidad*. https://bit.ly/3fLfXe1

Farhoud, N. (2021). Estimating the impact of financial accounting on risk management. [MBA Thesis, Lebanese International University].

Farr, D. (2008). *Sustainable urbanism—urban design with nature*. John Wiley & Sons.

Fatima, S. (2022). Impact of Covid-19 Pandemic on The Microfinance Sector in India. *Globus An International Journal of Management & IT, 14*(1), 25-27.

Fava, F., Gardossi, L., Brigidi, P., Morone, P., Carosi, D. A., & Lenzi, A. (2021). The bioeconomy in Italy and the new national strategy for a more competitive and sustainable country. *New Biotechnology*, *61*, 124–136. doi:10.1016/j.nbt.2020.11.009 PMID:33220517

Feldman, M. S. (2016). Past, Present, and future. In J. Howard-Grenville, C., Rerup, A., Langley, & H., Tsoukas (Eds.). Organizational Routines: How They are Created, Maintained, and Changed (pp. 23-46) Oxford: Oxford University Press

Fenequito, A. C., Robles, C. F., & Santos, A. F. & Valencia, K. C. (2023). Effects of Caregiver Burden to The Psychosocial Adjustment and Quality of Life of Caregivers of Older People with Debilitating Diseases. Globus An International Journal of Medical Science. *Engineering and Technology*, *12*(1), 41–53.

Fernandes, A. A. R., & Solimun. (2017). The mediating effect of strategic orientation and innovations on the effect of environmental uncertainties on the performance of business in the Indonesian aviation industry. *International Journal of Law and Management*, *59*(6), 1269–1278. doi:10.1108/IJLMA-10-2016-0087

Ferreira, C. (2017). The contested instruments of a new governance regime: Accounting for nature and building markets for biodiversity offsets. *Accounting, Auditing & Accountability Journal*, *30*(7), 1568–1591. doi:10.1108/AAAJ-12-2015-2336

Ferreira, V., Pié, L., Mainar-Causapé, A., & Terceño, A. (2023). The bioeconomy in Spain as a new economic paradigm: The role of key sectors with different approaches. *Environment, Development and Sustainability*, 1–25. doi:10.100710668-022-02830-5 PMID:36624734

Fialkiewicz, W., Burszta-Adamiak, E., Kolonko-Wiercik, A., Manzardo, A., Loss, A., Mikovits, C., & Scipioni, A. (2018). Simplified direct water footprint model to support urban water management. *Water (Basel)*, *10*(5), 630. doi:10.3390/w10050630

Figueiredo, B., & Scaraboto, D. (2016). The systemic creation of value through circulation in collaborative consumer networks. *The Journal of Consumer Research*, *43*(4), 509–533. doi:10.1093/jcr/ucw038

Fihlo, W., Brito, P., de Lima, I., & Pocivalisteanu, D. (2018). Towards a Sustainable Bioeconomy: Principles, Challenges and Perspectives. In *World sustainability series*. Springer International Publishing. doi:10.1007/978-3-319-73028-8

Filatov, V. V., Klimenko, V. A., Karelina, E. A., Ptitsyn, D. A., & Erpulev, V. A. (2022). Theoretical aspects of the creation of artificial intelligence systems in transport. *2022 Systems of Signals Generating and Processing in the Field of on Board Communications*, (pp. 1–5). IEEE.

Finucane, M. L. (2009). Why science alone won't solve the climate crisis: Managing climate risks in the Pacific. *Asia Pacific Issues*, *89*, 1–8.

Fiske, S. J., Crate, S. A., Crumley, C., Galvin, K., Lazrus, H., Lucero, L., Oliver-Smith, A., Orlove, B., Strauss, S., & Wilk, R. R. (2014). *Changing the atmosphere. Anthropology and climate change. Final report of the AAA global climate change task force*. American Anthropological Association.

Fitzpatrick, M. (2019). Pension reform and return-to-work policies. *Journal of Pension Economics and Finance*, *18*(4), 500–514. doi:10.1017/S1474747218000367

Fleddermann, C.-B. (2003). *Engineering Ethics* (2nd ed.). Pearson.

Folke, C., Hahn, T., Olsson, P., and Norberg, J. (2005). Adaptive Governance of Social–ecological Systems. *Annual Review of Environment and Resources*.

Folke, C., Carpenter, S., Elmqvist, T., Gunderson, L., Holling, C. S., & Walker, B. (2002). Resilience and Sustainable Development: Building Adaptive Capacity in a World of Transformations. *Ambio*, *31*(5), 437–440. doi:10.1579/0044-7447-31.5.437 PMID:12374053

Fong, C. M., & Chang, N. J. (2012). The impact of green learning orientation on proactive environmental innovation capability and firm performance. *African Journal of Business Management*, *6*(32), 727–735.

Food and Agriculture Organization. (2017). *Food and agriculture organization blue growth initiative: Partnering with countries to achieve the sustainable development goals*. United Nations.

Forman, J., & Sabin, M. (2015). Tontine Pensions. *University of Pennsylvania Law Review*, *163*(3), 755–831. doi:10.2139srn.2393152

Forttes, P. (2020). Envejecimiento y Atención a la Dependencia en Ecuador. https://bit.ly/3CjrV6k

Freeman, R. E. (1984). *Strategic Management: A Stakeholder Approach*. Pitman.

Frey, D. F. (2017). Economic growth, full employment and decent work: The means and ends in SDG 8. *International Journal of Human Rights*, *21*(8), 1164–1184. doi:10.1080/13642987.2017.1348709

Fu, Z., Cheng, J., Yang, M., & Batista, J. (2018). Prediction of industrial wastewater quality parameters based on wavelet de-noised ANFIS model. *2018 IEEE 8th Annual Computing and Communication Workshop and Conference (CCWC)*, (pp. 301–306). IEEE.

Fujii, H. (2016). Decomposition analysis of green chemical technology inventions from 1971 to 2010 in Japan. *Journal of Cleaner Production*, *112*, 4835–4843. doi:10.1016/j.jclepro.2015.07.123

Fujii, H., & Managi, S. (2016). Research and development strategy for environmental technology in Japan: A comparative study of the private and public sectors. *Technological Forecasting and Social Change*, *112*, 293–302. doi:10.1016/j.techfore.2016.02.012

Fullmer, R., & Sabin, M. (2019). Individual Tontine Accounts. *Journal of Accounting and Finance*, *8*(19), 2–46. doi:10.2139srn.3217551

Fung, K. L. (2018). *Expanding the green network on rooftops: A study of integrating green roofs as a part of urban green innovation space planning* (Order No. 10932095). ProQuest One Academic. (2124999351). http://wdg.biblio.udg.mx:2048/login?url=https://www.proquest.com/dissertations-theses/expanding-green-network-on-rooftops-study/docview/2124999351/se-2?accountid=28915

Fuso Nerini, F., Tomei, J., To, L. S., Bisaga, I., Parikh, P., Black, M., Borrion, A., Spataru, C., Castán Broto, V., Anandarajah, G., Milligan, B., & Mulugetta, Y. (2017). Mapping synergies and trade-offs between energy and the Sustainable Development Goals. *Nature Energy*, *3*(1), 10–15. doi:10.103841560-017-0036-5

Gabriel, N. (2014). Urban political ecology: Environmental imaginary, governance, and the non-human. *Geography Compass*, *8*(1), 38–48. doi:10.1111/gec3.12110

Gachet, I., Grijalva, D., Ponce, P., & Rodríguez, D. (2019). Vertical and Horizontal Inequality in Ecuador: The Lack of Sustainability. *Social Indicators Research*, *145*, 861–900. doi:10.100711205-017-1810-2

Gadenne, D. L., Kennedy, J., & McKeiver, C. (2009). An empirical study of environmental awareness and practices in SMEs. *Journal of Business Ethics*, *84*(1), 45–63. doi:10.100710551-008-9672-9

Gaikwad, A., Bhuta, N., Jadhav, T., Jangale, P., & Shinde, S. (2022). A review on forest fire prediction techniques. *2022 6th International Conference On Computing, Communication, Control And Automation (ICCUBEA)*, (pp. 1–5). IEEE.

Galdeano-Gómez, E., Céspedes-Lorente, J., & Martínez-del-Río, J. (2008). Environmental performance and spillover effects on productivity: Evidence from horticultural firms. *Journal of Environmental Management*, *88*(4), 1552–1561. doi:10.1016/j.jenvman.2007.07.028 PMID:17825476

Galvez, R. S. (2022). Learners' Psychological Well-Being and Academic Performance in The Midst of Health Crisis: A Correlational Study. *Cosmos An International Journal Art & Higher Education*, *11*(1), 45–52.

Gao, X., Lyu, Y., Shi, F., Zeng, J., & Liu, C. (2019). The impact of financial fac-tor market distortion on green innovation efficiency of high-tech industry. *Ekoloji*, *28*(107), 3449–3461.

Gao, Y., Sun, J., Li, F., He, S., Sandven, S., Yan, Q., Zhang, Z., Lohmann, K., Keenlyside, N., Furevik, T., & Suo, L. (2015). Arctic sea ice and Eurasian climate: A review. *Advances in Atmospheric Sciences*, *32*(1), 92–114. doi:10.100700376-014-0009-6

Garcia-Bernabeu, A., Hilario-Caballero, A., Pla-Santamaria, D., & Salas-Molina, F. (2020). A process oriented MCDM approach to construct a circular economy composite index. *Sustainability (Basel)*, *12*(2), 618. doi:10.3390u12020618

Gardossi, L., Philp, J., Brigidi, P., Morone, P., Carosi, D. A. R., & Lenzi, A. (2021). Is tourism really affected by logistical operations and environmental degradation? An empirical study from the perspective of Thailan. *New Biotechnology*, *61*, 124–136.

Gardossi, L., Philp, J., Fava, F., Winickoff, D., D'Aprile, L., Dell'Anno, B., Marvik, O. J., & Lenzi, A. (2023). Bioeconomy national strategies in the G20 and OECD countries: Sharing experiences and comparing existing policies. *EFB Bioeconomy Journal*, *3*, 100053. doi:10.1016/j.bioeco.2023.100053

Gasper, D., Shah, A., & Tankha, S. (2019). The framing of sustainable consumption and production in SDG 12. *Global Policy*, *10*(S1), 83–95. doi:10.1111/1758-5899.12592

Geddes, P. (1915). *Cities in Evolution*. Williams.

Geissdoerfer, M., Savaget, P., Bocken, N. M. P., & Hultink, E. J. (2017). The circular economy - a new sustainability paradigm? *Journal of Cleaner Production*, *143*, 757–768. doi:10.1016/j.jclepro.2016.12.048

Geng, Y., Sarkis, J., Ulgiati, S., & Zhang, P. (2013). Measuring China's circular economy. *Science*, *339*(6127), 1526–1527. doi:10.1126cience.1227059 PMID:23539584

Geraie, M. S., & Rad, F. M. (2015). Mediator role of the organizational identity green in relationship between total quality management and perceived innovation with sustainable competitive advantage. *International Journal of Biology, Pharmacy and Allied Sciences*, *4*, 266–276.

Gerges, F.-A. (2020). Will Beirut's blast be a catalyst for change? *Dev Discourse*. https://www.devdiscourse.com/article/politics/1159968-will-beiruts-blast-be-a-catalyst-for-change

Gesell, S. (1919). *The Monetary and Economic Problem*. Peter Owen Publishers.

Gesell, S. (1932). *Free Money*. Ernst Wasmuth Verlag.

Gesell, S. (1958). *The Natural Economic Order*. Peter Owen Publishers.

Ghatee, M. A., Taylor, W. R., & Karamian, M. (2020). The Geographical Distribution of Cutaneous Leishmaniasis Causative Agents in Iran and Its Neighboring Countries, A Review. *Frontiers in Public Health*, *8*, 111–113. doi:10.3389/fpubh.2020.00011 PMID:32391301

Ghosh, M., Sushil, R., & Ghosh, K. (2019). Detecting and reporting forest fire through deployment of three dimensional multi-sink wireless sensor network. *2019 4th International Conference on Internet of Things: Smart Innovation and Usages (IoT-SIU)*, (pp. 1–5). IEEE.

Ghosh, N., Kumar, B., Biswas, T., Patnaik, S., & Paul, R. (2021). IoT fog based framework to predict forest fire. *2021 Smart City Challenges & Outcomes for Urban Transformation (SCOUT)*, (pp. 256–259). IEEE.

Giblett, R. (2012). Nature is ordinary too: Raymond Williams as the founder of ecocultural studies. *Cultural Studies*, *26*(6), 922–933. doi:10.1080/09502386.2012.707221

Gibson-Graham, J. K. (1996). *The End of Capitalism as We Knew It: A Feminist Critique of Political Economy*. Blackwell Publishing.

Giesler, M. (2008). Conflict and compromise: Drama in marketplace evolution. *Journal of Consumer Research*, 34(April), Giesler, M. (2006). Consumer gift systems. *The Journal of Consumer Research*, *33*(September), 283–290. doi:10.1086/506309

Giguashvili, G., Makasarashvili, T., & Khorguashvili, T. (2021). The Main Challenges of Economic and Information Security of Business Companies in Georgia in The Conditions of COVID-19. *Globus An International Journal of Management & IT*, *13*(1), 19–23.

Gill, F. L., Viswanathan, K. K., & Karim, M. Z. A. (2018). The critical review of the pollution haven hypothesis. *International Journal of Energy Economics and Policy*, *8*(1), 167–174. http://hdl.handle.net/11159/1929

Giner, C. (2004). *J. Antropología urbana*. Barcelona. *Ariel.*

Gladkiy, Y. (2020). "Polar Silk road": project implementation and geoeconomic interests of Russia and China. *IOP Conf. Ser.: Earth Environ. Sci.* IOP Science. (Accessed 17.06.2023) doi:10.1088/1755-1315/434/1/012009

Glavič, P. (2021). Evolution and current challenges of sustainable consumption and production. *Sustainability (Basel)*, *13*(16), 9379. doi:10.3390u13169379

Glick, T. F., & Kirchner, H. (2000). *Hydraulic systems and technologies of Islamic Spain: History and archaeology. Working with water in Medieval Europe*. Brill.

Goda, Z., Foda, M. & Elsayyad, N. (2023). Using Green Roofs for Social Housing to Improve Energy Consumption in New Cities. (An Applied Study of Social Housing in Egypt's New Cairo City). *Future Cities and Environment*, *9*(1), 2, 1–11. doi:10.5334/fce.165

Gohari, A., Gohari, A., & Bin, A. (2023). Importance of green roof criteria for residential and governmental buildings: A multi-criteria decision analysis. *Environmental Science and Pollution Research International*, *30*(2), 3707–3725. doi:10.100711356-022-22472-4 PMID:35953748

Gökmenoğlu, K., & Taspinar, N. (2016). The relationship between CO2 emissions, energy consumption, economic growth and FDI: The case of Turkey. *The Journal of International Trade & Economic Development*, *25*(5), 706–723. doi:10.1080/09638199.2015.1119876

Golden, S. D., & Earp, J. A. L. (2012). Social Ecological Approaches to Individuals and Their Contexts: Twenty Years of Health Education & Behavior Health Promotion Interventions. *Health Education & Behavior*, *39*(3), 364–372. doi:10.1177/1090198111418634 PMID:22267868

Goldfarb, A., Greenstein, S., & Tucker, C. (2015). *Economic Analysis of the Digital Economy*. University of Chicago Press. doi:10.7208/chicago/9780226206981.001.0001

Goldfarb, K. E. (2016). 'Coming to Look Alike': Materializing Affinity in Japanese Foster and Adoptive Care. *Social Analysis*, *60*(2), 47–64. doi:10.3167a.2016.600204

Gómez Hinojosa, J. F. (1990). De la ecología a la ecofilía: Apuntes para una ecología liberadora'. *Pasos (El Sauzal)*, *30*(July–August).

Gong, Y., Liu, F., Liu, H., Zhao, H., Zhu, Y., Han, Q., Zhou, J., Jiang, T., & Li, S. (2022). Environment–Energy–Economy Analysis and Related Technical Transition Strategies for Spent Activated Carbon Regeneration in China. *ACS Sustainable Chemistry & Engineering*, *10*(35), 11512–11523. doi:10.1021/acssuschemeng.2c02855

Gong, Y., Li, X., & Li, Z. (2020). How to design green roofs to enhance urban green innovation space? Evidence from Beijing. *Habitat International*, *107*, 102–113. doi:10.1016/j.habitatint.2020.102113

González Álvarez, M. D. (1997). Processos de Planejamento nos Pólos Tecnológicos: um enfoque adaptativo. [Tese de Doutorado, Departamento de Engenharia Industrial, PUC-Rio, Rio de Janeiro].

Gonzalez-Vidal, A., Cuenca-Jara, J., & Skarmeta, A. F. (2019). IoT for water management: Towards intelligent anomaly detection. *2019 IEEE 5th World Forum on Internet of Things (WF-IoT)*, (pp. 858–863). IEEE.

Goosen, L. (2015). Educational Technologies for an ICT4D MOOC in the 21st Century. In D. Nwaozuzu, & S. Mnisi (Ed.), *Proceedings of the South Africa International Conference on Educational Technologies* (pp. 37 - 48). Pretoria: African Academic Research Forum.

Goosen, L. (2016a, February 18). *"We don't need no education"? Yes, they DO want e-learning in Basic and Higher Education!* UNISA. https://uir.unisa.ac.za/handle/10500/20999

Goosen, L. (2018a). Sustainable and Inclusive Quality Education Through Research Informed Practice on Information and Communication Technologies in Education. In L. Webb (Ed.), *Proceedings of the 26th Conference of the Southern African Association for Research in Mathematics, Science and Technology Education (SAARMSTE)* (pp. 215 - 228). Gabarone: University of Botswana.

Goosen, L., & Mukasa-Lwanga, T. (2017). Educational Technologies in Distance Education: Beyond the Horizon with Qualitative Perspectives. In U. I. Ogbonnaya, & S. Simelane-Mnisi (Ed.), *Proceedings of the South Africa International Conference on Educational Technologies* (pp. 41 - 54). Pretoria: African Academic Research Forum.

Goosen, L. (2004). *Criteria and Guidelines for the Selection and Implementation of a First Programming Language in High Schools*. Potchefstroom Campus: North West University. http://hdl.handle.net/10394/226

Goosen, L. (2008). A Brief History of Choosing First Programming Languages. In J. Impagliazzo (Ed.), *History of Computing and Education 3* (Vol. 269, pp. 167–170). Springer. doi:10.1007/978-0-387-09657-5_11

Goosen, L. (2018b). Trans-Disciplinary Approaches to Action Research for e-Schools, Community Engagement, and ICT4D. In T. A. Mapotse (Ed.), *Cross-Disciplinary Approaches to Action Research and Action Learning* (pp. 97–110). IGI Global. doi:10.4018/978-1-5225-2642-1.ch006

Goosen, L. (2018c). Students' Access to an ICT4D MOOC. In S. Kabanda, H. Suleman, & S. Jamieson (Ed.), *Proceedings of the 47th Annual Conference of the Southern African Computer Lectures' Association (SACLA 2018)* (pp. 183 - 201). Cape Town: University of Cape Town.

Government of Netherlands. (2016). *A Circular Economy in the Netherlands by 2050*. Government of Netherlands. https://circulareconomy.europa.eu/platform/sites/default/files/17037circulaireeconomie_en.pdf

Gragson, T. L., & Blount Ben, G. (Eds.). (1999). *Introduction Ethnoecology: Knowledge, Resources, and Rights*. University of Georgia.

Graham, B., Reilly, W., Beinecke, F., Boesch, D., Garcia, T., Murray, C., & Ulmer, F. (2011). *Deep water: The Gulf oil disaster and the future of offshore drilling. Report to the President*. National Commission on the BP Deepwater Horizon Oil Spill and Offshore Drilling.

Granda, M., & Jimenez, W. (2019). The evolution of socioeconomic health inequalities in Ecuador during a public health system reform (2006–2014). *International Journal for Equity in Health, 18*(31), 1–12. doi:10.118612939-018-0905-y PMID:30736808

Gravano, A. (2007b). Claves para la facilitación organizacional del proceso participativo en la planificación urbano-ambiental metropolitana. *Revista Regional de Trabajo Social, 21*(40), 9-22.

Gravano, A. (1992). Antropología Práctica, muestra y posibilidades de la Antropología Organizacional. *Publicar en Antropología y Ciencias Sociales, Buenos Aires*, (1), 95–126.

Gravano, A. (2005). *El barrio en la teoría social*. Espacio Editorial.

Gravano, A. (2006). Imaginarios regionales y circularidad en la planificación: El caso del TOAR. *Intersecciones*, (7), 305–323.

Greco, T. H. (2009). *The End of Money and the Future of Civilization*. Chelsea Green Publishing.

Greene, J., Caracelli, V., & Graham, W. (1989). Toward a conceptual framework for mixed-method evaluation designs. *Educational Evaluation and Policy Analysis, 11*(3), 255–274. doi:10.3102/01623737011003255

Greenhouse Gases . (2023). US EPA. https://www.epa.gov/report-environment/greenhouse-gases

Greer, J. M. (2012). *The Wealth of Nature: Economics as if Survival Mattered. Gabriola Island*. New Society Publishers.

Grether, J. M., Mathys, N. A., & De Melo, J. (2012). Unravelling the worldwide pollution haven effect. *The Journal of International Trade & Economic Development, 21*(1), 131–162. doi:10.1080/09638190903552040

Grimmer, J., Roberts, M., & Stewart, B. (2021). Machine learning for social science: An agnostic approach. *Annual Review of Political Science, 24*(1), 395–419. doi:10.1146/annurev-polisci-053119-015921

Grossman, G. M., & Krueger, A. B. (1991). *Environmental impacts of a North American free trade agreement (No. w3914)*. National Bureau of Economic Research., doi:10.3386/w3914

Grossman, G. M., & Krueger, A. B. (1995). Economic growth and the environment. *The Quarterly Journal of Economics, 110*(2), 353–377. doi:10.2307/2118443

Guedes, A. (2015). Guedes 2015, "The Arctic Council". *Janus*, 160–161. https://repositorio.ual.pt/bitstream/11144/2982/1/3.30_ArmandoMGuedes_Cons elhoArtico.pdf

Guiamalon, T. S. (2022). Social and Economic Development: State Universities and Colleges' (SUC's) Contribution Creativity Skills of The Students in Recycling. *Globus Journal of Progressive Education, 13*(1), 107–111.

Guiamalon, T. S. (2022). Stakeholders' Awareness and Acceptance of The Institution's Vision, Mission and Goals and Information Technology Program Objectives in A State University in The Philippines. *Globus Journal of Progressive Education, 12*(1), 51–60.

Guinnane, T., Jopp, T., & Streb, J. (2021). Bismarck's social security and its influence on Germany's demographic change. *Journal for Economic Policy, 10*(1), 262–265. doi:10.100710273-021-2893-6

Guiomar, N., Godinho, S., Pinto-Correia, T., Almeida, M., Bartolini, F., Bezak, P., Biró, M., Bjørkhaug, H., Bojnec, Š., Brunori, G., Corazzin, M., Czekaj, M., Davidova, S., Kania, J., Kristensen, S., Marraccini, E., Molnár, Z., Niedermayr, J., O'Rourke, E., & Wästfelt, A. (2018). Typology and distribution of small farms in Europe: Towards a better picture. *Land Use Policy*, *75*, 784–798. doi:10.1016/j.landusepol.2018.04.012

Gunderson, L. H., & Holling, C. S. (Eds.). (2002). Panarchy; Understanding Transformations in Human and Natural Systems, Island, Washington, DC.

Gunderson, L. (1999). Resilience, Flexibility and Adaptive Management: Antidotes for Spurious Certitude? *Conservation Ecology*, *3*(1), 7. http://www.consecol.org/vol3/iss1/art7. doi:10.5751/ES-00089-030107

Gunn, A., & Mintrom, M. (2016). Higher Education Policy Change in Europe: Academic Research Funding and the Impact Agenda. *European Education*, *48*(4), 241–257. doi:10.1080/10564934.2016.1237703

Gunn, P., Burak, H., Van Geel, P., & Baldwin, C. (2022). Development of resistor-capacitor and finite difference models to evaluate green roof thermal performance. *Building and Environment*, *226*, 1–15. doi:10.1016/j.buildenv.2022.109700

Gupeng, Z., & Xiangdong, C. (2012). The value of invention patents in China: Country origin and technology field differences. *China Economic Review*, *23*(2), 357–370. doi:10.1016/j.chieco.2012.02.002

Gupta, J., & Vegelin, C. (2016). Sustainable development goals and inclusive development. *International Environmental Agreement: Politics, Law and Economics*, *16*(3), 433–448. doi:10.100710784-016-9323-z

Gupta, T., & Khan, S. (2023). Education and International Collaboration in The Digital Age. *Globus Journal of Progressive Education*, *13*(1), 79–87.

Guria, P., & Bhattacharyya, A. (2021). Lightweight Cryptography in Cloud-Based IoT: An Analytical Approach. In P. Tomar (Ed.), Integration and Implementation of the Internet of Things Through Cloud Computing (pp. 190-216). IGI Global. doi:10.4018/978-1-7998-6981-8.ch010

Hadavi, S. (2017). Direct and indirect effects of the physical aspects of the environment on mental well-being. *Environment and Behavior*, *49*(10), 1071–1104. doi:10.1177/0013916516679876

Haines, S. (2019b). Managing expectations: Articulating expertise in climate services for agriculture in Belize. *Climatic Change*, *157*(1), 43–59. doi:10.100710584-018-2357-1

Halbritter, G., & Dorfleitner, G. (2015). The wages of social responsibility—Where are they? A critical review of ESG investing. *Review of Financial Economics*, *26*(1), 25–35. doi:10.1016/j.rfe.2015.03.004

Hamid, M. A., Isa, S. M., & Kiumarsi, S. (2020). Sustainable tourism practices and business performance from the tour operators' perspectives. *Anatolia*, *32*(1), 23–32. doi:10.1080/13032917.2020.1830135

Hamilton, J. T., & Kip Viscusi, W. (1999). Regulation of economic activity. In *Calculating risks? The spatial and political dimensions of hazardous waste policy* (Vol. 21). MIT Press.

Hamzaçebi, C., & Pekkaya, M. (2011). Determining of stock investments with grey relational analysis. *Expert Systems with Applications*, *38*(8), 9186–9195. doi:10.1016/j.eswa.2011.01.070

Hanks, A., & Dwoskin, L. (2019). The economic impact of social enterprises in the United States. *Journal of Entrepreneurship and Public Policy*, *8*(1), 1–17.

Hannan, M. (2005). Ecologies of Organizations: Diversity and Identity. *The Journal of Economic Perspectives*, *19*(1), 51–70. doi:10.1257/0895330053147985

Hannan, M. T., & Freeman, J. (1989). *Organizational Ecology*. Harvard University Press. doi:10.4159/9780674038288

Hannan, M. T., & Freeman, J. H. (1977). The Population Ecology of Organizations. *American Journal of Sociology*, *82*(5), 929–963. doi:10.1086/226424

Hannerz, U. (1986). *Exploración de la ciudad*. Fondo de Cultura Económica.

Hannerz, U. (2004). *Foreign New: Exploring the World of Foreign Correspondents*. University of Chicago Press.

Hartmut Resonanz, R. (2016). *Eine Soziologie der Weltbeziehung*. Suhrkamp.

Hashimoto, K. (2013). *En no Shakaigaku: Fukushi Shakaigaku no Shitenkara*. Harvest-sha.

Hatimtai, M. H., & Hassan, H. (2018). The relationship between the characteristics of innovation towards the effectiveness of ICT in Malaysia productivity corporation. *Malaysian Journal of Communication*, *34*(1), 253–269. doi:10.17576/JKMJC-2018-3401-15

Haward, M., Davidson, J., Lockwood, M., Hockings, M., Kriwoken, L., & Allchin, R. (2013). Climate change,scenarios and marine biodiversity conservation. *Marine Policy*, *38*, 438–446. doi:10.1016/j.marpol.2012.07.004

Heijman, W. (2016). How big is the bio-business? Notes on measuring the size of the Dutch bio-economy. *NJAS Wageningen Journal of Life Sciences*, *77*(1), 5–8. https://research.wur.nl/en/publications/how-big-is-the-bio-business-noteson-measuring-the-size-of-the-du(Accessed 13.06.2023). doi:10.1016/j.njas.2016.03.004

Heimann, T. (2019). Bioeconomy and SDGs: Does the bioeconomy support the achievement of the SDGs? *Earth's Future*, *7*(1), 43–57. doi:10.1029/2018EF001014

He, J. (2006). Pollution haven hypothesis and environmental impacts of foreign direct investment: The case of industrial emission of sulfur dioxide (SO2) in Chinese provinces. *Ecological Economics*, *60*(1), 228–245. doi:10.1016/j.ecolecon.2005.12.008

Hemsley-Brown, T., & Laing, A. (2019). Green roofs and sustainable urban development: An exploratory review of potential barriers to adoption. *Sustainable Cities and Society*, *50*, 101648. doi:10.1016/j.scs.2019.101648

Hendry, J. (2014). *Science and Sustainability: Learning from Indigenous Wisdom*. Palgrave Macmillan. doi:10.1057/9781137430069

Heyd, T. (1995). Indigenous knowledge, emancipation, and alienation. *Knowledge in Society*, *8*(1), 63–73. doi:10.1007/BF02698557

Hickel, J. (2020). *Less is More: How Degrowth Will Save The World*. Windmill.

Hicks, J. (2019). Campaigning for social security rights: Women in the informal economy and maternity benefits. *Agenda (Durban, South Africa)*, *33*(2), 32–41. doi:10.1080/10130950.2019.1609809

Hidayati, S., Hadi, S., Saputra, N., Iswanto, A. H., Fatriasari, W., Solihat, N. N., Antov, P., & Hua, L. S. (2023). Characterization of lignin isolated from oil palm empty fruit bunch using phosphoric acid. *Biomass Conversion and Biorefinery*. doi:10.100713399-023-04381-9

Hill, K. (2007). Urban ecological design and urban ecology: an assessment of the state of current knowledge and a suggested research agenda. V. Novotny, P. Brown (Eds.), Cities of the future: towards integrated sustainable water and landscape management. IWA Publishing, London.

Hilliard, A., & Jamieson, G. A. (2008). Winning solar races with interface design. *Ergonomics in Design*, *16*(2), 6–11. doi:10.1518/106480407X312374

Hindman, M. (2015). Building better models: Prediction, replication, and machine learning in the social sciences. *The Annals of the American Academy of Political and Social Science*, *659*(1), 48–62. doi:10.1177/0002716215570279

Hipp, L. (2020). Feeling secure vs. being secure? Qualitative evidence on the relationship between labour market institutions and employees' perceived job security from Germany and the U.S. *Contemporary Social Science*, *15*(4), 416–429. doi:10.1080/21582041.2019.1656816

Hirst, K. (2019). *Is anthropology a science?* ThoughtCo. https://www.thoughtco.com/isanthropology-a-science-3971060

Hjeij, Z. (2021). *Estimating the impact of the Lebanese Currency Depreciation on Small investors*. [MBA Thesis, Lebanese International University].

Hobeika, L. (2019). *High-interest rates mean a weaker currency*. Anna Har. https://en.annahar.com/article/1039542-highinterest-rates-mean-a-weaker-currency

Hodge, T. (1997). Toward a conceptual framework for assessing progress toward sustainability. *Social Indicators Research*, *40*(1/2), 5–98. doi:10.1023/A:1006847209030

Hodgson, A., Maxon, M., & Alper, J. (2022). The U.S. Bioeconomy: Charting a Course for a Resilient and Competitive Future. *Industrial Biotechnology (New Rochelle, N.Y.)*, *18*(3), 115–136. doi:10.1089/ind.2022.29283.aho

Hoeden, S. M., Soussana, J. F., & Tubiello, F. N. (2007). Adapting agriculture to climate change. *Agricultural Sciences*, *104*(50), 19691–19696. PMID:18077402

Hoffmann, J. A. (2020). Memory, Waterways, and Ecocultural Identity. In T. Milstein & J. Castro-Sotomayor (Eds.), *The Routledge Handbook of Ecocultural Identity*. Routledge. doi:10.4324/9781351068840-9

Hofman, J., Sharma, A., & Watts, D. (2017). Prediction and explanation in social systems. *Science*, *355*(6324), 486–488. doi:10.1126cience.aal3856 PMID:28154051

Hofman, J., Watts, D., Athey, S., Garip, F., Griffiths, T., Kleinberg, J., Margetts, H., Mullainathan, S., Salganik, M., Vazire, S., Vespignani, A., & Yarkoni, T. (2021). Integrating explanation and prediction in computational social science. *Nature*, *595*(7866), 181–188. doi:10.103841586-021-03659-0 PMID:34194044

Hoke, M. K., & Schell, L. M. (2020). Doing biocultural anthropology: Continuity and change. *American Journal of Human Biology*, *2020*(32), e23471. PMID:32681558

Holling, C. S. (1973). *Resilience and Stability of Ecological Systems. Annual Review of Ecology and Systematics* (Vol. 4). JSTOR. https://www.jstor.org/stable/2096802

Holmes, D. R., & Marcus, G. E. (2006). Fast capitalism: para-ethnography and the rise of the symbolic analyst. In M. Fisher & G. Downey (Eds.), *Frontiers of Capital: Ethnographic Reflections on the New Economy*. Duke University Press.

Holmes, D. R., & Marcus, G. E. (2021). How do we collaborate? An updated manifesto. In D. Boyer & G. Marcus (Eds.), *Collaborative Anthropology Today: A Collection of Exceptions*. Cornell University Press. doi:10.7591/cornell/9781501753343.003.0002

Höök, M., & Aleklett, K. (2009). Historical trends in American coal production and a possible future outlook. *International Journal of Coal Geology*, *78*(3), 201–216. doi:10.1016/j.coal.2009.03.002

Höök, M., & Aleklett, K. (2010). Trends in US recoverable coal supply estimates and future production outlooks. *Natural Resources Research*, *19*(3), 189–208. doi:10.100711053-010-9121-1

Hopwood, A. G. (2009). Accounting and the environment. *Accounting, Organizations and Society*, *34*(3-4), 433–439. doi:10.1016/j.aos.2009.03.002

Hopwood, A. G. (2010). In memoriam, Anthony G. Hopwood, 1944-2010. *Accounting, Organizations and Society*, *35*(4), 496–497. doi:10.1016/j.aos.2010.05.002

Horta Nogueira, L. A., Souza, G. M., Barbosa Cortez, L. A., & de Brito Cruz, C. H. (2020). 9 - Biofuels for Transport. *Future Energy*, *3*, 173–197. doi:10.1016/B978-0-08-102886-5.00009-8

Hoskisson, R. E., Gambeta, E., Green, C. D., & Li, T. X. (2018). Is my firm-specific investment protected? Overcoming the stakeholder investment dilemma in the resource-based view. *Academy of Management Review*, *43*(2), 284–306. doi:10.5465/amr.2015.0411

Hossain, A., Reza, I., Rahman, S., & Kayes, I. (2012). Climate Change and its Impacts on the Livelihoods of the Vulnerable People in the Southwestern Coastal Zone in Bangladesh. In W. Leal Filho (Ed.), *Climate Change and the Sustainable Use of Water Resources. Climate Change Management*. Springer. doi:10.1007/978-3-642-22266-5_15

House, W. (2022). Executive Order on Advancing Biotechnology and Biomanufacturing Innovation for a Sustainable, Safe, and Secure American Bioeconomy. *The White House*. https://www.whitehouse.gov/briefing-room/presidential-actions/2022/09/12/executive-order-on-advancing-biotechnology-and-biomanufacturing-innovation-for-a-sustainable-safe-and-secure-american-bioeconomy/

Howard-Grenville, J., Rerup, C., Langley, A., & Tsoukas, H. (2016). Introduction: Advancing a process perspective on routines by zooming out and zooming in. In J. Howard-Grenville, C., Rerup, A., Langley, & H., Tsoukas (Eds.). Organizational routines: how they are created, maintained, and changed (pp.1-22.) Oxford: Oxford University Press.

Howson, P. (2021). Distributed degrowth technology: Challenges for blockchain beyond the green economy. *Ecological Economics*, *184*, 107020. doi:10.1016/j.ecolecon.2021.107020

Hu, A. G. Z., Zhang, P., & Zhao, L. (2017). China as number one? Evidence from China's most recent patenting surge. *Journal of Development Economics*, *124*, 107–119. doi:10.1016/j.jdeveco.2016.09.004

Huan, Y., Kang, X., Wang, Y., & Wang, Y. (2022). IAVT: Anomalous vessel trajectory detection using AIS data. *2022 5th International Conference on Artificial Intelligence and Big Data (ICAIBD)*, 17–22.

Huang, J., Kong, F., Yin, H., Middel, A., Liu, H., & Meadows, M. E. (2023). Green roof effects on urban building surface processes and energy budgets. *Energy Conversion and Management*, *287*, 1–13. doi:10.1016/j.enconman.2023.117100

Huang, M., Ding, R., & Xin, C. (2021). Impact of technological innovation and industrial-structure upgrades on ecological efficiency in China in terms of spatial spillover and the threshold effect. *Integrated Environmental Assessment and Management*, *17*(4), 852–865. doi:10.1002/ieam.4381 PMID:33325155

Huang, Y., Liu, C., Lee, C., & Huang, Y. (2012). Designing a Personalized Guide Recommendation System to Mitigate Information Overload in Museum Learning. *Journal of Educational Technology & Society*, *15*(4), 150–166.

Hua, Y., Xie, R., & Su, Y. (2018). Fiscal spending and air pollution in Chinese cities: Identifying composition and technique effects. *China Economic Review*, *47*, 156–169. doi:10.1016/j.chieco.2017.09.007

Huerta-Carvajal, M., & Luna-Reyes, L. (2009). City Boosterism through Internet Marketing: An Institutional Perspective. In M. Gasco-Hernandez & T. Torres-Coronas (Eds.), *Information Communication Technologies and City Marketing: Digital Opportunities for Cities Around the World* (p. 148). IGI Global. doi:10.4018/978-1-60566-134-6.ch007

Hu, G., Zhao, L., Wu, T., Wu, X., Park, H., Fedorov, A., Wei, Y., Li, R., Zhu, X., Sun, Z., Ni, J., & Zou, D. (2021). Spatiotemporal variations and regional differences in air temperature in the permafrost regions in the Northern Hemisphere during 1980–2018. *The Science of the Total Environment*, *791*, 148358. doi:10.1016/j.scitotenv.2021.148358 PMID:34139490

Hui, L. (2018). China's Arctic Policy. *Xinhua*. http://www.xinhuanet.com/

Humpert, M. (2011). *The Future of the Northern Sea Route - A "Golden Waterway" or a Niche Trade Route*. The Artic Institute. https://www.thearcticinstitute.org/future-northern-sea-route-golden-waterway-niche/

Humpert, M. (2013). Putin and Xi Discuss Further Deepening of Arctic Partnership. *High North News*. https://www.highnorthnews.com/en/putin-and-xi-discuss-further-deepening-arctic-partnership

Hussain, S. A. I., & Mandal, U. K. (2016, July). Entropy based MCDM approach for Selection of material. In *National Level Conference on Engineering Problems and Application of Mathematics* (pp. 1–6).

Hyde, M., & Shand, R. (2017). *Retirement, Pensions and Justice: A Philosophical Analysis*. Palgrave Macmillan. doi:10.1057/978-1-137-60066-0

Ibarra, R. (2021). *Retos de los sistemas de pensiones y salud*. https://bit.ly/3EvnZlL

Ibarra-Jarrín, R., & Corral, C. (2017). IESS: riesgo de descapitalización y posibles medidas. *Coyuntura. 11*, 1-4. https://bit.ly/3fESHyb

Ihnatenko, M., & Novak, N. (2018). Development of regional programs for the development of agrarian enterprises with organic production based on the European and international experience. *Baltic Journal of Economic Studies*, *4*(4), 126–133. doi:10.30525/2256-0742/2018-4-4-126-133

Ilchenko, L. M., Sokolov, S. S., Egorova, K. V., Gubernatorov, O., & Knish, T. (2021). Analysis of the main types of automated information systems in the transport sector, functioning mainly in urban agglomerations. *2021 IEEE Conference of Russian Young Researchers in Electrical and Electronic Engineering (ElConRus)*, (pp. 401–406). IEEE. 10.1109/ElConRus51938.2021.9396630

Inter-American Development Bank. (2016). *Saving for Development*. Palgrave Macmillan. doi:10.1057/978-1-349-94929-8_7

International Energy Agency. (2021). *Key World Energy Statistics 2021*. IEA. https://www.iea.org/reports/key-world-energy-statistics-2021/supply

International Labour Organization. (2009). De Bismarck a Beveridge: seguridad social para todos. *Trabajo, 67*, 2-31. https://bit.ly/3rJwHVy

International Labour Organization. (2020). *Valuación actuarial del régimen de invalidez, vejez y muerte del Instituto Ecuatoriano de Seguridad (Primera ed.)*. Quito: OIT. https://bit.ly/3fJX1fz

International Labour Organization. (2022a). *Análisis de la legislación ecuatoriana de seguridad social a la luz del Convenio sobre la seguridad social (Primera ed.)*. Quito: OIT. https://bit.ly/3T3nL9s

International Labour Organization. (2022b). *Declaración de la OIT relativa a los principios y derechos fundamentales en el Trabajo*. ILO. https://www.ilo.org/declaration/lang--es/index.htm

International Labour Organization. (2022c). *Informe de análisis de la coherencia normativa del sistema de seguridad social administrado por el IESS*. Quito: OIT / Oficina de la OIT para los Países Andinos. https://bit.ly/3MgwvH9

International Labour Organization. (2022d). *Análisis de la afectación de la pandemia de la COVID-19 en el riesgo de informalidad laboral y pobreza en el Ecuador*. Quito: OIT /Oficina de la OIT para los Países Andinos. https://bit.ly/3SZMfQE

International Trade Center. (2023). World trade statistics. *Trade Map*. ITC. https://www.trademap.org/Index.aspx?nvpm=1%7c%7c%7c%7c%7c%7c%7c%7c%7c%7c%7c%7c%7c%7c%7c%7c%7c

International Union for Conservation of Nature (IUCN). (2021). *Guidelines for Planning and Monitoring Business Biodiversity Performance*. ICUN.

IRENA. (n.d.). *World Energy Transitions Outlook 2022*. IRENA. https://www.irena.org/Digital-Report/World-Energy-Transitions-Outlook-2022

Ishitha, S., Nagaraju, S., Mohan, H. A., Harshitha, M., & Gowda, G. R., & Jeevan. (2021). IoT based anti-poaching and fire alarm system for forest. *2021 IEEE Mysore Sub Section International Conference (MysuruCon)*, (pp. 711–715). IEEE. 10.1109/MysuruCon52639.2021.9641539

Issawi, C. (1978). The 1973 Oil Crisis and After. *Journal of Post Keynesian Economics*, *1*(2), 3–26. doi:10.1080/01603477.1978.11489099

Jabbour, P. (2018). *The challenges facing e-commerce businesses in Lebanon*. [MBA Thesis, Lebanese International University].

Jackson, J. B. C., Kirby, M. X., Berger, W. H., Bjorndal, K. A., Botsford, L. W., Bourque, B. J., Bradbury, R. H., Cooke, R., Erlandson, J., Estes, J. A., Hughes, T. P., Kidwell, S., Lange, C. B., Lenihan, H. S., Pandolfi, J. M., Peterson, C. H., Steneck, R. S., Tegner, M. J., & Warner, R. R. (2001). Historical overfishing and the recent collapse of coastal ecosystems. *Science*, *293*(5530), 629–638. doi:10.1126cience.1059199 PMID:11474098

Jaffe, A. B., Peterson, S. R., Portney, P. R., & Stavins, R. N. (1995). Environmental regulation and the competitiveness of US manufacturing: What does the evidence tell us? *Journal of Economic Literature*, *33*(1), 132–163. https://www.jstor.org/stable/2728912

Jain, S. K., & Kaur, G. (2004). Green marketing: An Indian perspective. *Decision (Washington, D.C.)*, *31*(2), 161–209.

Jakobson, L. (2010). *China Prepares for an Ice-free Artic*. SIPRIS- Stockholm International Peace Research Institute, march 2010 https://www.sipri.org/publications/2010/sipri-insights-peace-and-security/china

Jamei, E., Chau, H., Seyedmahmoudian, M., Mekhilef, S., & Alzahraa, F. (2023). Green roof and energy – role of climate and design elements in hot and temperate climates. *Heliyon*, *9*(5), 1–16. doi:10.1016/j.heliyon.2023.e15917 PMID:37215798

Janssen, M. A., Jager, W., Bakker, M., & Weijer, M. (2009). Towards a Theory of Gesell Currencies. *Ecological Economics*, *68*(7), 1812–1824.

Jensen, J., & Luchsinger, G. (2019). *Estado de la población mundial 2019*. New York: Fondo de población de las Naciones Unidas- Unfpa. https://bit.ly/3fQ6gLr

Jensen, C. B., & Miho, I. (2016). Attuning to the Webs of En: Ontography, Japanese Spirit Worlds, and the 'Tact' of Minakata Kumagusu. *HAU*, *6*(2), 149–172. doi:10.14318/hau6.2.012

Jevons, W. S. (1875). *Money and the Mechanism of Exchange*. D. Appleton and Company.

Jiang, Z., Wang, Z., and Lan, X. (2021). How Environmental Regulations Affect Corporate Innovation? the Coupling Mechanism of Mandatory Rules and Voluntary Management. *Technol. Soc.*, *65*, 101575. doi:. 2021.101575 doi:10.1016/j.techsoc

Jiang, Y. (2015). Foreign direct investment, pollution, and the environmental quality: A model with empirical evidence from the Chinese regions. *The International Trade Journal*, *29*(3), 1–16. doi:10.1080/08853908.2014.1001538

Jiao, N.Z., Chen, D.K., Luo, Y.M., Huang, X.P., Zhang, R., Zhang, H.B., Jıang, Z.J. & Zhang, F. (2015). Climate change and anthropogenic impacts on marine ecosystems and countermeasures in China. *Advances in Climate Change Research 6* (2015) 118e125

Jim, C. Y. (1999). A planning strategy to augment the diversity and biomass of roadside trees in urban Hong Kong. *Landscape and Urban Planning*, *44*(1), 13–32. doi:10.1016/S0169-2046(98)00113-3

Johnson, S. K. (2019) *Leaking Talent How People of Color are Pushed Out of Environmental Organizations How People of Color are Pushed Out of Environmental Organizations*. Diverse Green. diversegreen.org/research/leaking-talent/

Johnson, S. K. (2017, August 17th). What 11 CEOs have learned about championing diversity. *Harvard Business Review*.

Jones, M. J., & Solomon, J. F. (2013). Problematising accounting for biodiversity. *Accounting, Auditing & Accountability Journal*, *26*(5), 668–687. doi:10.1108/AAAJ-03-2013-1255

Jorgensen, D. (2015). Participant observation. In R. Scott & S. Kosslyn (Eds.), *Emerging trends in the social and behavioral sciences* (pp. 1–15). Wiley. doi:10.1002/9781118900772.etrds0247

Jozić, S., Bajić, D., & Celent, L. (2015). Application of compressed cold air cooling: Achieving multiple performance characteristics in end milling process. *Journal of Cleaner Production*, *100*, 325–332. doi:10.1016/j.jclepro.2015.03.095

Ju-Long, D. (1982). Control problems of grey systems. *Systems & Control Letters*, *1*(5), 288–294. doi:10.1016/S0167-6911(82)80025-X

Kadar, T., & Kadar, M. (2020, June). Sustainability Is Not Enough: Towards AI Supported Regenerative Design. In *2020 IEEE International Conference on Engineering, Technology and Innovation (ICE/ITMC)* (pp. 1-6). IEEE. 10.1109/ICE/ITMC49519.2020.9198554

Kahveci, Z., & Biçer, A. (2019). *Mavi dünya' için mavi ekonomi* (*Blue economy for the blue world)*, AA. https://www.aa.com.tr/tr/temiz-cevre-temiz-su/mavi-dunya-icin-mavi-ekonomi/1508543

Kalandides, A., & Grésillon, B. (2021). The Ambiguities of "Sustainable" Berlin. *Sustainability (Basel)*, *2021*(13), 1666. doi:10.3390u13041666

Kallinikos, J. (2001). 'Bureaucracy: Exploring organization forms in the age of contingency. Paper presented at the 17th EGOS Colloquium, The Odyssey of Organizing, 5–7 July, Lyon. (Organization, special issue on Bureaucracy, 2003). 616. *Organization Studies*, *24*(4).

Kaminska, M. (2022). Variations on Bismarck: Translations of Social Health Insurance in Post-Communist Healthcare Reforms in Central and Eastern Europe. In International Impacts on Social Policy (pp 449–462). Palgrave Macmillan. doi.org/ doi:10.1007/978-3-030-86645-7_35

Kam-Sing Wong, S. (2012). The influence of green product competitiveness on the success of green product innovation: Empirical evidence from the Chinese electrical and electronics industry. *European Journal of Innovation Management*, *15*(4), 468–490. doi:10.1108/14601061211272385

Kangra, K., & Singh, J. (2021). Internet of Things and Its Relevance. In P. Tomar (Ed.), Integration and Implementation of the Internet of Things Through Cloud Computing (pp. 19-36). IGI Global. doi:http://doi:10.4018/978-1-7998-6981-8.ch002

Kanuri, C., Venkat, K., Maiti, S., & Mulukutla, P. (2019). Leveraging innovation for last-mile connectivity to mass transit. *Transportation Research Procedia*, *41*, 655–669. doi:10.1016/j.trpro.2019.09.114

Kaplan, M., & Yildirim, S. (2023). The Sustainability of Seafood Market Under the Attack of Invasive Alien Species: A Review of Turkish Marines. In A. Ben Salem, L. Rhazi, & A. Karmaoui (Eds.), *Climatic and Environmental Significance of Wetlands: Case Studies from Eurasia and North Africa* (pp. 74–95). IGI Global.

Kaplan, M., Yildirim, S., & Yildirim, D. C. (2022). Pufferfish versus lionfish: Comparing risks for Turkish marine economics. *Marine Economics and Management*, *5*(2), 173–187. doi:10.1108/MAEM-09-2022-0008

Kaplan, O. (1940). Prediction in the social sciences. *Philosophy of Science*, *7*(4), 492–498. doi:10.1086/286658

Karamillo, R. N, Liezl Mae G. M., Bucao, T., Bueno, B., Bulatao, L., Cabrera, S., & Cacho, L. (2022). Projected Action Plan and Timeline in Pilot Testing of Simulation-Based Learning in The College of Nursing at Phinma-University of Pangasinan. Cosmos An International Journal Art & Higher Education, 11(2), 04-15.

Karimi-Maleh, H., Orooji, Y., Karimi, F., Karaman, C., Vasseghian, Y., Dragoi, E. N., & Karaman, O. (2023). Integrated approaches for waste to biohydrogen using nanobiomediated towards low carbon bioeconomy. *Advanced Composites and Hybrid Materials*, *6*(1), 29. doi:10.100742114-022-00597-x

Karstensen, K. H., Engelsen, C. J., & Saha, P. K. (2020). Circular economy initiatives in Norway. In S. Ghosh (Ed.), *Circular economy: Global perspective* (pp. 299–316). Springer. doi:10.1007/978-981-15-1052-6_16

Kaufmann, R., Hermann, J., & Watters, K. (1996). *Educational Planning: Strategic, Tactical, Operational*. Technomic Publishing Inc.

Kaufman, R., & Herman, J. (1991). *Strategic planning in education: Rethinking, Restructuring, Revitalizing*. Technomic.

Kawashima, N. (1998). Knowing the public. A review of museum marketing literature and research1. *Museum Management and Curatorship*, *17*(1), 21–39. doi:10.1080/09647779800301701

Kawulich, B. (2005). La observación participante como método de recolección de datos. *Forum Qualitative Social Research*, *68*(2), 43.

Kaya, G., Kayalica, M. O., Kumas, M., & Ulengin, B. (2017). The role of foreign direct investment and trade on carbon emissions in Turkey. *Environment and Ecology*, *8*(1), 8–17. doi:10.21511/ee.08(1).2017.01 PMID:28241803

Kaya, Y. (2012). Waste trade and the effectiviness of the legal arrengements for tranfortier movement of hazardous wastes. *Paradoks: The Journal of Economics. Social Politics*, *8*(2), 63–82.

Kellenberg, D. K. (2009). An empirical investigation of the pollution haven effect with strategic environment and trade policy. *Journal of International Economics*, *78*(2), 242–255. doi:10.1016/j.jinteco.2009.04.004

Kempera, J. A., & Ballantine, P. W. (2019). What do we mean by sustainability marketing? *Journal of Marketing Management*, *35*(3–4), 277–309. doi:10.1080/0267257X.2019.1573845

Kempiak, J., Hollywood, L., Bolan, P., & McMahon-Beattie, U. (2017). The Heritage Tourist: An understanding of the visitor experience at Heritage Attractions. *International Journal of Heritage Studies*, *23*(4), 375–392. doi:10.1080/13527258.2016.1277776

Kennedy, M. (1995). *Interest and Inflation Free Money: Creating an Exchange Medium That Works for Everybody and Protects the Earth*. New Society Publishers.

Keskin, I., & Soykan, G. (2019). Reliability improvement and energy saving at internet data center microgrids. *2019 IEEE International Conference on Environment and Electrical Engineering and 2019 IEEE Industrial and Commercial Power Systems Europe (EEEIC / I&CPS Europe)*, (pp. 1–5). IEEE. 10.1109/EEEIC.2019.8783781

Kessler, T. P., & Chavannes, N. (2018). Leveraging the social impact of small business in urban neighborhoods. *Journal of Small Business Management*, *56*(4), 602–616.

Khari, D., & Sharma, V. & Agarwal, N. (2020). Effect of pandemic COVID-19 on economic crisis and health issues globally. *Cosmos Journal of Engineering & Technology, 10*(1), 9- 15.

Kilbourne, W., McDonagh, P., & Prothero, A. (1997). Sustainable consumption and the quality of life: A macromarketing challenge to the dominant social paradigm. *Journal of Macromarketing*, *17*(1), 4–24. doi:10.1177/027614679701700103

Kilgour, J. (2017). Social Security Retirement Income Replacement Rates. *Compensation and Benefits Review, 48*(3), 1–9. doi:10.1177/0886368717736666

Kılıçarslan, Z., & Dumrul, Y. (2017). Foreign direct investments and CO2 emissions relationship: The case of Turkey. *Business and Economics Research Journal,* 8(4), 647-660. https://www.ceeol.com/search/article-detail?id=586046

Kircher, M. (2019). Bioeconomy: Markets, implications and investment opportunities. *Savings, 73*(3), 73. Advance online publication. doi:10.3390/economies7030073

Kircher, M. (2022). The bioeconomy needs economic, ecological and social sustainability. *AIMS Environmental Science, 9*(1), 33–50. doi:10.3934/environsci.2022003

Kirchherr, J., Reike, D., & Hekkert, M. (2017). Conceptualizing the circular economy: An analysis of 114 definitions. *Resources, Conservation and Recycling, 127*, 221–232. doi:10.1016/j.resconrec.2017.09.005

Kirezli, Ö. (2011). Museum marketing: Shift from traditional to experiential marketing. *International Journal of Management Cases, 13*(4), 173–184. doi:10.5848/APBJ.2011.00125

Kittinger, J. N., McClenachan, L., Gedan, K. B., & Blight, L. K. (2015). *Marine historical ecology in conservation: Applying the past to manage for the future.* University of California Press.

Klemm, W., Zinngrebe, Y., & Heidenreich, S. (2019). Green roofs as part of urban green infrastructure for innovative cities. *Environmental Innovation and Societal Transitions, 31*, 50–56. doi:10.1016/j.eist.2018.11.003

Klewes, J., Popp, D., & Rost-Hein, M. (2017a). Digital transformation and the challenges for organizational communications: an introduction. In J. Klewes, D. Popp, & M. Rost-Hein (Eds.), *Out-thinking Organizational Communications: The Impact of Digital Transformation* (pp. 1–6). Springer. doi:10.1007/978-3-319-41845-2_1

Klewes, J., Popp, D., & Rost-Hein, M. (2017b). Digital Transformation and Communication How Key Trends Will Transform the Way Companies Communicate. In *Out-thinking Organizational Communications: The Impact of Digital Transformation* (pp. 7–31). Springer. doi:10.1007/978-3-319-41845-2_2

Klewes, J., Popp, D., & Rost-Hein, M. (2017c). Managing the digital transformation: ten guidelines for communications professionals. In *Out-thinking Organizational Communications: The Impact of Digital Transformation* (pp. 187–195). Springer. doi:10.1007/978-3-319-41845-2_14

Knell, M. (2018). Increasing life expectancy and NDC pension systems. *Journal of Pension Economics and Finance, 17*(2), 170–199. doi:10.1017/S1474747216000226

KnioM. S. (2021). *Personal Communication.*

Knio, M. S., & Bellos, G. S. G. (2018). The Effective Valorization of Resources for the Economic Development of Tyre. *AJFM, 11*(11), 1794–1800. doi:10.31142/afmj/v3i11.04

Knowledge-Driven Developments in the Bioeconomy. (2017). *Economic complexity and evolution.* Springer International Publishing., doi:10.1007/978-3-319-58374-7

Ko, Y.-C., Zigan, K., & Liu, Y.-L. (2021). Carbon capture and storage in South Africa: A technological innovation system with a political economy focus. *Technological Forecasting and Social Change, 166*, 120633. . techfore. 2021. 120633 doi:10.1016/j

Koçak, E., & Şarkgüneşi, A. (2017). The impact of foreign direct investment on CO2 emissions in Turkey: New evidence from cointegration and bootstrap causality analysis. *Environmental Science and Pollution Research International, 25*(1), 790–804. . doi:10.100711356-017-0468-2 PMID:29063405

Köhler, M., & Poll, P. H. (2010). Long-term performance of selected old Berlin greenroofs in comparison to younger extensive green roofs in Berlin. *Ecological Engineering, 36*(5), 722–729. doi:10.1016/j.ecoleng.2009.12.019

Kolb, Ch. (1998). Review of Bilharz, Joy A., *The Allegany Senecas and Kinzua Dam: Forced Relocation through Two Generations*. H-AmIndian, H-Net Reviews. https://www.h-net.org/reviews/showrev.php?id=2366

Koliousis, I. G., Papadimitriou, S., Riza, E., Stavroulakis, P. J., & Tsioumas, V. (2018a). Scarcity theory and maritime clusters: From paradox to modelling. *Marine Policy, 93*, 40–46. doi:10.1016/j.marpol.2018.03.029

Koliousis, I. G., Papadimitriou, S., Stavroulakis, P. J., & Tsioumas, V. (2018b). The management of change within maritime clusters. *FME Transactions, 46*(3), 360–366. doi:10.5937/fmet1803360K

Kolomiets, T. V., & Tomashuk, I. V. (2021). Entrepreneurship and development of rural areas in Ukraine. *Colloquium-journal, 9*(96), 29–42.

Kompf, M. (2012). *Entreplexity = Entrepreneurship + Complexity: The writings and thoughts of Gene Luczkiw*. Sense Publishers. doi:10.1007/978-94-6209-064-4

Kopnina, H., Zhang, R., Anthony, S., Hassan, A., & Atkins, J. (2023). *How to account for extinction?: Discussion of Environmental, Social and Governance (ESG) in relation to biodiversity loss*. Working paper.

Korhonen, J., Honkasalo, A., & Seppälä, J. (2018). Circular economy: The concept and its limitations. *Ecological Economics, 143*, 37–46. doi:10.1016/j.ecolecon.2017.06.041

Korten, D. C. (2006). *The Great Turning: From Empire to Earth Community*. Berrett-Koehler Publishers.

Kosova, R., Qendraj, D.H., Xhafaj, E. (2022). Meta-Analysis ELECTRE III and AHP in Evaluating and Ranking the Urban Resilience. *Journal of Environmental Management* and Tourism, *3*(59): 756 - 768. doi:10.14505/jemt.v13.3(59).15

Kottack, C. P. (1999). The new ecological anthropology. *American Anthropologist, 101*(1), 23–35. doi:10.1525/aa.1999.101.1.23

Kousar, S., Sabri, P. S. U., Zafar, M., & Akhtar, A. (2017). Technological factors and adoption of green innovation-moderating role of government intervention: A case of SMEs in Pakistan. *Pakistan Journal of Commerce and Social Sciences, 11*(3), 833–861.

Kovačević, J., & Ramelli, S. (2020). Ecosystems for social impact: A review of literature and practice. *Journal of Social Entrepreneurship, 11*(2), 161–182.

Koverman, C. (2016). Next-Generation Connected Support in the Age of IoT: It's time to get proactive about customer support. *IEEE Consumer Electronics Magazine, 5*(1), 69–73. doi:10.1109/MCE.2015.2484698

Koyama, C. N., Watanabe, M., Hayashi, M., Shimada, M., Kobayashi, K., & Tadono, T. (2022). Advancements in global forest monitoring research founded on ALOS-2 long-term pantropical land observation. *IGARSS 2022 - 2022 IEEE International Geoscience and Remote Sensing Symposium*, (pp. 4445–4448). IEEE.

Koyuncu, H. (2021). İklim krizi: 1,5 ve 2 derecelik küresel ısınma arasındaki fark nedir, nasıl sonuçlara yol açabilir? *Euronews*. https://tr.euronews.com/2021/11/08/iklim-krizi-1-5-ve-2-derecelik-kuresel-s-nma-aras-ndaki-fark-nedir-nas-l-sonuclara-yol-aca

KPMG International. (2022). Big shifts, small steps. *Survey of Sustainability Reporting 2022*. KPMG International. https://kpmg.com/xx/en/home/insights/2022/09/survey-of-sustainability-reporting-2022.html

Kramer, J. M. (2021). From predator to parasite: On private property and our ecological disaster. *Journal of Economic Issues, 55*(June), 416–422. doi:10.1080/00213624.2021.1908804

Krauß, W. (2020). Narratives of change and the co-development of climate services for action. *Climate Risk Management, 28*, 100217. doi:10.1016/j.crm.2020.100217

Kremser, W., Pentland, B. T., & Brunswicker, S. (2017). *The Continuous Transformation of Interdependence in Networks of Routines*. Presented at European Group for Organizational.

Krishna, S. Kumaraswamy, M. S., & Nair, V. (2020). IoT based Water Parameter Monitoring System. *2020 5th International Conference on Communication and Electronics Systems (ICCES)*, (pp. 1299–1303). IEEE.

Krupar, S., & Ehlers, N. (2020). Biocultures: A critical approach to mundane biomedical governance. *Culture, Theory & Critique, 61*(4), 440–456. doi:10.1080/14735784.2020.1857810

Krysovatyy, A., Zvarych, I., & Zvarych, R. (2018). Circular economy in the context of alterglobalization. *Journal of International Students, 11*(4), 185–200. doi:10.14254/2071-8330.2018/11-4/13

KucherA. (2019). *Sustainable soil management in the formation of competitiveness of agricultural enterprises*. Academic Publishing House «Talent». https://doi.org/ doi:10.13140/RG.2.2.19554.07366

KucherL.HeldakM.OrlenkoA. (2018). Project management in organic agricultural production. *Agricultural and Resource Economics, 4*(3), 104–128. doi:10.22004/ag.econ.281753

Kuhn, T. S. (1962/1996). *The Structure of Scientific Revolutions* (3rd ed.). University of Chicago Press.

Kull, K. (1998). Semiotic Ecology: Different Natures in the Semiosphere. *Sign Systems Studies, 26*(1), 344–371. doi:10.12697/SSS.1998.26.15

Kumar, G., & Shashank, K. V. (2022). Smart farming based on AI, edge computing and IoT. *2022 4th International Conference on Inventive Research in Computing Applications (ICIRCA)*, (pp. 324–327). IEEE.

Kumar, M., Minai, A. F., Khan, A. A., & Kumar, S. (2020). IoT based energy management system for smart grid. *2020 International Conference on Advances in Computing, Communication & Materials (ICACCM)*, (pp. 121–125). IEEE. 10.1109/ICACCM50413.2020.9213061

Kumar, V., & Christodoulopoulou, A. (2014). Sustainability and branding: An integrated perspective. *Industrial Marketing Management, 43*(1), 6–15. doi:10.1016/j.indmarman.2013.06.008

Kuna, M., & Dreslerová, D. (2007). Landscape archaeology and "community areas" in the archaeology of central Europe. D. Hicks, L. McAtackney, G.J. Fairclough (Eds.), Envisioning landscape: Situations and standpoints in archaeology and heritage. Left Coast Press.

Kuosmanen, T. (2020). *How Big is the Bioeconomy? Reflections from an economic perspective*. Publications Office of the European Union, Luxembourg. , https://www.researchgate.net/publication/340916792_How_Big_is_the_Bioeconomy#fullTextFileContent doi:10.2760/144526

Kurth, T., Wübbels, G., Portafaix, A., Meyer zum Felde, A., & Zielcke, S. (2021). *The biodiversity crisis is a business crisis*. Boston Consulting Group.

Kurt, Ü., Kılıç, C., & Özekicioğlu, H. (2019). Doğrudan yabancı yatırımların co2 emisyonu üzerindeki etkisi: Türkiye için ardl sınır testi yaklaşımı. *Selçuk Üniversitesi Sosyal Bilimler Meslek Yüksekokulu Dergisi, 22*(1), 213–224. doi:10.29249elcuksbmyd.487139

KuveytTürk. (2022). *OECD (Ekonomik Kalkınma ve İşbirliği Örgütü) Nedir? Hangi Ülkeler OECD Kapsamında Yer Alır?* OECD. https://www.kuveytturk.com.tr/blog/finans/oecd-ekonomik-kalkinma-ve-isbirligi-orgutu-nedir-hangi-ulkeler-oecd-kapsaminda-yer-alir#:~:text=T%C3%BCrkiye%2C%20ABD%2C%20Fransa%2C%20Kanada,ve%-20L%C3%BCksemburg%20OECD%20kurucu%20%C3%BCyeleridir

Kwon, K., Lee, S., & Kim, S. (2022). AI-based home energy management system considering energy efficiency and resident satisfaction. *IEEE Internet of Things Journal*, *9*(2), 1608–1621. doi:10.1109/JIOT.2021.3104830

Ladrera, R.C. & Cagasan, U.A. (2022). A Review on the Impact of Climate Change on Agriculture Sector:Its Implication to Crop Production and Management. *Eurasian Journal of Agricultural Research*, *6*(1), 52-61.

Laibach, N., Börner, J., & Bröring, S. (2019). Exploring the future of the bioeconomy: An expert-based scoping study examining key enabling technology fields with potential to foster the transition toward a bio-based economy. *Technology in Society*, *58*, 101118. doi:10.1016/j.techsoc.2019.03.001

Lakatos, E. S., Yong, G., Szilagyi, A., Clinci, D. S., Georgescu, L., Iticescu, C., & Cioca, L. I. (2021). Conceptualizing core aspects on circular economy in cities. *Sustainability (Basel)*, *13*(14), 7549. doi:10.3390u13147549

Lake, I. R. (2012, November). Hooper, L., Abdelhamid, A., Bentham, G., Boxall, A.B.A., Draper, A., Fairweather-Tait, S., Hulme, M., Paul R. Hunter, P.R., Gordon, N., & Waldron, K.W. (2012). Climate Change and Food Security: Health Impacts in Developed Countries. *Environmental Health Perspectives*, *120*(11). PMID:23124134

Lambert, A. (2019). Psychotic, acritical and precarious? A Lacanian exploration of the neoliberal consumer subject. *Marketing Theory*, *19*(3), 329–346. doi:10.1177/1470593118796704

Lambooy, T. E., Maas, K. E. H., van't Foort, S., & Van Tilburg, R. (2018). Biodiversity and natural capital: Investor influence on company reporting and performance. *Journal of Sustainable Finance & Investment*, *8*(2), 158–184. doi:10.1080/20430795.2017.1409524

Langdon, J., & Claridge, J. (2011). Transport in medieval England. *History Compass*, *9*(11), 864–875. doi:10.1111/j.1478-0542.2011.00804.x

Language selection. (n.d.). European Commission. https://ec.europa.eu/info/files/communication-european-green-deal_en

Lan, J., Kakinaka, M., & Huang, X. (2012). Foreign direct investment, human capital and environmental pollution in China. *Environmental and Resource Economics*, *51*(2), 255–275. doi:10.100710640-011-9498-2

Lapassade, G. (1986). *Grupos, organizaciones e instituciones: la transformación de la burocracia*. Gedisa.

Larson, E. D. (2008). *Biofuel production technologies: status, prospects and implications for trade and development*. UNCTAD.

Lasserre, F., Huang, L., & Alexeeva, O. V. (2017). China's strategy in the Arctic: Threatening or opportunistic? *The Polar Record*, *53*(1), 31–34. doi:10.1017/S0032247415000765

Latour, B. (1987). *Science in Action: How to Follow Scientists and Engineers Through Society*. Harvard University Press.

Latour, B. (2005). *Reassembling the Social: An Introduction to Actor-Network Theory*. Oxford University Press.

Latour, B. (2013). *An inquiry into modes of existence: An anthropology of the moderns* (C. Porter, Trans.). Harvard University Press.

Leal Filho, W., Abubakar, I. R., Kotter, R., Grindsted, T. S., Balogun, A. L., Salvia, A. L., Aina, Y. A., & Wolf, F. (2021). Framing electric mobility for urban sustainability in a circular economy context: An overview of the literature. *Sustainability (Basel)*, *13*(14), 7786. doi:10.3390u13147786

Lee, C. B., Wan, J., Shi, W., & Li, K. (2014). A cross-country study of competitiveness of the shipping industry. *Transport Policy*, *35*, 366–376. doi:10.1016/j.tranpol.2014.04.010

Lee, C. G. (2009). Foreign direct investment, pollution and economic growth: Evidence from Malaysia. *Applied Economics*, *41*(13), 1709–1716. doi:10.1080/00036840701564376

Lee, P., Sims, E., Bertham, O., Symington, H., Bell, N., Pfaltzgraff, L., Sjögren, P., Wilts, H., & O'Brien, M. (2017). *Towards a circular economy: waste management in the EU*. European Parliament. doi:10.2861/978568

Lee, S., & Hamelin, L. (2023). Unravelling global future scenarios in the perspective of bioeconomy planning. *Biomass and Bioenergy*, *168*, 106670. Advance online publication. doi:10.1016/j.biombioe.2022.106670

Leff, E. (1995). *Green Production*. Guilford Press.

Lehmann, S. (2010). *The principles of green urbanism: transforming the city for sustainability*. Routledge.

Leite, S. (2022). Using the SDGs for global citizenship education: Definitions, challenges, and opportunities. *Globalisation, Societies and Education*, *20*(3), 401–413. doi:10.1080/14767724.2021.1882957

Lekander, J. (2017). How do institutional pension managers consider real estate. *Journal of Property Investment & Finance*, *35*(1), 26–43. doi:10.1108/JPIF-05-2016-0033

Lemon, K. N., & Verhoef, P. C. (2016). Understanding customer experience throughout the customer journey. *Journal of Marketing*, *80*(6), 69–96. doi:10.1509/jm.15.0420

Leonard, W. R. (2020). Re-examining Biocultural Approaches in Human Biology. *American Journal of Human Biology*, *32*(4), e23475. doi:10.1002/ajhb.23475 PMID:32885558

Lester, R. K. (2005). *Universities, Innovation, and the Competitiveness of Local Economies A Summary Report from the Local Innovation Systems Project – Phase I*. Working Paper, Industrial Performance Center Massachusetts Institute of Technology.

Levinson, A., & Taylor, M. S. (2008). Unmasking the pollution haven effect. *International Economic Review*, *49*(1), 223–254. doi:10.1111/j.1468-2354.2008.00478.x

Lévi-Strauss, C. (1962). *La Pensée Sauvage*. Plon.

Ley para la Justicia Laboral y el Reconocimiento del Trabajo en el Hogar. (2015). *Asamblea Nacional Constituyente*. https://bit.ly/3rEamZO

Liang, F. (2008). *Does foreign direct investment harm the host country's environment? Evidence from China*. Rutgers University.

Liang, Y., Tan, Q., Song, Q., & Li, J. (2021). An analysis of the plastic waste trade and management in Asia. *Waste Management (New York, N.Y.)*, *119*, 242–253. . doi:10.1016/j.wasman.2020.09.049 PMID:33091837

Liao, S. H., Chang, J. C., Cheng, S. C., & Kuo, C. M. (2004). Employee relationship and knowledge sharing: A case study of a Taiwanese finance and securities firm. *Knowledge Management Research and Practice*, *2*(1), 24–34. doi:10.1057/palgrave.kmrp.8500016

Lier, M. (2018). Synthesis on bioeconomy monitoring systems in the EU Member States. *Natural resources and bioeconomy studies, 38*. https://www.researchgate.net/publication/339325322_Synthesis_on_bioeconom y_monitoring_systems_in_the_EU_Member_States_indicators_for_monitoring_the_progress_of_bioeconomy_Synthesis_on_bioeconomy_monitoring_systems_ in_the_EU_Member_States_-indicators_#fullTextFileContent

Lietaer, B., & Arnsperger, C. (2005). Money and Sustainability: The Missing Link. *Ecological Economics*, *54*(2-3), 164–168.

Lietaer, B., & Belgin, S. (2011). *New Money for a New World.* Qiterra Press.

Lietaer, B., & DeMeulenaere, S. (2005). Complementary Currencies: A Global Typology. *International Journal of Community Currency Research*, *9*, 1–23.

Lietaer, B., & Dunne, J. (2018a). *The Future of Money: Creating New Wealth, Work and a Wiser World.* Chelsea Green Publishing.

Lietaer, B., & Dunne, J. (2018b). *The Money Alternative: Creating Wealth in the 21st Century.* Watkins Publishing.

Lietaer, B., & Dunne, J. (2020a). Money and Sustainability: The Missing Link - A New Paradigm for the 21st Century. *Sustainability*, *12*(1), 1–24.

Lietaer, B., & Hallsmith, G. (2012). *Community Currency Guide: Guidelines for Community Currency Designers.* New Economics Institute.

Lietaer, B., & Kennedy, M. (1999). Complementary Currencies: A Beginner's Guide. *International Journal of Community Currency Research*, *3*, 1–10.

Li, G., Wang, X., Su, S., & Su, Y. (2019). How green technological innovation ability influences enterprise competitiveness. *Technology in Society*, *59*, 101136. doi:10.1016/j.techsoc.2019.04.012

Li, G., Zhou, Y., Liu, F., & Tian, A. (2021). Regional difference and convergence analysis of marine science and technology innovation efficiency in China. *Ocean and Coastal Management*, *205*, 105581. doi:10.1016/j.ocecoaman.2021.105581

Li, H. R. (2016). Study on green transportation system of international metropolises. *Procedia Engineering*, *137*, 762–771. doi:10.1016/j.proeng.2016.01.314

Li, J., Pan, S. Y., Kim, H., Linn, J. H., & Chiang, P. C. (2015). Building green supply chains in eco-industrial parks towards a green economy: Barriers and strategies. *Journal of Environmental Management*, *162*, 158–170. doi:10.1016/j.jenvman.2015.07.030 PMID:26241931

Li, J., Yin, Z., Li, Y., Sun, S., Li, M., & Li, Y. (2021). The application of green roofs in urban green innovation spaces: A case study in Jinan City, China. *Environmental Science and Pollution Research International*, *28*(3), 3525–3537. doi:10.100711356-020-11544-2 PMID:32892283

Li, L., Lin, H., Lyu, Y., & Cheng, L. (2021). Invulnerability in innovation networks: The importance of technology cluster coupling and network centralization. *Technology Analysis and Strategic Management*, 1–13. doi:10.1080/09537325.2021.1979202

Li, M., & Luo, M. (2021). Review of existing studies on maritime clusters. *Maritime Policy & Management*, *48*(6), 795–810. doi:10.1080/03088839.2020.1802786

Lin, Y. X., Sha, K. C., & Wang, J. (2020). Practical Experience and Inspiration of Foreign Emission Permit System. *Environ. Impact. Assess.*, *42*(1), 14–16.

Li, N. B. (2019). Environmental Regulation and Corporate Green Technology Innovation - a Conditional Process Analysis. *Inn. Mong. Soc. Sci. Chin. Ed*, *40*(6), 109–115. doi:10.14137/j.cnki.issn1003-5281.2019.06.016

Lin, C. Y., & Ho, Y. H. (2010). The influences of environmental uncertainty on corporate green behavior: An empirical study with small and medium-size enterprises. *Social Behavior and Personality*, *38*(5), 691–696. doi:10.2224bp.2010.38.5.691

Lin, C., Wang, J., & He, Q. (2011). Internet of Things technology for energy saving of telecom operators. *IET International Conference on Communication Technology and Application (ICCTA 2011)*. IET.

Lindblom, C. E. (1959). The Science of Muddling through. *Public Administration Review*, *19*(2), 79–88. doi:10.2307/973677

Lines, E. R., Allen, M., Cabo, C., Calders, K., Debus, A., Grieve, S. W. D., Miltiadou, M., Noach, A., Owen, H. J. F., & Puliti, S. (2022). AI applications in forest monitoring need remote sensing benchmark datasets. *2022 IEEE International Conference on Big Data (Big Data)*, (pp.4528–4533). IEEE. 10.1109/BigData55660.2022.10020772

Lin, J. L., & Lin, C. L. (2002). The use of the orthogonal array with grey relational analysis to optimize the electrical discharge machining process with multiple performance characteristics. *International Journal of Machine Tools & Manufacture*, *42*(2), 237–244. doi:10.1016/S0890-6955(01)00107-9

Linser, S., & Lier, M. (2020). The contribution of sustainable development goals and forest-related indicators to national bioeconomy progress monitoring. *Sustainability (Basel)*, *12*(7), 2898. doi:10.3390u12072898

Lin, Y. H., & Chen, Y. S. (2017). Determinants of green competitive advantage: The roles of green knowledge sharing, green dynamic capabilities, and green service innovation. *Quality & Quantity*, *51*(4), 1663–1685. doi:10.100711135-016-0358-6

Li, R. Y. M., Meng, L., Leung, T. H., Zuo, J., Tang, B., & Wang, Y. (2018). Unmaking Waste in Construction in the EU and the Asian Circular Economy: A Formal Institutional Approach. In R. Crocker, C. Saint, G. Chen, & Y. Tong (Eds.), *Unmaking Waste in Production and Consumption: Towards the Circular Economy* (pp. 225–240). Emerald Publishing Limited. doi:10.1108/978-1-78714-619-820181018

Li, S., Song, X., & Wu, H. (2015). Political connection, ownership structure, and corporate philanthropy in China: A strategic-political perspective. *Journal of Business Ethics*, *129*(2), 399–411. doi:10.100710551-014-2167-y

Liu, C. (2018). China's role in Arctic governance 'cannot be ignored'. *Global Times*. https://www.globaltimes.cn/content/1128626.shtml

Liu, D., Jiang, K., Shalley, C. E., Keem, S., & Zhou, J. (2016). Motivational mechanisms of employee creativity: A meta-analytic examination and theoretical extension of the creativity literature. *Organizational Behavior and Human Decision Processes*, *137*, 236–263. doi:10.1016/j.obhdp.2016.08.001

Liu, J., Chen, J., & Tao, Y. (2015). Innovation Performance in New Product Development Teams in China's Technology Ventures: The Role of Behavioral Integration Dimensions and Collective Efficacy. *Journal of Product Innovation Management*, *32*(1), 29–44. doi:10.1111/jpim.12177

Liu, J., Zhao, M., & Wang, Y. (2020). Impacts of government subsidies and environmental regulations on green process innovation: A nonlinear approach. *Technology in Society*, *63*, 101417. doi:10.1016/j.techsoc.2020.101417

Liu, Q., Wang, S., Zhang, W., Zhan, D., & Li, J. (2018). Does foreign direct investment affect environmental pollution in China's cities? A spatial econometric perspective. *The Science of the Total Environment*, *613*, 521–529. . doi:10.1016/j.scitotenv.2017.09.110 PMID:28923755

Liu, X., Lei, T., Boré, A., Lou, Z., Abdouraman, B., & Ma, W. (2022). Evolution of global plastic waste trade flows from 2000 to 2020 and its predicted trade sinks in 2030. *Journal of Cleaner Production*, *376*, 134373. doi:10.1016/j.jclepro.2022.134373

Liu, Y., Wang, A., & Wu, Y. (2021). Environmental regulation and green innovation: Evi-dence from China's new environmental protection law. *Journal of Cleaner Production*, *297*, 126698. doi:10.1016/j.jclepro.2021.126698

Liu, Z., Liu, W., Walker, T. R., Adams, M., & Zhao, J. (2021). How does the global plastic waste trade contribute to environmental benefits: Implication for reductions of greenhouse gas emissions? *Journal of Environmental Management, 287*, 112283. . doi:10.1016/j.jenvman.2021.112283 PMID:33706087

Li, X., & Norford, L. (2016). Evaluation of cool roof and vegetations in mitigating urban heat island in a tropical city, Singapore. *Environmental Science and Pollution Research*, 30, 3707–3725. *Urban Climate, 16*, 59–74. doi:10.1016/j.uclim.2015.12.002

Li, Z., Liao, G., Wang, Z., & Huang, Z. (2018). Green loan and subsidy for promoting clean production innovation. *Journal of Cleaner Production, 187*, 421–431. doi:10.1016/j.jclepro.2018.03.066

Llanos, M. (2009). *Epistemología de las ciencias sociales*. UNMSM.

Llonch, P., Haskell, M. J., Dewhurst, R. J., & Turner, S. P. (2017). Current available strategies to mitigate greenhouse gas emissions in livestock systems: An animal welfare perspective. *Animal, 11*(2), 274–284. doi:10.1017/S1751731116001440 PMID:27406001

López, M., Hincapié, V., & Fuentes, A. (2015). *Personas mayores y solidaridad intergeneracional en la familia (Primera ed.)*. Madrid: Grupo editorial Cinca, S. A. https://core.ac.uk/download/pdf/200765085.pdf

López-Arranz, A., & Picatoste, X. (2022). Retirement Pension Financing to Guarantee Older Workers' Rights. *International Advances in Economic Research, 28*, 45–47. doi:10.100711294-022-09846-7

López-Gamero, M. D., Claver-Cortés, E., & Molina-Azorín, J. F. (2008). Complementary resources and capabilities for an ethical and environmental management: A qual/quan study. *Journal of Business Ethics, 82*(3), 701–732. doi:10.100710551-007-9587-x

López-Roldán, P., & Fachelli, S. (2021). *Towards a Comparative Analysis of Social Inequalities between Europe and Latin America*. Springer. doi:10.1007/978-3-030-48442-2

Lorenz, K., & Lal, R. (2023). Organic Agriculture and Climate Change. Springer eBooks. doi:10.1007/978-3-031-17215-1

Lourau, R. (1979). *El análisis institucional*. Amorrortu.

Loureiro, M., Pêgo, A., & Raposo, I. (2021). The Real Blockchain Game Changer: Protocols and DAOs for Coordinating Work to Provide Goods and Services. In Political and Economic Implications of Blockchain Technology in Business and Healthcare (pp. 160-172). IGI Global.

Lozier, J. (2022). Arctic: Toward the End of the Exception? Strategic, Nuclear and Maritime Issues in the Region. *Proliferation Papers, No. 64*, https://www.ifri.org/sites/default/files/atoms/files/lozier_arctic_end_of_exception_2022.pdf

Lu, I.-H. (2022). Research on energy-saving efficiency of the air-conditioning host in agricultural plant bacteria room. *2022 IEEE 5th Eurasian Conference on Educational Innovation (ECEI)*, (pp. 384–387). IEEE.

Lu, P., Liu, Y., & Zhou, J. (2022). AI intelligent control method for energy saving of HVAC equipment in smart parks based on deep learning. *2022 4th International Academic Exchange Conference on Science and Technology Innovation (IAECST)*, (pp. 1560–1564). IEEE.

Lucier, C. A., & Gareau, B. J. (2015). From waste to resources? Interrogating 'race to the bottom' in the global environmental governance of the hazardous waste trade. *Journal of World-systems Research, 21*(2), 495–520. doi:10.5195/jwsr.2015.11

Luck, L., Jackson, D., & Usher, K. (2006). Case study: A bridge across the paradigms. *Nursing Inquiry, 13*(2), 103–109. doi:10.1111/j.1440-1800.2006.00309.x PMID:16700753

Lu, F. (2023). Corporate green initiatives on real circular economy or window-dressing? In G. Donleavy & C. Noronha (Eds.), *Comparative CSR and sustainability: New Accounting for Social Consequences*. Routledge.

Lundvall, B.-Å. (1992). (ed.). National Systems of Innovation: towards a theory of innovation and interactive learning. Printer Publisher. Londres.

Luo, Y., Salman, M., & Lu, Z. (2021). Heterogeneous impacts of environmental regulations and foreign direct investment on green innovation across different regions in China. *The Science of the Total Environment*, *759*, 143744. doi:10.1016/j.scitotenv.2020.143744 PMID:33341514

Luque, A. (2022). Analysis of the concept of informal economy through 102 definitions: Legality or necessity. *Open Research Europe*, *2022*(1), 134. doi:10.12688/openreseurope.13990.2

Luque, A., & Herrero-García, N. (2019). How corporate social (ir)responsibility in the textile sector is defined, and its impact on ethical sustainability: An analysis of 133 concepts. *Corporate Social Responsibility and Environmental Management*, *26*(6), 1–22.

Lyle, J. T. (1985). *Design for human ecosystems: landscape, land use, and natural resources*. Van Nostrand.

Ma, J., Feng, S., Li, X., Zhang, X., & Zhang, D. (2019). Research on the internet of things architecture for intelligent passenger transportation services and its application. *2019 4th International Conference on Electromechanical Control Technology and Transportation (ICECTT)*, (pp. 194–197). IEEE.

Maas, C. J., & Hox, J. J. (2005). Sufficient sample sizes for multilevel modeling. *Methodology: European Journal of Research Methods for the Behavioral and Social Sciences*, *1*(3), 86–92. doi:10.1027/1614-2241.1.3.86

Mahanand, M., & Singh, D. S. (2019). A Study on Social and Economic Justice by Fundamental Rights. *Globus Journal of Progressive Education*, *9*(1), 38–40.

Maksymiv, Y., Yakubiv, V., Pylypiv, N., Hryhoruk, I., Piatnychuk, I., & Popadynets, N. (2021). Strategic challenges for sustainable governance of the bioeconomy: Preventing conflict between SDGs. *Sustainability (Basel)*, *13*(15), 8308. doi:10.3390u13158308

Malik, Akash and Gharpe, Dr PS. (2022). Rehabilitation Engineering and Its Technological Solutions to Problems Confronted by Engineers. Globus An International Journal of Medical Science. *Engineering and Technology*, *11*(2), 24–27.

Malinowski, B. (1961). *Argonauts of the Western Pacific*. Routledge.

Mandavilli, S., Rys, M. J., & Russell, E. R. (2008). Environmental impact of modern roundabouts. *International Journal of Industrial Ergonomics*, *38*(2), 135–142. doi:10.1016/j.ergon.2006.11.003

Manniche, J., Larsen, K. T., & Broegaard, R. B. (2021). The circular economy in tourism: Transition perspectives for business and research. *Scandinavian Journal of Hospitality and Tourism*, *21*(3), 247–264. doi:10.1080/15022250.2021.1921020

Manta, F. (2019). China's policies on Polar Silk Road and impacts on trade and energy use in Arctic region. *Chinese Journal of Population Resources and Environment*, *17*(4) 322-328. doi:10.1080/10042857.2019.1685360

Manunen, O. (2000). An activity-based costing model for logistics operations of manufacturers and wholesalers. *International Journal of Logistics*, *3*(1), 53–65. doi:10.1080/13675560050006673

Mao, Y., & Robinson, D. (2019). Three pillars of sustainability: In search of conceptual origins. *Sustainability Science*, *14*(3), 681–695. doi:10.100711625-018-0627-5

Maran, T. (2018). Two Decades of Ecosemiotics in Tartu. *Sign Systems Studies*, *46*(4), 630–639. doi:10.12697/SSS.2018.46.4.11

Maran, T., & Kull, K. (2014). Ecosemiotics: Main principles and current developments. *Geografiska Annaler. Series B, Human Geography*, *96*(1), 41–50. doi:10.1111/geob.12035

Marchi, V. D. (2012). Environmental innovation and R&D cooperation: Empirical evidence from Spanish manufacturing firms. *Research Policy*, *41*(3), 614–623. doi:10.1016/j.respol.2011.10.002

Marcone, R. D., Schmid, M., & Meylan, G. (2022). Closing the gap between EU-wide national bioeconomy monitoring frameworks and urban circular bioeconomy development. *Journal of Cleaner Production*, *379*, 134563. doi:10.1016/j.jclepro.2022.134563

Marcu, A.-E., Suciu, G., Olteanu, E., Miu, D., Drosu, A., & Marcu, I. (2019). IoT System for Forest Monitoring. *2019 42nd International Conference on Telecommunications and Signal Processing (TSP)*, (pp. 629–632). IEEE.

Marcuse, H. (1972). *Counterrevolution and Revolt*. Boston Beacon Press.

Markard, J. (2012). Sustainability transitions: An emerging field of research and its prospects. *Research Policy 41*(6). https://www.sciencedirect.com/science/article/abs/pii/S004873331200056X

Marković, M., Popović, Z., & Marjanović, I. (2023). Towards a Circular Economy: evaluation of Waste Management Performance in EU Countries. *Serbian Journal of Management. 18*(1). doi:10.5937/sjm18-40073

Marković, M. (2022). Has Serbia Reached the Quality of Education of EU Countries? Multi-Criteria Evaluation. *Oradea Journal of Business and Economics*, *7*(2), 17–25. doi:10.47535/1991ojbe153

Marković, M., & Marjanović, I. (2022). The Role of Polyculture in Sustainable Agricultural Development and Prevention of Land Degradation. In S. Milutinović & S. Živković (Eds.), *Prevention and Management of Soil Erosion and Torrential Floods* (pp. 132–153). IGI Global. doi:10.4018/978-1-7998-8459-0.ch007

Marković, M., Stanković, J. J., Marjanović, I., & Tsaples, G. (2023a). A non-compensatory approach to the creation of composite indices of agricultural sustainability of the European Union countries. *International Journal of Sustainable Agricultural Management and Informatics*, *9*(1), 45–57. doi:10.1504/IJSAMI.2023.127538

Marotta, G. (2019). Behind the success of dominated personal pension plans: Sales force and financial literacy factors. *Journal of Pension Economics and Finance*, *19*(4), 532–547. doii:10.1017/S1474747219000209

Marouni, H. (2021). *Estimating the impact of entrepreneurial ventures and startups on the Lebanese economy*. [MBA Thesis, Lebanese International University].

Martínez-García, A. (2022). Artificial Intelligence for Sustainable Complex Socio-Technical-Economic Ecosystems. *Computation (Basel, Switzerland)*, *10*(6), 95. doi:10.3390/computation10060095

Martín, M. (2012). Mercados de activos financieros. *Revista de Fomento Social*, *67*(267), 549–552. doi:10.32418/rfs.2012.267.1788

Maruani, T., & Amit-Cohen, I. (2007). Open Space planning models: A review of approaches and methods. *Landscape and Urban Planning*, *81*(1–2), 1–13. doi:10.1016/j.landurbplan.2007.01.003

Maryville University. (2022). Why Is Agriculture Important? Benefits and Its Role. *Maryville Online*. https://online.maryville.edu/blog/why-is-agriculture-important/

Masco, J. (2006). *The Nuclear Borderlands: The Manhattan Project in Post–Cold War New Mexico*. Princeton University Press. doi:10.1515/9781400849680

Maskus, K. E. (2004). Encouraging international technology transfer. In UNCTAD-ICTSD Project on IPRs and Sustainable Development, 7. https://unctad.org/system/files/official-document/ictsd2004ipd7_en.pdf

Mason, P. (2018). Accounting for social benefits: The search for a past event. *International Social Security Review*, *71*, 73–80. doi:10.1111/issr.12180

Massaro, A., Panarese, A., & Galiano, A. (2021). Technological platform for hydrogeological risk computation and water leakage detection based on a convolutional neural network. *2021 IEEE International Workshop on Metrology for Industry 4.0 & IoT (MetroInd4.0&IoT)*, (pp. 225–230). IEEE. 10.1109/MetroInd4.0IoT51437.2021.9488561

Massoud, F. (2021). *The Effects of the Consecutive Crises in Lebanon on Financial Markets Efficiency and Money Laundering.* [MBA Thesis, Lebanese International University].

Mateescu, I., Popescu, S., Paun, L., Roata, G., Bancila, A. , & Oancea, A. (2011). Bioeconomy. What is bioeconomy? How will bioeconomy develop the next two decades. *Studia Universitatis «Vasile Goldiş». Seria Ştiinţele Vieţii*, *21*(2), 451.

Mathers, A., & Deonandan, R. (2018). Are the Sustainable Development Goals Realistic and Effective: A Qualitative Analysis of Key Informant Opinions. *OIDA International Journal of Sustainable Development*, *11*(3), 11–16.

Mayo, E. (1945). *The social problems of an Industrial Civilization*. Harvard University Press.

Mazjoub, Z. (2018). Business Ethics (1st ed.). Open Stax.

Mazur, K. V., & Tomashuk, I. V. (2019). Governance, and regulation as an indispensable condition for developing the potential of rural areas. *Baltic Journal of Economic Studies*, *5*(5), 67–78. doi:10.30525/2256-0742/2019-5-5-67-78

McClenachan, L., & Kittinger, J. N. (2013). Multicentury trends and the sustainability of coral reef fisheries in Hawai'i and Florida. *Fish and Fisheries*, *14*(3), 239–255. doi:10.1111/j.1467-2979.2012.00465.x

McClintock, P. (2014, March). *Box office milestone: 'Frozen' becomes No. 1 animated film of all time.* Hollywood Reporter. https://www.hollywoodreporter.com/news/box-office-milestone-frozen-becomes-692156

McCool, W. C., Vernon, K. B., Yaworsky, P. M., & Codding, B. F. (2022). Subsistence strategy mediates ecological drivers of human violence. *PLoS One*, *17*(5), e0268257. doi:10.1371/journal.pone.0268257 PMID:35604917

McCormick, K., & Kautto, N. (2013). The Bioeconomy in Europe: An overview. *Sustainability (Basel)*, *5*(6), 2589–2608. doi:10.3390u5062589

McElroy, A. (1990). Biocultural Models in Studies of Human Health and Adaptation. *Medical Anthropology Quarterly*, *4*(3), 243–265. doi:10.1525/maq.1990.4.3.02a00010

McHarg, I. (1996). *Design with nature*. The Natural History Press.

McIntosh, R. J., Tainter, J. A., & McIntosh, S. K. (Eds.). (2000). *The Way the Wind Blows: Climate Change, History, and Human Activity*. Columbia University Press.

McKay. (2022). Exceeding 1.5C global warming could trigger multiple climate tipping points. *Science*, *377*(6611), eabn7950. doi:10.1126cience.abn7950 PMID:36074831

McKinnon, R. (2020). Introduction: Social security, inclusive growth and social cohesion. *International Social Security Review*, *73*(3), 5–12. doi:10.1111/issr.12242

Mclean, F. (1997). *Marketing the museum*. Routledge. doi:10.4324/9780203277744

McLeroy, K. R., Bibeau, D., Steckler, A., & Glanz, K. (1988, Winter). An ecological perspective on health promotion programs. *Health Education Quarterly*, *15*(4), 351–377. doi:10.1177/109019818801500401 PMID:3068205

McMillan, J. (2014). *The End of Banking: Money, Credit, and the Digital Revolution.* Zero/One Economics.

Medina, E. (2017). Reformas al Código del Trabajo introducidas por la Ley de Justicia Laboral y Reconocimiento del Trabajo en el Hogar. Foro. *Revista De Derecho*, (24), 5–22. https://revistas.uasb.edu.ec/index.php/foro/article/view/458

Meerow, S., & Newell, J. P. (2017). Detroit, Spatial planning for multifunctional green infrastructure: Growing resilience in. *Landscape and Urban Planning*, *72*, 62–75. doi:10.1016/j.landurbplan.2016.10.005

Meeus, J. H. A. (1995). Pan-European landscapes. *Landscape and Urban Planning*, *31*(1-3), 57–79. doi:10.1016/0169-2046(94)01036-8

Mehmetoglu, M., & Engen, M. (2011). Pine and Gilmore's concept of experience economy and its dimensions: An empirical examination in tourism. *Journal of Quality Assurance in Hospitality & Tourism*, *12*(4), 237–255. doi:10.1080/1528008X.2011.541847

Mell, I. C. (2010). *Green infrastructure: concepts, perceptions, and its use in spatial planning.* [Doctoral Thesis, School of Architecture, Planning and Landscape Newcastle University].

Mellor, M. (2010). *The Future of Money: From Financial Crisis to Public Resource.* Pluto Press.

Melnikov, V., Osipov, V. I., Brouchkov, A. V., Falaleeva, A. A., Badina, S. V., Zheleznyak, M. N., Sadurtdinov, M. R., Ostrakov, N. A., Drozdov, D. S., Osokin, A. B., Sergeev, D. O., Dubrovin, V. A., & Fedorov, R. Y. (2022). Climate warming and permafrost thaw in the Russian Arctic: Potential economic impacts on public infrastructure by 2050. *Natural Hazards*, *112*(1), 1–21. doi:10.100711069-021-05179-6

Mentens, J., Raes, D., & Hermy, M. (2006). Green roofs as a tool for solving the rainwater runoff problem in the urbanized 21st century? *Landscape and Urban Planning*, *77*(3), 217–226. doi:10.1016/j.landurbplan.2005.02.010

Merçon, J., Vetter, S., Tengö, M., Cocks, M., Balvanera, P., Rosell, J. A., & Ayala-Orozco, B. (2019). From local landscapes to international policy: Contributions of the biocultural paradigm to global sustainability. *Global Sustainability*, *2*(e7), 1–11. doi:10.1017us.2019.4

Mesa-Lago, C. (1985). Desarrollo de la seguridad social en América Latina. Santiago: Naciones Unidas. https://bit.ly/3CeMm49

Mesa-Lago, C., & Valero, D. (2020). The New Wave of Pension Reforms in Latin America. In *Economic Challenges of Pension Systems*. Springer. doi:10.1007/978-3-030-37912-4_12

Mesimaki, M., Hauru, K., Kotze, D. J., & Lehvavirta, S. (2017). Neo-spaces for urban livibility? Urbanities' versatile mental image of green roofs in the Helsinki metropolitan area, Finland. *Land Use Policy*, *61*, 587–600. doi:10.1016/j.landusepol.2016.11.021

Mestre, G., Matos-Carvalho, J. P., & Tavares, R. M. (2022). Irrigation management system using artificial intelligence algorithms. *2022 International Young Engineers Forum (YEF-ECE)*, (pp. 69–74). IEEE.

Mike, F. (2020). Kirlilik sığınağı hipotezi Türkiye için geçerli mi? ardl sınır testi yaklaşımından bulgular. *Dogus University Journal, 21(2).* 107-121. Retrieved from https://dergipark.org.tr/en/pub/doujournal/issue/66683/1043221

Milanović, T., Savić, G., Martić, M., Milanović, M., & Petrović, N. (2022). Development of the Waste Management Composite Index Using DEA Method as Circular Economy Indicator: The Case of European Union Countries. *Polish Journal of Environmental Studies*, *31*(1), 1–14. doi:10.15244/pjoes/139896

Miles, M. B., & Huberman, A. M. (1994). *Qualitative data analysis: An expanded sourcebook.* Sage Publications, Inc.

Milevsky, M. (2015). *King William's Tontine: Why the Retirement Annuity of the Future Should Resemble Its Past.* Cambridge University. doi:10.1017/CBO9781139879316

Milevsky, M., & Salisbury, T. (2015). Optimal retirement income tontines. *Insurance, Mathematics & Economics, 64*(1), 91–105. doi:10.1016/j.insmatheco.2015.05.002

Milici, R. C., Flores, R. M., & Stricker, G. D. (2009). Estimation of United States peak coal production. Paper presented at the *Geological Society of America Abstracts with Programs for the 2009 Annual Meeting and Exposition,* doi:10.1016/j.coal.2012.10.002

Miller, T., & Mejía, I. (2020). *El envejecimiento de la población en Ecuador: la revolución silenciosa.* https://bit.ly/3rCuqLR

Millimet, D. L., & List, J. A. (2004). The case of the missing pollution haven hypothesis. *Journal of Regulatory Economics, 26*(3), 239–262. doi:10.100711149-004-7550-7

Millington, N. (2013). Post-industrial imaginaries: Nature, representation and ruin in Detroit, Michigan. *International Journal of Urban and Regional Research, 37*(1), 279–296. doi:10.1111/j.1468-2427.2012.01206.x

Milstein, T., & Castro-Sotomayor, J. (Eds.). (2020). *Routledge Handbook of Ecocultural Identity.* Routledge. doi:10.4324/9781351068840

Miner, K. R., Turetsky, M. R., Malina, E., Bartsch, A., Tamminen, J., McGuire, A. D., Fix, A., Sweeney, C., Elder, C. D., & Miller, C. E. (2022). Permafrost carbon emissions in a changing Arctic. *Nature Reviews. Earth & Environment, 3*(1), 55–67. doi:10.103843017-021-00230-3

Ministry of Agriculture, Fisheries, and Food. (2023). *Encuesta sobre Superficies y Rendimientos de Cultivos.* MAPA. https://www.mapa.gob.es/es/estadistica/temas/estadisticas-agrarias/agricultura/esyrce/

Ministry Of Ecology And Environment of The People's Republic Of China. (2023). *Announcement on the Regulation of the Imported Waste Management Catalogue.* https://www.mee.gov.cn/xxgk2018/xxgk/xxgk01/201804/t20180419_629582.html

Ministry of Economic and Social Inclusion (MIES). (2013). *Ciudadanía Activa y Envejencimiento Activo: Diseño de una política pública a favor de los derechos de las personas adultas mayores.* Subsecretaría de Atención Intergeneracional. Quito: MIES. https://bit.ly/3SMOhEi

Ministry of Economy and Finance. (2021). *Boletín de deuda pública interna y externa. Resolución No. MEF-2018-0134.* Quito: Ministerio de Economía y Finanzas. https://bit.ly/3RIZQeq

Ministry of Economy and Finance. (2022). *Justificativo de la proforma del presupuesto general del Estado.* Ministry of Economy and Finance. https://bit.ly/3SMM1g7

Ministry of Finance. (2015). *Resumen ejecutivo justificativa proforma presupuesto general del Estado 2016.* Quito: Subsecretaría de presupuesto. https://bit.ly/3CIPTte

Mintzberg, H. (1979). *The structuring of organizations: A synthesis of the research.* Prentice Hall.

Mohd, H. A. & Norhidayah, S. (2016), Sustainable food production: insights of Malaysian halal small and medium sized enterprises. *International Journal of Production Economics, 181* (part B), 303–314. . doi:10.1016/j.ijpe.2016.06.003

Mohr, S. H., & Evans, G. M. (2009). Forecasting coal production until 2100. *Fuel, 88*(11), 2059–2067. doi:10.1016/j.fuel.2009.01.032

Molnar, C. (2015, May 14). *10.2 Biotechnology in Medicine and Agriculture.* Pressbooks. https://opentextbc.ca/biology/chapter/10-2-biotechnology-in-medicine-and-agriculture/

Monteiro, P. V., Noronha, T. D., & Neto, P. A. (2013). A Differentiation Framework for Maritime Clusters: Comparisons across Europe. *Sustainability (Basel)*, *5*(9), 4076–4105. doi:10.3390u5094076

Montgomery, E., Bennett, J. W., & Scudder, T. (1973). The impact of human activities on the physical and social environments: New directions in anthropological ecology. *Annual Review of Anthropology*, *2*(1), 27–61. doi:10.1146/annurev.an.02.100173.000331

Moore, S. A., Rosenfeld, H., Nost, E., Vincent, K., & Roth, R. E. (2018). Undermining methodological nationalism: Cosmopolitan analysis and visualization of the North American hazardous waste trade. *Environment and Planning A. Environment & Planning A*, *50*(8), 1558–1579. doi:10.1177/0308518X18784023

Morimoto, R. (2022). A wild boar chase: Ecology of Harm and Half-Life Politics in Coastal Fukushima. *Cultural Anthropology*, *37*(1), 69–98. doi:10.14506/ca37.1.08

Morseletto, P. (2020). Restorative and regenerative: Exploring the concepts in the circular economy. *Journal of Industrial Ecology*, *24*(4), 763–773. doi:10.1111/jiec.12987

Mostafavi, M., & Doherty, G. (Eds.), *Ecological urbanism*. Lars Müller Publisher.

Moustier, P., Holdsworth, M., Anh, D., Seck, P. A., Renting, H., Caron, P., & Bricas, N. (2023). Science and Innovations for Food Systems Transformation. In Springer eBooks. doi:10.1007/978-3-031-15703-5

Munara, M., Kumar, N., & Shanmugam, K. (2022). Recommending IoT based real-time water quality monitoring system in Malaysia. *2022 IEEE 2nd Mysore Sub Section International Conference (MysuruCon)*, (pp. 1–5). IEEE.

Murphy, M. (2022). *Gıda krizi: BM, Ukrayna'daki savaşın küresel gıda krizine yol açabileceği uyarısında bulundu*, BBC. https://www.bbc.com/turkce/haberler-dunya-61503829

Murphy, K. M., Shleifer, A., & Vishny, R. W. (1993). Why is rent-seeking so costly to growth? *The American Economic Review*, *83*(2), 409–414. https://www.jstor.org/stable/2117699

Murray, N. (2017). Urban disaster risk governance. A systemic review, PPI-Centre. UCL Institute of Education.

Muscat, A., de Olde, E. M., Ripoll-Bosch, R., Van Zanten, H. H., Metze, T. A., Termeer, C. J., van Ittersum, M. K., & de Boer, I. J. (2021). Principles, drivers and opportunities of a circular bioeconomy. *Nature Food*, *2*(8), 561–566. doi:10.103843016-021-00340-7 PMID:37118163

Mussatto, S. I., Dragone, G., Guimaraes, P. M. R., Silva, J. P. A., Carneiro, L. M., Roberto, I. C., Vicente, A., Domingues, L., & Teixeira, J. A. (2010). Technological trends, global market, and challenges of bio-ethanol production. *Biotechnology Advances*, *28*(6), 817–830. doi:10.1016/j.biotechadv.2010.07.001 PMID:20630488

Mutafoglu, T. H. (2012). Foreign direct investment, pollution, and economic growth: Evidence from Turkey. *Journal of Developing Societies*, *28*(3), 281–297. doi:10.1177/0169796X12453780

Mu, X., Fang, C., Yang, Z., & Guo, X. (2022). Impact of the COVID-19 Epidemic on Population Mobility Networks in the Beijing–Tianjin–Hebei Urban Agglomeration from a Resilience Perspective. *Land (Basel)*, *2022*(11), 675. doi:10.3390/land11050675

Mwongera, K., & Langat, K. (2014). Current developments in Voice over Internet protocol (VoIP) and emerging challenges. *Proceedings of the International Conference on Sustainable Research and Innovation*. IEEE.

Nabavi, S. A., Motlagh, N. H., Zaidan, M. A., Aslani, A., & Zakeri, B. (2021). Deep learning in energy modeling: Application in smart buildings with distributed energy generation. *IEEE Access : Practical Innovations, Open Solutions*, *9*, 125439–125461. doi:10.1109/ACCESS.2021.3110960

Namkung, Y., & Jang, S. (2013). Effects of restaurant green practices on brand equity formation: Do green practices really matter? *International Journal of Hospitality Management, 33*(2), 85–95. doi:10.1016/j.ijhm.2012.06.006

Naranjo Bonilla, M. (2014). *Social protection systems in Latin America and the Caribbean: Ecuador. ECLAC – Project Documents collection*. United Nations. https://bit.ly/3EmSUAw

Narula, N., Ali, R., & Lipton, A. (2018). *The Future of Money: How New Lending Technologies Will Transform the Global Economy*. Harvard Business Review Press.

NASA Global Climate Change. (n.d.). *The Causes of Climate Change*. NASA. https://climate.nasa.gov/causes/

NASA. (2021). Ice Persists in the North West Passage. *NASA Earth Observatory*. NASA. https://earthobservatory.nasa.gov/

National Secretariat of Bakery and Development. (2017). *Plan Nacional del Buen Vivir 2017-2021*. Quito: Senplades. https://bit.ly/3mhqTQx

Navadia, N., Kaur, G., Bhardwaj, H., Singh, T., Sakalle, A., Acharya, D., & Bhardwaj, A. (2021). Applications of Cloud-Based Internet of Things. In *Integration and Implementation of the Internet of Things Through Cloud Computing* (pp. 65–84). IGI Global. doi:10.4018/978-1-7998-6981-8.ch004

Ndubisi, F. (2002). *Ecological planning: a historical and comparative synthesis*. The Johns Hopkins University Press.

Ndubisi, F. (2014). *The ecological design and planning reader*. Island Press. doi:10.5822/978-1-61091-491-8

Nedopil, C. (2022). Integrating biodiversity into financial decision-making: Challenges and four principles. *Business Strategy and the Environment*. doi:10.1002/bse.2884

Nelson, F., & Combe, M. (2022). What is the taskforce on nature-related financial disclosures. *Equity, 36*(8), 20–21.

Netting, R. M. (1974). Agrarian ecology. *Annual Review of Anthropology, 3*(1), 21–56. doi:10.1146/annurev.an.03.100174.000321

Neumann, T. (2021). Does it pay for new firms to be green? An empirical analysis of when and how different greening strategies affect the performance of new firms. *Journal of Cleaner Production, 317*, 128403. doi:10.1016/j.jclepro.2021.128403

Ngugi, J., & Goosen, L. (2021). Innovation, Entrepreneurship, and Sustainability for ICT Students Towards the Post-COVID-19 Era. In Handbook of Research on Entrepreneurship, Innovation, Sustainability, and ICTs in the Post-COVID-19 Era (pp. 110-131). IGI Global.

Nicolini, D., & Monteiro, P. (2017). The practice approach: For a praxeology of organizational and management studies. In A. Langley & H. Tsoukas (Eds.), *The Sage Handbook of Process Organization Studies*. Sage.

Nießen, D., Danner, D., Spengler, M., & Lechner, C. (2020). Big Five personality traits predict successful transitions from school to vocational education and training: A large-scale study. *Frontiers in Psychology, 11*, 1827. doi:10.3389/fpsyg.2020.01827 PMID:32903700

Nikken, P. (2010). La protección de los derechos humanos: haciendo efectiva la progresividad de los derechos económicos, sociales y culturales. *Revista IIDH, 52*(1), 55-140. https://www.corteidh.or.cr/tablas/r25563.pdf

Noaman, S. B., Harguem, S., & Bellos, G. S. G. (2022). Estimating the Impact of the Lebanese Economic Crisis on Employee Turnover through Motivation among other mediating variables as impeded by COVID-19 [Retrived from: http://www.resmilitaris.net.]. *Res Militaris, 12*(6), 1351–1374.

Nogué, S., & Sala, M. (2018). The use of green roofs in urban green innovation spaces: A case study in Barcelona. *Urban Forestry & Urban Greening, 33*, 19–29. doi:10.1016/j.ufug.2018.02.008

Noronha, C., Zhang, R., & Donleavy, G. (2024). Developing the phenomenon *grande* in management research: The pandemic, lockdowns, and biodiversity extinction in the Anthropocene. In D. Crowther & S. Seifi (Eds.), *The Routledge Companion to the Future of Management Research*. Routledge.

Norsk, L. (2023), *The Artic Council*. Artic Council. https://arctic-council.org/

North, P. (2007). *Money and Liberation: The Micropolitics of Alternative Currency Movements*. University of Minnesota Press.

Norton, T. A., Parker, S. L., Zacher, H., & Ashkanasy, N. M. (2015). Employee green behavior: A theoretical framework, multilevel review, and future research agenda. *Organization & Environment*, *28*(1), 103–125. doi:10.1177/1086026615575773

Nost, E. (2019). Climate services for whom? The political economics of contextualizing climate data in Louisiana's coastal master plan. *Climatic Change*, *157*(1), 27–42. doi:10.100710584-019-02383-z

Nöth, W. (1998). Ecosemiotics. *Sign Systems Studies*, *26*(1), 332–343. doi:10.12697/SSS.1998.26.14

Nowacki, M., & Kruczek, Z. (2020). Experience marketing at Polish museums and visitor attractions: The co-creation of visitor experiences, emotions and satisfaction. *Museum Management and Curatorship*, *36*(1), 62–81. doi:10.1080/09647775.2020.1730228

Nowakowski, P., & Mrówczyńska, B. (2018). Towards sustainable WEEE collection and transportation methods in circular economy-Comparative study for rural and urban settlements. *Resources, Conservation and Recycling*, *135*, 93–107. doi:10.1016/j.resconrec.2017.12.016

Nozawa, S. (2015). Phatic Traces: Sociality in Contemporary Japan. *Anthropological Quarterly*, *88*(2), 373–400. https://muse.jhu.edu/article/582266. doi:10.1353/anq.2015.0014

Nsamenang, A. B. (2015). Ecocultural Theories of Development. In J. Wright (Ed.), *International Encyclopedia of the Social & Behavioral Sciences* (Vol. 6, pp. 838–844). Elsevier. doi:10.1016/B978-0-08-097086-8.23205-4

NSW Environment Protection Authority. (2021). *What are biodiversity credits?* Biodiversity Offsets Scheme. https://www.environment.nsw.gov.au/topics/animals-and-plants/biodiversity-offsets-scheme/about-the-biodiversity-offsets-scheme/what-are-biodiversity-credits

NTV. (2023). *Mavi ekonominin yeşil kalkınmadaki rolü* (*The role of the blue economy in green development*),. NTV. https://www.ntv.com.tr/turkiye/mavi-ekonominin-yesil-kalkinmadaki-rolu,5qPFiFcgh0WAdX-g_yj9RQ

O'Neil, D. (2018, September 16). The EU needs a stability and wellbeing pact, not more growth. *The Guardian*. https://www.theguardian.com/politics/2018/sep/16/the-eu-needs-a-stability-and-wellbeing-pact-not-more-growth

O'Neill, K. (2001). The changing nature of global waste management for the 21st century: A mixed blessing? *Global Environmental Politics*, *1*(1), 77–98. doi:10.1162/152638001570642

Ober-Heilig, N., Bekmeier-Feuerhahn, S., & Sikkenga, J. (2014). Enhancing museum brands with experiential design to attract low-involvement visitors. *Arts Marketing: An International Journal*, *4*(1/2), 67–86. doi:10.1108/AM-01-2014-0006

Odum, E. P. (1957). *Fundamentals of ecology*. W. B. Saunders Company.

Odum, E. P. (1959). *Fundamentals of Ecology* (2nd ed.). W. B. Saunders.

OECD, Inter-American Development Bank and The World Bank. (2014). *Pensions at a Glance: Latin America and the Caribbean*. Organization for Economic Co-operation and Development. doi:10.1787/pension_glance-2014-

OECD. (1997b). Government Venture Capital for Technology-Based Firms. Ducument OCDE/GD (97) 201. Paris: OECD, 1997b.

OECD. (2004). *Biotechnology for sustainable growth and development.* Organisation for Economic Co-operation and Development. https://www.oecd.org/science/emerging-tech/33784888.pdf

OECD. (2016), *The Ocean Economy in 2030, Directorate for Science, Technology and Innovation Policy Note. OECD.* https://www.oecd.org/sti/futures/Policy-Note-Ocean-Economy.pdf

OECD. (2018). *Meeting Policy Challenges for a Sustainable Bioeconomy.* OECD Publishing. doi:10.1787/9789264292345-

OECD. (2021), *Adapting to a changing climate in the management of coastal zones.* OECD. https://www.oecd-ilibrary.org/docserver/b21083c5-en.pdf?expires=1679401663&id=id&accname=guest&checksum=3C774E8FE5923025E42023931179DE73

OECD. (2022). *Sustainable Ocean Economy.* OECD. https://stats.oecd.org/Index.aspx?DataSetCode=OCEAN

OECD. (n.d.). *Home.* OECD. https://www.oecd-ilibrary.org/sites/bd16d851-en/index.html?itemId=/content/component/bd16d851-en

OECD. (n.d.). Ocean and climate change. *Agriculture and Fisheries, Fisheries and Aquaculture, Fisheries and Aquaculture statistics.* OECD. https://www.oecd.org/ocean/topics/climate-change-ocean/

OECD. Stat, (2022a). Employment in fisheries, aquaculture and processing. *Agriculture and Fisheries, Fisheries and Aquaculture, Fisheries and Aquaculture statistics.* OECD. https://stats.oecd.org/

OECD. Stat, (2022b). International trade of fisheries commodities. *Agriculture and Fisheries, Fisheries and Aquaculture, Fisheries and Aquaculture statistics.* OECD. https://stats.oecd.org/

OECD. Stat, (2022c). Inland Fisheries. *Agriculture and Fisheries, Fisheries and Aquaculture, Fisheries and Aquaculture statistics.* OECD. https://stats.oecd.org/

OECD. Stat, (2022d). Fishing fleet. *Agriculture and Fisheries, Fisheries and Aquaculture, Fisheries and Aquaculture statistics.* OECD. https://stats.oecd.org/

OECD. Stat, (2022e). Aquaculture production. *Agriculture and Fisheries, Fisheries and Aquaculture, Fisheries and Aquaculture statistics.* OECD. https://stats.oecd.org/

OECD. Stat, (2022f). *Marine Landings. Agriculture and Fisheries, Fisheries and Aquaculture, Fisheries and Aquaculture statistics.* OECD. https://stats.oecd.org/

OECD. Stat, (2022g). Fisheries And Aquaculture Indicators. *Agriculture and Fisheries, Fisheries and Aquaculture, Fisheries and Aquaculture statistics.* OECD. https://stats.oecd.org/

Okamoto, T. (2019). China's "Territorial Sovereignty" and Its Origins. *Review the Japan Institute of international affairs,* 3(2).

Olczyk, M. (2016). A systematic retrieval of international competitiveness literature: A bibliometric study. *Eurasian Economic Review,* *6*(3), 429–457. doi:10.100740822-016-0054-9

Olías, L. (2022). *El Gobierno prevé que las pensiones subirán alrededor del 8,5% en 2023.* https://bit.ly/3RJ0qc6

Olimid, A. P., Georgescu, C. M., & Gherghe, C. L. (2022). Influences of Covid-19 Crisis on Resilience Theories: An analysis of Community, Societal and Governance Resilience. *Revista de Științe Politice. Revue des Sciences Politiques,* *73*, 38–51.

Olivia, P. F. (2012). *Developing the Curriculum* (6th ed.).

Oller Alonso, M. (2021). *La Responsabilidad Social Corporativa de las empresas del mármol en la comarca del Almanzora, Almería (2019-2021): análisis de sus estrategias de comunicación integral.* [Doctorado en Gestión de la Información y de la Comunicación en las Organizaciones Universidad de Murcia Escuela Internacional de Doctorado].

One Earth. (2021). *Four Steps for Earth: Bringing the Post-2020 Global Biodiversity Framework into the Mainstream.* Elsevier. . doi:10.1016/j.oneear.2020.12.011

Ones, D. S., & Dilchert, S. (2012). Employee green behaviors. In D. S. S. E. Jackson (Ed.), *Managing human resource for environmental sustainability*. Jossey-Bass.

Orange, E., & Cohen, A. M. (2010). From eco-friendly to eco-intelligent. *The Futurist*, *44*(5), 24–32.

Oriade, A., & Cameron, H. (2016). Logistics and Supply Chain Management in Travel Industry. In Operations management in the travel industry. CAB INTERNATIONAL.

Orrell, D. (2014). *Money and Sustainability: The Missing Link*. Routledge.

Ortiz-Crespo, G. (2022). *18 millones y 8 mil millones*. https://bit.ly/3fT1qx3

OrtizI.DuranF.UrbanS.WoodsesV.YuZ. (2018). Reversing Pension Privatization: Rebuilding Public Pension Systems in Eastern European and Latin American Countries. doi.org/ doi:10.2139/ssrn.3275228

Ortner, S. (2010). Access: Reflections on studying up in Hollywood. *Ethnography*, *11*(2), 211–233. doi:10.1177/1466138110362006

Ortner, S. (2013). *Not Hollywood: Independent Film at the Twilight of the American Dream*. Duke University Press.

Ostreng, W. (2012). On the Geopolitical Significance of Arctic States. *Arctis Knowledge Hub*. http://www.arctis-search.com/

Ostrom, E. (2005). Understanding Institutional Diversity, Princeton University Press, Princeton New Jersey, USA.

Özkaya, G. (2022). Evaluation of regions and four income groups in terms of social protection through critic based grey relational analysis method. In C. A. Ceylan & M. C. U. O. Kaya (Eds.), *Research & Reviews in Social, Human and Administrative Sciences* (pp. 108–129). Gece Publishing.

Pagoropoulos, A., Pigosso, D. C. A., & McAloone, T. C. (2017). The emergent role of digital technologies in the Circular Economy: A Review. *Procedia CIRP*, *64*, 19–24. doi:10.1016/j.procir.2017.02.047

Palazzo, D. (1997). Sulle spalle di giganti. [On the Shoulders of Giants]. Franco Angeli, Milano.

Palazzo, D., & Steiner, F. R. (2011). *Urban ecological design—a process for regenerative places*. Island Press.

Palicki, M. (Ed.). (2021). (rep.). TEA/AECOM 2021 Index and Museum Index: The Global Attractions Attendance Report (pp. 1–89). Themed Entertainment Association (TEA).

Palmer, L. (2013). Melting Arctic Ice Will Make Way for More Ships—and More Species Invasions. *Scientific American*. https://www.scientificamerican.com/

Palomino, J. C., & Durán, P. (2020). The influence of green roofs on the urban heat island effect: A review of the current state of knowledge. *Urban Climate*, *34*, 100696. doi:10.1016/j.uclim.2020.100696

Palyvoda, O. M., & Seliverstova, L. S. (2017). Management of the Industry's Innovative Development in the European Union Based on Cluster Infrastructure Formation. *Scientific Bulletin of Polissia*, *1*(1), 185–191. doi:10.25140/2410-9576-2017-1-1(9)-185-191

Pandey, R., Srivastava, D., Dizayee, R. H., Lobanova, A., Rabbi, F., & Sutkar, P. R. (2022). Enhancing the accuracy of forest monitoring through AI to reduce carbon footprint. *2022 4th International Conference on Inventive Research in Computing Applications (ICIRCA)*, (pp. 1177–1182). IEEE.

Pandolfi, J. M., Bradbury, R. H., Sala, E., Hughes, T. P., Bjorndal, K. A., Cooke, R. G., McArdle, D., McClenachan, L., Newman, M. J. H., Paredes, G., Warner, R. R., & Jackson, J. B. C. (2003). Global trajectories of the long-term decline of coral reef ecosystems. *Science, 301*(5635), 955–958. doi:10.1126cience.1085706 PMID:12920296

Paramio, L., Tiago, M. T. B., Tiago, F. B., & Vieira, J. C. (2013). Contributions of Marine Cluster for Regional Sustainability: Economic Prospective for Azores. *Maria Scientia-Scientific e-Journal, 6*, 48-67.

Páramo, D., & Ramírez, E. (2017). *Etnomarketing, la dimensión cultural del marketing*. Klasse Editorial.

Paredes, P. (2022). Sistema de jubilación en crisis: Soluciones urgentes. *Dialoguemos*. https://dialoguemos.ec/2022/05/sistema-de-jubilacion-en-crisis-soluciones-urgentes/

Paris, B., Michas, D., Balafoutis, A., Nibbi, L., Skvaril, J., Li, H., Pimentel, D., da Silva, C., Athanasopoulou, E., Petropoulos, D., & Apostolopoulos, N. (2023). A Review of the Current Practices of Bioeconomy Education and Training in the EU. *Sustainability (Basel), 15*(2), 954. doi:10.3390u15020954

Patzek, T. W., & Croft, G. D. (2009). Potential for coal-to-liquids conversion in the United States: Fischer–Tropsch synthesis. *Natural Resources Research, 18*(3), 181–191. doi:10.100711053-009-9098-9

Paulson, S., D'Alisa, G., Demaria, F., & Kallis, G. (2020). *The Case for Degrowth*. Polity.

Pauly, D. (1995). Anecdotes and the Shifting Baseline Syndrome of Fisheries. *Trends in Ecology & Evolution, 10*(10), 430. doi:10.1016/S0169-5347(00)89171-5 PMID:21237093

Payri, F., & Desantes, J. M. (2017). *Motores de combustión interna alternativos*. Editorial Universitat Politècnica de València.

Peace, G. (2023, February, 4). *Data from, the global plastic waste trade 2016-2018 and the offshore impact of China's foreign waste import ban*. Greeen Peace. https://www.greenpeace.org/eastasia/publication/5907/data-from-the-global-plastics-waste-trade-2016-2018-and-the-offshore-impact-of-chinas-foreign-waste-import-ban/

Pearce, D. W., & Turner, R. K. (1990). *Economics of Natural Resources and the Environment*. Hemel Hempstead, Harvester Wheatsheaf.

Peet, R., & Watts, M. (1996) Liberation ecology: development, sustainability, and environment in an age of market triumphalism. in: R. Peet & M. Watts (eds.) Liberation Ecologies: Environment, Development, Social Movements. Routledge.

Pego, A. (2020a). New Challenges for the Tourism Sector in the Algarve Region Based on Evaluation of the Circular Economy. In Strategic Business Models to Support Demand, Supply, and Destination Management in the Tourism and Hospitality Industry (pp. 185-199). IGI Global. doi:10.4018/978-1-5225-9936-4.ch010

Pego, A. (2020b). Circularity in Portugal: Features of New Business Challenges. In Networked Business Models in the Circular Economy (pp. 224-240). IGI Global.

Pego, A., & de Matos, M. D. (2017). The impact of technology and information systems in rural tourism in the Algarve region. *Dos Algarves: a Multidisciplinary e-Journal*(25), 39-59.

Pego, A., Marques, M., Salvador, R., & Monteiro, A. (2016). The potential offshore energy cluster in Portugal. In Progress in renewable energy offshore (pp. 867-873). doi:10.1201/9781315229256-102

Pego, A., Raposo, I. G., & Loureiro, M. (2021). Blockchain and Clinical Data Economics: The Tokenization of Clinical Research in the EU. In Political and Economic Implications of Blockchain Technology in Business and Healthcare (pp. 269-291). IGI Global.

Pego, A. (2018). The influence of renewable energy in smart cities. the europen perspetive for new business. *Revista Lusófona de Economia e Gestão das Organizações*, *1*(7), 63–70.

Pego, A. (2019). The Portuguese Offshore Energy SWOT Analysis. *Journal of Physics: Conference Series*, *1222*(1), 012036. doi:10.1088/1742-6596/1222/1/012036

Pego, A. (2021). Business Model for a Circular Economy based on Portuguese Offshore energy cluster. *Technium Social Sciences Journal*, *19*, 499–505.

Pego, A., & Bernardo, M. (2018). Decision making in Rural Tourism Management: The Case of Algarve. In *Handbook of Research on Entrepreneurial Ecosystems and Social Dynamics in a Globalized World* (pp. 274–289). IGI Global. doi:10.4018/978-1-5225-3525-6.ch012

Pego, A., & Bernardo, M. (2019). The Role of Urban Living Labs in Entrepreneurship, Energy, and Governance of Smart Cities. In *Handbook of Research on Entrepreneurship and Marketing for Global Reach in the Digital Economy* (pp. 203–221). IGI Global. doi:10.4018/978-1-5225-6307-5.ch009

Peng, J., & Wegge, N. (2015). China's bilateral diplomacy in the Arctic, Polar Geography. *Polar Geography*, *38*(3), 233–249. doi:10.1080/1088937X.2015.1086445

Pentland, A. (2004, October). *Social Dynamics: Signals and Behavior. International Conference on Developmental Learning*, Salk Institute, San Diego, CA

Pentland, B. T., Hærem, T., & Hillison, D. (2011). The (N)Ever-Changing World: Stability and Change in Organizational Routines. *Organization Science*, *22*(6), 1369–1383. doi:10.1287/orsc.1110.0624

Pereira, Á., & Vence, X. (2012). Key business factors for eco-innovation: An overview of recent firm-level empirical studies. *Cuadernos de Gestión*, *12*, 73–103. doi:10.5295/cdg.110308ap

Pereira, P., Zhao, W., Symochko, L., Inacio, M., Bogunovic, I., & Barceo, D. (2022). The Russian-Ukrainian armed conflict will push back the sustainable development goals. *Geography and Sustainability*, *3*(3), 277–287. doi:10.1016/j.geosus.2022.09.003

Pereira, V., Munjal, S., & Ishizaka, A. (2019, October). Outsourcing and offshoring decision making and its implications for businesses- A synthesis of research pursuing five pertinent questions. *Journal of Business Research*, *103*, 348–355. doi:10.1016/j.jbusres.2019.07.009

Pessino, C., & Ter-Minassian, T. (2021). *Enfrentando los costos fiscales del envejecimiento poblacional en América Latina y el Caribe con lecciones de los países desarrollados.* Banco Interamericano de Desarrollo. doi.org/ doi:10.18235/0003242

Peter, C. (2021). Social innovation for sustainable urban developmental transitions in Sub-Saharan Africa: Leveraging economic ecosystems and the entrepreneurial state. *Sustainability (Basel)*, *13*(13), 7360. doi:10.3390u13137360

Peterson, N., & Broad, K. (2009). Climate and weather discourse in anthropology: From determinism to uncertain futures. In *Anthropology and climate change: From encounters to actions* (pp. 70–86). Left Coast Press.

Pethig, R. (1976). Pollution, welfare and environmental policy in the theory of comparative advantage. *Journal of Environmental Economics and Management*, *2*(3), 160–169. doi:10.1016/0095-0696(76)90031-0

Petridis, N. E., Petridis, K., & Stiakakis, E. (2020). Global e-waste trade network analysis. *Resources, Conservation and Recycling, 158*, 104742. doi:10.1016/j.resconrec.2020.104742

Petrie, M. (2021). *Environmental Governance and Greening Fiscal Policy: Government Accountability for Environmental Stewardship*. Palgrave Macmillan. doi:10.1007/978-3-030-83796-9

Petterson, M., Kensek, K., & Noble, D. (2017). Supple Skins: Considering the Relevance, Scalability, and Design Strategies for Façade System Resilience. *Journal of Architectural Education, 71*(1), 34–45. doi:10.1080/10464883.2017.1260919

Pfaffenbichler, P., Emberger, G., & Shepherd, S. A. (2010). System dynamics approach to land use transport interaction modelling: The strategic model MARS and its application. *System Dynamics Review, 26*(3), 262–282. doi:10.1002dr.451

Pfeffer, J., & Salancik, G. (1978). *The External Control of Organizations: A Resource Dependence Perspective*. Harper & Row.

Pfeifer, J. (2006). Laws of nature. In S. Sarkar & J. Pfeifer (Eds.), *The philosophy of science: An encyclopedia* (pp. 439–444).

Pieris, T. P. D., Kulasekera, A. L., & Chathuranga, K. V. D. S. (2022). Multi-AI based wireless sensor node for forest fire detection. *2022 7th International Conference on Information Technology Research (ICITR)*, 1–6.

Piffero, S. P. (2013). The Militarisation of the Arctic: Political, Economic and Climate Challenges. *UFRGS Model United Nations Journal 1*, 11–70, https://www.ufrgs.br/.

Pimienta-Bueno, J.A. (1999). *Parque de Inovação Tecnológica e Cultural da Gávea: A Visão da PUC-Rio*. Documento Interno.

Pincus, R. (2020). Three-Way Power Dynamics in the Arctic. *Strategic Studies Quarterly, 14*, 40–63. doi:10.2307/26891883

Pine, B. J. I. I., & Gilmore, J. H. (1998). Welcome to the Experience Economy. *Harvard Business Review, 76*(4), 97–105. PMID:10181589

Pineiro-Villaverde, G., & García-Álvarez, M. T. (2020). Sustainable consumption and production: Exploring the links with resources productivity in the EU-28. *Sustainability (Basel), 12*(21), 8760. doi:10.3390u12218760

Pinto, M. (2021). *La protección ante el desempleo: medidas aplicadas durante la crisis del COVID-19. Documentos de Proyectos*. Santiago: Comisión Económica para América Latina y el Caribe. https://bit.ly/3fOctra

Pizarro, C. (2014). La entrevista etnográfica como práctica discursiva: Análisis de caso sobre las pistas meta-discursivas y la emergencia de categorías nativas. *Revista de Antropologia, 57*(1), 461–496. doi:10.11606/2179-0892.ra.2014.87770

Plumwood, V. (1995). Nature, self and gender: feminism, environmental philosophy, and the critique of rationalism. C. Pierce & D. VanDeVeer (Eds) People, Penguins and Plastic Trees. Wadsworth.

Ponsignon, F., Durrieu, F., & Bouzdine-Chameeva, T. (2017). Customer experience design: A case study in the cultural sector. *Journal of Service Management, 28*(4), 763–787. doi:10.1108/JOSM-01-2017-0016

Pontarollo, N., Orellana, M., & Segovia, J. (2020). The Determinants of Subjective Well-Being in a Developing Country: The Ecuadorian Case. *Journal of Happiness Studies, 21*, 3007–3035. doi:10.100710902-019-00211-w

Popescu, G., Comanescu, M., & Sabie, O. (2016). The Role of Human Capital In The Knowledge-Networked Economy. *Psychosociological Issues in Human Resource Management, 4*(1), 168–174. doi:10.22381/PIHRM4120168

Porras, A. X. (2015). La seguridad social en Ecuador: Un necesario cambio de paradigmas. Foro. *Revista de Derecho*, (24), 89–116. https://repositorio.uasb.edu.ec/bitstream/10644/5285/1/08-TC-Porras.pdf

Porter, M. (2001). Clusters of Innovation: regional foundations of U.S. competitiveness. Council on Competitiveness. Washington, D.C: Monitor Group.

Postel, S. (1994). Carrying capacity: Earth's bottom line. In L. R. Brown, A. T. Durning, C. Flavin, H. F. French, N. Lenssen, M. D. Lowe, A. Misch, S. Postel, M. Renner, & P. Weber, (Eds.), *State of the World* (pp. 3–21). W.W. Norton.

Pozzato, A. (2017-2018). Smart cities: an innovative solution for the management of global risks. [MA thesis, Ca' Foscari University of Venice, Italy].

Pradhan, P., Pradhan, L., Rybski, D., Lucht, W., & Kropp, J. P. (2017). A Systematic Study of Sustainable Development Goal (SDG) Interactions. *Earth's Future*, *5*(11), 1169–1179. doi:10.1002/2017EF000632

Pranee, P., & Jirasuwankul, N. (2021). Experimental study and modeling of automatic home energy management system using AI. *2021 International Conference on Power, Energy and Innovations (ICPEI)*, (pp. 103–106). IEEE. 10.1109/ICPEI52436.2021.9690647

PRC. (2018). *China's Arctic Policy.* White paper by The State Council Information Office of the People's Republic of China. https://english.www.gov.cn/archive/white_paper/2018/01/26/content_2814760266 60336.htm

Press, M., & Arnould, E. J. (2014). Narrative transparency. *Journal of Marketing Management*, *30*(13–14), 1353–1376. doi:10.1080/0267257X.2014.925958

Pretty, J. (2011). Interdisciplinary progress in approaches to address social-ecological and ecocultural *systems. Environmental Conservation*, *38*(2), 127–139. doi:10.1017/S0376892910000937

Priscilla, D., & Pandi. (2023). Agriculture-based Automation with Recommendation Systems based on AI Models. *2023 Third International Conference on Artificial Intelligence and Smart Energy (ICAIS)*, (pp. 1582–1589). IEEE.

Prothero, A., Dobscha, S., Freund, J., Kilbourne, W. E., Luchs, M. G., Ozanne, L. K., & Thøgersen, J. (2011). Sustainable consumption: Opportunities for consumer research and public policy. *Journal of Public Policy & Marketing*, *30*(1), 31–38. doi:10.1509/jppm.30.1.31

Puneet, K., & Shagun, P. (2009). Operation Research Endorsing The Social Transformation with Management and Information Technology Convergent. *IIMT Business Review*, *1*(1), 47–52.

Punla, N., Regalado, D., & Navarro, R. M. (2022). Perceived Social Support as Moderator Between Self-Stigma and Help-Seeking Attitudes. *Globus An International Journal of Medical Science, Engineering and Technology*, *11*(1), 87-98.

Radder, L., & Han, X. (2015). An examination of the museum experience based on pine and Gilmore's experience economy realms. *Journal of Applied Business Research*, *31*(2), 455–470. doi:10.19030/jabr.v31i2.9129

Rahim, A., Eladwiah, R., & Abdul Rahman, A. (2013). Resource-based framework of green IT capability toward firms 'competitive advantage. *Resource (Saint Joseph, Mich.)*, *6*, 18–2013.

Ramaswamy, V. (2009). Co-Creation of Value: Towards an Expanded Paradigm of Value Creation. *Marketing Review St. Gallen*, *6*(6), 11–17. doi:10.100711621-009-0085-7

Rands, M. R., Adams, W. M., Bennun, L., Butchart, S. H., Clements, A., Coomes, D., Entwistle, A., Hodge, I., Kapos, V., Scharlemann, J. P. W., Sutherland, W. J., & Vira, B. (2010). Biodiversity conservation: Challenges beyond 2010. *Science*, *329*(5997), 1298–1303. doi:10.1126cience.1189138 PMID:20829476

Ranjan, V., Reddy, M. V., Irshad, M., & Joshi, N. (2020). The internet of things (IOT) based smart rain water harvesting system. *2020 6th International Conference on Signal Processing and Communication (ICSC)*, (pp. 302–305). IEEE.

Rappaport, R. A. (2000). *Pigs for the Ancestors: Ritual in the Ecology of a New Guinea People.* Waveland Press.

Raworth, K. (2012). A safe and just space for humanity: Can we live within the doughnut? Oxfam Policy and Practice. *Climate Change and Resilience*, *8*, 1–26.

Raza, S., Orooji, Y., Ghasali, E., Hayat, A., Karimi-Maleh, H., & Lin, H. (2023). Engineering approaches for CO2 converting to biomass coupled with nanobiomaterials as biomediated towards circular bioeconomy. *Journal of CO2 Utilization, 67*. doi:10.1016/j.jcou.2022.102295

Razavi, S. (2022). Making the Right to Social Security a Reality for All Workers. *The Indian Journal of Labour Economics : the Quarterly Journal of the Indian Society of Labour Economics*, *65*, 269–294. doi:10.100741027-022-00378-6 PMID:35911186

Redman, C. (2014). Should sustainability and resilience be combined or remain distinct pursuits? *Ecology and Society*, *19*(2), 190–202. doi:10.5751/ES-06390-190237

Reguera, A. (2009) Empresarios de ayer, de hoy y de siempre. Un recorrido latinoamericano por sus formas espacio-temporales *América Latina en la historia económica, 32*. https://www.scielo.org.mx/scielo.php?script=sci_arttext&pid=S1405- 22532009000200003&lng=es&nrm=iso

Reich, D. (2018). *Who We Are and How We Got Here: Ancient DNA and the New Science of the Human Past*. Oxford University Press.

Reid, H., Sahlén, L., Stage, J., & Macgregor, J. (2008). Climate change impacts on Namibia's natural resources and economy. *Climate Policy*, *8*(5), 452–466. doi:10.1080/14693062.2008.9685709

Reisinger, A., & Clark, H. (2018). How much do direct livestock emissions contribute to global warming? *Global Change Biology*, *24*(4), 1749–1761. doi:10.1111/gcb.13975 PMID:29105912

Renwick, D. W. S., Redman, T., & Maguire, S. (2008). Green human resource management: A review and research agenda. *International Journal of Management Reviews*, *10*(1), 1–18. doi:10.1111/j.1468-2370.2011.00328.x

Reynoso, C. (1995b). Hacia la perfección del consenso: Los lugares comunes de la antropología. *Intersecciones*, *1*, 51–72.

Ricketts, K. D., Palmer, J., Navarro-Garcia, J., Lee, C., Dominik, S., Barlow, R., Ridoutt, B., & Richards, A. (2023). Bridging organisational discourse and practice change: Exploring sustainable procurement for Australian beef. *Sustainability Accounting. Management and Policy Journal*, *14*(2), 265–288.

Ridley-Duff, R., & Bull, M. (2019). The social and economic impact of social enterprises: A systematic review of the evidence. *Social Enterprise Journal*, *15*(1), 1–25.

Riegel, E. C. (1949). *Flight from Inflation: The Monetary Alternative*. Devin-Adair Company.

Riemer, F. J., Lapan, S. D., & Quartaroli, M. T. (2012). *Qualitative research: an introduction to methods and designs*. Jossey-Bass.

Rmeity, R. (2021). *Estimating the impact of the Banking Industry on the Financial Crisis*. [MBA Thesis, Lebanese International University].

Roberts, C. M. (2007). *The unnatural history of the sea*. Island Press.

Roca, A. (2008). *La comprensión de la tecnología en la antropología*. Actas VII ESOCITE UFRJ, Jornadas Latino-Americanas de Estudos Sociais das Ciências e das Tecnologías.

Rodionova, M. V., Poudyal, R. S., Tiwari, I., Voloshin, R. A., Zharmukhamedov, S. K., Nam, H. G., Zayadan, B. K., Bruce, B. D., Hou, H. J., & Allakhverdiev, S. I. (2017). Biofuel production: Challenges and opportunities. *International Journal of Hydrogen Energy*, *42*(12), 8450–8461. doi:10.1016/j.ijhydene.2016.11.125

Rodríguez, Y. (2020). *Mitigación de la pandemia en los sistemas de seguridad social de Ecuador. Dirección de Proyectos e Investigación.* México D.F: Conferencia Interamericana de Seguridad Social (CISS). https://bit.ly/3ChO3ht

Rodriguez-Anton, J. M., Rubio-Andrada, L., Celemín-Pedroche, M. S., & Alonso-Almeida, M. D. M. (2019). Analysis of the relations between circular economy and sustainable development goals. *International Journal of Sustainable Development and World Ecology*, *26*(8), 708–720. doi:10.1080/13504509.2019.1666754

Rodríguez, C., Florido, C., & Jacob, M. (2020). Circular economy contributions to the tourism sector: A critical literature review. *Sustainability (Basel)*, *12*(11), 4338. doi:10.3390u12114338

Roethlisberger, F., & Dickson, W. (1939). *Management and the Worker*. Harvard Univ. Press.

Rofman, R., Apella, I., & Vezza, E. (2015). Expanding economic protection to the elderly in Latin America. *International Social Security Review*, *68*(1), 1–22. doi:10.1111/issr.12057

Rogers, E. M. (2003). Elements of diffusion. *Diffusion of Innovations*, *5*, 1–38.

Rogoff, K. (2020). *Mapping the COVID-19 Recession*. Project Syndicate. https://www.project-syndicate.org/commentary/mapping-covid19-global-recession-worst-in-150-years-by-kenneth-rogoff-2020-04

Roni, S., Merga, M., & Morris, J. (2020). *Conducting quantitative research in education*. Springer. doi:10.1007/978-981-13-9132-3

Røseth, T. (2014). Russia's China Policy in the Artic. *Strategic Analysis, 38*(6). doi:10.1080/09700161.2014.952942

Rowe, D. B. (2011). Green Roofs as a means of pollution abatement. *Environmental Pollution*, *159*(8-9), 2100–2110. doi:10.1016/j.envpol.2010.10.029 PMID:21074914

Rumeng, T., Tong, W., Ying, S., & Yanpu, H. (2021). Intelligent energy saving solution of 5G base station based on artificial intelligence technologies. *2021 IEEE International Joint EMC/SI/PI and EMC Europe Symposium*, (pp. 739–742). IEEE. 10.1109/EMC/SI/PI/EMCEurope52599.2021.9559261

Sabin, M. (2010). Fair Tontine Annuity. *The Social Science Research Network*, 3-67. doi.org/ doi:10.2139/ssrn.1579932

Saha, N., Sáha, T., & Sáha, P. (2022). Entrepreneurial University and Social Innovation Ecosystems: Do They Support HEIs' Knowledge-Based Economic Development? In Artificiality and Sustainability in Entrepreneurship (pp. 215-240). Springer.

Saidani, M., Yannou, B., Leroy, Y., Cluzel, F., & Kendall, A. (2019). A taxonomy of circular economy indicators. *Journal of Cleaner Production*, *207*, 542–559. doi:10.1016/j.jclepro.2018.10.014

Salame, R. (2022). Lebanese lira lost 82 percent of its purchasing power in two-year period, UN data shows. *L'Orientle Jour.* https://today.lorientlejour.com/article/1289780/lebanese-lira-lost-82-percent-of-its-purchasing-power-in-two-year-period-un-data-shows.html

Salimath, M., & Jones, R. III. (2011). Population ecology theory: Implications for sustainability. *Management Decision*, *49*(6), 874–910. doi:10.1108/00251741111143595

Salisbury, T., & Milevsky, A. (2016). Equitable Retirement Income Tontines: Mixing Cohorts Without Discriminating. Astin Bulletin. *The Journal of the IAA*, *46*(3), 571–604. doi:10.1017/asb.2016.19

Sangtani, R., Nogueira, R., Yadav, A. K., & Kiran, B. (2023). Systematizing Microbial Bioplastic Production for Developing Sustainable Bioeconomy: Metabolic Nexus Modeling, Economic and Environmental Technologies Assessment. *Journal of Polymers and the Environment*, *31*(7), 2741–2760. doi:10.100710924-023-02787-0 PMID:36811096

Sangwan, D. S. (2022). Impact of Russia Ukraine War on Indian Economy. Globus An International Journal of Medical Science. *Engineering and Technology*, *11*(1), 36–39.

Sanquist, T. F. (2008). Human factors and energy use. *HFES Bulletin*, *51*(11), 1–3.

Santiago, C. E. (1987). The impact of foreign direct investment on export structure and employment generation. *World Development*, *15*(3), 317–328. doi:10.1016/0305-750X(87)90016-7

Santoro, G., Vrontis, D., Thrassou, A., & Dezi, L. (2018). The Internet of Things: Building a knowledge management system for open innovation and knowledge management capacity. *Technological Forecasting and Social Change*, *136*, 347–354. doi:10.1016/j.techfore.2017.02.034

Santoro, V. (2021). Model of health and social inequalities in Ecuador: Progress and challenges. A systematic review. *Journal of Public Health*, *29*, 775–785. doi:10.100710389-019-01183-0

Sapkota, P., & Bastola, U. (2017). Foreign direct investment, income, and environmental pollution in developing countries: Panel data analysis of Latin America. *Energy Economics*, *64*, 206–212. doi:10.1016/j.eneco.2017.04.001

Satya, A., Cheah, W., Yazdi, S., Cheng, Y., Khoo, K., Vo, D., Bui, X. D., Vithanage, M., & Show, P. (2023). Progress on microalgae cultivation in wastewater for bioremediation and circular bioeconomy. *Environmental Research*, *218*, 114948. doi:10.1016/j.envres.2022.114948 PMID:36455634

Saud, S., Chen, S., Haseeb, A., Khan, K., & Imran, M. (2019). The Nexus between Financial Development, Income Level, and Environment in Central and Eastern European Countries: A Perspective on Belt and Road Initiative. *Environmental Science and Pollution Research International*, *26*(16), 16053–16075. doi:10.100711356-019-05004-5 PMID:30968296

Saule, M. (2004) *Rural Tourism as a tool for Local Development, Diagnosis of the French area of Haute-Corrèze.* [Master's thesis, the Bournemouth University, UK].

Saunders, M. (2019). *Research Methods for Business Students* (8th ed.). Pearson.

Sauvé, S., Bernard, S., & Sloan, P. (2016). Environmental sciences, sustainable development and circular economy: Alternative concepts for trans-disciplinary research. *Environmental Development*, *17*, 48–56. doi:10.1016/j.envdev.2015.09.002

Savickiene, J. & Miceikiene, A. (2018), Sustainable economic development assessment model for family farms. *AgriculturalEconomics – Czech*, *64*, 527–535. . doi:10.17221/310/2017-AGRICECON

Saxenian, A. (1994). *Regional Advantage: culture and competition in Silicon Valley and Route 128.* Harvard Business Press.

Schaefer, G. (2016). *The Future of Money: Bitcoin and Other Cryptocurrency Technologies Are a Way of Life in This Small Swiss Town.* St. Martin's Press.

Scherer, F. M. (1980). *Industrial Market Structure and Economic Performance*. Rand McNally.

Schieber, S. (2017). Generational Equity and Social Security Financing Reform. *The Journal of Retirement*, *5*(1), 12–31. doi:10.3905/jor.2017.5.1.012

Schlapfer, A., & Marinova, D. (2001). *Local Innovation Systems: nature, importance and role.* Conference Proceedings: International Summer Academy on Technological Studies: user involvement in technological innovation. Deutschlandsberg. Austria.

Schludi, M. (2005). *The Reform of Bismarckian: A Comparison of Pension Politics in Austria, France, Germany, Italy and Sweden* (Vol. 6). Amsterdam University Press. https://bit.ly/3rHymLz

SchmidtC.SchneiderY.SteffenS.StreitzD. (2020). Capital misal-location and innovation. SSRN 3489801. doi:10.2139/ssrn.3489801

Schmidt, U. (2016). Insurance Demand Under Prospect Theory: A Graphical Analysis. *The Journal of Risk and Insurance, 83*(1), 77–89. doi:10.1111/jori.12098

Schmitt, B. (1999). Experiential Marketing. *Journal of Marketing Management, 15*(1-3), 53–67. doi:10.1362/026725799784870496

Schnarch, B. (2004). Ownership, control, access, and possession (OCAP) or self-determination applied to research: A critical analysis of contemporary First Nations research and some options for First Nations communities. *Journal of Aboriginal Health, 1*, 80–97.

Schnegg, M., O'Brian, C. I., & Sievert, I. J. (2021). It's our fault: A global comparison of different ways of explaining climate change. *Human Ecology: an Interdisciplinary Journal, 49*(3), 327–339. doi:10.100710745-021-00229-w

Schröder, P., Lemille, A., & Desmond, P. (2020). Making the circular economy work for human development. *Resources, Conservation and Recycling, 156*, 104686. doi:10.1016/j.resconrec.2020.104686

Schroeder, P., Anggraeni, K., & Weber, U. (2019). The relevance of circular economy practices to the sustainable development goals. *Journal of Industrial Ecology, 23*(1), 77–95. doi:10.1111/jiec.12732

Schuster, K., Grob, K., Vossen, R., Richert, A., & Jeschke, S. (2016). Preparing for Industry 4.0 - Collaborative Virtual Learning Environments in Engineering Education. In Automation, Communication and Cybernetics in Science and Engineering 2015/1016 (p. 425). Springer.

Scudder, T. (1973). The Human Ecology of Big Projects: River Basin Development and Resettlement. [Section 3 in Montgomery, Bennett, and Scudder (1973)] In B. J. Siegel, A. R. Beals, & S. A. Tyler (Eds.), *Annual Review of Anthropology*. Annual Reviews Inc.

Scudder, T., & Colson, E. (1982). From Welfare to Development: A Conceptual Framework for the Analysis of Dislocated People in Involuntary. In A. Hansen & A. Oliver-Smith (Eds.), *Migration and Resettlement: The Problems and Responses of Dislocated People* (pp. 267–287). Westview Press, Westview Special Study.

Sebastian, I., Mocker, M., Ross, J., Moloney, K., Beath, C., & Fonstad, N. (2017). How Big Companies Navigate Digital Transformation. *MIS Quarterly Executive, 16*(3), 197–213.

Segismundo, M. & Dayao, E. P. (2021). Employability and Productivity of The 21st Century Workforce: A Tracer Study of LCUP-CBEA Graduates AY 2015 to 2018. *Cosmos Journal of Engineering & Technology, 11*(2), 39-47.

Sekaran, U. (2000). *Research Methods for Business: A skill-building approach* (4th ed.). Wiley.

Seker, F., Ertugrul, H. M., & Cetin, M. (2015). The impact of foreign direct investment on environmental quality: A bounds testing and causality analysis for Turkey. *Renewable & Sustainable Energy Reviews, 52*, 347–356. doi:10.1016/j.rser.2015.07.118

Sele, K., & Grand, S. (2016). Unpacking the dynamics of ecologies of routines: Mediators and their generative effects in routine interactions. *Organization Science, 27*(3), 722–738. doi:10.1287/orsc.2015.1031

Seroka-Stolka, O. (2016). Green initiatives in environmental management of logistics companies. *Transportation Research Procedia, 16*, 483–489. doi:10.1016/j.trpro.2016.11.045

Shafi Que, M., Kim, R. & Rafi Q, M. (2018). Green roof benefits, opportunities, and challenges– A review. *Renewable and Sustainable Energy Reviews, 90*, 757-773.

Shahrul, N. S., & Normah, M. (2015). Digital version newspaper: Implication towards printed newspaper circulation in Malaysia. *Malaysian Journal of Communication, 31*(2), 687–701.

Shahzad, F., Lu, J., & Fareed, Z. (2019). Does firm life cycle impact corporate risk taking and performance? *Journal of Multinational Financial Management, 51*, 23–44. doi:10.1016/j.mulfin.2019.05.001

Sharma, T. (2019). Melting Arctic Sea Ice Opens New Maritime Shipping Route. *Global Security Review*. https://globalsecurityreview.com/

Sharma, R., & Malaviya, P. (2023). Ecosystem services and climate action from a circular bioeconomy perspective. *Renewable & Sustainable Energy Reviews, 175*, 113164. Advance online publication. doi:10.1016/j.rser.2023.113164

Shashank, A., Vincent, R., Sivaraman, A. K., Balasundaram, A., Rajesh, M., & Ashokkumar, S. (2021). Power Analysis of Household Appliances using IoT. *2021 International Conference on System, Computation, Automation and Networking (ICSCAN)*, (pp. 1–5). 10.1109/ICSCAN53069.2021.9526428

Shaw, C., & Ivens, J. (2002). *Building Great Customer experiences*. Palgrave Macmillan. doi:10.1057/9780230554719

Shevchenko, A., & Petrenko, O. (2020). Current state of micro and small agribusiness in Ukraine. *Agricultural and Resource Economics, 6*(1), 146–160. doi:10.51599/are.2020.06.01.10

Shi, X., Jiang, H., Li, H., & Wang, Y. (2020). Upgrading port-originated maritime clusters: Insights from Shanghai's experience. *Transport Policy, 87*, 19–32. doi:10.1016/j.tranpol.2019.11.002

Shoeb, A., & Nisar, T. (2015) Green Human Resource Management: Policies and practices, *Cogent Business & Management, 2* (1), 1030817-103. https://www.tandfonline.com/doi/full/10.1080/23311975.2015.1030817 doi:10.1080/23311975.2015.1030817

Show, P., Ogino, C., & Ibrahim, M. (n.d.). *Biotechnology for Sustainability and Social Well Being*. Google Books. https://books.google.com.pk/books?id=YD1AEAAAQBAJ&pg=PA156&lpg=PA156&dq=Biotechnology+is+a+key+component+of+the+bioeconomy+and+has+numerous+applications+in+various+fields+such+as+medicine,+agriculture,+and+industry.+Biotechnology+involves+the+use+of+biological+systems+and+living+organisms+to+produce+new+products+or+improve+existing+ones.+Biotechnology+has+the+potential+to+transform+numerous+industries+by+enabling+the+development+of+new+bio-based+products+and+processes.+One+of+the+most+promising+applications+of+biotechnology+is+in+the+field+of+medicine,+enabling+the+development+of+new+therapies+and+treatments+for+various+diseases.+Biotechnology+also+has+numerous+applications+in+agriculture,+such+as+the+development+of+genetically+modified+crops+that+are+resistant+to+pests+and+diseases.+However,+biotechnology+also+poses+several+challenges+that+need+to+be+addressed+through+innovation,+such+as+the+potential+impact+of+GMOs+on+human+health+and+the+environment.&source=bl&ots=yvMWdS4WST&sig=ACfU3U0-rF_RwVp1dmiY8xmcmsN

Shukla, D. A. (2022). Urgency of Current Times – Peace and Non-Violence. *Globus Journal of Progressive Education, 12*(1), 61–63.

Shukla, S. N., & Champaneria, T. A. (2017). Survey of various data collection ways for smart transportation domain of smart city. *2017 International Conference on I-SMAC (IoT in Social, Mobile, Analytics and Cloud) (I-SMAC)*, (pp. 681–685). 10.1109/I-SMAC.2017.8058265

Shukman, D. (2021). *Dünya küresel ısınma eşiği olan 1.5 C dereceye 'gelecek beş yıl içinde ulaşabilir [The world may reach 1.5°C, the global warming threshold, within the next five years]*. BBC. https://www.bbc.com/turkce/haberler-dunya-57264316

Siddula, S. S., Babu, P., & Jain, P. C. (2018). Water level monitoring and management of dams using IoT. *2018 3rd International Conference On Internet of Things: Smart Innovation and Usages (IoT-SIU)*, (pp. 1–5). IEEE.

Sigman, H. (1996). The effects of hazardous waste taxes on waste generation and disposal. *Journal of Environmental Economics and Management, 30*(2), 199–217. doi:10.1006/jeem.1996.0014

Signorelli, A. (1999). Antropología Urbana. Barcelona: Anthropos: UAM.

Simberloff, D. (2018). Nature, Culture, and Natureculture: The Role of Nonnative Species in Biocultures. In R. Rozzi, R. H. May Jr, F. S. Chapin III, F. Massardo, M. C. Gavin, I. J. Klaver, A. Pauchard, M. A. Nunez, & D. Simberloff (Eds.), *From Biocultural Homogenization to Biocultural Conservation* (pp. 207–218). Nature. doi:10.1007/978-3-319-99513-7_12

Simpson, K. H., de Vries, F. P., Dallimer, M., Armsworth, P. R., & Hanley, N. (2022). Ecological and economic implications of alternative metrics in biodiversity offset markets. *Conservation Biology, 36*(5), e13906. doi:10.1111/cobi.13906 PMID:35288986

Singh, G., Gupta, R. and Agarwal, Nidhi (2020). Economic Re-Engineering: COVID-19. *International Journal on Emerging Technologies, 11*(4), 01-05.

Singh, M. (2019). Disaster Management: Managing The Risk of Environmental Calamity. *Cosmos An International Journal Art & Higher Education*, *8*(2), 25–27.

Singh, S. K., Del Giudice, M., Chierici, R., & Graziano, D. (2020). Green innovation and environmental performance: The role of green transformational leadership and green human resource management. *Technological Forecasting and Social Change, 150*, 119762. doi:10.1016/j.techfore.2019.119762

Singh, T., Singh, D., & Bedi, S. S. (2021). Monitoring and Detecting Plant Diseases Using Cloud-Based Internet of Things. In *Integration and Implementation of the Internet of Things Through Cloud Computing* (pp. 217–235). IGI Global. doi:10.4018/978-1-7998-6981-8.ch011

Siu, N. Y., Zhang, T. J., & Kwan, H. Y. (2021). Reference effects and customer engagement in a museum visit. *International Journal of Contemporary Hospitality Management, 34*(2), 482–508. doi:10.1108/IJCHM-02-2021-0208

Sjöstedt, G. (1977). The exercise of international civil power: A framework for analysis. *Cooperation and Conflict, 12*(1), 21–39. doi:10.1177/001083677701200102

Skydan, O., Nykolyuk, O., Pyvovar, P., & Martynchuk, I. (2020). Methodological approach to the evaluation of agricultural business system flexibility. *Management Theory and Studies for Rural Business and Infrastructure Development, 41*(4), 444–462. doi:10.15544/mts.2019.36

Slater, D., & Tonkiss, F. (2001). *Market society: Markets and modern social theory*. Polity Press.

Sneideriene, A., & Rugine, H. (2019). Theoretical approach on the green technologies development. *Regional Formation and Development Studies, 2*(28), 124–134.

Social Security Administration. (2020). *Social Security Programs Throughout the World: The Americas, 2019. Office of Retirement and Disability Policy*. Washington: SSA Publication. https://bit.ly/3rEYDtO

Social Security Law. (2001). *Asamblea Nacional Constituyente*. SSL. http://www.oas.org/juridico/pdfs/mesicic4_ecu_segu.pdf

Sole, R. V., & Bascompte, J. (1995). Spatial Self Organization Self-Organization in complex Ecosystems Princeton University Press.

Sonker, J., Kumar, S., & Pal, S. (2021). An Implementation of Energy Awareness Management in Heritage Hotels of India. *Turkish Online Journal of Qualitative Inquiry, 12*(7).

Sørensen, F., & Bærenholdt, J. O. (2020). Tourist practices in the circular economy. *Annals of Tourism Research, 85*, 103027. doi:10.1016/j.annals.2020.103027

Spee, P., Jarzabkowski, P., & Smets, M. (2016). The influence of routine interdependence and skillful accomplishment on the coordination of standardizing and customizing. *Organization Science, 27*(3), 759–781. doi:10.1287/orsc.2016.1050

Spirn, A. W. (2011). *Ecological urbanism: a framework for the design of resilient cities.* Anne Whistonspirn. http://annewhistonspirn.com/pdf/spirn_ecological_urbanism-2011.pdf

Srinivasan, A., Baskaran, K., & Yann, G. (2019). IoT based smart plug-load energy conservation and management system. *2019 IEEE 2nd International Conference on Power and Energy Applications (ICPEA)*, (pp. 155–158). IEEE.

Srividhya, S., & Sankaranarayanan, S. (2020). IoT–fog enabled framework for forest fire management system. *2020 Fourth World Conference on Smart Trends in Systems, Security and Sustainability (WorldS4)*, (pp. 273–276). IEEE. 10.1109/WorldS450073.2020.9210328

Staff, C. B. (2022). In-depth Q&A: The IPCC's sixth assessment on how to tackle climate change. *Carbon Brief.* https://www.carbonbrief.org/in-depth-qa-the-ipccs-sixth-assessment-on-how-to-tackle-climate-change/

Stamm, I. (2017). The Human Right to Social Security and Its Impact on Socio-Political Action in Germany and Finland. *Journal of Human Rights and Social Work, 2*(1), 25–33. doi:10.100741134-017-0030-7

Stankevičius, A., Novikovas, A., Bakaveckas, A., & Petryshyn, O. (2020). EU waste regulation in the context of the circular economy: Peculiarities of interaction. *Entrepreneurship and Sustainability Issues, 8*(2), 533–545. doi:10.9770/jesi.2020.8.2(32)

Stanković, J. J., Janković-Milić, V., Marjanović, I., & Janjić, J. (2021). An integrated approach of PCA and PROMETHEE in spatial assessment of circular economy indicators. *Waste Management (New York, N.Y.), 128*, 154–166. doi:10.1016/j.wasman.2021.04.057 PMID:33991754

Stanković, J. J., Popović, Ž., & Marjanović, I. (2021a). Assessing smartness and urban development of the European cities: An integrated approach of Entropy and VIKOR. In Y. I. Topcu, Ö. Özaydın, Ö. Kabak, & Ş. Önsel Ekici (Eds.), *Multiple Criteria Decision Making: Beyond the Information Age 25, (69–97).* Springer., doi:10.1007/978-3-030-52406-7_3

Stavroulakis, P. J., & Papadimitriou, S. (2017). Situation analysis forecasting: The case of European maritime clusters. *Maritime Policy & Management, 44*(6), 779–789. doi:10.1080/03088839.2017.1330560

Stavroulakis, P. J., Papadimitriou, S., Tsioumas, V., Koliousis, I. G., Rizad, E., & Tsirikou, F. (2020b). Exploratory spatial analysis of maritime clusters. *Marine Policy, 120*, 104125. Advance online publication. doi:10.1016/j.marpol.2020.104125

Stavroulakis, P. J., Papadimitriou, S., Tsioumas, V., Koliousis, I., Riza, E., & Kontolatou, E. O. (2020a). Strategic competitiveness in maritime clusters. *Case Studies on Transport Policy, 8*(2), 341–348. doi:10.1016/j.cstp.2019.10.008

Stavroulakis, P. J., Papadimitriou, S., & Tsirikou, F. (2021). Perceptions of competitiveness for maritime clusters. *Ocean and Coastal Management, 205*(1), 105546. Advance online publication. doi:10.1016/j.ocecoaman.2021.105546

Steiner, F. R. (2016). Human ecology: how nature and culture shape our world. The Island Press: Washington, D.C.

Steiner, F. R. (1991). *The living landscape: an ecological approach to landscape planning*. McGraw-Hill.

Steiner, F. R. (2001). *Human ecology: following nature's lead.* Island Press.

Steinzor, R. (2012). The case for abolishing centralized white house regulatory review. *Michigan Journal of Environmental & Administrative Law*, *1*(1.1), 209. doi:10.36640/mjeal.1.1.case

Stiglitz, J. (2015). *Globalization and its discontents*. Penguin.

Stokols, D. (1996). Translating social ecological theory into guidelines for community health promotion. *American Journal of Health Promotion, 10*(4), 282-98. doi:10.4278/0890-1171-10.4.282

Stokols, D. (1992). Establishing and maintaining healthy environments: Toward a social ecology of health promotion. *The American Psychologist*, *47*(1), 6–22. doi:10.1037/0003-066X.47.1.6 PMID:1539925

Stone, M. (2003). Is sustainability for development anthropologists? *Human Organization*, *62*(2), 93–99. doi:10.17730/humo.62.2.9cfam6hd6e04yac8

Stoner, A. M. (2021). Things are getting worse on our way to catastrophe: Neoliberal environmentalism, repressive desublimation, and the autonomous ecoconsumer. *Critical Sociology*, *47*(May), 491–506. doi:10.1177/0896920520958099

Stovin, V., Vesuviano, G., & Kasmin, H. (2012). The hydrological performance of a green roof test bed under UK climatic conditions. *Journal of Hydrology (Amsterdam)*, *414*, 148–161. doi:10.1016/j.jhydrol.2011.10.022

Strang, V. (2017). The Gaia complex: Ethical Challenges to an Anthropocentric 'Common Future. In B. Marc & L. Jerome (Eds.), *The Anthropology of Sustainability* (pp. 255–283). Palgrave. doi:10.1057/978-1-137-56636-2_12

Strohm, L. A. (1993). The environmental politics of the international waste trade. *Journal of Environment & Development*, *2*(2), 129–153. doi:10.1177/107049659300200209

Stubbs, W., & Cocklin, C., (2008). Conceptualizing a sustainability business model. *Organizational Environment, 21* (2), 103e127.

Studies: EGOS Le Boterf, G. (1999). *L'ingénierie des compétences*. París: Éditions d'Organisation.

Suchman, M. (1995). Managing legitimacy: Strategic and institutional approaches. *Academy of Management Review*, *20*(3), 571–610. doi:10.2307/258788

Sueyoshi, T., & Goto, M. (2012). Efficiency-based rank assessment for electric power industry: A combined use of data envelopment analysis (DEA) and DEA-discriminant analysis (DA). *Energy Economics*, *34*(3), 634–644. doi:10.1016/j.eneco.2011.04.001

Sundbo, J. (1998). *The Theory of Innovation: entrepreneurs, technology, and strategy*. Edward Elgar.

Sung, P. L., Hsiao, T. Y., Huang, L., & Morrison, A. M. (2021). The influence of green trust on travel agency intentions to promote low-carbon tours for the purpose of sustainable development. *Corporate Social Responsibility and Environmental Management*, *28*(4), 1185–1199. doi:10.1002/csr.2131

Sun, X., Zhou, X., Chen, Z., & Yang, Y. (2020b). Environmental efficiency of electric power industry, market segmentation and technologi-cal innovation: Empirical evidence from China. *The Science of the Total Environment*, *706*, 135749. doi:10.1016/j.scitotenv.2019.135749 PMID:31940733

Sustainability for all, (n.d.).*The Blue Economy: Benefits Of Investing In Ocean Conservation*, Active Sustainability. https://www.activesustainability.com/water/blue-economy/?_adin=02021864894

Sutton, R. (2014). Aesthetics for green roofs and green walls. *The Journal of Living Architecture*, *1*(2), 1–20. doi:10.46534/jliv.2014.01.02.001

Taagepera, R. (2008). *Making social sciences more scientific: The need for predictive models.* Oxford University Press. doi:10.1093/acprof:oso/9780199534661.001.0001

Tabak, J. (2009). *Biofuels.* Infobase Publishing.

Tainter, J. A. (1988). *The Collapse of Complex Societies.* Cambridge University Press.

Tainter, J. A. (2008). Collapse and Sustainability: Rome, the Maya, and the Modern World. *Archeological Papers of the American Anthropological Association, 24*(1), 201–214. doi:10.1111/apaa.12038

Taklo, S. K., & Tooranloo, H. S. & Shahabaldini parizi, Z. (. (2020). Green Innovation: A Systematic Literature Review. *Journal of Cleaner Production, 2020*(7). doi:10.1016/j.jclepro.2020.122474

Tang, C. F., & Tan, B. W. (2015). The impact of energy consumption, income and foreign direct investment on carbon dioxide emissions in Vietnam. *Energy, 79*, 447–454. doi:10.1016/j.energy.2014.11.033

Tan, K. S., & Eze, U. C. (2008). An empirical study of internet-based ICT adoption among Malaysian SMEs. *Communications of the IBIMA, 1*, 1–12.

Tan, P. Y., Wong, N. H., Tan, T. K., & Wong, K. W. (2019). Green roof innovation in Singapore: Policies, performance, and potential. *Journal of Environmental Management, 233*, 128–135. doi:10.1016/j.jenvman.2018.12.060

Tan, R., Shi, Y., Fan, Y., Zhu, W., & Wu, T. (2022). Energy saving technologies and best practices for 5G radio access network. *IEEE Access : Practical Innovations, Open Solutions, 10*, 1–1. doi:10.1109/ACCESS.2022.3174089

Tarantino-Curseri, S. (2018). Antropología, etología, sociología, y psicología, 4 disciplinas útiles en el ámbito empresarial. *Management Review, 3*(3). doi:10.18583/umr.v3i3.126

Teece, D. J. (2016). Dynamic capabilities and entrepreneurial management in large organizations: Toward a theory of the (entrepreneurial) firm. *European Economic Review, 86*, 202–216. doi:10.1016/j.euroecorev.2015.11.006

Temesgen, A., Storsletten, V., & Jakobsen, O. (2021). Circular economy – reducing symptoms or radical change? *Philosophy of Management, 20*(March), 37–56. doi:10.100740926-019-00112-1

Thatte, A. A. (2007). *Competitive advantage of a firm through supply chain responsiveness and SCM practices* [Doctoral dissertation, The University of Toledo].

The Commonwealth. (n.d.). *Action Group on Sustainable Blue Economy.* The Commonwealth. https://thecommonwealth.org/bluecharter/sustainable-blue-economy#:~:text=The%20blue%20economy%20embraces%20economic,of%20poverty%20and%20climate%20change

Thompson, L. (1998). *Older and Wiser- The Economics of Public Pensions.* The Urban Institute Press. http://biblioteca.ciess.org/adiss/g528/f3

Thompson, G. F., & Steiner, F. R. (Eds.). (1997). *Ecological design and planning.* John Wiley & Sons, Inc.

Tillman, H. (2018). The Polar Silk Road: China's New Frontier of International Cooperation. *China Quarterly of International Strategic Studies 4*(3), 345–62. https://www.worldscientific.com/

TNFD. (2023). *TNFD Integrated Framework Version 0.4.* TNFD. https://framework.tnfd.global/wp-content/uploads/2023/03/23-23882-TNFD_v0.4_Integrated_Framework_v7.pdf

Tobey, J. A. (1990). The Effects of Domestic Environmental Policies on Patterns of World Trade: An Empirical Test. *Kyklos, 43*(2), 191–209. doi:10.1111/j.1467-6435.1990.tb00207.x

Toffler, A. (1970). *Future shock.* Random House.

Toledo, A. (2016). Seguro de desempleo para Ecuador: características y sostenibilidad. *Analítika, 12*, 7-45. https://bit.ly/3fVkydS

Toledo, V. M. (1997) Sustainable development at the village community level: a Third World perspective in: Fraser Smith (Ed.) Environmental Sustainability. St. Lucie Press. doi:10.1201/9781420048742.ch11

Tolliver, C., Fujii, H., Keeley, A. R., & Managi, S. (2021). Green innovation and finance in Asia. *Asian Economic Policy Review*, *16*(1), 67–87. doi:10.1111/aepr.12320

Tolstykh, T., Gamidullaeva, L., Shmeleva, N., Gromov, S., & Ermolenko, A. (2022). Megapolis as a Symbiosis of Socio-Economic Ecosystems: The Role of Collaboration. *Journal of Open Innovation*, *8*(3), 126. Advance online publication. doi:10.3390/joitmc8030126

Tomas, M. (2021). (pp. 176–195). The Era of the Dragon: How China Is Preparing to be the World's Top Economy in 21 Century. IGI Global. doi:10.4018/978-1-7998-7568-0.ch009

TomashukI. V.BaldynyukV. M. (2021). Identification of problems and prospects of rural infrastructure development of Ukraine *Economic sciences*, *13* (100), 58-70. doi:10.24412/2520-6990-2021-13100-58-70

Tornatzky, L. G., & Klein, K. J. (1982). Innovation characteristics and innovation adoption-implementation: A meta-analysis of findings. *IEEE Transactions on Engineering Management*, *29*(1), 28–45. doi:10.1109/TEM.1982.6447463

Torres, W. (2021). *El IESS da más beneficios en pensiones a las personas de altos ingresos*. https://bit.ly/3ysXysT

Tran, T., Goto, H., & Matsuda, T. (2021). The impact of China's tightening environmental regulations on international waste trade and logistics. *Sustainability (Basel)*, *13*(2), 987. doi:10.3390u13020987

Trawick, P., & Alf, H. (2015). Revisiting the Image of Limited Good: On Sustainability, Thermodynamics, and the Illusion of Creating Wealth. *Current Anthropology*, *56*(1), 1–27. doi:10.1086/679593

Trigedga, H. (2013). Biodiversity offsetting: Problematisation of of an emerging governance regime. *Accounting, Auditing & Accountability Journal*, *26*(5), 806–832. doi:10.1108/AAAJ-02-2013-1234

Trist, E. L. (1976). *A Concept of Organizational Ecology: an invited address to the three Melbourne universities.*

Trujillo, M. L. N. (2021). Notes for a critical and ecological view of patriarchal capitalism in the web of life. *Capital and Class*, *45*(March), 21–32. doi:10.1177/0309816820929115

Truong, Y., Mazloomi, H., & Berrone, P. (2021). Understanding the impact of symbolic and substantive environmental actions on organizational reputation. *Industrial Marketing Management*, *92*, 307–320. doi:10.1016/j.indmarman.2020.05.006

Tsing, A. L. (2015). *The mushroom at the end of the world.* Princeton University Press.

Tuesta, D. (2017). Retirement and labour markets under the context of pension reform in Latin America. *Economic and Political Studies*, *5*(4), 475–500. doi:10.1080/20954816.2017.1384622

Turrentine, J. (2022). *What Are the Causes of Climate Change?* NRDC. https://www.nrdc.org/stories/what-are-causes-climate-change

Tushingham, S., Hopt, J., Christiansen, C., Bommelyn, M. L., Green, J., Peterson, M. R., Steinruck, S., & Stewart, C. (2020). In the footsteps of Amelia Brown: Collaborative historical ecology at Shin-yvslh-sri~, a Tolowa village on the north coast of California. *Journal of Island & Coastal Archaeology*, *15*(1), 3–27. doi:10.1080/15564894.2018.1539789

Ugai, T. (2016). Evaluation of sustainable roof from various aspects and benefits of agriculture roofing in urban core. *Procedia: Social and Behavioral Sciences*, *216*, 850–860. doi:10.1016/j.sbspro.2015.12.082

Ullah, Z., Naveed, R. T., Rehman, A. U., Ahmad, N., Scholz, M., Adnan, M., & Han, H. (2021). Towards the development of sustainable tourism in pakistan: A study of the role of tour operators. *Sustainability (Basel), 13*(9), 4902. doi:10.3390u13094902

UNCTAD. (2022). *5 global actions needed to build a sustainable ocean economy.* UNCTAD. https://unctad.org/news/5-global-actions-needed-build-sustainable-ocean-economy

Ungolo, F., Christiansen, M., Kleinow, T., & Macdonald, A. (2019). Survival analysis of pension scheme mortality when data are missing. *Scandinavian Actuarial Journal, 19*(6), 523–547. doi:10.1080/03461238.2019.1580610

United Nations Development Programme. (2023). Human Development Report 2021/2022. [Data set]. https://hdr.undp.org/system/files/documents/global-report-document/hdr2021-22overviewenpdf.pdf

United Nations Environment Program. (2023). *Ozone layer recovery is on track, helping avoid global warming by 0.5°C.* UNEP. https://www.unep.org/news-and-stories/press-release/ozone-layer-recovery-track-helping-avoid-global-warming-05degc

United Nations Environment Programme. (2012). *Green economy in a blue world: Synthesis report.* United Nations.

United Nations General Assembly. (1948). *Declaración Universal de los Derechos Humanos.* ONU. https://www.un.org/es/about-us/universal-declaration-of-human-rights

United Nations General Assembly. (2017). Resolution 71/312 Our ocean, our future: Call for action. In *Resolution adopted by the general assembly on 6 July 2017.* United Nations.

United Nations. (1987). *Our Common Future. Report of the World Commission on Environment and Development.* Oxford University Press.

United Nations. (2020). *World Population Prospects. Department of Economic and Social Affairs.* UN. https://population.un.org/wpp/Graphs/DemographicProfiles/Pyramid/218

United Nations. (2022). *World Population Prospects 2022: Summary of Results.* UN. https://www.un.org/development/desa/pd/sites/www.un.org.development.desa.pd/files/wpp2022_summary_of_results.pdf

United Nations. (2023). Guterres urges countries to recommit to achieving SDGs by 2030 deadline. *UN New: Global perspective Human stories.* UN. https://news.un.org/en/story/2023/04/1136017

US DoA. (n.d.). *Biotechnology Frequently Asked Questions.* USDA. https://www.usda.gov/topics/biotechnology/biotechnology-frequently-asked-questions-faqs

Ūsas, J., Balezentis, T., & Streimikiene, D. (2021). Development and integrated assessment of the circular economy in the European Union: the outranking approach. *Journal of Enterprise Information Management.* doi:10.1108/JEIM-11-2020-0440

Uthoff, A. (2001). *La reforma del sistema de pensiones en Chile: desafíos pendientes.* Santiago: Naciones Unidas ONU. https://bit.ly/3ecSfai

Vaid, S. K., Vaid, G., Kaur, S., Kumar, R., & Sidhu, M. S. (2022). Application of multi-criteria decision-making theory with VIKOR-WASPAS-Entropy methods: A case study of silent Genset. *Materials Today: Proceedings, 50*, 2416–2423. doi:10.1016/j.matpr.2021.10.259

Valdez, J. & Nunag, R. (2023). The Research and Creative Works Colloquium and Its Effect on Research Performance. *Cosmos Journal of Engineering & Technology, 13*(1), 57–69.

Vale, J., Branco, M., & Ribeiro, J. (2011). *Intellectual Capital in a Seaport Context.* ECIC.

Valenzuela, K. (2021). Transnational Social Protection and the Role of Countries of Origin: The Cases of Mexico, Guatemala, Bolivia, and Ecuador. In Latin American Diasporas in Public Diplomacy (pp 27–51). Palgrave Macmillan Series in Global Public Diplomacy. Cham: Palgrave Macmillan. doi.org/ doi:10.1007/978-3-030-74564-6_2

Van der Ryn, S., & Cowan, S. (2007). *Ecological design: tenth anniversary edition.* Island Press.

Van Heerden, D., & Goosen, L. (2021). Students' Perceptions of e-Assessment in the Context of Covid-19: The Case of UNISA. In *Proceedings of the 29th Conference of the Southern African Association for Research in Mathematics, Science and Technology Education (SAARMSTE)* (pp. 291-305). SAARMSTE.

Van Heerden, D., & Goosen, L. (2012). Using Vodcasts to Teach Programming in an ODL Environment. *Progressio, 34*(3), 144–160.

Van Holt, T., Statler, M., Atz, U., Whelan, T., van Loggerenberg, M., & Cebulla, J. (2020). The cultural consensus of sustainability-driven innovation: Strategies for success. *Business Strategy and the Environment, 29*(2), 3399–3409. doi:10.1002/bse.2584

Van Regenmortel, S., De Donder, L., Dury, S., Smetcoren, A., De Witte, N., & Verté, D. (2016). Social exclusion in later life: A systematic review of the literature. *Journal of Population Ageing, 9*(4), 315–344. doi:10.100712062-016-9145-3

Vargas-Sánchez, A. (2018). The unavoidable disruption of the circular economy in tourism. *Worldwide Hospitality and Tourism Themes, 10*(6), 652–661. doi:10.1108/WHATT-08-2018-0056

Velho, G. (1978). Observando o familiar. In: Nunez, E. de O. (Org.). An aventura sociológica: objetividade, paixao, improviso e método na pesquisa social. Rio de Janeiro: Zahar.

Vermesan, O., & Friess, P. (2015). Building the Hyperconnected Society: Internet of Things Research and Innovation Value Chains, Ecosystems and Markets. Gistrup: River Publishers. doi:10.13052/rp-9788793237988

Vesci, M., Conti, E., Rossato, C., & Castellani, P. (2021). The mediating role of visitor satisfaction in the relationship between museum experience and word of mouth: Evidence from Italy. *The TQM Journal, 33*(1), 141–162. doi:10.1108/TQM-02-2020-0022

Vial, G. (2019). Comprensión de la transformación digital: Una revisión y una agenda de investigación. *The Journal of Strategic Information Systems, 28*(2), 118–144. doi:10.1016/j.jsis.2019.01.003

Vijayaraghavan, K. (2016). Green roofs: A critical review on the role of components, benefits, limitations, and trends. *Renewable & Sustainable Energy Reviews, 57*, 740–752. doi:10.1016/j.rser.2015.12.119

Villasmil, M., Alvarado-Peña, L., Socorro-González, C., & Gamboa, N. (2018). Contribuciones parafiscales: Factor determinante para la seguridad social. [RCS]. *Revista de Ciencias Sociales, XXIV*(3), 21–35. https://bit.ly/3T9B4Fn

Vincent, K., Roth, R. E., Moore, S. A., Huang, Q., Lally, N., Sack, C. M., Nost, E., & Rosenfeld, H. (2019). Improving spatial decision making using interactive maps: An empirical study on interface complexity and decision complexity in the North American hazardous waste trade. *Environment and Planning. B, Urban Analytics and City Science, 46*(9), 1706–1723. doi:10.1177/2399808318764122

Visser, M., & Togt, K. (2016). Learning in Public Sector Organizations: A Theory of Action, Approach. *Public Organization Review, 16*(2), 235–249. doi:10.100711115-015-0303-5

Vogelpohl, T. (2023). Understanding the bioeconomy through its instruments: Standardizing sustainability, neoliberalizing bioeconomies? *Sustainability Science, 18*(2), 1–15. doi:10.100711625-022-01256-2 PMID:36628104

Von Braun, J. (2015). Bioeconomy. Oxford University Press eBooks. doi:10.1093/acprof:oso/9780198733201.003.0011

Von Mises, L. (1912). *The Theory of Money and Credit*. Yale University Press.

Vranjanac, Ž., Rađenović, Ž., Rađenović, T., & Živković, S. (2023). Modeling circular economy innovation and performance indicators in European Union countries. *Environmental Science and Pollution Research International*, 1–12. doi:10.100711356-023-26431-5 PMID:36918486

W. S. (2007). The Anthropocene: Are Human Beings Now Overwhelming the Forces of Nature? *Ambio*, *36*, 614–621. doi:10.1579/0044-7447(2007)36[614:TAAHNO]2.0.CO;2 PMID:18240674

Walker, P. (2005). Political ecology: Where is the ecology? *Progress in Human Geography*, *29*(1), 73–82. doi:10.1191/0309132505ph530pr

Wallace-Wells, D. (2019). *The uninhabitable earth: Life after warming*. Tim Duggan Books.

Walter, I., & Ugelow, J. L. (1979) Environmental policies in developing countries. *Technology, Development and Environmental Impact 8(2)*:102–109. https://www.jstor.org/stable/4312437

Wandrey, J. (2008). Circum-Arctic Resource Appraisal: Estimates of Undiscovered Oil and Gas North of the Arctic Circle. *Fact Sheet*. USGS. https://pubs.usgs.gov/fs/2008/3049/

Wang, F. Z., Jiang, T., & Guo, X. C. (2018). Government Quality, Environmental Regulation and Corporate Green Technology Innovation. *Sci. Res. Manage, 39*(1), 26–33.

Wang, C., Zhao, L., Lim, M. K., Chen, W. Q., & Sutherland, J. W. (2020). Structure of the global plastic waste trade network and the impact of China's import Ban. *Resources, Conservation and Recycling*, *153*, 104591. doi:10.1016/j.resconrec.2019.104591

Wang, K., & Jiang, W. (2021). State ownership and green innovation in China: The con-tingent roles of environmental and organizational factors. *Journal of Cleaner Production*, *314*, 128029. doi:10.1016/j.jclepro.2021.128029

Wang, Q., Xia, C., Alagumalai, K., Le, T., Yuan, Y., Khademi, T., & Lu, H. (2023). Biogas generation from biomass as a cleaner alternative towards a circular bioeconomy: Artificial intelligence, challenges, and future insights. *Fuel*, *333*, 126456. doi:10.1016/j.fuel.2022.126456

Wang, T., Yu, Z., Ahmad, R., Riaz, S. S., Khan, K. U., Siyal, S., Chaudhry, M. A., & Zhang, T. (2022). Transition of bioeconomy as a key concept for the agriculture and agribusiness development: An extensive review on ASEAN countries. *Frontiers in Sustainable Food Systems*, *6*, 998594. doi:10.3389/fsufs.2022.998594

Wang, X., Shuster, W., Pal, C., Buchberger, S., Bonta, J., & Avadhanula, K. (2011). Low impact development design—Integrating suitability analysis and site planning for reduction of post-development stormwater quantity. *Sustainability (Basel)*, *2*(8), 2467–2482. doi:10.3390u2082467

Wang, Y., Hu, H., Dai, W., & Burns, K. (2021). Evaluation of industrial green development and industrial green competitiveness: Evidence from Chinese urban agglomerations. *Ecological Indicators*, *124*, 107371. doi:10.1016/j.ecolind.2021.107371

Wang, Y., Sun, X., & Guo, X. (2019). Environmental regulation and green productivity growth: Empirical evidence on the Porter Hypothesis from OECD industrial sectors. *Energy Policy*, *132*, 611–619. doi:10.1016/j.enpol.2019.06.016

Watt, J. (1880). *Steam-Hammers*. Alamy.

Webber, S. (2019). Putting climate services in contexts: Advancing multi-disciplinary understandings: Introduction to the special issue. *Climatic Change*, *157*(1), 1–8. doi:10.100710584-019-02600-9

Weick, K. E. (1969). *The Social Psychology of Organizing*. McGraw Hill.

Weinert, J., & Gründl, H. (2021). The modern tontine. *European Actuarial Journal, 11*(1), 49–86. doi:10.100713385-020-00253-y

Wei, X., Lo, C. K. Y., Jung, S., & Choi, T.-M. (2021). From co-consumption to co-production: A systematic review and research synthesis of collaborative consumption practices. *Journal of Business Research, 129*(May), 282–294. doi:10.1016/j.jbusres.2021.02.027

Weng, M. H., & Lin, C. Y. (2011). Determinants of green innovation adoption for small and medium-size enterprises (SMES). *African Journal of Business Management, 5*(22), 9154–9163.

Wen, Z., Xie, Y., Chen, M., & Dinga, C. D. (2021). China's plastic import ban increases prospects of environmental impact mitigation of plastic waste trade flow worldwide. *Nature Communications, 12*(1), 425. . doi:10.103841467-020-20741-9 PMID:33462243

Weston, A. (2009). *The incompleat eco-philosopher: essays from the edges of environmental ethics*. State University of New York Press.

Westwood, R., & Low, D. R. (2003). The multicultural muse: Culture, creativity, and innovation. *International Journal of Cross Cultural Management, 3*(2), 235–259. doi:10.1177/14705958030032006

Wiering, M., Liefferink, D., Boezeman, D., Kaufmann, M., Crabbé, A., & Kurstjens, N. (2020). The Wicked Problem the Water Framework Directive Cannot Solve. The Governance Approach in Dealing with Pollution of Nutrients in Surface Water in the Netherlands, Flanders, Lower Saxony, Denmark, and Ireland. *Water (Basel), 12*(5), 1240. doi:10.3390/w12051240

Wigneron, J.-P., Dayan, S., Kruszewski, A., & Aluome, C. AI-Yaari, M. G.-E. A., Fan, L., Guven, S., Chipeaux, C., Moisy, C., Guyon, D., & Loustau, D. (2018). The aqui network: Soil moisture sites in the "Les Landes" forest and graves vineyards (Bordeaux Aquitaine region, France). *IGARSS 2018 - 2018 IEEE International Geoscience and Remote Sensing Symposium*, (pp. 3739–3742). IEEE.

Wild, S. (n.d.). What is the bioeconomy and how could it help fight climate change? *Horizon Magazine*. https://ec.europa.eu/research-and-innovation/en/horizon-magazine/what-bioeconomy-and-how-could-it-help-fight-climate-change

Wiley, A. S., & Cullin, J. M. (2016). What Do Anthropologists Mean When They Use the Term Biocultural? *American Anthropologist, 2016*(118), 554–569. doi:10.1111/aman.12608

Wilk, R. R., & Wilhite, H. L. (1985). Why don't people weatherize their homes? An ethnographic solution. *Energy, 10*(May), 621–629. doi:10.1016/0360-5442(85)90093-3

Willermet, C. (2016). Biological Anthropology in 2015: Open Access, Biocultural Interactions, and Social Change. *American Anthropologist, 118*(2), 118. doi:10.1111/aman.12529

Winston, A. S. (2006). *Green to Gold*. Yale University Press.

Wirtenberg, J., Harmon, K. D., Russell, W. G., & Fairfield, K. D. (2007). HR's role in building a sustainable enterprise. *Human Resource Planning, 30*, 10–20.

WMO, 2021, *Arctic assessment report shows faster rate of warming,* World Meteorological Organization (wmo.int)

WMO. (2016). *NASA, analysis confirms July 2016 as warmest month in 136 years.* World Meteorological Organization (wmo.int).

Wolf, G. (2017). New Challenges of the Digital Transformation: The Comeback of the Vision Mission System. In Out-thinking Organizational Communications: The Impact of Digital Transformation (pp. 113 - 128). Springer.

Wolf, E. (1972). Ownership and political ecology. *Anthropological Quarterly*, *45*(3), 201–205. doi:10.2307/3316532

Wolf, I. (2021). Political Stress and the Sustainability of Funded Pension Schemes: Introduction of a Financial Theory. *Journal of Risk and Financial Management*, *14*(11), 1–12. doi:10.3390/jrfm14110525

Wolf, I. (2022). Studying the equilibrium of pension designs when shifting to funded pension schemes: Economic theory and links to political factors. *SN Business & Economics*, *2*(76), 1–21. doi:10.100743546-022-00236-z PMID:35789707

Wolf, M. (2016). Rethinking urban epidemiology: Natures, networks and materialities. *International Journal of Urban and Regional Research*, *40*(5), 958–982. doi:10.1111/1468-2427.12381 PMID:32336869

World Bank Group. (2022). *Urgent Climate Action Crucial for Bangladesh to Sustain Strong Growth.* World Bank. https://www.worldbank.org/en/news/press-release/2022/10/31/urgent-climate-action-crucial-for-bangladesh-to-sustain-strong-growth

World Bank. (2020b). *Total fisheries production (metric tons) - OECD members*. World Bank. https://data.worldbank.org/indicator/ER.FSH.PROD.MT?locations

World Bank. (2020c). *Captured Fisheries Production (Metric Tons)*. World Bank. https://data.worldbank.org/indicator/ER.FSH.CAPT.MT?locations=OE

World Bank. (2023). *World GDP statistics,* https://data.worldbank.org/

World Bank. (n.d.). *Overview*. World Bank. https://www.worldbank.org/en/topic/agriculture/overview

World Commission on Environment and Development. (1987). *Our common future*. Oxford University Press.

World Economic Forum (WEF). (2022a). *Biodiversity credits as an economic driver for nature can deliver net positive biodiversity gains*. WEF. https://www.weforum.org/agenda/2022/12/biodiversity-credits-nature-cop15/#:~:text=Biodiversity%20credits%20as%20an%20economic,%20deliver%20net%20positive%20biodiversity%20gains

World Economic Forum (WEF). (2022b). *Biodiversity Credits: Unlocking Financial Markets for Nature-Positive Outcomes*. Geneva, Switzerland: World Economic Forum.

World Economic Forum (WEF). (2023). *The European Union's Artificial Intelligence Act, explained.* WEF. https://www.weforum.org/agenda/2023/03/the-european-union-s-ai-act-explained/

World Wildlife Fund (WWF). (2022). Living. *The Planetary Report*. https://wwflpr.awsassets.panda.org/downloads/lpr_2022_full_report.pdf

Wu, H. H. (2006). A comparative study of using grey relational analysis in multiple attribute decision making problems. *Quality Engineering*, *15*(2), 209–217. doi:10.1081/QEN-120015853

Wu, H. Y., Tsai, A., & Wu, H. S. (2019). A HYBRID MULTI-CRITERIA DECISION ANALYSIS APPROACH FOR ENVIRONMENTAL PERFORMANCE EVALUATION: AN EXAMPLE OF THE TFT-LCD MANUFACTURERS IN TAIWAN. [EEMJ]. *Environmental Engineering and Management Journal*, *18*(3), 597–616. doi:10.30638/eemj.2019.056

Wurlod, J. D., & Noailly, J. (2018). The impact of green innovation on energy intensity: An empirical analysis for 14 industrial sectors in OECD countries. *Energy Economics*, *71*, 47–61. doi:10.1016/j.eneco.2017.12.012

Wu, W., & Jim, C. Y. (2021). Urban green infrastructure and sustainable urbanization: A comprehensive review of the roles of green roofs. *Landscape and Urban Planning*, *212*, 104097. doi:10.1016/j.landurbplan.2021.104097

Wysocki, J. (2021). Innovative green initiatives in the manufacturing SME sector in Poland. *Sustainability (Basel)*, *13*(4), 2386. doi:10.3390u13042386

Xiao, Z., Kerr, W.A. (2022). Biotechnology in China-regulation, investment, and delayed commercialization. *GM Crops Food 13*(1), 86–96. (Accessed 18.06. 2023) doi:10.1080/21645698.2022.2068336

Xu, W., Chen, W. Q., Jiang, D., Zhang, C., Ma, Z., Ren, Y., & Shi, L. (2020). Evolution of the global polyethylene waste trade system. *Ecosystem Health and Sustainability*, *6*(1), 1756925. doi:10.1080/20964129.2020.1756925

Xu, Z., Frankwick, G. L., & Ramirez, E. (2016). Effects of big data analytics and traditional marketing analytics on new product success: A knowledge fusion perspective. *Journal of Business Research*, *69*(5), 1562–1566. doi:10.1016/j.jbusres.2015.10.017

Yang, Y., Lili, W., & Hongchi, Z. (2022). *Bibliometric research on the evolution of resilience theme from the perspective of Geographical Science*. 2022 29th International Conference on Geoinformatics, Beijing, China. 10.1109/Geoinformatics57846.2022.9963870

Yang, M., Chen, L., Wang, J., Msigwa, G., Osman, A. I., Fawzy, S., Rooney, D. W., & Yap, P. S. (2022). Circular economy strategies for combating climate change and other environmental issues. *Environmental Chemistry Letters*, *21*(1), 55–80. doi:10.100710311-022-01499-6

Yıldırım, M., & Destek, M. A. ve Özsoy F. N. (2017). Doğrudan yabancı yatırımlar ve kirlilik sığınağı hipotezi. *C. Ü. İktisadi ve İdari Bilimler Dergisi, 18(2),* 99-111. Retrieved from http://esjournal.cumhuriyet.edu.tr/en/download/article-file/370376

Yildirim, S. (2023), Greenwashing: a rapid escape from sustainability or a slow transition?, *LBS Journal of Management & Research*. https://doi.org/ doi:10.1108/LBSJMR-11-2022-0077

Yıldırım, S., & Yıldırım, D. Ç. (2022). İklim Değişikliği ve Gıda Güvenliği İlişkisine Kısa Bir Bakış [A Brief Overview of the Relationship between Climate Change and Food Security]. Dirençli Şehir Yolunda Toplum Politikaları, Ed. Lehimler, H.M., Nobel yayıncılık.

Yildirim, S., & Kaplan, M. (2020). Seafood security and sustainability through sustainable development: a review of Turkish seafood market. In A. Obayelu & O. Obayelu (Eds.), *Developing Sustainable Food Systems, Policies, and Securities* (pp. 16–36). IGI Global. doi:10.4018/978-1-7998-2599-9.ch002

Yildirim, S., & Kaplan, M. (2022). The threat of invasive alien marine species to the blue economy: the mediterranean case. In L. Raimi & J. Kah (Eds.), *Implications for Entrepreneurship and Enterprise Development in the Blue Economy* (pp. 50–80). IGI Global.

Yıldırım, S., & Yıldırım, D. C. (2021). Achieving seafood security in the Mediterranean region: a case of Turkey. In R. Castanho & J. Martın Gallardo (Eds.), *Management and Conservation of Mediterranean Environments* (pp. 175–195). IGI Global. doi:10.4018/978-1-7998-7391-4.ch011

Ying, Z., Ai-jun, Y., Jian-wei, J., & Ji, L. (2011). Study on measurement model for forest risk assessment in Beijing. *2011 International Conference on Computer Distributed Control and Intelligent Environmental Monitoring*, (pp. 1750–1755). IEEE. 10.1109/CDCIEM.2011.494

Yin, R. K. (2003). *Case study research: Design and methods*. SAGE Publications.

Yongming, Z. (2010). How Our Village Becomes “Ecocultural.” In Cultural Change and Persistence. Palgrave Macmillan US.

Youness, H. (2017). *Trends of Business* (1st ed.).

Yu, Z., Khan, S. A. R., Ponce, P., & Jabbour, A. B. L. (2022). Factors affecting carbon emissions in emerging economies in the context of a green recovery: Implications for sustainable development goals. *Technological Forecasting and Social Change, 176*, 121417. doi:10.1016/j.techfore.2021.121417

Zailani, S., Iranmanesh, M., Nikbin, D., & Jumadi, H. B. (2014). Determinants and environmental outcome of green technology innovation adoption in the transportation industry in Malaysia. *Asian Journal of Technology Innovation, 22*(2), 286–301. doi:10.1080/19761597.2014.973167

Zan, L., & Lusiani, M. (2011). Managing Change and Master Plans: Machu Picchu Between Conservation and Exploitation. *Archaeologies, 7*(2), 329–371. doi:10.100711759-011-9167-7

Zedler, J. B., & Stevens, M. L. (2018). Western and Traditional Ecological Knowledge in Ecocultural Restoration. *San Francisco Estuary and Watershed Science, 16*(3), 1–18. doi:10.15447fews.2018v16iss3art2

Zhang, Z. Zeng, Y., Liu, H., Zhao, C., Wang, F., & Chen, Y. (2022). Smart DC: An AI and digital twin-based energy-saving solution for data centers. *NOMS 2022-2022 IEEE/IFIP Network Operations and Management Symposium*, (pp. 1–6). IEEE.

Zhang, Z., Wei, M., Xu, X., Hu, C., & Xie, W. (2022). Intelligent energy saving technology and strategy of 5G RAN. *2022 IEEE International Symposium on Broadband Multimedia Systems and Broadcasting (BMSB)*, (pp. 1–6). IEEE. 10.1109/BMSB55706.2022.9828681

Zhang, D. G., & Lu, Y. Q. (2017). Impact of market segmentation on energy efficiency. *China Popul Resour Environ, 27*(1), 65–72.

Zhang, L., Liu, Z., Zhang, Y., & Ai, J. (2018). Intelligent path planning and following for UAVs in forest surveillance and fire fighting missions. *2018 IEEE CSAA Guidance, Navigation and Control Conference (CGNCC)*, (pp. 1–6). IEEE.

Zhang, R., & Noronha, C. (2023). Assessing the nexus between cross-boarder infrastructure projects and extinction accounting - from the belt and road initiative. *Social and Environmental Accountability Journal, 43*(1), 30–55. doi:10.1080/0969160X.2022.2132969

Zhang, W., He, L., & Yuan, H. (2021). Enterprises' decisions on adopting low-carbon technology by considering consumer perception disparity. *Technovation, 102238*. Advance online publication. doi:10.1016/j.technovation.2021.102238

Zhang, X., Zhao, C., Shao, M., Chen, Y., Liu, P., & Chen, G. (2022). The roadmap of bioeconomy in China. *Engineering Biology, 6*(4), 71–81. doi:10.1049/enb2.12026 PMID:36968339

Zhang, Z., & Sundaresan, S. (2010). Knowledge markets in firms: Knowledge sharing with trust and signalling. *Knowledge Management Research and Practice, 2010*(8), 322–339. doi:10.1057/kmrp.2010.22

Zhao, C., Liu, M., Du, H., & Gong, Y. (2021). The evolutionary trend and impact of global plastic waste trade network. *Sustainability (Basel), 13*(7), 3662. doi:10.3390u13073662

Zhao, C., Qi, X., Wang, J., Du, F., & Shi, X. (2022). Predicting possible new links to future global plastic waste trade networks. *Sustainability (Basel), 14*(8), 4692. doi:10.3390u14084692

Zhao, X., Shang, Y., & Song, M. (2020). Industrial structure distortion and urban ecological efficiency from the perspective of green entre-preneurial ecosystems. *Socio-Economic Planning Sciences, 72*, 100757. doi:10.1016/j.seps.2019.100757

Zhou, K. Z., Gao, G. Y., & Zhao, H. (2017). State Ownership and firm innovation in China: An integrated view of institutional and efficiency logics. *Administrative Science Quarterly, 62*(2), 375–404. doi:10.1177/0001839216674457

Zhou, M., Chen, J., Ge, Q., Huang, X., & Liu, Y. (2013). AIS data based identification of systematic collision risk for maritime intelligent transport system. *2013 IEEE International Conference on Communications (ICC)*, (pp. 6158–6162). IEEE. 10.1109/ICC.2013.6655590

Zhou, P., Ang, B. W., & Poh, K. L. (2008). Measuring environmental performance under different environmental DEA technologies. *Energy Economics*, *30*(1), 1–14. doi:10.1016/j.eneco.2006.05.001

Zhou, Y., Yuen, K. F., Tan, B., & Thai, V. V. (2021a). Maritime knowledge clusters: A conceptual model and empirical evidence. *Marine Policy*, *123*, 104299. doi:10.1016/j.marpol.2020.104299

Zhou, Y., Yuen, K. F., Tan, B., & Thai, V. V. (2021b). The effect of maritime knowledge clusters on maritime firms' performance: An organizational learning perspective. *Marine Policy*, *128*, 104472. doi:10.1016/j.marpol.2021.104472

Zhu, F., Lv, Y., Chen, Y., Wang, X., Xiong, G., & Wang, F.-Y. (2020). Parallel transportation systems: Toward IoT-enabled smart urban traffic control and management. *IEEE transactions on intelligent transportation systems: a publication of the IEEE Intelligent Transportation Systems Council, 21*(10), 4063–4071. doi:10.1109/TITS.2019.2934991

Zhu, J., & Liu, W. (2020). A tale of two databases: The use of Web of Science and Scopus in academic papers. *Scientometrics*, *123*(1), 321–335. doi:10.100711192-020-03387-8

Zhu, Y., Tian, D., & Yan, F. (2020). Effectiveness of Entropy Weight Method in Decision Making. *Mathematical Problems in Engineering*, *1–5*, 1–5. doi:10.1155/2020/3564835

Zogopoulos, E. (2020). Why Blue Economy Is So Important. *Energy Industry Review*. https://energyindustryreview.com/analysis/why-blue-economy-is-so-important/

Zreik, Z. (2021). *The transition of the Lebanese Economy from a service economy to a productive economy,* [MBA Thesis, Lebanese International University].

Zupic, I., & Čater, T. (2015). Bibliometric methods in management and organization. *Organizational Research Methods*, *18*(3), 429–472. doi:10.1177/1094428114562629

About the Contributors

Ana Pego holds a Ph.D. in Geography and Territorial Planning (Universidade Nova de Lisboa, Portugal), an MBA (Open University, Portugal) and a BSC in Economics (Universidade de Évora, Portugal). She is an integrated researcher at CICS Nova- Research Center in Social Sciences; in her research, she focuses on the circular economy, offshore energy, smart cities and tourism. She is a coordinator of five European projects as Erasmus and Horizon Europe. She is an author of several articles about offshore energy clusters, tourism, and the circular economy. She also collaborates with Open University in Macroeconomics and Business Management classes.

***

Georges Bellos currently is an Educational Leadership and Management Graduate at the Lebanese International University, Mount Lebanon Campus. His educational background is varied. He currently is a Masters' Degree Holder in Geological and Archaeological Sciences and Management, with emphasis on Economic Development and Tourism Management, and holding a Teaching Diploma in Entrepreneurial Skills Training. His current aspirations are to join a PhD program and pursue a doctoral degree in a hybrid PhD program combining Geosciences, Archeology, tourism and education into one PhD research in a multi-disciplinary track in order to improve the tourism situation in Lebanon, through education. There are other academic functions Mr. Bellos aspires to hold, such as a resident faculty in a higher education institution as well as an active member in the Civil society whereby enhancing entrepreneurial opportunities for youths willing to improve the country's situation, through contributing in becoming better quality and more responsible citizens. Aside from Academic aspirations, Mr. Bellos also seeks to become a professional researcher, and become also active in publications and help contribute in relationships with academia and the private sector.

Anthony Bolton is a PhD candidate in the School of Computing at the University of South Africa. He is also Chief Information Officer (CIO) and Chief Technical Officer (CTO) Global Telecom, End User Infrastructure & Services, Immersive Technology Development and Engineering at General Motors in Ireland.

S. Castañeda-Burciaga has a PhD in Strategic Business Management; PhD in Education.

Marcelo Centeno Ponce is a Junior Researcher at the Technical University of Manabí (Ecuador) within the Department of Social Sciences.

Ana Corrales Mendoza is a Junior Researcher at the Technical University of Manabí (Ecuador) within the Department of Social Sciences.

João Pedro Almeida Couto is Associated Professor at University of Azores, Ph.D. In International Business, and has the MBA by the University of Porto. Presides the Department of Economics and Management Council and coordinates the management section. Teaches Management Introduction, Strategic Management and International Business. He is also Research Fellow at the Centre of Applied Economics Studies of the Atlantic (CEEAplA - Portugal). His main research interests include Strategic Management, Entrepreneurship, International Business and Marketing. He has published several articles in international journals such as the Business Horizins, Tourism Management, Environmental Research, Journal of Business and Economic Management, Multinational Business Review, Journal of Comparative International Management among others.

Hasan Hüseyin Erdoğan received his PhD in Business Administration (Marketing) from Yıldız Technical University. He also holds a bachelor's degree in Tourism Administration and a master's degree in Business Administration. He worked as a tourism lecturer at Okan University and Piri Reis University. Also, he worked as tour guide at a French travel agency. He is currently working as a tourism lecturer at Pamukkale University. He teaches vocational foreign language, general tourism, tourism geography, and front office services. He also assesses European Union projects as an independent expert for the Turkish National Agency. His research interests are tourism marketing and museum marketing.

Rinat Galiautdinov is a Principal Software Developer and Software Architect having the expertise in Information Technology and Computer Science. Mr. Galiautdinov is also an expert in Banking/Financial industry as well as in Neurobiological sphere. Mr. Rinat Galiautdinov works on the number of highly important researches (including such the spheres as: Drones, Financial Systems and Private Money, Brain interface and neuro-biology) as an independent researcher and projects holder and developer.

Leila Goosen is a full professor in the Department of Science and Technology Education of the University of South Africa. Prof. Goosen was an Associate Professor in the School of Computing, and the module leader and head designer of the fully online signature module for the College for Science, Engineering and Technology, rolled out to over 92,000 registered students since the first semester of 2013. She also supervises ten Masters and Doctoral students, and has successfully completed supervision of 43 students at postgraduate level. Previously, she was a Deputy Director at the South African national Department of Education. In this capacity, she was required to develop ICT strategies for implementation. She also promoted, coordinated, managed, monitored and evaluated ICT policies and strategies, and drove the research agenda in this area. Before that, she had been a lecturer of Information Technology (IT) in the Department for Science, Mathematics and Technology Education in the Faculty of Education of the University of Pretoria. Her research interests have included cooperative work in IT, effective teaching and learning of programming and teacher professional development.

María Fernanda Higuera-Cota is a research assistan.t

Sejana Jose V. is pursuing PhD in Tourism Management from Christ University, Bangalore. She has successfully completed her Mphil and Masters in Tourism Administration from Christ University, Bangalore. She is currently working as Lecturer at Mount Carmel College, Bangalore. She has one year of industry experience working in a tour operation company and teaching experience for four years. Her research interest includes Green Transportation, Sustainable Tourism, Organizational behaviour, and Tourism Logistics.

Merve Kaplan was born in 1986. She was graduated her master's degree from Kocaeli University in Turkey. The subject of her thesis was "Effects of shucking and cooking of striped venus clam (Chamelea gallina L.,1753) collected from kandira beach with different thermal methods on meat yield chemical composition". She is an aquaculture engineer and she has been working in Kocaeli Provincial Directorate of Agriculture and Forestry, Turkey. Also she is a phd student at Canakkale Onsekiz Mart University in Turkey. Also she is preparing her phd thesis.

Mazen Kotob is an Educator recognized as a passionate, student-centered instructor continually seeking to improve learning and higher education culture. He earned B.S. degree in Chemistry, MA in education from American University of Beirut and a Ph.D. in Education from the Lebanese University. Dr. Kotob is an expert in curriculum design and construction, focusing on topics including Inquiry Based Learning, Integrating technology into the curriculum and authentic assessment. He worked as Head of department and as action research designer for private educational institutions. He is a Certified Senior trainer at Intel-Education and a research consultant at Parkville Global Advisory.

Elmarie Kritzinger joined the University of South Africa's College of Science, Engineering and Technology (CSET) in 2000 and currently holds the position of Professor in the School of Computing. Prof Kritzinger completed her PhD in 2006 and Post Graduate Certificate in Education in 2012. She is currently enrolled for her Master's in Education (Online Technology). Her research primarily focuses on Cyber Safety awareness, training and education for school learners, teachers and schools. The main aim is to establish and promote social responsibilities within communities to establish and grow a cyber-safety culture within South Africa. Prof Kritzinger has established herself as a mature researcher and has published in accredited national and international journals, contributed to a chapter in a book and presented at peer-reviewed conferences across the globe. Prof Kritzinger currently hold a NRF C3 rating within her research field.

Arturo Luque is full professor at the Technical University of Manabí, Ecuador, and member of Euro-Mediterranean Observatory on Public Policies and Democratic Quality at the University King Juan Carlos (Spain) and visiting Professor at the University of Rosario, Colombia. He holds a Ph.D. in Social Sciences and Law and a Master of two years about Labor Relations. He supervises several doctoral theses (Ph.D.) in Córdoba (Spain) and in the Basque Country (GEZKI Institute) like a director and co-director. It has 81 indexed articles and his latest publications includes topics about digitalization (Practicality, support or premeditated calculation in the digital age: the case of Ecuador), Globalization (Analysis of the concept of informal economy through 102 definitions: legality or necessity), corporate social (Ir)responsibility (How corporate social (ir)responsibility in the textile sector is defined, and its impact on ethical sustainability: An analysis of 133 concepts), transnational companies (The transnational textile companies relationship with environment: a Delphi analysis approach) and their relations with ethics (Corruption in the transnational textile industry: an exception or the rule?), social economy and cooperative system like economic alternative (Socially responsible public management: case spinning development in Ecuador), media analysis and social conflicts (Analysis of the indigenous uprising of Ecuador, 2019: between a measured legal response and lawfare, 2019) and sustainability (Public Strategy and Eco-Social Engagement in Latin American States: An Analysis of Complex Networks Arising from Their Constitutions).

Bartolomé Marco-Lajara is Professor of Business Organization at the University of Alicante (Spain). He holds a PhD in Economics and his doctoral dissertation focused in strategic alliances management. His research interests are on Strategic Management and Tourism Management. He is author of several books, book chapters and international articles related to teaching methodology and the areas above mentioned. On the other hand, he is member of the Tourism Research Institute at the University of Alicante since its foundation. He has participated in many public- and private-funded projects, being the main researcher in many of them. He was assistant dean of the Economics Faculty at the University of Alicante during eight years, holding now the position of Head of the Department of Business Organization at this University.

Milan Marković, PhD in Economics, is a Research Associate at the Innovation Centre of the University of Niš. From 2014 to 2019 he was engaged in the project "Improving Competitiveness of Public and Private Sector by Networking Competences in the Process of European Integrations of Serbia". He is a member of the editorial board of the Journal "Economics of Sustainable Development", a member of the Association of Economists and Managers of the Balkans and the Balkan Scientific Association of Agricultural Economists, and a reviewer of many scientific publications. His key areas of interest are sustainable development, economics of agriculture and international finance.

Javier Martínez Falcó (PhD) is Assistant Professor in the Department of Management at the University of Alicante. In the field of research, he focuses his interest on issues related to the Strategic Management of the Company, specifically in the areas of Knowledge Management and Corporate Sustainability of wine companies, on which he has written several publications in the form of articles, book chapters and contributions to conferences. He has also participated in several national research projects and teaches Strategic Management on the ADE, TADE and DADE degree courses.

Luis Antonio Millán Tudela is Assistant Professor in the Department of Management at University of Alicante. MBA and Bachelor of Mechanical Engineering, his research focuses on business longevity, and more specifically on the factors that influence the survival of companies. This activity is developed within the research group "Dirección Estratégica, conocimiento e innovación en una economía global (DECI-GLOBAL)". He also teaches strategic management in the Degree of Business Administration.

Maria Manuela Santos Natário is Coordinating Professor at Polytechnics of Guarda, Portugal, Phd. in Economics of Innovation and Competitiveness by University of Évora, and as the MBA by the University of Beira Interior. Teaches Innovation and Creativity, Innovation and Competitiveness, International Management and Economy. She is member of the Research Unit on Governance, Competitiveness and Public Policies (GOVCOPP) and of the Research Unit for Inland Development (UDI/IPG) (Portugal). Her main interests include Economy of Innovation, Innovation and Competitiveness. She has published articles in international journals such as European Journal of Innovation Management, Tékhne Studies Review, Journal of Business Economics and Management, Journal of the American Academy of Business, Journal of Comparative International Management, Journal of Knowledge-based Innovation in China. She has published articles in international books.

Carlos Noronha is an Associate Professor of Accounting at the Faculty of Business Administration, University of Macau, Macau SAR, People's Republic of China.

Elisabeth T. Pereira is an Associate Professor at the Department of Economics, Management, Industrial Engineering and Tourism (DEGEIT) - University of Aveiro (Portugal), where she is Vice-Diretor of the Doctoral Programme in Business and Economic Sciences, member of the Council of the Department and coordinates the International Mobility and Erasmus+ programs for Economics studies. She is a full senior researcher at GOVCOPP research unit. She is a former Chair of the Subject Committee of Economics and Business of Campus Europae – European University Foundation (2012-2016). She has been a Member and Chair of the Economics and Business panel of the Foundation for Science and Technology (FCT). She holds an MSc on Economics (University of Coimbra) and a PhD in Business and Economics Competitiveness (University of Aveiro). She teaches since 1995 at the University of Aveiro at the undergraduate and graduate levels on the programs of the 1st, 2nd and 3rd cycles in the fields of Economics and Business management. She has been the supervisor of several dozen national and international M.Sc., PhD and Post-doctoral theses. She is the author of several dozen articles published in international scientific journals indexed, books, book chapters, and institutional reports. She is also a reviewer of national and international journals. She is involved in several national and international projects, as coordinator and researcher, on competitiveness, labor market, poverty and innovation fields. Her current research interests are in Macroeconomics, Competitiveness, Gender Research, Labour Markets, Poverty and Social Innovation, Real State Economics, Monetary Economics and cryptocurrencies, Circular Economy and Tourism. She has been involved as coordinator and lecturer in several international summer schools, as well as chair and member of scientific committees at international conferences.

Carlos Rodríguez-Maillard research assistant.

S M Nazmuz Sakib was born in Dinajpur, Bangladesh on the 16th of April in 2001. He achieved his Primary School Certificate (PSC) from Cotton Research Station School, Sreepur, Gazipur and after several months of studying Class 6 in Alhaz Dhanai Bepari Memorial High School, Sreepur, Gazipur, he relocated to Jashore. He accomplished his Junior School Certificate (JSC) from Jagodishpur Mirzapur Ismail Secondary School, Jagodishour, Chaowgacha, Jashore and his Secondary School Certificate (SSC) from A. K. High School And College, Dhania, Dhaka-1236. He obtained his Higher Secondary Certificate (HSC) from Chowgacha Government College, Chowgacha, Jashore. He received a BSc in Business Studies, earning a CGPA of 4 out of 4 (1st class 1st) with 97.06% marks from School of Business and Trade, an online business school situated in Switzerland. He completed an MBA in Human Resources from International MBA Institute, an online business institute in Switzerland. He studied Bachelor of Laws (Hon's) or LLB(Hon's) at Dhaka International University, Dhaka, Bangladesh. He has accomplished CPD verified advanced diploma and diploma programs i.e. Advanced Diploma in Tissue Engineering, Advanced Diploma in Genetic Engineering - Theory and Application, Diploma in Fashion Design, Diploma in Nutrition, Therapeutics and Health, Diploma in Biology, Diploma in

Pharmacy Technician, Software Test Engineer Diploma (Advanced), Advanced Diploma in Soil Science and Technology, Diploma in Energy Economics, Energy Systems and Environmental Impact, Advanced Diploma in Political Ideologies, Diploma in Community Psychology, Diploma in Training of Trainers, Advance Diploma in Principles of Industrial Engineering, Advanced Diploma in Production & Operation Management, Advanced Diploma in Modelling and Analytics for Supply Chain Management, Business Administration Diploma (Foundation), Diploma in Effective Human Resource Administration, Diploma in Audio System Engineering, Diploma in ISO Standards - Integrated Management System (IMS), Diploma in Lean Manufacturing - Productive Management with Fundamental Tools, Advanced Diploma in Tourism and Hospitality Management, Diploma in Web Design, Diploma in Human Resources, and Diploma in Basic English Grammar, Diploma in River Engineering and has obtained certifications such as TESOL from Arizona State University, TEFL from Teacher Record, Scrum Master Professional Certificate (SMPC®), EF SET Certificate (C2 Level), Google IT Support Professional Certificate, Google Data Analytics Professional Certificate, Google Project Management Professional Certificate, Google Business Intelligence Professional Certificate, Google Digital Marketing & E-commerce Professional Certificate and IBM New Collar: Customer Engagement Specialist credential etc. As a multidisciplinary researcher, he has published several articles in the Asian Pacific Journal of Environment and Cancer, Simulacra, Journal of Innovation Information Technology and Application (JINITA), Waste Technology, and The IUP Journal of Electrical and Electronics Engineering. Several of his articles are published by Cambridge Open Engage, a publication of Cambridge University Press. Some of the titles of his published research works include "The Impact of Oil and Gas Development on the Landscape and Surface in Nigeria," "Assessing the Impact of Arctic Melting in the Predominantly Multilateral World System," "LiDAR Technology - An Overview," "Exploring the Intersection of Software Engineering and Mobile Technology from 2010 to 2021: A Review of Recent Research" "Comparing the sociology of culture in Bangladesh and India: Similarities and differences in Bangladeshi and Indian cultures" "Electrochemical Waste Water Treatment.". Published chapters i.e. "Chapter 3: Artificial Intelligence Model for Analyzing the Buying Patterns of Customers" and "Chapter 11: Restaurant Sales Prediction Using Machine Learning" of the book titled "Handbook of Research on AI and Machine Learning Applications in Customer Support and Analytics".

Eduardo Sánchez García (PhD) is assistant professor in the Department of Business Organization at the University of Alicante (Spain). His research is in the field of strategic management of the company. Specifically, he focuses on the study of the territorial agglomeration of companies in clusters and the absorptive capacity, and its influence on the innovative performance and entrepreneurial orientation of companies. He is co-author of several papers presented at conferences, both national and international, as well as several articles and book chapters related to his field of research.

Ahmed Yusuf SARIHAN is currently an Assistant Proffessor in Bandirma Onyedi Eylul Universtiy/ Türkiye. His work focus on international trade, import-export management, economic integrations and global supply chain. His educational background includes Bachelor, Master and PhD in international trade. He also worked in the international trade department of Okan University/Türkiye before in his academic career.

Maria João Tomás is an Associate Professor at Universidade Autónoma de Lisboa, Integrated Researcher Observare. PhD History- FCSH-UNL, MD-FL-UL,, Intenrships -University of Basel, UCLA.

Omar Vargas-González is a professor and Head of Systems and Computing Department at Tecnologico Nacional de Mexico Campus Ciudad Guzman, professor at Telematic Engineering at Centro Universitario del Sur Universidad de Guadalajara with a master degree in Computer Systems. Has been trained in Innovation and Multidisciplinary Entrepreneurship at Arizona State University (2018) and a Generation of Ecosystems of Innovation, Entrepreneurship and Sustainability for Jalisco course by Harvard University T.H. Chan School of Health. At present conduct research on diverse fields such as Entrepreneurship, Economy, Statistics, Mathematics and Information and Computer Sciences. Has colaborated in the publication of over 20 scientific articles and conducted diverse Innovation and Technological Development projects

José Vargas-Hernandez is a research professor.

Bindi Varghese is a Doctorate in Commerce,specializing inTourism. As an academician and tourism professional, she has over 18 years of Academic and one year of Industrial experience. Currently, she is affiliated with Christ University, as an Associate Professor and is the Research Coordinator at School of Business Studies and Social Sciences. She has served many educational institutions in South India and has served as a national and international expert, for a decade among the educational institutions of India. Currently, she is actively associated with Indian Tourism Congress (ITC) and Kerala Development Society (KDS), New Delhi. The active researches undertaken include Impact Assessment Studies, Medical Tourism, Destination Management Organization and Ecological Studies. Dr. Bindi completed a major research project on the title "Strategic Intervention of Destination Management Organizations to Enhance Competitiveness of Tourism Destinations– A Model for Karnataka" funded by Christ University. Along with her academic expertise, she is also an Section editor for 'ATNA- Journal of Tourism Studies', published by Christ University, Bengaluru. She has authored one book; on Medical Tourism in India: by an international publisher in Germany, and has also contributed chapters to edited books and has published several articles in areas of Destination Management, Governance, Medical Tourism, E-Tourism etc. To her credit, she has edited a book on "Evolving Paradigms in Tourism and Hospitality in Developing Countries: A Case Study of India". The book is published by CRC Press, Taylor and Francis group - an international publisher in US and released in 2018

Seda Yıldırım currently works at Tekirdag Namık Kemal University in Turkey. She studies on sustainable development, 2030 sustainable development goals, sustainable consumption patterns, Islamic finance and banking, consumption patterns, environmental quality and food security. She is married and has two children.

Durmuş Çağrı Yıldırım was born in 1981. After he graduated from Department of Economics in Kocaeli University in 2005, he started his master's education in Economics in the same year. He got the degree of Masters in Economic Policy in 2007. In 2011, he got his PhD degree in Economics. While he was a doctorate student, he studied one semester in Almeria University-Spain. Between 2006-2012, he worked as a research assistant at University of Kocaeli. He works at Tekirdag Namık Kemal University since April 2015. He studies on sustainable development, carbon emission, health and education issues, employment, innovation.

Ruopiao Zhang is an Assistant Professor of Accounting at the School of Business,Macau University of Science and Technology, Macau SAR, People's Republic of China.

Index

D

E

F

G

H

I

L

M

N

O

P

R

S

T

U

V

W

Ensure Quality Research is Introduced to the Academic Community

Become an Evaluator for IGI Global Authored Book Projects

The overall success of an authored book project is dependent on quality and timely manuscript evaluations.

Applications and Inquiries may be sent to:
development@igi-global.com

Applicants must have a doctorate (or equivalent degree) as well as publishing, research, and reviewing experience. Authored Book Evaluators are appointed for one-year terms and are expected to complete at least three evaluations per term. Upon successful completion of this term, evaluators can be considered for an additional term.

If you have a colleague that may be interested in this opportunity, we encourage you to share this information with them.

www.ingramcontent.com/pod-product-compliance
Lightning Source LLC
LaVergne TN
LVHW060350300926
849162LV00006B/1233